Frontiers of Optical Spectros

Investigating Extreme Physical (
Optical Techniques

NATO Science Series

A Series presenting the results of scientific meetings supported under the NATO Science Programme.

The Series is published by IOS Press, Amsterdam, and Kluwer Academic Publishers in conjunction with the NATO Scientific Affairs Division

Sub-Series

I. Life and Behavioural Sciences	IOS Press
II. Mathematics, Physics and Chemistry	Kluwer Academic Publishers
III. Computer and Systems Science	IOS Press
IV. Earth and Environmental Sciences	Kluwer Academic Publishers
V. Science and Technology Policy	IOS Press

The NATO Science Series continues the series of books published formerly as the NATO ASI Series.

The NATO Science Programme offers support for collaboration in civil science between scientists of countries of the Euro-Atlantic Partnership Council. The types of scientific meeting generally supported are "Advanced Study Institutes" and "Advanced Research Workshops", although other types of meeting are supported from time to time. The NATO Science Series collects together the results of these meetings. The meetings are co-organized bij scientists from NATO countries and scientists from NATO's Partner countries – countries of the CIS and Central and Eastern Europe.

Advanced Study Institutes are high-level tutorial courses offering in-depth study of latest advances in a field.
Advanced Research Workshops are expert meetings aimed at critical assessment of a field, and identification of directions for future action.

As a consequence of the restructuring of the NATO Science Programme in 1999, the NATO Science Series has been re-organised and there are currently Five Sub-series as noted above. Please consult the following web sites for information on previous volumes published in the Series, as well as details of earlier Sub-series.

http://www.nato.int/science
http://www.wkap.nl
http://www.iospress.nl
http://www.wtv-books.de/nato-pco.htm

Series II: Mathematics, Physics and Chemistry – Vol. 168

Frontiers of Optical Spectroscopy
Investigating Extreme Physical Conditions with Advanced Optical Techniques

edited by

Baldassare Di Bartolo
Boston College,
Chestnut Hill, MA, U.S.A.

and

Ottavio Forte
Raytheon Command, Control, Communications and Information Systems,
Sudbury, MA, U.S.A.

Kluwer Academic Publishers

Dordrecht / Boston / London

Published in cooperation with NATO Scientific Affairs Division

Proceedings of the NATO Advanced Study Institute on
Frontiers of Optical Spectroscopy
Investigating Extreme Physical Conditions with Advanced Optical Techniques
Erice, Italy
16 May–1 June 2003

A C.I.P. Catalogue record for this book is available from the Library of Congress.

ISBN 1-4020-2750-8 (PB)
ISBN 1-4020-2749-4 (HB)
ISBN 1-4020-2751-6 (e-book)

Published by Kluwer Academic Publishers,
P.O. Box 17, 3300 AA Dordrecht, The Netherlands.

Sold and distributed in North, Central and South America
by Kluwer Academic Publishers,
101 Philip Drive, Norwell, MA 02061, U.S.A.

In all other countries, sold and distributed
by Kluwer Academic Publishers,
P.O. Box 322, 3300 AH Dordrecht, The Netherlands.

Printed on acid-free paper

Printed in the Netherlands.

CONTENTS

THE PARTICIPANTS

La nature de notre esprit nous porte à chercher l'essence ou le 'pourquois' des choses. L'experience nous apprend bientôt que nous ne devons pas aller au delà du 'comment'.

Claude Bernard

THE DIRECTOR AND HIS TEAM

PREFACE

This book presents an account of the course "Frontiers of Optical Spectroscopy," held in Erice, Sicily, Italy, from May 16 to June 1, 2003. This meeting was organized by the International School of Atomic and Molecular Spectroscopy of the "Ettore Majorana" Centre for Scientific Culture.

Advanced spectroscopic techniques allow the probing of very small systems and very fast phenomena, conditions that can be considered "extreme" at the present status of our experimentation and knowledge. Quantum dots, nanocrystals, and single molecules are examples of the former and events on the femtosecond scale examples of the latter. The purpose of this institute was to examine the realm of phenomena of such extreme type and the techniques that permit their investigations.

The technical advances have enabled the observation of new phenomena such as high-harmonic generation in atoms, metrology using femtosecond lasers, observation of entangled states in semiconductor quantum dots (a prerequisite for quantum computing), Bose-Einstein condensation, light as slow as a bicyclist, and many more. The fruitful cross-fertilization of optical techniques and phenomena in physical and chemical systems was the motivation of this Institute.

Each lecturer aimed at developing a coherent section of the program starting at a somewhat fundamental level and ultimately reaching the frontier of knowledge in the field in a systematic and didactic fashion. The formal lectures were complemented and illustrated by additional seminars and discussions. The course was addressed to workers in spectroscopy-related fields from universities, laboratories, and industries. Senior scientists were encouraged to participate.

The Institute provided the participants with an opportunity to present their research work in the form of short seminars or posters.

The participants came from 15 different countries: Belarus, Bulgaria, Denmark, France, Germany, Israel, Italy, The Netherlands, Romania, Russia, Spain, Sweden, Switzerland, Turkey, and United States.

There were 18 formal lectures, one interdisciplinary lecture, and 5 long seminars. In addition, 11 short seminars and 16 posters were presented. Two round-table discussions were held. The first round-table discussion took place during the first week of the school in order to evaluate the work done and consider suggestions and proposals regarding the organization, format and presentation of the lectures. The second round-table discussion was held at the conclusion of the course, so that the participants could comment on the work done during the entire meeting and discuss various proposals for the next course of the International School of Atomic and Molecular Spectroscopy.

The full-text lectures/long seminars and the abstracts of short seminars and posters are reported in this book.

The secretary of the course was Ottavio Forte.

I wish to acknowledge the sponsorship of the meeting by NATO, NASA, the ENEA Organization, Boston College, the Italian Ministry of University and Scientific Research and Technology, the USA National Science Foundation, and the Sicilian Regional Government.

I would like to thank the Co-Director of the Course, Academian Alexander Voitovich, the members of the organizing committee (Prof. Martin Wegener, Dr. Giuseppe Baldacchini, Prof. Claus Klingshirn, Dr. Cees Ronda, Prof. Eric Mazur, Dr. James Barnes, Dr. Norman Barnes, Prof. Ralph von Baltz, and Prof. Steve Arnold), the secretary of the course (Mr. Ottavio Forte) and Prof. Xuesheng Chen for their help in organizing and running the course.

A special thank you goes to my brother Francesco who received all the participants in his house for a social gathering.

I am looking forward to our activities at the Majorana Centre in years to come, including the next 2005 meeting of the International School of Atomic and Molecular Spectroscopy.

Baldassare (Rino) Di Bartolo
Director of the International School of
Atomic and Molecular Spectroscopy of
the "Ettore Majorana" Center

NOTE: During the preparation of this volume we received the sad news of the untimely death of Dr. James Barnes. The paper whose abstract appears on page 687 will be his last contribution to our schools. We shall always remember his enthusiastic participation in our meetings, his gentlemanly courtesy and his warm friendship.

LIST OF PAST INSTITUTES

Advanced Study Institutes Held at the "Ettore Majorana" Centre in Erice, Sicily, Italy, Organizes by the International School of Atomic and Molecular Spectroscopy

1974 – Optical Properties of Ions in Solids
1975 – The Spectroscopy of the Excited State
1977 – Luminescence of Inorganic Solids
1979 – Radiationless Processes
1981 – Collective Excitations in Solids
1983 – Energy Transfer Processes in Condensed Matter
1985 – Spectroscopy of Solid-State Laser Type Materials
1987 – Disordered Solids: Structures and Processes
1989 – Advances in Nonradiative Processes
1991 – Optical Properties of Excited State in Solids
1993 – Nonlinear Spectroscopy of Solids: Advances and Applications
1995 – Spectroscopy and Dynamics of Collective Excitations in Solids
1996 – Workshop on Luminescence Spectroscopy
1997 – Ultrafast Dynamics of Quantum Systems: Physical Processes and Spectroscopic Techniques
1998 – Workshop on Advances in Solid State in Luminescence Spectroscopy
1999 – Advances in Energy Transfer Processes
2000 – Workshop on Advanced Topics in Luminescence Spectroscopy
2001 – Spectroscopy of Systems with Spatially Confined Structures
2002 – Workshop on the Status and Prospects of Luminescence Research

1. INVESTIGATING PHYSICAL SYSTEMS WITH OPTICAL SPECTROSCOPY

B. DI BARTOLO
Department of Physics, Boston College
Chestnut Hill, MA 02467, USA

Abstract

The article is based on the lectures that I delivered at the beginning of the course "Frontiers of Optical Spectroscopy," a NATO Advanced Study Institute that took place at the Ettore Majorana Center in Erice, Italy, May 16 - June 1, 2003.

The purpose of this contribution is to present some background material useful to deal with the application of optical spectroscopy to the study of physical systems.

In the introductory lecture we differentiate between two cases of "extreme physical conditions":

i) extreme conditions that predate experimentation, having been produced artificially and objectively different from more common ones, and
ii) extreme conditions created by an experimenter who employs some technical procedure to vary or modify the status of some systems and bring them into conditions different from their natural ones.

In the second lecture we treat the interaction of radiation with atoms and molecules. We introduce the concept of transition rate. In addition, we deal with the optical Bloch equations, the Rabi oscillations, and the mechanisms responsible for the broadening of spectral lines.

1. Introduction

At the beginning of the NATO Advanced Study Institute on "Frontiers of Optical Spectroscopy – Investigating *Extreme* Physical Conditions with Advanced Optical Techniques," I thought it was appropriate to present to the participants some considerations regarding the nature and purpose of such conditions. I want to report in this introduction these considerations, incorporating in them some of the input from the audience.

As suggested by a participant, "extreme" is a relative term, and conditions that may seem extreme today may, at a later time, be considered normal. We shall then at this point appraise the situation in terms of today's experimental reality.

Extreme physical conditions are sought or prepared for in several human endeavors. An engineer, when building a bridge, will make himself sure that it will withstand much stronger pressure that it is ever likely to experience. For a scientist extreme physical conditions of a system under study provide an appropriate situation in which the relevance of a certain parameter is enhanced and therefore made more amenable to be studied and understood. Examples that come to mind are the medical observation of patients walking on a treadmill and the study of orthophrenic children.

Claude Bernard (1813-1878) in his treatise on experimental medicine [1] makes a distinction between the observer and the experimenter:

B. Di Bartolo and O. Forte (eds.), Frontiers of Optical Spectroscopy, 1-28.

"We give the name of the observer to somebody who applies the procedures of investigation, that may be simple or complex, to the study of phenomena that he does not influence and who collects the data as nature provides them.

We give the name of the experimenter to somebody who employs the procedures of investigation in order to vary or modify in some way the natural phenomena and make them appear in circumstances or conditions that are different from those in which nature will ever present them."

In this scheme of things astronomy is a science of observation, because an astronomer cannot act on the celestial bodies and chemistry is a science of experimentation, because chemists act on nature and modify it. What about physics and what about the subjects of our course? We can make the following observations:

1. For certain systems the extreme physical condition precedes the measurement. The act of measurement may interfere with them, but has no influence on their "extreme" condition. We may assign to this category nanostructures, atoms in microcavities, and media that slow down light propagation.
2. For other systems the act of measurement brings about the "extreme" condition. We may assign to this category spectroscopy of solids at high temperature, and spectroscopy of atoms under intense radiation.
3. Other systems may present a more complex behavior. In femtospectroscopy short light pulses may be used to measure events on a femtosecond scale or may be used light to create the extreme condition of very intense radiation that may produce a light emission with a very wide light spectrum.

Having made this attempt at creating a framework for the various topics of our course, I moved to the consideration of a subject of fundamental importance to spectroscopy, the interaction of radiation with atoms and molecules.

2. Interaction of Radiation with Atoms and Molecules

2.1. TWO-LEVEL SYSTEM

Let us consider a system with a time-independent Hamiltonian H_0. The time-dependent Schroedinger equation gives

$$H_0 \psi = i\hbar \frac{\partial \psi}{\partial t} \tag{1}$$

If the system is in a stationary state labeled i

$$\psi(t) = \psi_i(t) = e^{-i(E_i/\hbar)t} \psi_i(0) \tag{2}$$

where the energy values are given by

$$H_0 \psi_i(0) = E_i \psi_i(0) \tag{3}$$

We shall assume that the wavefunctions $\psi_i(t)$ are orthonormal.

Let us now suppose that the system is subjected to a time-dependent perturbation represented by $H'(t)$. The system will be represented by a wavefunction $\psi(t)$ such that

$$H\psi(t)=\left(H_0+H^{'}\right)\psi(t)=i\hbar\frac{\partial\psi(t)}{\partial t} \tag{4}$$

We can expand $\psi(t)$ in terms of the complete set $\psi_i(t)$

$$\psi(t)=\sum_i c_i(t)\psi_i(t) \tag{5}$$

If $H^{'}=0$, the coefficients $c_i's$ are time-independent. Replacing eq. (5) in eq. (4),

$$\left(H_0+H^{'}\right)\sum_i c_i(t)\psi_i(t)=i\hbar\left[\sum_i c_i(t)\frac{\partial\psi_i(t)}{\partial t}+\sum_i\frac{\partial c_i(t)}{\partial t}\psi_i(t)\right] \tag{6}$$

Then

$$\sum_i c_i(t)H^{'}\psi_i(t)=i\hbar\sum_i\frac{\partial c_i(t)}{\partial t}\psi_i(t) \tag{7}$$

where we have taken advantage of eqs. (2) and (3). Multiplying by $\psi_k^*(t)$ and integrating over all space we obtain

$$i\hbar\frac{\partial c_k(t)}{\partial t}=\sum_i c_i(t)\left\langle\psi_k(t)\right|H^{'}\left|\psi_i(t)\right\rangle=\sum_i c_i(t)M_{ki}e^{i\omega_{ki}t} \tag{8}$$

where

$$\omega_{ki}=\frac{E_k-E_i}{\hbar};\quad M_{ki}=\left\langle\psi_i(0)\right|H^{'}\left|\psi_i(0)\right\rangle \tag{9}$$

We shall now make the following simplifying assumptions:

(a) The system has only two energy levels, say *1* and *2*, and

(b) the diagonal matrix0 elements of $H^{'}$ are zero.

The coupled equations (8) become

$$\begin{cases} i\dot{c}_2(t)=c_1V_{21}e^{i\omega_0 t} \\ i\dot{c}_1(t)=c_2V_{12}e^{-i\omega_0 t} \end{cases} \tag{10}$$

where

$$\omega_0=\frac{E_2-E_1}{\hbar};\quad V_{ij}=\frac{M_{ij}}{\hbar} \tag{11}$$

2.2. THE HAMILTONIAN OF THE INTERACTION WITH RADIATION

Let a polarized electromagnetic wave interact with an atom. Let the nucleus of the atom be at the origin of a system of coordinates as in Fig. 1 and let the electrons have coordinates $\vec{r}_i$; let also be

x the direction of polarization of the $\vec{E}$ field. The size of an atom is of the order of the Bohr radius

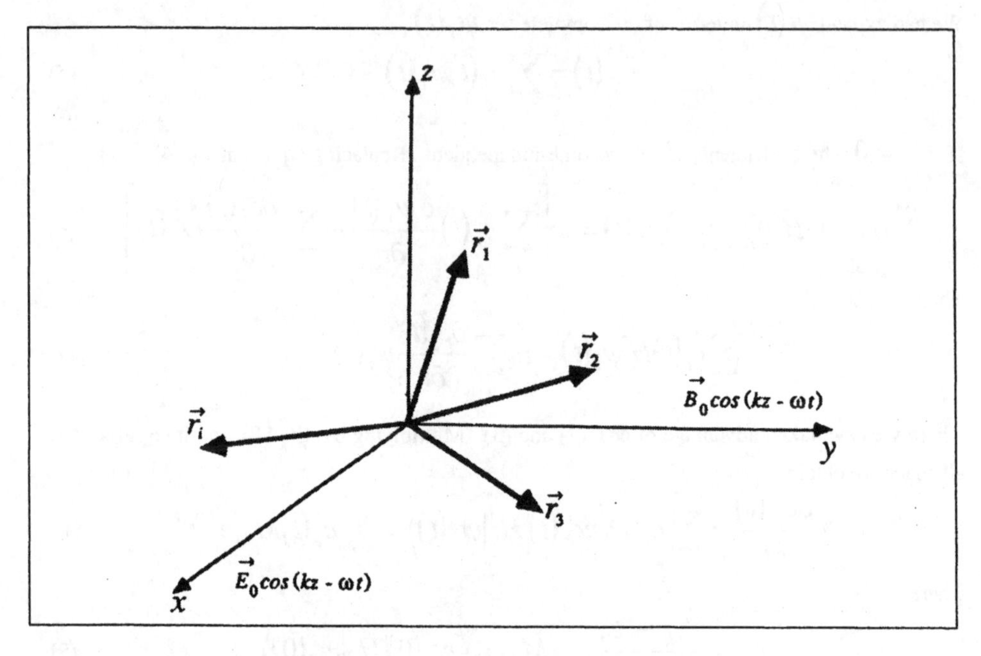

Figure 1: An Atom Interacting with Radiation.

$$a_0 = \frac{4\pi\varepsilon_0 \hbar^2}{me^2} = 5 \times 10^{-9}\, cm \quad \langle\langle \quad \lambda \text{ radiation} \tag{12}$$

Then

$$\frac{a_0}{\lambda} \langle\langle 1 \tag{13}$$

and the field acting on the atom is given by

$$\vec{E} = \vec{E}_0 \cos \omega t \tag{14}$$

with no variation of $\vec{E}$ across the atom. The total electric dipole moment is given by

$$e\vec{p} = e\sum_{i=1}^{Z} \vec{r}_i \tag{15}$$

where Z = number of electrons in the atom. The interaction Hamiltonian can be written as follows

$$H' = -e\vec{p} \cdot \vec{E}_0 \cos \omega t \tag{16}$$

$\vec{p}$, and therefore H', are odd operators: then

$$V_{11} = V_{22} = 0 \tag{17}$$

But

$$\begin{aligned} V_{12} &= \frac{1}{\hbar}\langle \psi_1(0)|H'|\psi_1(0)\rangle = \frac{1}{\hbar}\langle \psi_1(0)| - e\vec{p}\cdot\vec{E}_0 \cos\omega t|\psi_2(0)\rangle \\ &= -\frac{eE_0}{\hbar}\cos\omega t\langle \psi_1(0)|p^x|\psi_2(0)\rangle = -\frac{eE_0 p_{12}^x}{\hbar}\cos\omega t \\ &= V\cos\omega t \end{aligned} \tag{18}$$

where

$$V = -\frac{eE_0 p_{12}^x}{\hbar} = \frac{eE_0}{\hbar}\langle \psi_1(0)|p^x|\psi_2(0)\rangle \tag{19}$$

and

$$p^x = \sum_{i=1}^{Z} e x_i \tag{20}$$

Example: Atom of Hydrogen [1]
Let

$$|\psi_1\rangle = 1s \text{ state}$$
$$|\psi_2\rangle = 2p_x \text{ state}$$

We can write:

$$\psi_1 = \frac{1}{\sqrt{\pi a_0^3}} e^{-r/a_0}$$

$$\psi_2 = \frac{1}{\sqrt{2^5 \pi a_0^5}} e^{-r/2a_0} x$$

The energies of the various levels are given by

$$E_n = -\frac{me^4}{32\pi^2\varepsilon_0^2\hbar^2 n^2}$$

Then

$$V = -\frac{eE_0}{\hbar}\langle \psi_1|x|\psi_0\rangle = 5.93\times 10^4\, E_0 \sec^{-1}$$

But

$$\hbar\omega_0 = E_2 - E_1 = -\frac{me^4}{32\pi^2\varepsilon_0^2\hbar^2}\left(\frac{1}{4} - 1\right) = \hbar\left(1.63\times 10^{16}\right)$$

In order to make $V \approx \omega_0$ we need a field strength of $\sim 3\times 10^{11}\ V/m$. For light beams produced by conventional (non-laser) sources

$$V \ll \omega_0$$

The upper limit for conventional spectroscopic sources is represented by the field $E_0 = 10^3\ V/m$ produced by a mercury lamp with its emission line at 2537 A .

2.3. TRANSITION RATES

Using the expression (18) for the matrix element V_{12}, we can rewrite the eqs. (10) as follows:

$$\begin{cases} V \cos \omega t e^{-i\omega_0 t} c_2 = i\dot{c}_1 \\ V^* \cos \omega t e^{i\omega_0 t} c_1 = i\dot{c}_2 \end{cases} \tag{21}$$

$|c_2(t)|^2$ represents the probability of finding the system in its upper state ψ_2 at time t , and $|c_2(t)|^2 / t$ the rate at which this probability increases.

We set the initial conditions:

$$\begin{cases} c_1(0) = 1 \\ c_2(0) = 0 \end{cases} \tag{22}$$

and use the approximation

$$c_1(t) \approx 1 \tag{23}$$

We obtain

$$c_2(t) = \frac{V^*}{2}\left[\frac{1 - e^{i(\omega_0 + \omega)t}}{\omega_0 + \omega} + \frac{1 - e^{i(\omega_0 - \omega)t}}{\omega_0 - \omega}\right] \tag{24}$$

We use an additional approximation by neglecting the first term in the square brackets in (24). When $\omega \approx \omega_0$ the second term in [ ] is much larger than the first and this is the reason for such an approximation. Then

$$c_2(t) = \frac{V^*}{2}\frac{1 - e^{i(\omega_0 - \omega)t}}{\omega_0 - \omega} \tag{25}$$

$$\begin{aligned} |c_2(t)|^2 &= \frac{|V|^2}{4(\omega_0 - \omega)^2}\left(1 - e^{i(\omega_0 - \omega)t}\right)\left(1 - e^{-i(\omega_0 - \omega)t}\right) \\ &= |V|^2 \frac{\sin^2 \frac{\omega_0 - \omega}{2} t}{(\omega_0 - \omega)^2} \end{aligned} \tag{26}$$

When $\omega = \omega_0$

$$|c_2(t)|^2 = \frac{1}{4}|V|^2 t^2 \tag{27}$$

namely the probability of excitation is proportional to t^2. For $\omega \neq \omega_0$ such probability has an oscillatory behavior. Fig. 2 shows the dependence of the probability of excitation at a certain time t on the angular frequency ω.

Integrating over a range $\Delta\omega$ we obtain

$$\begin{aligned} |c_2(t)|^2 &= \int_{\omega_0-\frac{\Delta\omega}{2}}^{\omega_0+\frac{\Delta\omega}{2}} |V|^2 \frac{\sin^2[(\omega_0-\omega)t/2]}{(\omega_0-\omega)^2} d\omega \\ &= \frac{2e^2|p_{12}^x|^2 W(\omega_0)}{\varepsilon_0 \hbar^2} \int_{\omega_0-\frac{\Delta\omega}{2}}^{\omega_0+\frac{\Delta\omega}{2}} \frac{\sin^2[(\omega_0-\omega)t/2]}{(\omega_0-\omega)^2} d\omega \end{aligned} \tag{28}$$

where we have made use of the relations

$$V = \frac{eE_0 p_{12}^x}{\hbar} \tag{29}$$

$$\frac{1}{2}\varepsilon_0 E_0^2 = W(\omega)d\omega \tag{30}$$

$W(\omega)d\omega$ being the energy density of the wave for frequencies in $(\omega, \omega + d\omega)$.

The value of the integral is $\frac{t^2}{4}\Delta\omega$ for $t\,\Delta\omega \langle\langle 1$ and $\frac{1}{2}\pi t$ for $t\,\Delta\omega \rangle\rangle 1$. The latter condition, which we shall hold true, leads to a transition probability $|c_2(t)|^2$ proportional to t:

$$|c_2(t)|^2 = \frac{2e^2|p_{12}^x|^2 W(\omega_0)}{\varepsilon_0 \hbar^2}\frac{1}{2}\pi t = \frac{\pi e^2|p_{12}^x|^2 W(\omega_0)t}{\varepsilon_0 \hbar^2} \tag{31}$$

The linear approximation expressed above breaks down for t long enough to make $|c_2(t)|^2 \rangle 1$, contrary to the normalization condition. However, as long as $t \rangle\rangle \tau_R$ = radiative lifetime, the linear dependence of the transition probability on time is valid.

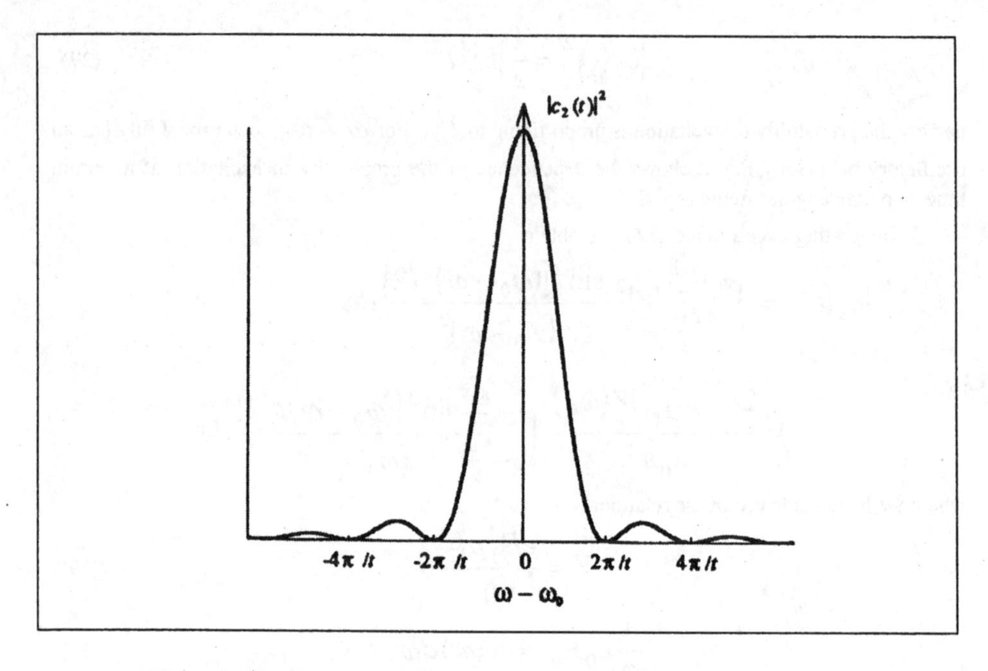

Figure 2: The Probability of Finding the System in State ψ_2 at Time t.

2.4. OPTICAL BLOCH EQUATIONS

In the presence of a perturbation $H'(t)$ the wavefunction of a two-level system is expressed by

$$\psi(\vec{r},t) = c_1(t)\psi_1(\vec{r},t) + c_2(t)\psi_2(\vec{r},t) \tag{32}$$

where $c_1(t)$ and $c_2(t)$ are such that

$$|c_1(t)|^2 + |c_2(t)|^2 = 1 \tag{33}$$

are solutions of

$$\begin{cases} V\cos\omega t\, e^{-i\omega_0 t} c_2 = i\dot{c}_1 \\ V^* \cos\omega t\, e^{i\omega_0 t} c_1 = i\dot{c}_2 \end{cases} \tag{34}$$

We now look for more general solutions of these equations, by doing the following:

1) We use the same approximation that reduced eq. (24) to the eq. (25) form*,
2) we retain terms in all orders in V, and
3) we assume the electromagnetic radiation interacting with the two-level system monochromatic.

We define an *atomic density matrix* as follows:

* In what follows we shall call this approximation the *usual* approximation.

$$\begin{cases} D_{11} = c_1 c_1^* = |c_1|^2 = \dfrac{N_1}{N} \\ D_{22} = c_2 c_2^* = |c_2|^2 = \dfrac{N_2}{N} \\ D_{12} = c_1 c_2^* \\ D_{21} = c_2 c_1^* \end{cases} \tag{35}$$

where N_1, N_2 and N are the populations in level 1, level 2 and total, respectively. Note that

$$\begin{cases} D_{11} + D_{22} = 1 \\ D_{12} = D_{21}^* \end{cases} \tag{36}$$

Then

$$\begin{aligned} \dot{D}_{22} &= c_2^* \dot{c}_2 + \dot{c}_2 c_2^* = c_2^* \frac{V^*}{i} \cos \omega t \, e^{i\omega_0 t} c_1 + c.c. \\ &= -i \cos \omega t \left[V^* \, e^{i\omega_0 t} D_{12} - V e^{-i\omega t} D_{21} \right] = -\dot{D}_{11} \end{aligned} \tag{37}$$

and

$$\begin{aligned} \dot{D}_{12} &= c_1 \dot{c}_2^* + c_2^* \dot{c}_1 = c_1 \left(\frac{V^*}{i} \cos \omega \, t \, e^{i\omega_0 t} c_1 \right)^* + c_2^* \left(\frac{V}{i} \cos \omega \, t \, e^{-i\omega_0 t} c_2 \right) \\ &= iV \cos \omega t \, e^{-i\omega_0 t} \left(D_{11} - D_{22} \right) = \dot{D}_{21}^* \end{aligned} \tag{38}$$

We now use the *usual* approximation; accordingly, we consider the terms oscillating with frequency $\omega_0 + \omega$ negligible with respect to the terms oscillating at frequency $\omega_0 - \omega$. We write

$$\begin{aligned} \dot{D}_{22} &= -\dot{D}_{11} = -\frac{i}{2} V^* e^{i(\omega_0 - \omega)t} D_{12} + \frac{i}{2} V e^{-i(\omega_0 - \omega)t} D_{21} \\ \dot{D}_{12} &= \dot{D}_{21}^* = \frac{i}{2} V e^{-i(\omega_0 - \omega)t} \left(D_{11} - D_{22} \right) \end{aligned} \tag{39}$$

These equations are called the *optical Bloch equations* [3].

2.5. RABI OSCILLATIONS

The optical Bloch equations can be written as follows:

$$\begin{cases} \dot{D}_{11} = \frac{i}{2} V^* e^{i(\omega_0-\omega)t} D_{12} - \frac{i}{2} V e^{i(\omega_0-\omega)t} D_{21} \\ \dot{D}_{22} = -\frac{i}{2} V e^{-i(\omega_0-\omega)t} D_{12} + \frac{i}{2} V e^{i(\omega_0-\omega)t} D_{21} \\ \dot{D}_{12} = \frac{i}{2} V e^{-i(\omega_0-\omega)t} D_{11} - \frac{i}{2} V e^{-i(\omega_0-\omega)t} D_{22} \\ \dot{D}_{21} = -\frac{i}{2} V^* e^{i(\omega_0-\omega)t} D_{11} + \frac{i}{2} V^* e^{i(\omega_0-\omega)t} D_{22} \end{cases} \tag{40}$$

We shall introduce the following trial solutions

$$\begin{cases} D_{11} = D_{11}^0 e^{\alpha t} \\ D_{22} = D_{22}^0 e^{\alpha t} \\ D_{12} = D_{12}^0 e^{-i(\omega_0-\omega)t} e^{\alpha t} \\ D_{21} = D_{21}^0 e^{i(\omega_0-\omega)t} e^{\alpha t} \end{cases} \tag{41}$$

with the quantities D_{ij}^0 and α independent of time. Using the relations (41) in the eq. (40) we obtain a system of homogeneous equations in the unknown quantities D_{ij}^0:

$$\begin{pmatrix} -\alpha & 0 & \frac{i}{2}V^* & -\frac{i}{2}V \\ 0 & -\alpha & -\frac{i}{2}V^* & \frac{i}{2}V \\ \frac{i}{2}V & -\frac{i}{2}V & i(\omega_0-\omega)-\alpha & 0 \\ -\frac{i}{2}V^* & \frac{i}{2}V^* & 0 & -i(\omega_0-\omega)-\alpha \end{pmatrix} \begin{pmatrix} D_{11}^0 \\ D_{22}^0 \\ D_{12}^0 \\ D_{21}^0 \end{pmatrix} = 0 \tag{42}$$

These equations admit solutions if the determinant of the coefficients is equal to zero:

$$\alpha^2 \left[\alpha^2 + (\omega_0-\omega)^2 + |V|^2\right] = 0 \tag{43}$$

The possible values of α are

$$\begin{cases} \alpha_1 = 0 \\ \alpha_2 = i\Omega \\ \alpha_3 = -i\Omega \end{cases} \tag{44}$$

where

$$\Omega = \sqrt{(\omega_0 - \omega)^2 + |V|^2} \tag{45}$$

We note here that the dependence of Ω on $|V|^2$ indicates a change in the frequency of oscillations of the "coupled" system which consist of atom and light beam: this effect is called the *dynamic Stark effect.*

The most general solution is then

$$D_{ij} = D_{ij}^{(1)} + D_{ij}^{(2)} e^{i\Omega t} + D_{ij}^{(3)} e^{-j\Omega t} \tag{46}$$

Additional oscillatory exponentials are present in the off-diagonal elements. By using the initial conditions and the optical Bloch equations we can obtain the constant coefficients.

Example

$$\begin{aligned} c_1(0) = 1 &\longrightarrow D_{22}(0) = 0 \\ c_2(0) = 0 &\longrightarrow D_{12}(0) = 0 \end{aligned} \tag{47}$$

The density matrix elements are

$$\begin{cases} D_{22} = \dfrac{|V|^2}{\Omega^2} \sin^2 \dfrac{1}{2}\Omega t \\ D_{11} = 1 - D_{22} = 1 - \dfrac{|V|^2}{\Omega^2} \sin^2 \dfrac{1}{2}\Omega t \\ D_{12} = e^{-i(\omega_0 - \omega)t} \dfrac{V}{\Omega^2} \sin \dfrac{1}{2}\Omega t \left[-(\omega_0 - \omega) \sin\left(\dfrac{1}{2}\Omega t\right) + i\Omega \cos\left(\dfrac{1}{2}\Omega t\right) \right] \\ D_{21} = e^{i(\omega_0 - \omega)t} \dfrac{V^*}{\Omega^2} \sin \dfrac{1}{2}\Omega t \left[-(\omega_0 - \omega) \sin\left(\dfrac{1}{2}\Omega t\right) - i\Omega \cos\left(\dfrac{1}{2}\Omega t\right) \right] \end{cases} \tag{48}$$

For zero detuning $(\omega = \omega_0)$: $\Omega = |V|$ and

$$\begin{cases} D_{22} = \sin^2 \dfrac{1}{2}|V|t \\ D_{12} = i\dfrac{V}{|V|} \sin\left(\dfrac{1}{2}|V|t\right) \cos\left(\dfrac{1}{2}|V|t\right) \end{cases} \tag{49}$$

The behavior of the quantity D_{22} for different values of detuning

$$d = \frac{\omega_0 - \omega}{|V|} \tag{50}$$

is represented in Fig. 3. We shall make the following observations:

1) For zero detuning $(\omega = \omega_0)$ the atom oscillates between ground and excited states: these oscillations are called *Rabi oscillations* and their frequency $|V|$ is called *Rabi frequency*. Solutions for the similar problem of a spin system in an oscillatory magnetic field were obtained by Rabi [4]. When

$$\frac{|V|}{2}t = \frac{eE_0|p_{12}^x|}{2\hbar}t = \pi \tag{51}$$

$D_{22} = 1$, all the atoms are in the upper state, i.e. the population is completely *inverted.* When

$$\frac{|V|}{2}t = \frac{eE_0|p_{12}^x|}{2\hbar}t = 2\pi \tag{52}$$

the original situation, with all the atoms in the ground state, is restored.

2) Since the *usual* approximation was originally used in deriving the optical Bloch equations the solutions (48) are valid only if $(\omega_0 - \omega) \ll (\omega_0 + \omega)$.

3) The solutions for D_{22} and D_{12} refer to monochromatic radiation. In effect the oscillations can be seen experimentally when the frequency spread of the electromagnetic radiation is much smaller than the linewidth of the transition.

4) We have to dedicate some attention to the fact that the processes that broaden the linewidth of the transition introduce modifications in the optical Bloch equations. We shall return on this point.

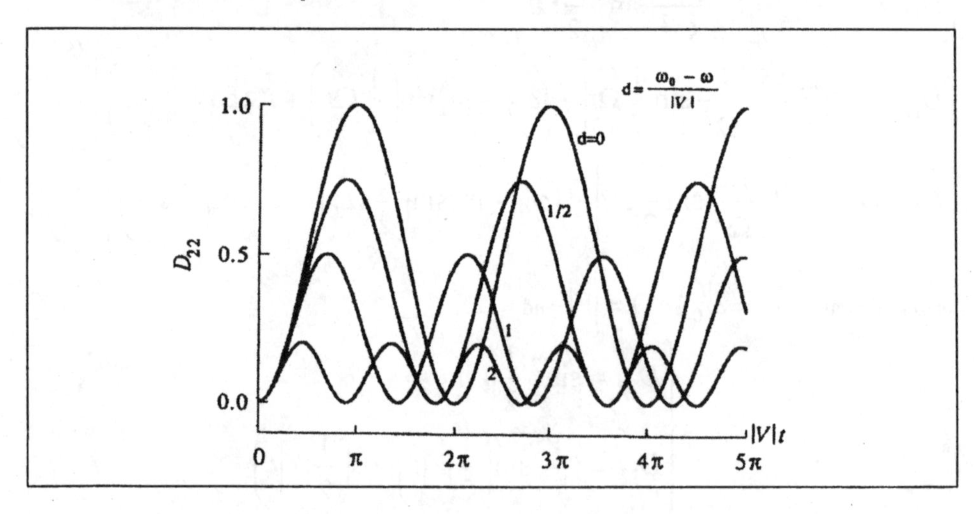

Figure 3: Probability of Excitation as Function of Time.

2.6. BROADENING OF SPECTRAL LINES

2.6.1. *Definition of Susceptibility*

Consider a gas of atoms and apply it to an electric field

$$E(t) = \int_{-\infty}^{+\infty} E(\omega) e^{-i\omega t} d\omega \tag{53}$$

where

$$E(\omega) = \frac{1}{2\pi} \int_{-\infty}^{+\infty} E(t) e^{i\omega t} dt \tag{54}$$

Since $E(t)$ is real

$$E(-\omega) = E^*(\omega) \tag{55}$$

Let $P(t)$ = atomic polarization:

$$P(t) = \int_{-\infty}^{+\infty} P(\omega) e^{i\omega t} d\omega \tag{56}$$

where

$$P(\omega) = \varepsilon_0 \chi(\omega) E(\omega) \tag{57}$$

and $\chi(\omega)$ = electric susceptibility. Then

$$P(t) = \varepsilon_0 \int_{-\infty}^{+\infty} \chi(\omega) E(\omega) e^{-i\omega t} d\omega \tag{58}$$

A "real" stimulus $E(t)$ must produce a "real" response $P(t)$: then

$$\chi(-\omega) = \chi^*(\omega) \tag{59}$$

Therefore, if we write in general

$$\chi(\omega) = \chi'(\omega) + i\chi''(\omega) \tag{60}$$

eq. (59) gives

$$\begin{cases} \chi'(-\omega) = \chi'(\omega) \\ \chi''(-\omega) = -\chi''(\omega) \end{cases} \tag{61}$$

The above relations are called *crossing relations* for the real and imaginary parts of the susceptibility.

2.6.2. *Electric Dipole Moment*

Consider a gas of Z -electron, 2-level atoms, and apply to it an electric field along the x direction:

$$E(t) = \frac{1}{2} E_0 \left(e^{-i\omega t} + e^{i\omega t} \right) \tag{62}$$

A polarization will set in:

$$P(t) = \frac{1}{2} \varepsilon_0 E_0 \left[\chi(\omega) e^{-i\omega t} + \chi(-\omega) e^{i\omega t} \right] \tag{63}$$

The electric dipole moment of one atom in the x direction will be

$$d(t) = -\langle \psi(t) | e p^x | \psi(t) \rangle \tag{64}$$

where

$$ep^x = e\sum_{i=1}^{Z} x_i \tag{65}$$

But

$$\begin{aligned}\psi(t) &= c_1(t)\psi_1(\vec{r},t) + c_2(t)\psi_2(\vec{r},t) \\ &= c_1(t)e^{-i(E_1/\hbar)t}\psi_1(\vec{r}) + c_2(t)e^{-i(E_2/\hbar)t}\psi_2(\vec{r})\end{aligned} \tag{66}$$

Then

$$\begin{aligned}d(t) = -\langle c_1 e^{-i(E_1/\hbar)t}\psi_1(\vec{r}) + c_2 e^{-i(E_2/\hbar)t}\psi_2(\vec{r})|ep^x|c_1(t)e^{-i(E_1/\hbar)t}\psi_1(\vec{r}) \\ + c_2(t)e^{-i(E_2/\hbar)t}\psi_2(\vec{r})\rangle = -e\left[c_1^* c_2 p_{12}^x e^{-i\omega_0 t} + c_1 c_2^* p_{21}^x e^{i\omega_0 t}\right]\end{aligned} \tag{67}$$

where c_1 and c_2 are solutions of the equations

$$\begin{cases} V\cos\omega t\, e^{-i\omega_0 t} c_2 = i\dot{c}_1 \\ V^* \cos\omega t\, e^{i\omega_0 t} c_1 = i\dot{c}_2 \end{cases} \tag{68}$$

In the absence of electromagnetic field $V = 0$ and $\dot{c}_1 = \dot{c}_2 = 0$: the probabilities of occupancy of the two levels are constant and no "spontaneous" emission is possible. To remedy the situation we introduce an additional damping term in the second equation (68) and write

$$i\dot{c}_2 = V^* \cos\omega t\, e^{i\omega_0 t} c_1 - i\gamma c_2 \tag{69}$$

In absence of a perturbing field

$$\dot{c}_2 = -\gamma c_2 \tag{70}$$

and

$$c_2(t) = c_2(0)e^{-\gamma t} \tag{71}$$

For a gas of atoms with a population $N_2(0)$ at time $t = 0$ in the upper level:

$$N_2(t) = N_2(0)e^{-2\gamma t} \tag{72}$$

We identify the quantity 2γ with the Einstein's A coefficient:

$$2\gamma = A_{21} \tag{73}$$

We now take eq. (69) and set in it $c_1 = 1$: we then have

$$i\dot{c}_2 = V^* \cos\omega t\, e^{i\omega_0 t} - i\gamma c_2 \tag{74}$$

and

$$c_2(t) = -\frac{1}{2}V^*\left[\frac{e^{i(\omega_0+\omega)t}}{\omega_0 + \omega - i\gamma} + \frac{e^{i(\omega_0-\omega)t}}{\omega_0 - \omega - i\gamma}\right] \tag{75}$$

$|c_2(t)|^2$ is of the order $|V|^2$; this makes $|c_1(t)|^2$ differing from 1 by a term of the order $|V|^2$: then

$$d(t) \approx -e\left[c_2 p_{12}^x e^{-i\omega_0 t} + c_2^* p_{21}^x e^{i\omega_0 t}\right]$$
$$= \frac{e^2 |p_{12}^x|^2 E_0}{2\hbar}\left[\frac{e^{i\omega t}}{\omega_0 + \omega - i\gamma} + \frac{e^{-i\omega t}}{\omega_0 - \omega - i\gamma} + \frac{e^{-i\omega t}}{\omega_0 + \omega + i\gamma} + \frac{e^{i\omega t}}{\omega_0 - \omega + i\gamma}\right] \tag{76}$$

We average over the random orientations of the atoms by introducing a factor of 1/3 and writing

$$d(t) = \frac{e^2 |p_{12}|^2 E_0}{6\hbar}\left[\frac{e^{i\omega t}}{\omega_0 + \omega - i\gamma} + \frac{e^{-i\omega t}}{\omega_0 - \omega - i\gamma} + \frac{e^{-i\omega t}}{\omega_0 + \omega + i\gamma} + \frac{e^{i\omega t}}{\omega_0 - \omega + i\gamma}\right] \tag{77}$$

where p_{12} = magnitude of $\vec{p}_{12}$. Then the polarization is given by

$$P(t) = \frac{N}{V} d(t)$$
$$= \frac{Ne^2 |p_{12}|^2 E_0}{6\hbar V}\left[\frac{e^{i\omega t}}{\omega_0 + \omega - i\gamma} + \frac{e^{-i\omega t}}{\omega_0 - \omega - i\gamma} + \frac{e^{-i\omega t}}{\omega_0 + \omega + i\gamma} + \frac{e^{i\omega t}}{\omega_0 - \omega + i\gamma}\right]$$
$$= \frac{1}{2}\varepsilon_0 E_0\left[\chi(\omega) e^{-i\omega t} + \chi(-\omega) e^{-i\omega t}\right] \tag{78}$$

Then

$$\chi(\omega) = \frac{Ne^2 |p_{12}|^2}{3\varepsilon_0 \hbar V}\left(\frac{1}{\omega_0 - \omega - i\gamma} + \frac{1}{\omega_0 + \omega + i\gamma}\right) \tag{79}$$

and

$$\chi(-\omega) = \frac{Ne^2 |p_{12}|^2}{3\varepsilon_0 \hbar V}\left(\frac{1}{\omega_0 + \omega - i\gamma} + \frac{1}{\omega_0 - \omega + i\gamma}\right) \tag{80}$$

We note that

$$\chi(-\omega) = \chi^*(\omega) \tag{81}$$

2.6.3. *Radiative Broadening*

We shall first seek a relation between the atomic absorption coefficient and the imaginary part of the susceptibility.

A gas of atoms can be considered as a dielectric medium in which the polarization is related to the electric field through the electric susceptibility χ:

$$\vec{P} = \varepsilon_0 \chi \vec{E} \tag{82}$$

The susceptibility is frequency-dependent. The dielectric constant K is related to the susceptibility as follows:

$$K = \left(\frac{kc}{\omega}\right)^2 = 1 + \chi = n^2 \tag{83}$$

where $k = \frac{2\pi}{\lambda}$ = magnitude of the wave vector and n = index of refraction. If we let

$$n = n_r + in_i \tag{84}$$

we obtain

$$k = \frac{\omega}{c} n_r + i\frac{\omega}{c} n_i \tag{85}$$

We can also write

$$(n_r + in_i)^2 = n_r^2 - n_i^2 + 2in_r n_i = \left(\frac{kc}{\omega}\right)^2 = 1 + \chi' + i\chi'' \tag{86}$$

and

$$\begin{cases} n_r^2 - n_i^2 = 1 + \chi', \\ 2n_r n_i = \chi'' \end{cases} \tag{87}$$

A traveling wave moving in, say, the z-direction would have time and space dependences given by

$$e^{i(kz - \omega t)} = e^{i\left(\frac{\omega}{c} n_r + i\frac{\omega}{c} n_i\right) z - i\omega t} = e^{i\omega\left(\frac{n_r}{c} z - t\right)} e^{-\frac{\omega}{c} n_i z} \tag{88}$$

The intensity of the wave will go down as

$$e^{-\alpha z} = e^{-\frac{2\omega n_i}{c} z} \tag{89}$$

where

$$e^{-\alpha z} = \text{absorption coefficient} = \frac{2\omega n_i}{c} = \frac{\omega \chi''}{n_r c} \tag{90}$$

We now do the following:

a) we consider the gas dilute and set $n_r = 1$, and

b) we use the *usual* approximation:

$$\chi(\omega)=\chi'(\omega)+i\chi''(\omega)=\frac{Ne^2|p_{12}|^2}{3\varepsilon_0\hbar V}\frac{1}{\omega_0-\omega-i\gamma}$$
$$=\frac{Ne^2|p_{12}|^2}{3\varepsilon_0\hbar V}\frac{\omega_0-\omega}{(\omega_0-\omega)^2+\gamma^2}+i\frac{Ne^2|p_{12}|^2\pi}{3\varepsilon_0\hbar V}\frac{\gamma/\pi}{(\omega_0-\omega)^2+\gamma^2} \tag{91}$$

Then

$$\chi''(\omega)=\frac{Ne^2|p_{12}|^2\pi}{3\varepsilon_0\hbar V}\frac{\gamma/\pi}{(\omega_0-\omega)^2+\gamma^2} \tag{92}$$

and

$$\alpha(\omega)=\frac{\pi Ne^2|p_{12}|^2\omega_0}{3\varepsilon_0 c\hbar V}f_L(\omega) \tag{93}$$

where

$$f_L(\omega)=\text{Lorentzian lineshape}=\frac{\gamma/\pi}{(\omega_0-\omega)^2+\gamma^2} \tag{94}$$

The function $f_L(\omega)$ is represented in Fig. 4.

2.6.4. *Power Broadening*

We have obtained the following result for the susceptibility correct to second order in p_{12}:

$$\chi(\omega)=\frac{Ne^2|p_{12}|^2}{3\varepsilon_0\hbar V}\left(\frac{1}{\omega_0-\omega-i\gamma}+\frac{1}{\omega_0+\omega+i\gamma}\right) \tag{95}$$

This result is consistent with a linear response of the atoms to the electric field of the light beam. If we want to include higher terms in $\chi(\omega)$ we have to use the optical Bloch equations:

$$\begin{cases}\dot{D}_{22}=-\dot{D}_{11}=-\frac{i}{2}V^*e^{i(\omega_0-\omega)t}D_{12}+\frac{i}{2}Ve^{-i(\omega_0-\omega)t}D_{21}\\ \dot{D}_{12}=\dot{D}_{21}^*=\frac{i}{2}Ve^{-i(\omega_0-\omega)t}(D_{11}-D_{22})\end{cases} \tag{96}$$

but we have to include spontaneous emission. We do so by considering the two equations

$$\begin{aligned}\dot{c}_2&=-iV^*\cos\omega t\,e^{i\omega_0 t}c_1-\gamma c_2\\ \dot{c}_1&=-iV\cos\omega t\,e^{-i\omega_0 t}c_2\end{aligned} \tag{97}$$

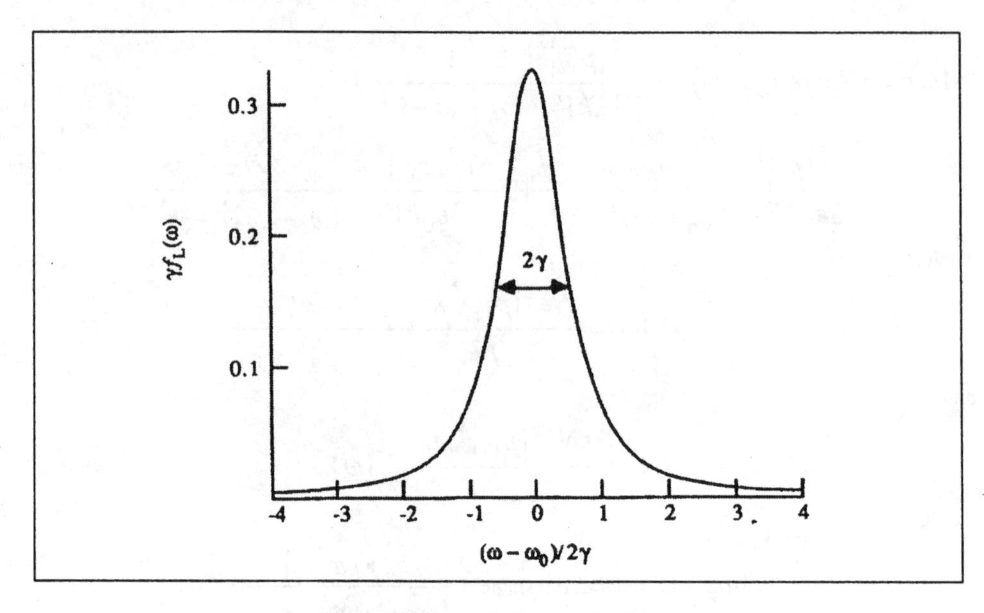

Figure 4: Shape of a Lorentzian Line

Then

$$\begin{aligned}\dot{D}_{22} &= c_2^* \frac{dc_2}{dt} + c_2 \frac{dc_2^*}{dt} \\ &= c_2^*\left(-iV^* \cos\omega t\, e^{i\omega_0 t} c_1 - \gamma c_2\right) + c_2\left(iV \cos\omega t\, e^{-i\omega_0 t} c_1^* - \gamma c_2^*\right) \\ &= -iV^* \cos\omega t\, e^{i\omega_0 t} D_{22} - \gamma D_{22} + iV \cos\omega t\, e^{-i\omega_0 t} D_{21} - \gamma D_{22}\end{aligned}$$

$$\approx -\frac{i}{2} V^* e^{i(\omega_0-\omega)t} D_{12} + \frac{i}{2} V e^{-i(\omega_0-\omega)t} D_{21} - 2\gamma D_{22}$$

$$\dot{D}_{12} = c_1 \frac{dc_2^*}{dt} + c_2^* \frac{dc_1}{dt} = iV \cos\omega t\, e^{-i\omega_0 t}\left(D_{11} - D_{22}\right) - \gamma D_{12}$$

$$\approx \frac{i}{2} V e^{-i(\omega_0-\omega)t}\left(D_{11} - D_{22}\right) - \gamma D_{12}$$

where we have used the *usual* approximation. In summary

$$\begin{cases} \dot{D}_{22} = -\dot{D}_{11} - \frac{i}{2} V^* e^{i(\omega_0-\omega)t} D_{12} + \frac{i}{2} V e^{-i(\omega_0-\omega)t} D_{21} - 2\gamma D_{22} \\ \dot{D}_{12} = \dot{D}_{21}^* - \frac{i}{2} V e^{-i(\omega_0-\omega)t}\left(D_{11} - D_{22}\right) - \gamma D_{12} \end{cases} \tag{98}$$

The solutions of these equations are no longer purely oscillatory; after a certain time we have a steady state situation

$$\begin{cases} D_{22} = \dfrac{|V|^2/4}{(\omega_0-\omega)^2+\gamma^2+|V|^2/2} \\ D_{12} = -e^{-i(\omega_0-\omega)t}\dfrac{\dfrac{|V|^2}{2}(\omega_0-\omega-i\gamma)}{(\omega_0-\omega)^2+\gamma^2+|V|^2/2} \end{cases} \tag{99}$$

The atomic dipole moment is given, according to eq. (67), by

$$\begin{aligned} d(t) &= -e\left[D_{21}p_{12}^x e^{-i\omega_0 t} + D_{12}p_{21}^x e^{i\omega_0 t}\right] \\ &= e\left[e^{i(\omega_0-\omega)t}\frac{\dfrac{V^*}{2}(\omega_0-\omega+i\gamma)p_{12}^x}{(\omega_0-\omega)^2+\gamma^2+|V|^2/2}e^{i\omega_0 t} + c.c.\right] \\ &= e\left[e^{i\omega t}\frac{\dfrac{V^*}{2}(\omega_0-\omega+i\gamma)p_{12}^x}{(\omega_0-\omega)^2+\gamma^2+|V|^2/2} + c.c.\right] \end{aligned} \tag{100}$$

Then

$$\begin{aligned} P(t) = \frac{N}{V}d(t) &= e^{-i\omega t}\frac{N}{V}\frac{e^2|p_{12}^x|^2 E_0}{2\hbar}\frac{\omega_0-\omega+i\gamma}{(\omega_0-\omega)^2+\gamma^2+|V|^2/2} \\ &+ e^{i\omega t}\frac{N}{V}\frac{e^2|p_{12}^x|^2 E_0}{2\hbar}\frac{\omega_0-\omega-i\gamma}{(\omega_0-\omega)^2+\gamma^2+|V|^2/2} \\ &= \frac{1}{2}\varepsilon_0 E_0\chi(\omega)e^{-i\omega t} + \frac{1}{2}\varepsilon_0 E_0\chi(-\omega)e^{i\omega t} \end{aligned} \tag{101}$$

Then

$$\chi(\omega) = \frac{Ne^2|p_{12}|^2}{3\varepsilon_0\hbar V}\frac{(\omega_0-\omega)+i\gamma}{(\omega_0-\omega)^2+\gamma^2+|V|^2/2} \tag{102}$$

This expression is not "complete" because we have used the *usual* approximation; the complete expression can be written as follows:

$$\chi(\omega) \propto \left[\frac{\omega_0 - \omega + i\gamma}{(\omega_0 - \omega)^2 + \gamma^2 + |V|^2/2} + \frac{\omega_0 + \omega - i\gamma}{(\omega_0 - \omega)^2 + \gamma^2 + |V|^2/2} \right] \tag{103}$$

If we use the complete expression we obtain

$$\chi(-\omega) = \chi^*(\omega) \tag{104}$$

a condition that is not verified for the simpler expression (102).

Now we can write

$$\chi''(\omega) = \frac{Ne^2 |p_{12}|^2}{3\varepsilon_0 \hbar V} \frac{\gamma}{(\omega_0 - \omega)^2 + \gamma^2 + |V|^2/2} \tag{105}$$

Because of the presence of the term $|V|^2/2$ in the denominator the susceptibility is dependent on the field strength, the rate of absorption of incident light is reduced and the linewidth of the atomic transition is given by

$$2\sqrt{\gamma^2 + \frac{1}{2}|V|^2} \tag{106}$$

The additional contribution to the linewidth is called *power broadening* or *saturation broadening*.

2.6.5. *Damped Rabi Oscillations*

The diagonal elements of the atomic density matrix give us, when multiplied by the total atomic population N the populations of atoms in states ψ_1 and ψ_2

$$\begin{cases} N_2 = ND_{22} = \dfrac{N|V|^2/4}{(\omega_0 - \omega)^2 + \gamma^2 + |V|^2/2} \\ N_1 = ND_{11} = N - N_2 \end{cases} \tag{107}$$

These expressions apply in the steady state. In the limit of weak intensity of incident light the value of N_2 is proportional to the intensity of light. We note also that the steady state value above is independent of the initial conditions.

For studying the transients we need, however, the initial conditions. The equations to consider are eqs. (98); let us deal with the simple case in which $\omega_0 = \omega$ (zero detuning) and

$$\begin{cases} D_{12}(0) = 0 \quad (c_2 = 0) \\ D_{12}(0) = 0 \quad (c_1 = 1) \end{cases} \tag{108}$$

R. A. Smith [5] has given and C. Yang [6] has worked out in great detail the solutions for D_{22} in three different situations for this case of zero detuning:

$\underline{|V| \rangle \frac{1}{2}\gamma}$

$$D_{22} = \frac{|V|^2/2}{2\gamma^2 + |V|^2}\left[1 - \left(\cos at + \frac{3\gamma}{2a}\sin at\right)\exp\left(-\frac{3}{2}\gamma t\right)\right] \tag{109}$$

where

$$a = \left(V^2 - \frac{\gamma^2}{4}\right)^{1/2} \tag{110}$$

$|V| = \frac{1}{2}\gamma$

$$D_{22} = \frac{1}{18}\left[1 - \left(\frac{3}{2}\gamma t + 1\right)\exp\left(-\frac{3}{2}\gamma t\right)\right] \tag{111}$$

$|V| \langle \frac{1}{2}\gamma$

$$D_{22} = \frac{|V|^2/2}{2\gamma^2 + |V|^2}\left[1 - \left(\cosh\frac{1}{2}a't + \frac{3\gamma}{a'}\sinh\frac{1}{2}a't\right)\exp\left(-\frac{3}{2}\gamma t\right)\right] \tag{112}$$

where

$$a' = \left(\gamma^2 - 4|V|^2\right)^{1/2} \tag{113}$$

For $\gamma = 0$, $a = |V| = \Omega$, and we recover the expression (49)

$$D_{22} = \frac{1}{2}\left(1 - \cos|V|t\right) = \sin^2\frac{1}{2}|V|t \tag{114}$$

In Fig. 5 we represent the variations of D_{22} as given by (109) for the case of zero detuning. We note the following:

1) The greater is γ the more damped are the oscillations. For $\eta = \frac{\gamma}{|V|} = \frac{1}{3}$ a single maximum remains.
2) The incident light must be of large enough intensity in order to make

$$\eta \langle\langle \frac{1}{3}, \text{i.e.} |V| \rangle\rangle 3\gamma$$

in order to generate significant oscillations in the populations of atoms in the two states.

3) The steady state value of D_{22}

$$\frac{|V|^2/2}{2\gamma^2 + |V|^2} = \frac{1}{2 + 4\eta^2} \tag{115}$$

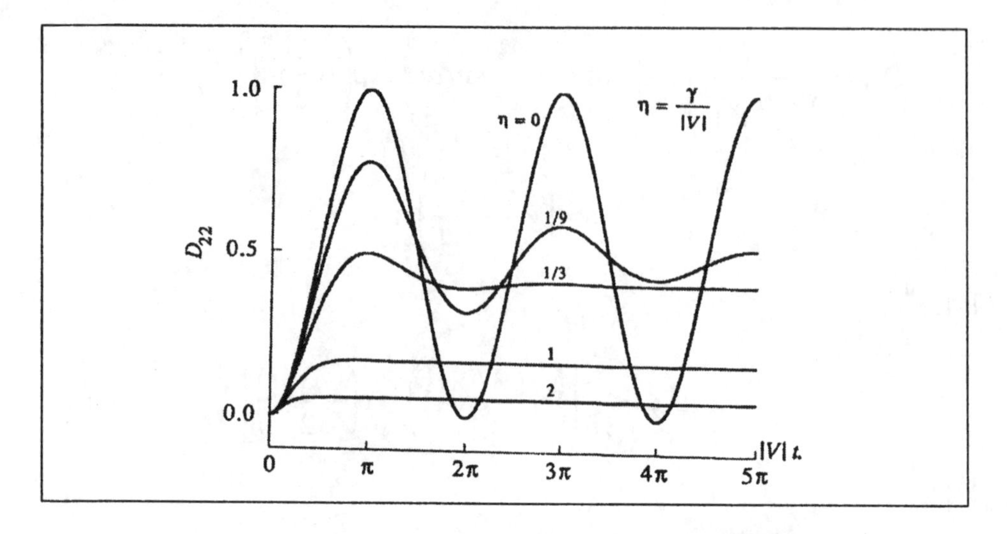

Figure 5: D_{22} for Zero Detuning and Various Values of $\eta = \dfrac{\gamma}{|V|}$

decreases with increasing η.

4) The oscillatory behavior of the populations is called *optical nutation.* The nutation frequency depends on $|V|$, detuning, and radiation damping.

5) The effect of a beam of light on atomic excitation can be summarized as follows:
   a) The atoms are initially in the ground state. The beam is attenuated as energy is transferred from the beam to the atoms.
   b) If the intensity of the beam is large enough so that

$$|V| \rangle (\omega_0 - \omega) \text{and} |V| \rangle\rangle \gamma$$

   the excited state component of the wavefunction will exceed the ground state component and energy is transferred back from the atom into the beam, increasing its initial intensity.
   c) This "cycle" repeats itself producing optical nutation. The oscillations in the atomic populations are accompanied by oscillations in the intensity of the transmitted light. Experimental observations of this type were made by MacGillivray and co-workers [18] who observed the effects of optical nutation in the D_2 transition of sodium atoms. A detailed theory of experiments of this type is given in [8].

2.6.6. *Collision Broadening*

Consider a gas of molecules and single out a molecule with velocity $\vec{v}$. We call $p(t)$ the probability that such a molecule does *not* collide in a time t. We call also wdt the probability that the molecule collides in a time in the interval $(t, t + dt)$: we shall assume $w = w(v)$ independent of past collisions, and then independent of time. We can write

$$p(t+dt)=p(t)(1-wdt) \tag{116}$$

or

$$p(t)+\frac{dp}{dt}dt=p(t)-p(t)wdt$$

$$\frac{1}{p}\frac{dp}{dt}=-w \tag{117}$$

Since v does not change in a time w^{-1} between collisions we can integrate and obtain

$$p(t)=e^{-wt} \tag{118}$$

The probability that a molecule, not having collided in time t, collides in $(t,t+dt)$ is

$$\mathsf{P}(t)dt=p(t)wdt=e^{-wt}wdt \tag{119}$$

We verify that

$$\int_0^\infty \mathsf{P}(t)dt=\int_0^\infty e^{-wt}wdt=1 \tag{120}$$

The mean time between collisions is

$$\tau=\bar{t}=\int_0^\infty p(t)tdt=\int_0^\infty e^{-wt}wtdt=\frac{1}{w} \tag{121}$$

Then we can write

$$p(t)dt=\frac{e^{-t/\tau}}{\tau}dt \tag{122}$$

Since $w=w(v)$, $\tau=\tau(v)$. We can then characterize a gas of atoms by the average collision time of the atoms traveling with the mean speed $\bar{v}$:

$$\tau=\tau(\bar{v}) \tag{123}$$

For a gas of N atoms of mass m and radius $\frac{d}{2}$ in a volume V we find [9]

$$\tau^{-1}=\frac{4d^2N}{V}\left(\frac{\pi kT}{m}\right)^{1/2} \tag{124}$$

In a gas of N_2 molecules at room temperature and atmospheric pressure, $\tau=6\times10^{-10}$ sec.

Collisions among atoms or molecules can be *elastic* or *inelastic*. An elastic collision leaves the atom in the same quantum state, but changes the phase of the atomic wavefunction. An inelastic collision produces a change in the state of the colliding atom: inelastic collisions are taken into account in the Bloch equations by suitably increasing the rate of decay γ.

We shall consider only the elastic collisions because it is found experimentally that they represent the dominant line-broadening mechanism in a very large number of cases and conditions. The presence of collisions of this type changes the off-diagonal terms D_{12} and D_{21}, but does not change D_{11} and D_{22}:

$$\begin{cases} \dot{D}_{22} = -\frac{i}{2}V^* e^{i(\omega_0-\omega)t} D_{12} + \frac{i}{2} V e^{-i(\omega_0-\omega)t} D_{21} - 2\gamma D_{22} \\ \dot{D}_{12} = \frac{i}{2} V e^{-i(\omega_0-\omega)t} (D_{11} - D_{22}) - \gamma' D_{12} \end{cases} \tag{125}$$

The steady state solutions of these equations are

$$\begin{cases} D_{22} = \dfrac{\left(|V|^2/4\right)\gamma'/\gamma}{(\omega_0-\omega)^2 + \gamma'^2 + \frac{1}{2}\left(\frac{\gamma'}{\gamma}\right)|V|^2} \\ D_{12} = e^{-i(\omega_0-\omega)t} \dfrac{(V/2)(\omega_0-\omega-i\gamma')}{(\omega_0-\omega)^2 + \gamma'^2 + \frac{1}{2}\left(\frac{\gamma'}{\gamma}\right)|V|^2} \end{cases} \tag{126}$$

The susceptibility can be rederived as follows:

$$\chi(\omega) = \frac{Ne^2|p_{12}|^2}{3\varepsilon_0 \hbar V} \frac{\omega_0 - \omega + i\gamma'}{(\omega_0-\omega)^2 + \gamma'^2 + \frac{\gamma'}{\gamma}\frac{|V|^2}{2}} \tag{127}$$

The linewidth is now given by

$$2\left[\gamma' + \frac{1}{2}\left(\frac{\gamma'}{\gamma}\right)|V|^2\right]^{1/2} \tag{128}$$

where

$$\gamma' = \gamma + \gamma_{coll} = \gamma + \frac{1}{\tau} \tag{129}$$

The linewidth (128) contains the effects of radiative broadening (γ), power broadening $(|V|)$ and collision broadening (τ).

2.6.7. *Doppler Broadening*

A photon of energy $\hbar\omega$ carries a momentum

$$\hbar\vec{k} = \hbar\frac{\omega}{c}\frac{\vec{k}}{|\vec{k}|} \tag{130}$$

An atom residing originally in the ground level absorbs this photon and moves to an excited level. Let E_1 and E_2 be the energies of the ground and excited level of the atom, respectively. Let also $\vec{v}_1$ and $\vec{v}_2$ be the velocities of the atom before and after the absorption of the photon.

Conservation of momentum and conservation of energy are expressed by the relations

$$\begin{cases} m\vec{v} = \hbar\vec{k} = m\vec{v}_2 \\ E_1 = \frac{1}{2}mv_1^2 + \hbar\omega = E_2 + \frac{1}{2}mv_2^2 \end{cases} \tag{131}$$

We shall call

$$\omega_0 = \frac{E_2 - E_1}{\hbar} \tag{132}$$

and we shall take the direction in which the proton travels as the x-direction. The above relations give us

$$\omega_0 = \omega - \frac{\omega v_{1x}}{c} - \frac{\hbar\omega^2}{2mc^2} \tag{133}$$

Typical values of some of the above quantities are

$$\frac{\omega v_{1x}}{c} \approx 10^{-8}$$

$$\frac{\hbar\omega^2}{2mc^2} \approx 10^{-9}$$

Then the third term in (133) is negligible with respect to the second term and we can write, dropping the subscript 1,

$$\omega = \frac{\omega_0}{1 - \frac{v_x}{c}} \approx \omega_0\left(1 + \frac{v_x}{c}\right) \tag{134}$$

A photon can be absorbed by an atom which has an x-component of the velocity v_x if its frequency ω is related with the frequency ω_0 of the atom by the relation (134).

The probability that an atom in a gas at temperature T has the x-component of its velocity in $(v_x, v_x + dv_x)$ is proportional to

$$\exp\left(-\frac{mv_x^2}{2kT}\right)dv_x = \exp\left(-\frac{mc^2(\omega-\omega_0)^2}{2\omega_0^2 kT}\right)dv_x \tag{135}$$

Let $g(\omega)$ be the profile of the spectral line. We can write

$$g(\omega)d\omega = g(v_x)dv_x \tag{136}$$

and, if we use (134),

$$g(\omega) = g(v_x)\frac{dv_x}{d\omega} = g(v_x)\frac{c}{\omega_0} \tag{137}$$

Then

$$\exp\left(-\frac{mv_x^2}{2kT}\right)dv_x = \frac{c}{\omega_0}\exp\left(-\frac{mc^2(\omega-\omega_0)^2}{2\omega_0^2 kT}\right)d\omega \qquad (138)$$

The profile represented by (138) is called a *Gaussian lineshape*. The FWHM (full width at half maximum height) of this line is

$$2\Delta\omega = 2\omega_0\left(\frac{2kT\ln 2}{mc^2}\right)^{1/2} \qquad (139)$$

The mean square spread is

$$\sigma = \left[\overline{(\omega-\omega_0)^2}\right]^{1/2} = \omega_0\left(\frac{kT}{mc^2}\right)^{1/2} = \frac{\Delta\omega}{(2\ln 2)^{1/2}} \qquad (140)$$

We can then write for a normalized Gaussian line:

$$f_G(\omega) = \frac{1}{\sqrt{2\pi\sigma^2}}\exp\left(-\frac{(\omega-\omega_0)^2}{2\sigma^2}\right) \qquad (141)$$

2.6.8. Composite Lineshape

Both the collision broadening and the Doppler broadening are proportional to $T^{1/2}$ and $m^{-1/2}$ and may be changed by changing the temperature of the gas. But the collision broadening is also proportional to the density N/V and, for this reason, is also called *pressure broadening*.

A *composite* lineshape may arise when two different mechanisms contribute to the broadening of a spectral line. If these two mechanisms produce, say, the profiles $F_1(\omega)$ and $F_2(\omega)$, then the profile of the composite line is the convolution of the two spectral functions:

$$f(\omega) = \int_{-\infty}^{+\infty} F_1(s)F_2(\omega-\omega_0-s)ds \qquad (142)$$

where ω_0 = common central frequency.

We can now make the following statements:

1) Any number of line-broadening mechanisms can be combined by repeated convolutions. The resultant lineshape is independent of the order in which the convolutions are performed.

2) For two Lorentzian broadening mechanisms giving width $2\gamma_1$ and $2\gamma_2$,

$$2\gamma = 2\gamma_1 + 2\gamma_2 \qquad (143)$$

3) For two Gaussian broadening mechanisms giving widths 2Δ, and $2\Delta_2$,

$$\Delta^2 = \Delta_1^2 + \Delta_2^2 \qquad (144)$$

4) The convolution of a Lorentzian and a Gaussian profiles is a *Voigt Profile* [10]. If $\Delta\nu_L$ and $\Delta\nu_G$ are the widths given by the Lorentzian and Gaussian mechanisms, respectively, the Voigt profile is given by

$$r_v(v) = \frac{2\ln 2}{\pi\sqrt{\pi}} \frac{\Delta v_L}{\Delta v_G} \int_{-\infty}^{+\infty} \frac{e^{-y^2}}{a^2 + (\Omega - y)^2} dy \tag{145}$$

where

$$\Omega = \frac{2v\sqrt{\ln 2}}{\Delta v_G} \tag{146}$$

$$a = \frac{\Delta v_L}{\Delta v_G} \ln 2 \tag{147}$$

Note that

$$r_v(v) \xrightarrow[\Delta v_L \to 0]{} \text{Gaussian lineshape}$$

$$r_v(v) \xrightarrow[\Delta v_G \to 0]{} \text{Lorentzian lineshape}$$

The Voigt profile is important because it occurs in the case when collision broadening and Doppler broadening are both significant and are affecting the spectral line independently. In Fig. 6 we report the three shapes of a Gaussian, a Lorentzian and a Voigt profiles with the same half width.

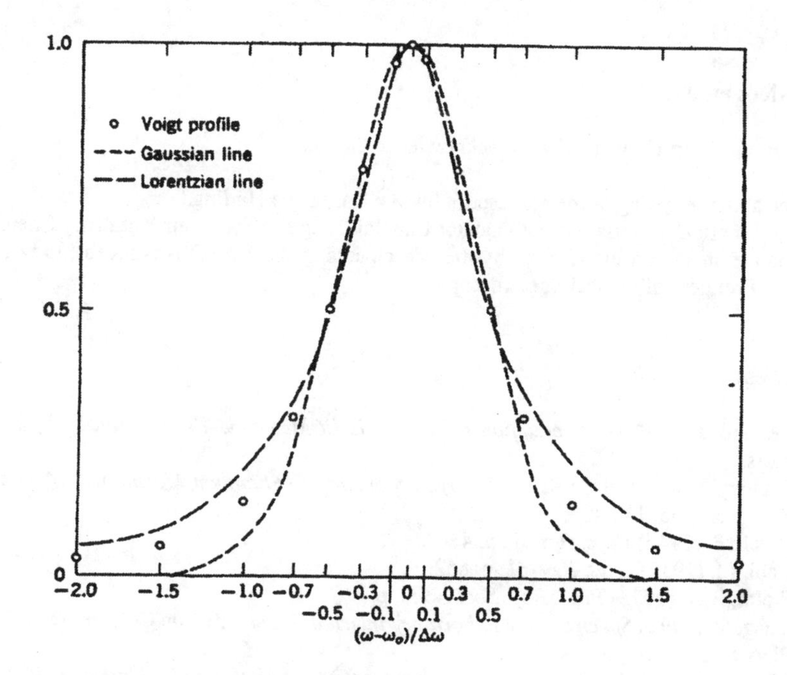

Figure 6: Lorentzian, Gaussian and Voigt Line Shapes

The mechanisms which broaden spectral atomic lines fall into two broad categories and situations:

1) Different atoms absorb or emit radiation at somewhat different ω 's. The profile is in this case Gaussian and the broadening is called *inhomogeneous*. Examples: Doppler broadening in gases and broadening of lines of laser ions in solids at very low temperatures.
2) Each atom absorbs or emits in the same way, with no particular frequency associated to a particular atom. The width is related to the time interval Δt in which the atom is left undisturbed:

$$\Delta\omega \, \Delta t \geq 1$$

The profile in this case is Lorentzian, and the broadening is called *homogeneous*. Examples: radiative broadening, collision broadening, broadening of lines of laser ions in solids at $T \rangle 77\text{K}$.

We shall report now the example of Neon atoms at $p = 0.5$ torr and room temperature. The two pressure independent mechanisms give the following contributions to the linewidth:

Doppler broadening:	1.7 GHz
Radiative broadening:	20 MHz

The pressure-dependent mechanism gives:

Collision broadening:	0.64 MHz

At sufficiently high pressures, the collision broadening predominates over the Doppler broadening.

Acknowledgements

The author of this article would like to acknowledge the following:

- the accurate typing of the manuscript by Ms. Kashawna Harling,
- the benefit of discussions with Doctor Chunlai Yang and Professor Xuesheng Chen,
- the comments on his lectures by the participants in the NATO Advanced Study Institute on "Frontiers of Optical Spectroscopy."

References

1. Bernard, C. (1984) *Introduction à l'Etude de la Médecine Expérimentale*, Flammarion, Paris.
2. Pauling, L. and Wilson, E.B. (1935) *Introduction to Quantum Mechanics*, McGraw Hill, New York, pp. 133-136.
3. Bloch, F. (1946) *Phys. Rev.* **70**, p. 460.
4. Rabi, I.I. (1937) *Phys. Rev.* **51**, p. 652.
5. Smith, R.A. (1978) *Proc. Roy. Soc.* **A362**, p. 1.
6. Yang, C. (1992) *Spectroscopy Laboratory Internal Report*, Boston College, Department of Physics.
7. MacGillivray, W.R., Pegg, D.T., and Standage, M.C. (1978) *Optics Commun.* **25**, p. 355.
8. Brewer, R.G. in *Nonlinear Spectroscopy*, Bloembergen N. editor (1977) North-Holland, Amsterdam, p. 355.
9. Reif, F. (1965) *Fundamentals of Statistical and Thermal Physics*, McGraw Hill, New York, p. 470.
10. Posener, D.W. (1959) *Austral J. Phys.* **12**, p. 184.

2. LIGHT–MATTER INTERACTIONS ON THE FEMTOSECOND TIME SCALE

C. A. D. Roeser and E. Mazur

Department of Physics and Division of Engineering and Applied Sciences
Harvard University
9 Oxford St.
Cambridge, MA 02138 USA

Abstract The subject of electromagnetism in the presence of matter is both extensively studied and rich in diverse phenomena. It spans such topics as the quantization of the electromagnetic field to the semiclassical treatment of light–matter interactions to the derivation of the Fresnel reflectivity formulas. Interest in femtosecond optics is rooted in nonlinear optical phenomena and in the complex electron and lattice dynamics that occur in a material following intense ultrashort-pulse irradiation. The experiments we discuss in this paper are concerned mainly with the latter and lie at the intersection between femtosecond optics and materials science.

1. Light–matter interactions

Fundamental to any description of light–matter interactions are Maxwell's equations [1],

$$\nabla \times \mathbf{E} = -\frac{\partial \mathbf{B}}{\partial t} \tag{1}$$

$$\nabla \times \mathbf{H} = -\frac{\partial \mathbf{D}}{\partial t} + \mathbf{J} \tag{2}$$

$$\nabla \cdot \mathbf{D} = \rho \tag{3}$$

$$\nabla \cdot \mathbf{B} = 0 \tag{4}$$

where, along with the usual field terms, $\mathbf{E}, \mathbf{D}, \mathbf{B}$, and $\mathbf{H}$, are the source terms of charge ρ and current $\mathbf{J}$. The influence of matter is cast in terms

B. Di Bartolo and O. Forte (eds.), Frontiers of Optical Spectroscopy, 29-54.

of constitutive relations among the fields,

$$\mathbf{B} = \mu_0 \mu \mathbf{H} \tag{5}$$
$$\mathbf{D} = \epsilon_0 \epsilon \mathbf{E} \tag{6}$$
$$\mathbf{J} = \sigma \mathbf{E} \tag{7}$$

for which the vacuum (matter-less) conditions are $\mu \rightarrow 1$, $\epsilon \rightarrow 1$, and $\sigma \rightarrow 0$. As written, the equations are essentially linear, in that an applied **E** field of frequency ω generates a **D** field in the bulk of a material at ω and no other frequency. To isolate the response of the material, we introduce the polarization **P**,

$$\mathbf{D} = \epsilon_0 \mathbf{E} + \mathbf{P} \tag{8}$$
$$\mathbf{P} = \epsilon_0 \chi^{(1)} \mathbf{E} \tag{9}$$

where the (linear) susceptibility $\chi^{(1)}$ is related to the dielectric constant by $\epsilon = \left(1 + \chi^{(1)}\right)$. Linear optical properties are fully described by either ϵ or $\chi^{(1)}$, which are complex, or by the complex index of refraction $\eta + i\kappa$.[1]

1.1 Relationship between linear optical properties and band structure

For a detailed description of light-matter interactions we refer the reader to Refs. [2] and [3]. Here we only highlight the aspects of the semiclassical desciption of light-matter interactions that are of particular relevance for this article.

While the electromagnetic field is treated classically, the electrons are governed by the Hamiltonian [2]

$$\mathcal{H} = \frac{\hat{\mathbf{p}}^2}{2m} + V(\hat{\mathbf{r}}) + \frac{e}{mc} \mathbf{A} \cdot \hat{\mathbf{p}} \tag{10}$$

where the first term is a kinetic energy term involving the momentum operator $\hat{\mathbf{p}}$, the second term is the electron–ion Coulomb interaction, and the third term encompasses the coupling between the applied field (represented by the vector potential **A**) and the electrons.[2] The eigenstates of the above system in the absence of the perturbing field **A** are the Bloch wavefunctions $|n, \mathbf{k}\rangle$, which in the position representation take the form [4]

$$\langle \mathbf{r} | n, \mathbf{k} \rangle = u_{n,\mathbf{k}}(\mathbf{r}) e^{i(\mathbf{k} \cdot \mathbf{r})}. \tag{11}$$

Here, $u_{n,\mathbf{k}}(\mathbf{r})$ is a function with the periodicity of the lattice potential $V(\mathbf{r})$, and n and **k** correspond to the band index and crystal momentum, respectively, in the reduced-zone scheme [4]. The energy eigenvalues

$E_n(\mathbf{k})$ constitute the band structure of the crystal.[3] The difference between the band structures of different materials arises from differences in their lattice potentials, due to variations in composition, lattice configuration, or both. Of particular interest to the experiments described in Section 4 is the fact that a lattice potential that is changing in time gives rise to a time-varying band structure. Ultrashort laser pulses allow one to track the major features of the band structure via their manifestation in the linear optical properties of the material.

To investigate the interaction of light with the system described by Eq. (10), we consider the situation where the applied field excites electrons from an occupied (valence band) state to an unoccupied (conduction band) state. The number and energy distribution of such transitions give rise to the optical properties of a solid. Specifically, the imaginary part of the dielectric tensor can be written as [2, 3]

$$\mathrm{Im}\left[\epsilon_i(\omega)\right] \sim \frac{1}{\omega^2} \sum_{n_c, n_v, \mathbf{k}} \delta\left(E_{n_c}(\mathbf{k}) - E_{n_v}(\mathbf{k}) - \hbar\omega\right) \left|\langle n_c, \mathbf{k}|\hat{\mathrm{p}}_i|n_v, \mathbf{k}\rangle\right|^2. \tag{12}$$

The momentum matrix element quantifies the strength of the coupling for vertical transitions between various conduction and valence band states.[4] The dependence of the momentum matrix element on the direction of the applied field, denoted by the subscript $i = x, y, z$, allows for an anisotropic optical response (*e.g.*, birefringence). The term common to all elements of the dielectic tensor is the joint density of states (JDOS)

$$\mathrm{JDOS} = \sum_{n_c, n_v, \mathbf{k}} \delta\left(E_{n_c}(\mathbf{k}) - E_{n_v}(\mathbf{k}) - \hbar\omega\right), \tag{13}$$

which depends solely on the shape of the band structure. The JDOS peaks at photon energies equal to the most common transition energy between different states in $\mathbf{k}$ space. The fact that parallel conduction and valence bands produce a large peak in $\mathrm{Im}[\epsilon(\omega)]$ is a direct consequence of the form of the JDOS in Eq. (13). In fact, the linear optical response of many solids is dominated by only a few peaks in their JDOS — that is, by resonances at a small number of photon energies produced by only a few regions of parallel bands.

It is important to note that the correspondence between band structure and dielectric function is not one-to-one. Many band structures can produce the same dielectric function,[5] which means that the interpretation of optical properties must be done cautiously. Changes in the linear optical properties can be used to make general statements about the changes in band structure, but additional information is often required to localize the dynamics in $\mathbf{k}$ space.

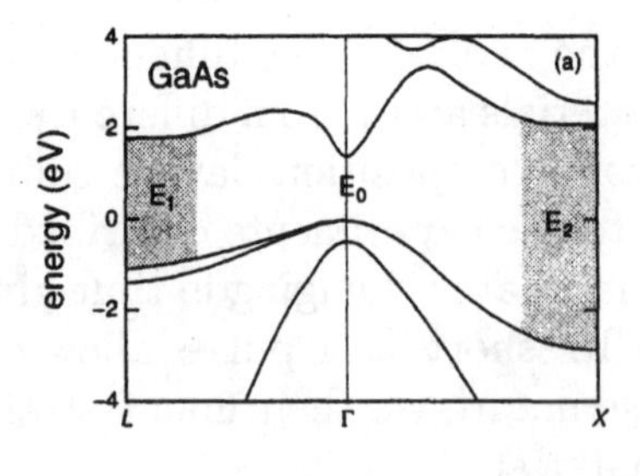

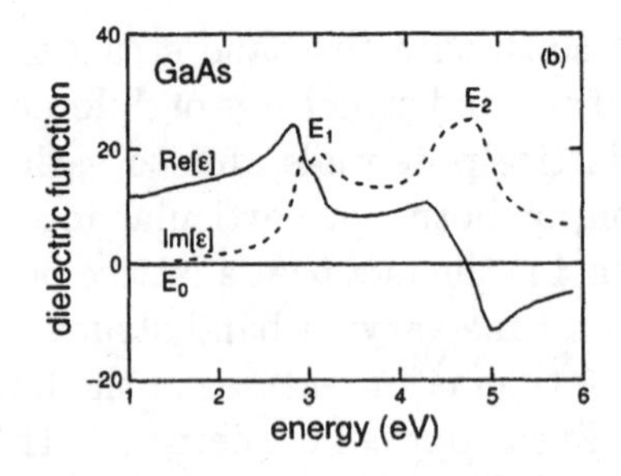

Figure 1. (a) Band structure [3] and (b) dielectric function [5] of GaAs.

As an example of the direct relationship between band structure and dielectric function, Figure 1 shows the band structure and the dielectric function of GaAs. The characteristic absorption peaks in $\mathrm{Im}[\epsilon(\omega)]$ at 3.1 eV (E_1) and 4.7 eV (E_2) are due in part to a large joint densities of states around the L and X valleys, as indicated by the shaded regions in Figure 1(a). The real part shows the characteristic dispersive structure for each absorption peak, in agreement with the Kramers–Kronig relations.

1.2 The Drude–Lorentz model

The Drude–Lorentz model, also known as the Lorentz oscillator model when applied to semiconductors and as the Drude model when applied to metals, attempts to describe the optical response of a material as that of a damped classical harmonic oscillator. While simple, involving only a few free parameters, the Drude–Lorentz model is surprisingly good at describing the optical properties of many semiconductors and metals.

The Lorentz model describes, in a phenomenological way, the polarization induced in a material by the applied $\mathbf{E}$ field. The situation we consider is that of an electron in a solid that is described by its displacement x from its equilibrium position.[6] The equation of motion for the displacement x is taken to be that of a harmonic oscillator,

$$\frac{d^2}{dt^2}x + \Gamma\frac{d}{dt}x + \omega_0^2 x = F(t) \tag{14}$$

where Γ is a phenomenological damping coefficient, ω_0 is the resonance frequency of the oscillator (a real resonance in the material), and the driving force is due to the applied field,

$$F(t) = \frac{e}{m}\left[Ee^{-i\omega t} + E^* e^{i\omega t}\right]. \tag{15}$$

Without loss of generality, the equation of motion for $x(t)$ can be solved by neglecting the second driving term above and considering a trial solution of the form $x(t) = Ce^{-i\omega t}$...

$$
\begin{aligned}
C[-\omega^2 - i\Gamma\omega + \omega_0^2]e^{-i\omega t} &= \frac{e}{m}Ee^{-i\omega t} && (16)\\
\Rightarrow C &= \frac{e}{m}E\frac{1}{\omega_0^2 - \omega^2 - i\Gamma\omega} && (17)\\
\Rightarrow x(t) &= \frac{e}{m}E\frac{e^{-i\omega t}}{\omega_0^2 - \omega^2 - i\Gamma\omega}\,. && (18)
\end{aligned}
$$

While $x(t)$ describes the motion of a single electron, it is often the case that many electrons in a solid respond in the same fashion. Thus, if N electrons respond as $x(t)$, then the total polarization is given by

$$P(t) = Nex(t) = \epsilon_0\chi^{(1)}E(t), \qquad (19)$$

where Eq. (9) is used to relate the applied field $E(t)$ to the polarization $P(t)$ and to introduce the linear susceptibility. Hence,

$$\chi^{(1)} = \frac{Ne}{\epsilon_0}\frac{x(t)}{E(t)} = \frac{Ne^2}{\epsilon_0 m}\frac{1}{\omega_0^2 - \omega^2 - i\Gamma\omega}\,. \qquad (20)$$

For our purposes, the dielectric function is more useful than the susceptibility, and takes the form,

$$
\begin{aligned}
\mathrm{Re}[\epsilon(\omega)] &= 1 + f\frac{E_{\mathrm{res}}^2 - (\hbar\omega)^2}{(E_{\mathrm{res}}^2 - (\hbar\omega)^2)^2 + (\Gamma\cdot\hbar\omega)^2} && (21)\\
\mathrm{Im}[\epsilon(\omega)] &= f\frac{\Gamma\cdot h\omega}{(E_{\mathrm{res}}^2 - (\hbar\omega)^2)^2 + (\Gamma\cdot\hbar\omega)^2}\,. && (22)
\end{aligned}
$$

There are two methods by which the Lorentz model is applied to real absorbing materials. First, because the Lorentzian shape of $\mathrm{Im}[\epsilon_{\mathrm{Lorentz}}(\omega)]$ is similar to a δ-function, materials can be modeled by a distribution of Lorentz oscillators, analogous to the distribution of δ-function contributions to the JDOS in Eq. (13). The sum of many oscillators would produce a "single resonance" in the material (*e.g.*, the E_2 peak in GaAs). However, for modeling changes in the dielectric function on a femtosecond time scale, this technique is numerically challenging to implement (due to noise in the data) as well as physically unsatisfying in the interpretation of its results.

A second method of applying the Lorentz model to real materials is to describe an entire resonance by the three free parameters of a single oscillator; the resonant frequency ω_0, the linewidth Γ, and the oscillator strength $f = Ne^2/\epsilon_0 m$. Each parameter is connected to features of the

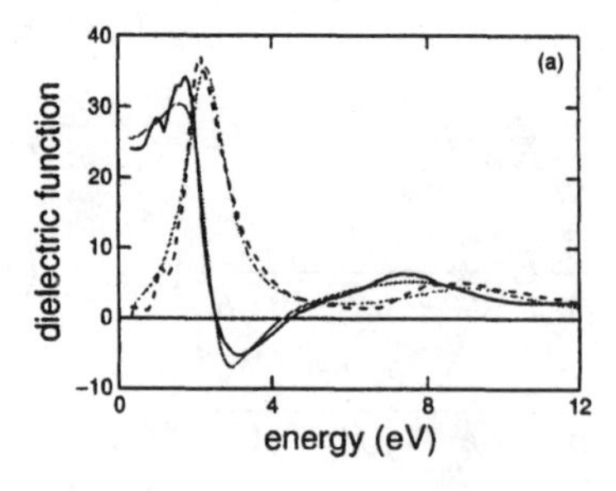

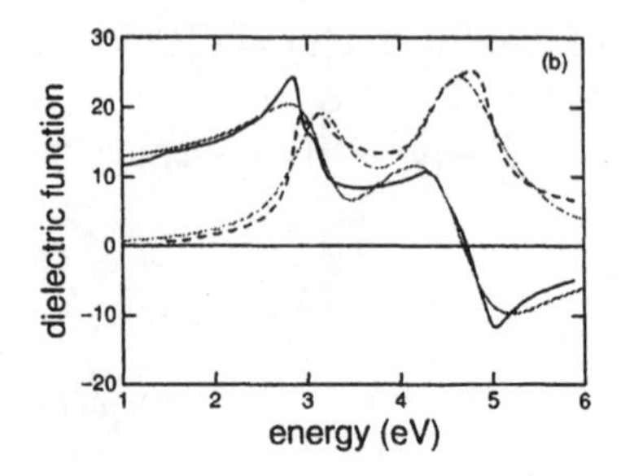

Figure 2. Lorentz oscillator model fits to (a) Te and (b) GaAs. Black lines represent literature values of Re[ϵ] (solid) and Im[ϵ] (dashed) [5], while gray lines represent the best-fit values of Re[$\epsilon_{\text{Lorentz}}$] (solid) and Im[$\epsilon_{\text{Lorentz}}$] (dash-dotted).

band structure. The resonant frequency ω_0 corresponds to the position of the peak in the JDOS. The linewidth Γ is related to the distribution of energy levels around the resonant frequency — sharper absorption lines correspond to smaller values of Γ, arising from regions of parallel bands. Lastly, the oscillator strength f carries information about the number of states contributing to the resonance at ω_0.

As an example of the success of the Lorentz model in describing real materials, Figure 2 shows fits to Te and to GaAs. In each case, the fit is the sum of two Lorentz oscillators with different values of ω_0, Γ, and f for each term. This two-oscillator model follows the major features of the literature optical properties in each case, but fails to capture smaller features. For example, Im[$\epsilon_{\text{Lorentz}}(\omega)$] does not vanish for photon energies below the band gap, nor is it sensitive to sharp features near other critical points. Nevertheless, the Lorentz model is sensitive to large resonances in a material — the parameters of the fit to GaAs in Figure 2(b) indicate resonances at 3.18 eV (E_1 peak) and 4.67 eV (E_2 peak).[7]

The form of the Drude model for metals is

$$\epsilon_{\text{Drude}}(\omega) = 1 + \frac{Ne^2}{\epsilon_0 m}\left(\frac{i\tau}{\omega(1 - i\omega\tau)}\right), \tag{23}$$

which is equivalent to Eqs. (21) and (22) with $\Gamma \to 1/\tau$ and $\omega_0 \to 0$. By convention, a plasma frequency $\omega_p^2 = Ne^2/\epsilon_0 m$ is defined to play the role of the oscillator strength above. The classical derivation of $\epsilon_{\text{Drude}}(\omega)$ is analogous to the above derivation for $\epsilon_{\text{Lorentz}}(\omega)$, except that an induced current $\mathbf{J}$ rather than an induced polarization $\mathbf{P}$ results in a differential equation that lacks a harmonic potential term [4]. The optical properties of many metals consist of a Drude (intraband) contribution from "free" electrons in half-filled bands in addition to Lorentz oscillator (interband) contributions from available vertical transitions. That is, even good

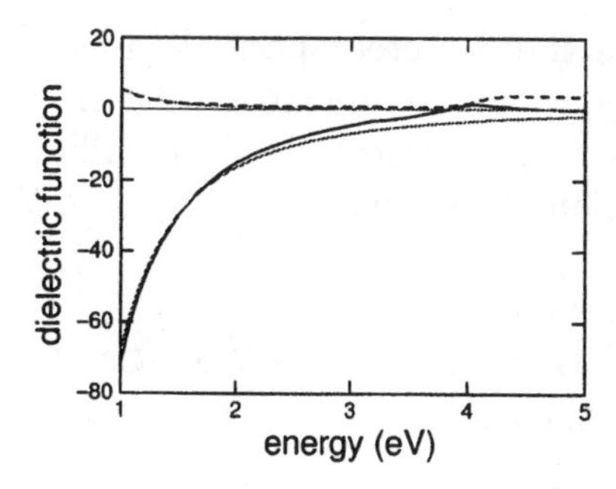

Figure 3. Drude model fit to Ag. Black lines represent literature values of Re[ϵ] (solid) and Im[ϵ] (dashed) [5], while gray lines represent the best-fit values of Re[$\epsilon_{\mathrm{Drude}}$] (solid) and Im[$\epsilon_{\mathrm{Drude}}$] (dash-dotted).

metals are rarely described by the Drude model alone. To illustrate this fact, Figure 3 shows a Drude model fit to Ag, with $\omega_{\mathrm{p}} = 8.3$ eV and $\tau = 50$ fs. While the fit describes low-photon-energy behavior well, it is not accurate near 4 eV due to resonant contributions to $\epsilon(\omega)$.

1.3 The Kramers–Kronig relations

Thus far we have discussed both the real and the imaginary part of the dielectric function as if the two quantities were independent. In reality, $\mathrm{Re}[\epsilon(\omega)]$ and $\mathrm{Im}[\epsilon(\omega)]$ are linked through the Kramers–Kronig relations [6]

$$\mathrm{Re}[\epsilon(\omega)] = 1 + \frac{2}{\pi}\mathcal{P}\int_0^\infty \mathrm{d}\nu\, \frac{\nu\,\mathrm{Im}[\epsilon(\nu)]}{\nu^2 - \omega^2} \tag{24}$$

$$\mathrm{Im}[\epsilon(\omega)] = \frac{2\omega}{\pi}\mathcal{P}\int_0^\infty \mathrm{d}\nu\, \frac{1 - \mathrm{Re}[\epsilon(\nu)]}{\nu^2 - \omega^2} \tag{25}$$

where $\mathcal{P}$ represents the principal value of the integral. It is worth noting that the Kramers–Kronig relations follow from the fact that the electric field $\mathbf{E}$ drives the material polarization $\mathbf{P}$, as in Eq. (9) [6]. Interestingly, the availability of transitions at a single photon energy contributes locally to $\mathrm{Im}[\epsilon(\omega)]$ (see Eq. (12)) but affects $\mathrm{Re}[\epsilon(\omega)]$ globally according to the Kramers–Kronig relations. For fitting dielectric function data to the Lorentz model when some of the material resonances lie outside the measured spectal range, the expression for the imaginary part will fit the data correctly due to the local contribution of transitions, but the real part will have unaccounted-for global contributions. It is often the case that an additive constant to the real part can dramatically improve the fit by playing the role of these Kramers–Kronig-type contributions from resonances outside the detected spectral range. Because the real part of the Lorentz model is mostly constant far from a resonance,[8] a single additive-constant free parameter is often sufficient to capture all the resonance contributions to $\mathrm{Re}[\epsilon(\omega)]$ from outside the spectral range of the data.

2. Ultrafast dynamics of solids under intense photoexcitation

Despite the broad array of topics that fall under the title "ultrafast dynamics of solids," the discussion below is of limited range. For instance, we do not discuss the myriad of electronic phenomena that have been observed at low excitation densities of 10^{14} to 10^{18} cm^{-3}, a review of which can be found in Ref. [7]. The reason for this omission is that such phenomena are rarely observed in experiments where the excited carrier density is on the order of 10^{22} cm^{-3} as in the experiments described here. Although excited carrier effects are present and are more pronounced than at lower densities, the material dynamics are often dominated by the ionic motion that results from excitation of a significant fraction of the valence electrons. The text that follows is an attempt to present the framework in which these dynamics are understood.

2.1 Molecular dynamics and coherent control

The idea that solid dynamics are determined by ionic motion is rooted in the microscopic picture of molecular electronic transitions and molecular dynamics. In fact, the molecular case is even more extreme than the solid one — photoexcitation in molecules often results in dissociation, whereas the sharing of electrons in a solid leads to only a partial weakening of bonds under photoexcitation.[9] Consider a diatomic sodium molecule, where the energy of the system as a function of nuclear displacement is as shown in Figure 4. The curves in this figure show the energy of the system in different electronic configurations as a function of ionic separation. Often, the excited state potentials have a shape that results in dissociation (no minimum at finite separations) or bond stretching (a minimum at a different separation than the ground state). When this is the case, electronic transitions are coupled to molecular vibrational transitions [8], where our intuition predicts that the ensuing nuclear motion is determined by the new potential in a classical way — an excitation from the ground $\mathrm{X}\,^1\Sigma_g^+$ state to the $2\,^1\Pi_g$ state of Figure 4 would leave the nuclei displaced from equilibrium and they thus begin to oscillate. The idea that the nuclei remain fixed during the electronic

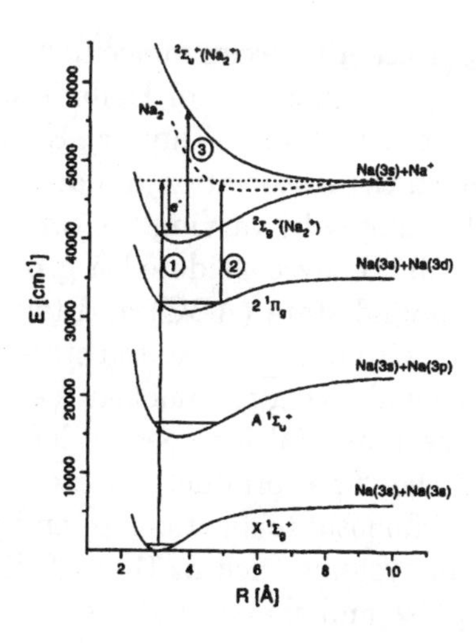

Figure 4. Potential energy curves for a diatomic sodium molecule, showing bond stretching or dissociation for different excited state potentials. Numbers 1, 2 and 3, indicate the possible transitions to products in a two-pulse excitation scheme after the first pulse excites the electronic system to $2\,^1\Pi_g$. After Ref. [9].

transition[10] is equivalent to the approximation of vertical transitions in crystals and is known as the Franck-Condon principle [8].

The quantum mechanical derivation of molecular excited state dynamics was first provided by Heller [10, 11]. To summarize, immediately after photoexcitation, the ground nuclear eigenstate evolves on the excited state potential surface following the classical trajectory. The anharmonicity of the excited state potential determines the rate at which the nuclear wavepacket spreads and leads to deviations from the classical trajectory. Of primary interest in our case is understanding the way in which the resulting nuclear dynamics can affect properties such as the dielectric function. Heller points out the lack of such a description at the time, stating [11]

> After the electrons have made a transition, the nuclei experience new forces; they find themselves displaced relative to the equilibrium geometry of the new potential surface, and interesting dynamics should ensue. Unfortunately, most discussions of electronic transitions cut short any allusions to dynamics and explain the absorption spectrum in terms of Franck-Condon overlaps of the initial nuclear wavefunction with a time-independent vibrational *eigenfunction* of the upper electronic potential surface. We (and the nuclear wave function) are left hanging; we are given no explanation of the time evolution of the hapless nuclei which, once the photon is absorbed, are ready to move in ways that *determine* the spectra.

Before making the connection between molecular and crystal dynamics, we discuss an important application of Heller's work — the coherent control of molecular dissociation. Tannor, Kosloff, and Rice devised a scheme under which the dissociation dynamics of a hypothetical molecule can be controlled simply by varying the time delay between two femtosecond pulses [12, 13]. They considered a ground state potential energy surface with one bound state (ABC) and two dissociated states (AB + C, A + BC), along with different excited state potential surfaces. They demonstrated that by allowing the nuclear wave function to propagate on the excited state potential for specific lengths of time before the second pulse arrives, the "final product" of the two-pulse excitation can be controlled.[11] Experimental realizations of end-product control in molecular dissociation (in systems such as that of Figure 4) have been achieved with multiple-pulse and shaped-pulse excitations [9, 14].

2.2 The molecular picture of crystal dynamics

Many of the features of the dynamics of solids can be understood within the framework described above, albeit with some extensions. Photoexcitation of a large density of electrons establishes a new potential energy surface on which the ions move. The new potential may have no minimum near the initial lattice configuration, resulting in large nuclear displacements, disordering, and often "damage." If a new potential minimum is established, then the ions can respond to the new potential in a more controlled fashion. Nuclear motion on the new potential energy surface leads to commensurate changes in the band structure, and, in turn, in the optical properties of the solid. That is, the available transitions for the electrons are determined by the lattice configuration, the dynamics of which are determined by the excited electrons.

Additional considerations when discussing solids concern the treatment of the "excited electronic state." First, the manifold of excited energy states is virtually a continuous function of the excited electron density. In general, the excited electron density cannot specify a unique potential because of the possible permutations of transitions among the 10^{23} cm^{-3} valence electrons. The idea that material dynamics depend on the excited electron density alone is an approximation that holds when the carriers can thermalize before any significant nuclear motion occurs. Excited electrons (holes) thermalize within 10 fs at densities of 10^{21} cm^{-3} or more,[12] leading to a Fermi-Dirac distribution within the conduction (valence) band and a loss of memory of the initial excited carrier configuration. Because the ions spend most of their time (all but 10 fs) evolving on a potential determined by a Fermi-Dirac distribution of carriers, the excited electron density is often sufficient to specify the "excited electronic state." A second concern is that the electronic state and the nuclear state do not evolve independently. A particular excited electron distribution establishes a potential surface to which the lattice responds by deformation. This deformation results in a new band structure, resulting in a redistribution of electrons and, in general, a modified potential. In essence, the excited electron potential becomes a dynamic quantity which depends on the nuclear coordinates. These many-body interactions are more pronounced in solids than in molecules and can dynamically modify the potential energy surface and further perturb the semiclassical trajectories of the ions. Electron-electron interactions of exchange and correlation, electron-phonon interactions, and other many-body interactions offer avenues by which the electron distribution within a band can exert a force on the ions.

Even with the further complications of dealing with solids, nuclear dynamics can still be treated with excited state potential surfaces. The accuracy of "exact" calculated results for solids is less than for molecules, simply due to the increased complexity of a condensed system. Usually, approximations to reduce the complexity of the system make the problem tractable and model a specific experimental situation. We present an example of such a treatment in the following section.

2.3 Ultrafast disordering of zincblende semiconductors

The observation of laser-induced disordering on a time scale shorter than the thermalization time between excited carriers and the lattice was observed by a number of groups working with semiconductors [15, 16, 17, 18, 19]. Although the disordering of solids via thermal processes (*i.e.*, melting) has been known for a long time, a theoretical description of lattice instability as a result of photoexcitation was first provided by Stampfli and Bennemann [20, 21]. Apparently derived independently from the work described above, their treatment of zincblende semiconductors shares many features with the molecular description of nuclear motion on an excited state potential.

Stampfli and Bennemann consider a tight-binding Hamiltonian that includes nearest-neighbor interactions only. A calculation of the potential surface for the lattice in the ground electronic state configuration is shown for silicon in Figure 5(a). A clear minimum exists as a function of transverse acoustic (δ_t) and longitudinal optical (δ_l) lattice distortions, indicating a stable lattice. The band structure and optical properties of the calculated ground state configuration agree well with the known

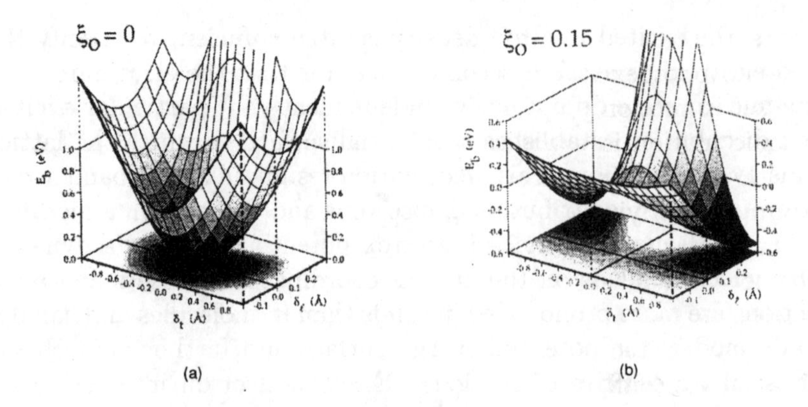

Figure 5. **Potential energy surface as a function of transverse acoustic (δ_t) and longitudinal optical (δ_l) lattice displacements for silicon (a) in the ground state electronic configuration and (b) when 15% of valence electrons are excited to the conduction band. The stable minimum at $\delta_t = \delta_l = 0$ becomes unstable for sufficient excitation densities, as shown here. After Ref. [21].**

properties of Si. After photoexcitation, the excited electrons rapidly take on a Fermi-Dirac distribution[13] and the lattice potential is recalculated with the different band-filling. As shown in Figure 5(b), when 15% of valence electrons are excited to the conduction band,[14] the lattice potential no longer displays a stable minimum at the ground state lattice configuration. The calculated trajectory of the ions is shown in Figure 6, displaying oscillations and translation in one direction and pure translation in another, following the shape of the potential in Figure 5(b). The significant motion of ions on a time scale of 100 fs agrees with the experimental results of many groups [15, 16, 17, 18, 19], and the onset of metallic behavior on this time scale has also been observed [22, 23].

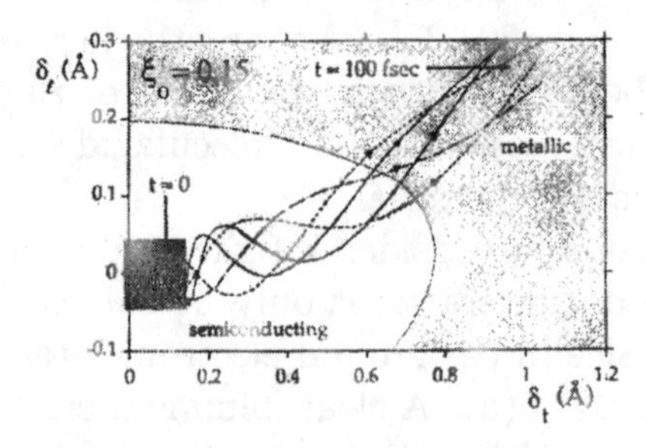

Figure 6. Trajectory of ionic motion on the potential shown in Figure 5(b). Significant nuclear displacements are predicted to occur within 100 fs. After Ref. [21].

The main difference between the response of silicon and that of Na_2 molecules to photoexcitation is that an ensemble of ions, rather than two, are set in motion on a new potential surface. As shown in Figure 6, the ionic motion is confined to a certain path for all initial conditions (within a certain range). As discussed in Section 4, the physical mechanisms underlying the response of tellurium to intense photoexcitation are similar to those that govern the response of silicon. Whereas nuclear motion in silicon develops along two directions, the photoinitiated nuclear motion in tellurium affects only a single lattice parameter. Moreover, the excited state potential in tellurium has a minimum near the initial lattice configuration, which leads to lattice vibrations that are analogous to the vibrations observed in Na_2.

3. Nonlinear optical properties

Although the majority of this article concerns linear optics in solids, nonlinear optical processes are encountered as well. A nonlinear optical response occurs when fields of frequency different than that of the applied field are generated. The nonlinear response is usually isolated from the linear one, and Eq. (9) becomes

$$\mathbf{P} = \epsilon_0 \chi^{(1)} \mathbf{E} + \mathbf{P}^{NL} \tag{26}$$

$$\mathbf{P}^{NL} = \epsilon_0 \chi^{(2)} : \mathbf{E} \cdot \mathbf{E} + \epsilon_0 \chi^{(3)} :: \mathbf{E} \cdot \mathbf{E} \cdot \mathbf{E} + \cdots . \tag{27}$$

Essentially, the material response is expanded in powers of the applied $\mathbf{E}$ field. Note that the total field is involved in driving the polarization and this total field may involve contributions from laser pulses of different frequency travelling in different directions with different field polarizations. The total polarization then acts as a driving term, as in Eq. (8), to radiate fields at the fundamental frequency[15] as well as at the frequency of the nonlinear polarization.

Broadly speaking, there are two criteria which must be satisfied in order to generate nonlinear radiation. For a nonlinear process involving n fields of the form[16]

$$E_j(t) = \frac{1}{2} \left(E_j e^{i(\vec{k}_j \cdot \mathbf{r} - \omega_j t)} + E_j^* e^{-i(\vec{k}_j \cdot \mathbf{r} - \omega_j t)} \right), \tag{28}$$

for $j = 1 \ldots n$, these criteria are

$$\omega_{NL} = \sum_{i=1}^{n} \pm \omega_i \tag{29}$$

$$\vec{k}_{NL} = \sum_{i=1}^{n} \pm \vec{k}_i \tag{30}$$

where the choice of sign indicates whether the "ω" or the "$-\omega$" term of a particular field contributes, as determined by the particular nonlinear process considered and the experimental arrangement. Equation (29) is always satisfied in a nonlinear process and specifies the nonlinear frequency generated. Equation (30) is the "phase-matching" condition and determines the direction of the radiated nonlinear field.[17] In the remainder of this section, we present an overview of second- and third-order nonlinearities and refer the interested reader to Ref. [24] for further details on optical nonlinearities.

3.1 Second-order nonlinearities

The type of second-order nonlinear process most often encountered in the lab is second harmonic generation (SHG), which involves two degenerate driving fields ($\omega_1 = \omega_2$) of the form in Eq. (28). The nonlinear polarization generated by SHG is

$$P^{(2)}(2\omega_1) = \epsilon_0 \chi^{(2)}(2\omega_1 : \omega_1, \omega_1) E_1 E_1 e^{i\left(\vec{k}(2\omega_1)\cdot \mathbf{r} - 2\omega_1 t\right)} \tag{31}$$

where the criteria of Eqs. (29) and (30) involve only positive terms (*e.g.*, $\omega_{NL} = \omega_1 + \omega_1$). We have allowed for less-than-perfect phase matching because the dispersion of the material determines whether $\vec{k}(2\omega_1) = 2\vec{k}(\omega_1)$. When the phase matching is not perfect, the nonlinear field generated at different positions in the crystal intefere somewhat destructively, reducing the total radiated nonlinear field.

The most common application of SHG is in measuring the duration of a femtosecond pulse via autocorrelation. In practice, two copies of the same pulse are overlapped in a nonlinear crystal with a controllable delay τ between the two pulses. Because the nonlinear field depends on the total intensity, the SHG signal S varies with the temporal overlap of the pulses. One can extract the pulse duration from the shape of $S(\tau)$.[18]

3.2 Third-order nonlinearities

In general, third-order nonlinearities require three driving fields. In practice, two or even all three fields are degenerate. One situation of particular interest is the following interaction between two fields of frequency ω_1 and ω_2

$$\begin{aligned} P^{(3)}(\omega_{NL}) &= \frac{1}{2}\epsilon_0 \chi^{(3)}(\omega_{NL} : \omega_1, -\omega_1, \omega_2) E_1 E_1^* E_2 e^{i\left(\vec{k}_2\cdot \mathbf{r} - \omega_2 t\right)} && (32) \\ \omega_{NL} &= \omega_1 - \omega_1 + \omega_2 && (33) \\ \vec{k}_{NL} &= \vec{k}_1 - \vec{k}_1 + \vec{k}_2 && (34) \end{aligned}$$

This particular nonlinear mixing can result in intensity-dependent effects. For $\chi^{(3)}$ real, an intensity-dependent index of refraction leads to self-focussing of a gaussian beam.[19] An imaginary $\chi^{(3)}$ produces intensity-dependent absorption (*i.e.* two-photon absorption). Both phenomena are widely applied in the field of optics. Self-focussing, or Kerr lensing, is used to mode-lock oscillators and contributes to the generation of white-light femtosecond pulses. Two-photon absorption (TPA) is commonly used for cross-correlation of ultrashort pulses because it is automatically phase-matched and produces a field at the fundamental.

Of particular interest to researchers is how nonlinear susceptibilities are related to (and can reveal) material properties. For instance, the process of two-photon absorption described in Eq. (32) does not occur unless a photon of energy $\hbar(\omega_1 + \omega_2)$ is absorbed linearly. This is not to say that $\text{Im}[\chi^{(3)}]$ depends on $\text{Im}[\chi^{(1)}]$. Rather, both depend on the availability and distribution of, in this case, vertical electronic transitions.

In addition to electronic transitions, the existence of lattice vibrations (phonons) serves to enhance nonlinear susceptibilities, in particular $\chi^{(3)}$, via the change in linear optical properties with lattice distortion. When the interaction in Eq. (32) is used to probe (or excite) a phonon ω_v of a solid where $\omega_1 - \omega_2 = \omega_\text{v}$, it is called a Raman interaction. The form of the phonon contribution to the nonlinear susceptibility is given by [24]

$$\chi^{(3)}_\text{Raman}(\omega_2 : \omega_1, -\omega_1, \omega_2) \sim \left(\frac{\partial \alpha}{\partial q}\right)^2 \frac{1}{\omega_\text{v}^2 - (\omega_1 - \omega_2)^2 + 2i\,(\omega_1 - \omega_2)\,\gamma}, \tag{35}$$

where q is a displacement of the lattice associated with the phonon and γ is the associated damping constant. By convention, ω_2 is referred to as the Stokes frequency when $\omega_2 < \omega_1$ and as the anti-Stokes frequency otherwise. The quantity α is the polarizability of the material, which changes as the lattice is distorted. Note that $\chi^{(3)}_\text{Raman}$ contains a resonance denominator of similar form to the linear susceptibility in Eq. (20), however the "strength" of the Raman process depends on the sensitivity of the polarizability to lattice displacement. Analogous to the electronic case, where an applied field induces an oscillating electronic polarization that leads to excitation of an electron, Raman interactions lead to the excitation of phonons.

4. Ultrafast Materials Science

Early investigations into ultrafast materials science relied on intense femtosecond laser pulses to initiate and probe dynamics that follow from photoinduced lattice instabilities [15, 25]. Recently, the focus has shifted from photoinduced instabilities that lead to a disordered state [18, 17, 19] to those that result in an altered lattice configuration [26, 27, 28] and to the methods by which lattice dynamics can be controlled [29, 30, 31, 32]. As discussed in Section 2, the nuclear motion is believed to follow a trajectory dictated by the potential surface of the electronic excited state [21], much like the semiclassical picture of nuclear dynamics in molecules [10, 12, 9, 14].

In this section, we discuss the electron and lattice dynamics of a variety of semiconductors following excitation by an intense femtosecond laser pulse. We observe these dynamics by measuring the dielectric tensor of the material with femtosecond time resolution. The linear optical properties of a material provide a view of the underlying band structure and lattice configuration that the reflectivity of a sample alone cannot. Consequently, measurements of the femtosecond time-resolved dielectric tensor provide a greater amount of information about electron and lattice dynamics and about the nature of ultrafast phase transitions than other optical probes.

We performed pump–probe experiments on commercially available GaAs, on *a*-GeSb thin films, and on single-crystal tellurium using 800-nm pulses from a multipass amplified Ti:sapphire laser, producing 0.5-mJ, 35-fs pulses at a repetition rate of 1 kHz [33]. In each case, *s* polarized pump pulses excite the sample while the *p* polarized transient reflectivity is measured using a white-light pulse (1.65 – 3.2 eV). Two-photon absorption measurements [34] indicate that the time-resolution of the pump–probe setup is better than 50 fs, while calculations based on measurements of the spectrum and chirp of the white-light probe indicate that the time resolution of the probe varies from 20 fs near 1.7 eV to 60 fs near 3.2 eV [35]. The entire system is calibrated to obtain absolute reflectivity. Measurements of the absolute reflectivity at two angles of incidence allow for determination of the linear optical properties by numerical inversion of the Fresnel formulas. Further details of this experimental technique can be found in Ref. [36].

4.1 Ultrafast carrier and lattice dynamics in GaAs

Shortly after the introduction of femtosecond laser sources, numerous experiments were conducted on semiconductors where a transition to a metallic state is observed upon laser irradiation. Experimental techniques included pump–probe reflectivity measurements [15], both reflectivity and second harmonic measurements [25, 18, 16, 17], and pump–probe microscopy [37, 38]. While each experiment reveals a laser-induced phase transition with high precision, the nature of the resulting phase and the changes in the band structure are difficult to determine. This difficulty is due to the fact that many different values of $\epsilon(\omega)$, and hence many different band structures and material phases, can yield the same reflectivity at a particular angle of incidence.

We performed single-shot femtosecond time-resolved dielectric function measurements of GaAs to investigate carrier and lattice dynamics associated with its ultrafast semiconductor-to-metal transition under intense photoexcitation [22, 39]. Figure 7 shows dielectric function measurements of GaAs. Without excitation of the sample, $\epsilon(\omega)$ matches literature values of the dielectric function [5], confirming that our technique measures the dielectric function correctly. Figure 7(b) shows $\epsilon(\omega)$ 500 fs after excitation below the threshold for permanent damage ($F_{\mathrm{th}} = 1.0$ kJ/m^2). Shortly after excitation, before the ions of the lattice can move, changes in $\epsilon(\omega)$ are due to the presence of excited carriers in the conduction band. The decrease of $\mathrm{Im}[\epsilon(\omega)]$ around the E_1 critical point (near 3 eV) is likely due to Pauli blocking of the transition by electrons in the conduction band. At higher excitation fluences, a transition to a metallic state is observed, an example of which is shown in Figure 7(c). This data is well fit by the Drude model, which describes free-electron (metallic) behavior. The parameters of the fit (a plasma frequency of 13 eV and a relaxation time of 0.18 fs) reveal that virtually all of the valence electrons are free and that the band gap has completely collapsed. Theoretical calculations of the evolution of the dielectric function of GaAs after femtosecond-pulse excitation agree with our experimental results [40, 41, 42].

4.2 Ultrafast phase changes in *a*-GeSb

The speed of ultrafast phase transitions and the large reflectivity variations associated with them make materials that display such transitions good candidates for optical switches and high speed optical data storage. Thin films of *a*-GeSb allow optically induced, optically reversible amorphous-to-crystalline transitions. In 1998, Sokolowski-Tinten and co-workers presented normal-incidence reflectivity measurements which suggested that femtosecond pulses above the threshold for permanent crystallization can induce an ultrafast *disorder-to-order* transition in amorphous $Ge_{0.06}Sb_{0.94}$ films within 200 fs [43]. The suggestion of a subpicosecond amorphous-to-crystalline phase transition raises an important question: how can lattice ordering occur in less time than it takes to establish thermal equilibrium between the laser-excited electrons and the lattice?

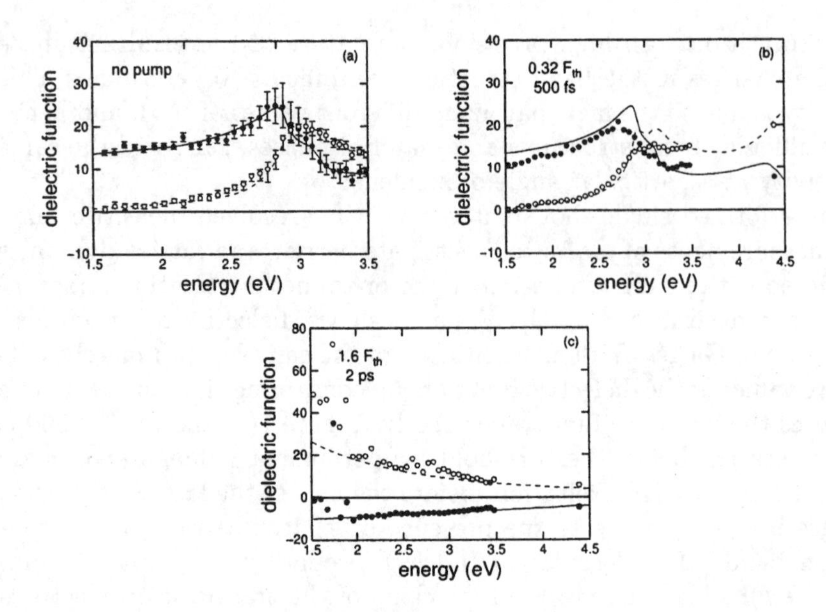

Figure 7. Dielectric function data for GaAs — • = Re[ϵ], ∘ = Im[ϵ]. (a) Under no excitation, $\epsilon(\omega)$ matches literature values of the dielectric function, represented by the solid and dashed curves [5]. An example of changes in $\epsilon(\omega)$ due to the presence of excited carriers is shown in (b). (c) At sufficiently high pump fluences, a semiconductor-to-metal transition is observed, as evidenced by the fit to the Drude model (ω_p = 13.0 eV and τ = 0.18 fs).

We performed single-shot dielectric function measurements of a 50-nm thin film of a-$Ge_{0.06}Sb_{0.94}$ to determine the nature of the phase during its ultrafast phase transition [44]. Figure 8(a) shows the agreement between $\epsilon(\omega)$ obtained at a time delay of −1 ps and literature values of the dielectric function [5]. As a reference, the dielectric function of the crystalline phase is also shown.[20] Because the film is optically thin and covered by a 1.25-nm SbO_2 oxide layer [45], this sample is considered a four-medium system: air, oxide, a-GeSb thin film, and fused silica substrate.

Figure 8(b) shows the response of the dielectric function 200 fs after arrival of a pump pulse of fluence $F = 320\ \mathrm{J/m^2}$, which is 60% above the threshold for permanent crystallization (F_{cr}). At this excitation fluence, the dielectric function remains unchanged from 200 fs to 475 ps. The same dielectric function is observed on subpicosecond time scales for all fluences above F_{cr}, indicating the existence of a nonthermal phase after femtosecond-pulse excitation. The existence of a new phase at ultrashort time delays for all fluences above F_{cr} was correctly identified by the authors of Ref. [43], however, the material is not crystalline, as evidenced

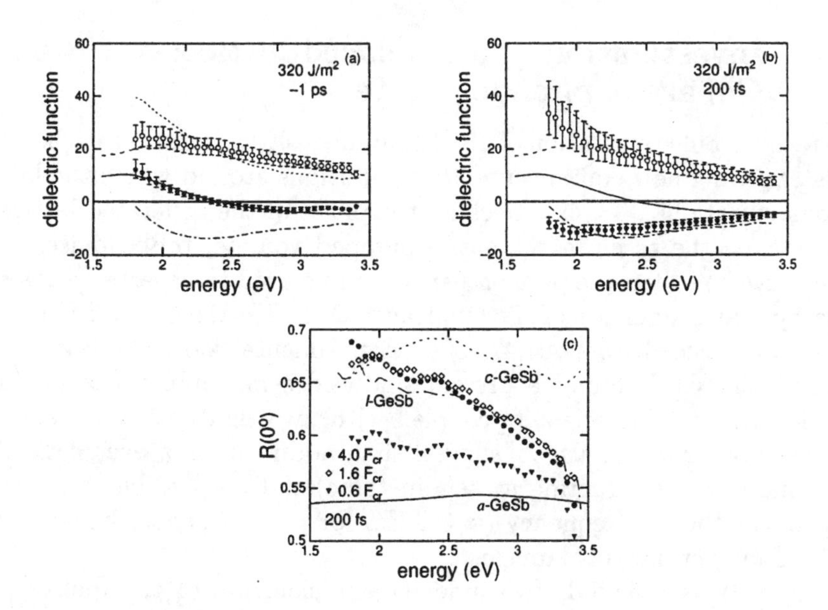

Figure 8. (a),(b) Dielectric function data for a-GeSb thin films — $\bullet = \mathrm{Re}[\epsilon(\omega)]$, $\circ = \mathrm{Im}[\epsilon(\omega)]$: (a) $\epsilon(\omega)$ under no excitation (-1 ps time delay), and (b) $\epsilon(\omega)$ 200 fs after excitation of 320 $\mathrm{J/m^2}$. In both plots, the solid and dashed curves show the real and imaginary parts of $\epsilon(\omega)$ for the amorphous phase from previous measurements, [46] and the dotted and dash-dotted curves show the real and imaginary parts of $\epsilon(\omega)$ of the crystalline phase. (c) Normal-incidence reflectivity calculated from the time-resolved $\epsilon(\omega)$ data. The reflectivity of the amorphous, crystalline, and liquid phases are shown for reference.

by the discrepancy between the measured dielectric function and that of the crystalline phase (see Figure 8(b)). This discrepancy is brought out by Figure 8(c), which shows the normal-incidence reflectivity as calculated from our time-resolved dielectric function measurements. Only at the 2.01-eV photon energy of the experiments in Ref. [43] does the reflectivity at 200 fs after excitation above F_{cr} match that of the crystalline phase. Furthermore, even at 2.01 eV, we find that for angles of incidence near or above the pseudo-Brewster, the reflectivity does not go to the crystalline level for pump fluences above F_{cr}. Our measurements thus show that broadband measurements of $\epsilon(\omega)$ enable one to distinguish phases that may appear the same based on reflectivity or transmission for a single photon energy at a single angle of incidence.

4.3 Investigation of a displaced lattice: Coherent phonons in Te

Ultrashort-pulse excitation of Te instantaneously weakens lattice bonding, establishing new equilibrium lattice positions around which the lattice ions vibrate [47, 48, 49]. Because the phase of the generated lattice oscillations is the same in the entire pumped volume, probe pulses of shorter duration than the phonon period can be used to observe changes in the optical properties of Te (typically, $\Delta R/R \sim 10\%$) at different degrees of lattice distortion [47, 50]. Experimental work by Bardeen [51] and others [52] found a pressure-induced semiconductor-to-metal transition in Te. These results coupled with investigations of coherent phonons in other materials [53] suggest that modification or even control of the phase (semiconducting *vs.* semimetallic) of Te is possible at a rate equal to the phonon frequency (≈ 3 THz) for pump fluences below the threshold for permanent damage.

Because Te is uniaxial, two independent elements of the dielectric tensor must be measured to fully characterize the ultrafast material response, as described in Ref. [36]. Figure 9 shows the excellent agreement between measured and literature values [5] of both the ordinary and extraordinary dielectric functions. This agreement not only validates the technique for uniaxial materials, it also shows that no cumulative irradiation effects arise from operating in a configuration where the sample is not translated between laser pulses.

The dynamics of $\epsilon_{\mathrm{ord}}(\omega)$ are shown in Figure 10. Within the error of the measurement, dielectric function values remain constant at all times before the pump arrives. After excitation, the oscillatory behavior of the

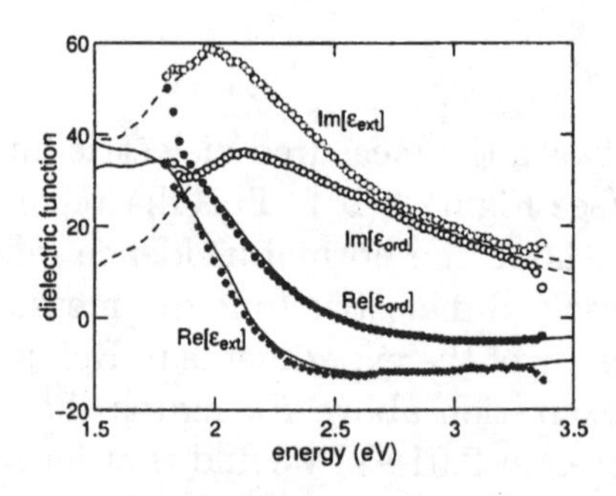

Figure 9. Dielectric tensor data for Te — • = Re[ϵ], ∘ = Im[ϵ]. At −500 fs time delay, both the (a) ordinary and (b) extraordinary dielectric function agree with literature values for the dielectric tensor, represented by the solid and dashed curves. [5]

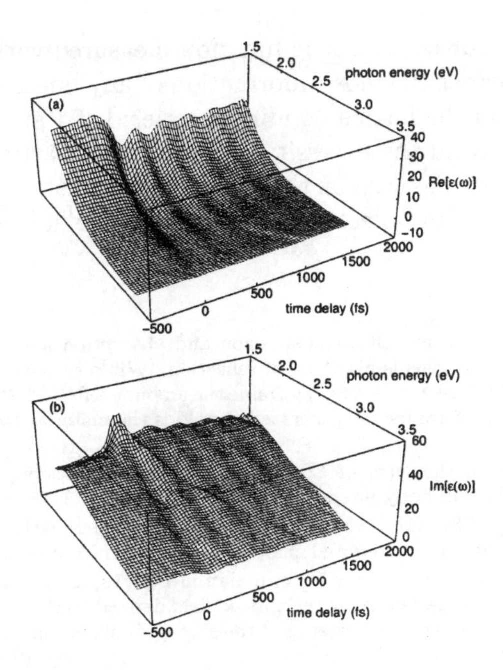

Figure 10. Dynamics of the ordinary dielectric function of Te for excitation with a fluence of 120 J/m^2. Both (a) the real part and (b) the imaginary part show oscillatory behavior due the excitation of coherent phonons.

optical properties indicate the presense of coherent phonons. A decaying offset from the initial values, separate from the oscillation, represents the relaxation of the equilibrium lattice spacing as electrons diffuse from the probed region. In contrast to reflectivity-only studies of coherent phonons in materials, the dielectric function data clearly indicate a shift of absorption resonances to lower photon energies. The broad resonance near 2 eV has moved to lower energies, as indicated by the shift in the peak of $\mathrm{Im}[\epsilon_{\mathrm{ord}}(\omega)]$ and the zero of $\mathrm{Re}[\epsilon_{\mathrm{ord}}(\omega)]$. The magnitude and direction of the shift suggest that the lattice may be sufficiently displaced at the peak of the phonon oscillation to cross the conduction and valence bands, but that the short duration of the crossing could prohibit any metallic character from emerging [54]. In addition, we observe larger changes in $\epsilon_{\mathrm{ord}}(\omega)$ than in $\epsilon_{\mathrm{ext}}(\omega)$, which may be attributed to the fact that the motion of the coherent phonons is confined to the *ab* plane.

5. Summary

The availability of electronic transitions, the existence of vibrational modes, and the dynamics of nuclei all influence the optical properties of solids. The time-resolved dielectric function measured with a reflectometry technique provides the most information of any linear optical probe, revealing changes in the lattice bonding, carrier distribution, and phase of a material. We avoid the necessity of assuming a particular model of the material dynamics as well as the potential pitfalls of other methods that measure changes in reflectivity at a single photon energy.

Notes

1. Although it seems obvious that transmission and absorption are determined by the "bulk" properties of a solid, this is also true of reflection. While surface quality affects the amount of scattered light, bulk properties determine the amount reflected. This is emphasized in Eq. (8), where a driving term for the generated **D** field is the material polarization created by the applied **E** field.

2. It is apparent from the form of Eq. (10) that electron–electron, electron–phonon, electron–hole, and other multibody interactions are not considered here.

3. The Hamiltonian of Eq. (10), while ignoring important contributions from multibody interactions, captures many of the essential characteristics of semiconductor band structures.

4. Although the coupling term of the Hamiltonian in Eq. (10) is proportional to the field **A**, the material properties experienced by the applied field depend on its direction rather than its magnitude. Any field-strength dependence of the susceptibility results in a fundamentally nonlinear system.

5. The dependence of the dielectric function on the JDOS (Eq. (12)) illustrates how the essential distinction among band structures — their k-dependence — is lost in the summation.

6. We take the ions to be fixed in this derivation.

7. For completeness, the parameters of the fit to Te are $\omega_1 = 2.37$ eV, $\Gamma_1 = 1.28$ eV, $f_1 = 105$ eV^2, $\omega_2 = 9.39$ eV, $\Gamma_2 = 4.43$ eV, $f_2 = 170$ eV^2, and an additive constant to the real part of 3.66. The parameters of the fit to GaAs are $\omega_1 = 3.18$ eV, $\Gamma_1 = 0.75$ eV, $f_1 = 38.7$ eV^2, $\omega_2 = 4.67$ eV, $\Gamma_2 = 1.14$ eV, $f_2 = 125$ eV^2, and an additive constant to the real part of 1.85. The real additive constant arises from Kramers–Kronig-type contributions from resonances outside the spectral range of the fit, as discussed in Section 1.3.

8. The real part varies significantly within a linewidth Γ of the resonance frequency ω_0.

9. It is rarely the case that a pump pulse is intense enough to provide one photon for every valence electron in the pumped volume of the solid.

10. This is also representative of the Born-Oppenheimer approximation, where the electronic quantum numbers are the so-called "fast variables" and the nuclear positions are the "slow variables."

11. For an anharmonic excited state potential, the ability to control the end products is reduced, essentially because the projection of a spread nuclear wave function onto the ground state potential can "split" the wave function between the two end products.

12. At densities of 10^{18} cm^{-3}, carrier thermalization occurs within hundreds of femtoseconds [55]. Extrapolation of the results of Becker *et al.* [55] to 10^{21} cm^{-3} gives a thermalization time on the order of 10 fs.

13. Immediately after photoexcitation, the excited electrons are distributed in the conduction band according to the pump spectrum.

14. In practice, the excitation of 15% of valence electrons to the conduction band is rarely achieved in pump–probe experiments. Excitation of a few percent of valence electrons is both more common and often sufficient to initiate phase transitions.

15. The "fundamental" frequency refers to the center or carrier frequency of the applied **E** field.

16. Here and for the remainder of the discussion we ignore the polarization of the applied **E** field.

17. Many details of phase-matching are omitted here because they are beyond the scope of this article. See Refs. [24] and [56].

18. The ease of such a measurement makes it attractive, but it does not fully characterize the temporal profile of the laser pulse [57].

19. For self-focussing, and any other self-action effects, $\omega_1 = \omega_2$.

20. Literature values of $\epsilon(\omega)$ for c-$Ge_{0.06}Sb_{0.94}$ are not available. The data presented are measurements taken in our apparatus of a region of the sample that was permanently crystallized by laser irradiation.

References

[1] H. A. Haus, *Waves and Fields in Optoelectronics.* Englewood Cliffs, NJ: Prentice-Hall, 1st ed., 1984.

[2] P. Y. Yu and M. Cardona, *Fundamentals of Semiconductors.* Berlin: Springer-Verlag, 1996.

[3] M. L. Cohen and J. Chelikowsky, *Electronic Structure and Optical Properties of Semiconductors.* Berlin: Springer Verlag, 2nd ed., 1989.

[4] N. W. Ashcroft and N. D. Mermin, *Solid State Physics.* Philadelphia: Saunders College, 1976.

[5] E. D. Palik, *Handbook of Optical Constants of Solids.* New York: Academic Press, 1985.

[6] J. D. Jackson, *Classical Electrodynamics.* New York: Wiley, 2nd ed., 1975.

[7] J. Shah, *Ultrafast Spectroscopy of Semiconductors and Semiconductor Nanostructures.* Berlin: Springer-Verlag, 1996.

[8] P. W. Atkins, *Molecular Quantum Mechanics.* New York: Oxford University Press, 1983.

[9] A. Assion, T. Baumert, J. Helbing, V. Seyfried, and G. Gerber, "Coherent control by a single phase shaped femtosecond laser pulse," *Chem. Phys. Lett.*, vol. 259, p. 488, 1996.

[10] E. J. Heller, "Quantum corrections to classical photodissociation models," *J. Chem. Phys.*, vol. 68, p. 2066, 1978.

[11] E. J. Heller, "The semiclassical way to molecular spectroscopy," *Acc. Chem. Res.*, vol. 14, p. 368, 1981.

[12] D. J. Tannor, R. Kosloff, and S. A. Rice, "Coherent pulse sequence induced control of selectivity of reactions: Exact quantum mechanical calculations," *J. Chem. Phys.*, vol. 85, p. 5805, 2086.

[13] S. A. Rice and M. Zhao, *Optical Control of Molecular Dynamics.* New York: John Wiley and Sons, 2000.

[14] T. Brixner, N. H. Damrauer, P. Niklaus, and G. Gerber, "Photoselective adaptive femtosecond quantum control in the liquid phase," *Nature*, vol. 414, p. 57, 2001.

[15] C. V. Shank, R. Yen, and C. Hirlimann, "Time-resolved reflectivity measurements of femtosecond-optical-pulse-induced phase transitions in silicon," *Phys. Rev. Lett.*, vol. 50, p. 454, 1983.

[16] H. W. K. Tom, G. D. Aumiller, and C. H. Brito-Cruz, "Time-resolved study of laser-induced disorder of Si surfaces," *Phys. Rev. Lett.*, vol. 60, p. 1438, 1988.

[17] S. V. Govorkov, I. L. Shumay, W. Rudolph, and T. Schröder, "Time-resolved second-harmonic study of femtosecond laser-induced disordering of GaAs surfaces," *Opt. Lett.*, vol. 16, p. 1013, 1991.

[18] P. Saeta, J.-K. Wang, Y. Siegal, N. Bloembergen, and E. Mazur, "Ultrafast electronic disordering during femtosecond laser melting of GaAs," *Phys. Rev. Lett.*, vol. 67, p. 1023, 1991.

[19] K. Sokolowski-Tinten, H. Schulz, J. Bialkowski, and D. von der Linde, "Two distinct transitions in ultrafast solid-liquid phase transformations of GaAs," *Appl. Phys. A*, vol. 53, p. 227, 1991.

[20] P. Stampfli and K. H. Bennemann, "Theory for the instability of the diamond structure of Si, Ge, and C induced by a dense electron-hole plasma," *Phys. Rev. B*, vol. 42, p. 7163, 1990.

[21] P. Stampfli and K. H. Bennemann, "Time dependence of the laser-induced femtosecond lattice instability of Si and GaAs: Role of longitudinal optical distortions," *Phys. Rev. B*, vol. 49, p. 7299, 1994.

[22] J. P. Callan, A. M.-T. Kim, L. Huang, and E. Mazur, "Ultrafast electron and lattice dynamics in semiconductors at high excited carrier densities," *Chem. Phys.*, vol. 251, p. 167, 1998.

[23] J. P. Callan, A. M.-T. Kim, C. A. D. Roeser, and E. Mazur, "Universal dynamics during and after ultrafast laser-induced semiconductor-to-metal transitions," *Phys. Rev. B*, vol. 64, p. 073201, 2001.

[24] R. W. Boyd, *Nonlinear Optics*. New York: Academic Press, 2nd ed., 2003.

[25] C. V. Shank, R. Yen, and C. Hirlimann, "Femtosecond-time-resolved surface structural dynamics of optically excited silicon," *Phys. Rev. Lett.*, vol. 51, p. 900, 1983.

[26] O. V. Misochko, M. Tani, K. Sakai, S. Nakashima, V. N. Andreev, and F. A. Chudnovsky, "Phonons in V_2O_3 above and below the mott transition: a comparison of time- and frequency-domain spectroscopy results," *Physica B*, vol. 263-264, p. 57, 1999.

[27] A. Cavalleri, C. Tóth, C. W. Siders, J. A. Squier, F. Ráski, P. Forget, and J. C. Kieffer, "Femtosecond structural dynamics in VO_2 during an ultrafast solid-solid phase transition," *Phys. Rev. Lett.*, vol. 87, p. 237401, 2001.

[28] K. Sokolowski-Tinten, C. Blome, J. Blums, A. Cavalleri, C. Dietrich, A. Tarasevitch, I. Uschmann, E. Förster, M. Kammler, M. Horn-von-Hoegon, and D. von der Linde, "Femtosecond x-ray measurement of coherent lattice vibrations near the Lindemann stability limit," *Nature*, vol. 81, p. 3679, 1998.

[29] M. W. Wefers, H. Kawashima, and K. A. Nelson, "Optical control over femtosecond polarization dynamics," *J. Phys. Chem. Solids*, vol. 57, p. 1425, 1995.

[30] M. Hase, K. Mizoguchi, H. Harima, S. Nakashima, M. Tani, K. Sakai, and M. Hangyo, "Optical control of coherent optical phonons in bismuth films," *Appl. Phys. Lett.*, vol. 69, p. 2474, 1996.

[31] A. Bartels, T. Dekorsy, H. Kurz, and K. Köhler, "Coherent control of acoustic phonons in semiconductor superlattices," *Appl. Phys. Lett.*, vol. 72, p. 2844, 1998.

[32] M. F. DeCamp, D. A. Reis, P. H. Bucksbaum, and R. Merlin, "Dynamics and coherent control of high-amplitude optical phonons in bismuth," *Phys. Rev. B*, vol. 74, p. 738, 2001.

[33] S. Backus, J. Peatross, C. P. Huang, M. M. Murnane, and H. C. Kapteyn, "Ti:sapphire amplifier producing millijoule-level, 21-fs pulses at 1 kHz," *Opt. Lett.*, vol. 20, p. 2000, 1995.

[34] T. F. Albrecht, K. Seibert, and H. Kurz, "Chirp measurement of large-bandwidth femtosecond optical pulses using two-photon absorption," *Opt. Commun.*, vol. 84, p. 223, 1991.

[35] S. A. Kovalenko, A. L. Dobryakov, J. Ruthmann, and N. P. Ernsting, "Femtosecond spectroscopy of condensed phases with chirped supercontinuum probing," *Phys. Rev. A*, vol. 59, p. 2369, 1999.

[36] C. A. D. Roeser, A. M.-T. Kim, J. P. Callan, L. Huang, E. N. Glezer, Y. Siegal, and E. Mazur, "Femtosecond time-resolved dielectric function measurements by dual-angle reflectometry," *Rev. Sci. Instrum.*, vol. 74, p. 3413, 2003.

[37] M. C. Downer, R. L. Fork, and C. V. Shank, "Femtosecond imaging of melting and evaporation at a photoexcited silicon surface," *J. Opt. Soc. Am. B*, vol. 2, p. 595, 1985.

[38] K. Sokolowski-Tinten, J. Bialkowski, A. Cavalleri, D. von der Linde, A. Oparin, J. Meyer-ter-Vehn, and S. I. Anisimov, "Transient states of matter during short pulse laser ablation," *Phys. Rev. Lett.*, vol. 81, p. 224, 1998.

[39] L. Huang, J. P. Callan, E. N. Glezer, and E. Mazur, "GaAs under ultrafast excitation: Response of the dielectric function," *Phys. Rev. Lett.*, vol. 80, p. 185, 1998.

[40] J. S. Graves and R. E. Allen, "Response of GaAs to fast intense laser pulses," *Phys. Rev. B*, vol. 58, p. 13627, 1998.

[41] R. E. Allen, T. Dumitrica, and B. Torralva in *Ultrafast Processes in Semiconductors* (K.-T. Tsen, ed.), New York: Academic Press, 2000.

[42] L. X. Benedict, "Dielectric function for a model of laser-excited GaAs," *Phys. Rev. B*, vol. 63, pp. 075202–1, 2001.

[43] K. Sokolowski-Tinten, J. Solís, J. Bialkowski, J. Siegel, C. N. Afonso, and D. von der Linde, "Dynamics of ultrafast phase changes in amorphous GeSb films," *Phys. Rev. Lett.*, vol. 81, p. 3679, 1998.

[44] J. P. Callan, A. M.-T. Kim, C. A. D. Roeser, E. Mazur, J. Solis, J. Siegel, C. N. Afonso, and J. C. G. de Sande, "Ultrafast laser-induced phase transitions in amorphous GeSb films," *Phys. Rev. Lett.*, vol. 86, p. 3650, 2001.

[45] J. C. G. de Sande, F. Vega, C. N. Afonso, C. Ortega, and J. Siejka, "Optical properties of Sb and SbO_x films," *Thin Solid Films*, vol. 249, p. 195, 1994.

[46] J. Solís 2001. private communication.

[47] T. K. Cheng, S. D. Brorson, A. S. Kazeroonian, J. S. Moodera, G. Dresselhaus, M. S. Dresselhaus, and E. P. Ippen, "Impulsive excitation of coherent phonons observed in reflection in bismuth and antimony," *Appl. Phys. Lett.*, vol. 57, p. 1004, 1990.

[48] H. J. Zeiger, J. Vidal, T. K. Cheng, E. P. Ippen, G. Dresselhaus, and M. S. Dresselhaus, "Theory for displacive excitation of coherent phonons," *Phys. Rev. B*, vol. 45, p. 768, 1992.

[49] P. Tangney and S. Fahy, "Density-functional theory approach to ultrafast laser excitation of semiconductors: Application to the a_1 phonon in tellurium," *Phys. Rev. B*, vol. 65, p. 054302, 2002.

[50] S. Hunsche, K. Wienecke, T. Dekorsy, and H. Kurz, "Impulsive softening of coherent phonons in tellurium," *Phys. Rev. Lett.*, vol. 75, p. 1815, 1995.

[51] J. Bardeen, "Pressure change of resistance of tellurium," *Phys. Rev.*, vol. 75, p. 1777, 1949.

[52] F. A. Blum and B. C. Deaton, "Properties of the group VI B elements under pressure. II. Semiconductor-to-metal transition of tellurium," *Phys. Rev.*, vol. 137, p. A1410, 1965.

[53] T. K. Cheng, L. H. Acioli, J. Vidal, H. J. Zeiger, G. Dresselhaus, M. S. Dresselhaus, and E. P. Ippen, "Modulation of a semiconductor-to-semimetal transition at 7 THz via coherent lattice vibrations," *Appl. Phys. Lett.*, vol. 62, p. 1901, 1993.

[54] A. M.-T. Kim, C. A. D. Roeser, and E. Mazur, "Modulation of the bonding-antibonding splitting in Te by coherent phonons," *Phys. Rev. B*, vol. 68, p. 012301, 2003.

[55] P. C. Becker, H. L. Fragnito, C. H. Brito-Cruz, R. L. Fork, J. E. Cunningham, J. E. Henry, and C. V. Shank, "Femtosecond photon echoes from band-to-band transitions in GaAs," *Phys. Rev. Lett.*, vol. 61, p. 1647, 1988.

[56] A. Yariv, *Quantum Electronics*. New York: Wiley, 3rd ed., 1989.

[57] R. Trebino, K. W. DeLong, D. N. Fittinghoff, J. N. Sweetser, M. A. Krumbügel, B. A. Richman, and D. J. Kane, "Measuring ultrashort laser pulses in the time-frequency domain using frequency-resolved optical gating," *Rev. Sci. Instrum.*, vol. 68, p. 3277, 1997.

3. PHOTONS AND PHOTON STATISTICS: FROM INCANDESCENT LIGHT TO LASERS

RALPH v. BALTZ
Institut für Theorie der Kondensierten Materie
Universität Karlsruhe, D–76128 Karlsruhe, Germany *

1. Introduction

The majority of experiments in optics can be understood on the basis of Classical Electrodynamics. Maxwell's theory is perfectly adequate for understanding diffraction, interference, image formation, and even nonlinear phenomena such as frequency doubling or mixing. However, many fascinating quantum effects like correlations between photons are not captured, e.g., the photons in a single mode laser well above the threshold photons are completely uncorrelated, whereas photons in thermal light have a tendency to "arrive" in pairs.

This contribution addresses the following questions:

- What is a photon?
- Description and examples of relevant photon states.
- Discussion of basic optical devices and measurements.

2. Nature of light

2.1. HISTORICAL MILESTONES

Isaac Newton (1643 – 1727):	Founder of the corpuscular theory.
Christian Huyghens (1629 – 1695):	Founder of the wave theory of light.
Thomas Young (1773 – 1829):	Independent pioneering work about
Augustin Fresnel (1788 – 1827):	waves & interference.
James C. Maxwell (1831 – 1879):	Theory of the Electromagnetic Field.
Heinrich Hertz 1888:	Discovery of electromagnetic waves.

* http://www-tkm.uni-karlsruhe.de (e.g, for previous Erice contributions).

B. Di Bartolo and O. Forte (eds.), Frontiers of Optical Spectroscopy, 55-92.

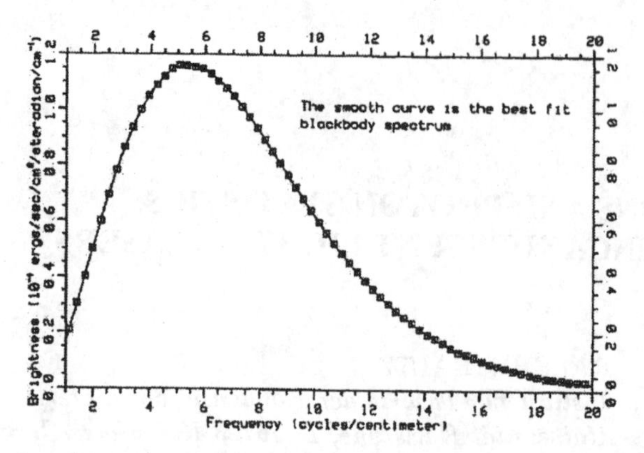

Figure 1. A perfect Planckian: The $T = 2.7$K cosmic background radiation. From[14].

Maxwell's theory of the Electromagnetic Field (EMF) did not only give a beautiful unification of electric and magnetic phenomena but it also predicted the existence of electromagnetic waves as undulations of electric and magnetic fields propagating through space. Wave theory had won a glorious victory! But then the incomprehensible happened:

Max Planck 1900:	Quantization of mode–oscillators.
Phillip Lenard 1902:	Difficulties with photoelectric effect.
Albert Einstein 1905:	Postulate of light quanta of energy $\hbar\omega$.
G. I. Taylor 1909:	Interference with "single photons".
A. H. Compton 1923:	Photons have $E(\mathbf{p}) = c\lvert\mathbf{p}\rvert$.
E.O.Lawrence, J.W.Beams 1927:	No photoelectron delay time.
P. A. M. Dirac 1927	Quantum Theory of the EMF.

The era of quantum physics started with Planck's[1] derivation of the black body spectrum in terms of quantized mode oscillators which is depicted in Fig. 1. The spectral energy density (energy per volume, energy $\hbar\omega$, polarization degree, and solid angle $d\Omega$) reads

$$e(\hbar\omega, T)d(\hbar\omega)\,d\Omega = \frac{(\hbar\omega)^2}{(2\pi c\hbar)^3}\,\frac{\hbar\omega}{e^{\frac{\hbar\omega}{k_B T}} - 1}\,d(\hbar\omega)\,d\Omega. \tag{1}$$

Lenard's[2] observations on the photoelectric effect were incompatible with the predictions of the Maxwell–Theory where energy is distributed continuously in space. The photocurrent was found to be proportional to the intenstity I of the light, however the energy of individual electrons did not depend on I, yet it increased with light frequency. Nevertheless, Lenard thought that the light does not supply the energy which is necessary to release an electron but merely triggers the photoelectric emission.

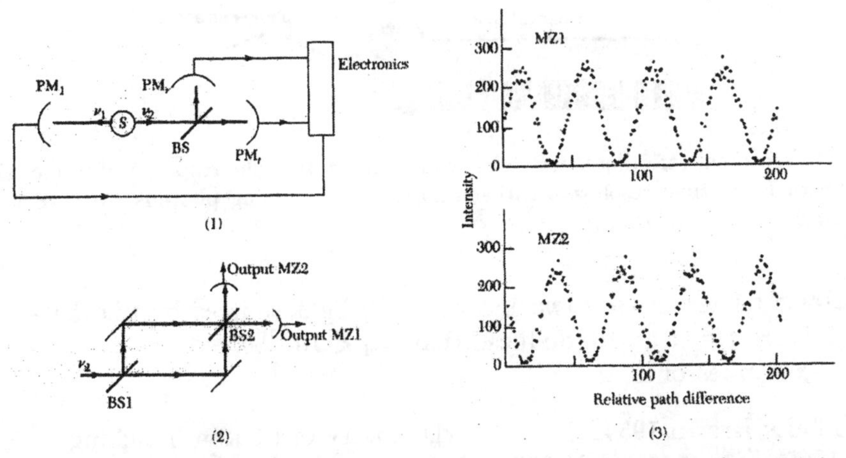

Figure 2. (1) Triggered photon cascade experiment to produce single photon states. (2) Mach–Zehnder Interferometer. (3) Number counts in the outputs of the photodetector MZ1 and MZ2 as a function of the path difference between the arms of the interferometer. (One channel corresponds to a variation of $\delta = \lambda/50$). According to Grangier et al.[11].

In 1905 Einstein published three seminal contributions: About Brownian motion, special relativity, and the photoelectric effect. For the latter contribution entitled *Über einen die Erzeugung und Verwandlung des Lichts betreffenden heuristischen Gesichtspunkt*[3] he was awarded the Nobel prize in 1921. Guided by an ingenious thermodynamic approach of the black body radiation he got the inspiration that the energy transported by light is distributed in a granular rather than in a continuous fashion in space ("darts of energy"[4]). The most direct evidence of the particle nature of light quanta is provided by the Compton–effect[5].

The existence of light quanta is in apparent contradiction with typical wave properties like interference fringes. It was expected that such fringes fade out if the intensity of the incident light becomes smaller and smaller so that the probability of having more than a single photon in the spectrometer becomes negligible. Interference experiments at very low intensity were carried out in 1909 by Taylor[6] and later, by Dempster and Batho[7], and by Janossy et al.[9]. With great disappointment all these investigators reported a null result. This discovery lead Dirac to the famous statement "a photon interferes (only) with itself". Since 1985, coherence experiments with genuine single photon states of light are possible, see Fig. 2.

With a simple but ingenious method Lawrence and Beams[8] studied in 1927 the time variation of the photoelectric emission from a metal surface illuminated by light flashes of 10^{-8}s duration. It was found that photolectric emission starts in less than 3×10^{-9}s after the beginning of the illumination of a potassium surface. From another experiment which was designed to

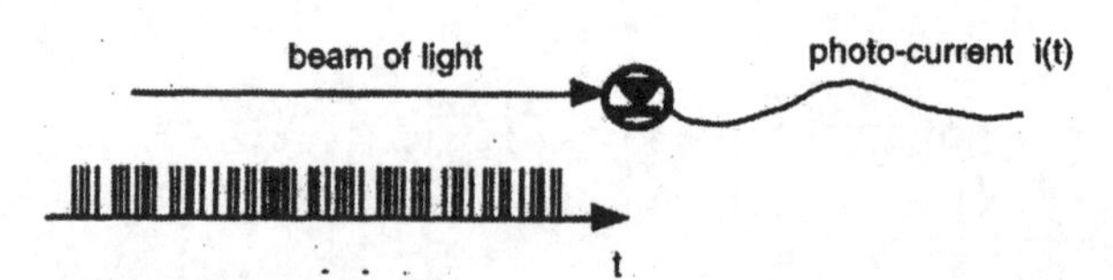

Figure 3. Photoelectric detection. The "comb" refers to a the response of a photodetector with hight time resolution rather than to the incoming photons. According to Bachor[38].

investigate beats between two incoherent(!) light sources by photoelectric mixing Forrester et al.[10] deduced that $t_{del} < 10^{-10}$s.

Further milestones are:

R. Hanbury Brown 1954	Discovery of photon bunching.
Kimble, Dagenais, Mandel 1977	Generation of nonclassical light.
R. E. Slusher, B. Yurke 1985	Generation of squeezed states

2.2. WHAT IS A PHOTON?

The photon hypothesis relies on four basic facts:

- Point–like localization of energy.
- Transport of energy and momentum through space with $E(\mathbf{p}) = c|\mathbf{p}|$.
- Absence of a delay time for the emission of photoelectrons.
- A photon interferes with itself.

Evidences for the particle nature of the excitations of the EMF are the photoelectric effect[2] and the Compton effect[5], respectively. The non-existence of a delay time[8, 10] is less frequently mentioned although it is equally important as the other properties.

Today our interpretation of photons differs substantially from the original idea of small energy "bullets" or "darts"[4]

- Photons are just the energy eigenstates of the EMF.
- The localization arises as the outcome of a measurement which causes the state of the EMF to "collapse" into an eigenstate of the device as a result of a position measurement, e.g., the absorption of a photon by an atom or in a pixel of a CCD–camera.

Although photons (in free space) have a definite energy–momentum relation, photons are not "objects" in the sense of individual, localizable classical particles. By contrast, they are indistinguishable, nonlocalizable and obey Bose statistics. A figure like Fig. 3 is dangerous as it pretends that photons in a light beam have well defined positions. The notion of classical and quantum particles is intrinsically very different. In particular the collapse of the quantum state is foreign to all classical (wave–) theories, in particular, a localization smaller than the wavelength is not possible. The nonexistence of a waiting time for the photoelectrons is simply the consequence of statistics, some come promptly, others come later. This will be discussed in Section 5.3.

It was left to Dirac[12] to combine the wave– and particle like aspects of light so that this description is capable of explaining all interference and particle phenomena of the EMF. We shall follow his traces in Section 4.

For an excellent survey on photons see Paul's book[32]. Alternative theories are, e.g., discussed in the Rochester Proceedings from 1972[47].

3. Classical Description of the EMF: Waves

3.1. MAXWELL–EQUATIONS

For our purposes a detailed knowledge how to calculate field configurations for specific systems in not required. However, we have to know the relevant dynamical variables of the EMF.

The state of the EMF is described by two (mathematical) vector fields $\mathcal{E}, \mathcal{B}$ which are coupled to the charge and current density of matter $\rho, \mathbf{j}$ by the Maxwell–Equations [1]

$$\frac{\partial \mathcal{E}}{\partial t} - c^2 \text{curl}\, \mathcal{B} = -\frac{1}{\epsilon_0}\mathrm{j}(\mathbf{r},t), \qquad \text{div}\, \mathcal{E} = \frac{1}{\epsilon_0}\rho(\mathbf{r},t), \tag{2}$$

$$\frac{\partial \mathcal{B}}{\partial t} + \text{curl}\, \mathcal{E} = 0, \qquad \text{div}\, \mathcal{B} = 0. \tag{3}$$

Analogous to the interpretation of Classical Mechanics one may view the two differential equations (lhs) with respect to time as equations of motion of the Maxwell field, whereas the rhs–set represents "constraints". Hence, from the 6 components of $\mathcal{E}, \mathcal{B}$ at most 6-2=4 components are independent dynamical variables at each space point. Therefore, potentials $\Phi, \mathcal{A}$ are more appropriate than $\mathcal{E}, \mathcal{B}$

$$\mathcal{E} = -\dot{\mathcal{A}}(\mathbf{r},t) - \text{grad}\, \Phi(\mathbf{r},t), \tag{4}$$

$$\mathcal{B} = \text{curl}\, \mathcal{A}(\mathbf{r},t). \tag{5}$$

In contrast to $\mathcal{E}, \mathcal{B}$ the potentials $\mathcal{A}$, Φ are not uniquely determined, rather $\mathcal{A} \to \mathcal{A}' = \mathcal{A} + \text{grad}\, \Lambda(\mathbf{r},t), \Phi \to \Phi' = \Phi - \dot{\Lambda}(\mathbf{r},t)$ lead to the same $\mathcal{E}, \mathcal{B}$–fields and, hence, contain the "same physics". $\Lambda(\mathbf{r},t)$ is an arbitrary gauge function. This property is called gauge invariance or gauge symmetry and it is considered as a fundamental principle of nature.

---

[1] Vectors are set in boldface, electromagnetic fields in caligraphic style.

In Quantum Optics (and in solid state physics as well), the *Coulomb gauge*, div $\mathcal{A} = 0$, is particularily convenient, as the equations for Φ and $\mathcal{A}$ decouple

$$\Delta\Phi(\mathbf{r},t) = -\frac{1}{\epsilon_0}\rho(\mathbf{r},t), \tag{6}$$

$$\Delta\mathcal{A}(\mathbf{r},t) - \frac{1}{c^2}\frac{\partial^2\mathcal{A}(\mathbf{r},t)}{\partial t^2} = -\mu_0\mathbf{j}_{tr}(\mathbf{r},t), \tag{7}$$

where

$$\mathbf{j}_{tr}(\mathbf{r},t) = \mathbf{j}(\mathbf{r},t) - \epsilon_0 \mathrm{grad}\,\frac{\partial\Phi(\mathbf{r},t)}{\partial t} \tag{8}$$

denotes the socalled transverse component of the current, $\mathrm{div}\,\mathbf{j}_{tr} = 0$. Some advantages of the Coulomb gauge are:

- Φ is not a dynamical system, i.e. there is no differential equation with respect to time, hence, $\Phi(\mathbf{r},t)$ follows $\rho(\mathbf{r},t)$ without retardation!
- As $\mathrm{div}\,\mathbf{j}_{tr} = 0$ only 2 of the 3 components of $\mathcal{A}$ are independent variables of the EMF.

3.2. MODES AND DYNAMICAL VARIABLES

In order to extract the dynamical variables of the EMF from Eq. (7) we decompose the vector potential in terms of modes $\mathbf{u}_\ell(\mathbf{r})$

$$\mathcal{A}(\mathbf{r},t) = \sum_\ell A_\ell(t)\,\mathbf{u}_\ell(\mathbf{r}), \tag{9}$$

$$\Delta\mathbf{u}_\ell(\mathbf{r}) + (\frac{\omega_\ell}{c})^2\,\mathbf{u}_\ell(\mathbf{r}) = 0, \quad \mathrm{div}\,\mathbf{u}_\ell(\mathbf{r}) = 0, \tag{10}$$

$$\int \mathbf{u}^*_\ell(\mathbf{r})\,\mathbf{u}_{\ell'}(\mathbf{r})\,d^3\mathbf{r} = \delta_{\ell,\ell'}. \tag{11}$$

In addition, there will be boundary conditions for $\mathcal{E}, \mathcal{B}$ which fix the eigenfrequencies ω_ℓ of the modes labelled by ℓ. (To lighten the notation we omit the index "tr" from now on). The set of $\mathcal{A}_\ell(t)$ represents the generalized coordinates or dynamical variables of the EMF which obeys the equation of motion

$$\ddot{A}_\ell(t) + \omega_\ell^2\,A_\ell(t) = \frac{1}{\epsilon_0}j_\ell(t). \tag{12}$$

$\mathbf{j}_\ell(t)$ is defined in the same way as in Eq. (9) and it can be obtained by using the orthogonality relations Eq. (11). Note, each mode is equivalent to a driven harmonic oscillator. A state of the EMF is, thus, specified by the set of mode amplitudes $A_\ell(t_0)$ and their velocities $\dot{A}_\ell(t_0)$ at a given instant of time t_0.

$\mathcal{A}(\mathbf{r},t)$ is a real field so that $\mathbf{u}_\ell(\mathbf{r})$ as well as $A_\ell(t)$ ought to be real. Nevertheless, the choice of complex modes may be convenient. In particular, in free space we will use "running plane waves".

$$\mathbf{u}_{\mathbf{k},\sigma}(\mathbf{r},t) = \frac{1}{\sqrt{V}} \boldsymbol{\epsilon}_{\mathbf{k},\sigma}\, e^{\imath \mathbf{k}\mathbf{r}}, \tag{13}$$

$$\boldsymbol{\epsilon}^*_{\mathbf{k},\sigma'} \cdot \boldsymbol{\epsilon}_{\mathbf{k},\sigma} = \delta_{\sigma,\sigma'}. \tag{14}$$

$\boldsymbol{\epsilon}_{\mathbf{k},\sigma}$ denotes the polarization vector which, by $\operatorname{div}\mathbf{u} = \imath\mathbf{k}\cdot\boldsymbol{\epsilon}_{\mathbf{k},\sigma} = 0$, is orthogonal to the wave vector $\mathbf{k}$. (Here the notation "transversal" becomes manifest). The two independent polarization vectors will be labelled by $\sigma = 1, 2$. V denotes the normalization volume and, as usual, periodic boundary conditions implied. As a result, we obtain

$$\mathcal{A}(\mathbf{r},t) = \sum_{\mathbf{k},\sigma} A_{\mathbf{k},\sigma}(t)\, \boldsymbol{\epsilon}_{\mathbf{k},\sigma}\, e^{\imath\mathbf{k}\mathbf{r}} = \sum_{\mathbf{k},\sigma}{}' \left(A_{\mathbf{k},\sigma}(t)\, \boldsymbol{\epsilon}_{\mathbf{k},\sigma}\, e^{\imath\mathbf{k}\mathbf{r}} + cc\right). \tag{15}$$

As $\mathcal{A}(\mathbf{r},t)$ is a real field we must have $\mathcal{A}_{-\mathbf{k},\sigma} = \mathcal{A}^*_{\mathbf{k},\sigma}$, i.e. the amplitudes for $\mathbf{k}$ and $-\mathbf{k}$ are not independent, only those amplitudes with $k_z > 0$ are independent dynamical variables. (This is the meaning of the prime in $\sum'$. $k_z = 0$ requires additional investigation). Fortunately, this problem can be circumvented by a redefinition of the $A_{\mathbf{k},\sigma}$ and as a result we have[39, 48]

$$\mathcal{A}(\mathbf{r},t) = \sum_{\mathbf{k},\sigma} \sqrt{\frac{\hbar}{2\epsilon_0\omega_{\mathbf{k}} V}} \left(a_{\mathbf{k},\sigma}(t)\, \boldsymbol{\epsilon}_{\mathbf{k},\sigma}\, e^{\imath\mathbf{k}\mathbf{r}} + a^*_{\mathbf{k},\sigma}(t)\, \boldsymbol{\epsilon}^*_{\mathbf{k},\sigma}\, e^{-\imath\mathbf{k}\mathbf{r}}\right), \tag{16}$$

$$\begin{aligned}
\mathcal{E}(\mathbf{r},t) &= -\frac{\partial \mathcal{A}(\mathbf{r},t)}{\partial t} \\
&= \sum_{\mathbf{k},\sigma} \sqrt{\frac{\hbar}{2\epsilon_0\omega_{\mathbf{k}} V}} \left(\imath\omega_{\mathbf{k}}\, a_{\mathbf{k},\sigma}(t)\, \boldsymbol{\epsilon}_{\mathbf{k},\sigma}\, e^{\imath\mathbf{k}\mathbf{r}} + cc\right), \qquad (17)\\
\mathcal{B}(\mathbf{r},t) &= \operatorname{curl} \mathcal{A}(\mathbf{r},t) \\
&= \sum_{\mathbf{k},\sigma} \sqrt{\frac{\hbar}{2\epsilon_0\omega_{\mathbf{k}} V}} \left(\imath(\mathbf{k}\times\boldsymbol{\epsilon}_{\mathbf{k},\sigma})\, a_{\mathbf{k},\sigma}(t)\, e^{\imath\mathbf{k}\mathbf{r}} + cc\right). \qquad (18)
\end{aligned}$$

In order to make the $a_{\mathbf{k},\sigma}$ dimensionless and in anticipation of the quantum treatment $\hbar$ has been "smuggelt in". In contrast to Eq. (12) $a_{\mathbf{k},\sigma}(t)$ obeys the first order differential equation

$$\frac{da_{\mathbf{k},\sigma}(t)}{dt} + \imath\omega_{\mathbf{k}}\, a_{\mathbf{k},\sigma}(t) = \imath\sqrt{\frac{1}{2\epsilon_0\hbar\omega_{\mathbf{k}}}}\, j_{\mathbf{k},\sigma}(t), \tag{19}$$

which has been already used performing the time–derivative of $\mathcal{A}(\mathbf{r},t)$ in Eq. (17). (The contribution from $\mathbf{j}_{tr}$ drops out in the final result).

From the fields we obtain the energy (Hamiltonian) of the radiation field (including the interaction with the current) and the momentum

$$\begin{aligned} H &= \int \left(\frac{\epsilon_0}{2} \mathcal{E}_{tr}^2(\mathbf{r},t) + \frac{1}{2\mu_0} \mathcal{B}^2(\mathbf{r},t) - \mathbf{j}_{tr}(\mathbf{r},t)\, \mathcal{A}(\mathbf{r},t) \right) d^3\mathbf{r} \\ &= \sum_{\mathbf{k},\sigma} \hbar\omega_{\mathbf{k}}\, a_{\mathbf{k},\sigma}^* a_{\mathbf{k},\sigma} - \sqrt{\frac{\hbar}{2\epsilon_0\omega_{\mathbf{k}}}} \left(j_{\mathbf{k},\sigma}^*(t)\, a_{\mathbf{k},\sigma} + cc \right), \end{aligned} \tag{20}$$

$$P = \int \left(\frac{1}{\mu_0} \mathcal{E}_{tr}(\mathbf{r},t) \times \mathcal{B}(\mathbf{r},t) \right) d^3\mathbf{r} = \sum_{\mathbf{k},\sigma} \hbar\mathbf{k} a_{\mathbf{k},\sigma}^* a_{\mathbf{k},\sigma}. \tag{21}$$

(In contrast to most treatments of the subject no efforts have been made to preserve the "natural sequence" of the amplitudes $a_{\mathbf{k},\sigma}, a_{\mathbf{k},\sigma}^*$. Potential energy/momentum contributions from the scalar potential to Eqs. (20-21) have been omitted, see Kroll's article[48].)

The complex amplitudes $a_{\mathbf{k},\sigma}$ represent the dynamical variables of the EMF. Its real and imaginary parts are called *quadrature amplitudes*

$$a_{\mathbf{k},\sigma} = X_{\mathbf{k},\sigma}^{(1)} + \imath X_{\mathbf{k},\sigma}^{(2)}, \tag{22}$$

which (apart from numerical factors) are the analogues of position and momentum of a mechanical oscillator.

Example:
In free space (without current source) the complex amplitudes of a single mode with the initial condition $a(t=0) = a_0$ reads

$$a(t) = a_0\, e^{-\imath\omega t}, \tag{23}$$

$$X_1(t) = +\Re a_0 \cos(\omega t) + \Im a_0 \sin(\omega t), \tag{24}$$

$$X_2(t) = -\Re a_0 \sin(\omega t) + \Im a_0 \cos(\omega t). \tag{25}$$

Together with Eqs. (17,18) and $\mathbf{k} = (k,0,0)$, $\epsilon_{\mathbf{k},\sigma} = (0,1,0)$ this field describes a linearly polarized plane wave propagating along the x–direction.

3.3. SPECIAL STATES OF THE EMF

There are two classes of states of the classical EMF:

- Deterministic states: $a_{\mathbf{k},\sigma}(t)$ are specified for all modes at time t_0.
- Random (stochastic) states with a probability distribution $P(\{a_{\mathbf{k},\sigma}\},t)$.

In radio physics these states are termed *signals* and *noise*. Prior to the advent of the laser, the only possibility to produce radiation which is correlated over some space–time domain was to filter black body radiation with respect to frequency and spatial directions.

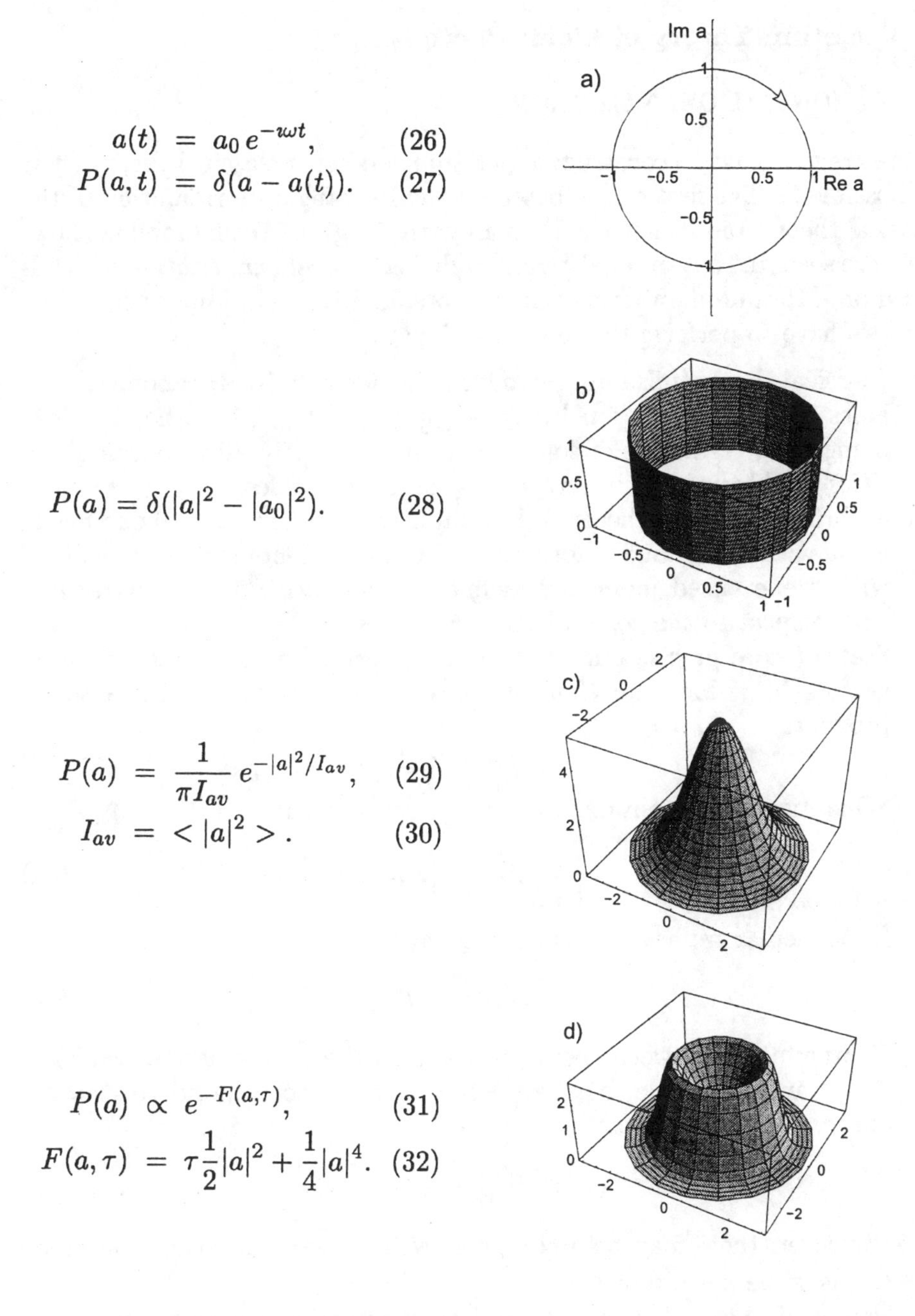

$$a(t) = a_0 \, e^{-\imath\omega t}, \qquad (26)$$
$$P(a,t) = \delta(a - a(t)). \qquad (27)$$

$$P(a) = \delta(|a|^2 - |a_0|^2). \qquad (28)$$

$$P(a) = \frac{1}{\pi I_{av}} \, e^{-|a|^2/I_{av}}, \qquad (29)$$
$$I_{av} = < |a|^2 > . \qquad (30)$$

$$P(a) \propto e^{-F(a,\tau)}, \qquad (31)$$
$$F(a,\tau) = \tau \frac{1}{2}|a|^2 + \frac{1}{4}|a|^4. \qquad (32)$$

Figure 4. Model probability distributions for the complex field amplitude a of a single mode. Ideal Laser (a), amplidute stabilized laser with phase fluctuations (b), Gaussian (=thermal) noise (c). (d) Laser model which includes both amplitude and phase fluctuations according to a "phase transition" far from equilibrium. $F(a,\tau)$ is a "Ginzburg–Landau free energy" and τ is the pump parameter[33].

4. Quantum Theory of Light: Photons

4.1. CANONICAL QUANTIZATION

There are two ways to construct a quantum version of classical theory (if at all possible !). The first one is based on the Lagrangian formulation of the classical theory and it uses the Feyman path integral. We shall follow, however, the second, conventional "trail" called *canonical quantization* which is based on a Hamiltonian form, state vectors, and the Schrödinger equation. Here we have to perform the following steps:

- Classical theory in Hamiltonian form, i.e. identify (real) canonical variables p_j, q_k with Poisson–brackets $\{p_j, q_k\} = \delta_{j,k}$. (All other brackets being zero, regardless of components j, k). Rewrite all physicall quantities (=observables) in terms of canonical variables.
  To bring a classical theory in Hamiltonian form (if possible) one has to begin with a Lagrangian formulation. Canonical momenta are identical with "generalized momenta" defined as derivative of the Lagrangian with respect to the generalized coordinates.
- States (more precise pure states) are described by (normalized) state vectors $|\psi\rangle$ which are elements of a Hilbert space $\mathcal{H}$ with a scalar product.
$$\langle\psi_1|\psi_2\rangle = \langle\psi_2|\psi_1\rangle^*. \tag{33}$$
- "Quantization" is obtained by the translation rule
$$\{A, B\} \rightarrow \frac{\imath}{\hbar}\left[\hat{A}, \hat{B}\right], \quad \left[\hat{A}, \hat{B}\right] = \hat{A}\hat{B} - \hat{B}\hat{A}. \tag{34}$$
  In particular we have for canonical variables
$$[\hat{p}_j, \hat{q}_k] = -\imath\hbar\delta_{j,k}. \tag{35}$$
  Commutators between the p's or between the q's themselves vanish.
- The translation rule for the operators which correspond to classical observables $G(p, q, t)$ are
$$\hat{G} = G^{cl}(p = \hat{p}, q = \hat{q}, t). \tag{36}$$
  (However, there may be ambiguities with noncommutating operators. $\hat{G}$ has to be hermitian.)
- Values of observables are defined as expectation values of the corresponding operators:
$$\langle G\rangle = \langle\psi|\hat{G}|\psi\rangle = \langle\psi|\psi'\rangle, \tag{37}$$
  where $|\psi'\rangle = \hat{G}|\psi\rangle$.

- Dynamics: Initially, the system is supposed to be in state $|\psi_0\rangle = |\psi(t_0)\rangle$. Then, sequence of states $|\psi(t)\rangle$ which the system runs through as a function of time is governed by the *Schrödinger equation*

$$\imath\hbar \frac{\partial|\psi(t)\rangle}{\partial t} = \hat{H}|\psi(t)\rangle, \tag{38}$$

where $\hat{H}$ denotes the *Hamiltonian* (=energy) of the system. Note, Eq. (38) holds not only for non–relativistic particles but also for photons!

- States vectors $|\psi\rangle$ describe *pure states* with zero entropy. They are the analoga of the ideal mechanical states with fixed q, p or the ideal states of the classical EMF with fixed $a_{\mathbf{k},\sigma}$ ("signals").
The counterparts of the classical statistical states, e.g., thermal radiation Eq (29), are called *mixed states.* They have nonzero entropy and are described by a *density operator* $\hat{\rho}$

$$\langle G\rangle = \text{trace}\,(\hat{\rho}\,\hat{G}). \tag{39}$$

4.2. QUANTUM OPTICS

The quantum version of the EMF together with a (nonrelativistic) theory of matter is called *Quantum Optics.* To follow the scheme outlined in the previous section we have to bring the Maxwell–Theory into a Hamiltonian form. This is, however, almost trivial because the EMF is dynamically equivalent to a system of uncoupled harmonic oscillators with generalized "coordinates" A_ℓ, see Eq. (12). We guess the Lagrangian

$$L \propto \sum_\ell \frac{1}{2}\dot{A}_\ell^2 - \frac{\omega_\ell^2}{2}A_\ell^2 + \frac{1}{\epsilon_0}j_\ell(t)\,A_\ell \tag{40}$$

$$= \int\left(\frac{\epsilon_0}{2}\mathcal{E}_{tr}^2(\mathbf{r},t) - \frac{1}{2\mu_0}\mathcal{B}^2(\mathbf{r},t) + \mathbf{j}_{tr}(\mathbf{r},t)\,\mathcal{A}(\mathbf{r},t)\right)d^3\mathbf{r}. \tag{41}$$

The canonical variables are $Q_\ell = A_\ell, P_\ell = \dot{A}_\ell$ and the Hamiltonian is obtained from $H = P\dot{Q} - L$.

For complex modes, as used in Eqs. (16–18), each amplitude contains two real variables X_1, X_2 which (apart from numerical factors) correspond to canonical momentum and position, ($p = \sqrt{2m\hbar\omega}X_2, x = \sqrt{2\hbar/m\omega}X_1$, $\{p, x\} = 1$). The complex amplitudes obey the Poisson bracket relations

$$\{a_{\mathbf{k},\sigma}\,,\,a^*_{\mathbf{k}',\sigma'}\} = \frac{\imath}{\hbar}\,\delta_{\mathbf{k},\mathbf{k}'}\,\delta_{\sigma,\sigma'}\,. \tag{42}$$

Quantization is obtained by replacing the classical amplitudes $a_{\mathbf{k},\sigma}, a^*_{\mathbf{k},\sigma}$ by "ladder operators" $\hat{a}_{\mathbf{k},\sigma}, \hat{a}^\dagger_{\mathbf{k},\sigma}$ with commutation relations

$$\left[\hat{a}_{\mathbf{k},\sigma}, \hat{a}^\dagger_{\mathbf{k}',\sigma'}\right] = \delta_{\mathbf{k},\mathbf{k}'}\, \delta_{\sigma,\sigma'}. \tag{43}$$

Field operators, Hamiltonian, and the momentum operator are (in the Schrödinger picture)

$$\hat{\boldsymbol{\mathcal{A}}}(\mathbf{r}) = \sum_{\mathbf{k},\sigma} \sqrt{\frac{\hbar}{2\epsilon_0 \omega_{\mathbf{k}} V}} \left(\hat{a}_{\mathbf{k},\sigma}\, \boldsymbol{\epsilon}_{\mathbf{k},\sigma}\, e^{\imath \mathbf{k}\mathbf{r}} + \hat{a}^\dagger_{\mathbf{k},\sigma}\, \boldsymbol{\epsilon}^*_{\mathbf{k},\sigma}\, e^{-\imath \mathbf{k}\mathbf{r}}\right), \tag{44}$$

$$\hat{\boldsymbol{\mathcal{E}}}(\mathbf{r}) = \sum_{\mathbf{k},\sigma} \sqrt{\frac{\hbar}{2\epsilon_0 \omega_{\mathbf{k}} V}} \left(\imath \omega_{\mathbf{k}}\, \hat{a}_{\mathbf{k},\sigma}\, \boldsymbol{\epsilon}_{\mathbf{k},\sigma}\, e^{\imath \mathbf{k}\mathbf{r}} + cc\right), \tag{45}$$

$$\hat{\boldsymbol{\mathcal{B}}}(\mathbf{r}) = \sum_{\mathbf{k},\sigma} \sqrt{\frac{\hbar}{2\epsilon_0 \omega_{\mathbf{k}} V}} \left(\imath(\mathbf{k} \times \boldsymbol{\epsilon}_{\mathbf{k},\sigma})\, \hat{a}_{\mathbf{k},\sigma}\, e^{\imath \mathbf{k}\mathbf{r}} + cc\right). \tag{46}$$

$$\hat{H} = \sum_{\mathbf{k},\sigma} \hbar\omega_{\mathbf{k}}\, \hat{a}^\dagger_{\mathbf{k},\sigma}\hat{a}_{\mathbf{k},\sigma} - \sqrt{\frac{\hbar}{2\epsilon_0 \omega_{\mathbf{k}}}} \left(\jmath^*_{\mathbf{k},\sigma}(t)\, \hat{a}_{\mathbf{k},\sigma} + hc\right), \tag{47}$$

$$\hat{P} = \sum_{\mathbf{k},\sigma} \hbar\mathbf{k}\, \hat{a}^\dagger_{\mathbf{k},\sigma}\hat{a}_{\mathbf{k},\sigma}, \tag{48}$$

The (infinite) zero point energy which arises from the noncommutativity of the $\hat{a}_{\mathbf{k},\sigma}, \hat{a}^\dagger_{\mathbf{k},\sigma}$ operators has been omitted in the Hamiltonian as this has no influence on the dynamics of the EMF. The zero point fluctuations of the EMF, however, are still present in the fields, as we shall see later.

As it is well known, the stationary states of a (free) harmonic oscillator, $|n\rangle$, $n = 0, 1, 2 \ldots$, are the eigenstates of the number operator $\hat{N} = \hat{a}^\dagger\hat{a}$. In addition, they are are nondegenerate, orthogonal, and normalizable, $\langle m|n\rangle = \delta_{m,n}$. The action of the $\hat{a}, \hat{a}^\dagger$–operators on these states is

$$\hat{a}^\dagger\hat{a}|n\rangle = n|n\rangle, \tag{49}$$

$$\hat{a}|n\rangle = \sqrt{n}|n-1\rangle, \tag{50}$$

$$\hat{a}^\dagger|n\rangle = \sqrt{n+1}|n+1\rangle. \tag{51}$$

These operators are also called "ladder operators" because repeated operation on a particular energy eigenstate creates the "ladder" of all other states, with $\hat{a}^\dagger$ we climb up, whereas with $\hat{a}$ we climb down the ladder.

The states of the infinite set of mode oscillators of the EMF is, thus, labelled by the (infinite set of) quantum numbers $\{n_{\mathbf{k},\sigma}\}$ which individually can take on different nonnegative integers. These states describe the stationary states of the free EMF $|\{n_{\mathbf{k},\sigma}\}\rangle$ and their time dependence is,

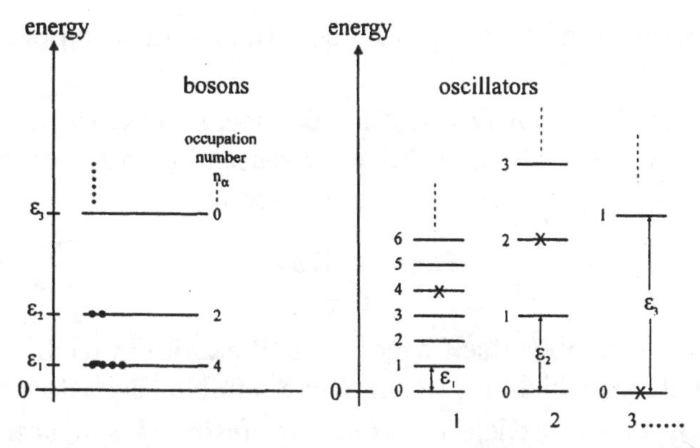

Figure 5. Equivalence of a system of N (noninteracting) bosons with single–particle energies ϵ_ℓ and occupation numbers n_ℓ and an infinite (uncoupled) set of harmonic oscillators with frequencies $\omega_\ell = \epsilon_\ell/\hbar$. Note that the zero–point energies of the oscillators are omitted. Dots symbolize particles, crosses excited states, respectively. $N = 6$. According to Ref.[49]

as usual, obtained by an exponential factor $\exp(-\imath n\omega t)$ for each mode. All other states can be represented as a superposition of these *number states*, which, therefore, represent a natural basis for the description of the quantum states of the EMF.

For a wide–range introduction see Scully and Zubairy's book on *Quantum Optics*[35].

4.3. OSCILLATORS AND PHOTONS

Dirac[12] has made the important observation that

> *...a system of noninteracting bosons with single particle energies ϵ_ℓ is dynamically equivalent to a system of uncoupled oscillators and vice versa. The two systems are just the same looked at from two different points of view.*

Here, *dynamic equivalence* implies that all states of an N–boson system which are conventionally described by a symmetric wave function are equally well described in the "oscillator picture", where each single particle state with energy ϵ_ℓ correponds to an oscillator with frequency $\omega_\ell = \epsilon_\ell/\hbar$. Remarkably, the commutation relations $[\hat{a}_\ell, \hat{a}^\dagger_{\ell'}] = \delta_{\ell,\ell'}$ between the ladder operators are fully equivalent to the permutation symmetry of the boson wavefunction, and, fortunately, a great deal of notational redundancy in the description of a many–body system is removed. In addition, all operators in the particle picture (lhs of Fig. 5) can be translated into operators acting on the oscillator states. These operators are conveniently expressed in terms of ladder operators and contain products with an equal number of $\hat{a}_{\mathbf{k},\sigma}$ and $\hat{a}^{\dagger}_{\mathbf{k}',\sigma'}$ operators.

The union of all sets of $N = 1, 2\ldots$ particle subspaces plus the $N = 0$ "no particle" vacuum state is called *Fock space.* The number states $|\{n_\ell\}\rangle$ are the eigenstates of the particle number operator

$$\hat{N} = \sum_{\ell} \hat{a}^{\dagger}_{\ell}\,\hat{a}_{\ell}. \tag{52}$$

Now, the particle number itself becomes a dynamical variable and we can even describe states which are not particle number eigenstates of the system. $\hat{a}^{\dagger}_{\ell}$, $\hat{a}_{\ell}$ are called particle *creation and destruction operators* because they change the number of particles by one. The Fock representation is also called *occupation number representation* or "second quantization". It is much more flexible than the original formulation with a fixed particle number.

The bosons corresponding to the quantized oscillators with $\ell = (\mathbf{k}, \sigma)$ are called *photons.* Photons behave very different from massive particles like electrons. They can be created and annihilated arbitrarily and they are not localizable. This will become obvious by the discussion of various examples in the next sections.

Example:
The momentum operator (in one dimension) is translated according to

$$\sum_{j=1}^{N} \hat{p}_j \rightarrow \sum_{\ell,\ell'} \langle \ell' | \hat{p} | \ell \rangle\, \hat{a}^{\dagger}_{\ell'}\,\hat{a}_{\ell}. \tag{53}$$

ℓ labels any set of single particle states. Evaluate the expectation value of a single particle operator for both the wave–function and occupation number representation for $N = 3$ particles! (The full advantage of the occupation number representation will really show up if interaction of particles are included.)

4.4. SPECIAL PHOTON STATES

In the following we shall discuss some selected states of the EMF with respect to the expectation values of the fields, energy, and momentum. The physically relevant states cannot be eigenstates of the electrical field operator $\hat{\mathcal{E}}$ as these have infinite energy. ($\hat{\mathcal{E}}$ corresponds to the position (or momentum) of a mechanical oscillator).

The "quantum unit" of the electrical field is $\mathcal{E}_0 = \sqrt{\hbar\omega/2\epsilon_0 V}$. For green light, $\lambda = 500\mathrm{nm}$, and a quantization volume of $V = 1\mathrm{cm}^3$, $\mathcal{E}_0 \approx 0.075\mathrm{V/m}$, whereas, in a microresonator of linear dimension $1\mu\mathrm{m}$, $\mathcal{E}_0 \approx 7.5 \times 10^4\mathrm{V/m}$!

4.4.1. *n-photons in a single mode*

We consider n photons in a single mode with $\mathbf{k} = (k, 0, 0)$ and linear polarization along y–direction $\epsilon_{\mathbf{k},\sigma} = (0, 1, 0)$. $|n\rangle$ is, of course, an eigenstate of the photon number operator Eq. (52). (Mode indices $\mathbf{k}, \sigma$ are omitted, for brevity).

Without a driving current source this state is an eigenstate of the Hamiltonian with energy $\hbar\omega n$ and it evolves in time according to

$$|n, t\rangle = |n\rangle \, e^{-\imath\omega n t}. \tag{54}$$

In addition, this state is also a momentum eigenstate with eigenvalue $n\hbar\mathbf{k}$, Eq. (48). However, $|n\rangle$ is not an eigenstate of the electrical field operator, Eq. (45), because $\hat{a}_{\mathbf{k},\sigma}, \hat{a}^{\dagger}_{\mathbf{k},\sigma}$ changes the number of photons by ± 1. The expectation value of the electrical field operator reads

$$\langle n, t|\hat{\mathcal{E}}(\mathbf{r})|n, t\rangle = 0, \tag{55}$$

$$\langle n, t|\hat{\mathcal{E}}^2(\mathbf{r})|n, t\rangle = \mathcal{E}_0^2 \, (2n + 1). \tag{56}$$

Certainly, such a state does not correspond to a classical sinusoidal wave, instead it is pure "quantum noise". Note, even in the vacuum state, $|0\rangle$, *zero point fluctuations* are present.

4.4.2. *Single photon wave packet*

We consider a superposition of single photon states refering to different modes (but with the same polarization).

$$| \, \phi_\sigma(\mathbf{k}), t >= \sum_{\mathbf{k}} \phi_\sigma(\mathbf{k}) \, e^{-\imath\omega_{\mathbf{k}} t} \, |1_{\mathbf{k},\sigma}\rangle, \tag{57}$$

where $\phi_\sigma(\mathbf{k})$ is an arbitrary normalizable function which, with some care, may be interpreted as a wave function of a photon (-wave packet) in momentum space. However, there is no reasonable photon position representation[40]. The question of localization of photons is discussed,e.g., by Clauser[50].

A special case is the superposition of just two modes. Such a "two colour state" can be created by a simultaneous excitation of two almost degenerate (atomic) states by a short laser pulse. If the bandwidth of the laser pulse embraces both components a coherent superposition of the two atomic states is created, which decays spontaneousely in a single photon "wave packet" state. Experimentally, this phenomenon shows up in the form of "quantum beats", see Fig. 6.

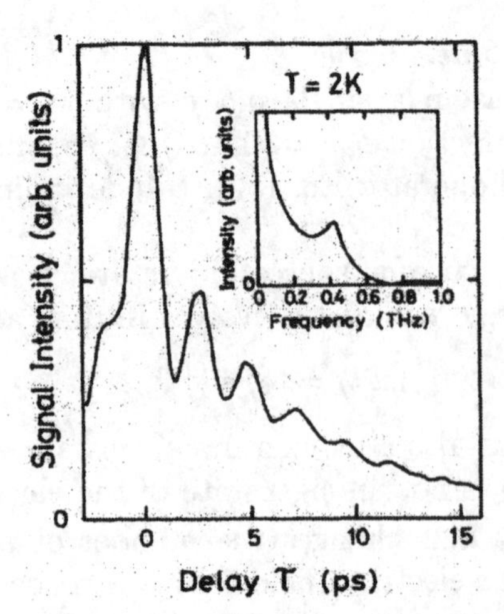

Figure 6. Quantum beats between a coherently prepared population of excitons and biexcitons in AlGaAs quantum well. According to Pandtke et al.[43].

4.4.3. *Coherent states (ideal single mode laser)*

We are searching for a state, in which the fields vary sinusoidally in space and time with $\mathcal{E}$–uncertaincy as small as possible, i.e., have time independent uncertaincies in the quadrature amplitudes with $\Delta X_1 = \Delta X_2 = 1/2$ and $\Delta X_1 \Delta X_2 = 1/4$. These states were already known by Schrödinger[13] and they correspond to a displaced Gaussian ground state wavefunction whose time evolution is depicted in Fig 7. In dimensionless quantities, we have

$$\psi(x,t) = (\pi)^{-1/4} \exp\left[-\frac{(x - x_c(t))^2}{2}\right] e^{\imath x p_c(t)} \, e^{\imath\varphi(t)}. \tag{58}$$

$x_c(t)$ and $p_c(t)$ are the solutions of the classical equations of motion of the oscillator and $\varphi(t)$ is a time dependent phase.

In number representation, these states are given by (without the $\varphi(t)$–term)

$$|\alpha\rangle = e^{-\frac{1}{2}|\alpha|^2} \sum_{n=0}^{\infty} \frac{\alpha^n}{\sqrt{n!}} |n\rangle, \quad \alpha = x_c + \imath p_c. \tag{59}$$

Nowadays, these states are called *coherent states, Glauber states*, or just *α–states.* Glauber[15] was the first who recognized their fundamental role for the description of laser radiation and coherence phenomena.

The α– states have a number of interesting properties:

– $|\alpha\rangle$ is an eigenstate of the destruction operator

$$\hat{a}|\alpha\rangle = \alpha|\alpha\rangle, \tag{60}$$

where α is an arbitrary complex number.

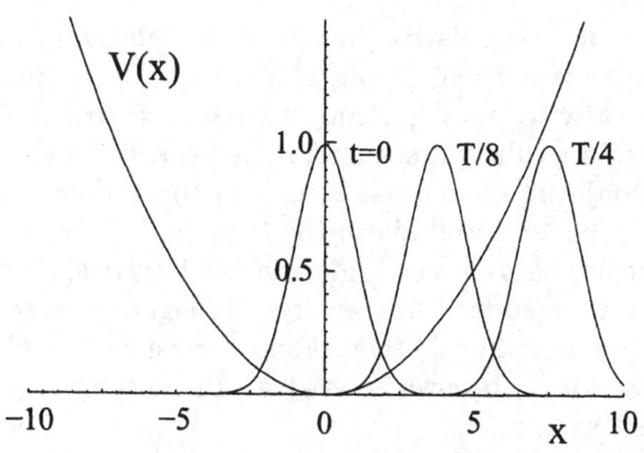

Figure 7. Time development of the coherent state wave function.

– α–states can be generated by the unitary "displacement" operator $\hat{D}$

$$|\alpha\rangle = \hat{D}(\alpha)|0\rangle, \tag{61}$$
$$\hat{D}(\alpha) = e^{\alpha\hat{a}^\dagger - \alpha^*\hat{a}} = e^{-\frac{1}{2}|\alpha^2|}e^{\alpha\hat{a}^\dagger}e^{-\alpha\hat{a}}, \tag{62}$$
$$\hat{D}^\dagger\,\hat{a}\,\hat{D} = \hat{a} + \alpha. \tag{63}$$

– The time dependence (even with a classical external current source) is obtained just by replacing $\alpha \to \alpha(t)$, where

$$|\,\alpha, t\rangle = e^{\imath\varphi(t)}\;|\,\alpha(t)\rangle, \tag{64}$$

Real and imaginary parts of $\alpha(t) = x_c(t) + \imath p_c(t)$ correspond to the position and momentum of a classical oscillator. For a free oscillator $\alpha(t) = \alpha\, e^{-\imath\omega t}$. (Phase $\varphi(t)$ has no influence on the "physics").

– Although the α–eigenvalues form a continuous spectrum, the $|\alpha\rangle$ states are normalizable but they are not orthogonal. Moreover, the set of $|\alpha\rangle$ states is complete (overcomplete) and forms a convenient basis for an "almost classical description" of laser physics.

For further properties see e.g., Scully and Zubairy[35] or Louissell[36].

Expectation values and uncertaincies of the electrical field and photon number and the probability to measure n photons are (omitting the polarization index)

$$\mathcal{E}(\mathbf{r}, t) = \langle\alpha(t)|\hat{\mathcal{E}}(\mathbf{r})|\alpha(t)\rangle = -2\mathcal{E}_0|\alpha|\,\sin(\mathbf{kr} - \omega_{\mathbf{k}}t + \phi), \tag{65}$$
$$(\Delta\mathcal{E})^2 = \mathcal{E}_0^2, \tag{66}$$
$$\langle\hat{N}\rangle = |\,\alpha\,|^2 = \bar{n}, \quad (\Delta\hat{N})^2 = \langle\hat{N}\rangle, \tag{67}$$
$$p_n = |\langle n|\alpha\rangle|^2 = e^{-\bar{n}}\,\frac{\bar{n}^n}{n!}. \tag{68}$$

$\alpha = |\alpha| \exp(\imath\phi)$. Note, the relative amount of fluctuations in the electrical field decreases with increasing amplitude, see Fig. 9. p_n denotes a *Poissonian distribution* with mean photon number $\bar{n} = |\alpha|^2$ and uncertainty $(\Delta n)^2 = \bar{n}$. Thus, in a coherent state photons behave like they were uncorrelated classical objects! In contrast to naive expectations, the photons in a (single mode) laser (and well above the threshold) "arrive" in a random fashion, in particular they do not "ride" on the electrical field maxima.

How to generate α–states? As α–states are eigenstates of the (nonhermitian) destruction operator $\hat{a}$, there is no corresponding observable and "measuring apparatus"! However, α–states can be simply generated from a classical current source

$$\hat{H} = \hbar\omega\hat{a}^\dagger\hat{a} - \left[f(t)\hat{a}^\dagger + f^*(t)\hat{a}\right], \tag{69}$$

where $f(t) \propto j(t)$. Nevertheless, it was a great surprise that the light–matter interaction in a laser (well above the threshold) could be modelled in such a simple way.

The amplitude of the electrical field of a laser may be well stabilized by saturation effects, but there is no possibility to control the phase, i.e., a more realistic laser state would be described by the density operator

$$\hat{\rho} = \int \frac{d\phi}{2\pi} \mid \alpha\rangle\langle\alpha \mid = \sum_n p_n \, |n\rangle\langle n|. \tag{70}$$

This state is made up of an (incoherent) superposition of n–photon states with a *Poissonian distribution.* A model of laser light with a finite linewidth which is caused by phase diffusion has been given by Jacobs[16].

Problem:

P1: Verify that Eqs. (58,64) are solutions of the time–dependent Schrödinger equation of the (driven) harmonic oscillator.

4.4.4. *Squeezed states*

Squeezed states correspond to wave functions which have an uncertainty in one of the quadrature amplitudes smaller than for the groundstate. A harmonic oscillator has the pecularity that any wave function will reproduce itself after the classical oscillation time $T = 2\pi/\omega$, moreover, there is an exact mirror image at $t = T/2$, see Fig. 8. In particular, we will study Gaussian wave packets which initially are minimum uncertainty wave packets with $\Delta X_1 \Delta X_2 = 1/4$, but $\Delta X_1 < 1/2$, $\Delta X_2 > 1/2$ (or vice versa).

$$\Psi(x,t) = \exp\left(-\frac{x^2}{2}\,w(t) + xv(t) + u(t)\right). \tag{71}$$

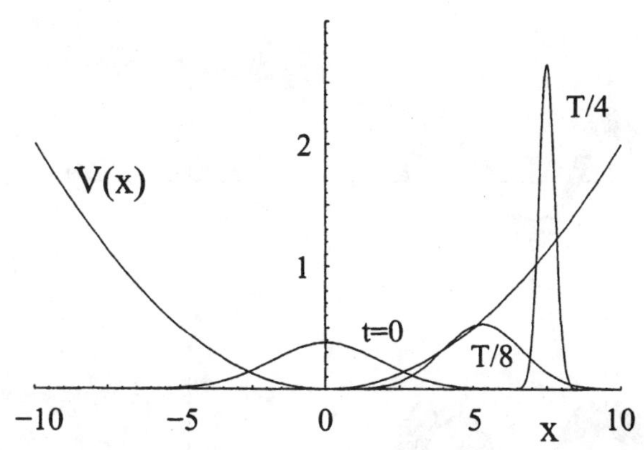

Figure 8. **Time development of a squeezed state wave function.**

$u(t)$, $v(t)$, and $w(t)$ can be complex. These parameters follow from an insertion in the time dependent Schrödinger equation. We leave that as an exercise.

Squeezed states are characterized by two complex variables, usually termed α and ξ and they can be constructed by first operating with the unitary squeezing operator $\hat{S}(\xi)$ on the vacuum and then by shifting with $\hat{D}(\alpha)$.

$$\begin{aligned} |\alpha,\xi\rangle &= \hat{D}(\alpha)\,\hat{S}(\xi)\,|0\rangle, & (72)\\ \hat{S}(\xi) &= e^{\frac{1}{2}(\xi^*\hat{a}^2-\xi(\hat{a}^\dagger)^2)}, & (73)\\ \hat{S}^\dagger(\xi)\,\hat{a}\,\hat{S}(\xi) &= \cosh(r)\hat{a} - e^{-2\imath\theta}\sinh(r)\hat{a}^\dagger, & (74) \end{aligned}$$

where $\xi = r\exp(\imath\theta)$. (Some authors prefer a reversed sequence of $\hat{D}(\alpha)$ and $\hat{S}(\xi)$). Squeezed states are also the eigenstates of a transformed destruction operator $\hat{b}$[18]

$$\begin{aligned} \hat{b} &= \mu\hat{a} + \nu\hat{a}^\dagger, \quad \mu = \cosh(r), \quad \nu = e^{-2\imath\theta}\sinh(r,) & (75)\\ \hat{b}|\beta\rangle &= \beta|\beta\rangle, & (76)\\ \beta &= \alpha\cosh(r) + \alpha^* e^{-2\imath\theta}\sinh(r), & (77)\\ \langle\hat{N}\rangle &= |\alpha|^2 + \sinh^2(r), & (78)\\ (\Delta N)^2 &= |\,\alpha\cosh(r) - \alpha^* e^{\imath\theta}\sinh(r)\,|^2 + \frac{1}{2}\sinh^2(2r). & (79) \end{aligned}$$

For moderate squeezing the photon distribution function is similar to a Poissonian, but with a narrower width, see e.g., Bachor[38] (p. 234).

Squeezed states can be generated by various nonlinear processes, e.g. degenerate parametric amplification of an initial coherent (=laser) state.

$$\hat{H}_{par} = \hbar\omega\hat{a}^\dagger\hat{a} - k\left[e^{\imath(\omega t+\phi_p)}\,\hat{a}^2 + hc\right]. \qquad (80)$$

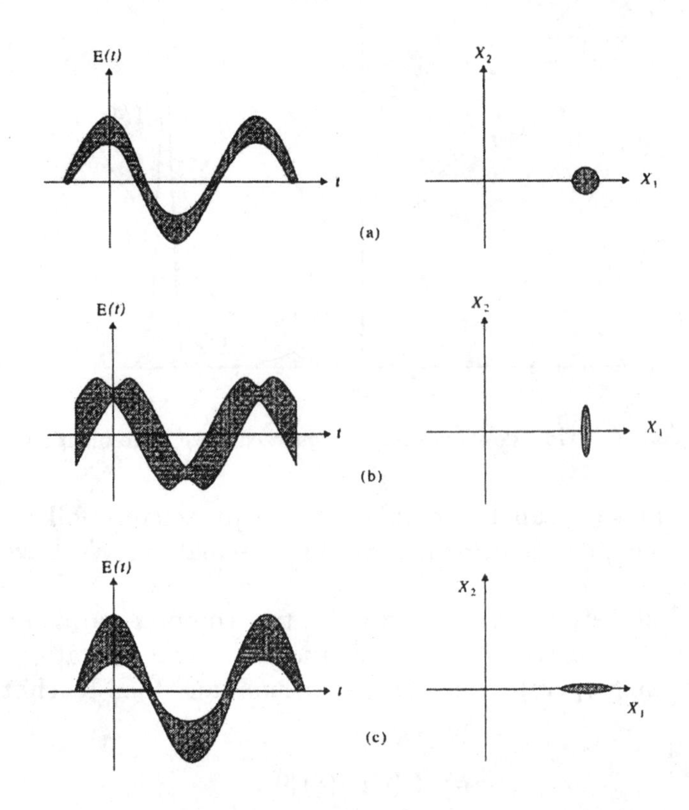

Figure 9. Contours and time dependence of the electrical field for (a coherent state), squeezed state with reduced quantum noise in X_1, and (c) a sqeezed state with reduced noise in X_2. According to Caves[19].

In this model, the pump at $\omega_p = 2\omega$ is treated as a classical source; the coupling constant k is proportional to the second order susceptibility χ_2 of the nonlinear crystal. An initial α–state evolves with $\alpha(t) = \alpha \exp(-\imath\omega t)$ and $\xi(t) = g \exp[-\imath(\omega t + \phi_p - \pi/2)]$. Note, squeezing is sensitive on the phase of the pump.

A nice review of squeezed states with applications has been given by Walls[21], for details see the articles by Stoler[17], Yuen[18], and Caves[19]. Communication by squeezed light has been discussed by Giacobino et al.[22].

Problems:

P2: Calculate the functions $w(t), v(t), u(t)$ in Eq. (71).

P3: Show that $\hat{b}, \hat{b}^\dagger$ are Bose operators, i.e., $[\hat{b}, \hat{b}^\dagger] = 1$.

P4: Which property must the energy spectrum of a system have that its state reproduces after a finite time? (See also Chergui's contribution in this book and Ref.[44]).

P5: Are there photons in a <u>static</u> magnetic field? (Groundstate of Eq. (47)).

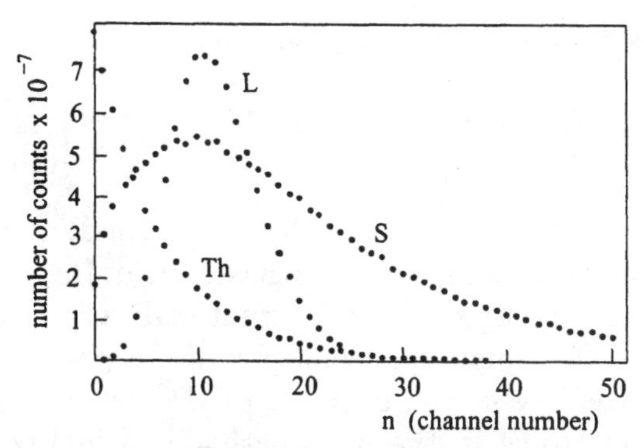

Figure 10. Photon count distribution for a single mode laser (L), thermal light (G), and a superposition of both (S). According to Arecchi[51].

4.4.5. *Thermal (chaotic) photon states*

A single mode of thermal (black body) radiation is described by a statistical operator

$$\hat{\rho} = \frac{1}{Z} e^{-\beta \hat{H}} = \sum_{n=0}^{\infty} b_n \mid n\rangle\langle n \mid, \tag{81}$$

$$b_n = \frac{\bar{n}^n}{(\bar{n}+1)^{n+1}} = \frac{1}{1+\bar{n}} \left(1 + \frac{1}{\bar{n}}\right)^{-n}, \tag{82}$$

$$\langle \hat{N} \rangle = \mathrm{tr}(\hat{\rho}\,\hat{N}) = \frac{1}{e^{\beta\hbar\omega} - 1} = \bar{n}, \tag{83}$$

$$(\Delta \hat{N})^2 = \bar{n}(\bar{n}+1). \tag{84}$$

$\beta = 1/(k_B T)$, $Z = 1/(1 - \exp(-\beta\hbar\omega))$ is the partition function, b_n the Bose–Einstein photon distribution (geometric sequence), and $\bar{n}$ is the mean photon number in the mode. In contrast to a coherent state $\Delta\hat{N}/\bar{n} \to 1$ for $\bar{n} \to \infty$, see Fig. 10.

5. Optical devices and measurements

5.1. PHOTODETECTORS

In a classical description the electrical current of a photoelectric device like a photocell, a photomultiplier, or a photodiode is proportional to the light intensity (energy density) averaged over a cycle of oscillation

$$J_{PD} = \zeta I(t), \tag{85}$$

$$I(t) = < \mathcal{E}(t)^2 >_{cycl} = \mathcal{E}^{(-)}(t)\,\mathcal{E}^{(+)}(t). \tag{86}$$

(A factor 1/2 has been included in the definition of $< \ldots >_{cycl}$). $\mathcal{E}^{(\pm)}$ denote the positive(negative) frequency parts of the electric field (polarization properties and space variables omitted for simplicity)

$$\mathcal{E}(t) = \int_0^\infty \left(\mathcal{E}^{(+)}(\omega)\, e^{-\imath\omega t} + \mathcal{E}^{(-)}(\omega)\, e^{+\imath\omega t} \right) \frac{d\omega}{2\pi}. \tag{87}$$

For a free field $\mathcal{E}^{(\pm)}$ are identical with the first(second) terms of Eq. (17). (Note also the sign convention.) For a stochastic field, the product $\mathcal{E}^{(-)}\mathcal{E}^{(+)}$ has to be additionally averaged on the different realizations of the ensemble. In praxis ζ includes the (quantum) efficiency of the photodetector, too.

In a quantum treatment, the response of the detector arises from the ground state of the atoms in the photocathode to highly excited quasi–free states by absorption of photons. Initially, we have for the combined system "atom plus EMF" $|i>=|a,\{n\}>$. The electrical dipole interaction $\hat{H}_{dip} = -e\hat{\mathcal{E}}\mathbf{r}$ induces transitions to final states $|f>=|b,\{n'\}>$. With the golden rule and summing over all possible final states, we have for the total transition rate

$$p(t) \;=\; \zeta\langle\hat{\mathcal{E}}^{(-)}(\mathbf{r},t)\hat{\mathcal{E}}^{(+)}(\mathbf{r},t)\rangle, \tag{88}$$

where $\hat{\mathcal{E}}^{(\pm)}$ denote the creation/destruction parts of the electrical field operator (in the Heisenberg picture) and ζ includes the atomic dipole transition matrix element. Implicitly, we shall assume a perfect photocathode with unit quantum efficiency so that each absorbed photon causes an atom in the phototube to emit an electron and register a single count during times $t, t+dt$.

A good presentation of the quantum theory of a photodetector has been given by Glauber in the Proceedings of the Les Houches[45] and Fermi summerschools[46].

5.2. INTERFEROMETERS

Interferometers are devices to measure the correlation of the EMF between different space–time points. The prototype is the *Young double slit interference experiment* which is depicted in Fig. 11. Thermal light from a point source is rendered parallel by a lens, passes through a wavelength filter, and then falls on a screen which contains two slits or pinholes (as we assume for simplicity). Interference fringes show up on a second screen placed on the right of the first screen, many wavelengths apart. In the following discussion we shall ignore complications arising from the finite source diameter and consequent lack of perfect parallelism in the illuminating beam, diffraction effects at the pinholes (or slits), etc., in order that attention be focused on the properties of the incident EMF rather than on details of the measuring device.

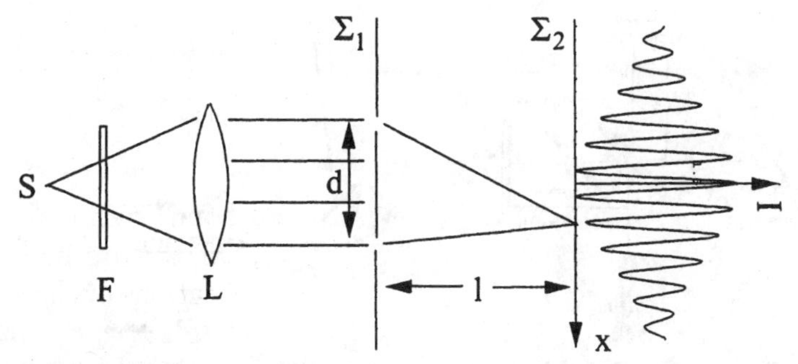

Figure 11. **Arrangement of components for an idealized Young's interference experiment. Interferogram shown in the limit of infinitely small slits and $\lambda \ll d \ll L$. Gaussian spectral filter F.**

Let $\mathcal{E}(\mathbf{r}, t)$ be the electrical field of the radiation at point $\mathbf{r}$ on the observation screen at time t. This field is a superposition of the incident field at points $\mathbf{r}_1, \mathbf{r}_2$ at earlier times t_1, t_2,

$$\mathcal{E}(\mathbf{r}, t) = u_1 \mathcal{E}^{in}(\mathbf{r}_1, t_1) + u_2 \mathcal{E}^{in}(\mathbf{r}_2, t_2), \tag{89}$$

$$t_1 = t - s_1/c, \quad t_2 = t - s_2/c. \tag{90}$$

Coefficients u_1, u_2 depend on the geometry and they are purely imaginary since the secondary waves radiated by the pinholes (or slits) are $\pi/2$ out of phase with the primary beam. For simplicity, we consider identical pinholes and approximate $u_1 = u_2 = u_0 =$const.

The (cycle averaged) intensity of light at position $\mathbf{r}$ can be expressed in terms of the (first order) correlation function $G_1(\mathbf{r}_2, t_2; \mathbf{r}_1, t_1)$

$$I(t) \propto G_1(1,1) + G_1(2,2) + 2\Re G_1(2,1), \tag{91}$$

$$G_1(\mathbf{r}_2, t_2; \mathbf{r}_1, t_1) = \langle \hat{\mathcal{E}}^{(-)}(\mathbf{r}_2, t_2) \hat{\mathcal{E}}^{(+)}(\mathbf{r}_1, t_1) \rangle, \tag{92}$$

where $G_1(2,1)$ is short for $G_1(\mathbf{r}_2, t_2; \mathbf{r}_1, t_1)$ etc. It is seen from Eq. (91) that the intensity on the second screen consists of three contributions: First two terms represent the intensities caused by each of the pinholes in the absence of the other, whereas the third term gives rise to interference effects.

For a symmetric configuration (equal slit width, homogeneous illumination, $G_1(1,1) = G_1(2,2)$)) the visibility of the fringes is given by the magnitude of the normalized correlation function $g_1(1,2)$

$$\mathcal{V} = \frac{I_{max} - I_{min}}{I_{max} + I_{min}} = |g(2,1)|^2, \tag{93}$$

$$g_1(\mathbf{r}_2, t_2; \mathbf{r}_1, t_1) = \frac{G_1(\mathbf{r}_2, t_2; \mathbf{r}_1, t_1)}{\sqrt{G_1(1,1) G_1(2,2)}}. \tag{94}$$

Coherence (i.e. the possibility of interference) of light as measured with the double slit experiment is therefore a measure of correlation in the EMF.

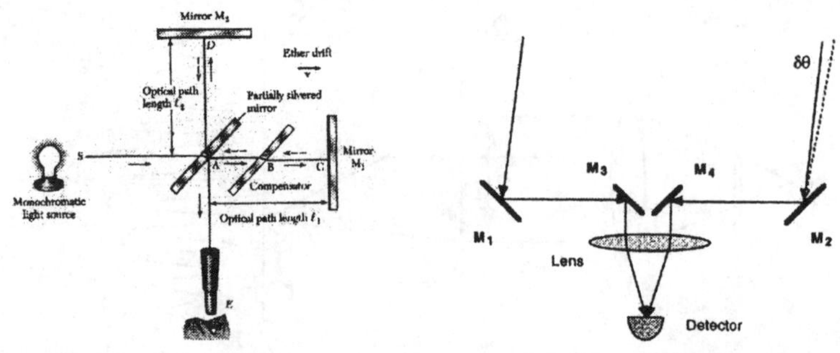

Figure 12. Sketch of the Michelson interferometer (left) and Michelson stellar interferometer (right). With these instruments the temporal and spatial correlation of the EMF can be measured independently. According to Bachor[38].

The Michelson interferometers – as depicted in Figs. 12 – are even better suited to investigate coherence properties as these instruments separately measure the temporal and spatial dependencies of $G_1(2, 1)$.

Prior to the invention of the laser, coherent light was made out of "chaotic" radiation by (wave length) filters and apertures. Here, the visibility in the Michelson interferometers vanishes (and remains zero!) if the distance between the arms becomes larger than the (longitudinal) coherence length $l_{coh} = ct_{coh}$ (point like source of spectral width $\Delta\lambda$) or larger than the (transversal) coherence diameter d_{coh} (monochromatic source of angular diameter $\Delta\theta$)

Coherence time: $t_{coh} = \frac{2\pi}{\Delta\omega} = \frac{\lambda_0^2}{c\Delta\lambda}$

Coherence diameter: $d_{coh} = \frac{2\pi}{\Delta k} = 1.22\frac{\lambda}{\Delta\theta}$

The numerical factor of 1.22 holds for a circular light source. Note, the vanishing of the interference pattern is not the result of interference but of the shift of individual patterns which are produced independently by each frequency component or each volume element of an extended source[41].

The first star which was measured by Michelson and Pease with the 20ft (=6m armlength) stellar interferometer at Mt. Wilson observatory was the supergigant Betelgeuze in the stellar configuration of Orion ($\Delta\theta = 43\times10^{-3}$ seconds of an arc). Turbulence in the atmosphere severely affected these measurements and all efforts to increase the armlength proved to be unsuccessful. Hence, in total only 6 stars could be studied.

Examples:

red Cd–line: $\lambda_0 = 643.8$nm, $\Delta\lambda = 0.0013$nm, $l_{coh} = 32$cm.
sun: $\lambda_0 = 2.4$m, $\Delta\theta = 32'$, $d_{coh} = 316$m.

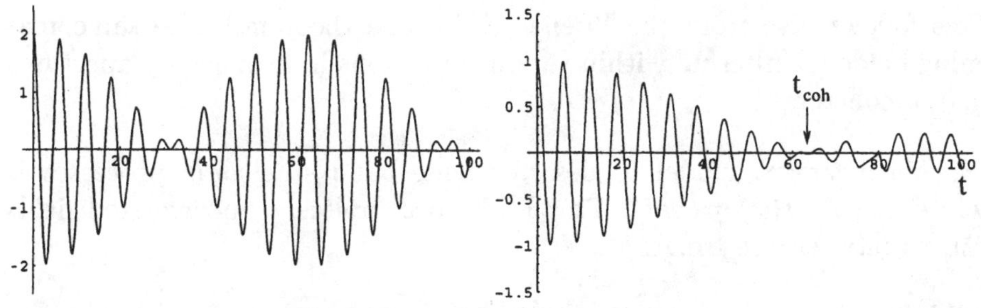

Figure 13. First order coherence functions as a function of time. (Left:) Two modes with $\omega_j = (1 \pm 0.05)\omega_0$ (left) and (right:) many uncorrelated modes with a "box" spectrum centered at ω_0 and full width $\Delta\omega = 0.1\omega_0$.

5.2.1. *Examples for* G_1

We study two examples for the (first order) classical coherence functions.

a) Single mode (deterministic/stochastic) field.

$$G_1(\mathbf{r}_2, t_2; \mathbf{r}_1, t_1) = < |a_{\mathbf{k},\sigma}|^2 > \exp\left(-\imath\left[\mathbf{k}(\mathbf{r}_2 - \mathbf{r}_1) - \omega_{\mathbf{k}}(t_2 - t_1)\right]\right) \quad (95)$$

Obviously, this correlation function is (apart from the numerical value of $< |a_{\mathbf{k},\sigma}|^2 >$) the same for a deterministic and a stochastic single mode field and displays maximum contrast.

b) Many statistically independent modes (of equal polarization) and intensity profile: $I(\omega_{\mathbf{k}}) = |\mathcal{E}(\omega)|^2$, $< \hat{a}^*_{\mathbf{k}',\sigma'} a_{\mathbf{k},\sigma} >= I(\omega_{\mathbf{k}})\delta_{\mathbf{k},\mathbf{k}'}\delta_{\sigma,\sigma'}$.

$$\mathcal{E}^{(+)}(\mathbf{r}, t) = \sum_{\mathbf{k}} a_{\mathbf{k},\sigma}\, e^{\imath(\mathbf{k}\mathbf{r} - \omega_{\mathbf{k}} t)}, \quad (96)$$

$$G_1(\mathbf{r}_2, t_2; \mathbf{r}_1, t_1) = \sum_{\mathbf{k}} I(\omega_{\mathbf{k}})\, e^{-\imath[\mathbf{k}(\mathbf{r}_2 - \mathbf{r}_1) - \omega_{\mathbf{k}}(t_2 - t_1)]}. \quad (97)$$

In particular, we have at the same space point, $\mathbf{r}_2 = \mathbf{r}_1 = \mathbf{r}$, (as measured by a Michelson interferometer)

$$G_1(t_2 - t_1) = G_1(\mathbf{r}, t_2; \mathbf{r}, t_1) = \int_0^\infty I(\omega)\, e^{\imath\omega(t_2 - t_1)} \frac{d\omega}{2\pi} \quad (98)$$

For a Gaussian line centered at $\omega_0 > 0$

$$I(\omega) = I_0\, e^{-\frac{(\omega - \omega_0)^2}{2(\Delta\omega)^2}}, \quad (99)$$

$$G_1(t) = \sqrt{2} I_0 \Delta\omega\, e^{-\frac{(\Delta\omega t)^2}{2}}\, e^{\imath\omega(t)}. \quad (100)$$

Some examples are depicted in Figs. 11 and 13.

Concerning their coherence properties, the filtered many mode field is virtually indistinguishable from the single mode field provided $t < t_{coh}$.

(This follows also from the Wiener–Khinchine theorem.) The same reasoning holds for filtering within various directions in **k**–space by apertures (spatial coherence).

NB: A convenient model to describe wide–band stochastic processes is “white noise”, $I(\omega)$ =const. This leads to completely uncorrelated fields with no interference fringes.

Problem:
P6: Study the case of two statistically independent modes of equal intensity and discuss the interference pattern for the Na-D doublett $\lambda_0 = 589$nm, $\Delta\lambda = 0.6$nm (Fizeau 1862).

5.3. INTENSITY CORRELATIONS: HANBURY–BROWN & TWISS EFFECT

In the previous section we considered first order field correlation. For fields with identical spectral properties the classical and quantum treatments leads to the same result. This can be different when studying intensity (=photon) correlations.

A bit of history: The intrinsic problems of the Michelson stellar interferometer – mechanical instability at armlength longer than 6m, and the effect of atmospheric turbulence, were overcome by the invention of the *intensity interferometer*. The idea dates back to 1949 where R. Hanbury Brown, a radio astronomer at Jodrell Bank, was trying to design a radio interferometer which would solve the intriguing problem of measuring the angular size of the most prominent radio sources: Cygnus A and Cassiopeia A. If, as some people thought at that time their angular size is as small as the largest visible stars, then global base lines would be needed and a coherent superposition of the signals would be impossible in praxis (around 1950!).

The new and unconventional idea was to correlate (low frequency) intensities instead of superimposing (high frequency) amplitudes. First, a pilot model was built in 1950 and was tested by measuring the angular diameter of the sun at 2.4m wavelength, and, subsequently, the radio sources Cygnus A and Cassiopeia A. The intermediate–frequency outputs of the completely independent superheterodyne receivers were rectified in square law detectors and bandpass filtered (1…2.5KHz). Then, the LF outputs were brought together by radio links (or telefone!). After analogue multiplication of the LF signals and integration, the correlator output

$$G_2^{cl}(\mathbf{r}_2, t_2; \mathbf{r}_1, t_1) = \langle I(\mathbf{r}_2, t_2)\, I(\mathbf{r}_1, t_1) \rangle, \tag{101}$$

displayed the expected correlations, see Fig. 14. (A constant term has been subtracted by LF bandpass filtering so that $G_2(2,1) \to 0$ for large

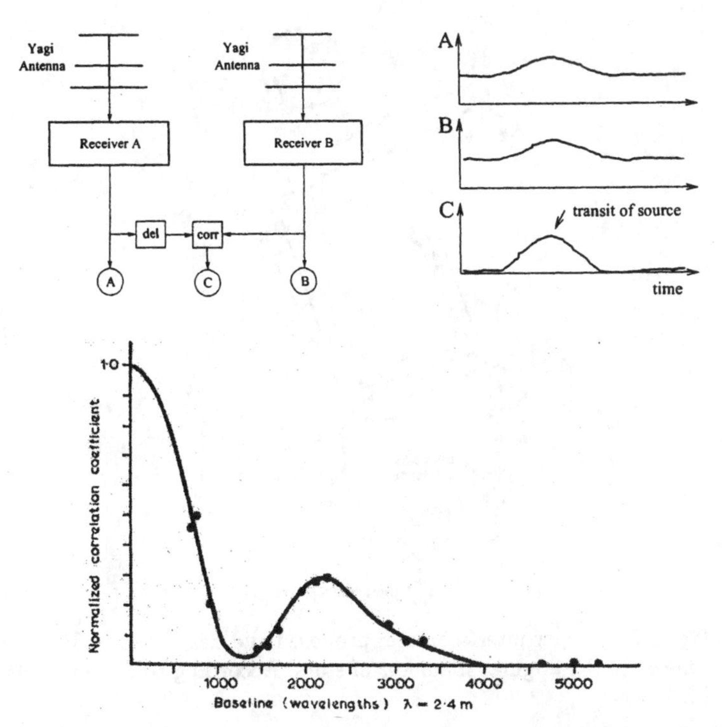

Figure 14. (a) Sketch of the RF interferometer at $\lambda = 2.4$m wavelength. (b) Output of the individual receivers A,B and correlation C showing the transit of a radio source through the arial beam. (c) Normalized correlation function for the radio source Cygnus A which consists of two almost equal components with an angular diameter of 45" and a separation of 1'25". According to Hanbury Brown[42].

separation of $\mathbf{r}_2$ and $\mathbf{r}_1$)). However, to the great disappointment of the investigators, the adventure was over at a separation of less than 5km. This experiment could have been done by conventional technique!

For many independent modes, Eq. (101) can be evaluated in the same way as for G_1,

$$G_2^{cl}(\mathbf{r}_2, t_2; \mathbf{r}_1, t_1) = \sum_{\mathbf{k}} \left(\langle |a_{\mathbf{k}}|^4 \rangle - 2 \langle |a_{\mathbf{k}}|^2 \rangle^2 \right) + |G_1(0,0;0,0)|^2 + |G_1(\mathbf{r}_2, t_2; \mathbf{r}_1, t_1)|^2. \tag{102}$$

Moreover, for Gaussian thermal light $\langle |a_{\mathbf{k}}|^4 \rangle = 2\langle |a_{\mathbf{k}}|^2 \rangle^2$, so that the first term of Eq. (102) drops out. In all other cases this contribution is negative so that the "contrast" in $G_2(2,1)$ for adjacent $(\mathbf{r}_2, t_2)$, $(\mathbf{r}_1, t_1)$ and distant arguments is smaller than for thermal radiation. For thermal radiation, intensity correlation measurements yield the same information as conventional first order coherence experiments, e.g. using the Michelson interferometers.

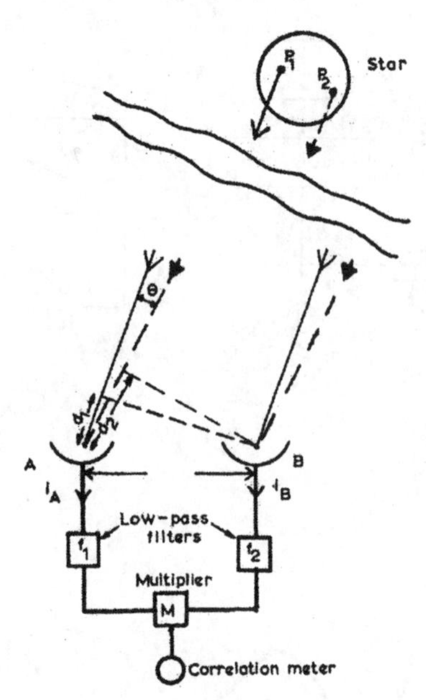

Figure 15. Optical intensity interferometer proposed and developed by Hanbury Brown and Twiss to measure the angular diameter of stars. According to Hanbury Brown[42].

The optical analogue of the intensity interferometer seemed to be straightforward: Antennas and receivers will respectively be replaced by mirrors and photodetectors, as sketched in Fig. 15. In principle, the theory is the same for all wavelengths but the trouble of course was worrying about photons. In the RF spectrum the energy flows rather smoothly whereas in the optical region energy comes in "photon–bursts", see Fig. 3. A correlator (or coincidence counter) measures the combined absorption of photons at different space time points $(\mathbf{r}_2, t_2)$ and $(\mathbf{r}_1, t_1)$. As a result, we have

$$G_2(\mathbf{r}_2, t_2; \mathbf{r}_1, t_1) = \langle \hat{\mathcal{E}}^{(-)}(\mathbf{r}_2, t_2)\hat{\mathcal{E}}^{(-)}(\mathbf{r}_1, t_1)\hat{\mathcal{E}}^{(+)}(\mathbf{r}_2, t_2)\hat{\mathcal{E}}^{(+)}(\mathbf{r}_1, t_1)\rangle. \quad (103)$$

Note, the sequence of operators matters, creation and destruction operators are in "normal order" (creation operators left to the destruction operators). Nevertheless, for thermal radiation, the classical result given by Eq. (102) remains valid.

If one thinks in terms of photons one must accept that thermal photons at two well separated detectors are correlated – they tend to to "arrive" in pairs ("photon bunching")! But how, if the photons are emitted at random in a thermal source, can they appear in pairs at two well separated detectors? What about the sacred number–phase uncertainty relation?

$$\Delta\hat{N}\Delta\hat{\Phi} \geq 1. \quad (104)$$

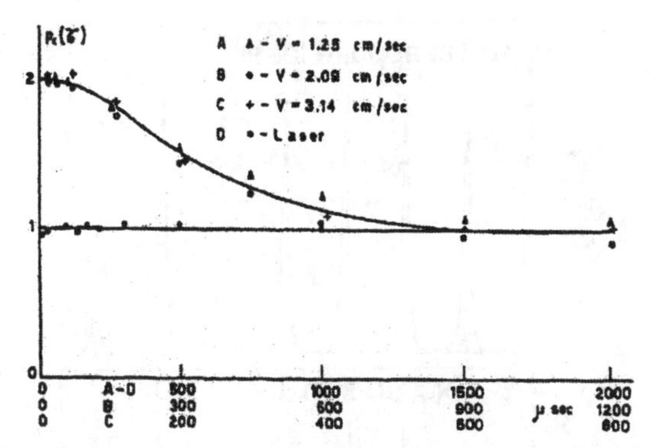

Figure 16. Photon coincidences for single mode chaotic light and laser light. Coherence time of the chaotic light depends on the speed v of the rotating ground glass disk. According to Arecchi et al.[20].

Wouldn't a photon number measurement destroy phase relations and, hence, all interference phenomena? However, for an (intensity) interferometer the absolute phase is not relevant, only the phase difference matters and this difference is not touched by Eq. (104).

Eventually, this problem was settled by experiment which clearly shows that photon bunching exists in thermal radiation [23, 24]. Later on, a well functioning stellar interferometer was built in Australia. More on the exciting scientific story about this instrument and its history can be found in Hanbury Brown's book[42].

Today, the photon bunching effect can be simply demonstrated with an artificial "chaotic" source which synthesizes pseudothermal light by passing laser radiation through a rotating ground glass disk with long adjustable coherence times ("Martienssen lamp")[25], see Fig. 16.

It is instructive to define a coincidence ratio

$$R = \frac{C - C_{rand}}{C_{rand}} = \frac{(\Delta n)^2 - < n >}{(< n >)^2}, \tag{105}$$

where $C \propto G_2(1,1) =< \hat{N}(\hat{N}-1) >$. The number of random coincidences is proportional to $C_{rand} = G_2(2,1) =< \hat{N} >^2$ when the separation of $\mathbf{r}_2, t_2$, $\mathbf{r}_1, t_1$ is much larger than the coherence area/time.

Coherent states:	$(\Delta N)^2 = \bar{n}$	$R = 0,$
thermal states:	$(\Delta N)^2 = \bar{n}(\bar{n}+1)$	$R = 1,$
number states:	$(\Delta N)^2 = 0$	$R = -\frac{1}{n}.$

Classical states have photon number distributions which are always broader than a Poissonian, i.e., $(\Delta N)^2 \geq \bar{n}$, hence, the correlation ratio is positive: Classical states always show "photon bunching". The α states as generated

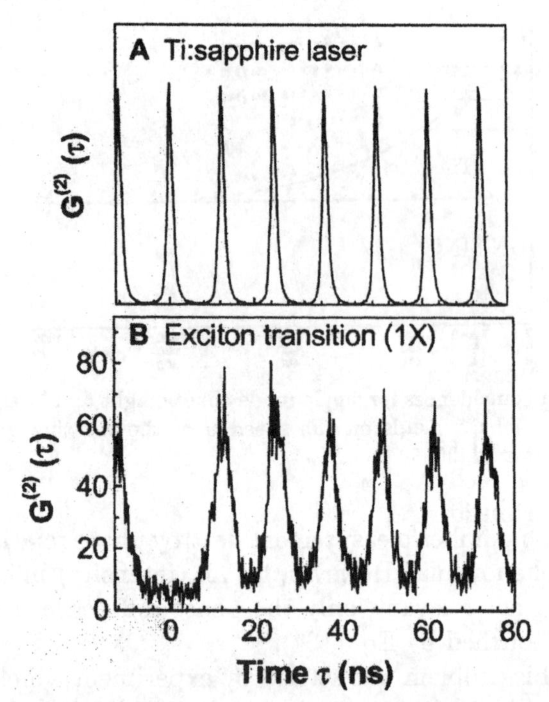

Figure 17. Single photon turnstile device. (A) Unnormalized second order correlation function of a mode locked Ti:sapphire laser (FWHM=250 fs) and (B) a single quantum dot excitonic ground state (1X) emission under pulsed excitation conditions (82MHz). According to Michler et al.[30].

by an amplitude stabilized laser, represent the optimum classical state with respect to photon fluctuations.

On the other hand, states which have less photon number fluctuations than a Poissonian, e.g., the number states, show "antibunching", i.e., the photons prefer to "come" not too close, see Fig. 22. In particular, the single photon state $|n = 1 >$ is the most nonclassical state one can think of! Obviously, photon bunching is not a "typical Bose property".

The generation of nonclassical light (which still showed "photon bunching") was first demonstrated in 1977 by Kimble, Dagenais, and Mandel[27]; the first clear evidence for antibunching, $R \approx 0$, was presented by Diedrich and Walther[28] in 1987, using a single Mg–Ion in a Paul–trap.

For a review about photon antibunching, see the article by Paul[29]. Presently, nonclassical photon states became attractive in semiconductor optics in connection with quantum communication. For instance Michler et al.[30] have developed a "Quantum dot single photon turnstile device", see Fig. 17.

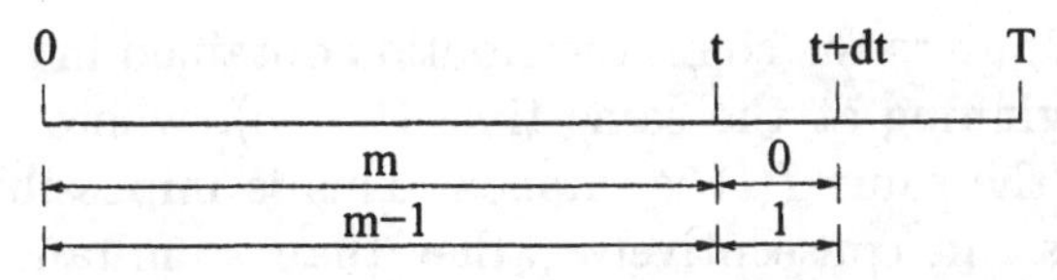

Figure 18. Time intervals used for the derivation of $P_m(T)$.

5.4. PHOTON COUNTING

The number of photons which a counter records in any interval of time fluctuates randomly. In a simple counting experiment we may imagine that the counter is exposed to the radiation field for a fixed time interval T. After a time delay T_{del}, which is long compared to the coherence time of the light, the original experiment is repeated, and a second number of counted photons is recorded and so on. The results can be expressed by a probability distribution $P_m(T)$ for the counting of m photons during an observation time T.

We consider a particular period of counting $[t, t+T]$ as shown in Fig. 18. There are two ways in which m photons can be counted in the periods between times t and $t + dt$. The probability that more than one photon being counted during the time interval dt is proportional to $(dt)^2$ and, thus, neglegible.

$$P_m(t + dt) = P_m(t)\,[1 - p(t)\,dt] + P_{m-1}(t)\,p(t)\,dt \tag{106}$$

Rearranging terms and using $P_m(t+dt) - P_m(t) = \dot{P}_m(t)dt + O(dt)^2$ we obtain a chain of coupled differential equations

$$\frac{dP_m(t)}{dt} = \zeta I(t)\,[P_{m-1}(t) - P_m(t)]\,, \tag{107}$$

$$P_{-1} = 0, \quad P_0(0) = 1, \quad P_m(0) = 0 \quad (m > 1), \tag{108}$$

which can be solved by recursion.

The probability that no photon to be recorded during time t becomes

$$P_0'(t) = -\zeta\, I(t)\, P_0(t), \tag{109}$$

$$P_0(T) = e^{-\zeta\, \bar{I}(T)\, T}, \tag{110}$$

$$\bar{I}(T) = \frac{1}{T}\int_0^T I(t')\,dt'. \tag{111}$$

$\bar{I}(T)$ is the mean intensity during the observation time T. The remaining counting functions $P_m(T)$ can be obtained from Eq. (107), beginning with $m = 1$ and proceeding to higher values. As a result, we obtain

$$P_m(T) = \frac{[\zeta \bar{I}(T) T]^m}{m!}\, e^{-\zeta\, \bar{I}(T)\, T}. \tag{112}$$

Equation (112) gives the count distribution obtained in a series of measurements all beginning at the same time ($t = 0$), where the same time duration T, and the same $I(t)$ is implied. This is impossible in practice: Counting periods run consecutively rather than simultaneously. Photon measurements imply the absorption of photons so that a single photon can be counted only once! The intensity $\bar{I}(T)$ in general fluctuates for different members of the ensemble and the measured photon count distribution is an average of $P_m(T)$ over a large number of different starting times (which are separated in time by a period much larger than the coherence time).

$$\bar{P}_m(T) = \langle P_m(T) \rangle. \tag{113}$$

A nice introduction to photon counting is still Loudon's book[34].

5.4.1. *Examples*

a) Constant intensity (amplitude stabilized single mode laser).
$I(t) = I_0$ is constant so that the quantity to be averaged is time–idependent.

$$\bar{P}_m(T) = \exp(-\bar{m})\,\frac{\bar{m}^m}{m!}, \quad \bar{m} = \zeta I_0 T. \tag{114}$$

This is a *Poissonian distribution* with a mean photon count number $\bar{m}$, see Fig. 19. This distribution has been already discussed in Section 4.4.3.

The fluctuations which occur for a beam of constant intensity are called *particle fluctuations*. They are due to the discrete nature of the photoelectric process in which energy can be removed from the light beam only in whole quanta $\hbar\omega$.

b) Chaotic (thermal) light, long time limit ($T \gg t_{coh}$).
Another important case for which the Poisson–distribution Eq. (114) holds, follows from the fact that $\bar{I}(T)$ can be constant even if $I(t)$ is a fluctuating quantity. This case holds for chaotic light (of arbitrary type) if the time of measurement is much larger than the coherence time of the light, so that all fluctuations are averaged out during a long time period.

c) Chaotic (thermal) light, short time limit ($T \ll t_{coh}$).
The probability distribution for the instantaneous intensity of thermal light is given by Eq. (29). With the usual ergodic hypothesis the time average in Eq. (113) is converted to an ensemble average over the distribution $p(I) = \exp(-I/I_{av})/(\pi I_{av})$, Eq. (29), leading to

$$\bar{P}_m(T) = \int_0^\infty p(I)\,P_m(T)\,dI = \frac{\bar{m}^m}{[\bar{m}+1]^{m+1}} = b_m. \tag{115}$$

b_m is the Bose–Einstein distribution function, see Fig. 20, which we have already met in Section 4.4.5.

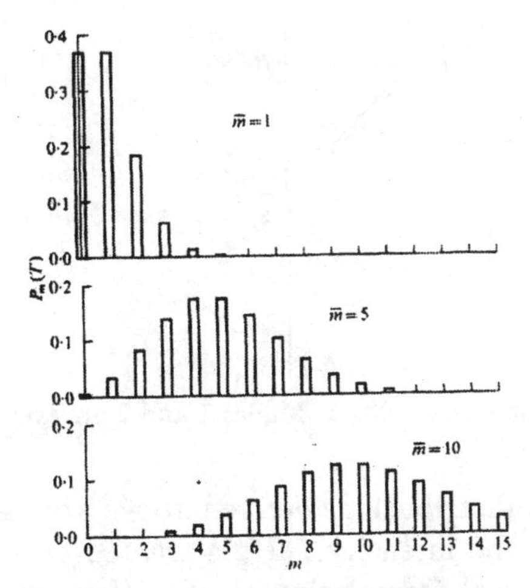

Figure 19. **Poisson form of the photon count distribution for light beams of constant intensity (Single mode laser well above threshhold). According to Loudon[34].**

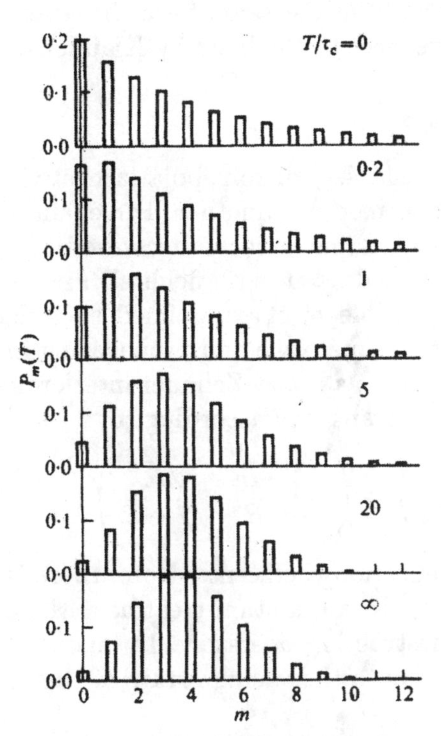

Figure 20. **Photon count distribution for chaotic (thermal) single mode light for $\bar{m} = 4$ and different counting times T. According to Loudon[34].**

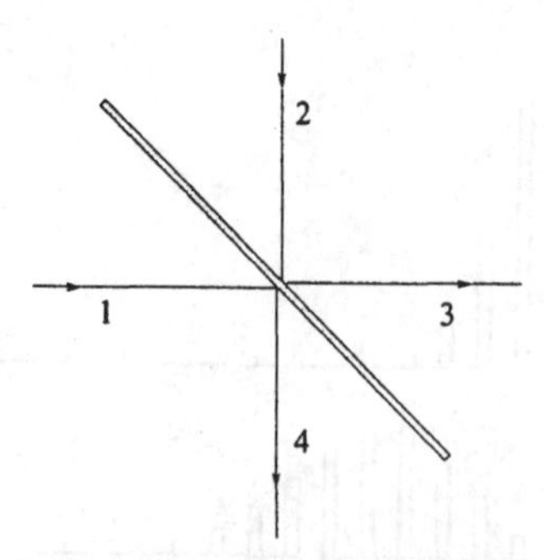

Figure 21. Sketch of a beam slitter. Modes 1 and 2 are transformed into 3, 4.

The photon count distributions derived above are based on a semiclassical approach where the intensity $\bar{I}(T)$ is treated as a classical quantity. The quantum mechanical formulation (in the Heisenberg picture) has to take into account that the operators $\hat{\mathcal{E}}^{(-)}(\mathbf{r},t)$, $\hat{\mathcal{E}}^{(+)}(\mathbf{r},t)$ do not commute. Formally, the result is similar to Eq. (112), but the evaluation is much more laborious than in the semiclassical case. For coherent and thermal states the count distributions have the same form, however, this is not true in general. For details we refer to the book by Klauder and Sudarshan[37].

5.5. BEAM SPLITTERS

Optical components like lenses, mirrors, polarizers etc., are used in quantum optics to transform one mode in another. For example, let us consider a beam splitter which transforms an incoming wave beam with field $\mathcal{E}_1$ into a transmitted and a reflected beam with fields $\mathcal{E}_3, \mathcal{E}_4$, respectively. However, there is also a second possible input axis defined, such that $\mathcal{E}_2$ would produce the output waves in the same place and propagating in the same direction as $\mathcal{E}_3$ and $\mathcal{E}_4$ (e.g. as used in a Mach–Zehnder interferometer). The complex amplitudes of the EMF transform according to

$$\begin{pmatrix} a_3 \\ a_4 \end{pmatrix} = \begin{pmatrix} \sqrt{1-R} & \sqrt{R} \\ -\sqrt{R} & \sqrt{1-R} \end{pmatrix} \begin{pmatrix} a_1 \\ a_2 \end{pmatrix} . \tag{116}$$

R is the (intensity) reflection coefficent. (Note, there is a phase jump of π of the reflected beams). In a quantum treatment the complex amplitudes will be replaced by destruction operators. In quantum optics, the case of "no incident wave" refers to the vacuum state of that mode rather than to "zero field".

In quantum optics amplitudes a_j become operators $\hat{a}_j$ and Eq. (116) represents a unitary basis transformation. A beam splitter does not "split" photons, rather it acts as a random selector which divides the incident flow of photons in a reflected and a transmitted one. As a consequence, the

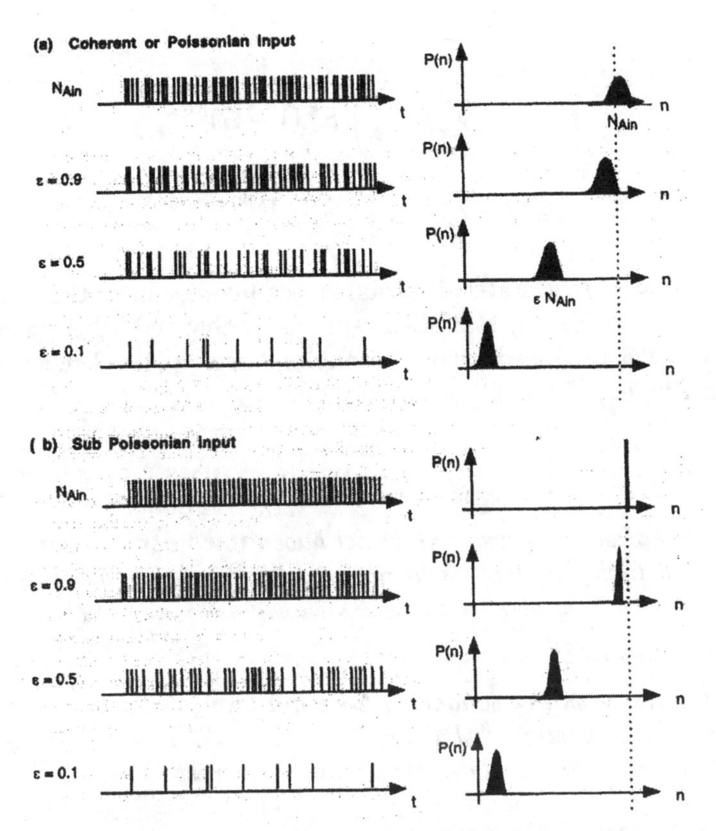

Figure 22. Change of photon statistics of an n photon state after passing a beam splitter.(a) Coherent state, (b) n-photon state. According to Bachor[38].

photon statistics of the reflected/transmitted beams correspond to that of the input beam after a random selection process has taken place. For a coherent state with a Poissonian distribution, a random selection yields again a Poissonian, hence a coherent state remains a coherent state after reflection or transmission through a mirror, yet with a reduced value of α.

The situation may be different for states with a non–Poissonian photon distribution. For an incident $|n>$–photon state the probability that $k(\leq n)$ photons are reflected ($n-k$ being transmitted) becomes

$$p_k^{(n)} = \binom{n}{k} R^k \, (1-R)^{n-k}, \quad k = 0, 1, \ldots n. \tag{117}$$

The binominal coefficient arises from the indistinguishability of the photons. Hence, an incident photon state with distribution p_n transforms according to

$$p_k^{ref} = \sum_n p_n \binom{n}{k} R^k (1-R)^{n-k}, \tag{118}$$

$$p_k^{tr} = \sum_n p_n \binom{n}{k} (1-R)^k R^{n-k}. \tag{119}$$

In addition to coherent states, thermal states likewise have the remarkable property that the photon statistics remains unchanged when passing the splitter (yet with a reduced mean photon number $\bar{k} = (1-R)\bar{n}$, see Fig. 22). For details see, e.g., Paul[32] or Bachor[38].

6. Outlook

Die ganzen Jahre bewusster Grübelei haben mich der Antwort der Frage "Was sind Lichtquanten" nicht näher gebracht. Heute glaubt zwar jeder Lump, er wisse es, aber er täuscht sich...

Literal translation:

All the years of willful pondering have not brought me any closer to the answer to the question "what are light quanta". Today every good-for-nothing believes he should know it, but he is mistaken...

ALBERT EINSTEIN[2]

But, in contrast to Einstein, most of us have given up any hope for objective realism...

For a discussion on conceptual difficulties and different interpretations of Quantum Mechanics see Costa's interdisciplinary article in this book.

7. Acknowledgement

Thanks to Prof. Di Bartolo and his team, the staff of the Majorana Center, and all the participants which again provided a wonderful time in a stimulating atmosphere.

[2] In a letter to M. Besso, 1951. See first page of Ref.[32].

References

1. M. Planck, Verh. Dtsche. Phys. Ges. **2**, 202, 237 (1900), *Theorie der Wärmestrahlung*, 6. Auflage, J. A. Barth, Leizpig (1966).
2. P. Lenard, *Ueber die lichtelektrische Wirkung*, Annalen der Physik, **8**, 149–198 (1902).
3. A. Einstein, *Über einen die Erzeugung und Verwandlung des Lichts betreffenden heuristischen Gesichtspunkt*, Annalen der Physik, **17**, 132 (1905).
4. G. Breit, *Are Quanta Unidirectional?*, Phys. Rev. **22**, 314 (1923).
5. A. H. Compton, *The Spectrum of Scattered X-Rays*, Phys. Rev **22**, 409 (1923).
6. G. I. Taylor, Proc. Cambridge Phil. Soc. **15**, 114 (1909).
7. A. J. Dempster and H. F. Batho, *Light Quanta and Interference*, Phys. Rev. **30**, 644 (1927).
8. E. O. Lawrence and J. W. Beams, *The element of time in the photoelectric effect*, Phys. Rev. **32**, 478 (1928).
9. L. Janossy, *Experiments and Theoretical Considerations Concerning The Dual Nature of Light*, in H. Haken and M. Wagner (eds.), Cooperative Phenomena, Springer Verlag, Berlin (1973).
10. A. T. Forrester, R. A. Gudmundsen, and P. O. Johnson, *Photoelectric Mixing of Incoherent Light*, Phys. Rev. **99**, 1691 (1955).
11. P. Grangier, G. Roger, and A. Aspect,*Experimental Evidence for Photon Anticorrelation Effects on a Beam Splitter: A New Light on Single-Photon Interferences.* Eur. Phys. Lett. **1**, 173 (1986). See also Physics World, Feb. (2003).
12. P. A. M. Dirac, *The Quantum Theory of the Emission and Absorption of Radiation*, Proc. Roy. Soc. A **114**, 243 (1927), see also *The Principles of Quantum Mechanics*, fourth edition, Oxford, At the Clarendon Press (1958).
13. E. Schrödinger, *Die Naturwissenschaften*, **28**, 664 (1926), reprinted in: *Collected Papers on Wave Mechanics by E. Schrödinger*, Chelsea, New York (1982).
14. J. C. Mather et al., *A preliminary Measurement of the Cosmic Microwave Background Radiation by the Cosmic Backgroung Explorer (COBE) Satellite*, The Astrophysical Journal, **354**, L37-L40, (1990).
15. R. J. Glauber, *The Quantum Theory of Optical Coherence*, Phys. Rev. **130**, 2529 (1963), *Coherent and Incoherent States of the Radiation Field*, Phys. Rev. **131**, 2766 (1964), and in Ref.[45, 46].
16. S. F. Jacobs, *How Monochromatic is Laser Light*, Am. J. Phys. **47**, 597 (1979).
17. D. Stoler, *Equivalence Class of Minimum Uncertainty Packets*, Phys. Rev. D**1**, 3217 (1970).
18. H. P. Yuen, *Two-photon coherent states of the radiation field*, Phys. Rev. A**13**, 2226 (1976).
19. C. M. Caves, *Quantum-mechanical noise in an interferometer*, Phys. Rev. D**23**, 1693 (1981).
20. F. T. Arecchi, E. Gatti, and A. Sona, *Time distribution of Photons From Coherent and Gaussian Sources*, Phys. Lett. **20**, 27 (1966).
21. D. F. Walls, *Squeezed states of light*, Nature **306**, 141 (1983).
22. E. Giacobino, C. Fabre, and G. Leuchs, Physics World **2**, Feb., p.31 (1989).
23. R. Hanbury Brown and R. Q. Twiss, *Correlations ...*, Nature **177**, 27 (1956).
24. G. A. Rebka and R. V. Pound, *Time-Correlated Photons*, Nature **180**, 1035 (1957).
25. W. Martiensen and E. Spiller, *Coherence and Fluctuations in Light Beams*, Am. J. Phys. **32**, 919 (1964).

26. H. J. Kimble, M. Dagenais, and L. Mandel, *Photon Antibunching in Resonance Fluorescence*, Phys. Rev. Lett **39**, 691 (1977).
27. M. Dagenais and L. Mandel, *Investigations of two-time correlations in photon emissions from a single atom*, Phys. Rev. A**18**, 2217 (1978).
28. F. Diedrich and H. Walther, *Nonclassical Radiation of a Single Stored Ion*, Phys. Rev. Lett. **58**, 203 (1987).
29. H. Paul, *Photon Antibunching*, Rev. Mod. Phys. **54**, 1061 (1982).
30. P. Michler et al., *A Quantum Dot Single-Photon Turnstile Device*, Science **290**, 2282 (2000).
31. S. Strauf, P. Michler, M. Kluder, D. Hommel, G. Bacher, and A. Forchel, Phys. Rev. Lett. **89**, 177403-1 (2002).
32. H. Paul, *Photonen. Eine Einführung in die Quantenoptik*, B. G. Teubner Stuttgart, Leipzig (1999).
33. H. Haken, *Light*, Vols. I and II, North Holland (1981).
34. R. Loudon, *The Quantum Theory of Light*, Clarendon Press, Oxford (1973).
35. M. O. Scully and S. S. Zubairy, *Quantum Optics*, Cambridge University Press (1999).
36. W. H. Louissell, *Quantum Statistical Properties of Radiation*, Wiley (1973).
37. Klauder and E. G. C Sudarshan, *Quantum Optics*, W. A. Benjamin (1968).
38. H. A. Bachor, *A Guide to Experiments in Quantum Optics*, Wiley–VCH (1998).
39. E. G. Harris, *A Pedestrian Approach to Quantum Field Theory*, Wiley (1972).
40. L. D. Landau and E. M. Lifshitz, Vol. IVA, *Relativistische Quantenfeldtheorie*, Akademie Verlag, Berlin (1970).
41. M. Born and E. Wolf, *Principles of Optics*, Pergamon Press, 2nd ed., Oxford (1964).
42. R. Hanbury Brown, *The Intensity Interferometer*, Taylor and Francis (1974).
43. K.-H. Patke, D. Oberhauser, V. G. Lyssenko, J. M. Hvam, and G. Weimann, *Coherent generation and interference of excitons and biexcitons in* $GaAs/Al_xGa_{1-x}As$ *quantum wells*, Phys. Rev. B **45**, 2413 (1993).
44. K. Razi Naqvi and S. Waldenstrøm, *Revival, Mirror Revival and Collapse may occur even in a harmonic Oscillator Wavepacket*, Physica Scripta **62**, 12 (2000).
45. C. de Witt, A. Blandin, and C. Cohen–Tannoudji (eds), *Quantum Optics and Quantum Electronics*, Gordon and Breach (1965).
46. R. J. Glauber (ed.), *Quantum Optics*, Academic Press, New York (1969).
47. L. Mandel and E. Wolf, (eds.), *Coherence and Quantum Optics*, Plenum (1973).
48. N. Kroll, *Quantum Theory of Radiation*, in Ref.[45].
49. R.v.Baltz and C.F. Klingshirn, *Quasiparticles and Quasimomentum*, in: *Ultrafast Dynamics of Quantum Systems: Physical Processes and Spectroscopic Techniques*, NATO ASI series B Physics: Vol 372, edited by B. Di Bartolo, Plenum (1999).
50. J. F. Clauser *Localization of Photons*, in Ref.[47].
51. F. T. Arecchi, *Photocount distributions and field statistics*, in Ref.[46].

Solutions of the Problems

P1: Time dependent Schrödinger equation: $\dot{x}_c = p_c$, $\dot{p}_c = -\omega_0^2 x_x + f(t)$.

P2: $w(t) = [1 - \kappa \exp(-2\imath t)] \,/\, [1 + \kappa \exp(-2\imath t)]$, $u(t)$ from normalization.
$v(t) = \nu / [\exp(\imath t) + \kappa \exp(-\imath t)]$, κ, ν are constants.

P3: Use $\hat{b} = \mu \hat{a} + \nu \hat{a}^\dagger$, Eq. (43), and $\cosh^2 x - sinh^2 x = 1$.

P4: Energies can be non-equidistant but must be multiples of a unit.

P5: Yes! Eq. (47) is a displaced oscillator. $\langle \hat{N} \rangle = \sum_{\mathbf{k},\sigma} |\jmath_{\mathbf{k},\sigma}|^2 / (2\epsilon_0 \hbar \omega_{\mathbf{k}}^3)$.

P6: See Born and Wolf[41], p. 320.

4. CARRIER-WAVE NONLINEAR OPTICS

MARTIN WEGENER
Institut für Angewandte Physik, Universität Karlsruhe (TH), Wolfgang-Gaede-Straße 1, 76131 Karlsruhe, Germany

"The dream of intensifying light is as old as civilization. Legend has it that Archimedes focused the sun's rays[1] with a giant mirror to set the Roman fleet afire at Syracuse in 218 B.C. Although that story is a myth, it is true that around 200 B.C. another Greek, Diocles, had invented the first ideal focusing optic, a parabolic mirror. Two millenia later mirrors and quantum mechanics were put together to make the most versatile of high-intensity light sources: the laser."[2]

1. Introduction

Today, laser pulses directly out of a mode-locked oscillator as short as five femtoseconds ($1\,\text{fs} = 10^{-15}\,\text{s}$) – equivalent to merely two cycles of light – are available with focused intensities around $10^{13}\,\text{W/cm}^2$. This enormous progress in the generation of **short <u>and</u> intense** femtosecond laser pulses has given access to a completely new regime of light-matter interaction and to a previously unexplored branch of nonlinear optics. In this series of lectures we want to give some general background in the spirit of a tutorial before we review some of our recent corresponding work on semiconductors.

This series of lectures is organized as follows: Section two is concerned with the properties of **short** – i.e., few optical cycles long – laser pulses derived directly from a mode-locked laser. In particular, we want to introduce the carrier wave and the pulse envelope and explain the meaning of the so called carrier-envelope offset (CEO) phase which turns out to be important for nonlinear optics as well as for frequency metrology. Indeed, we will see that the knowledge from the "ultrafast frontier" of spectroscopy can also be used at the "ultraslow frontier", i.e., for ultraprecise frequency domain spectroscopy. In the third section, we will define what we actually mean by **intense**. Quoting the intensity in units of W/cm^2 is precise, however, not really helpful for actual problems. One rather would like to compare for example an energy corresponding to the laser intensity with some other characteristic energy of the system. In this section

[1] The sun's light intensity on the earth's surface on a sunny day near the equator is $\approx 10^{-1}\,\text{W/cm}^2$.
[2] Taken from: Extreme Light, G. Mourou and D. Umstadter, Scientific American, May 2002.

B. Di Bartolo and O. Forte (eds.), Frontiers of Optical Spectroscopy, 93-186.

we introduce the **Rabi energy**, the **ponderomotive energy**, and the **Bloch energy** ($\hbar$ times the Bloch frequency[3]) – which all have to be compared with the **carrier photon energy** of the laser pulses for the case of solids. For atoms, the ponderomotive energy, the Rydberg energy and the rest energy of the electron are the relevant scales. In section four, we summarize our recent work on resonant excitation of optical transitions in semiconductors in the regime where the Rabi energy becomes comparable to the photon energy, which leads to carrier-wave Rabi flopping and to a dependence on the CEO phase. In section five we address third-harmonic generation in disguise of second-harmonic generation: In contrast to traditional nonlinear optics, an inversion symmetric material excited by intense few-cycle pulses can show a peak at the spectral position of the second harmonic of the laser photon energy. In section six we briefly discuss corresponding effects in atoms at amplified laser intensities around $10^{15}\,\mathrm{W/cm^2}$ where attosecond x-ray pulses can be generated. Finally, we sketch out the physics at colossal laser intensities up to $10^{28}\,\mathrm{W/cm^2}$. We will see that an electron in vacuum can lead to (true) second-harmonic generation, that even the vacuum itself gives rise to optical nonlinearities and that the electron acceleration can approach values comparable to those at the edge of a black hole ...

Exercise 1.1.: What is the light intensity in a "dark" room held at room temperature?

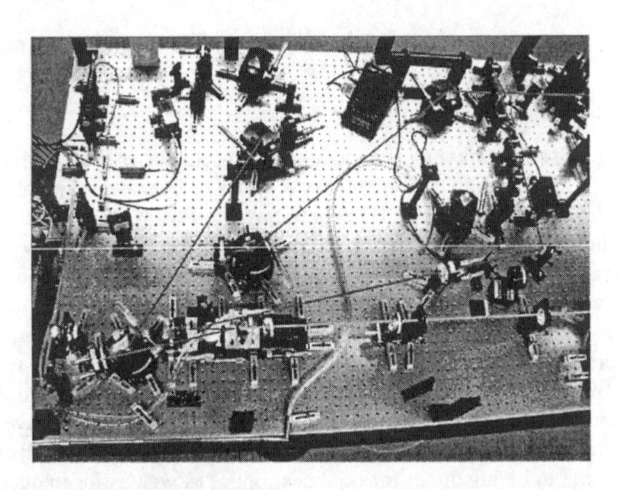

Figure 1: Photograph of the 5 fs mode-locked laser oscillator used for the carrier-wave nonlinear optics experiments in solids discussed in this article. Note the simplicity of the setup which is optimized for flexibility rather than for compactness. The depth of the optical table is 1.2 m. The pump beam (light grey lines) enters the picture on the lower RHS, the dispersion compensated output of the laser (dark grey lines) leaves the photograph on the upper RHS. The Ti:sapphire gain crystal is in the middle of the "z" on the lower LHS.

[3] $\hbar = 0.658\,\mathrm{eV\,fs}$, equivalent to $\hbar = 1.054 \times 10^{-34}$ Js, will be useful throughout this article.

2. Some Aspects of Few-Cycle Laser Pulses From Mode-Locked Oscillators

The basic principle of a laser is simple: It consists of a resonator (see Figs. 1 and 2) and a light amplifier – the gain medium. For the purpose of these lectures, the quantum optical aspects of the light field are not important, hence, it is sufficient to consider the well-known Maxwell equations of electrodynamics [1].

2.1. *MAXWELL EQUATIONS*

In S.I. units the Maxwell equations are given by

$$\vec{\nabla}\cdot\vec{D} = \rho \tag{1}$$

$$\vec{\nabla}\times\vec{E} = -\frac{\partial\vec{B}}{\partial t} \tag{2}$$

$$\vec{\nabla}\cdot\vec{B} = 0 \tag{3}$$

$$\vec{\nabla}\times\vec{H} = +\frac{\partial\vec{D}}{\partial t} + \vec{j}\,. \tag{4}$$

ρ is the electric charge density and $\vec{j}$ the electric current density. In media, the relation between the $\vec{E}$-field and the $\vec{D}$-field is given by[4]

$$\vec{E} = \frac{1}{\varepsilon_0}\left(\vec{D} - \vec{P}\right), \tag{5}$$

with the macroscopic polarization $\vec{P}$, and, similarly, for the $\vec{B}$-field and the $\vec{H}$-field the relation

$$\vec{B} = \mu_0\left(\vec{H} + \vec{M}\right), \tag{6}$$

with the magnetization $\vec{M}$. For the materials relevant in the context of this article, $\vec{M} = 0$ holds, and Eq. (6) simplifies to

$$\vec{B} = \mu_0\,\vec{H}\,. \tag{7}$$

In linear optics, one has

$$\vec{P} = \varepsilon_0\chi\vec{E}\,, \tag{8}$$

with the linear optical susceptibility χ. In this case, Eq. (5) simplifies to

$$\vec{D} = \varepsilon_0\,\varepsilon\,\vec{E}\,, \tag{9}$$

with the relative dielectric function $\varepsilon = 1 + \chi$. The Maxwell equations can be rewritten into the known wave equation[5] for the $\vec{E}$-field

$$\Delta\vec{E} - \frac{1}{c_0^2}\frac{\partial^2\vec{E}}{\partial t^2} = +\mu_0\,\frac{\partial^2\vec{P}}{\partial t^2}\,, \tag{10}$$

---

[4] $\varepsilon_0 = 8.8542\times10^{-12}\,\mathrm{AsV^{-1}m^{-1}}$ and $\mu_0 = 4\pi\times10^{-7}\,\mathrm{VsA^{-1}m^{-1}}$.

[5] Coming from Karlsruhe, we just have to remind you that it was Karlsruhe where Heinrich Hertz found the first experimental evidence for electromagnetic waves in the year 1887.

or, using Eq. (8), into

$$\Delta\vec{E} - \frac{1}{c^2}\frac{\partial^2\vec{E}}{\partial t^2} = 0\,, \tag{11}$$

with the medium velocity of light $c = c_0/n$, which is slower than the vacuum velocity of light $c_0 = 1/\sqrt{\epsilon_0\,\mu_0} = 2.998 \times 10^8\,\mathrm{m/s}$ by a factor identical to the (generally complex) refractive index n with

$$n = \sqrt{\varepsilon}\,. \tag{12}$$

2.2. *THE LIGHT INTENSITY*

Our eyes and most detectors are not sensitive to the electric field itself but to the number of photons which hit the detector per unit time. In other words, classically speaking: They are sensitive to the cycle-average of the modulus of the Poynting vector $\vec{S} = \vec{E} \times \vec{H}$. For plane waves in vacuum one has $|\vec{B}| = |\vec{E}|/c_0$ or equivalently $|\vec{E}| = |\vec{H}|\sqrt{\frac{\mu_0}{\varepsilon_0}}$, with the vacuum impedance

$$\sqrt{\frac{\mu_0}{\varepsilon_0}} = 376.7301\,\Omega\,, \tag{13}$$

leading to

$$S = |\vec{S}| = \sqrt{\frac{\varepsilon_0}{\mu_0}}\,|\vec{E}|^2\,, \tag{14}$$

which generally varies with time. For an electric field according to, e.g.,

$$|\vec{E}(t)|^2 = \tilde{E}_0^2\cos^2(\omega_0 t + \phi)\,, \tag{15}$$

the **light intensity** I, which is defined as the cycle-average[6] of the modulus of the **Poynting vector**, becomes[7]

$$\boxed{I = \langle S\rangle = \frac{1}{2}\sqrt{\frac{\varepsilon_0}{\mu_0}}\,\tilde{E}_0^2\,.} \tag{16}$$

Note that the intensity I does not depend on ϕ.

Example 2.1.: An electric field of $\tilde{E}_0 = 4 \times 10^9\,\mathrm{V/m}$ in vacuum corresponds to an intensity of $I = 2.1 \times 10^{12}\,\mathrm{W/cm^2}$. For comparison: This intensity corresponds to concentrating the power of thousand power plants with a power of 2 GW each to an area comparable to your finger tip – for a very short time. For the same electric field, the peak of the $\vec{B}$-field envelope is $\tilde{B}_0 = \mu_0\sqrt{\frac{\epsilon_0}{\mu_0}}\,\tilde{E}_0 = \tilde{E}_0/c_0 = 13.3\,\mathrm{T}$.

---

[6] Remember that $\langle\cos^2(\omega_0 t + \phi)\rangle = 1/2$.

[7] Within a dielectric, ϵ_0 has to be replaced by $\epsilon_0\,\epsilon$ in this relation.

2.3. ELECTRIC FIELD IN A LASER RESONATOR

Solutions of the wave equation (11) are, e.g., plane waves with

$$\vec{E}(\vec{r},t) = \vec{E}_0 \cos\left(\vec{K}\vec{r} - \omega t - \varphi\right) = \frac{\vec{E}_0}{2} \exp\left(\mathrm{i}\left(\vec{K}\vec{r} - \omega t - \varphi\right)\right) + \text{c.c.}\,, \tag{17}$$

which have to obey the **dispersion relation of light**

$$\boxed{\frac{\omega}{|\vec{K}|} = c = \frac{c_0}{n(\omega)}} \tag{18}$$

for the frequency ω and the wave vector of light $\vec{K}$.

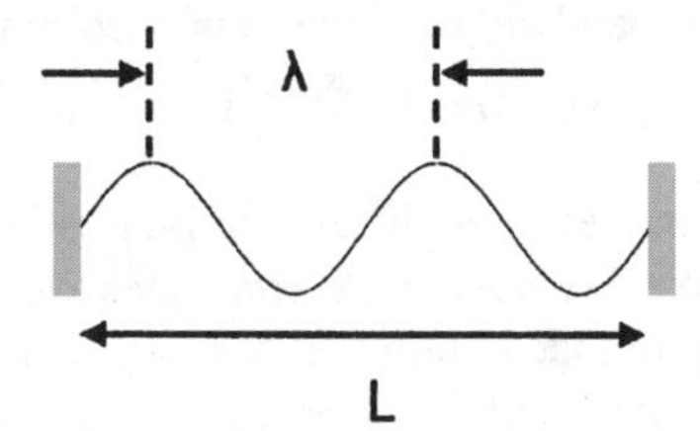

Figure 2: Scheme of a laser resonator, consisting of two mirrors separated by length L. A snapshot of a single mode ($N = 4$) of the electric field at wavelength λ is shown.

For the resonator shown in Fig. 2, we have the superposition of left-going and right-going waves, i.e., a standing wave, such that the electric field has nodes at the two mirrors, $\vec{E}(z = 0, t) = \vec{E}(z = L, t) = 0$. Thus, the **length L of the resonator** has to be an integer multiple, let us say N, of half the wavelength of light λ:

$$L = N\,\frac{\lambda}{2}\,. \tag{19}$$

With the dispersion relation of light Eq. (18) and with $|\vec{K}| = 2\pi/\lambda$, this can be rewritten into

$$\omega_N = N\,\Delta\omega \tag{20}$$

with the mode spacing

$$\boxed{\Delta\omega = c\,\frac{\pi}{L}\,.} \tag{21}$$

Example 2.2.: For a resonator with length $L = 1.5\,\mathrm{m}$ and with $c = c_0$ we obtain $\Delta\omega = 2\pi \times 100\,\mathrm{MHz}$, which is within the radio frequency (RF) regime.

The superposition principle tells us that any linear combination of these eigensolutions is also a solution of the resonator problem and we can write the general solution of standing waves in the resonator as

$$\boxed{\vec{E}(\vec{r},t) = \sum_{N=1}^{\infty} 2\,\vec{E}_0^N \sin(K_N z)\sin(\omega_N t + \varphi_N)\,.} \tag{22}$$

Details depend on the amplitudes $\vec{E}_0^N$, the phases φ_N of the modes with $N = 1...\infty$ and also on the dispersion relation Eq. (18) which connects the K_N and the ω_N. Let us consider three cases, (a) – (c) in Fig. 3, in which we study one component of the electric field vector, $E(t) = E_{x,y}(\vec{r} = \text{const.}, t)$, at a fixed point in space within the cavity. For the sake of simplicity we assume that the mode amplitudes are either constant or zero (which mimics the finite bandwidth of the gain medium), i.e., $\vec{E}_0^N = \vec{E}_0$ for all frequencies in the interval $[\omega_0 - \delta\omega/2, \omega_0 + \delta\omega/2]$ and $\vec{E}_0^N = 0$ elsewise. We choose $\delta\omega/\omega_0 = 0.6$.

In (a) we consider many modes N with random phases φ_N, $c =$ const. This leads to an electric field which looks like noise with some average intensity (Fig. 3(a)). This situation corresponds to a multimode continuous-wave (cw) laser – a bad cw laser. We conclude that a good cw laser must only work on a single mode.

In (b) all the phases φ_N are equal – they are locked –, in which case we can set them to zero, $c =$ const. A periodic train of identical pulses results (Fig. 3(b)). The duration of the individual pulses is inversely proportional to the width of the frequency interval $\delta\omega$. How can one realize this locking of the modes experimentally? By active or passive modulation of the resonator properties with frequency $\Delta\omega$, which is called **mode-locking**! Such modulation of the mode with frequency ω_N leads to sidebands at $\omega_N + \Delta\omega = \omega_{N+1}$ and $\omega_N - \Delta\omega = \omega_{N-1}$ for all N. This couples all the modes, hence it locks their phases φ_N, and it leads to a perfectly equidistant spacing of the modes in frequency space.

In (c) we give up the unrealistic assumption of a constant velocity of light c in the resonator, but the modes shall still be equidistant in frequency. The corresponding time-domain behavior of Eq. (22) is schematically shown in Figs. 3(c) and 4. The pulses are not identical under these conditions. In one round trip, a shift of the phase between the envelope and the carrier wave results from the fact that the **group velocity** v_{group} at frequency ω_0 (the velocity of the envelope)

$$v_{\text{group}} = \frac{\mathrm{d}\omega}{\mathrm{d}K} \tag{23}$$

with $K = |\vec{K}|$ and the **phase velocity** v_{phase} (the velocity of the carrier wave at frequency ω_0)

$$v_{\text{phase}} = c = \frac{\omega}{K} \tag{24}$$

are no longer identical. We can define a corresponding **carrier-envelope offset frequency** f_ϕ which is generally different from the repetition frequency $f_\text{r} = 1/t_\text{r} = \Delta\omega/(2\pi)$.

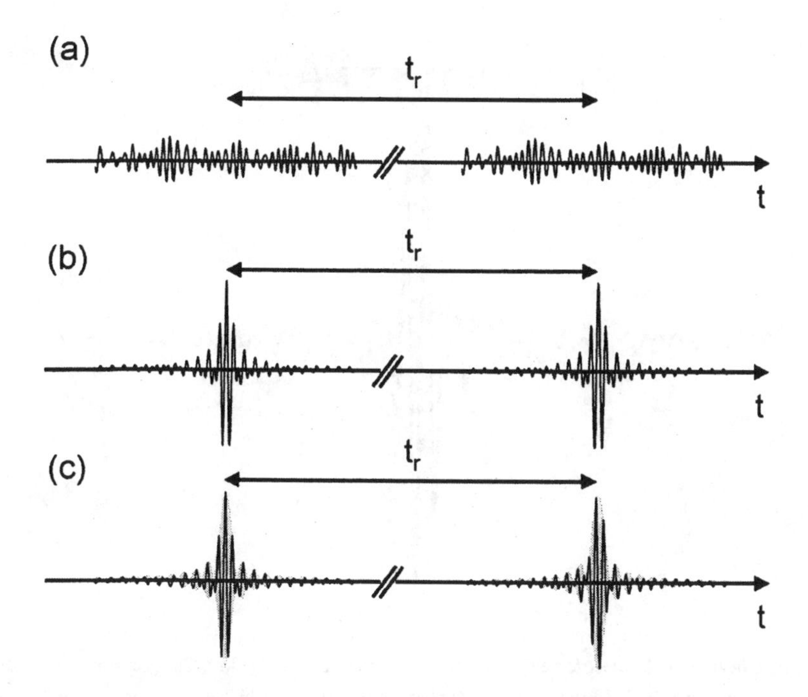

Figure 3: Electric field versus time t in the middle of the laser cavity according to Eq. (22). (a) Random phases φ_N, $c = \text{const.}$, (b) $\varphi_N = 0$ for all N, $c = \text{const.}$, (c) $\varphi_N = 0$ for all N, $c = c(\omega_N) \neq \text{const.}$ Note that (b) and (c) have been demagnified with respect to (a) in the vertical direction by a factor of about 10^6.

From Fig. 3(c) it becomes clear that the electric field according to Eq. (22) can alternatively be expressed[8] as

$$\boxed{E(t) = \sum_{N=-\infty}^{+\infty} \tilde{E}(t - N\,t_\mathrm{r}) \cos\Big(\omega_0(t - N\,t_\mathrm{r}) + N\,\Delta\phi + \phi\Big)\,.} \tag{25}$$

The cosine-term is the **carrier-wave oscillation** with carrier frequency ω_0, the prefactor $\tilde{E}$ is called the **envelope** of the pulse (grey areas in Fig. 3). t_r is the round-trip time, $\Delta\phi$ the pulse-to-pulse phase slip, and ϕ an overall phase. $(N\Delta\phi + \phi)$ is understood mod 2π, i.e. for all integers N, the term is element of the interval $[0, 2\pi]$. Later, we will only consider **one pulse out of the pulse train** according to Eq. (25), e.g., the one with $N = 0$, which leads to an electric field of

$$\boxed{E(t) = \tilde{E}(t) \cos(\omega_0 t + \phi)\,,} \tag{26}$$

[8]Note that the choice of the carrier frequency ω_0 is somewhat arbitrary, especially if the pulses are chirped. Often, one chooses ω_0 as the center of mass of the frequency spectrum of the laser pulses.

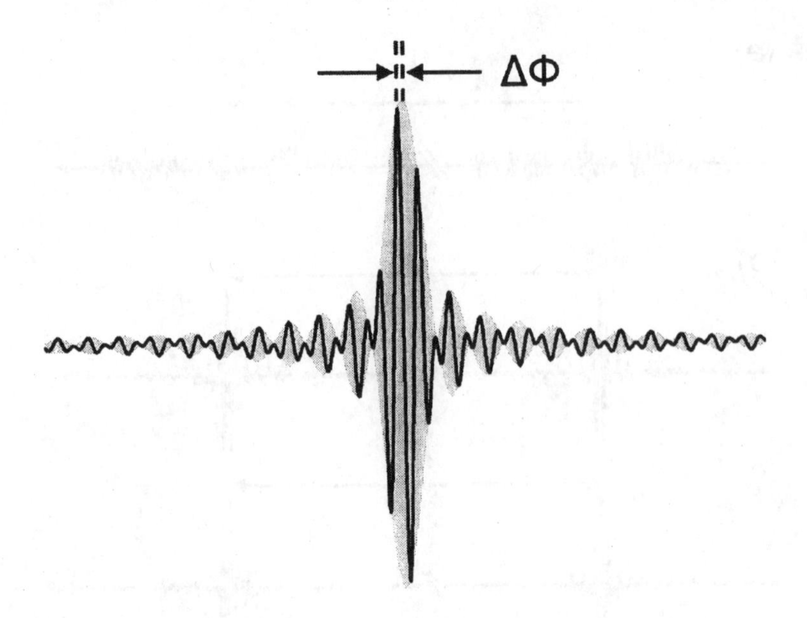

Figure 4: Magnification of the electric field versus time from the RHS of Fig. 3(c). The grey area corresponds to the electric field envelope. In this example, the CEO phase is $\phi = +\pi/2$, which corresponds to a pulse-to-pulse phase shift of $\Delta\phi = +\pi/2$ in Fig. 3(c).

with the so called **carrier-envelope offset (CEO) phase**[9] ϕ. The CEO phase of a single pulse has to be distinguished from the well known relative optical phase between two beams or pulses, e.g., in a Michelson interferometer.

What is the frequency-domain analogue of this behavior ? We compute the Fourier transform of the electric field, $E(\omega)$, via

$$
\begin{aligned}
E(\omega) &= \frac{1}{\sqrt{2\pi}} \int_{-\infty}^{+\infty} E(t)\, \mathrm{e}^{+\mathrm{i}\omega t}\, \mathrm{d}t \qquad (27) \\
&= \frac{1}{\sqrt{2\pi}} \int_{-\infty}^{+\infty} \sum_{N=-\infty}^{+\infty} \tilde{E}(t - N\, t_\mathrm{r}) \cos\Big(\omega_0 (t - N\, t_\mathrm{r}) + N\, \Delta\phi + \phi\Big)\, \mathrm{e}^{+\mathrm{i}\omega t}\, \mathrm{d}t
\end{aligned}
$$

The cosine-term can be written according to

$$
\cos(\omega_0 t + ...) = \frac{1}{2} \left(\mathrm{e}^{+\mathrm{i}(\omega_0 t + ...)} + \mathrm{e}^{-\mathrm{i}(\omega_0 t + ...)} \right) . \qquad (28)
$$

[9]The CEO phase is sometimes also called absolute optical phase.

The exponential with the "minus" sign leads to a peak in $E(\omega)$ at positive frequencies ω, the term with the "plus" sign to a peak at negative ω. The latter is omitted and we get

$$\begin{aligned}
E(\omega) &= \frac{1}{\sqrt{2\pi}} \int_{-\infty}^{+\infty} \sum_{N=-\infty}^{+\infty} \tilde{E}(t - N\,t_\mathrm{r})\,\frac{1}{2}\,\mathrm{e}^{-\mathrm{i}(\omega_0 (t-N\,t_\mathrm{r})+N\,\Delta\phi+\phi)}\,\mathrm{e}^{+\mathrm{i}\omega t}\,\mathrm{d}t \\
&= \frac{1}{2}\left(\sum_{N=-\infty}^{+\infty} \mathrm{e}^{-\mathrm{i}(N(\Delta\phi-\omega_0 t_\mathrm{r})+\phi)} \left(\frac{1}{\sqrt{2\pi}} \int_{-\infty}^{+\infty} \tilde{E}(t - Nt_\mathrm{r})\,\mathrm{e}^{+\mathrm{i}(\omega-\omega_0)t}\,\mathrm{d}t\right)\right) \\
&= \frac{1}{2}\left(\sum_{N=-\infty}^{+\infty} \mathrm{e}^{-\mathrm{i}(N(\Delta\phi-\omega_0 t_\mathrm{r})+\phi)}\,\mathrm{e}^{+\mathrm{i}N(\omega-\omega_0)t_\mathrm{r}}\right) \\
&\qquad\qquad \times \underbrace{\left(\frac{1}{\sqrt{2\pi}} \int_{-\infty}^{+\infty} \tilde{E}(t')\,\mathrm{e}^{+\mathrm{i}(\omega-\omega_0)t'}\,\mathrm{d}t'\right)}_{=:\,\tilde{E}_{\omega_0}(\omega-\omega_0)} \\
&= \frac{\mathrm{e}^{-\mathrm{i}\phi}}{2}\left(\sum_{N=-\infty}^{+\infty} \mathrm{e}^{-\mathrm{i}N(\Delta\phi-\omega t_\mathrm{r})}\right) \tilde{E}_{\omega_0}(\omega-\omega_0)\,. \qquad (29)
\end{aligned}$$

From the second to the third line we have substituted $t' = (t - Nt_\mathrm{r})$. $\tilde{E}_{\omega_0}(\omega - \omega_0)$ is the envelope of the optical spectrum[10]. With an optical spectrometer[11] one usually measures the intensity spectrum $\propto |\tilde{E}_{\omega_0}(\omega - \omega_0)|^2$ (which neither depends on ϕ nor on $\Delta\phi$). The spectrum is modulated by the sum over the exponentials in the last line of Eq. (29). The significant values of ω in this sum are the ones for which the terms for, e.g., N and $(N+1)$ add constructively, i.e. for which we have $(\omega t_\mathrm{r} - \Delta\phi) = M\,2\,\pi$ with integer M. This yields an **equidistant ladder of angular frequencies** ω_M with

$$\omega_M = M \times \frac{2\,\pi}{t_\mathrm{r}} + \frac{\Delta\phi}{t_\mathrm{r}}\,. \qquad (30)$$

Finally, we can convert from angular frequencies to frequencies via $\omega_M = 2\pi\, f_M$ and obtain

$$\boxed{f_M = M \times f_\mathrm{r} + f_\phi\,,} \qquad (31)$$

[10]In order to distinguish this envelope at carrier frequency ω_0 from other envelopes that will occur in this article, we have introduced the index ω_0.

[11]The full width at half maximum (FWHM), $\delta\omega$, of the intensity spectrum multiplied with the FWHM of the temporal intensity profile, δt, is the duration-bandwidth product $\delta\omega\,\delta t$. One obtains $\delta\omega\,\delta t \geq 2\pi \times 0.4413$ for a Gaussian, i.e., for a $\exp(-t^2)$ pulse, $\delta\omega\,\delta t \geq 2\pi \times 0.8859$ for a $\mathrm{sinc}^2(t) = (\sin(t)/t)^2$ pulse, $\delta\omega\,\delta t \geq 2\pi \times 0.3148$ for a $\mathrm{sech}^2(t) = 1/\cosh^2(t)$ pulse, and $\delta\omega\,\delta t \geq 2\pi \times 0.1103$ for a one-sided exponential, i.e., for a $\Theta(t)\exp(-t)$ pulse [2]. The latter is the absolute minimum of the product $\delta\omega\,\delta t$ for any pulse shape. For all these cases, the equality applies for zero chirp.

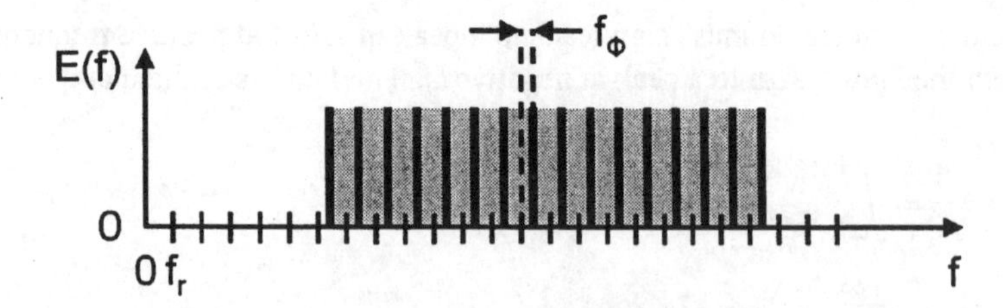

Figure 5: Scheme of the frequency domain analogue, $E(f)$, of the temporal behavior, $E(t)$, shown in Fig. 3(c). The spectrum exhibits peaks at the frequencies $f_M = M \times f_r + f_\phi$ with integer M, i.e., the equidistant frequency comb is up-shifted by the carrier-envelope offset frequency f_ϕ. The actual $E(f)$ corresponding to Fig. 3(c) contains more than 10^6 densely spaced peaks within the optical spectrum (indicated by the grey area).

with the **repetition frequency** $f_r = 1/t_r = \Delta\omega/(2\pi)$ (see Eq. (21))

$$\boxed{f_r = \frac{c}{2L}} \tag{32}$$

and the **carrier-envelope offset frequency**

$$\boxed{f_\phi = f_r \, \frac{\Delta\phi}{2\pi} \leq f_r \,.} \tag{33}$$

If $f_\phi \neq 0$, this frequency comb has a certain offset frequency [3, 4]. If one can arrange for $f_\phi = 0$, on the other hand, the eigenfrequencies form a ladder of equidistant frequencies starting at zero frequency with $M = 0$. These findings are summarized in Fig. 5.

Example 2.3.: Let us compute the carrier-envelope offset frequency f_ϕ for the following parameters: $L = 1.5\,\text{m}$, $v_{\text{group}} = 99.9999\%\,c_0$, $v_{\text{phase}} = c_0$, $\hbar\omega_0 = 1.5\,\text{eV} \Leftrightarrow 2\pi/\omega_0 = 2.8\,\text{fs}$. The difference, Δt, between the round-trip group delay time and the phase delay time is $\Delta t = 2L/v_{\text{group}} - 2L/v_{\text{phase}}$. With $1/(1-x) \approx (1+x)$ for $x \ll 1$ we get $\Delta t \approx 2L/c_0 \times ((1+10^{-6}) - 1) = 10\,\text{ns} \times 10^{-6} = 10\,\text{fs}$. This leads to $\Delta\phi = (10\,\text{fs}/(2.8\,\text{fs}) \times 2\pi) \bmod 2\pi = (3.57 \times 2\pi) \bmod 2\pi = 0.57 \times 2\pi$. With Eq. (33) and with $f_r = 100\,\text{MHz}$ from example 2.2. $\Rightarrow f_\phi = 57\,\text{MHz}$.

The time standard for one second within the S.I. system is related to a frequency of 9 192 631 770 Hz $\approx$ 9 GHz. This frequency can rather easily be locked to the repetition frequency f_r of the oscillator, which is typically around 100 MHz (see example 2.2.). If the pulse-to-pulse phase shift $\Delta\phi$ can be stabilized to $\Delta\phi = 0$ (section 2.4.), the frequency comb starts at zero frequency and looks **much like a ruler for frequencies** where one simply counts the number of millimeter tics to measure a length. This allows to connect the time standard for one second in the GHz regime to optical frequencies at hundreds of THz. Previously, this required a very (!) complicated procedure (see references given in Ref. [4]). The corresponding important implications for metrology (i.e., for using femtosecond lasers as frequency standards) are nicely discussed in a rather recent review article [4]. Some authors even speculate that this increased precision in measuring time might lead to experiments in which one could possibly observe the temporal variation of fundamental "constants" versus time on laboratory timescales [5, 6, 7]. If one could, e.g., measure the atomic Rydberg with increased frequency precision (relative precision of 10^{-15} or better), this value could be related to the fine structure constant. Performing such experiment today and comparing it to the result one year or several years later might reveal a difference. Strange but true: The ultrafast becomes useful for the ultraslow or ultraprecise, respectively ...

Let us finally note that one must be cautious with the choice of the electric field envelope $\tilde{E}(t)$ according to Eq. (26). In actual experiments, the optical spectrum does not contain zero-frequency (dc) components. According to the Maxwell equations, zero-frequency components are not radiated at all, low-frequency components are not efficiently radiated. Furthermore, they correspond to very large wavelengths, which do not propagate into the optical far-field because of diffraction. Zero dc component is equivalent to a vanishing time-average of the electric field, i.e., to the condition

$$\int_{-\infty}^{+\infty} E(t)\,\mathrm{d}t = 0 \tag{34}$$

for any value of the CEO phase ϕ. The $\tilde{E}(t) \propto \mathrm{sinc}(t)$ pulses (see, e.g., Fig. 5) we have discussed above and which we will frequently use below, do fulfill this condition for arbitrary values of ϕ. Generally, in the theory, however, one can get significant tails in the optical spectrum towards zero frequency for certain envelopes (which are well localized in time) and values of ϕ. In this case, the electric field envelope must not be assumed to be independent of ϕ. If one assumes a fixed envelope anyway, the light-matter interaction looses its gauge-invariance [8] and unphysical results are expected.

Exercise 2.1.: A light field with $\tilde{E}_0 = 4 \times 10^9\ \mathrm{V/m}$ propagates from air into a dielectric with $\epsilon = 10.9$. Reflections are completely suppressed via an ideal anti-reflection (AR) coating. What is $\tilde{E}_0$ inside the dielectric?

Exercise 2.2.: What – in principle – are the shortest *optical* pulses achievable?

2.4. PRINCIPLE OF MEASURING THE CARRIER-ENVELOPE OFFSET FREQUENCY

If one wants to stabilize or control f_ϕ, one obviously first needs to be able to measure f_ϕ, equivalent to determining the time derivative of the CEO phase ϕ. How can one measure ϕ of a laser pulse $E(t)$ according to Eq. (26)? It drops out when measuring the intensity (see subsection 2.1), it does not affect the optical spectrum, it does not show up in usual intensity autocorrelations or field autocorrelations. Generally, in order to observe a phase, one needs to compare the unknown field with a "reference".

The idea: If one had another field which would contain a phase 2ϕ rather than 1ϕ of the electric field itself, the interference of the two contributions (the beating) would oscillate with the difference, i.e. with $(2-1)\times\phi = \phi$.

Such "reference" can be generated by sending the electric field $E(t)$ onto a suitable nonlinear optical material, the optical polarization $P(t)$ of which can be expressed via

$$\boxed{P(t) = \epsilon_0 \left(\chi^{(1)} E(t) + \chi^{(2)} E^2(t) + \chi^{(3)} E^3(t) + \ldots\right),} \tag{35}$$

which is the generalization[12] of the linear optical polarization, Eq. (8), for large electric fields [1, 9]. Here the coefficients $\chi^{(N\neq 1)}$ are the **nonlinear optical susceptibilities** of order N, $\chi^{(1)} = \chi$ is the linear optical susceptibility. All even orders are only non-vanishing for lacking inversion symmetry of the medium[13]. The second derivative of the polarization (35) is the source term on the RHS of the wave equation (10). Consider the second-order contribution which – via the wave equation – gives rise to an electric field at carrier frequency $2\omega_0$, the so called second harmonic, which is given by

$$\begin{aligned} E_{2\omega_0}(t) &\propto E^2(t) \\ &= \tilde{E}^2(t)\cos^2(\omega_0 t + \phi) \\ &= \tilde{E}^2(t)\,\frac{1}{2}\left(1 + \underline{\cos(2\omega_0 t + 2\phi)}\right). \end{aligned} \tag{36}$$

The "1" in the last line reflects so called optical rectification and can (often but not always) be omitted because dc-components do not lead to propagating electromagnetic waves. Let us consider the resulting interference in frequency space. The Fourier transform of the cosine-terms have maxima at positive and at negative frequencies. As in section 2.3., we focus on the measurable positive frequency components (corresponding to the minus sign in the exponent). The resulting intensity from the interference can be written as

$$I_{\omega_0,2\omega_0}(\omega) \;\propto\; \left|\mathrm{e}^{-\mathrm{i}\phi}\tilde{E}_{\omega_0}(\omega) + \mathrm{e}^{-\mathrm{i}2\phi}\tilde{E}_{2\omega_0}(\omega)\right|^2 \tag{37}$$

---

[12] Vectors are omitted for simplicity. The range of validity of Eq. (35) is limited. It obviously assumes an instantaneous response of $P(t)$ with respect to $E(t)$, equivalent to no or negligible frequency dependence of the $\chi^{(N)}$. This is only justified "far away" from a resonance of the material. Also, Eq. (35) is only meaningful if the terms become rapidly smaller with increasing order N, i.e., if the electric field is not too large – if it is within the perturbative regime. In section 3.1. we evaluate the optical polarization microscopically for the case of semiconductors.

[13] Consider space inversion, i.e., $\vec{r} \to -\vec{r}$. Thus, $E(t) \to -E(t)$ and $P(t) \to -P(t)$. As $(-E(t))^2 = E^2(t)$, $(-E(t))^4 = E^4(t)$, ... it follows that $\chi^{(2)} = \chi^{(4)} = \ldots = 0$, while $\chi^{(1)}, \chi^{(3)}, \ldots$ can be nonzero.

$$= \left|\tilde{E}_{\omega_0}(\omega)\right|^2 + \left|\tilde{E}_{2\omega_0}(\omega)\right|^2 + 2\left|\tilde{E}_{\omega_0}(\omega)\,\tilde{E}_{2\omega_0}(\omega)\right| \times \underline{\cos(\phi)}\,.$$

The underlined $\cos(\phi)$-term delivers the anticipated dependence on ϕ. In order to actually observe this contribution in an experiment, at least the following two conditions have to be fulfilled.

- The amplitudes $\tilde{E}_{\omega_0}(\omega)$ and $\tilde{E}_{2\omega_0}(\omega)$ must be comparable in absolute value, otherwise the two constant, i.e. ϕ-independent, terms in Eq. (37) completely dominate the measured intensity $I_{\omega_0,2\omega_0}$. This condition can generally be fulfilled at some frequency ω in the optical frequency interval $[\omega_0, 2\omega_0]$.

- The product term must exhibit appreciable absolute strength in order not to be covered by noise in the experiment.

It turns out that the second condition is more difficult to fulfill than the first (see Fig. 6). From section 2.3. it is clear that a modulation according to

$$\boxed{I_{\omega_0,2\omega_0} = \dots + \dots \cos(\phi)} \tag{38}$$

leads to a **peak at frequency f_ϕ in the RF spectrum** [4, 10, 11, 12]. Similarly, an interference of a third-order contribution from Eq. (35), i.e.

$$\begin{aligned} E_{3\omega_0}(t) &\propto E^3(t) \\ &= \tilde{E}^3(t)\cos^3(\omega_0 t + \phi) \\ &= \tilde{E}^3(t)\,\frac{1}{4}\left(3\cos(\omega_0 t + \phi) + \underline{\cos(3\omega_0 t + 3\phi)}\right), \end{aligned} \tag{39}$$

with a contribution at frequency $3\omega_0$ with the fundamental at frequency ω_0 leads to a modulation with

$$\boxed{I_{\omega_0,3\omega_0} = \dots + \dots \cos(2\phi)\,,} \tag{40}$$

equivalent to a **peak at frequency $2f_\phi$ in the RF spectrum**, which is most prominent around the middle of the optical frequency interval $[\omega_0, 3\omega_0]$. We will come back to both types of interferences in sections 4.3. and 5.1.

When performing corresponding experiments, one often measures the RF power spectrum (as, e.g., in section 5.1.). Let us have a quick look at the details[14]. The beat signal $I(\phi)$, i.e., $I(\phi) = I_{\omega_0,2\omega_0}(\phi)$ from Eq. (38) or $I(\phi) = I_{\omega_0,3\omega_0}(\phi)$ from Eq. (40) or some more complicated general form, can be detected by a photomultiplier tube, which delivers a voltage signal $U(t)$. Assuming an integer ratio of repetition frequency and CEO

[14] For a first reading, the reader may want to continue with Chap. 3.

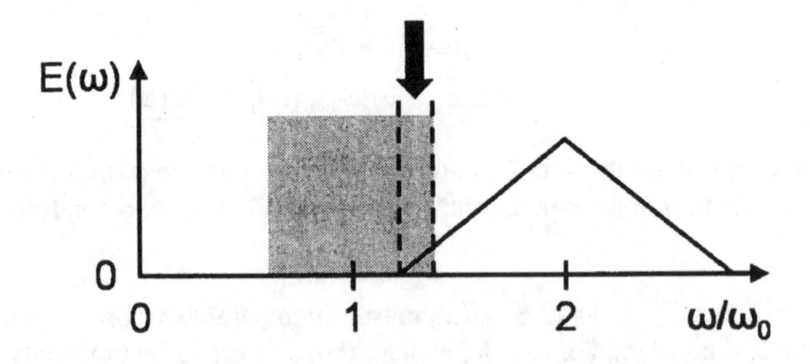

Figure 6: Scheme of the laser spectrum (grey area) of a $\text{sinc}^2(t)$ pulse covering slightly more than one octave in width ($\Leftrightarrow \delta\omega/\omega_0 = 2/3$, see Exercise 2.2.) and its second harmonic (not to scale). In the region of overlap (see arrow) the two contributions interfere and a dependence on the CEO phase ϕ results.

frequency for simplicity at this point, i.e., $f_\text{r}/f_\phi = r$ with integer r, the signal voltage (which is illustrated in Fig. 7) can be written as

$$U(t) = U_0 \sum_{N_\phi=-\infty}^{+\infty} \sum_{N_\text{r}=0}^{r-1} I_{N_\text{r}}\, \delta\left(t - [N_\phi t_\phi + N_\text{r} t_\text{r}]\right) , \tag{41}$$

with the integers N_ϕ and N_r, the abbreviation

$$I_{N_\text{r}} = I(\phi = N_\text{r} \frac{2\pi}{r}) , \tag{42}$$

the carrier-envelope offset period $t_\phi = 1/f_\phi$, and the (unimportant) prefactor U_0. Here we have approximated the actual temporal response of the photomultiplier by a δ-function. If this voltage signal is fed into an **RF spectrum analyzer**, the **RF power spectrum** $S_\text{RF}(f)$ versus RF frequency f is measured. It is defined by

$$S_\text{RF}(f) = \left| \frac{1}{\sqrt{2\pi}} \int_{-\infty}^{+\infty} \text{e}^{\text{i}2\pi f t}\, U(t)\, \text{d}t \right|^2 . \tag{43}$$

Inserting Eq. (41) into Eq. (43), the δ-functions in Eq. (41) select only contributions with $t = [N_\phi t_\phi + N_\text{r} t_\text{r}]$ from the integral and we obtain

$$\begin{aligned} S_\text{RF}(f) &= \frac{U_0^2}{2\pi} \left| \sum_{N_\phi=-\infty}^{+\infty} \sum_{N_\text{r}=0}^{r-1} I_{N_\text{r}}\, \text{e}^{\text{i}2\pi f\, [N_\phi t_\phi + N_\text{r} t_\text{r}]} \right|^2 \\ &= \frac{U_0^2}{2\pi} \left| \sum_{N_\text{r}=0}^{r-1} I_{N_\text{r}}\, \text{e}^{\text{i}2\pi f N_\text{r} t_\text{r}} \right|^2 \times \left| \sum_{N_\phi=-\infty}^{+\infty} \text{e}^{\text{i}2\pi f N_\phi t_\phi} \right|^2 . \end{aligned} \tag{44}$$

The nonvanishing values of the last sum correspond to those frequencies f, for which the terms for, e.g., N_ϕ and $(N_\phi + 1)$ add constructively, i.e., for which we have $2\pi f N_\phi t_\phi = M\, 2\pi$, thus

$$f = M/t_\phi = M \times f_\phi , \tag{45}$$

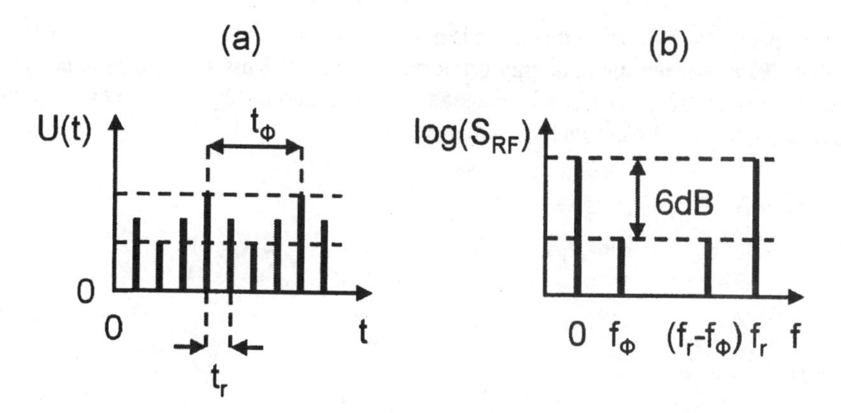

Figure 7: (a) Scheme of the output voltage $U(t)$ (linear scale) of a photomultiplier detecting an optical beat signal $I(\phi) = I_{\omega_0,2\omega_0}(\phi) = ... + ... \cos(\phi)$ according to Eq. (38). The ratio of repetition frequency to carrier-envelope offset frequency is $f_r/f_\phi = r = 4$ ($\Leftrightarrow \Delta\phi = \pi/2$, see Fig. 4). (b) Corresponding RF power spectrum $S_{RF}(f)$ on a logarithmic scale. The 50% modulation of $U(t)$ due to ϕ in (a) corresponds to peaks in the RF power spectrum at frequencies f_ϕ and $(f_r - f_\phi)$ which are $2 \times 3\,\mathrm{dB} = 6\,\mathrm{dB}$ smaller than the f_r peak and the $f = 0$ peak.

with integer M. This means that the RF power spectrum consists of a series of δ-peaks at integer multiples of the CEO frequency f_ϕ. The height of these peaks is given by the value of the first term in the second line of Eq. (44), i.e., by

$$\left| \sum_{N_r=0}^{r-1} I_{N_r}\, e^{i2\pi f N_r t_r} \right|^2 = \left| \sum_{N_r=0}^{r-1} I_{N_r}\, e^{i2\pi M f_\phi N_r t_r} \right|^2 = \left| \sum_{N_r=0}^{r-1} I_{N_r}\, e^{i2\pi M N_r / r} \right|^2 . \quad (46)$$

In general, some of the peaks with label M may not occur because they have zero height. It is obvious that replacing M by $(M+r)$ on the RHS of Eq. (46) delivers the same value, i.e., the height of the peak at, e.g., frequency f_ϕ is exactly the same as the height of the peak at frequency $(f_r + f_\phi)$. In other words: All relevant information of the RF power spectrum is contained in the frequency interval $[0, f_r]$. Along the same lines, the peaks at f_ϕ and $(f_r - f_\phi)$, respectively, have the same height as well (replace M by $(r - M)$ on the RHS of Eq. (46)). **The mixing products $(f_r - f_\phi)$, $(f_r - 2f_\phi)$ or $(f_r + f_\phi)$ ... essentially originate from the fact that the light intensity – and not the electric field itself – is subject to a Fourier transformation in an RF spectrum analyzer**. The Fourier transform of the laser electric field itself has been discussed in Eq. (29).

Eqs. (45) and (46) allow to compute the RF power spectrum from a known beat signal $I(\phi)$. An example is given in Fig. 7. In a real experiment, the photomultiplier does not exhibit a δ-response. In this case, the actual voltage signal can be written as the convolution of Eq. (41) with the response function of the photomultiplier. In the frequency domain, this convolution translates into the product of the "ideal" result with the power spectrum of the photomultiplier response function, i.e., there is an overall decay towards large RF frequencies f. Finally, real laser systems have noise, which shows up as a pedestal in the

RF power spectrum. Typically, this noise is not white-noise (which would be a constant in frequency space) but is rather roughly proportional to $1/f$. This sometimes makes, e.g., the frequency interval $[2f_r, 3f_r]$ advantageous as compared to $[0, f_r]$ in the experiment – although the intervals are equivalent in theory.

3. How Intense is the Light Field? Rabi Energy, Ponderomotive Energy and Bloch Energy

We now introduce three quantities: The Rabi energy, the ponderomotive energy and the Bloch energy. The ponderomotive energy is proportional to the intensity of light $I \propto \tilde{E}_0^2$ (see Eq. (16)), the Rabi energy as well as the Bloch energy ($\hbar$ times the Bloch frequency) are proportional to the associated strength of the electric field $\tilde{E}_0$. Interesting new physics will occur (sections 4 and 5) if these quantities become comparable to or even larger than the carrier photon energy $\hbar\omega_0$. The Rabi energy is related to **interband** optical transitions, the ponderomotive energy as well as the Bloch energy relate to **intraband** processes.

3.1. RABI ENERGY

If a two-level system is excited by a resonant light field, electrons absorb photons which pumps them from the ground state into the excited state (see Fig. 8). It is sometimes stated that one cannot reach inversion by optical pumping in a two-level system. This statement is, however, only true in the incoherent case, where one can only reach transparency indeed, i.e. 50% of the electrons are in the ground state, 50% are in the excited state. In contrast to this, if the system remains fully coherent in the quantum mechanical sense, complete inversion can be reached. If the light field remains switched on, stimulated emission brings the electrons back into the ground state. This oscillation of the inversion is known as Rabi oscillation or Rabi flopping [13, 1, 14].

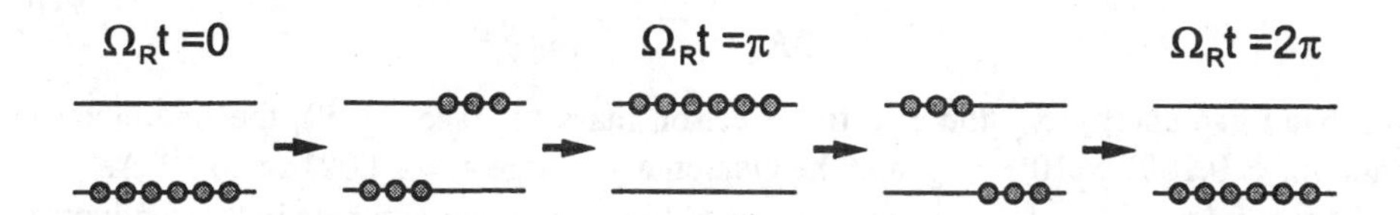

Figure 8: Scheme of a Rabi oscillation in a two-level system versus time t. The lower horizontal line represents the ground state, the upper line the excited state. The dots symbolize the electron occupation numbers.

To couple to the Maxwell equations, we have to compute the polarization P microscopically from a Hamiltonian $\mathcal{H}$ of the semiconductor. Neglecting the Coulomb interaction of carriers, any type of intraband optical processes at this point, phonons and their coupling to the carriers, suppressing spin indices and using the dipole approximation for the optical transitions from the valence (v) to the conduction (c) band at electron wave vector $\vec{k}$ we have [1]

$$\begin{aligned} \mathcal{H} &= \sum_{\vec{k}} E_{\mathrm{c}}(\vec{k})\, c^{\dagger}_{\mathrm{c}\vec{k}} c_{\mathrm{c}\vec{k}} + \sum_{\vec{k}} E_{\mathrm{v}}(\vec{k})\, c^{\dagger}_{\mathrm{v}\vec{k}} c_{\mathrm{v}\vec{k}} \\ &- \sum_{\vec{k}} d_{\mathrm{cv}}(\vec{k}) E(\vec{r},t) \left(c^{\dagger}_{\mathrm{c}\vec{k}} c_{\mathrm{v}\vec{k}} + c^{\dagger}_{\mathrm{v}\vec{k}} c_{\mathrm{c}\vec{k}} \right) . \end{aligned} \tag{47}$$

Here $E_{c,v}(\vec{k})$ are the single particle energies of electrons in the conduction and valence band, respectively (the band structure), which are schematically shown in Fig. 9. $d_{cv}(\vec{k})$ is the (real) dipole matrix element for an optical transition at electron wave vector $\vec{k}$. The creation $c^\dagger$ and annihilation c operators create and annihilate crystal electrons in the indicated band (c,v) at the indicated momentum ($\vec{k}$). The optical polarization is given by

$$P(\vec{r},t) = \frac{1}{V}\sum_{\vec{k}} d_{cv}(\vec{k})\left(p_{vc}(\vec{k}) + \text{c.c.}\right) + P_b(\vec{r},t), \tag{48}$$

where the optical transition amplitudes

$$p_{vc}(\vec{k}) = \langle c^\dagger_{v\vec{k}} c_{c\vec{k}} \rangle \tag{49}$$

depend on time t as well as parametrically on the spatial coordinate $\vec{r}$. As usual, the sum in Eq. (48) can be expressed via the combined density of states $D_{cv}(E)$ as $\sum_{\vec{k}} \ldots \rightarrow \int D_{cv}(E)\ldots \mathrm{d}E \rightarrow \sum_n D_{cv}(E_n)\ldots\Delta E$, which neglects all anisotropies. Sometimes, the background polarization $P_b(\vec{r},t) = \varepsilon_0\chi_b(z)E(\vec{r},t) = \varepsilon_0(\varepsilon_b(z)-1)E(\vec{r},t)$ is employed[15], which approximately accounts for all "very" high-energy optical transitions not explicitly accounted for in Eq. (47). It can be expressed in terms of the background dielectric constant $\varepsilon_b(\vec{r})$.

The dipole matrix element d_{cv} is approximately $\vec{k}$-independent and can be estimated on the basis of known material parameters by the following "rule of thumb" from $\vec{k}\cdot\vec{p}$ perturbation theory [1]:

$$|d_{cv}|^2 = \frac{\hbar^2 e^2}{2E_g}\left(\frac{1}{m_e} - \frac{1}{m_0}\right), \tag{50}$$

with band gap energy E_g, the effective electron mass m_e (see Fig. 9), the free electron mass $m_0 = 9.1091 \times 10^{-31}$ kg, and the elementary charge e $= 1.6021 \times 10^{-19}$ As.

The dynamics of $p_{vc}(\vec{k})$, as well as those of the occupation numbers in the conduction band

$$f_c(\vec{k}) = \langle c^\dagger_{c\vec{k}} c_{c\vec{k}} \rangle \tag{51}$$

and in the valence band

$$f_v(\vec{k}) = \langle c^\dagger_{v\vec{k}} c_{v\vec{k}} \rangle \tag{52}$$

are easily calculated from the Heisenberg equation of motion for any operator $\mathcal{O}$ according to

$$-\mathrm{i}\hbar\frac{\partial}{\partial t}\mathcal{O} = [\mathcal{H}, \mathcal{O}]. \tag{53}$$

Employing the usual anticommutation rules, i.e.

$$[c_{c\vec{k}}, c^\dagger_{c\vec{k}'}]_+ = \delta_{\vec{k}\vec{k}'}, \qquad [c_{v\vec{k}}, c^\dagger_{v\vec{k}'}]_+ = \delta_{\vec{k}\vec{k}'}, \tag{54}$$

---

[15] We will employ this approximation in section 4.2., but we will *not* use this approximation in section 4.3., i.e., we will explicitly account for all relevant optical transitions via the Bloch equations.

and that all other anticommutators are zero, this leads us to the known **Bloch equations for the transition amplitude**

$$\begin{aligned} \left(\frac{\partial}{\partial t} \right. &+ \left. \mathrm{i}\,\Omega(\vec{k})\right) p_{\mathrm{vc}}(\vec{k}) + \left(\frac{\partial}{\partial t} p_{\mathrm{vc}}(\vec{k})\right)_{\mathrm{rel}} \\ &= \mathrm{i}\hbar^{-1}\, d_{\mathrm{cv}}(\vec{k}) E(\vec{r},t) \left(f_{\mathrm{v}}(\vec{k}) - f_{\mathrm{c}}(\vec{k})\right), \end{aligned} \tag{55}$$

with the **optical transition energy**

$$\hbar\Omega(\vec{k}) = E_{\mathrm{c}}(\vec{k}) - E_{\mathrm{v}}(\vec{k}), \tag{56}$$

and **for the occupation in the conduction band**

$$\frac{\partial}{\partial t} f_{\mathrm{c}}(\vec{k}) + \left(\frac{\partial}{\partial t} f_{\mathrm{c}}(\vec{k})\right)_{\mathrm{rel}} = 2\hbar^{-1}\, d_{\mathrm{cv}}(\vec{k}) E(\vec{r},t)\,\mathrm{Im}\left(p_{\mathrm{vc}}(\vec{k})\right). \tag{57}$$

Here we have assumed a real dipole matrix element. $(1 - f_{\mathrm{v}}(\vec{k}))$ can be interpreted as the occupation of holes and obeys an equation similar to $f_{\mathrm{c}}(\vec{k})$. The terms with subscript "rel" have been added phenomenologically and describe dephasing and relaxation, respectively[16]. For very short time scales, they are not too important. Note that the transition amplitude $p_{\mathrm{vc}}(\vec{k})$ and the occupation factors $f_{\mathrm{c}}(\vec{k})$ and $f_{\mathrm{v}}(\vec{k})$ are easily connected to the components of the Bloch vector $(u, v, w)^{\mathrm{T}}$ via

$$\begin{pmatrix} u \\ v \\ w \end{pmatrix} := \begin{pmatrix} 2\,\mathrm{Re}(p_{\mathrm{vc}}(\vec{k})) \\ 2\,\mathrm{Im}(p_{\mathrm{vc}}(\vec{k})) \\ f_{\mathrm{c}}(\vec{k}) - f_{\mathrm{v}}(\vec{k}) \end{pmatrix}, \tag{58}$$

with equation of motion (relaxation omitted)

$$\boxed{\begin{pmatrix} \dot{u} \\ \dot{v} \\ \dot{w} \end{pmatrix} = \begin{pmatrix} 0 & +\Omega & 0 \\ -\Omega & 0 & -2\,\Omega_{\mathrm{R}}(t) \\ 0 & +2\,\Omega_{\mathrm{R}}(t) & 0 \end{pmatrix} \begin{pmatrix} u \\ v \\ w \end{pmatrix}.} \tag{59}$$

Here we have introduced the **Rabi frequency** $\Omega_{\mathrm{R}}(t)$ with

$$\boxed{\hbar\Omega_{\mathrm{R}}(t) = d_{\mathrm{cv}} E(\vec{r},t)} \tag{60}$$

and have suppressed the wave vector dependences for clarity. The Rabi frequency is proportional to the electric field and proportional to the dipole matrix element d_{cv}. **Note that the Rabi frequency itself oscillates with the frequency of light** ω_0, i.e., $\Omega_{\mathrm{R}}(t)$ periodically changes sign.

---

[16]For a state-of-the-art description of scattering processes see Refs. [1, 15, 16, 17, 18, 19, 20, 21].

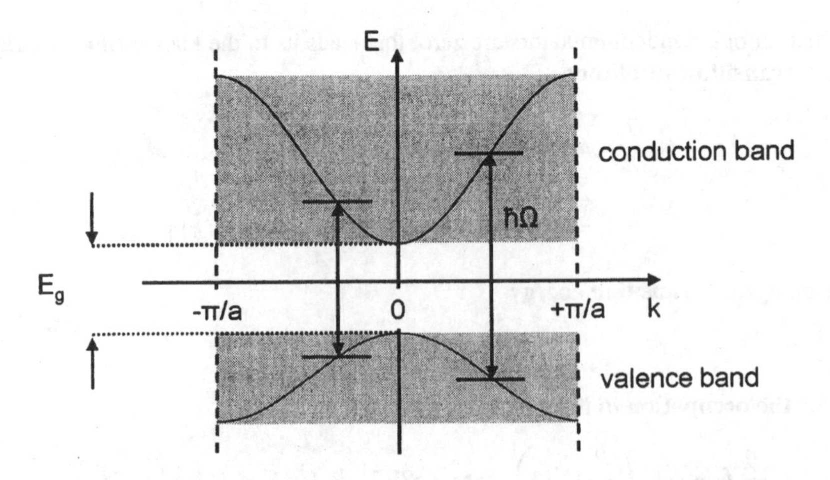

Figure 9: Scheme of the valence and the conduction band of a direct gap semiconductor in the first Brillouin zone, i.e. for wave numbers, k, in the interval $[-\pi/a, +\pi/a]$, a is the lattice constant. At each k, the optical interband transition resembles that of a two-level system with transition energy $\hbar\Omega = E_c(k) - E_v(k)$. Close to the center of the Brillouin zone, the bands are nearly parabolic and the effective mass approximation is justified. E_g is the band gap energy.

What is the meaning of the three components of the Bloch vector? With the above (and generally false) statement *that it is not possible to invert a two-level system by resonant excitation* one implies that electrons are *either* in the ground state *or* in the excited state. In quantum mechanics, however, they can be *both* in the ground state and in the excited state – i.e., they can be in a superposition state. The complex amplitude of this superposition state is encoded in the real and the imaginary part of the transition amplitude, i.e., it is encoded in the components u and v of the Bloch vector. The component w is simply the inversion of the two-level system, i.e., it is equal to -1 if all electrons are in the ground state, and it is +1 for complete inversion. It is easy to show that [1], for vanishing relaxation, the length of the Bloch vector is constant (the matrix in Eq. (59) is unitary) and equal to one, i.e.

$$\sqrt{u^2 + v^2 + w^2} = 1 . \tag{61}$$

Hence, all the physics can be represented as rotations of the Bloch vector on a sphere with radius unity, the Bloch sphere. For vanishing electric field, the Bloch vector rotates in the uv-plane with a frequency given by the optical transition frequency Ω, for very large fields one gets a rotation in the vw-plane with frequency $2\,\Omega_R(t)$. This oscillation is the **Rabi oscillation**. The resulting temporal evolution of the inversion w has already been shown in Fig. 8. If, for example, during the action of the electric field pulse, the Bloch vector performs one complete rotation in the vw-plane, the **pulse area** Θ is equal to 2π. There is, however, no simple mathematical expression for Θ. For finite Ω and Ω_R, the dynamics of the Bloch vector is a combination of both rotations, one in the uv-plane and one in the vw-plane. Graphical examples are shown in Fig. 10.

Example 3.1.: Consider zero light field, in which case we have $\Omega_R = 0$. From Eq. (59) we obtain the two coupled equations: $\dot{u} = +\Omega v$ and $\dot{v} = -\Omega u$, leading to $\ddot{u} + \Omega^2 u = 0$, which is nothing but the harmonic oscillator equation.

Example 3.2.: For an electric field of $E(\vec{r}, t) = \tilde{E}_0 = 4 \times 10^9$ V/m and GaAs parameters ($d_{cv} = 0.5\,$e nm, with the elementary charge e $= 1.6021 \times 10^{-19}$ As) we obtain $\hbar\Omega_R = 2\,$eV which is comparable to the photon energy of $\hbar\omega_0 = 1.42\,$eV corresponding to the GaAs band gap (see section 4.1.).

For semiconductors in general, one gets additional contributions to the Rabi frequency as a result of the Coulomb interaction of electrons and holes [1, 22, 23, 24, 25, 26, 27, 28, 29]. This modified or renormalized Rabi frequency can be interpreted as an internal field, which adds to the external laser field. We will neglect this aspect throughout this article.

Often (but not in this article), the Bloch equations are modified by a transformation to a frame, which rotates in the uv-plane with frequency ω_0. In addition, one neglects off-resonant contributions. This procedure is called the **rotating wave approximation**. Details can be found in Ref. [1]. Within this approximation, the transition frequency Ω in the Bloch equations has to be replaced by the detuning $\tilde{\Omega} = \Omega - \omega_0$, and one gets an **envelope Rabi frequency** $\tilde{\Omega}_R$ given by $\hbar\tilde{\Omega}_R = d_{cv}\tilde{E}(t)$, which varies in time according to the electric field envelope $\tilde{E}(t)$ but no longer oscillates with the frequency of light. The corresponding **envelope pulse area** $\tilde{\Theta}$ is defined according to

$$\tilde{\Theta} = \int_{-\infty}^{+\infty} \tilde{\Omega}_R(t)\,\mathrm{d}t\,. \tag{62}$$

Exercise 3.1.: Consider resonant excitation of a two-level system in the *incoherent limit*. (This brings us back to the qualitative discussion in the introduction of section 3.1.) Specifically, compute the steady-state inversion, w, of the two-level system via the optical Bloch equations (Eq. (59)) for the Bloch vector $(u, v, w)^{\mathrm{T}}$ and account for dephasing according to $\left(\frac{\partial}{\partial t} p_{vc}\right)_{\mathrm{rel}} = \frac{p_{vc}}{T_2}$ in the limit $T_2 \rightarrow 0$. T_2 is called the dephasing time or transverse relaxation time. The latter notion originates from nuclear magnetic resonance (NMR), where the components u and v of the Bloch vector correspond to the real space x and y-components of the magnetization. x and y are perpendicular (transverse) to the static magnetic field, which is usually oriented along the z-direction.

3.2. PONDEROMOTIVE ENERGY

We have seen in section 3.1. that the electric field of a laser pulse can promote electrons from the ground state (valence band) into the excited state (conduction band). In addition to this, the electric field can also accelerate electrons within the bands, classically speaking, or, quantum mechanically, it can modify the states, it can modify the band structure – an aspect, which is not contained in the Hamiltonian Eq. (47). This leads to "dressed electrons" within the effective mass approximation (section 3.2.) and to the "Wannier-Stark ladder" beyond the effective mass approximation (section 3.3.).

Let us start by assuming that the effective mass approximation applies. In this case, classically, Newton's second law[17] for the electron displacement, x, gives

$$m_{\rm e}\,\ddot{x}(t) = \hbar\dot{k}_x(t) = -{\rm e}\,E(t)\,, \tag{63}$$

with the (effective) electron mass $m_{\rm e}$. Inserting the electric field[18] $E(t)$ polarized along the x-direction

$$E(t) = \tilde{E}_0\cos(\omega_0 t + \phi)\,, \tag{64}$$

we obtain the solution

$$x(t) = \frac{{\rm e}\tilde{E}_0}{m_{\rm e}\omega_0^2}\cos(\omega_0 t + \phi)\,. \tag{65}$$

For the velocity $v(t)$ this results in

$$v(t) = -\frac{{\rm e}\tilde{E}_0}{m_{\rm e}\omega_0}\sin(\omega_0 t + \phi)\,. \tag{66}$$

Averaging[19] the kinetic energy $E_{\rm kin}(t) = \frac{m_{\rm e}}{2}v^2(t)$ over an optical cycle $2\pi/\omega_0$ we obtain the **ponderomotive energy** (or quiver energy)

$$\boxed{\langle E_{\rm kin}\rangle = \frac{1}{4}\frac{{\rm e}^2\tilde{E}_0^2}{m_{\rm e}\omega_0^2}\,.} \tag{67}$$

Obviously, the ponderomotive energy is directly proportional to the light intensity I ($\propto \tilde{E}_0^2$, see Eq. (16)). The peak kinetic energy is twice the ponderomotive energy.

---

[17] Here we have already omitted the term $-{\rm e}\,\vec{v}\times\vec{B}$ from the Lorentz force. In a medium, $|\vec{B}| = |\vec{E}|/c$ holds. Thus, $|\vec{v}\times\vec{B}|$ becomes comparable to $|\vec{E}|$ if $|\vec{v}| \approx c$, i.e., for relativistic velocities. For the parameters relevant for solids in this article, the electron velocities $|\vec{v}|$ are small compared to the medium light velocity c. However, for gases and extremely large intensities, the force $-{\rm e}\,\vec{v}\times\vec{B}$ can become important (see example 3.4. and section 6.).

[18] For clarity, we suppress the spatial dependence as well as the vector character of the electric field.

[19] Remember that $\langle\sin^2(\omega_0 t + \phi)\rangle = 1/2$.

Example 3.3.: For $\tilde{E}_0 = 4 \times 10^9\,\mathrm{V/m}$ and GaAs parameters ($m_e = 0.07 \times m_0$, with the free electron mass $m_0 = 9.1091 \times 10^{-31}$ kg and $\hbar\omega_0 = E_g^{\mathrm{GaAs}} = 1.42\,\mathrm{eV}$) one obtains $\langle E_{\mathrm{kin}} \rangle = 2.16\,\mathrm{eV}$. For $\tilde{E}_0 = 6 \times 10^9\,\mathrm{V/m}$ and ZnO parameters ($m_e = 0.24 \times m_0$ and $\hbar\omega_0 = 1.5\,\mathrm{eV}$) we get $\langle E_{\mathrm{kin}} \rangle = 1.27\,\mathrm{eV}$.

By the way: For $\tilde{E}_0 = 4 \times 10^9\,\mathrm{V/m}$ and GaAs parameters, the peak acceleration of the crystal electron, a_e^0, is given by $|a_e^0| = \mathrm{e}/m_e\,\tilde{E}_0 = 1.0 \times 10^{22}\,\mathrm{m/s^2} = 10^{21} \times \mathrm{g}$ with the gravitational acceleration constant near the earth's surface $\mathrm{g} = 9.81\,\mathrm{m/s^2}$. Compared with this, the maximum acceleration of a *Formula–1* race car, which is on the order of $10^1 \times \mathrm{g}$, is really negligible ...

Quantum mechanically but still within the effective mass approximation, the problem of a crystal electron in a laser field (oscillating in time) is somewhat analogous to that of the light field with carrier frequency ω_0 in a mode-locked laser oscillator (see section 2.3.). There, the electromagnetic wave packet (the laser pulse) periodically oscillates back and forth between the laser mirrors (with the round trip frequency f_r). This leads to sidebands of ω_0 – the frequency comb. These sidebands are rigidly shifted by the carrier-envelope offset frequency f_ϕ as a result of the phase slip $\Delta\phi$ of the electromagnetic wave packet from one round trip to the next according to Eq. (33). In analogy to this, semiclassically speaking, the electron wave packet in a periodic laser field acquires a quantum phase in one optical cycle, $\Delta\phi_e$, which is given by the cycle-average[20]

$$\Delta\phi_e = \langle 2\pi \frac{(v_{\mathrm{phase}} - v_{\mathrm{group}})\,\frac{2\pi}{\omega_0}}{\lambda_e} \rangle \,, \tag{68}$$

with the electron de Broglie wavelength $\lambda_e = 2\pi/k_x$ and the period of light $2\pi/\omega_0$. With the dispersion relation of, e.g., vacuum electrons or conduction band electrons in a semiconductor within the effective mass approximation

$$E_c(k_x) = \hbar\omega_e(k_x) = \frac{\hbar^2\,k_x^2}{2\,m_e}\,, \tag{69}$$

and with $v_{\mathrm{phase}} = \omega_e/k_x$ and $v_{\mathrm{group}} = \mathrm{d}\omega_e/\mathrm{d}k_x$, we obtain the phase slip of the oscillating electron wave packet from one optical cycle to the next

$$\begin{aligned}
\Delta\phi_e &= \langle 2\pi \frac{\left(\frac{\hbar k_x}{2\,m_e} - 2\,\frac{\hbar k_x}{2\,m_e}\right)\frac{2\pi}{\omega_0}}{2\pi/k_x} \rangle \\
&= -2\pi \frac{\langle \frac{\hbar^2 k_x^2(t)}{2\,m_e} \rangle}{\hbar\omega_0}
\end{aligned}$$

---

[20]Note that λ_e, v_{phase}, and v_{group} vary in time via $k_x = k_x(t)$.

$$= \; -2\pi \, \frac{\langle E_{\rm kin} \rangle}{\hbar\omega_0} \, . \qquad (70)$$

Note that the minus sign is due the fact that the electron group velocity is larger than its phase velocity, while for photons the situation is usually reversed, i.e., their group velocity is smaller than their phase velocity.

According to Eq. (70), **the phase slip becomes appreciable in magnitude if the ponderomotive energy approaches the carrier photon energy**, i.e. $\langle E_{\rm kin} \rangle/(\hbar\omega_0) \approx 1$. Furthermore, in analogy to the light field in a laser cavity, we expect that the density of states of the combined system *electron and light field*, i.e., of the so called **dressed electron**, exhibits **photon sidebands** of the electron density of states at $\pm N \times \hbar\omega_0$ (with integer N), which – in analogy to Eq. (33) – are shifted according to

$$-\hbar\omega_0 \, \frac{\Delta\phi_{\rm e}}{2\pi} = \langle E_{\rm kin} \rangle \, . \qquad (71)$$

Thus, the entire spectrum is shifted towards higher energy by the ponderomotive energy $\langle E_{\rm kin} \rangle$ [30, 31].

> Thus, $\langle E_{\rm kin} \rangle$ for electrons is analogous to f_ϕ for photons.

For semiconductors, however, the concept of the ponderomotive energy is only meaningful within the range of validity of the effective mass approximation, which obviously fails for large values of $\langle E_{\rm kin} \rangle$, typically already above several 0.1 eV (see Fig. 9). This limits the importance of the ponderomotive energy for semiconductors under extreme conditions.

For real electrons in atoms, on the other hand, it is always a valuable quantity. There, however, for the same laser intensity I, the ponderomotive energy is substantially smaller than in solids, because the free electron mass m_0 is larger than typical effective electron masses by about an order of magnitude. Also, for atoms, the ponderomotive energy has to be compared with the Rydberg energy (13.6 eV for H-atoms), which is at least one order of magnitude larger than typical transition energies in solids. Thus, with the combined effect of both aspects, the laser intensities in atoms have to be two to three orders of magnitude larger than in solids to make the ponderomotive energy comparable to a characteristic energy scale.

For yet larger intensities, the ponderomotive energy reaches the relativistic rest energy of the free electron $m_0 \, c_0^2$. We will come back to this aspect in section 6.

Example 3.4.: For $\tilde{E}_0 = 8 \times 10^{12}$ V/m, $\hbar\omega_0 = 1.5$ eV and free electrons ($m_{\rm e} = m_0 = 9.1091 \times 10^{-31}$ kg) we get $\langle E_{\rm kin} \rangle = 540$ keV, which is comparable to the relativistic rest energy $m_0 \, c_0^2 = 512$ keV of the electron. Thus, the non-relativistic expression of the kinetic energy Eq. (67) no longer applies. The corresponding intensity is 9×10^{18} W/cm^2. Thus, we anticipate an appreciable influence of relativistic effects already at intensities around 10^{18} W/cm^2 (see section 6.2.).

3.3. BLOCH ENERGY

The failure of the concept of the ponderomotive energy in solids for large laser intensities asks for a more general quantity which reflects the kinetic energy of the electrons within the bands without employing the effective mass approximation. Let us first consider a static electric field $\tilde{E}_0$ (again parallel to the x-direction). If the electrons are accelerated so much that their wave number k_x reaches the end of the first Brillouin zone, i.e. $k_x = \pm\pi/a$ with the lattice constant a, they are Bragg-reflected to $k_x = \mp\pi/a$, i.e., their momentum changes sign (Fig. 9). This leads to a real space oscillation of the electron position, known as Bloch oscillation (for a recent review see, e.g., chapter 8 of Ref. [1]). From Eq. (63) for a constant electric field $\tilde{E}_0$ starting from electron wave number $k_x(0) = 0$ we get

$$k_x(t) = -\frac{\mathrm{e}}{\hbar}\,\tilde{E}_0\,t\,. \tag{72}$$

For half a Bloch oscillation period $\pi/\Omega_{\text{Bloch}}$, the electron wave number hits the end of the Brillouin zone, i.e., $k_x(\pi/\Omega_{\text{Bloch}}) = -\pi/a$ and we obtain for the **Bloch frequency** Ω_{Bloch} from this semiclassical reasoning

$$\boxed{\hbar\Omega_{\text{Bloch}} = a\,\mathrm{e}\,\tilde{E}_0\,.} \tag{73}$$

As in our case the electric field of the laser pulse oscillates in time, also the Bloch frequency oscillates with the frequency of light ω_0 – in analogy to the Rabi frequency (Eq. (60) in section 3.1.).

What is the appropriate quantum mechanical picture? Without electric field, the electron wave functions of the atoms forming the solid overlap, which lifts their degeneracy, leading to the bands (see Fig. 9) describing delocalized electron wave functions. In the presence of a strong electric field, i.e., for $a\,\mathrm{e}\,\tilde{E}_0$ large compared with the width of the band (typically a few electron Volts), the potential drop over one lattice constant, $a\,\mathrm{e}\,\tilde{E}_0$, lifts the degeneracy and the wave functions become localized again. The corresponding eigenenergies, E_M, are evenly separated in energy according to the **Wannier-Stark ladder**

$$\boxed{E_M = M \times a\,\mathrm{e}\,\tilde{E}_0\,,} \tag{74}$$

with integer $M = -\infty, ..., -1,\ 0,\ 1, +\infty$. An electronic wave packet is a superposition of these Wannier-Stark states and leads to a quantum beating between these states in time. This quantum beating is the quantum mechanical analogue of the Bloch oscillations. Thus, the **Bloch frequency** Ω_{Bloch} is given by

$$\boxed{\hbar\Omega_{\text{Bloch}} = E_{M+1} - E_M = a\,\mathrm{e}\,\tilde{E}_0\,.} \tag{75}$$

This quantum mechanical result, Eq. (75), is identical to that of the semiclassical reasoning, Eq. (73). Note that Eq. (74) is analogous to the frequency comb of mode-locked laser oscillators (see section 2.3.), were $\Omega_{\text{Bloch}}/2\pi$ plays the role of the repetition frequency f_r (see Eq. (31)).

Example 3.5.: For $\tilde{E}_0 = 4 \times 10^9\,\mathrm{V/m}$ and $a = 0.5\,\mathrm{nm}$ (ZnO along the $\vec{c}$-axis or GaAs) this leads to a Bloch energy of $\hbar\Omega_{\mathrm{Bloch}} = 2\,\mathrm{eV}$, equivalent to a Bloch period of $2\pi/\Omega_{\mathrm{Bloch}} = 2\,\mathrm{fs}$. As this is already shorter than one cycle of light (e.g., $2\pi/\omega_0 = 2.8\,\mathrm{fs}$ for $\hbar\omega_0 = 1.5\,\mathrm{eV}$), one approaches the point at which the carriers experience Bragg reflections within an optical cycle, i.e., the electron kinetic energy reaches the top of the conduction band or the end of the first Brillouin zone.

In general, both **interband transitions** as well as **intraband transitions**, are important. Moreover, one generally also obtains a contribution to the optical polarization associated to the intraband processes. For the experiments on GaAs, ZnO and CdS discussed in this article, however, this contribution turns out to be much smaller than the interband polarization. Hence, one has to solve the problem discussed in sections 3.2. and 3.3. so far together with the optical Bloch equations (section 3.1.). This task has not been fully solved yet in the literature. Within the acceleration theorem approximation (see, e.g., Refs. [1, 32]), it can be considerably simplified by a transformation to the accelerated frame[21]. In essence, this leads to the same Bloch equations as described in section 3.1., however, with a time-dependent optical (interband) transition frequency Ω. For the electron wave vector, $\vec{k} = (k_x, k_y, k_z)^{\mathrm{T}}$, Newton's law (63) for an electric field polarized along the x-direction delivers $\hbar\dot{k}_x(t) = -\mathrm{e}E(t)$, the components k_y and k_z are constant in time. This leads to

$$\Omega(\vec{k}) \to \Omega(k_x(t), k_y, k_z)\,, \tag{76}$$

with

$$k_x(t) = k_x^0 - \frac{\mathrm{e}}{\hbar}\int_{-\infty}^{t} E(t')\,\mathrm{d}t'\,. \tag{77}$$

In Eq. (76), the actual dispersion relations of the bands enter according to $\Omega(\vec{k}) = (E_{\mathrm{c}}(\vec{k}) - E_{\mathrm{v}}(\vec{k}))/\hbar$ (see Fig. 9 and Eq. (56)). Thus, the transition frequency $\Omega = \Omega(t)$ itself oscillates in time with an amplitude which obviously increases with increasing peak field amplitude $\tilde{E}_0$. For symmetric dispersions $\Omega(\vec{k})$ with respect to k_x it oscillates with frequency $2\,\omega_0$, for a non-symmetric dispersion with frequency ω_0. In general, the oscillation of $\Omega(t)$ is not harmonic. After the pulse is over, i.e., for $t \to +\infty$, the transition frequency comes back to its original value, because of $\int_{-\infty}^{+\infty} E(t')\,\mathrm{d}t' = 0$ (see discussion at the end of section 2.3.).

Exercise 3.2.: What are the peak (classical) charge displacements, x_0, for an electric field of $\tilde{E}_0 = 4 \times 10^9\ \mathrm{V/m}$ for a) Rabi oscillations, b) free motion of a particle (ponderomotive energy) and c) Bloch oscillations ? How does x_0 scale with $\tilde{E}_0$ for the three cases a) – c) ?

---

[21] This semiclassical treatment of intraband processes is really only justified for weak electric fields, i.e., as long as the band structure remains intact. In principle, one has to calculate the time-dependent "band structure" under the influence of the strong laser electric field.

4. Carrier-Wave Rabi Flopping of Electrons in Semiconductors

The notion *carrier-wave Rabi flopping* was first used by S. Hughes, who discussed an ensemble of identical uncoupled two-level systems [33, 34] (also see Refs. [35, 36, 37, 38]). It refers to Rabi flopping under the condition that the Rabi frequency is approximately equal to the light frequency [39]. As one is interested in the system's dynamics on a timescale of one period of light or less, both, the rotating wave approximation and the slowly varying envelope approximation [1, 14] must obviously *not* be used. His theoretical work as well as that of others [40] is based on the theoretical framework of Ref. [41], i.e., on numerical solutions of the coupled Maxwell-Bloch equations (see sections 2.1. and 3.1.) in one dimension.

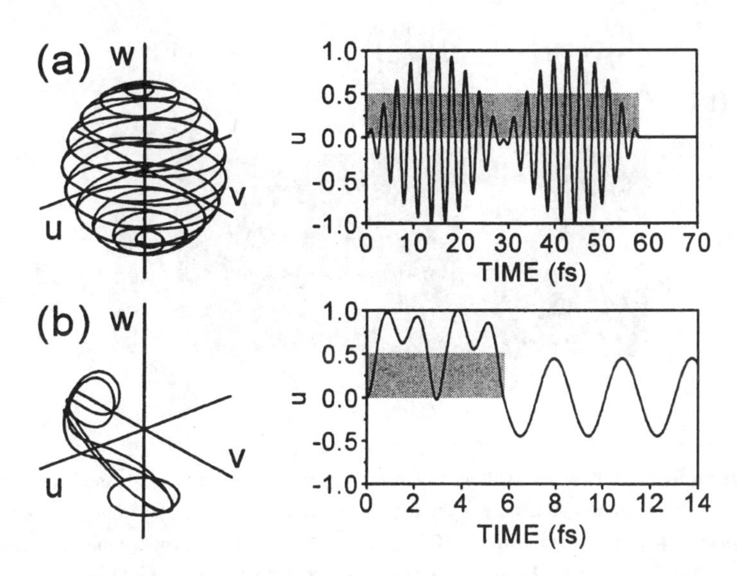

Figure 10: (a) Scheme of the trace of the Bloch vector for conventional Rabi flopping. The rotating wave approximation is not used. Pulse duration is 20 optical cycles, envelope pulse area is $\tilde{\Theta} = 2\pi$. (b) Same for carrier-wave Rabi flopping. Pulse duration is 2 optical cycles, $\tilde{\Theta} = 4\pi$. The optical pulse envelopes are indicated by the grey areas on the RHS.

What are the anticipated signatures of carrier-wave Rabi flopping? The condition Rabi period equal to the light period corresponds to a huge intensity (see section 3.1.). While it might be possible to reach this condition with pulses of several tens of femtoseconds in duration, it is not very likely that the samples will survive the large deposited energy (= intensity $\times$ duration). Thus, it seems favorable to study excitation with very short pulses, ideally with only one or two cycles of light in duration with minimum deposited energy. Remember that, for GaAs parameters, the period of light corresponding to the room temperature band gap energy is 2.9 fs. To highlight the general aspects of carrier-wave Rabi flopping, let us first review the behavior for an ensemble of uncoupled and identical two-level systems, which is the level of sophistication of Refs. [?, 39, 40, 41]. For reference, Fig. 10(a) schematically depicts conventional Rabi flopping plotted on the Bloch sphere,

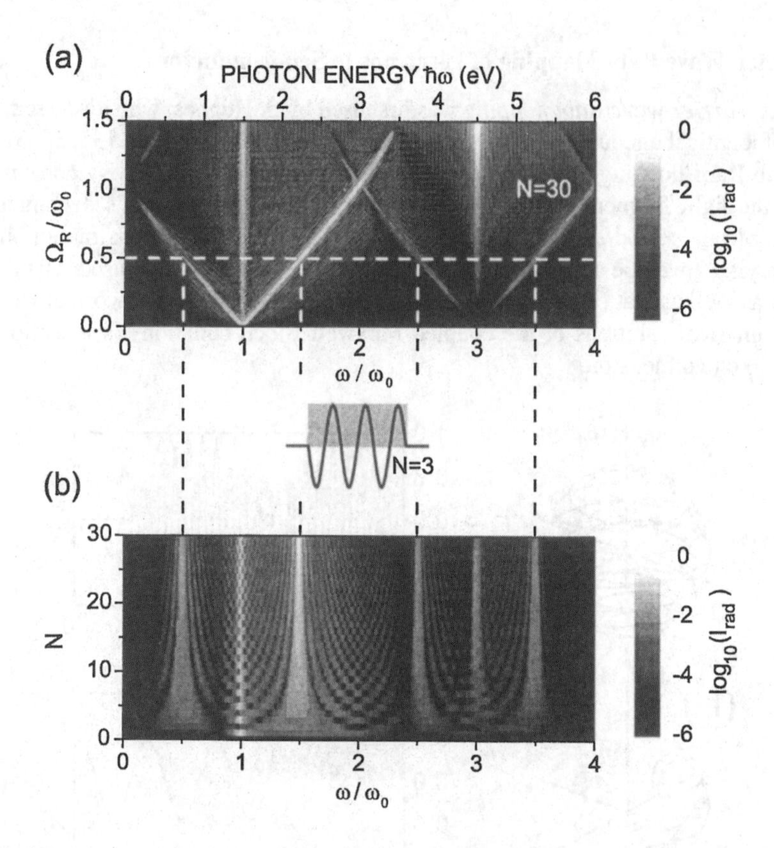

Figure 11: Grey-scale plots of the radiated light intensity I_{rad} (i.e., normalized square modulus of the Fourier transform of the second temporal derivative of the optical polarization $P(t) \propto u(t)$ from the Bloch equations) versus spectrometer frequency ω. (a) Position of the peaks of the fundamental and the third harmonic Mollow triplet versus peak Rabi frequency Ω_{R} in units of the laser carrier frequency ω_0 for $\Omega/\omega_0 = 1$ with $\hbar\omega_0 =$ 1.5 eV and for a $N = 30$ cycle long box-shaped optical pulse (see center inset for $N = 3$). (b) Dependence on the integer number of cycles $N = 1, 2, ...30$ in the pulse for fixed $\Omega_{\mathrm{R}}/\omega_0 = 0.5$. Note the occurrence of additional side maxima for few-cycle pulses.

i.e., the Rabi period is much larger than the light period. For clarity, we neglect any damping at this point. The components u and v of the Bloch vector $(u, v, w)^{\mathrm{T}}$ have been explained in section 3.1. In this representation, the optical oscillation corresponds to an orbiting of the Bloch vector parallel to the equatorial plane (uv-plane) with the optical transition frequency Ω (here $\Omega = \omega_0 = 2\pi/2.9\,\mathrm{fs}$), the oscillation of the inversion to a motion in the vw-plane. For a square-shaped pulse with envelope pulse area $\tilde{\Theta} = 2\pi$ starting from the south pole, i.e., all electrons are in the ground state (valence band), the Bloch vector spirals up to the north pole, i.e., all electrons are in the excited state (conduction band) and back to the south pole. This up and down leads to a temporal modulation of the real part of the optical transition amplitude $u(t)$ (Fig. 10(a)). Fig. 10(b) shows results for $\tilde{\Theta} = 4\pi$ and for a much shorter pulse, *such that the Rabi period equals the light period.* Two related aspects are obvious. First, although $\tilde{\Theta} = 4\pi$, the Bloch vector does not come back to the south pole. In this sense, the usual definition of the envelope pulse area $\tilde{\Theta}$ fails. Hence, also the area theorem of nonlinear optics [14], which is based on this definition, fails. Despite this failure, we quote $\tilde{\Theta}$ for reference at some points in this article. Second, it is obvious that the optical polarization becomes strongly distorted during the two cycles of the optical pulse (see u versus time in Fig. 10(b)). Thus, harmonics are being generated, the most prominent of which, for an inversion symmetric medium, is the third harmonic. For low intensities, this is nothing but the resonantly enhanced third-harmonic generation (see section 2.4.). Note that absolutely no harmonics are generated *after* the two cycles of the optical pulse (see Fig. 10(b)). Here one merely has a free (harmonic) oscillation of the optical polarization with the optical transition frequency Ω of the two-level system.

In the frequency domain (see Fig. 11), pairs of sidebands around the fundamental transition frequency, i.e, at $\Omega \pm \Omega_R$, and around its third harmonic, i.e., at $3\,\Omega \pm \Omega_R$ result from the temporal modulation of the optical transition with the Rabi frequency. Together with the corresponding central peaks, these pairs of sidebands form the fundamental **Mollow triplet** [42] and the **carrier-wave Mollow triplet** at the third harmonic, respectively.

4.1. EXPERIMENT: GALLIUM ARSENIDE

The experiments are performed with 5 fs linearly polarized (p-polarization) optical pulses at $f_r = 81$ MHz ($= 1/12$ ns) repetition frequency, which have recently become available [43]. Our home-built copy of this laser system very nearly reproduces the pulse properties described in Ref. [43]. The typical average output power of the laser lies in the range 120 – 230 mW.

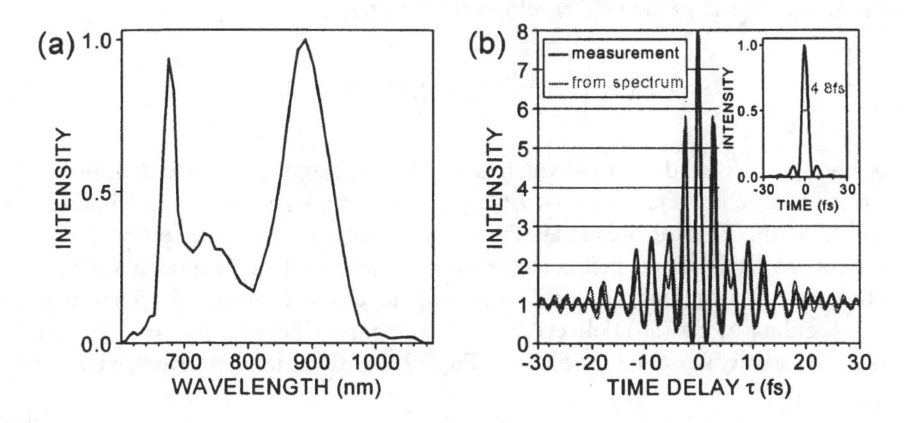

Figure 12: Experiment: (a) measured laser spectrum, (b) measured interferometric autocorrelation. The grey curve in (b) is the autocorrelation computed from the laser spectrum under the assumption of a constant spectral phase (no chirp). The inset in (b) depicts a 4.8 fs full width at half maximum real time intensity profile computed under the same assumption.

Fig. 12(a) shows a typical laser spectrum, which has been obtained via Fourier-transform of an interferogram taken with a pyroelectric detector, which is spectrally extremely flat. The Michelson interferometer used at this point and for all results throughout this article is carefully balanced and employs home-made beam splitters fabricated by evaporating a thin film of silver on a 100μm thin glass substrate. The Michelson interferometer is actively stabilized by means of the Pancharatnam screw [44], which allows for continuous scanning of the time delay while maintaining active stabilization. The remaining fluctuations in the time delay between the two arms of the interferometer are around ± 0.05 fs. The spectral wings which can be seen in Fig. 12(a) result from the spectral characteristics of the output coupler. The measured interferometric autocorrelation[22] depicted in Fig. 12(b) is very nearly identical to the one computed from the spectrum (Fig. 12(a)) under the assumption of a constant spectral phase. This shows that the pulses are nearly transform-limited. The intensity profile computed under the same assumption is shown as an inset in Fig. 12(b) and reveals a duration of about 5 fs. As a result of the strongly structured spectrum (a square-function to zeroth order), the intensity envelope versus time shows satellites (a sinc^2-function to zeroth order[23]). Using a high numerical aperture reflective microscope objective [45], we can tightly focus these pulses to a profile which is very roughly Gaussian with 1μm radius. This value has carefully been measured by a knife-edge technique at the sample position (see Fig. 17(a)). This sample position is equivalent to that of the second-harmonic generation (SHG) crystal used for the autocorrelation in terms of group delay dispersion. In front of the sample, each arm of the interferometer typically has an average power of about 8 mW. The resulting peak intensity of one arm can be roughly estimated as

$$I_0 = \frac{8\,\mathrm{mW}}{\pi\,(10^{-4}\,\mathrm{cm})^2}\;\frac{12\,\mathrm{ns}}{5\,\mathrm{fs}} = 0.6 \times 10^{12}\,\mathrm{W/cm}^2\,. \tag{78}$$

Following our discussion in the introduction (section 2.2. and Eq. (16)), this intensity corresponds to a peak of the field envelope (in vacuum)

$$\tilde{E}_0 = \sqrt{2\sqrt{\frac{\mu_0}{\varepsilon_0}}\,I_0} = 2.1 \times 10^9\,\mathrm{V/m} \tag{79}$$

for one arm, or 4.2×10^9 V/m – which is roughly the number we have used in several examples in sections 2 and 3 – for two constructively interfering arms of the interferometer ($\Leftrightarrow \tau = 0$). To further estimate the envelope pulse area $\tilde{\Theta}$ (see section 3.1.), one furthermore needs the dipole matrix element d_{cv} of the optical dipole transition. From the literature for GaAs we find $d_{cv} = 0.3\,e\,\mathrm{nm}$ [46] and $d_{cv} = 0.6\,e\,\mathrm{nm}$. $\vec{k}\cdot\vec{p}$ perturbation theory according to Eq. (50) delivers $d_{cv} = 0.65\,e\,\mathrm{nm}$. Choosing $d_{cv} = 0.5\,e\,\mathrm{nm}$ for GaAs in this article, the value for $\tilde{E}_0$ from Eq. (79) translates into an envelope pulse area of

$$\tilde{\Theta} = \hbar^{-1}\,d_{cv}\,\tilde{E}_0 \times 5\,\mathrm{fs} = 8.1 > 2\pi \tag{80}$$

[22] In an interferometric autocorrelation, the output of the Michelson interferometer is sent onto a $\chi^{(2)}$ medium. The resulting second harmonic is recorded as a function of the time delay τ of the interferometer.

[23] Remember that the sinc-function is defined as $\mathrm{sinc}(x) = \frac{\sin(x)}{x}$.

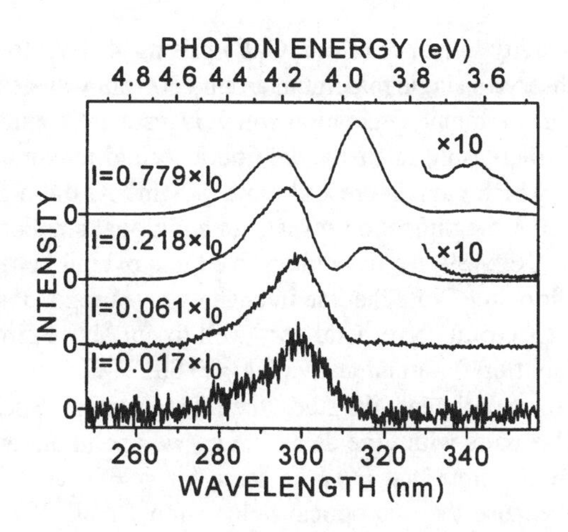

Figure 13: Experiment: Spectra of light emitted into the forward direction around the third harmonic of the GaAs band gap frequency. The spectra are shown on a linear scale, vertically displaced and individually normalized (from top to bottom: maxima correspond to 5664, 439, 34 and 4 counts/s). Excitation with 5 fs pulses. The intensity I of the pulses is indicated.

for one arm ($I = 0.601 \times I_0$ corresponds to 2π pulse area), and $> 4\pi$ (two Rabi periods) for two constructively interfering arms of the interferometer. For a 5 fs pulse and a 2.9 fs band gap period, this corresponds to a Rabi frequency which even slightly exceeds the light frequency. It is also interesting to give a very rough estimate for the excited carrier density under these conditions. The GaAs band-to-band absorption coefficient is $\alpha = 10^4\,\text{cm}^{-1}$. If all the light was absorbed according to this number – certainly an upper limit – one arrives at a carrier density of

$$n_{\text{eh}} = \alpha\, I_0 \times 5\,\text{fs}/1.42\text{eV} = 1.3 \times 10^{20}\,\text{cm}^{-3}\,. \tag{81}$$

For constructive interference of the two arms of the interferometer, this number needs to be multiplied by a factor of four. Thus, we can safely conclude that the highest carrier densities approach $10^{20}\,\text{cm}^{-3}$. In the experiment, we use a 0.6 μm thin film of GaAs clad between $Al_{0.3}Ga_{0.7}As$ barriers, grown by metal-organic vapor phase epitaxy on a GaAs substrate. The sample is glued onto a 1 mm thick sapphire disk and the GaAs substrate is removed. Finally, a $\lambda/4$-antireflection coating is evaporated. The light emitted by this sample, held under ambient conditions, is collected by a second reflective microscope objective [45], is spectrally pre-filtered by a sequence of four fused-silica prisms, and is sent into a 0.25 m focal length grating spectrometer connected to a liquid-nitrogen cooled, back-illuminated, UV-enhanced charge-coupled-device (CCD) camera. For a second set of experiments the transmitted light is dispersed in a miniature spectrometer which allows to simultaneously cover the wavelength range from 500 nm to 1100 nm.

Let us first discuss results for single pulses only, i.e., we block one arm of the interferometer. Fig. 13 shows spectra at the third harmonic for different pulse intensities I in multiples of I_0, as defined above. For the attenuation we have used metallic beam splitters on 100 μm thin fused silica substrates, the dispersion of which has carefully been compensated for by the extra-cavity sequence of four CaF_2 prisms [43]. At low intensity, i.e., for $I = 0.017 \times I_0$, we observe a single maximum around 300 nm wavelength which is interpreted as the usual third-harmonic generation which is resonantly enhanced by the GaAs band edge here. With increasing intensity, we find a second maximum emerging at the long wavelength side, which gains more and more weight. At the highest intensity, i.e., $I = 0.779 \times I_0$, the 10× magnification reveals an additional smaller maximum around 340 nm wavelength. In Ref. [39] we have interpreted this overall behavior as a signature of carrier-wave Rabi flopping. Note that the intensities revealing a double-peak structure in the third-harmonic spectrum correspond very well to our above simple estimates, i.e., we estimated a full Rabi flop for an intensity of $I = 0.601 \times I_0$.

In the second set of experiments we study the third-harmonic spectra for excitation with phase-locked pulse pairs with time delay τ, i.e., we open both arms of the interferometer. It is interesting to note that $\tilde{\Theta}$ is the same for $\tau = 0$ and for, e.g., τ equal to two optical cycles – because the two optical fields simply add. Yet, the corresponding Rabi frequency is larger for $\tau = 0$. For low intensities (Fig. 14(a)), i.e., for small Rabi frequency as compared to the light frequency, the third-harmonic spectrum is simply modulated as a function of τ due to interference of the laser pulses within the sample leading to a period of about 2.9 fs. In contrast to this, for higher intensities (Fig. 14(b)–(d)), where the Rabi frequency becomes comparable to the light frequency, the shape of the spectra changes dramatically with time delay τ. For, e.g., $\tau = 0$ in Fig. 14(b), the two pulses simply interfere constructively and we find the same spectral double-maximum structure as in the single pulse experiments (Fig. 13). For larger τ, i.e., after one or two optical cycles, this double maximum disappears and is replaced by one prominent and much larger maximum. For the highest intensity, i.e., for Fig. 14(d) – which corresponds to an envelope pulse area $\tilde{\Theta}$ of more than $4\,\pi$ – the behavior is quite involved with additional fine structure for $|\tau| < 1$ fs. Note that the spectra for $\tau = 0$ nicely reproduce the behavior seen in Fig. 13.

We have also deliberately introduced positive or negative group velocity dispersion by moving one of the extra-cavity CaF_2 prisms in or out of the beam with respect to the optimum position (Fig. 15) [39]. Obviously, this leaves the amplitude spectrum of the laser pulses unaffected. We find that one quickly gets out of the regime of carrier-wave Rabi flopping, i.e. both, the splitting at $\tau = 0$ as well as the dependence of the shape on the time delay τ, quickly disappear with increasing pulse chirp. This demonstrates that it is not just the large bandwidth of the pulses but the fact that they are short – two optical cycles – which is important for the observation of carrier-wave Rabi flopping.

Apart from the interference of the laser pulses in the sample, at larger time delays $|\tau|$ one additionally observes interference of the third-harmonic signals corresponding to the two phase-locked pulses on the detector leading to periods around one femtosecond in Fig. 16. It can also be seen from Fig. 16 that the splitting in the spectra gradually approaches zero for large time delays. In independent experiments we have verified that the pulses are really 5 fs in duration at the sample position, which rules out that the tails at large time delay are an artifact of our experiment.

It is clear that one also expects large induced transmission at the GaAs band gap as a result of the Rabi flopping, which brings us to the third set of experiments we have performed on GaAs. Fig. 17(a) schematically shows the geometry. To vary the excitation

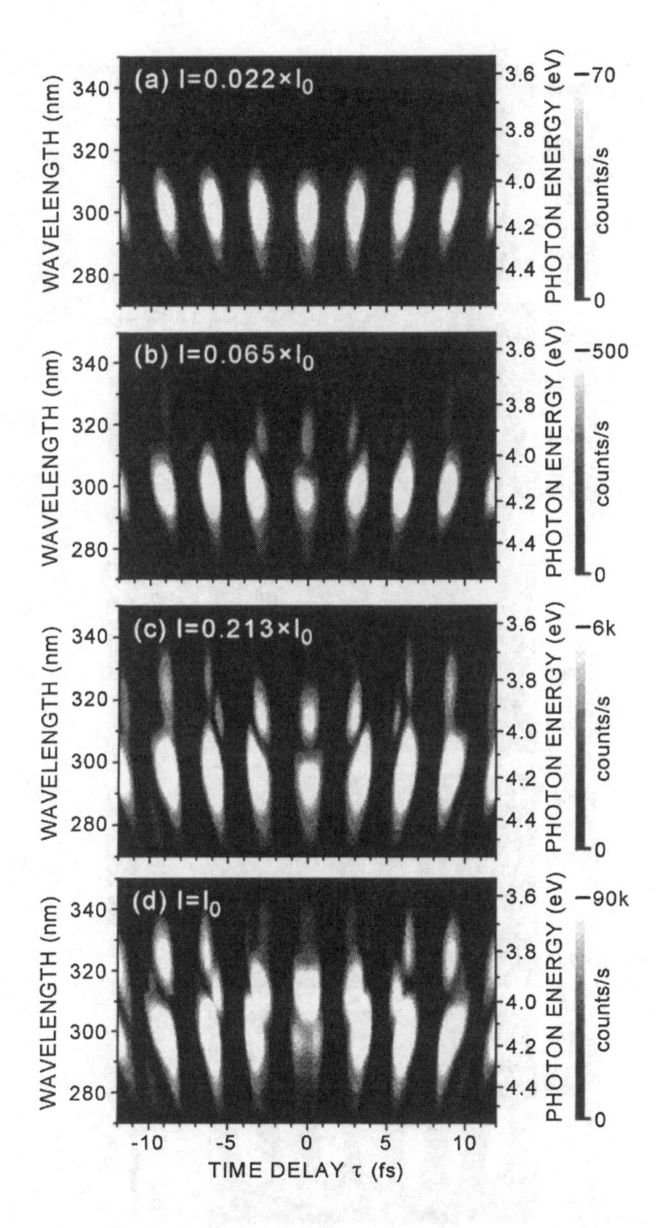

Figure 14: Experiment: Same as Fig. 13, however, using pairs of phase-locked 5 fs pulses. The signal around the third harmonic of the band gap is depicted versus time delay τ in a grey-scale plot (note the saturated grey scale on the right hand side). (a)–(d) correspond to different intensities I as indicated. I refers to one arm of the interferometer.

intensity without having to introduce filters (which would definitely require to change the dispersion compensation), we simply move the sample in the z-direction through the fixed

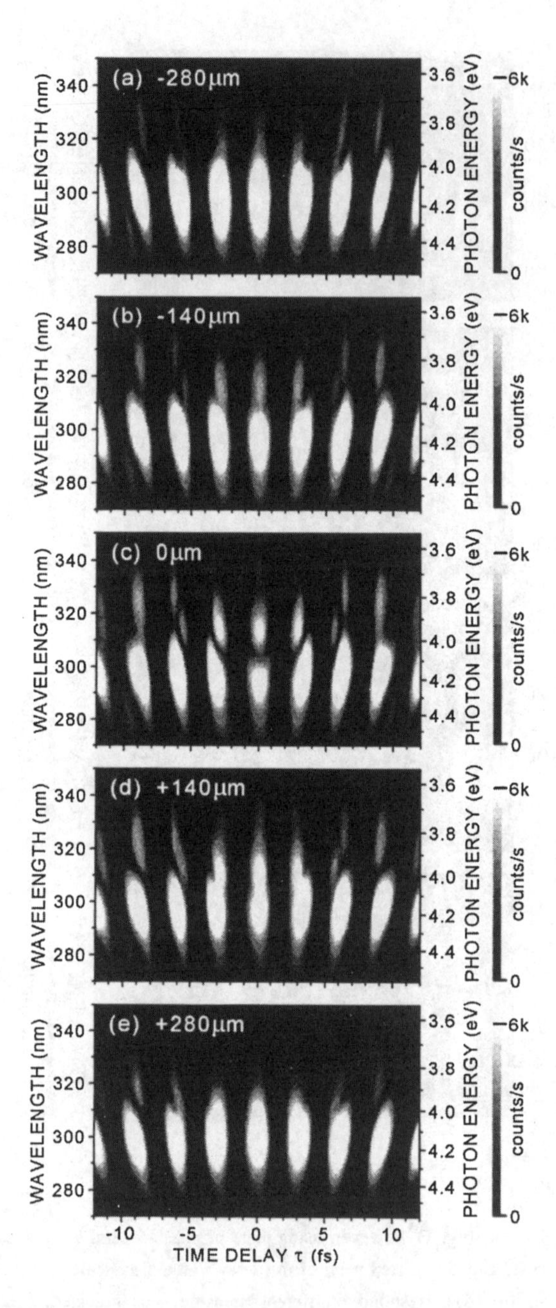

Figure 15: Experiment: As Fig. 14(c), however, the additional/removed amount of CaF_2 material is indicated. This changes the chirp of the pulses but not their amplitude spectrum. (c) corresponds to Fig. 14(c), i.e., to $I = 0.213 \times I_0$.

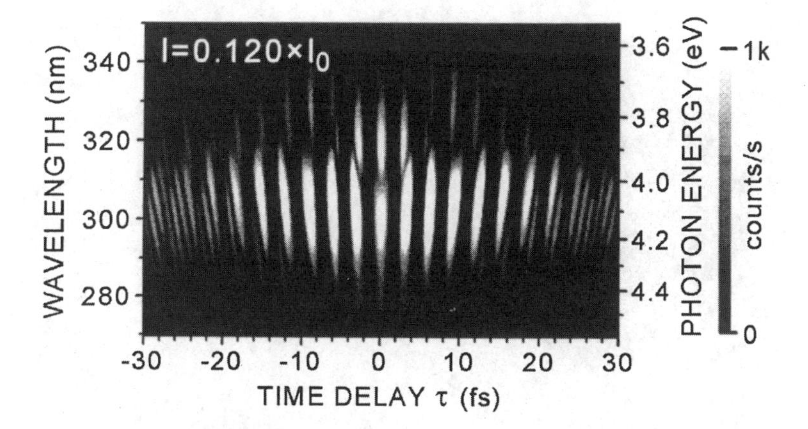

Figure 16: Experiment: Similar to Fig. 14(c) with $I = 0.120 \times I_0$, however, for a larger range of the time delay τ. Around $\tau = 0$ the interference of the laser pulses within the sample dominates, while we additionally observe interference of the third-harmonic signals corresponding to the two phase-locked pulses on the detector for larger $|\tau|$.

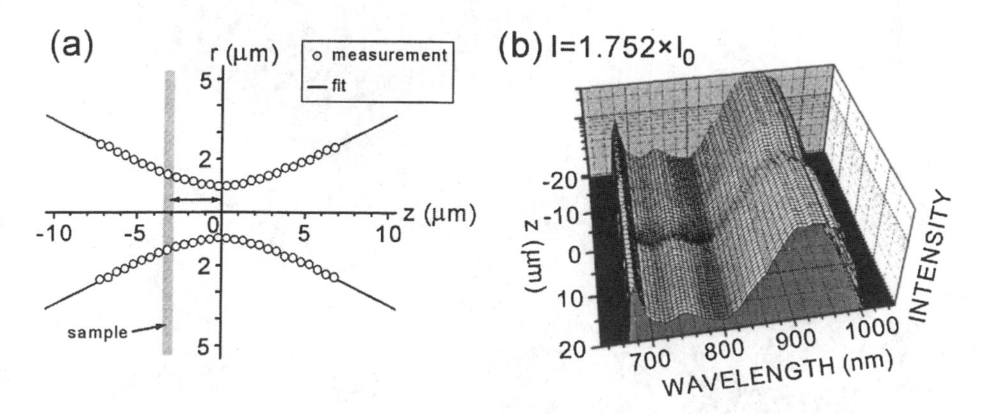

Figure 17: (a) Scheme of the z-scan experiment (the measured radii are fitted with the formula $r(z) = 0.97\,\mu\text{m}\sqrt{1 + 0.12 \cdot (z/\mu\text{m})^2}$, (b) transmitted light intensity on a logarithmic scale versus sample position z for an intensity of $I = 1.752 \times I_0$ (referring to $z = 0$).

focus ($z = 0$) of the microscope objective and collect the transmitted light with the fixed second microscope objective. To enhance the visibility of the changes in transmission shown in Fig. 17, we define a differential transmission, $\Delta T/T$, as

$$\frac{\Delta T}{T} = \frac{I_t(z) - I_t(z = -\infty)}{I_t(z = -\infty)}, \tag{82}$$

where $I_t(z)$ is the transmitted light intensity at sample position z. The condition $z = -\infty$ actually corresponds to $z = -20\,\mu\text{m}$ in the experiment, where the profile is so large that

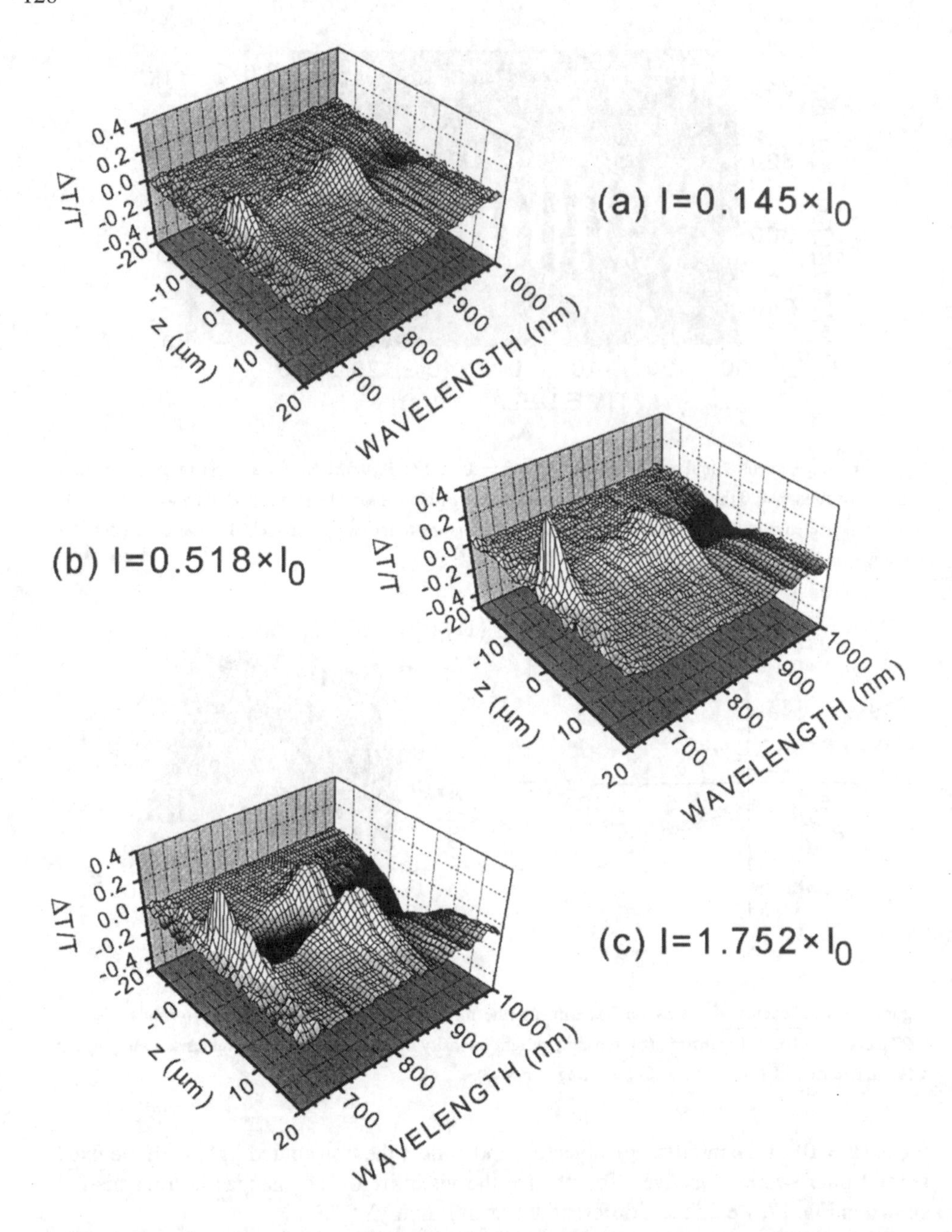

Figure 18: Differential transmission $\Delta T/T$ as a function of the sample coordinate z for three different incident intensities I (referring to $z = 0$). (a) $I = 0.145 \times I_0$, (b) $I = 0.518 \times I_0$, (c) $I = 1.752 \times I_0$.

we can safely assume that linear optics applies.

Fig. 18 shows corresponding results for three different incident light intensities I in units of I_0 as defined above. First, all results are closely symmetric around $z = 0$, which indicates that changes in absorption dominate. Changes in the refractive index might lead to focusing or defocusing of the beam which would result in an asymmetric dependence on z (similar to the known so called z-scan technique, e.g., described in Ref. [48]). Second, one can see a large increase in transmission for wavelengths shorter than the GaAs band edge (approximately 870 nm) around $z = 0$ (Fig. 18(a)). $z = 0$ corresponds to the highest intensity in each plot. The maximum around 670 nm wavelength results from bleaching of the band gap of the $Al_{0.3}Ga_{0.7}As$ barriers of the GaAs double heterostructure which accidentally coincides with the pronounced maximum in the laser spectrum (Fig. 12(a)) also around 680 nm. For larger intensity, Fig. 18(b), the transmission maximum around $z = 0$ flattens and we observe pronounced induced absorption for wavelengths longer than the GaAs band edge. For the highest intensity (Fig. 18(c)), this increased absorption becomes the dominating feature throughout most of the spectral range. Note that little if any induced transparency is observed for wavelengths between 780 nm (170 meV above the unrenormalized band gap $E_g = 1.42\,\text{eV}$) and 700 nm (350 meV above the unrenormalized band gap) while the laser spectrum (Fig. 12) still has significant amplitude there. This indicates that these states high up in the band-to-band continuum of GaAs must experience a much stronger damping (phase relaxation) and/or energy relaxation than those states near the band gap.

4.2. *THEORY*

In the introduction to this section we have already discussed the behavior for one two-level system resonant with the center frequency of the laser pulses. In the following we discuss the result of only the Bloch equations as a function of the detuning, which is relevant for states in the continuum of states of the semiconductor bands. Corresponding results[24] are depicted in Fig. 19. Here, ω denotes the (spectrometer) photon frequency, ω_0 the laser carrier frequency and $\Omega = \Omega(\vec{k}) = \hbar^{-1}(E_c(\vec{k}) - E_v(\vec{k}))$ the transition frequency of one transition within the band. All states are assumed to have the same dipole matrix element and the same phenomenological relaxation in Eq. (55) according to

$$\left(\frac{\partial}{\partial t} p_{vc}(\vec{k})\right)_{\text{rel}} = \frac{p_{vc}(\vec{k})}{T_2} \tag{83}$$

with the dephasing time $T_2 = 50\,\text{fs}$. Relaxation for the occupation numbers can be neglected. Without band gap renormalization, it is clear that there are no states below the band gap energy (dashed horizontal line); nevertheless, we depict these data. Again, the laser carrier frequency is centered at the band gap energy, i.e., we have $\hbar\omega_0 = E_g$. The laser spectrum is shown on the right hand side lower corner as the grey-shaded area. The spectrum for $\hbar\omega_0 = \hbar\Omega = E_g$ is also depicted by the white line. It corresponds to the result of a single resonantly excited two-level system. For small envelope pulse area, $\tilde{\Theta} = 0.5\pi$, we find a single rather narrow maximum around $\omega/\omega_0 = 3$ and $\Omega/\omega_0 = 1$. Its width correlates with the width of the laser spectrum. This single maximum is nothing but the usual, yet resonantly enhanced, third-harmonic generation. It experiences a

[24] Remember that the sech–function is defined as $\text{sech}(x) = 1/\cosh(x) = 2/\big(\exp(+x) + \exp(-x)\big)$.

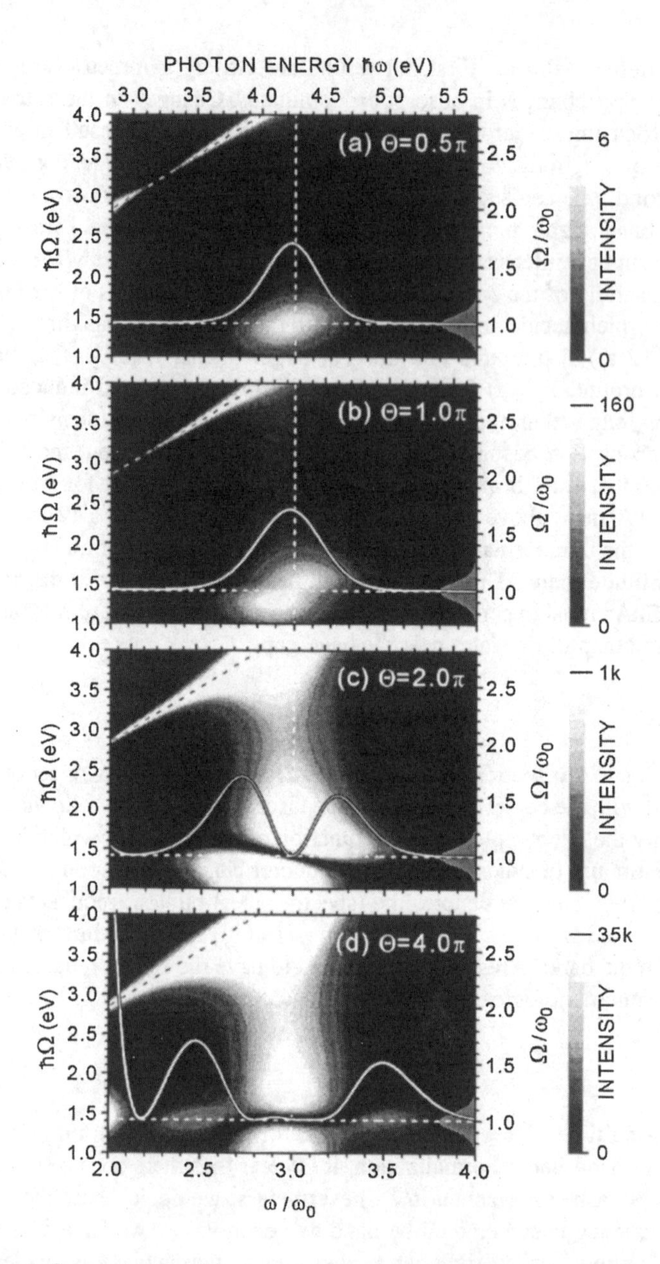

Figure 19: Theory: Grey-scale plot of the intensity (square modulus of p_{vc}) as a function of spectrometer frequency ω and transition frequency Ω (see Fig. 9). The carrier frequency ω_0 of the optical pulses (see grey areas on the RHS) is centered at the band gap frequency, i.e., $\hbar\omega_0 = E_g$. The spectrum for a transition right at the band gap, i.e. $\hbar\Omega = E_g$, is highlighted by the white curve. The diagonal dashed line corresponds to $\Omega = \omega$. Excitation with sech^2-shaped 5 fs pulses. The envelope pulse area $\tilde{\Theta}$ increases from (a) to (d).

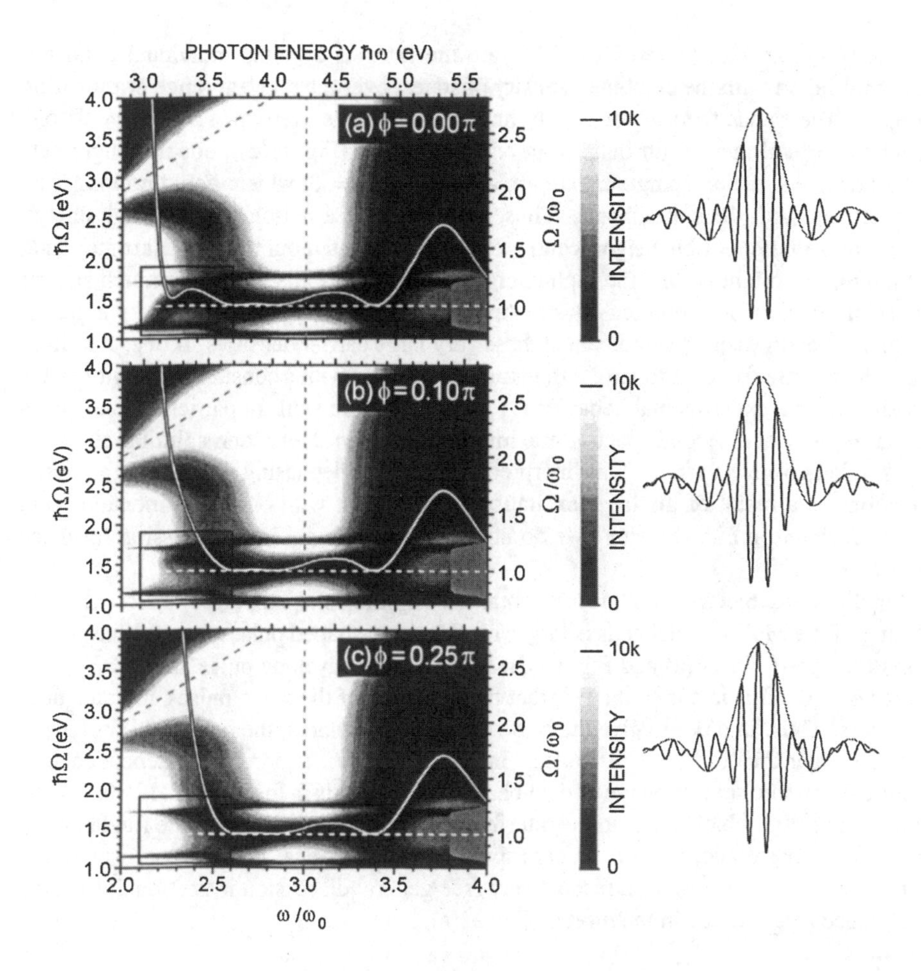

Figure 20: Theory: As Fig. 19, however for sinc^2-shaped $t_{\mathrm{FWHM}} = 5.6\,\mathrm{fs}$ pulses (spectrum see grey areas on the RHS) with envelope pulse area $\tilde{\Theta} = 4\pi$. The CEO phase ϕ, i.e., the phase between the electric field envelope and the carrier wave, is parameter. Note that significant changes occur within the black rectangles. (a) $\phi = 0.00\pi$, (b) $\phi = 0.10\pi$, and (c) $\phi = 0.25\pi$. The corresponding electric fields versus time are depicted on the RHS. The grey curves on the RHS are the field envelopes.

constriction for $\tilde{\Theta} = 1.0\pi$, which evolves into a shape that resembles an anticrossing for $\tilde{\Theta} = 2.0\pi$. Here, two separate peaks are only observed in a rather narrow region around $\hbar\Omega = \hbar\omega_0 = E_{\mathrm{g}}$, while for larger $\hbar\Omega$ only a single maximum occurs. Also, we find that the contribution of larger frequency transitions is by no means small. For, e.g., $\hbar\Omega = 2\,\mathrm{eV}$ transition energy, the signal is actually larger than for the band gap, i.e., for $\hbar\Omega = 1.42\,\mathrm{eV}$. This trend continues for yet larger pulse areas (see $\tilde{\Theta} = 4.0\pi$ in Fig. 19(d)). While there is considerable resonant enhancement (as can be seen from Fig. 19(a)), this enhancement becomes less important at large pulse areas because the resonant transitions are completely saturated.

The actual spectra (see section 3.1.) are the integral over the individual contributions, multiplied with the combined density of states, over the relevant range of transition energies. The bands themselves clearly have contributions even at $\hbar\Omega = 5\,\text{eV}$. If one would sum up all these contributions at, e.g., $\tilde{\Theta} = 4.0\pi$ (Fig. 19(d)), one no longer gets two maxima but rather a single maximum around $\omega/\omega_0 = 3$, which would no longer be in agreement with the experiments. Thus, there must be a reason why the high-energy transitions do not contribute significantly. It was first pointed out to us by Hartmut Haug that the reason might be that the high-energy transitions are likely to have much shorter dephasing times which significantly suppresses their contribution. Also, band gap renormalization becomes quite significant at these very large carrier densities. If one, e.g., integrates the spectra from 1.2 to 1.6 eV transition energy $\hbar\Omega$ with a constant density of states (not shown), the experimental behavior is reproduced quite well. In particular, one gets a gradual growth of a second spectral maximum rather than the sudden splitting observed for a single two-level system. This interpretation of short dephasing times of high-energy transitions as a result of the large excitation is consistent with our experimental observations depicted in Fig. 18, where we do not observe any induced transmission for these states either.

Finally, we depict in Fig. 20 results obtained for sinc^2-shaped pulses[25]. Note that the splitting of the Mollow sidebands is larger for the sinc^2-shaped pulses than for the sech^2-pulses (compare Fig. 20(a) and Fig. 19(d)) for the same envelope pulse area of $\tilde{\Theta} = 4\pi$ and for $\phi = 0$. This is due to the fact that the envelope of the sinc^2 pulses contains negative parts. Thus, in order to give the same envelope pulse area, the peak Rabi frequency has to be larger than for the sech^2-pulses. In Fig. 20 we furthermore find a dependence of the third-harmonic spectra on the optical phase ϕ (see signals in the lower LHS black rectangles in Fig. 20), which is due to the interference of the contributions of the fundamental and the third harmonic, i.e., of the contributions of the optical polarization originating from $\omega/\omega_0 = 1$ and $\omega/\omega_0 = 3$, respectively (see general discussion in section 2.4.). This interference is illustrated in the inset of Fig. 22(a).

4.3. ROLE OF THE CARRIER-ENVELOPE OFFSET PHASE

Theory: In order to actually observe this interesting interference and the corresponding dependence on the CEO phase, one must take care that propagation effects do not obscure the behavior anticipated from the Bloch equations. In order to account for propagation effects in a more realistic manner than in the previous section, where we have used the concept of a background dielectric constant, we employ an ensemble of two-level systems, which fits the measured shape of the linear dielectric function of GaAs according to Ref. [49] (reproduced in Fig. 21(d)).

The linear dielectric function of GaAs exhibits two strong resonances, the so called E_1 and E_2 resonance, which are due to the particular shape of the band structure. Following our discussion on high-energy transitions in section 4.2., only the optical nonlinearities of the band edge transitions are accounted for, the other transitions are assumed to be

[25] The long-time tails (posing significant numerical problems) have been suppressed by a Gaussian, i.e., $E(t) = \tilde{E}_0 \,\text{sinc}(t/t_0) \exp[-t^2/(2\tau_{\text{Gauss}}^2)] \cos(\omega_0 t + \phi)$ with $t_0 = t_{\text{FWHM}}/2.7831$ and $\tau_{\text{Gauss}} = 20$ fs.

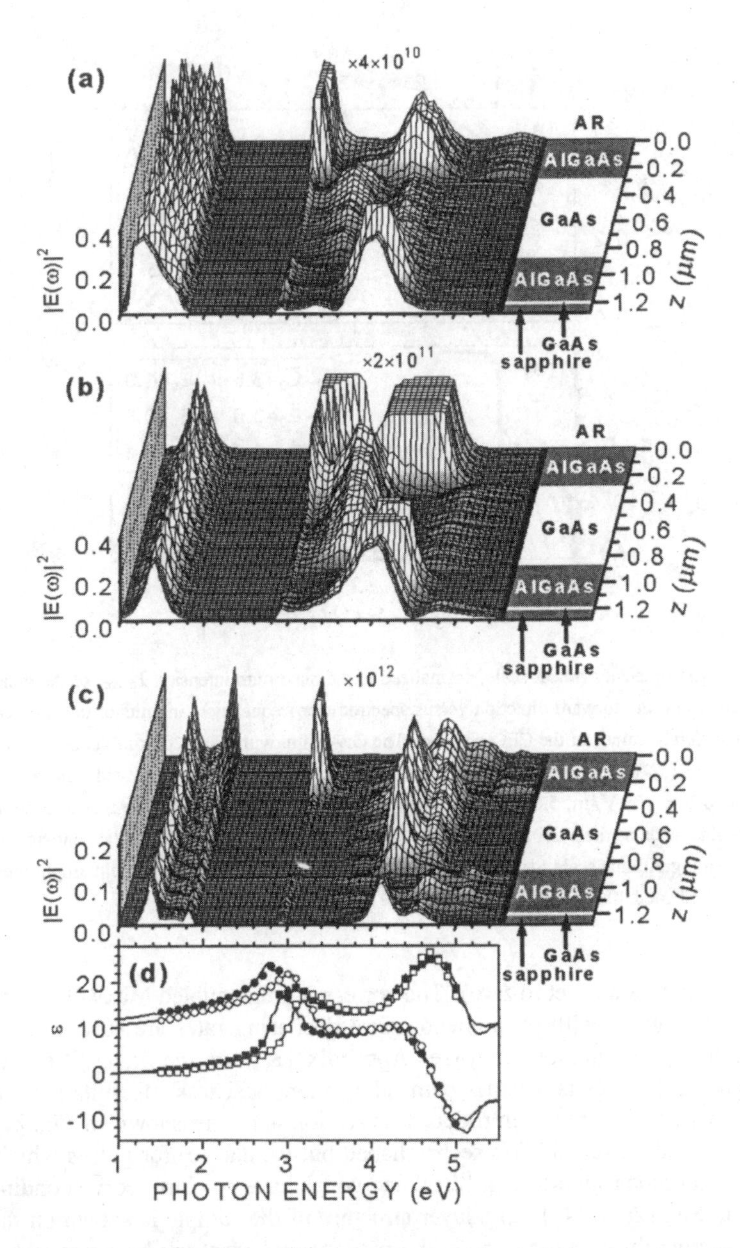

Figure 21: $|E(\omega)|^2$ (normalized to the maximum of the incident electric field spectrum) as a function of $\hbar\omega$ and z, $\phi = 0$. Note the strong variation as a function of z. (a) sinc^2-shaped 5.6 fs pulses, $\tilde{E}_0 = 3.5 \times 10^9$ V/m, the GaAs cap layer thickness is $d_{\text{cap}} = 30$ nm. (b) As (a), but for sech^2-shaped 5 fs incident optical pulses with $\tilde{E}_0 = 3.5 \times 10^9$ V/m, and $d_{\text{cap}} = 10$ nm. (c) As (a), but for incident pulses which are fitted [47] to the experiment (see Fig. 12(a)), $\tilde{E}_0 = 2.5 \times 10^9$ V/m, $d_{\text{cap}} = 10$ nm. (d) The real (circles) and imaginary (squares) part of the linear dielectric function of GaAs (full) and $Al_{0.3}Ga_{0.7}As$ (open), respectively, are shown for comparison. The symbols are the experimental data from Ref. [49], the full curves correspond to our modeling.

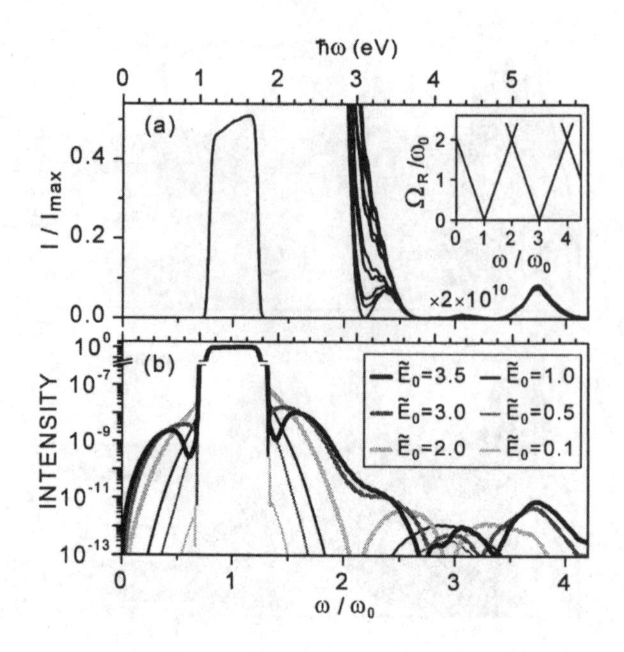

Figure 22: (a) Signal intensity (linear scale, normalized to the maximum intensity, I_{max}, of the incident laser spectrum) emitted into the forward direction versus spectrometer frequency ω in units of the laser carrier frequency ω_0 for different values of the CEO phase ϕ. The GaAs film with $L = 20\,\mathrm{nm}$ thickness on a substrate with $\epsilon_s = \mathrm{const} = (1.76)^2$ has no $Al_{0.3}Ga_{0.7}As$ barriers on either side, but a front-side antireflection (AR) coating, $\tilde{E}_0 = 3.5 \times 10^9\,\mathrm{V/m}$, 5.6 fs sinc^2-shaped pulses. The inset illustrates the interference of the different Mollow sidebands as the Rabi frequency $\Omega_{\mathrm{R}} = \hbar^{-1}\, d_{\mathrm{cv}}\, \tilde{E}$ increases. (b) As (a), but signal intensity (normalized) on a logarithmic scale for fixed $\phi = 0$ and for different incident electric field amplitudes (in units of $10^9\,\mathrm{V/m}$) as indicated. Taken from Ref. [50].

linear (their occupation is set to zero). The corresponding coupled Maxwell-Bloch equations in one dimension with phenomenological dephasing rates are solved numerically without further approximations [50] (see Appendix 13.1. on the finite-difference time-domain algorithm). This is a rather demanding numerical task. Results for the GaAs double heterostructure already introduced in section 4.1. are shown in Fig. 21 for (a) 5.6 fs sinc^2-shaped pulses, (b) 5 fs sech^2-shaped pulses, and (c) for pulses which match the experimental spectrum (see Fig. 12(a)) and which have no chirp (corresponding to the grey curve in Fig. 12(b)) [47]. The layer structure of the sample is shown on the RHS. While propagating through the sample, the fundamental spectrum becomes significantly distorted, which leads to a lengthening of the pulse in time, and, thus to a reduction of the field amplitude and the Rabi frequency. This effect is most pronounced for pulses corresponding to the experiment (see Fig. 21(c)), where especially the sharp high-energy peak of the laser spectrum is largely affected. The effect is close to negligible for sech^2-shaped pulses (Fig. 21(b)). The dispersive effects due to the linear dielectric function (Fig. 21(d)) further enhance the pulse distortions. Both effects lead to a reduction of the splitting of the Mollow sidebands around the third harmonic (compare Figs. 21(c) and 19). This explains

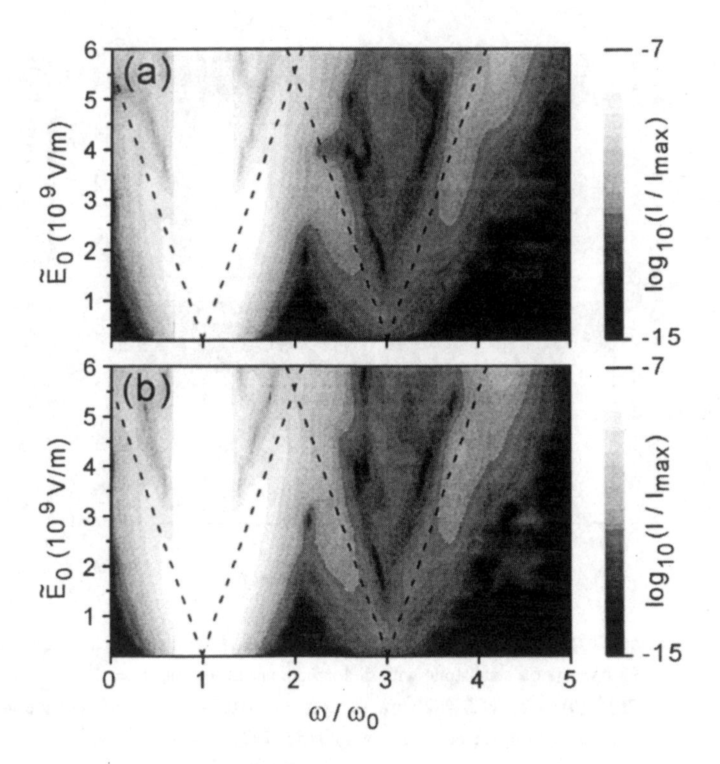

Figure 23: Grey-scale image of the emitted light intensity into the forward direction versus field amplitude of the incident pulses, corresponding to the inset in Fig. 22. Parameters as in Fig. 22(a). (a) $\phi = 0$, (b) $\phi = \pi/2$. Taken from Ref. [50].

the much (factor of two) smaller splitting seen in Fig. 13 as compared to the simple modeling (see white curves in Fig. 19) as well as the tails for large time delays $|\tau|$ in Fig. 16. This reduced splitting between the Mollow sidebands obviously largely suppresses the anticipated interference term. Thus, the GaAs double heterostructure discussed in section 4.1. is not suitable for observing a dependence on the CEO phase. Also, it becomes obvious from Fig. 21 that the signal around the third harmonic varies very strongly with propagation coordinate z. This is mainly due to the fact that the absorption coefficients (for the third harmonic) of both, GaAs and $Al_{0.3}Ga_{0.7}As$, are on the order of 1/(10 nm) which can easily be estimated from the corresponding linear dielectric functions shown in Fig. 21(d). As a dramatic result, the detected signal does not stem from the 600 nm thick GaAs layer sandwiched between $Al_{0.3}Ga_{0.7}As$ barriers – which we have initially believed [39] – but rather from the thin GaAs cap layer initially employed as an antioxidation layer. The band edge of the $Al_{0.3}Ga_{0.7}As$ barriers leads to an (almost) off-resonant nonlinear signal which is expected from the dependence on detuning shown in Fig. 19.

Fig. 22(a) shows the spectra of light emitted into the forward direction of a thin layer of GaAs which has no $Al_{0.3}Ga_{0.7}As$ barriers for various values of the CEO phase ϕ. Samples as such could be produced by molecular-beam epitaxy on GaAs substrates and

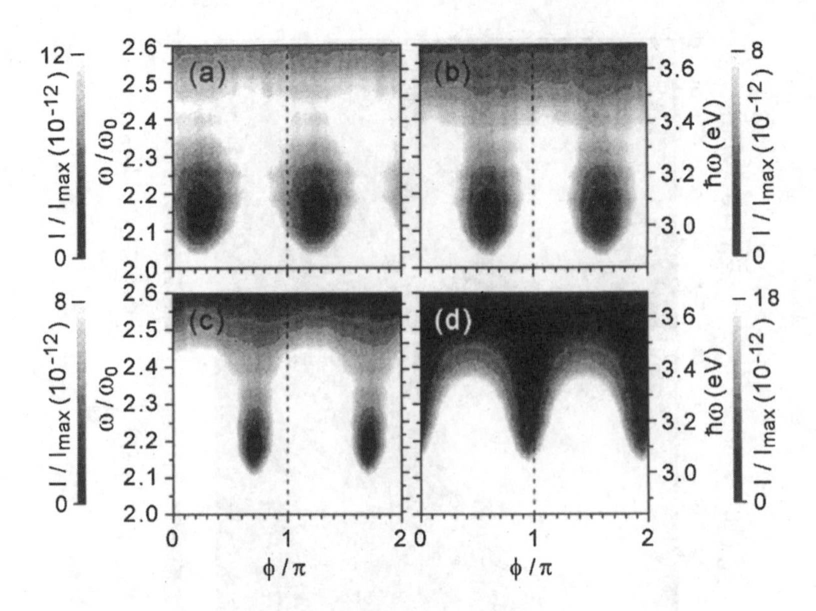

Figure 24: Grey-scale image of the emitted intensity as a function of ω and ϕ for a thin GaAs film with thickness L without $Al_{0.3}Ga_{0.7}As$ barriers on a substrate with dielectric constant ϵ_s. (a) $L = 100\,\text{nm}$, $\tilde{E}_0 = 3.5 \times 10^9\,\text{V/m}$, and $\epsilon_s = (1.76)^2$, (b) as (a) but $L = 20\,\text{nm}$, (c) as (b) but with an additional front-side antireflection coating (as in Fig. 22), (d) as (c) but for an electric field amplitude of $\tilde{E}_0 = 4.0 \times 10^9\,\text{V/m}$.

subsequent selective etching, first of the GaAs substrate and then of the $Al_{0.3}Ga_{0.7}As$ etch stop layer. The meeting of the different Mollow sidebands (see inset) can also be seen in the actual calculations (Fig. 22). Furthermore, it becomes clear from the intensity dependence shown in Fig. 22(b) that it is **not** simply the interference of the tail of the laser spectrum itself (which is roughly equal to a box-function) with the third-harmonic signal – which would be similar to the approach outlined in section 2.4. – but rather the interference of different Mollow sidebands. Fig. 23 exhibits the same data as Fig. 22, but represented as a grey-scale image.

Fig. 24(a) depicts the intensity spectra of light emitted into the forward direction versus CEO phase ϕ for a $L = 100\,\text{nm}$ thin GaAs film on a substrate with $\epsilon_s = (1.76)^2$ (e.g., sapphire). Note the dependence on ϕ with large visibility around $\omega/\omega_0 = 2.05-2.25$ (this is a 284 meV or 38 nm broad interval) and the period of π according to Eq. (40) (rather than 2π according to Eq. (38)) resulting from the inversion symmetry of the problem. In other words: The signal does not depend on the sign of the electric field. (b) Same for $L = 20\,\text{nm}$, indicating that one already has some distortions in (a) due to the finite thickness of the sample as a result of different group and phase velocities – the high-energy transitions do **not** react instantaneously as would be the case for the background dielectric constant. (c) As (b), but introducing a front-side $\lambda/4$-antireflection (AR) coating designed for the fundamental laser frequency ω_0. Note that (b) and (c) are shifted with respect to each other horizontally, because the incident optical pulses, and thus also ϕ, are distorted as a result of multiple reflections. (d) As (c), but for a different incident electric field

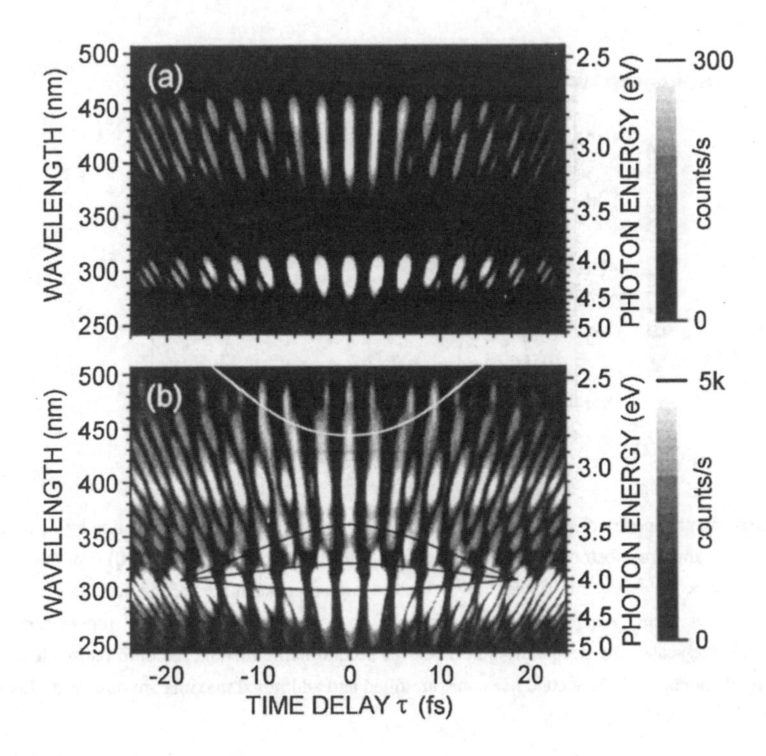

Figure 25: Experiment: Spectra of light (linear scale) emitted into the forward direction by a $l = 100\,\mathrm{nm}$ thin GaAs film on sapphire substrate resonantly excited by a pair of 5 fs pulses with time delay τ. The CEO phase ϕ of the laser pulses is not stabilized. (a) Excitation intensity $I = 0.24 \times 10^{12}\,\mathrm{W/cm^2}$, (b) $I = 2.8 \times 10^{12}\,\mathrm{W/cm^2}$ (both arms at $\tau = 0$). The contribution centered around $\lambda = 425\,\mathrm{nm}$ wavelength is due to surface SHG. The single peak in (a) centered around $\lambda = 300\,\mathrm{nm}$ wavelength (the third harmonic of the GaAs band gap) evolves into three peaks in (b), which are attributed to the carrier-wave Mollow triplet. The corresponding three black lines are a guide to the eye. The white curve at the top of (b) (another guide to the eye) indicates the position of the high-energy peak of the fundamental Mollow triplet. For (b) we estimate that the peak Rabi energy *within* the GaAs film (and accounting for reflection losses at the air-GaAs interface) is given by $\Omega_R/\omega_0 = 0.76$. Taken from Ref. [51].

amplitude $\tilde{E}_0$. This variation also leads to a horizontal shift, which is both interesting as well as disturbing. It is interesting on the one hand because no such intensity dependence occurs in off-resonant perturbative nonlinear optics (see section 2.4.) – pointing out the distinct difference between the two scenarios. It is disturbing on the other hand, because in order to use the effect to determine the CEO phase, one needs to calibrate the incident electric field amplitude, or, more precisely, the Rabi frequency. This is, however, possible in principle via the measured splitting of the Mollow sidebands.

Experiment: How can one actually get to *very thin* GaAs films with *high damage thresholds* and *without* $\mathrm{Al_{0.3}Ga_{0.7}As}$ *barriers*? This problem has led us to epitaxially

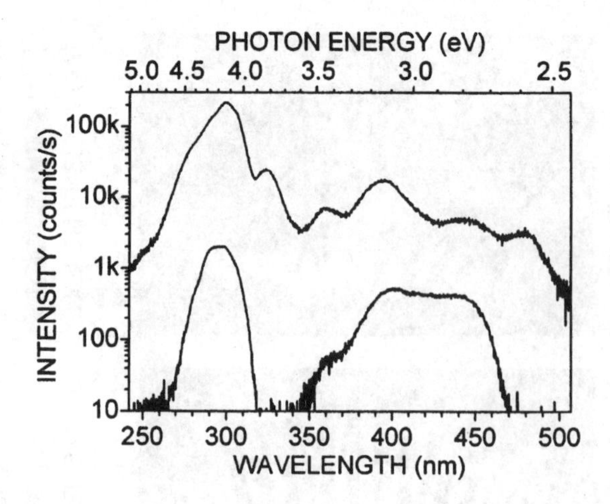

Figure 26: Experiment: Spectra of light (logarithmic scale) emitted into the forward direction by a $l = 100\,\text{nm}$ thin GaAs film on sapphire substrate excited by 5 fs pulses (cuts through Fig. 25 at $\tau = 0$). For low excitation intensity $I = 0.24 \times 10^{12}\,\text{W/cm}^2$ (lower curve), well separated peaks at the second harmonic and the third harmonic of the laser occur. When increasing the intensity to $I = 2.8 \times 10^{12}\,\text{W/cm}^2$ (upper curve, shown on the same absolute scale), the deep valleys between the second harmonic and the third harmonic as well as between the fundamental and the second harmonic are filled and additional maxima are observed. Taken from Ref. [51].

growing a GaAs layer of thickness l directly on sapphire substrate in a molecular-beam epitaxy machine [51]. The growth of the thin GaAs layer on the 0.43-mm-thick 51-mm-diameter epi-ready sapphire substrate (University Wafer) mounted in an In-free moly-block was performed the same as for a GaAs wafer with a growth rate of 0.25 nm/s. We have investigated four samples with GaAs film thicknesses of $l = 25\,\text{nm}$ (two samples), $l = 50\,\text{nm}$, and $l = 100\,\text{nm}$. Only the latter two samples turn out to have a damage threshold sufficiently large to actually perform the experiments at large Rabi energies. The experimental observation that thicker samples have a higher damage threshold is in contrast to what one might expect intuitively based on considering the areal energy density deposited in the film. The actual damage threshold might rather be determined by growth details or by different surface-to-volume ratios. Although these samples are not comparable in linewidth with state-of-the-art GaAs/$Al_{0.3}Ga_{0.7}As$ double heterostructures grown on GaAs substrate (see section 4.1), the $l = 100\,\text{nm}$ thin GaAs film on sapphire substrate does exhibit a band edge in linear optical transmission experiments at room temperature (not shown). As the relevant energy scales in our experiments are larger than 0.1eV anyway, linewidth is not much of an issue.

Fig. 25 shows the spectra of light emitted into the forward direction for the $l = 100\,\text{nm}$ sample. (a) corresponds to low excitation intensity, (b) to high excitation. The cuts through these data sets at $\tau = 0$ are depicted in Fig. 26. At high excitation (Fig. 25(b)), the emitted light intensity around the third harmonic of the GaAs band gap splits and overlaps with the second-harmonic generation (SHG) signal. From the dependence on l (not shown) we conclude that the SHG has a large surface contribution (or is even com-

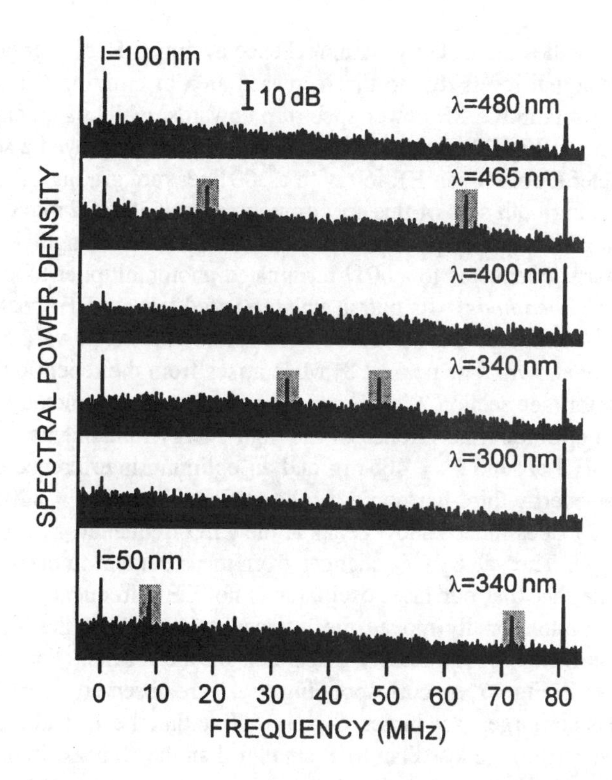

Figure 27: Experiment: Radio-frequency power spectra (logarithmic scale), 10 kHz resolution and video bandwidth, for various optical detection wavelengths λ and two GaAs film thicknesses l as indicated. The spectra are vertically displaced for clarity. The excitation intensity is comparable to that in Fig. 25(b), $\tau = 0$. The peaks at the CEO frequency f_ϕ and at $(f_r - f_\phi)$ are highlighted by grey areas. Taken from Ref. [51].

pletely generated at the two GaAs surfaces), while the third harmonic is consistent with a bulk effect. At $\tau = 0$, the spectrum exhibits three peaks around the third harmonic which evolve with time delay τ . The solid lines are guides to the eye and indicate that the splitting decreases with increasing $|\tau|$. These three peaks are interpreted as the carrier-wave Mollow triplet. Note also that a contribution from the fundamental moves into the picture from the top. Following the above theory, this is expected to be the high-energy peak of the fundamental Mollow triplet. The data of the $l = 50$ nm sample (not shown) are compatible with the $l = 100$ nm sample data (Fig. 25(a) and (b)), however - as already discussed above - the second-harmonic contribution is more prominent with respect to the third harmonic in the $l = 50$ nm case as compared to the $l = 100$ nm case due to a larger surface contribution. This significantly reduces the visibility of the low-energy peak of the third-harmonic Mollow triplet.

Fig. 27 shows measured radio-frequency (RF) power spectra of the signals corresponding to Fig. 25(b), $\tau = 0$. To enhance the signal levels, we have removed the interferometer, leading to a larger average laser power of about 45 mW in front of the sample. Compensating this increased power by slightly moving the GaAs film out of focus results in similar light intensities on the GaAs sample, hence in similar Rabi energies – but leads to larger absolute signal levels due to the increased area of emitting GaAs. This trick boosts the signal levels in the RF power spectrum upwards, which is essential considering the peak heights in Fig. 27. In these experiments we have employed a second optical grating spectrometer (Jobin Yvon HR460 with a 300 lines/mm grating blazed at 250 nm wavelength). Opening both slits of this spectrometer to a width of 2 mm corresponds to the detection of a 20 nm broad spectral interval with center wavelength λ. The exit slit of the spectrometer is connected to a 50 Ω-terminated photomultiplier tube (*Hamamatsu R 4332, Bialkali photocathode*). Its output voltage is fed into an RF spectrum analyzer (*Agilent PSA E4440A*) operated at 10 kHz resolution and video bandwidth. In Fig. 27 selected examples are shown. The peak at 81 MHz arises from the repetition frequency f_r of the laser oscillator (see section 2.3.). From the optical spectra depicted in Figs. 25 and 26 one expects an optimum interference of the high-energy fundamental Mollow triplet with the surface SHG around $\lambda = 465$ nm and an optimum interference of the surface SHG with the low-energy third-harmonic Mollow triplet around $\lambda = 340$ nm. At these wavelengths, Fig. 27 does indeed show peaks at the CEO frequency f_ϕ and at difference frequency $(f_r - f_\phi)$. The value of f_ϕ changes from measurement to measurement. This is partly due to the fact that our laser oscillator is not CEO-frequency stabilized. Furthermore, we have intentionally moved the intracavity prism near the high-reflector to demonstrate the influence of intracavity dispersion on the results. For other detection wavelengths shown in Fig. 27, no corresponding peaks are observed, even though the absolute signal levels are larger (see larger f_r-peak). Note that the f_ϕ and $(f_r - f_\phi)$ peaks in the RF power spectrum are less than 8 dB smaller than the f_r-peak, indicating that the relative modulation depth of the beat signal versus time is as large as 40%. Similar results are observed for the $l = 50$ nm thin sample (see lowest data set in Fig. 27). As we have discussed in the theory section, for such sample thicknesses, only small changes of the CEO phase within the GaAs film are expected. As argued above, this might allow for measuring the CEO phase itself. However, we will see in sections 5.1. and 5.3. that other semiconductor choices and excitation conditions are likely to be even better suited for this task than resonant excitation of GaAs.

5. "Off-Resonant" Carrier-Wave Nonlinear Optics of Electrons in Semiconductors

In section 4, we have concentrated on resonant excitation of the semiconductor band gap, exemplified by the III-V semiconductor GaAs. Let us now discuss two examples for "off-resonant" excitation, using the II-VI semiconductors ZnO and CdS.

5.1. *EXPERIMENT: ZINC OXIDE*

The direct gap semiconductor ZnO has a room temperature band gap energy of $E_{\rm g} = 3.3\,\rm eV$. Amazingly, the precise value of the band gap energy of ZnO was subject of scientific discussions until rather recently [52]. ZnO has a $\vec{c}$-axis without inversion symmetry and is birefringent ($\vec{E}||\vec{c}$ and $\vec{E} \perp \vec{c}$ are inequivalent). For ZnO single crystal platelets (here about $100\,\mu\rm m$ thick), the $\vec{c}$-axis lies within the plane of the platelet, while for the 350 nm thin ZnO epitaxial film discussed in this article, the $\vec{c}$-axis is perpendicular to the film. The ZnO interband dipole matrix element is smaller than for GaAs. From $\vec{k}\cdot\vec{p}$ perturbation theory according to Eq. (50) we obtain $d_{\rm cv} = 0.19\,\rm e\,nm$ (remember that we have used $d_{\rm cv} = 0.5\,\rm e\,nm$ for GaAs).

In the ZnO experiments [53] we use the same 5 fs laser system with $\hbar\omega_0 \approx 1.5\,\rm eV$, the same stabilized Michelson interferometer and the same focusing optics as in the GaAs experiments (see section 4.1.). Also, the reference intensity $I_0 = 0.6\times10^{12}\,\rm W/cm^2$ (one arm of the interferometer) has the same definition and the same value. The light emitted by the samples into the forward direction is spectrally filtered (3 mm *Schott BG 39*) to remove the prominent fundamental laser spectrum and is sent into a 0.25 m focal length grating spectrometer connected to a charge-coupled-device (CCD) camera. Alternatively, we send the light through a combination of filters (3 mm *Schott BG 39*, 3 mm *Schott GG 455*, and *Coherent 35-5263-000* 480 nm cut-off interference filter for the $100\,\mu\rm m$ ZnO single crystal and *Coherent 35-5289-000* 500 nm cut-off interference filter for the 350 nm epitaxial film, respectively) onto a $50\,\Omega$-terminated photomultiplier tube (*Hamamatsu R 4332, Bialkali photocathode*), connected to an RF spectrum analyzer (*Agilent PSA E4440A*). No kind of intentional spatial filtering of the emitted light is performed.

Fig. 28 shows measured optical spectra in the spectral region energetically above the laser spectrum (which has its short wavelength cut-off above 650 nm) and below the band gap of ZnO of $E_{\rm g} = 3.3\,\rm eV$ for (a) low excitation and (b), (c) large excitation intensity versus time delay τ of the interferometer. Note that (b) and (c) are the same data plotted with different levels of saturation in order to reveal details. The measured intensity is given in actual counts per second (one count corresponds to about two photons). All spectral components shown in Fig. 28 are also easily visible with the naked eye. If the ZnO sample is moved out of the focus by some tens of microns, all the spectral components shown in Fig. 28 completely disappear, indicating that none of them comes directly out of the laser. Polarization dependent experiments under these conditions show that all these spectral components have the same linear polarization as the laser pulses. In (a), the light around 390-470 nm wavelength is due to second-harmonic generation (SHG), the

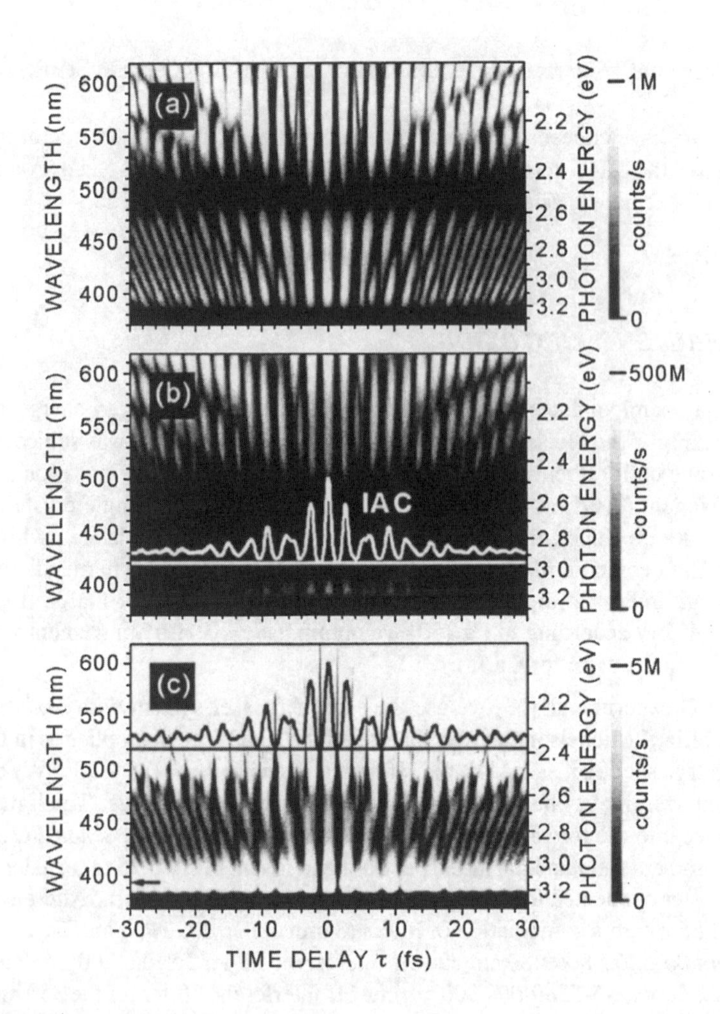

Figure 28: Experiment: Spectra of light emitted by the 100 μm thick ZnO single crystal into the forward direction versus time delay τ of the Michelson interferometer, $\vec{E}||\vec{c}$. (a) $I = 0.15 \times I_0$, (b) $I = 2.04 \times I_0$, and (c) as (b), but different saturation of the grey-scale. The light intensities decay near the ZnO band gap of $E_g = 3.3$ eV. The white curve in (b) labeled IAC is the independently measured interferometric autocorrelation using a BBO crystal, the black curve in (c) is a cut through the ZnO data at 395 nm wavelength (see arrow).

components above 500 nm are due to self-phase modulation[26] (SPM). Interestingly, the independently measured interferometric autocorrelation of the laser pulses (using a thin

[26] As already pointed out in section 2.4., the $\epsilon_0\, \chi^{(3)}\, E^3(t)$ term in Eq. (35) also contains a contribution at the fundamental laser frequency ω_0 (see Eq. (39)). This contribution can be rewritten [9] in terms of a nonlinear refractive index according to $n(t) = n_0 + n_2\, I(t)$, with the linear optical refractive index n_0 and the nonlinear contribution given by the coefficient n_2 and the laser intensity $I(t)$. For a wave propagating over length l, this nonlinear index leads to a phase shift – called self-phase modulation (SPM) – of $n_2\, I(t)\, 2\pi/\lambda_0\, l$, with the vacuum laser center wavelength λ_0. The derivative of this time-dependent phase shift gives rise to a time-dependent shift of the instantaneous frequency, i.e., to a broadening of the laser spectrum.

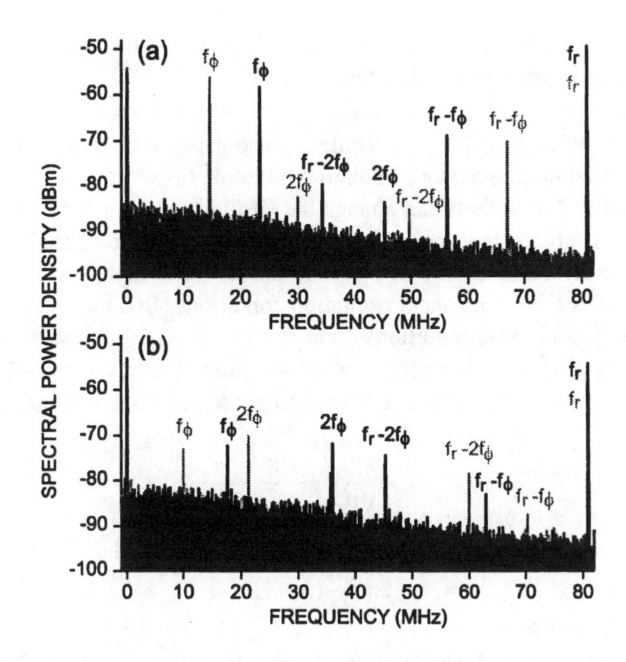

Figure 29: Experiment: RF spectra, 10 kHz resolution and video bandwidth, $\tau = 0$ fs (equivalent to an average total power of 64 mW in front of the sample). (a) 100 μm thick ZnO single crystal, $\vec{E}||\vec{c}$, corresponding to Figs. 28 (b), (c), optical filter roughly corresponds to 455 nm – 480 nm, (b) 350 nm thin ZnO epitaxial layer, $\vec{E} \perp \vec{c}$, optical filter roughly corresponds to 455 nm – 500 nm. The peaks at the repetition frequency f_r, the carrier-envelope offset frequency f_ϕ, its second harmonic $2f_\phi$ and the mixing products $(f_r - f_\phi)$ and $(f_r - 2f_\phi)$ are labeled. The black and grey data correspond to slightly different laser end mirror positions. When removing intracavity prism material (CaF_2), the f_ϕ and $2f_\phi$ (($f_r - f_\phi$) and ($f_r - 2f_\phi$)) peaks shift to the left (right).

beta barium-borate (BBO) SHG crystal, see curve labeled IAC in Fig. 28(b)), is closely reproduced by a cut at 395 nm wavelength (see white curve in Fig. 28(c)). This indicates that the pulses are not severely broadened in the ZnO crystal due to, e.g., group velocity dispersion. Furthermore, the spectral width of the SHG contribution indicates that phase-matching effects do not play a major role, which is not surprising considering the short Rayleigh range of the microscope objective of only several microns (see Fig. 17(a)). For higher intensities, the spectral overlap of SPM and SHG becomes immediately obvious from the spectra in Fig. 28(c) and a rich fine structure as a function of τ appears in this spectral region. Feeding the spectral components of this interference region into an RF spectrum analyzer, we find clear evidence for a peak at the carrier-envelope offset frequency f_ϕ (Fig. 29) which arises because ϕ changes from pulse to pulse of the mode-locked laser oscillator due to different group delay and phase delay times per round trip of the laser cavity (see section 2.3.). To further check this assignment, we also depict the RF spectrum for a slightly different laser end mirror mirror position, which shifts the f_ϕ peak as well as the mixing product $(f_r - f_\phi)$ (see labels in Fig. 29). In the 350 nm thin ZnO film (Fig. 29(b)), both, the f_ϕ peak as well as the $2f_\phi$ peak are still visible. Interestingly, the $2f_\phi$ peak is even larger than the f_ϕ peak.

If one is interested not only in determining a dependence on the CEO phase ϕ but rather interested in measuring ϕ of the incident pulse itself, it obviously relevant to ask whether ϕ changes while propagating through the sample (as we have done for the GaAs case in section 4.3.). Two effects can change the CEO phase within the sample: Linear optical propagation effects and nonlinear optical effects. The magnitude of the linear optical propagation effects is much easier to estimate for the ZnO case as compared to the GaAs case because of the off-resonant excitation conditions. In other words: Absorption plays no role here. Under these conditions, for a carrier frequency ω_0 and a center vacuum wavelength $\lambda_0 = 2\pi c_0/\omega_0$, the change of ϕ of the pulse, $\delta\phi$, as a result of propagation over a length l results from the different group and phase velocities according to

$$\begin{aligned} \delta\phi &= 2\pi \frac{\left(\dfrac{l}{v_{\text{group}}} - \dfrac{l}{v_{\text{phase}}}\right)}{2\pi/\omega_0} \\ &= -2\pi \frac{\mathrm{d}n}{\mathrm{d}\lambda}(\lambda_0)\, l\,, \end{aligned} \tag{84}$$

with the vacuum-wavelength dependent refractive index $n(\lambda)$. From the first to the second line in Eq. (84) a few straightforward mathematical manipulations are necessary. For many materials, the dependence $n(\lambda)$ is parametrized using the so called Sellmeier formula[27] according to

$$n^2(\tilde{\lambda}) = \epsilon(\tilde{\lambda}) = \mathcal{A} + \frac{\mathcal{B}\,\tilde{\lambda}^2}{\tilde{\lambda}^2 - \mathcal{C}^2} + \frac{\mathcal{D}\,\tilde{\lambda}^2}{\tilde{\lambda}^2 - \mathcal{E}^2} \tag{85}$$

with $\tilde{\lambda} = \lambda/(0.1\,\text{nm})$. For the relevant wavelengths λ, the refractive index according to Eq. (85) is real. Ref. [54] determined the fit parameters $\mathcal{A} = 2.0065$, $\mathcal{B} = 1.5748 \times 10^6$, $\mathcal{C} = 10^8$, $\mathcal{D} = 1.5868$, and $\mathcal{E} = 2606.3$ for ZnO, $\vec{E} \perp \vec{c}$. This fit is applicable to the visible part of the spectrum only. With Eq. (84) for $\lambda_0 = 826\,\text{nm}$ ($\Leftrightarrow \hbar\omega_0 = 1.5\,\text{eV}$) and after some tedious mathematics, this finally delivers

$$\boxed{\delta\phi_{\text{ZnO},\,\lambda_0=826\,\text{nm}} = 0.013 \times 2\pi\, l/100\,\text{nm}\,.} \tag{86}$$

For the above experiments with 100 μm thick ZnO single crystals, the effective interaction length l is given by the depth of focus (see Fig. 17(a)) which is on the order of five microns, in which case we have $\delta\phi = 0.7 \times 2\pi$ – a significant change of the CEO phase within the sample. For the $l = 350$ nm thin ZnO epitaxial film, $\delta\phi$ is merely 4.6% of 2π, which might be sufficiently small for many applications.

---

[27] This is nothing but the dielectric function of the sum of two resonances as, e.g., from the optical Bloch equations, plus a constant.

5.2. THEORY: "THG IN DISGUISE OF SHG"

From our discussion in section 2.4. it is clear that the peaks in the RF spectra at f_ϕ originate from an interference of the fundamental with the second harmonic, while the peaks at $2f_\phi$ and the mixing products $(f_r - 2f_\phi)$ in Fig. 29 stem from an interference of the fundamental and the third harmonic. *This is really amazing!* The centers of the fundamental and the third harmonic are separated by twice the laser carrier frequency – yet, they do interfere. Is a simple description of the nonlinear optical polarization in terms of nonlinear optical susceptibilities according to Eq. (35) going to work at all? To see inasmuch it works, we solve Eq. (35) together with the one-dimensional Maxwell equations (see section 4.2.) numerically, i.e., we do not employ the slowly varying envelope approximation[28] and work with the real electric laser field $E(z, t)$ and with the actual layer structure of the sample, i.e., a $l = 350\,\text{nm}$ thin layer of ZnO on a semi-infinite sapphire substrate. The latter has a dielectric constant of $\epsilon_s = (1.76)^2$ (compare sections 4.2. and 4.3.), i.e., its optical nonlinearities are neglected[29]. For the description of the ZnO layer we use the nonlinear suceptibilities $\chi^{(2)} = 4 \times 10^{-12}\,\text{m/V}$ [57] and $\chi^{(3)} = 7 \times 10^{-21}\,\text{m}^2/\text{V}^2$ [58]. Terms of order four and higher are neglected. On this level of modelling, the optical spectra of the experiment are nicely reproduced (not shown). Inspection of the corresponding calculated RF power spectrum (calculated on the basis of Eqs. (44) and (46)) shows a peak at frequency f_ϕ but no contribution at frequency $2f_\phi$ (not shown), while in the experiment the $2f_\phi$ peak is even larger than the f_ϕ peak (see Fig. 29(b)). Thus, we conclude that **the $2f_\phi$ peak in the RF power spectra of ZnO can definitely not be explained on the basis of perturbative off-resonant nonlinear optics.**

The solution [59] to the "riddle of the $2f_\phi$ peak" was already visible in section 4.2. and is related to the signal contribution labeled $\Omega = \omega$ in Fig. 19. In section 4.2. we have focused on the behavior around $\Omega/\omega_0 = 1$ on the vertical axis of Fig. 30, which is due to carrier-wave Rabi flopping. Remember that these calculations are based on the optical Bloch equations of two-level systems introduced in section 3.1., i.e, **they correspond to an inversion-symmetric material.** In order to make the $\Omega = \omega$ contribution in Fig. 19 more visible, we depict similar calculations in Fig. 30, where we have chosen broader frequency intervals for the vertical as well as for the horizontal axes, have plotted the intensity on a logarithmic rather than on a linear scale to enhance the dynamic range, and have chosen ZnO parameters ($\forall$ transition frequencies Ω: $d_{cv} = 0.19\,e\,\text{nm}$, $T_1 = \infty$, $T_2 = 50\,\text{fs}$) as well as sinc^2-shaped pulses of $t_{\text{FWHM}} = 5.0\,\text{fs}$ in duration with $\hbar\omega_0 = 1.5\,\text{eV}$.

The signal contribution around the $\Omega = \omega$ line in Fig. 30 [30] exhibits a constriction, the exact position of which depends on the Rabi frequency. Let us start the discussion with the part above this constriction and consider a transition frequency Ω on the vertical axis

---

[28] This is in contrast to, e.g., Refs. [55, 56] which address similar phenomena in optical fibers. There, the slowly varying envelope approximation is necessary because one needs to integrate the equations for propagation lengths on the order of millimeters.

[29] Indeed, no measurable nonlinear signal of the sapphire substrate itself occurs in independent additional experiments.

[30] Note that under these conditions, the square modulus of the optical polarization P is not simply proportional to that of the optical transition amplitude p_{vc}. This difference arises from a tail of the contributions in the spectrum at negative frequencies which reaches up to the positive frequencies of interest shown in Fig. 30.

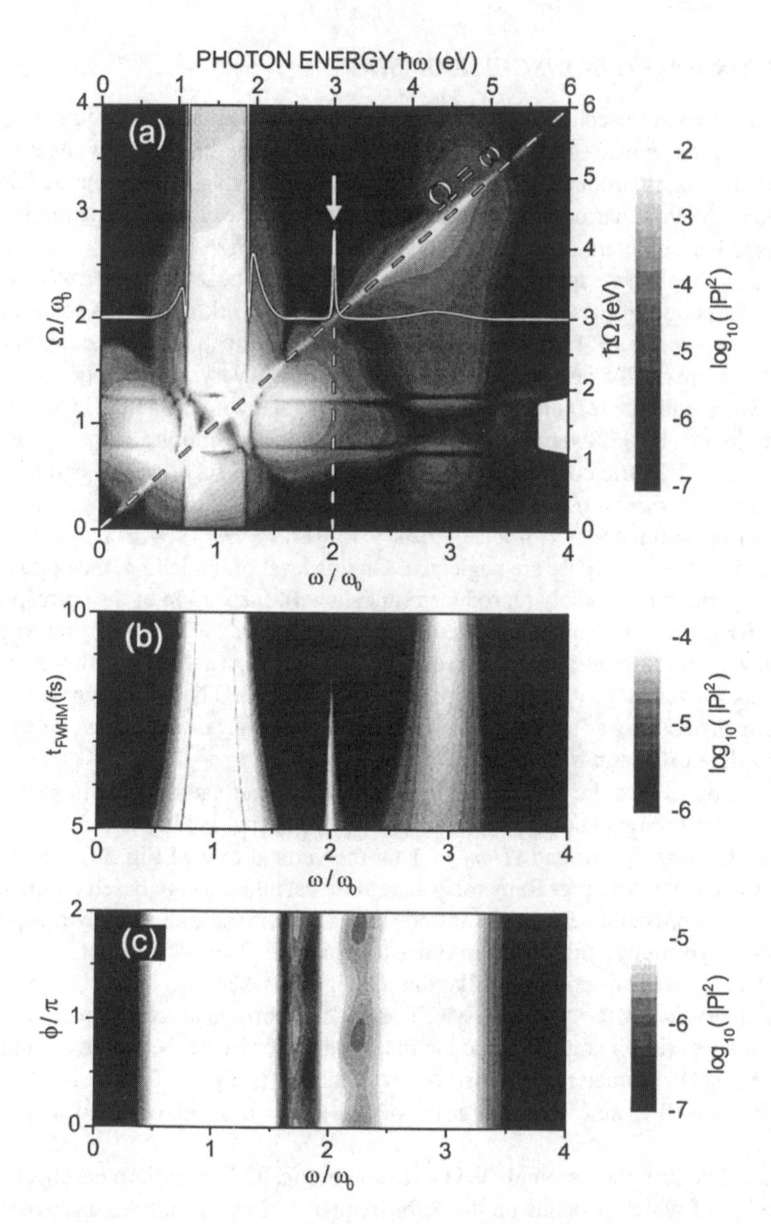

Figure 30: Grey-scale image of the square modulus of the optical polarization P (normalized) versus spectrometer frequency ω in units of the laser carrier frequency ω_0 with $\hbar\omega_0 = 1.5\,\text{eV}$. The peak Rabi frequency of the exciting $\text{sinc}^2(t)$ pulses with duration t_{FWHM} is $\Omega_{\text{R}}/\omega_0 = 0.76$. (a) $|P(\omega)|^2$ versus transition frequency Ω for a fixed CEO phase $\phi = 0$ and $t_{\text{FWHM}} = 5\,\text{fs}$. The white curve is a cut though the data at $\Omega/\omega_0 = 2$ (linear scale). The laser pulse spectrum is shown as the grey area on the RHS. (b) $|P(\omega)|^2$ versus pulse duration t_{FWHM} for fixed $\Omega/\omega_0 = 2$ and $\phi = 0$. (c) $|P(\omega)|^2$ versus ϕ for fixed $\Omega/\omega_0 = 2$ and $t_{\text{FWHM}} = 5\,\text{fs}$.

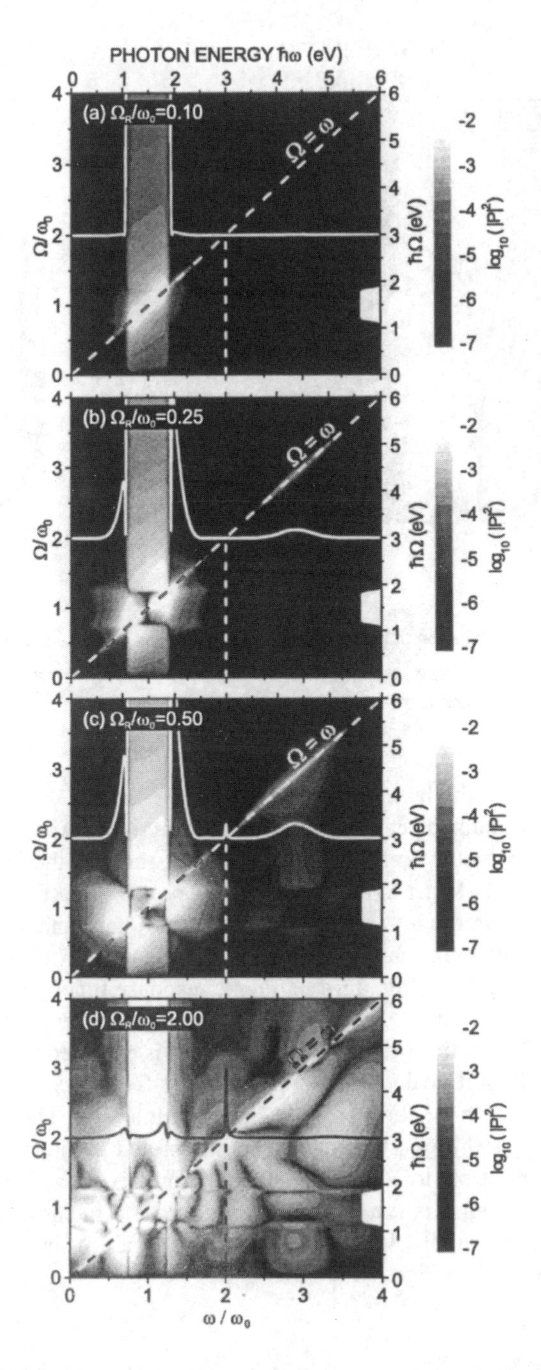

Figure 31: As Fig. 30, but for different Rabi energies. (a) $\Omega_R/\omega_0 = 0.10$, (b) $\Omega_R/\omega_0 = 0.25$, (c) $\Omega_R/\omega_0 = 0.50$, and (d) $\Omega_R/\omega_0 = 2.0$.

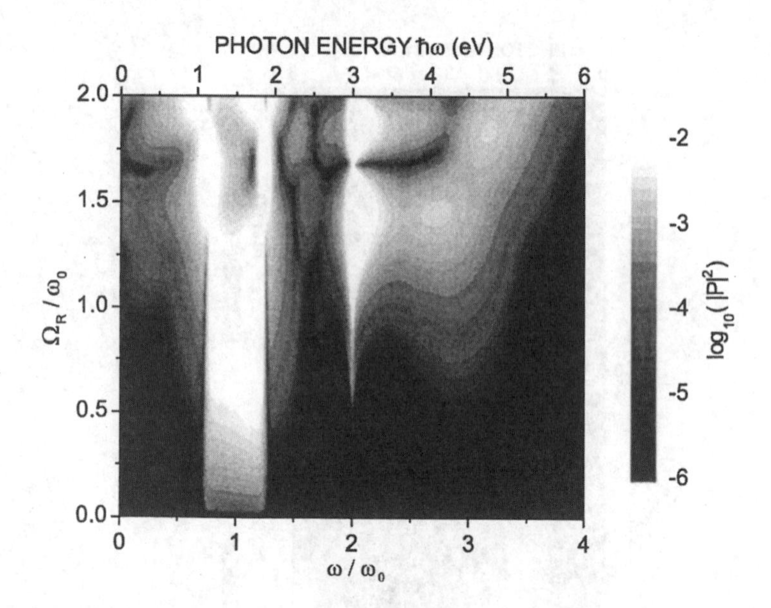

Figure 32: As Fig. 30, but as a function of peak Rabi energy Ω_R in units of the laser carrier frequency ω_0 for fixed $\Omega/\omega_0 = 2$, $\phi = 0$ and $t_{\mathrm{FWHM}} = 5\,\mathrm{fs}$. Around $\Omega_R/\omega_0 \approx 1.7$ on the vertical axis, a Rabi flop is completed even though the excitation is off-resonant for one-photon absorption.

at the second harmonic of the laser carrier frequency ω_0, i.e., at $\Omega/\omega_0 = 2$ in Fig. 30(a). Here we observe a well-defined peak in the optical spectra right at spectrometer frequency $\omega = 2\omega_0$ (see white curve in Fig. 30(a) which is a cut through the data at $\Omega/\omega_0 = 2$ plotted on a *linear* scale). What is the origin of this peak? A part of it is the resonant enhancement way down in the low-energy tail of the third harmonic of the laser photon energy. For pulses containing many cycles of light, this contribution would disappear because of negligible overlap of the third-harmonic response function and the resonance – it is specific for the regime of few-cycle pulses (see dependence on pulse duration depicted in Fig. 30(b)). This third-harmonic contribution is expected to be associated with a phase 3ϕ – even though it peaks right at spectrometer frequency $\omega = 2\omega_0$. Its signal strength roughly scales with the third power of the laser intensity. The part of the $\Omega = \omega$ signal contribution below this constriction can be interpreted as the resonantly enhanced SPM due to absorption of photons from the high-energy tail of the laser spectrum with phase 1ϕ. Its signal strength is roughly proportional to the intensity itself. Thus, the upper part gains relative weight with respect to the lower part for increasing intensity or increasing Rabi energy Ω_R in Figs. 31 and 32.

Exercise 5.1.: Consider a sinc^2-pulse with $\hbar\omega_0 = 1.5\,\mathrm{eV}$ exciting a narrow two-level resonance at $\Omega/\omega_0 = 2$. What is the maximum pulse duration t_{FWHM} which leads to third-harmonic generation in disguise of second-harmonic generation in the $\chi^{(3)}$-limit ?

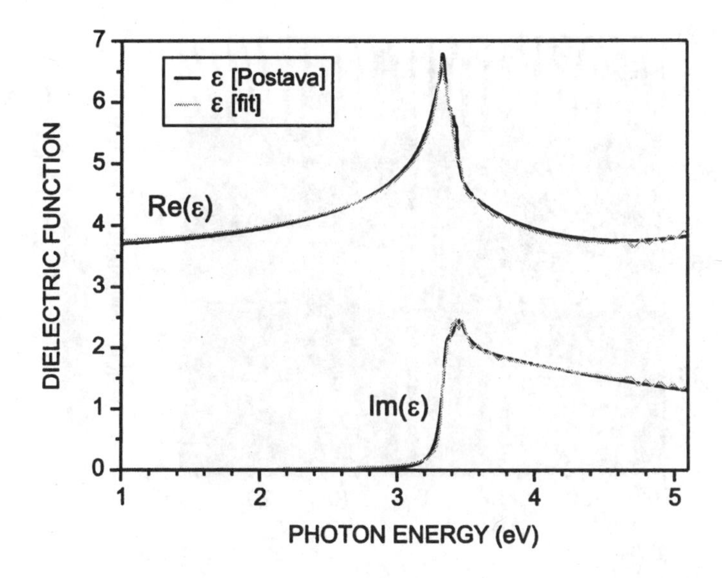

Figure 33: Dielectric function of ZnO versus (spectrometer) photon energy $\hbar\omega$. The black curve represents the data measured in Ref. [60], the grey curve is a fit using an ensemble of 45 two-level systems with different transition frequencies Ω. This ensemble is used for the calculations presented in Fig. 34.

Fig. 30(c) shows the dependence on the CEO phase ϕ for a selected transition frequency of $\Omega/\omega_0 = 2$. All other parameters are as in Fig. 30(a). It becomes obvious that a part of the interference occurs in between the fundamental, i.e. $\omega/\omega_0 = 1$, and $\omega/\omega_0 = 2$. Note that the period of the signal versus ϕ is π rather than 2π, equivalent to a peak at frequency $2f_\phi$ in the RF spectrum. The usual SHG would appear in the same region in the optical spectra as the third harmonic in disguise of a second harmonic, but its phase is 2ϕ rather than 3ϕ, thus, it leads to a peak at frequency f_ϕ in the RF spectra when beating with the fundamental. Another part of the interference in Fig. 30(c) occurs in between $\omega/\omega_0 = 2$ and $\omega/\omega_0 = 3$. This shows that the peak around $\omega/\omega_0 = 2$ in Fig. 30(a) is indeed a mixture of resonantly enhanced SPM and resonantly enhanced THG.

Fig. 31 illustrates the dependence on Rabi energy. For large Rabi energies (d), the *"THG in disguise of SHG"* becomes the dominating feature in the optical spectrum (black curve at $\Omega/\omega_0 = 2$). The unusually small contribution of P around the laser carrier frequency, i.e., at $\omega/\omega_0 = 1$ in Fig. 31(d) is due to carrier-wave Rabi flopping (see section 4.). Indeed, the inversion w starts at -1, reaches values near +1 in the maximum and comes almost back to -1 after the pulse – even though the excitation is "off-resonant" with $\Omega/\omega_0 = 2$. This illustrates the fact that the detuning of the carrier frequency from resonance becomes negligible if the Rabi energy is larger than the detuning.

If one interpretes the transition energy $\hbar\Omega$ in Fig. 30(a) as the band gap energy E_g of a semiconductor, the lower RHS triangle formed by the $\Omega = \omega$ line experiences strong reabsorption in the semiconductor band-to-band continuum, while the upper LHS triangle is in the transparency regime of the semiconductor. *"THG in disguise of SHG"* overlaps with

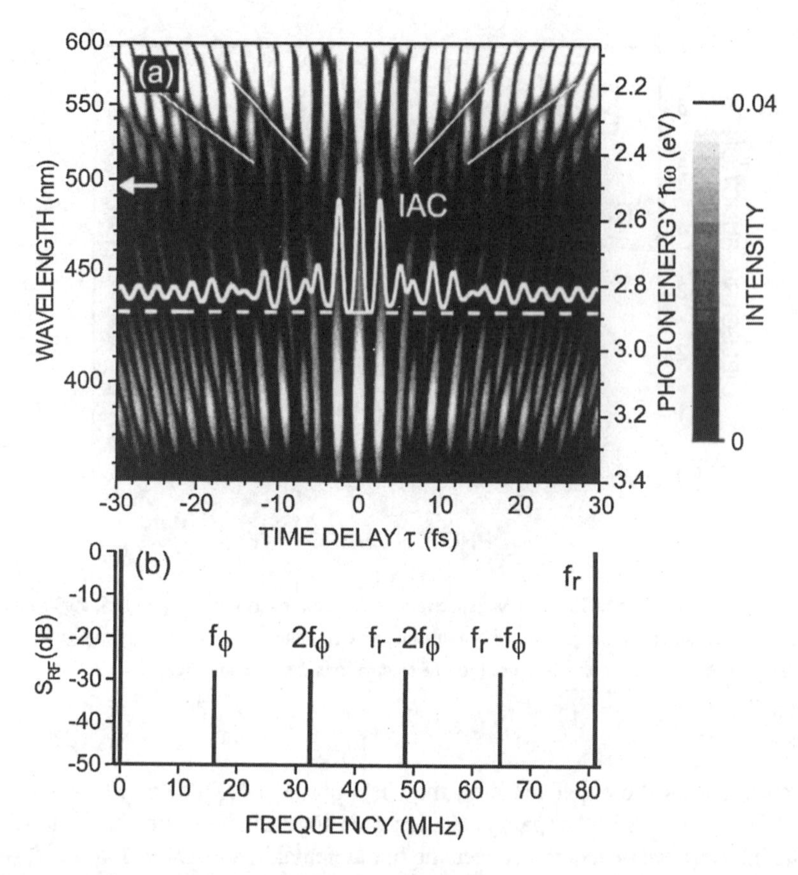

Figure 34: Theory, $\tilde{E}_0 = 6 \times 10^9$ V/m at $\tau = 0$. (a) Grey-scale image of the light intensity (normalized) emitted into the forward direction versus spectrometer photon energy $\hbar\omega$ and time delay τ. The data are averaged over $\phi = 0...2\pi$. The white thin lines are a guide to the eye. The white curve labelled IAC is the interferometric autocorrelation of the laser pulses. (b) Radio-frequency power spectrum S_{RF} of the intensity at the spectral position indicated by the arrow on the LHS in (a), $\tau = 0$.

this line. In order to study corresponding reabsorption and phase-matching effects, we now present numerical solutions of the coupled Maxwell-Bloch equations in one dimension without using the rotating wave approximation and without using the slowly varying envelope approximation and accounting for the actual sample geometry, i.e., we model a 350 nm thin film of ZnO on a sapphire substrate with dielectric constant $\epsilon_s = (1.76)^2$. Furthermore, for a semiconductor, one does not have a single nonlinear two-level system but rather a band continuum, i.e., one needs to integrate P along the vertical axis in Fig. 30(a). To be close to the experiment, we fit an ensemble of 45 two-level systems (3.3 eV – 7.9 eV) to the known measured linear dielectric function of ZnO over a broad frequency regime [60] (see Fig. 33). Thus, linear and nonlinear propagation effects as well as multi-photon absorption into high-energy states are accounted for exactly within this model. In addition to this, ZnO has no inversion symmetry and shows a nonvanishing

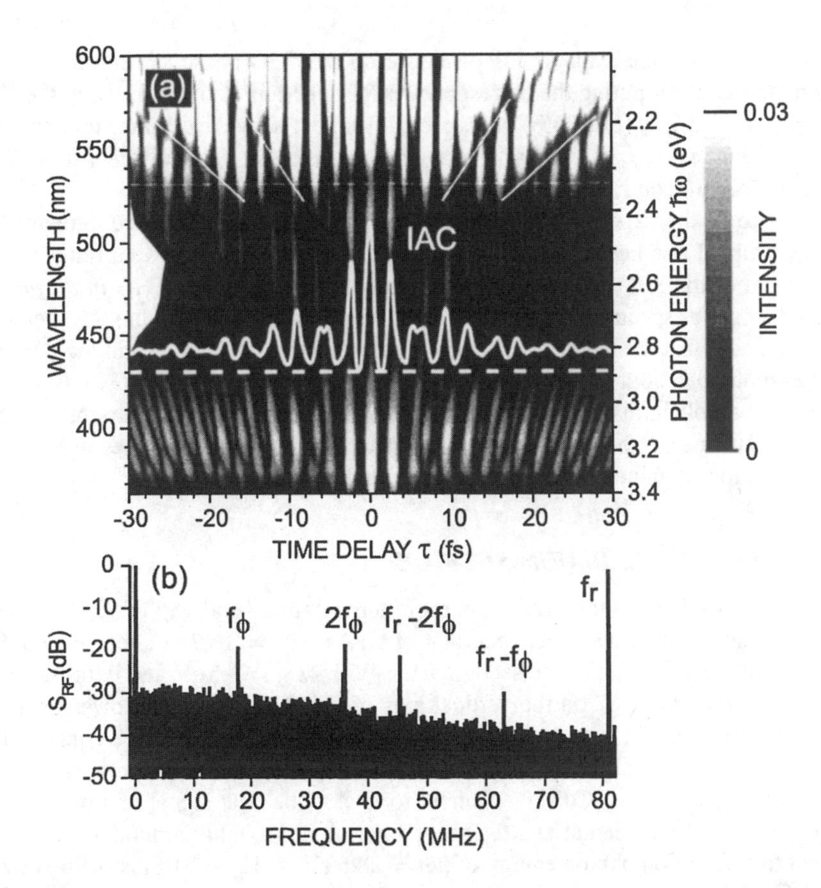

Figure 35: Experiment on the 350 nm ZnO film, $\tilde{E}_0 = 6 \times 10^9$ V/m at $\tau = 0$. (a) Grey-scale image of the light intensity (normalized) emitted into the forward direction versus spectrometer photon energy $\hbar\omega$ and time delay τ. ϕ is not stabilized. The white thin lines are a guide to the eye. The white curve labelled IAC is the interferometric autocorrelation of the laser pulses, obtained from an independent measurement using a β-barium borate SHG crystal. (b) Radio-frequency power spectrum S_{RF} of the intensity in the spectral interval indicated by the grey area on the LHS in (a), $\tau = 0$. Note the good agreement with the theory calculated under the same conditions (Fig. 34).

$\chi^{(2)}$–susceptibility. For simplicity, we describe this aspect by a frequency-independent $\chi^{(2)} = 4 \times 10^{-12}$ m/V [57]. Furthermore, we employ excitation with a pair of collinearly propagating identical pulses with time delay τ. The pulses are taken directly from the experiment [47] (see section 4.3.). The ratio of the pulse repetition frequency f_{r} and the CEO frequency f_ϕ is set to $f_{\mathrm{r}}/f_\phi = 5$, with $f_{\mathrm{r}} = 81$ MHz.

In Fig. 34(a) we depict the calculated optical spectra versus time delay τ. Note that none of the spectral components originates from the incident pulses directly, all of them are rather generated in the 350 nm thin ZnO layer. The spectral components above 530 nm wavelength are due to SPM, those in the range from 365 nm to 455 nm wavelength are due to a combination of conventional SHG and *"THG in disguise of SHG"*. In between the two regions, interference leads to a dependence on the CEO phase ϕ. Indeed, filtering out this region and computing the corresponding RF spectrum delivers peaks at the CEO frequency f_ϕ and at $2\,f_\phi$ (Fig. 34(b)) as expected from our above reasoning. Interestingly, the peak at frequency $2\,f_\phi$ is comparable in strength to that at f_ϕ, equivalent to a *"THG in disguise of SHG"* contribution comparable to traditional SHG.

Fig. 35 shows the experimental results corresponding to Fig. 34. The agreement is good, especially if one keeps in mind that there are no fit parameters to arrange for the relative height of the SPM component, the traditional SHG and the *"THG in disguise of SHG"* in the optical spectrum, or for the height of any of the peaks in the RF spectrum. In particular, note that the heights of the peaks at frequencies f_ϕ and $2\,f_\phi$ in the RF spectrum are comparable, both for theory and experiment, indicating that *"THG in disguise of SHG"* is comparable in magnitude to conventional SHG under these conditions. Furthermore, the nodal lines in the optical spectra indicated by the white thin lines in Fig. 35(a) nicely match those obtained for the complete modelling (see Fig. 34(a).

5.3. EXPERIMENT: CADMIUM SULFIDE

The direct gap semiconductor CdS has a room temperature band gap of $E_g = 2.5\,\text{eV}$, which lies roughly half way in between that of GaAs ($E_g = 1.42\,\text{eV}$, see section 4.1.) and that of ZnO ($E_g = 3.3\,\text{eV}$, see section 5.1.). With $\hbar\omega_0 = 1.5\,\text{eV}$ and with $\hbar\Omega = E_g$ one arrives at $\Omega/\omega_0 = 1.67$ on the vertical axis of Fig. 31. Due to the lower band gap of CdS compared to ZnO, larger nonlinearities are anticipated for CdS (see contributions around $\omega/\omega_0 = 1.5$ and $\Omega/\omega_0 = 1.67$ in Fig. 31). Although the nonlinearities of GaAs are yet larger than those of CdS, in contrast to GaAs, the beat signal arising from the interference of the fundamental and the second harmonic (roughly around $\omega/\omega_0 = 1.5$, equivalent to a detection photon energy of $\hbar\omega = 2.25\,\text{eV} \approx \text{E}_g - 10\%$) is still below the band gap for CdS, while reabsorption is a major problem in the GaAs case (see sections 4.2. and 4.3.). These two aspects combined should result in a CEO-phase beat signal larger than that of both GaAs and ZnO. This has been our main motivation to study this material system. Similar to ZnO, CdS has no inversion symmetry and the CdS single crystal platelets have a crystallographic $\vec{c}$-axis which lies within the platelet plane. From $\vec{k} \cdot \vec{p}$ perturbation theory, we estimate a CdS dipole matrix element of $d_{cv} = 0.23\,\text{e}\,\text{nm}$ (using Eq. (50) with $m_e = 0.2 \times m_0$ and $E_g = 2.5\,\text{eV}$). The free-standing CdS single crystal platelet discussed here has a thickness of about $l = 50\,\mu\text{m}$. The 5 fs laser system as well as the other experimental conditions are identical to sections 4.1. and 5.1. Indeed, when focussing the 5 fs pulses onto the sample, the light emitted by the CdS platelet looks much brighter as compared to ZnO (for GaAs, hardly anything could be seen with the naked eye). Due to the reabsorption at photon energies above $E_g = 2.5\,\text{eV}$, the emission appears greenish for CdS rather than white for ZnO. Suppressing the laser spectrum itself with a short-pass filter (3 mm *Schott BG 39*) and with no further filtering leads to f_ϕ peaks in the RF spectrum as large as 35 dB at 100 kHz resolution and video bandwidth – yet larger than for ZnO – under conditions comparable to those in the ZnO experiments. Unfortunately, however, the CdS crystals deteriorate on a timescale of several seconds, which makes this observation useless for applications.

6. Attosecond Pulses and Interaction of Intense Laser Fields With Atoms, Electrons and the Vacuum

So far, we have discussed the interaction of electrons bound in solids and light up to light intensities around $10^{13}\,\mathrm{W/cm^2}$. Let us now make a journey up to intensities of $10^{28}\,\mathrm{W/cm^2}$.

6.1. ELECTRONS BOUND IN ATOMS

How can electromagnetic pulses be generated which are yet shorter than a few femtoseconds? From exercise 2.2. it has become clear that the carrier photon energy of such pulses would move out of the visible range towards larger photon energies leading to extreme ultraviolet (EUV) or soft x-ray pulses. The general idea of mode-locking, which we have introduced in section 2.3., is still expected to work. However, it has not been possible so far to realize an x-ray laser or even a mode-locked x-ray laser oscillator. What is *the* essential ingredient of mode-locking? It is to consider a superposition of electromagnetic waves with equidistant frequencies and deterministic phases leading to a periodic train of pulses in the time domain. A comb of equidistant frequencies can also be obtained outside a laser cavity, namely by generation of harmonics, e.g., according to Eq. (35). For example, the 21st, 23rd, 25th, ... harmonics are evenly separated by twice the laser carrier frequency ω_0. If one was able to generate all this harmonics at the same time in a medium, if all of their phases would be fixed and favorable, a **train of attosecond pulses** ($1\,\mathrm{as} = 10^{-18}\,\mathrm{s}$) would arise.

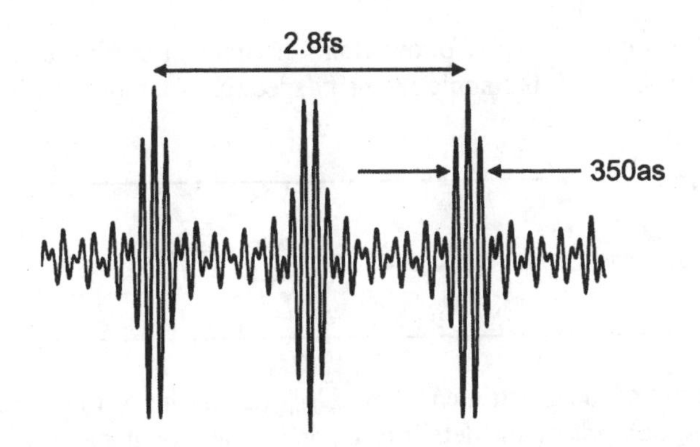

Figure 36: Scheme of the electric field versus time of a train of attosecond pulses. It arises from the superposition of the 21st, 23rd, 25th, ... 31st harmonic of the fundamental wave with carrier photon energy $\hbar\omega_0 = 1.5\,\mathrm{eV}$, equivalent to a fundamental light period of $2\pi/\omega_0 = 2.8$ fs. All harmonics are assumed to have equal amplitude and phase, the envelope of the fundamental is taken as constant in time. Note that the period of the electric field of the pulse train is 2.8 fs, that of the intensity would be 1.4 fs.

This scenario was first predicted in Refs. [61, 62]. Recently, attosecond beating as a result of the superposition of high harmonics, obtained by focusing a femtosecond laser pulse in a gas jet has been observed experimentally indeed [63, 64]. To make these extreme harmonics appreciable in strength, one clearly needs very large electric field amplitudes - yet larger than the ones we have discussed for solids. Indeed, at these intensities, solids usually vaporize, leaving gases of atoms as an option. As gases have inversion symmetry (on a macroscopic scale), all even harmonics are absent in perturbative nonlinear optics (see section 2.4.) – as already accounted for in the above example. A graphical example is shown in Fig. 36. A periodic train of pulses with a width of the individual pulses around 350 attoseconds and with a period corresponding to the period of light of the fundamental results. In section 2.4. we have seen that the interference of the fundamental with the second harmonic or the interference of the fundamental with the third harmonic can lead to a dependence on the CEO phase ϕ. In analogy to this, we also expect an influence of the CEO phase [65, 66] on the interference of, e.g., the ..., 79st, 81rd, 83th, ... harmonic and thus also on the shape of the train of attosecond pulses. This has recently been demonstrated experimentally [67] for soft x-ray photon energies in the range from 120 eV to 130 eV. In this work [67], the authors have used a CEO phase stabilized mode-locked laser oscillator, have amplified the pulses, generated a "white-light" continuum via self-phase modulation in a hollow-core waveguide filled with neon, have re-compressed the resulting pulses and used these 5 fs pulses for excitation of a 2 mm long sample of neon gas at intensities around $I = 7 \times 10^{14}\,\mathrm{W/cm^2}$ (in vacuum this intensity corresponds to a peak of the electric field envelope of $\tilde{E}_0 = 7.3 \times 10^{10}\,\mathrm{V/m}$ and with $\hbar\omega_0 = 1.5\,\mathrm{eV}$ to an electron ponderomotive energy of $\langle E_{\mathrm{kin}} \rangle = 46.7\,\mathrm{eV}$). Amazingly, the CEO phase of these excitation pulses turns out to be also fixed if the CEO phase of the mode-locked oscillator is fixed with a relative jitter of merely 50 mrad [67], equivalent to less than 1% of 2π.

Mathematically, we can closely follow along the lines of section 2.4. This leads to the general form for the high harmonic intensity spectrum (compare with $I_{\omega_0,2\omega_0}(\omega)$ in Eq. (37))

$$\boxed{I_{\omega_0,\,3\omega_0,\,\ldots,79\omega_0,\,81\omega_0,\,83\omega_0,\ldots}(\omega) \propto \left| \sum_{N,\,\mathrm{odd}} \mathrm{e}^{-\mathrm{i}N\phi}\, \tilde{E}_{N\omega_0}(\omega) \right|^2 .} \tag{87}$$

The shape and height of the spectral envelope $\tilde{E}_{N\omega_0}(\omega)$ of the N-th harmonic with carrier frequency $N\omega_0$ depends on the details (e.g., on the electron dynamics, pulse duration and shape). In order to get a feeling for the overall qualitative behavior, let us consider the simplest possible case and assume an instantaneous response according to the nonlinear optical susceptibilities in Eq. (35), such that the polarization $P(t)$ is a sum over terms $\propto E^N(t)$ with $E(t) = \tilde{E}(t)\cos(\omega_0 t + \phi)$. For, e.g., a Gaussian envelope with $\tilde{E}(t) = \tilde{E}_0 \exp(-(t/t_0)^2)$, the Fourier transform of any power N of the envelope is again a Gaussian, i.e.

$$\tilde{E}_{N\omega_0}(\omega) = \omega^2\, \eta_N\, \mathrm{e}^{-\left(\frac{\omega - N\omega_0}{\sigma_{N\omega_0}}\right)^2} . \tag{88}$$

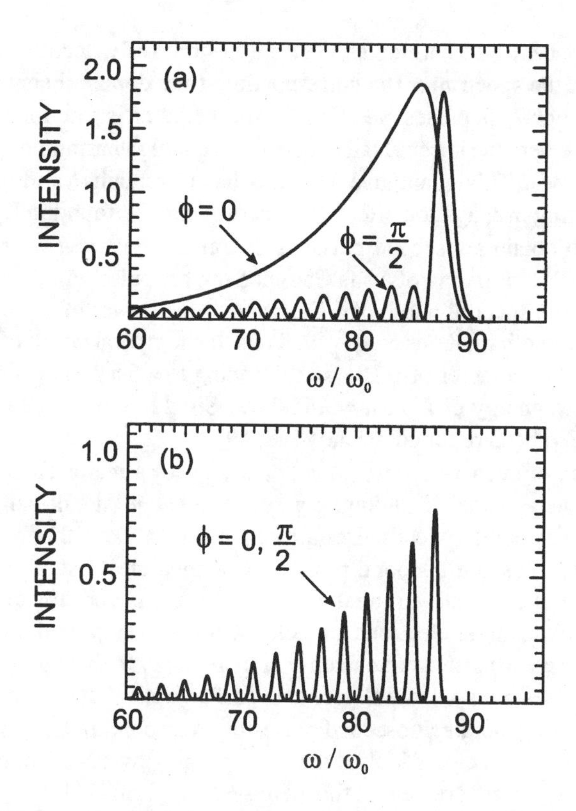

Figure 37: Scheme of the intensity spectrum of high harmonics and its dependence on CEO phase ϕ according to Eqs. (87) and (88). Here we have set η_N = const. for $N = 1, 3, 5, ..., 87 = N_{\text{cut-off}}$ and 0 elsewise, $\hbar\omega_0 = 1.5\,\text{eV}$. The pulse duration of the incident Gaussian pulses is $t_{\text{FWHM}} = t_0\, 2\sqrt{\ln\sqrt{2}}$. (a) $t_{\text{FWHM}} = 5\,\text{fs}$, (b) $t_{\text{FWHM}} = 20\,\text{fs}$. The two curves in each figure correspond to $\phi = 0, \pi, 2\pi, ...$ and $\phi = \pi/2, 3\pi/2, ...$, respectively. The latter has been stretched vertically by factor 4 for clarity in (a). Note that the various peaks at odd harmonics in (a) merge for "cos" 5 fs pulses, whereas the individual peaks are clearly separated for "sin" 5 fs pulses. No influence of the CEO phase is visible on this scale for 20 fs pulses, see (b).

The coefficients $\sigma_{N\omega_0}$ in the denominator of the exponent are given by

$$\sigma_{N\omega_0} = 2\sqrt{N}/t_0\,. \tag{89}$$

Note that the spectral width is proportional to $\sigma_{N\omega_0}$ and scales as $\propto \sqrt{N}$, which substantially increases the spectral overlap of adjacent hight harmonics (see section 2.4.). Our reasoning implies an optically thin medium, the ω^2 factor stems from the Fourier transform of the second temporal derivative on the RHS of Eq. (10). The prefactors η_N depend on intensity and can be calculated in principle. The resulting dependence on the CEO phase ϕ is illustrated in Fig. 37(a) for 5 fs excitation pulses. For CEO phase $\phi = 0, \pi, 2\pi, ...$, the tails of the different odd harmonics add up constructively, leading to a smooth total spectrum. The Fourier transform of this smooth spectrum is a single attosecond pulse in the center of the optical pulse. For CEO phase $\phi = \pi/2, 3\pi/2, ...$, the tails of two adjacent odd harmonics interfere destructively, leading to deep valleys in between them in the spectrum. The corresponding time domain behavior is a train of attosecond pulses. For 20 fs pulses (see Fig. 37(b)), the latter is true for any value of the CEO phase, because here the spectral tails of the different high harmonics hardly interfere at all (also see Fig. 36). This qualitative overall behavior remains valid in more realistic calculations, including propagation effects, which can, e.g., be found in Ref. [67].

Another way to obtain attosecond pulses is to consider just *one very high harmonic* as, e.g., the $N = 101$st harmonic of a 5 fs Gaussian optical pulse with $\hbar\omega_0 = 1.5\,\text{eV}$. We have just seen that the spectral width of a Gaussian to the power of N scales as $\propto \sqrt{N}$, hence, the temporal width scales as $\propto 1/\sqrt{N}$. Thus, the temporal width of the $N = 101$st harmonic is reduced by a factor of $\sqrt{101} \approx 10$, leading to a $5\,\text{fs}/10 = 500$ as short pulse with a carrier photon energy of $N\,\hbar\omega_0 = 151.5\,\text{eV}$. **Single x-ray pulses** with a duration of 650 as have indeed been reported recently [68, 69].

Characterization of such x-ray attosecond pulses poses a major challenge. Recently, several proposals were made, including, e.g., an attosecond streak camera [70]. In a conventional streak camera, optical pulses are sent onto a photocathode, which emits an electron bunch with the same temporal profile. A ramped electrostatic field deflects the electrons and converts time into a spatial variation along one coordinate on a screen. In the attosecond streak camera, electrons are generated by x-ray photoionization of atoms and deflected by the electric field of an intense optical pulse. In this approach, only single attosecond x-ray pulses with a duration less than half an optical cycle can be measured.

This possibility to generate attosecond pulses or attosecond pulse trains shall be sufficient motivation to have a closer look at the microscopic physics of high harmonic generation in atoms. For a recent review of this field see, e.g., Ref. [71].

High harmonics from two-level systems: It is tempting to simply continue along the lines of sections 4 and 5 where we have successfully used the Bloch equations of a two-level system (see section 3.1.) with transition frequency Ω. Here, the two levels could be two characteristic energy levels in an atom. Indeed, it has been shown that high harmonics ($N \approx 77$) and attosecond pulses are generated if a two-level system is resonantly excited ($\Omega/\omega_0 = 1$) with 18 fs optical pulses and pulse areas up to $200\,\pi$ [34] (with a dipole matrix element of $0.26\,\text{e}\,\text{nm}$, this is equivalent to a peak electric field of $5 \times 10^{10}\,\text{V/m}$ [34]). A more systematic overview is given in Figs. 38 and 39 [72]. Here we have employed $N = 30$ cycle long, box-shaped optical pulses, identical to the ones in Fig. 11 (see illustration there). For large Ω but not too large peak Rabi frequencies Ω_R in Fig. 38, well separated high harmonics are observed, as expected from our above heuristic reasoning based on nonlinear optical susceptibilities. On the diagonal, where $\omega = \Omega$, very large resonant enhancement effects are observed. This is also true for the adjacent harmonics at spectrometer frequencies $\omega = \Omega \pm 2M\omega_0$ with integer M, which altogether leads to a band of enhancement around the diagonal in Fig. 38. Especially note that large contributions can occur at the spectral position of even harmonics, which is the obvious generalization of what we have said about *"THG in disguise of SHG"* in section 5. For resonant excitation, i.e., for $\Omega/\omega_0 = 1$ on the vertical axis in Fig. 38, the various Mollow sidebands (see section 4.3.), centered at odd integer values ω/ω_0, give rise to a complex structure because the behavior of the Mollow sidebands deviates strongly from simple

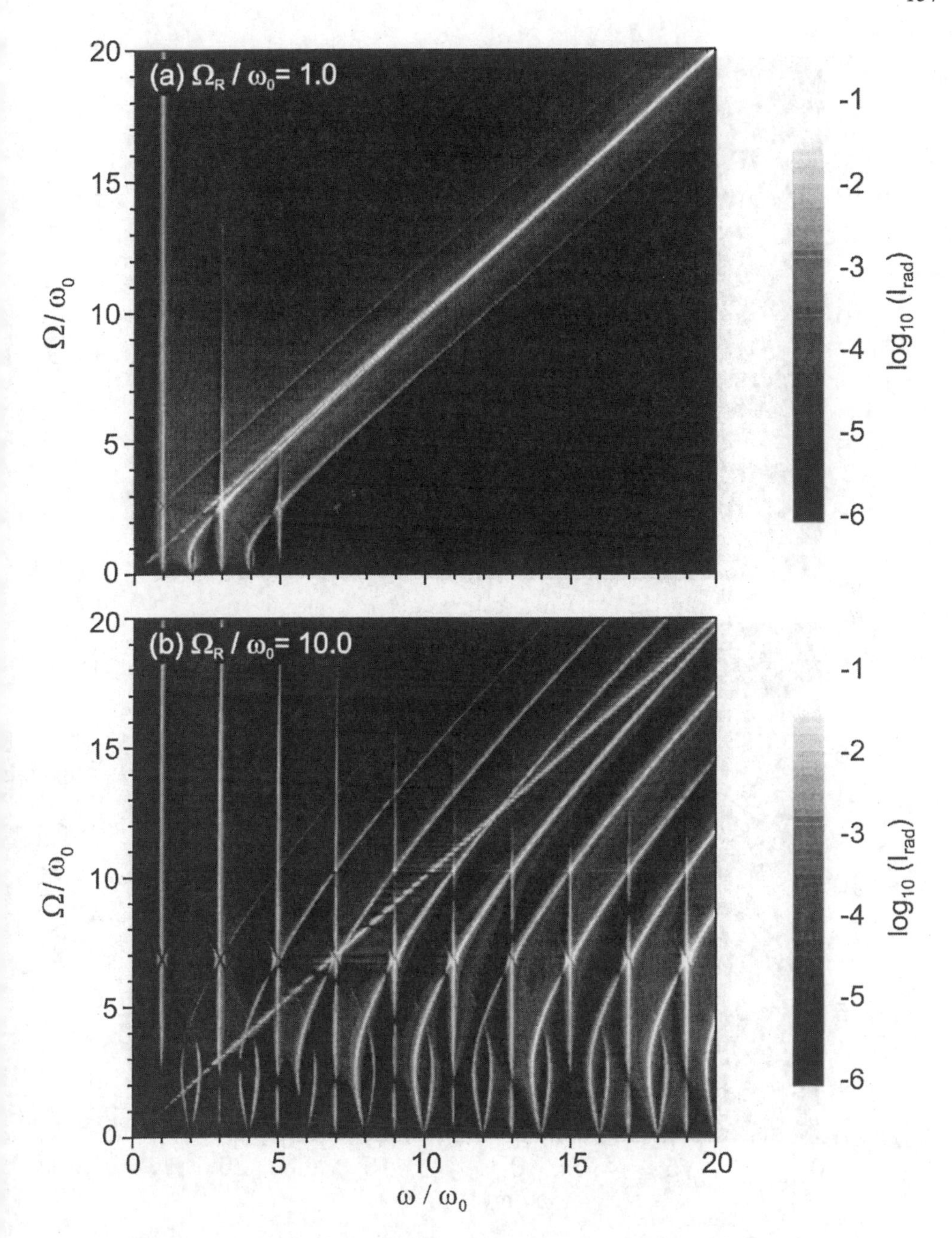

Figure 38: Grey-scale image of the radiated intensity $I_{\mathrm{rad}} \propto |\omega^2 \, u(\omega)|^2$ from exact numerical solutions of the two-level system Bloch equations (59) versus spectrometer frequency ω and transition frequency Ω in units of the laser carrier frequency ω_0. The peak Rabi frequency Ω_{R} of the exciting $N = 30$ cycle long box-shaped optical pulses is parameter (pulses identical to Fig. 11). (a) $\Omega_{\mathrm{R}}/\omega_0 = 1$, (b) $\Omega_{\mathrm{R}}/\omega_0 = 10$. Compare with Figs. 30 and 31. Taken from Ref. [72].

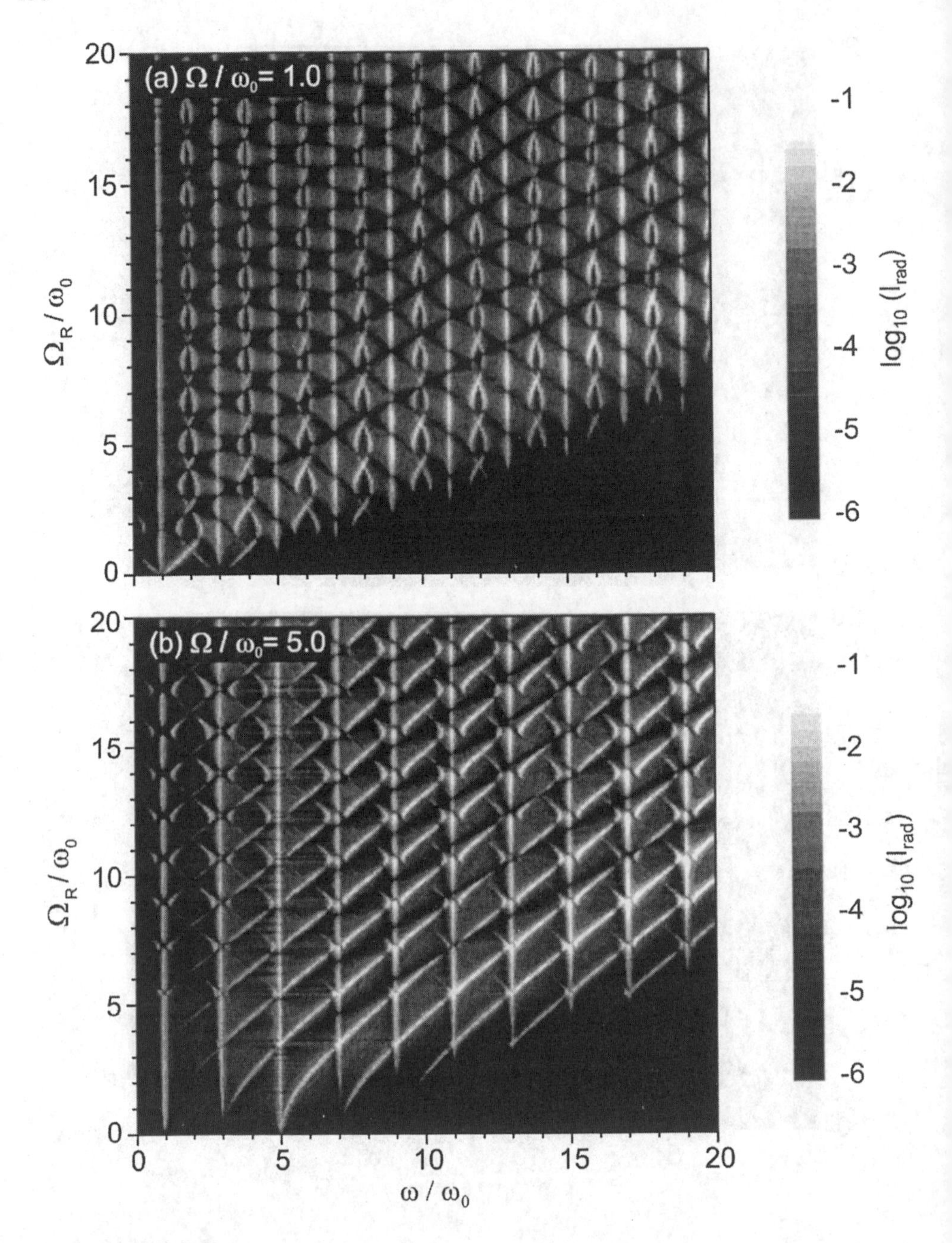

Figure 39: As Fig. 38, but versus spectrometer frequency ω and peak Rabi frequency Ω_R for two fixed values of the transition frequency Ω. (a) $\Omega/\omega_0 = 1$, (b) $\Omega/\omega_0 = 5$. Compare with Fig. 11. Taken from Ref. [72].

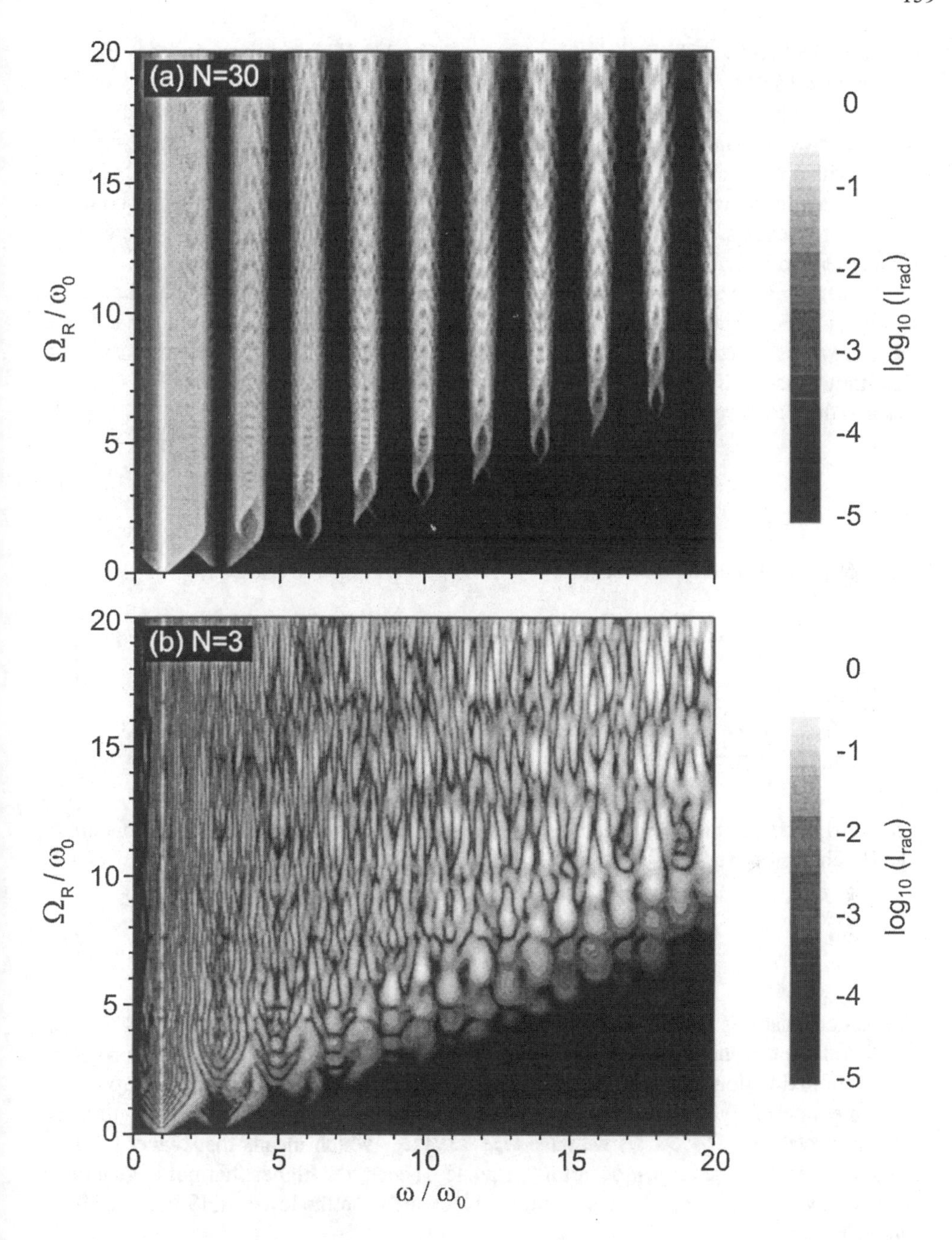

Figure 40: As Fig. 39(a), but for Gaussian optical pulses with CEO phase $\phi = 0$ and with a full width at half maximum of $t_{\mathrm{FWHM}} = N\ 2\pi/\omega_0$. (a) $N = 30$, (b) $N = 3$. Taken from Ref. [72].

sidebands shifted by $\pm\Omega_R$ (see section 4.1., e.g., Fig. 11). This can be seen from Fig. 39. The various Mollow sidebands are "repelled" from the odd harmonics. Whenever two Mollow sidebands cross each other, a set of peaks at even harmonics occurs in the spectrum. The corresponding period versus Ω_R/ω_0 is $\pi/2$ for $\Omega_R/\omega_0 \gg 1$. For excitation pulses with a non-constant envelope within the pulse as, e.g., a Gaussian, one effectively averages along the vertical axis in Fig. 39 (see Fig. 40(a)). Moreover, for pulses containing only few cycles of light, many of these intricate and beautiful structures merge into each other and interfere, giving rise to rather "messy" spectra (see Fig. 40(b)).

Is there anything that can easily be evaluated on a piece of paper rather than on a computer? For times much shorter than a cycle of light, $2\pi/\omega_0$, we can employ the "frozen-wave approximation", i.e., we approximate $\Omega_R(t) = \Omega_R$ as constant in time. In this limit, it is straightforward to solve the Bloch equations (59) analytically [1]. This leads to the Bloch vector

$$\begin{pmatrix} u(t) \\ v(t) \\ w(t) \end{pmatrix} = \mathcal{M}(t) \begin{pmatrix} u(0) \\ v(0) \\ w(0) \end{pmatrix}, \tag{90}$$

with the (3×3) rotation matrix

$$\mathcal{M}(t) = \begin{pmatrix} \frac{4\Omega_R^2 + \Omega^2 \cos(\Omega_{\text{eff}} t)}{\Omega_{\text{eff}}^2} & \frac{\Omega}{\Omega_{\text{eff}}} \sin(\Omega_{\text{eff}} t) & \frac{2\Omega\Omega_R}{\Omega_{\text{eff}}^2}(\cos(\Omega_{\text{eff}} t) - 1) \\ -\frac{\Omega}{\Omega_{\text{eff}}} \sin(\Omega_{\text{eff}} t) & \cos(\Omega_{\text{eff}} t) & -\frac{2\Omega_R}{\Omega_{\text{eff}}} \sin(\Omega_{\text{eff}} t) \\ \frac{2\Omega\Omega_R}{\Omega_{\text{eff}}^2}(\cos(\Omega_{\text{eff}} t) - 1) & \frac{2\Omega_R}{\Omega_{\text{eff}}} \sin(\Omega_{\text{eff}} t) & \frac{\Omega^2 + 4\Omega_R^2 \cos(\Omega_{\text{eff}} t)}{\Omega_{\text{eff}}^2} \end{pmatrix}. \tag{91}$$

Obviously, the optical polarization $P(t) \propto u(t)$ as well as the other two components of the Bloch vector oscillate with the effective frequency Ω_{eff}, which is given by

$$\boxed{\Omega_{\text{eff}} = \sqrt{4\,\Omega_R^2 + \Omega^2}\,.} \tag{92}$$

Remember that this **frozen-wave approximation** is only justified for times $t \ll 2\pi/\omega_0$, hence relevant in the limit $\Omega_{\text{eff}} \gg \omega_0$. It can be viewed as the opposite of the rotating wave approximation (see section 3.1.). There, almost nothing is supposed to happen on the timescale of light, whereas here all the significant dynamics takes place within an optical cycle. For $\Omega_R \gg \Omega$, we have $\Omega_{\text{eff}} \approx 2\,\Omega_R$, which means that twice the peak Rabi frequency is the largest occuring frequency, hence, the highest harmonic generated is given by $N_{\text{cut-off}} \approx 2\,\Omega_R/\omega_0$ (compare black areas on the lower RHS in Figs. 39 (a) and (b)).

Starting from the ground state, i.e., from Bloch vector $(0, 0, -1)^T$ at time $t = 0$, the inversion according to Eqs. (90) and (91) is given by

$$w(t) = -\frac{\Omega^2 + 4\,\Omega_R^2 \cos(\Omega_{\text{eff}} t)}{\Omega_{\text{eff}}^2}\,. \tag{93}$$

Thus, the two-level system can even perform Rabi flopping for far off-resonant conditions, i.e. for $\Omega \gg \omega_0$, if the intensity is so large that it roughly corresponds to a Rabi frequency of $\Omega_R = \Omega$, which leads to $w(t) = -1/5 - 4/5 \times \cos(\sqrt{5}\,\Omega_R t)$ with maximum inversion $w = +3/5$ ($\Leftrightarrow$ 80% maximum occupation of the excited state).

Within the "square-wave approximation" (see Appendix 13.2.), which can be viewed as an extension of the frozen-wave approximation, one can analytically show that the constrictions formed by the crossing Mollow triplets in Fig. 39 can be interpreted as points of commensurability of the three frequencies ω_0, Ω, and Ω_R. Specifically, commensurability occurs for Rabi frequencies with

$$\boxed{\frac{\Omega_R}{\omega_0} = \frac{\pi}{2}\sqrt{M^2 - \frac{1}{4}\left(\frac{\Omega}{\omega_0}\right)^2}\,,} \tag{94}$$

with integer $M = 1, 2, 3, \ldots$ For these Rabi freqiencies, an integer number of Rabi flops is completed after half an optical cycle and, thus, peaks at even integers ω/ω_0 occur in the optical spectrum.

Photo-ionization of atoms: However, such two-level-system approach completely ignores one of the most crucial aspects of the problem, i.e., that the atom can be ionized (in which case the two-level system effectively "disappears"). To see this, let us compare the optical field with the electric field which attracts the electron to the nucleus. The simplest example is the 1s state of a hydrogen atom, in which case the field is given by

$$|\vec{E}|_{\mathrm{H,\,1s}} = \frac{\mathrm{e}^2}{4\pi\epsilon_0\, r_\mathrm{B}^2}\,. \tag{95}$$

With the hydrogen Bohr radius of $r_\mathrm{B} = 0.053\,\mathrm{nm}$, we obtain an electric field of

$$|\vec{E}|_{\mathrm{H,\,1s}} = 5.17 \times 10^{11}\,\mathrm{V/m}\,, \tag{96}$$

which is comparable to the laser fields under discussion here. The corresponding electric potential experienced by the electron in the presence of a strong laser field is illustrated schematically in Fig. 41: The light field introduces a large oscillating tilt in the potential, which can sweep the electron out of its bound state, i.e., ionize the atom. In this process, the timescale of the light period (a few femtoseconds) has to be compared with the classical orbit time of the electron in the 1s state given by the Rydberg energy of $\hbar\Omega_{\mathrm{Rydberg}} = 13.6\,\mathrm{eV}$, equivalent to a period of $2\pi/\Omega_{\mathrm{Rydberg}} = 304$ attoseconds. Under these conditions, it is certainly not justified to compute the harmonics on the basis of the nonlinear optical susceptibilities according to Eq. (35) (as we have done in the above intuitive discussion), because significant electron dynamics takes place on the timescale of the fundamental period of light and, hence, the nonlinear polarization does not follow the driving electric field instantaneously.

What is the highest harmonic, $N_{\mathrm{cut-off}}$, which is generated? Ref. [73] argues that the harmonics are mostly generated by the ionized electron during the first optical cycle after

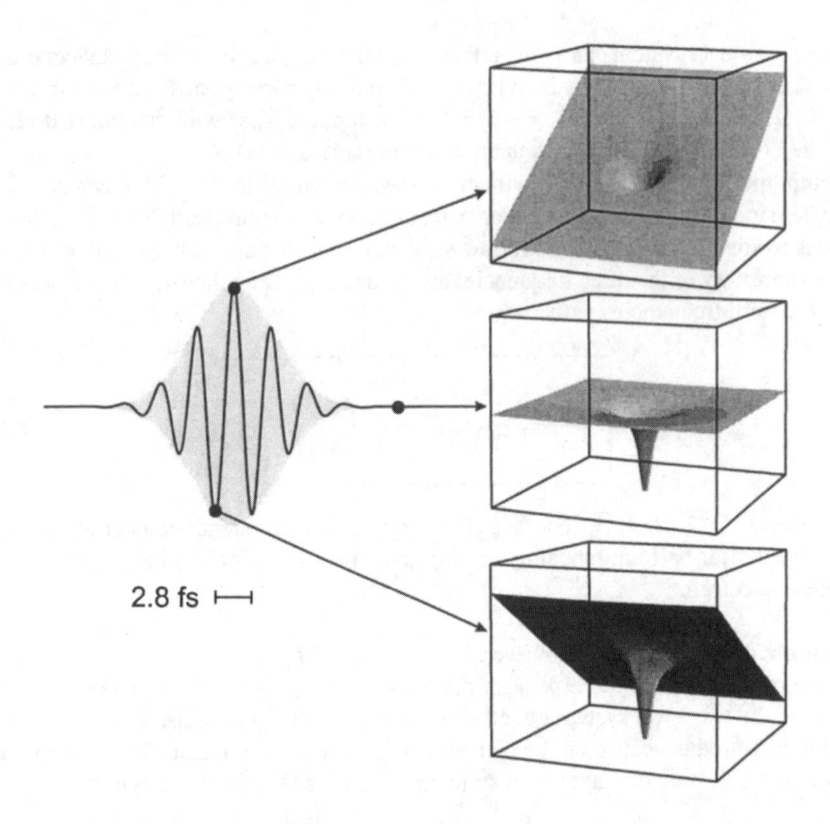

Figure 41: Electric field $E(t)$ of a Gaussian $t_{\mathrm{FWHM}} = 5\,\mathrm{fs}$ laser pulse with carrier photon energy $\hbar\omega_0 = 1.5\,\mathrm{eV}$ and $\phi = 0$ versus time and (two-dimensional) scheme of the resulting electric potential experienced by an electron initially bound in an atom at three characteristic points in time. The large "tilt" along the electric field vector axis in the center of the pulse leads to tunnelling of the electron. At very large fields, the electron can even (classically) be swept out of its binding potential. Both processes lead to ionization of the atom.

it is ionized due to the strong laser field. The highest harmonic $N_{\mathrm{cut-off}}$ is then given by the simple, semi-empirical formula

$$\boxed{N_{\mathrm{cut-off}} \times \hbar\omega_0 = E_{\mathrm{ion}} + 3.17 \times \langle E_{\mathrm{kin}} \rangle_{\mathrm{1st}}\,,} \tag{97}$$

where E_{ion} is the ionization potential of the atom, and $\langle E_{\mathrm{kin}} \rangle_{\mathrm{1st}}$ is the electron ponderomotive energy in the first optical cycle after ionization. This seems plausible: Before ionization, the electron more or less harmonically orbits around the nucleus (classically speaking), long after ionization, the electron is free and again no harmonics are generated (the electron velocities are not yet relativistic at intensities around $I = 10^{15} \ldots 10^{16}\,\mathrm{W/cm^2}$). Thus, most of the anharmonicities in the electron motion occur in the transition regime. Eq. (97) indirectly introduces a dependence on the pulse duration. For long and weak pulses, the atom is ionized in the center of the pulse and the ponderomotive energy is small, hence the cut-off is low. For long and intense pulses the atom is ionized far before the maximum of the electric field of the pulse is reached and, again, the ponderomotive energy is low. For short and intense pulses, the atom is ionized near the center of the pulse where the intensity and the ponderomotive energy are large, thus, the cut-off is shifted towards high harmonics. In Ref. [73], the authors used 26 fs pulses and generated, e.g., in He resolved harmonics up to $N = 221$ at intensities of $I = 6 \times 10^{15}\,\mathrm{W/cm^2}$ (and unresolved harmonics up to $N = 297$). One possible application of these high harmonics could be using 13 nm wavelength radiation ($\Leftrightarrow N \approx 61$) for EUV nanometer lithography in which case large EUV fluences are desirable. It has been demonstrated that quasi phase-matching in modulated, centimeter-long, gas-filled hollow waveguides can increase the EUV fluences significantly [74].

6.2. *ELECTRONS IN VACUUM*

When a gas is excited with yet larger laser intensities than the ones we have just discussed, the atoms are rapidly ionized and the electromagnetic field essentially acts on free (i.e., unbound) electrons. At some point, the laser intensity becomes so large that the magnetic component of the Lorentz force on a free electron with charge $-\mathrm{e}$,

$$\vec{F}(\vec{r},t) = \underbrace{-\,\mathrm{e}\,\vec{E}(\vec{r},t)}_{\text{electric}} \underbrace{-\,\mathrm{e}\,\vec{v}(t) \times \vec{B}(\vec{r},t)}_{\text{magnetic}} = \frac{\mathrm{d}(m_\mathrm{e}\vec{v})}{\mathrm{d}t}\,, \tag{98}$$

associated with the light field becomes comparable to the electric component [75] (see example 3.4.). Using $\tilde{E}_0/\tilde{B}_0 = c_0$ (see section 2.2.), this is equivalent to the condition $v_0/c_0 \approx 1$. With the peak velocity $v_0 = \mathrm{e}\tilde{E}_0/(m_0\omega_0)$ from Newton's second law (Eq. (98) with $m_\mathrm{e} \to m_0$ and $\tilde{B}_0 \approx 0$), this is furthermore equivalent to stating that the **dimensionless parameter**[31] $\mathcal{E}$, which is given by

$$\boxed{\mathcal{E} = \frac{\mathrm{e}\,\tilde{E}_0}{m_0\,\omega_0\,c_0}}\,, \tag{99}$$

becomes comparable to unity.

Example 6.1.: In vacuum, for electrons with rest mass $m_0 = 9.1091 \times 10^{-31}$ kg, carrier photon energy $\hbar\omega_0 = 1.5\,\mathrm{eV}$ and with the fundamental constants e and c_0, we have

$$\begin{aligned} \mathcal{E} &= 1 \qquad (100) \\ &\Leftrightarrow \\ \tilde{E}_0 &= 3.8 \times 10^{12}\,\mathrm{V/m} \\ &\Leftrightarrow \end{aligned}$$

[31] In the literature, the parameter $\mathcal{E}$ is often rather called a. In this article, the symbol a has already been used for the lattice constant and will later also be used for the acceleration.

$$\tilde{B}_0 = 1.3 \times 10^4\,\mathrm{T}$$
$$\Leftrightarrow$$
$$I = 1.9 \times 10^{18}\,\mathrm{W/cm^2}\,.$$

If we had just the magnetic field component of the laser field and if it were a static field, a non-relativistic electron would simply orbit in circles around the magnetic field axis with the cyclotron frequency ω_c given by

$$\boxed{\omega_c = e\tilde{B}_0/m_0\,.} \tag{101}$$

Introducing Eq. (101) into Eq. (99) we get

$$\boxed{\mathcal{E} = \frac{\hbar\omega_c}{\hbar\omega_0}\,.} \tag{102}$$

Thus, we can equivalently say that **something special is expected to happen when the cyclotron energy $\hbar\omega_c$ becomes comparable to the carrier photon energy $\hbar\omega_0$**. Obviously, the cyclotron energy is just another entry into the list of energies which are associated to the laser intensity I (section 3.: Rabi energy, ponderomotive energy and Bloch energy).

Fields on the order of $\mathcal{E} = 1$ can lead to **relativistic nonlinear Thomson scattering** on free electrons[32] [76, 77, 78]. Let us first give an intuitive explanation and then look at the mathematics. If the electric field propagating along z is polarized along the x–direction, the electron harmonically oscillates along x with frequency ω_0 according to Eq. (65) for low intensities (the electron velocity components in the y and z–direction are constant). The periodic acceleration (see RHS of Eq. (10)) of the charged electron is the source of waves, i.e., light can be scattered elastically into directions other than the z–direction. For larger intensities, the magnetic field component along the y–direction leads to a force comparable in amplitude to the electric component and the electron can, e.g., move in figure-eight patterns (i) in the xz-plane with an additional drift in the z–direction (ii) superimposed. Let us have a closer look at aspects (i) and (ii). (i) The figure-eight pattern results from the fact that the z-component of the force, $F_z = -e\,v_x\,B_y$, oscillates with frequency $2\omega_0$ because v_x as well as B_y oscillate with ω_0, while the x-component of the force oscillates with frequency ω_0 as argued above. This, e.g., leads to second-harmonic generation – even though an electron in vacuum has inversion symmetry (see Fig. 42 RHS column). (ii) A drift motion perpendicular to $\vec{E}$ and perpendicular to $\vec{B}$ also occurs for orthogonal *static* electric and magnetic fields in the non-relativistic regime – as described in many textbooks. The Hall effect [1] also stems from this drift motion.

---

[32] Non-relativistic linear Thomson scattering on an individual charge is analogous to Rayleigh scattering on an electric dipole – which makes the sky blue (Hertz dipole).

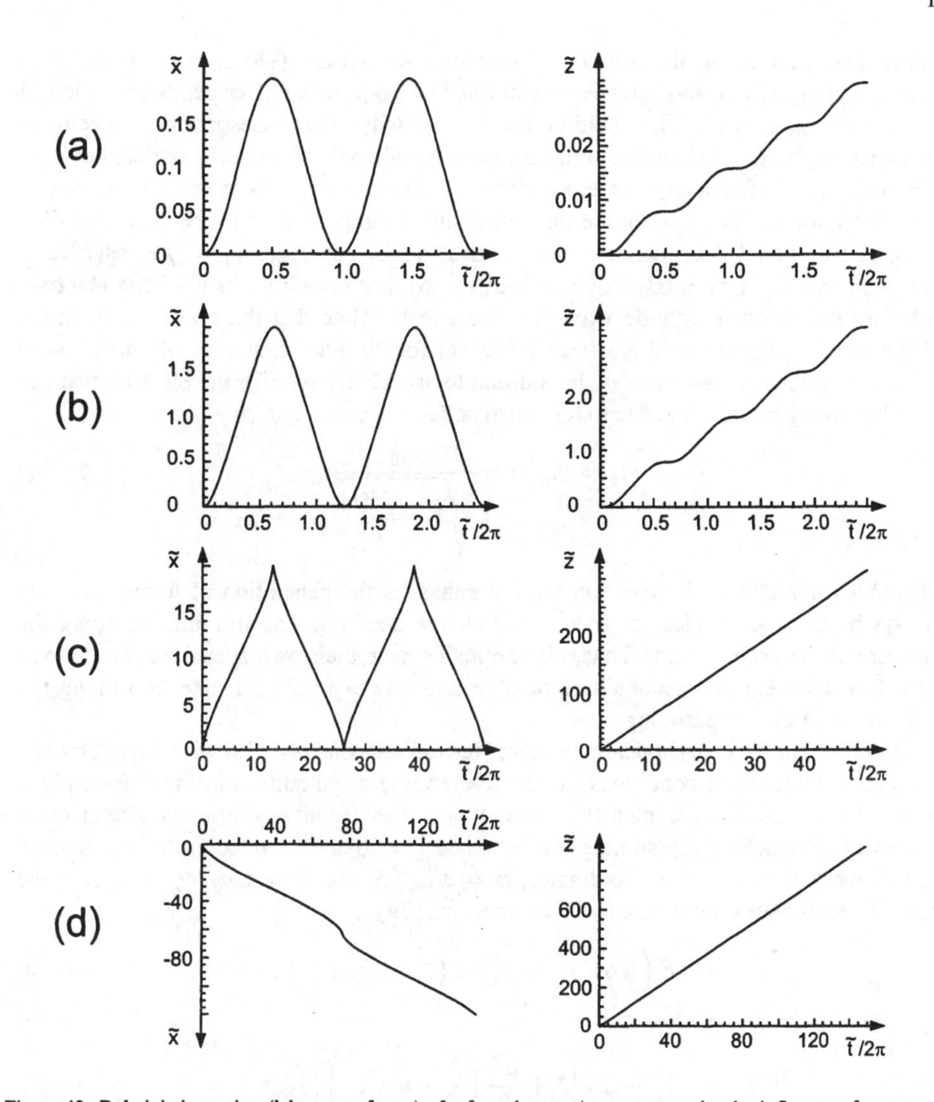

Figure 42: Relativistic motion (laboratory frame) of a free electron in vacuum under the influence of a strong laser field according to Eqs. (104)–(106). The light propagates along z with $\vec{E}$ and $\vec{B}$ being polarized along x and y, respectively. $\mathcal{E}$ is the normalized electric field strength given by $\mathcal{E} = \omega_c/\omega_0$. (a) $\mathcal{E} = 0.1$, $\zeta_0 = 0$; (b) $\mathcal{E} = 1$, $\zeta_0 = 0$; (c) $\mathcal{E} = 10$, $\zeta_0 = 0$; and (d) $\mathcal{E} = 10$, $\zeta_0 = \pi/2$. Note the different vertical and horizontal scales. In (a), the oscillation period of $\tilde{z}(\tilde{t})$ is π, equivalent to a motion of $z(t)$ with frequency $2\omega_0$, which leads to second-harmonic generation. In (b) and (c), the period of $\tilde{x}(\tilde{t})$ becomes larger than 2π, equivalent to an oscillation frequency of $x(t)$ smaller than ω_0 due to the relativistic Doppler red-shift. Indeed, $\tilde{z}(\tilde{t}) \approx \tilde{t}$ in (c) is equivalent to a drift velocity along z close to the speed of light c_0. The trajectories as well as the periods also depend on the initial condition of the electron. For example, the period of $\tilde{x}(\tilde{t})$ in (c) is $26 \times 2\pi$, while it is about $76 \times 2\pi$ in (d) – although the laser intensity is the same.

Under these conditions, the solution of Newton's second law (98) is given by $v_x(t) = v_0 \cos(\omega_c t)$, $v_y(t) = 0$ and $v_z(t) = -v_0 \sin(\omega_c t) + v_{\text{drift}}$, where v_0 depends on the initial conditions. $v_{\text{drift}} = E_x/B_y$ would be the drift velocity. The corresponding trajectories are called trochoids. A roughly similar behavior is anticipated for intense optical fields, in which case the drift velocity can approach the speed of light c_0. The acceleration towards this drift velocity occurs when the laser intensity is ramped up. As a result of the drift along z, the spatial dependence of the incident fields, i.e., $E_x(z,t) = \tilde{E}_0 \cos(K_z z - \omega_0 t - \phi)$ and $B_y(z,t) = \tilde{B}_0 \cos(K_z z - \omega_0 t - \phi)$, becomes important – **"the electron rides on the electromagnetic wave like a surfer"**. Note that the spatial dependence of the electromagnetic field has been irrelevant for all other examples of carrier-wave nonlinear optics discussed so far. In addition to aspects (i) and (ii), the electron mass in Eq. (98) changes with time due to relativistic effects according to

$$m_e = m_e(t) = \frac{m_0}{\sqrt{1 - \frac{\vec{v}^2(t)}{c_0^2}}}. \tag{103}$$

This anharmonicity in the electron motion enhances the generation of harmonics. As always in electrodynamics, an accelerated charge does not lead to radiation along the direction of its acceleration. Thus, all harmonics have their own specific emission patterns [78] and, e.g., no second-harmonic emission is expected for detection along the z-direction from our reasoning.

At first sight, it seems hopeless to solve Newton's second law (98) with the relativistic mass (103) under these conditions exactly. Nevertheless and quite amazingly, for a plane wave with a constant light intensity, an exact analytical implicit solution can be given in terms of a parameter ζ. Assuming a CEO phase $\phi = 0$ and introducing the normalized and dimensionless x and z coordinates, $\tilde{x} = x\,\omega_0/c_0$ and $\tilde{z} = z\,\omega_0/c_0$ as well as the normalized dimensionless time $\tilde{t} = t\,\omega_0$, one gets [79]

$$\tilde{x}(\zeta) = \mathcal{E}\left((\cos\zeta_0 - \cos\zeta) - (\zeta - \zeta_0)\sin\zeta_0\right), \tag{104}$$

$$\tilde{z}(\zeta) = \tilde{t} - \zeta, \tag{105}$$

$$\begin{aligned} \tilde{t}(\zeta) &= (\zeta - \zeta_0)\left[1 + \frac{\mathcal{E}^2}{2}\left(\frac{1}{2} + \sin^2\zeta_0\right)\right] \\ &\quad + \frac{\mathcal{E}^2}{2}\left[-\frac{\sin 2\zeta}{4} + 2\cos\zeta\sin\zeta_0 - \frac{3\sin 2\zeta_0}{4}\right]. \end{aligned} \tag{106}$$

The initial electron velocity $\vec{v}$ is assumed to be zero at time $t = 0$. The parameter ζ_0 results from the initial position of the electron at $t = 0$ and is given by $\tilde{z}(\tilde{t} = 0) = -\zeta_0$ (see Eq. (105)). It can be interpreted as the **phase of the electron** and is **equal to the CEO phase ϕ at this point**, but not identical to it in general: For a cloud of electrons with an extent along the z-direction comparable to or even larger than the wavelength of light $2\pi c_0/\omega_0$, one has a distribution of values for ζ_0. A similar effect can occur in the actual focus of a lens where the phase fronts are not plane everywhere. Selected examples of electron trajectories are given in Fig. 42. In each of the plots, ζ runs from 0 to 4π.

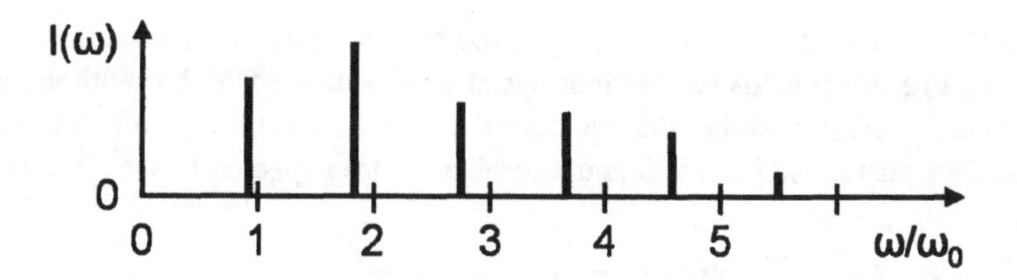

Figure 43: Scheme of the intensity spectrum $I(\omega)$ of light emitted by an electron in vacuum driven by a strong laser field at carrier frequency ω_0 according to Eq. (107) with $\tilde{\omega}_0/\omega_0 \approx 0.9$ corresponding to $\mathcal{E} \approx 0.6$ and $\zeta_0 = 0$. Note that the ratio $\tilde{\omega}_0/\omega_0$ depends on $\mathcal{E}$ as well as on ζ_0, see Fig. 42.

In the limit $\mathcal{E}^2 \ll 1$, one has $\tilde{t} = \zeta - \zeta_0$, thus $\tilde{z} = -\zeta_0 = \text{const.}$, while the x-coordinate oscillates harmonically with time, i.e., $x(t) \propto \cos(\omega_0 t)$ – as explained above (Fig. 42(a)). In Fig. 42(b) where $\mathcal{E} = 1$, the excursion along the x-direction is already on the order of $\tilde{x} = 1$, corresponding to an actual value[33] of about $x = 0.1\,\mu\text{m}$ for $\hbar\omega_0 = 1.5\,\text{eV}$. Deep in the relativistic regime, i.e. for $\mathcal{E}^2 \gg 1$, a modification arises which we have not discussed thus far. In Fig. 42(c) where $\mathcal{E} = 10$, the oscillation frequency $\tilde{\omega}_0$ is smaller than ω_0 by factor 26. This means that, e.g., for a laser carrier frequency of $\hbar\omega_0 = 1.5\,\text{eV}$, photons with photon energies as small as 58 meV are emitted. This modification stems from the fact that the "source" of the incident electromagnetic wave (the laser) and the "observer" (the electron) move away from each other due to the relativistic drift velocity of the electron (which is parallel to the light wave vector). This leads to a relativistic Doppler red-shift. Thus, the electron "feels" a driving frequency $\tilde{\omega}_0$ which is smaller than ω_0. Note, however, that the ratio $\tilde{\omega}_0/\omega_0$ cannot simply be calculated from the usual textbook (longitudinal) Doppler effect formulae which would only apply if the electron was a system of inertia – which it is not.

The peak heights I_N in the intensity spectrum $I(\omega)$ of the emitted light depend sensitively on ζ_0, CEO phase ϕ and also on the detection direction. It should, however, be clear from Fig. 42 that the general form is always given by

$$\boxed{I(\omega) = \sum_{N=1}^{\infty} I_N\,\delta(\omega - N\,\tilde{\omega}_0)\,,} \tag{107}$$

where the order N covers **odd and even harmonics of the Doppler red-shifted laser carrier frequency** $\tilde{\omega}_0$ **with**

$$\boxed{0 < \tilde{\omega}_0 = \tilde{\omega}_0(\mathcal{E}, \zeta_0) \leq \omega_0\,.} \tag{108}$$

[33] In order to make our model of isolated electrons in vacuum realistic, residual scatterers (e.g. ionized atoms) in the vacuum should have a number N_{atom} per volume V of less than $(1/x)^3 = 10^{15}\,\text{cm}^{-3} = 10^{21}\,\text{m}^{-3}$. With the equation of state of ideal gases from thermodynamics, i.e., $PV = N_{\text{atom}} k_B T$ with Boltzmann's constant $k_B = 1.3804 \times 10^{-23}$ J/K, this density translates into a maximum residual pressure of $P = 10^{21}\,\text{m}^{-3}\,k_B T = 4\,\text{Pa}$ at $T = 300\,\text{K}$ – a "bad vacuum".

Fig. 43 illustrates this behavior. For example, the highly anharmonic motion of $x(t)$ corresponding to Fig. 42(c) leads to contributions at large values of N, but without any even contributions. Finally, remember that we have discussed a light field with constant intensity in time. For pulsed excitation, $\mathcal{E}$ and thus also $\tilde{\omega}_0$ are expected to vary in time.

6.3. TOWARDS COLOSSAL LASER INTENSITIES

Multi-Terawatt laser pulses ($I = 10^{19} \dots 10^{20}\,\mathrm{W/cm^2}$) can even initiate **photonuclear fission** in solid targets [80, 81] – a phenomenon, which was predicted theoretically more than ten years earlier [82]. The physics of this laser-induced fission of nuclei is roughly similar to the laser-induced ionization of atoms we have discussed in section 6.1.

The peak focused laser intensities starting to become available today or in the near future are around $I = 10^{23}\,\mathrm{W/cm^2}$ [83]. Some authors [83, 84] already speculate about future physics, as, e.g., **light-induced** (i.e., $\hbar\omega_0 \approx 1\,\mathrm{eV}$) **electron-positron generation** in vacuum: According to relativistic quantum mechanics, the vacuum, i.e., the fully occupied Dirac sea, is analogous to the fully occupied valence band of a semiconductor. Positrons correspond to holes and the real electrons are analogous to conduction band electrons. Hence, electron-positron generation in vacuum is roughly similar to electron-hole generation in a semiconductor with a band gap energy of "E_g"$= 1.024\,\mathrm{MeV} = 2\,m_0 c_0^2$ under far off-resonant excitation conditions ("E_g"$\approx 10^6 \times \hbar\omega_0$). Note that this would be **nonlinear optics in vacuum,** i.e., such effects would violate the superposition principle following from the Maxwell equations in vacuum.

However, in contrast to electron-hole pair generation in semiconductors, a photon corresponding to a single plane wave in vacuum cannot generate an electron-positron pair ! This is due to the fact that it is not possible to simultaneously fulfill energy (E) and momentum (p) conservation in this process with the photon dispersion

$$E = p\,c_0 \tag{109}$$

and the relativistic electron dispersion

$$E = \pm\sqrt{(m_0 c_0^2)^2 + (p\,c_0)^2} = \pm\, m_e c_0^2 \tag{110}$$

– the photon momentum is too large (see Fig. 44). This problem can be circumvent by using counterpropagating laser beams with an effective photon momentum equal to zero (or by considering other four-particle processes). Recently, even **Rabi oscillations of the Dirac sea** have been predicted theoretically [85] for oppositely directed laser beams [34].

Zetta[35]- and Exawatt lasers with focused intensities around $I = 10^{26} \dots 10^{28}\,\mathrm{W/cm^2}$ might become accessible around the year 2010 and beyond. At these intensities, the peak

[34] Also see our discussion on Rabi flopping in the two-level model within the "frozen-wave approximation" in section 6.1. which also holds for a transition frequency $\Omega/\omega_0 \approx 10^6$.

[35] One Zettawatt=1 ZW=10^{21} W, one Exawatt=1 EW=10^{18} W.

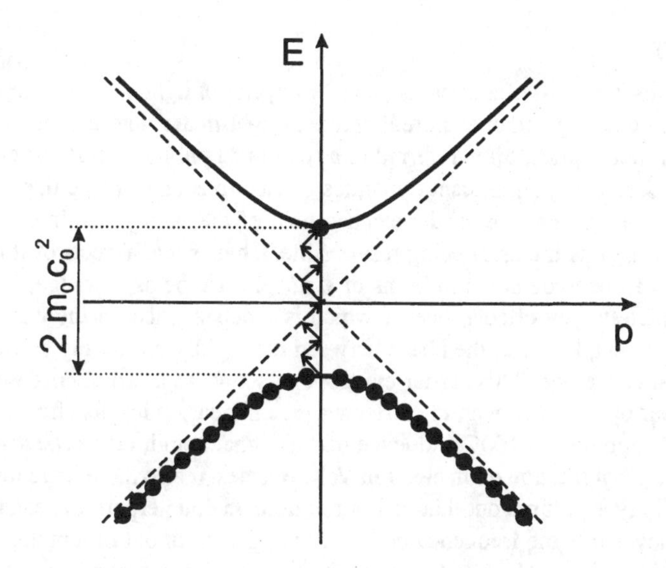

Figure 44: Relativistic quantum mechanical dispersion relation of electrons in vacuum, i.e., $E = \pm\sqrt{(m_0 c_0^2)^2 + (p\, c_0)^2}$, (solid curves) and dispersion relation of photons (dashed straight lines with $E = \pm p\, c_0$). The vacuum corresponds to a fully occupied lower band and an empty upper band as indicated by the full dots. An unoccupied state in the lower band is called positron, an occupied state in the upper band is called electron. Compare with Fig. 9. The arrows indicate the generation of an electron-positron pair by many photons from two counterpropagating light beams (not to scale). Actually, around 10^6 laser photons with $\hbar\omega_0 \approx 1\,\text{eV}$ are necessary as $2\, m_0 c_0^2 = 1.024\,\text{MeV}$.

electron acceleration a_e^0 becomes truly colossal (also see example 3.3.). For example at an intensity of $I = 10^{28}\,\text{W/cm}^2$, $\Leftrightarrow \tilde{E}_0 = 2.7 \times 10^{17}\,\text{V/m}$ and $\tilde{B}_0 = 1.7 \times 10^6\,\text{T}$ in vacuum, one gets a classical, i.e., non-relativistic acceleration of $|a_e^0| = e/m_0\, \tilde{E}_0 = 4.7 \times 10^{28}\,\text{m/s}^2 = 4.8 \times 10^{27} \times \text{g}$. This acceleration would be comparable to the gravitational acceleration near the edge of a black hole. In the latter case, the large gravitational acceleration is the origin of the so called Hawking radiation [86], the theoretically predicted energy loss channel of a black hole. The **Unruh radiation** [87, 88, 89] would be the analogue of that for acceleration by the laser electric field [83]. If the acceleration a_e was constant, **thermal radiation** would result, with a temperature T given by the relation

$$\boxed{k_B T = \frac{\hbar}{2\pi}\frac{a_e}{c}\,.} \tag{111}$$

For, e.g., $a_e = 10^{28}\,\text{m/s}^2$, $c = c_0$ and Boltzmann's constant $k_B = 1.3804 \times 10^{-23}\,\text{J/K}$, one arrives at a temperature of $T = 42 \times 10^6\,\text{K}$. Obviously, this Unruh radiation has to be distinguished from Bremsstrahlung, as usual originating from accelerated charges in the Maxwell equations (4).

7. Summary

Nonlinear optics with pulses as short as just a few cycles of light and with large intensities takes us into an exciting and unexplored regime of light-matter interaction and into a new regime of nonlinear optics, often referred to as extreme nonlinear optics or **carrier-wave nonlinear optics**. Here, significant dynamics of electrons takes place during an optical cycle and one has to think about the electric field of the laser pulse interacting with a material rather than its intensity being responsible. Thus, some aspects of this *ultrashort time* behavior can be understood in terms of *static* electric fields - strange, but true. This leads to completely new effects, one of which is a dependence on the carrier-envelope offset phase of a single pulse, the phase between the rapidly oscillating carrier wave and the electric field envelope. This carrier-envelope offset phase is strictly irrelevant in traditional nonlinear optics. However, in carrier-wave nonlinear optics, its effects can become quite large and approach a 100% influence on the signal amplitude in spectral regions as large as a significant fraction of an electron Volt. Besides, it also has a large impact on frequency metrology as it turns out that its Fourier domain counterpart, the carrier-envelope offset frequency, shifts the frequency comb of the pulse train out of a mode-locked laser oscillator. If this shift can be fixed to zero, the frequency comb can be used for measuring frequencies much like a ruler for measuring millimeters. In particular, the microwave time standard of the S.I. unit system can conveniently be related to optical frequencies. **In solids**, the strength of the laser field can be expressed by the Rabi energy for interband transitions and by the ponderomotive energy or the Bloch energy for the case of intraband transitions, which all have to be compared with the carrier photon energy of the exciting laser pulses. Interband transitions tend to be more pronounced for resonant excitation, intraband effects take over in the off-resonant limit. For resonant excitation, *carrier-wave Rabi oscillations* do occur. This regime is potentially relevant for applications because already today, semiconductor saturable absorbers do play a role in passive mode-locking of femtosecond lasers. On the scale of just few cycles of light, coherent effects eventually take over. The combination of saturable absorption, coherence and times on the order of a cycle of light leads directly to carrier-wave Rabi oscillations. Their specific signature is the occurrence of Mollow sidebands around the third harmonic of the optical transition photon energy. If these sidebands interfere with those of the fundamental, a *dependence on the carrier-envelope offset phase* results. For off-resonant excitation, we have not only observed a peak at f_ϕ in the RF spectra but also one at $2f_\phi$, as evidence of interference of the fundamental with the second harmonic and the third harmonic, respectively. Using, e.g., ZnO, allows for measuring f_ϕ and hence controlling the carrier-envelope offset phase with powers as low as just some tens of mW directly from a mode-locked femtosecond laser oscillator, which opens a whole new field of possible experiments. Furthermore, the ZnO experiments offer the perspective to determine the carrier-envelope offset phase itself – the ultimate complete characterization of laser pulses. **In atoms**, the laser intensities have to be yet larger than in solids in order to reach characteristic energies of the atoms. When the ponderomotive energy reaches the Rydberg energy, atoms are ionized by the light field on a timescale of an optical cycle, which results in the generation of harmonics beyond the 100th harmonic of the laser frequency. These harmonics can be used to generate single attosecond pulses or trains of attosecond pulses in the EUV.

8. Acknowledgements

Thanks to Prof. Dr. B. Di Bartolo and all the participants of the school for another wonderful time in Erice. After being in Erice as a student once and as a lecturer twice, I am always looking forward to come again to this beautiful and pieceful place and to its scientifically stimulating atmosphere.

The scientific results (both experimental and theoretical) described in this article have been obtained by the two graduate students Oliver D. Mücke and Thorsten Tritschler in my research group. Technical support by Thorsten Kuhn and Lars Walter is acknowledged. Werner A. Hügel frequently provided valuable help with the spectrometer software in the laboratory. We thank Uwe Morgner (Electrical Engineering, Universität Karlsruhe (TH), Germany and presently Max Planck Institut für Kernphysik, Heidelberg, Germany) and Franz X. Kärtner (Electrical Engineering, Universität Karlsruhe (TH), Germany and later Massachusetts Institute of Technology, Boston, U.S.A.) for intense support concerning the building of the five femtosecond laser system in the initial phase of the experiments and for stimulating discussions, especially on the role of the carrier-envelope offset phase and corresponding measurements.

We thank Wolfgang Stolz (Universität Marburg, Germany) as well as Galina Khitrova and Hyatt Gibbs (Optical Sciences Center, Tucson, U.S.A.) for several high quality GaAs samples specifically made for our experiments, Claus Klingshirn (Universität Karlsruhe (TH), Germany) for providing the ZnO and CdS single crystal platelets and Heinz Kalt (Universität Karlsruhe (TH), Germany) for the thin, epitaxially grown ZnO films used in our early experiments.

The theoretical part of this work has largely benefited from discussions with the group of Hartmut Haug (Universität Frankfurt, Germany) who shared their theoretical results concerning state-dependent damping in carrier-wave Rabi flopping with us prior to publication of their work. Also, Wilfried Schäfer (Forschungszentrum Jülich) helped us concerning band structure and dipole matrix element calculations of semiconductors. During his stay in January/February 2002 in Karlsruhe, we had many stimulating discussions with Renbao Liu (Tsinghua University, China) on intraband effects in general and on the ZnO experiments in their early stage as well as on the formulation of Bloch equations without inversion symmetry in particular. This stay was financed by the Deutsche Forschungsgemeinschaft (DFG) graduate school *Collective Phenomena in Solids* GRK 284. Furthermore, Leonid Keldysh (Lebedev Physics Institute, Moscow) visited us in February 2003 which led to many stimulating discussions as well. This stay also became possible through support by the graduate school GRK 284.

This work is financially supported by the DFG via the Leibniz Award 2000 (project DFG-We 1497/9) and by project DFG-We 1497/11-1. The latter is a cooperation of the group of M.W. and the group of Franz X. Kärtner (Electrical Engineering, Universität Karlsruhe (TH), Germany). A lot of the equipment was funded by the Alfried Krupp von Bohlen und Halbach foundation via the Krupp-Award 1993 and by the Teaching Award of the State of Baden-Württemberg 1998.

Finally, I thank Ursula Bolz, Werner Hügel, Oliver D. Mücke and Thorsten Tritschler for help in preparing special figures for this manuscript and Karin Gottschalk-Wegener for a critical reading of this manuscript and for much more ...

9. References

1. W. Schäfer and M. Wegener, *Semiconductor Optics And Transport Phenomena*, Advanced Texts in Physics, Springer (2002)
2. K. L. Sala, G. A. Kenney-Wallace, and G. E. Hall, IEEE Journal of Quantum Electron. **16**, 990 (1980)
3. E. V. Baklanov, and V. P. Chebotaev, Sov. J. Quantum Electron. **7**, 1252 (1977)
4. S. T. Cundiff, J. Ye, and J. L. Hall, Rev. of Sci. Instrum. **72**, 3749 (2001); S. T. Cundiff and J. Ye, Rev. Mod. Phys. **75**, 325 (2003)
5. S. Karshenboim, Can. J. Phys. **78**, 639 (2000)
6. Th. Udem, S. A. Diddams, K. R. Vogel, C. W. Oates, E. A. Curtis, W. D. Lee, W. M. Itano, R. E. Drullinger, J. C. Bergquist, and L. Hollberg, Phys. Rev. Lett. **86**, 4996 (2001)
7. Th. Damour, F. Piazza, and G. Veneziano, Phys. Rev. Lett. **89**, 081601 (2002)
8. L. B. Madsen, Phys. Rev. A **65**, 053417 (2002)
9. Y. R. Shen, *The Principles of Nonlinear Optics*, John Wiley & Sons (1984)
10. A. Apolonski, A. Poppe, G. Tempea, Ch. Spielmann, Th. Udem, R. Holzwarth, T. W. Hänsch, and F. Krausz, Phys. Rev. Lett. **85**, 740 (2000)
11. D. J. Jones, S. A. Diddams, J. K. Ranka, A. Stentz, R. S. Windeler, J. L. Hall, and S. T. Cundiff, Science **288**, 635 (2000)
12. U. Morgner, R. Ell, G. Metzler, T. R. Schibli, F. X. Kärtner, J. G. Fujimoto, H. A. Haus, and E. P. Ippen, Phys. Rev. Lett. **86**, 5462 (2001)
13. I. I. Rabi, Phys. Rev. **51**, 652 (1937)
14. P. Meystre and M. Sargent III, *Elements of Quantum Optics*, second edition, Springer (1991)
15. L. Bányai, D. B. Tran Thoai, E. Reitsamer, H. Haug, D. Steinbach, M. U. Wehner, M. Wegener, T. Marschner, and W. Stolz, Phys. Rev. Lett. **75**, 2188 (1995)
16. M. U. Wehner, M. H. Ulm, D. S. Chemla, and M. Wegener, Phys. Rev. Lett. **80**, 1992 (1998)
17. W. A. Hügel, M. F. Heinrich, and M. Wegener, Q. T. Vu, L. Bányai, and H. Haug, Phys. Rev. Lett. **83**, 3313 (1999)
18. Q. T. Vu, H. Haug, W. A. Hügel, S. Chatterjee, and M. Wegener, Phys. Rev. Lett. **85**, 3508 (2000)
19. H. Haug, Nature **414**, 261 (2001)
20. R. Huber, F. Tauser, A. Brodschelm, M. Bichler, G. Abstreiter, and A. Leitenstorfer, Nature **414**, 286 (2001)
21. W. A. Hügel, M. Wegener, Q. T. Vu, L. Bányai, H. Haug, F. Tinjod, and H. Mariette, Phys. Rev. B **66**, 153203 (2002)
22. K. Leo, M. Wegener, J. Shah, D. S. Chemla, E. O. Göbel, T. C. Damen, S. Schmitt-Rink, and W. Schäfer, Phys. Rev. Lett. **65**, 1340 (1990)
23. M. Wegener, D. S. Chemla, S. Schmitt-Rink, and W. Schäfer, Phys. Rev. A **42**, 5675 (1990)
24. R. Binder, S. W. Koch, M. Lindberg, N. Peyghambarian, and W. Schäfer, Phys. Rev. Lett. **65**, 899 (1990)
25. S. T. Cundiff, A. Knorr, J. Feldmann, S. W. Koch, E. O. Göbel, and H. Nickel, Phys. Rev. Lett. **73**, 1178 (1994)

26. H. Giessen, A. Knorr, S. Haas, S.W. Koch, S. Linden, J. Kuhl, M. Hetterich, M. Grün, and C. Klingshirn, Phys. Rev. Lett. **81**, 4260 (1998)
27. A. Schülzgen, R. Binder, M.E. Donovan, M. Lindberg, K. Wundke, H.M. Gibbs, G. Khitrova, and N. Peyghambarian, Phys. Rev. Lett. **82**, 2346 (1999)
28. L. Bányai, Q.T. Vu, B. Mieck, and H. Haug, Phys. Rev. Lett. **81**, 882 (1998)
29. C. Ciuti, C. Piermarocchi, V. Savona, P.E. Selbmann, P. Schwendimann, and A. Quattropani, Phys. Rev. Lett. **84**, 1752 (2000)
30. A.P. Jauho and K. Johnsen, Phys. Rev. Lett. **76**, 4576 (1996); K. Johnsen and A.P. Jauho, Phys. Rev. B **57**, 8860 (1998)
31. A.H. Chin, J.M. Bakker, and J. Kono, Phys. Rev. Lett. **85**, 3293 (2000)
32. T. Meier, G. von Plessen, P. Thomas, and S.W. Koch, Phys. Rev. Lett. **73**, 902 (1994)
33. S. Hughes, Phys. Rev. Lett. **81**, 3363 (1998)
34. S. Hughes, Phys. Rev. A **62**, 055401 (2000)
35. R. Bavli and H. Metiu, Phys. Rev. Lett. **69**, 1986 (1992)
36. M. Yu. Ivanov, P.B. Corkum, and P. Dietrich, Laser Phys. **3**, 375 (1993)
37. A. Levinson, M. Segev, G. Almogy, and A. Yariv, Phys. Rev. B **49**, R661 (1994)
38. T. Zuo, S. Chelkowski, and A.D. Bandrauk, Phys. Rev. A **49**, 3943 (1994)
39. O.D. Mücke, T. Tritschler, M. Wegener, U. Morgner, and F.X. Kärtner, Phys. Rev. Lett. **87**, 057401 (2001)
40. V.P. Kalosha and J. Herrmann, Phys. Rev. Lett. **83**, 544 (1999)
41. R.W. Ziolkowski, J.M. Arnold, and D.M. Gogny, Phys. Rev. A **52**, 3082 (1995)
42. B.R. Mollow, Phys. Rev. **188**, 1969 (1969); B.R. Mollow, Phys. Rev. A **2**, 76 (1970); B.R. Mollow, Phys. Rev. A **5**, 2217 (1972)
43. U. Morgner, F.X. Kärtner, S.H. Cho, Y. Chen, H.A. Haus, J.G. Fujimoto, E.P. Ippen, V. Scheuer, G. Angelow, and T. Tschudi, Opt. Lett. **24**, 411 (1999)
44. M.U. Wehner, M.H. Ulm, and M. Wegener, Opt. Lett. **22**, 1455 (1997)
45. The first microscope objective has a focal length of 5.41 mm and a numerical aperture of NA=0.5 (*Coherent 25-0522*), the second one 13.41 mm and NA=0.5 (*Coherent 25-0555*). In our experiments, we lose about a factor of two in average power on the first microscope objective. This is due to the fact that we chose the beam diameter to be larger than the objective aperture in order to get to the minimum spot radius. Thus, all relevant powers have been quoted *in front of the sample* consistently throughout this article.
46. N. Peyghambarian, S.W. Koch, and A. Mysyrowicz, *Introduction to semiconductor optics* (Prentice Hall, Englewood Cliffs, NJ, 1993)
47. The fit formula for the envelope of the electric field spectrum is given by the sum of three Gaussians, i.e. by $|\tilde{E}_{\omega_0}(\omega)| = \sum_{n=1}^{3} E_n \exp(-(\omega-\omega_n)^2/\sigma_n^2)$, with the parameters $E_2/E_1 = 0.72$, $E_3/E_1 = 1.16$ and E_1 being determined by $\tilde{E}_0$; $\hbar\omega_1 = 1.38\,\text{eV}$, $\hbar\omega_2 = 1.68\,\text{eV}$, and $\hbar\omega_3 = 1.82\,\text{eV}$; $\hbar\sigma_1 = 0.10\,\text{eV}$, $\hbar\sigma_2 = 0.19\,\text{eV}$, and $\hbar\sigma_3 = 0.03\,\text{eV}$. The real electric field $E(t)$ of an individual pulse results from the real part of the Fourier transform of $|\tilde{E}_{\omega_0}(\omega)|$. Note that the CEO phase ϕ of $E(t)$ can be modified without explicitely decomposing it into carrier wave $\cos(\omega_0 t + \phi)$ and envelope $\tilde{E}(t)$ – which would not be possible analytically anyway. One can rather replace $|\tilde{E}_{\omega_0}(\omega)| \rightarrow |\tilde{E}_{\omega_0}(\omega)| \exp(-\mathrm{i}\phi)$ and proceed as above.

48. M. Sheik-Bahae, A. A. Said, T.-H. Wei, D. J. Hagan, and E. W. van Stryland, IEEE J. Quantum Electron. **26**, 760 (1990)
49. D. E. Aspnes, S. M. Kelso, R. A. Logan, and R. Bhat, J. Appl. Phys. **60**, 754 (1986)
50. O. D. Mücke, T. Tritschler, M. Wegener, U. Morgner, and F. X. Kärtner, Phys. Rev. Lett. **89**, 127401 (2002)
51. O. D. Mücke, T. Tritschler, M. Wegener, U. Morgner, F. X. Kärtner, G. Khitrova, and H. M. Gibbs, Phys. Rev. Lett., submitted (2003)
52. V. Skrikand, and D. R. Clarke, J. Appl. Phys. **83**, 5447 (1998)
53. O. D. Mücke, T. Tritschler, M. Wegener, U. Morgner, and F. X. Kärtner, Opt. Lett. **27**, 2127 (2002)
54. X. W. Sun and H. S. Kwok, J. of Appl. Phys. **86**, 408 (1999)
55. V. P. Kalosha and J. Herrmann, Phys. Rev. A, **R 62**, 011804 (2000)
56. A. L. Gaeta, Opt. Lett. **27**, 924 (2002)
57. H. Cao, J. Y. Wu, H. C. Ong, J. Y. Dai, and R. P. H. Chang, Appl. Phys. Lett. **73**, 572 (1998)
58. $\chi^{(3)} = 4/3\, c_0 n_0^2 \epsilon_0 n_2$ with $n_0 = 1.99$ and $n_2 = 5 \times 10^{-19}\,\mathrm{m^2/W}$ from: R. Adair, L. L. Chase, and S. A. Payne, Phys. Rev. B **39**, 3337 (1989)
59. T. Tritschler, O. D. Mücke, M. Wegener, U. Morgner, and F. X. Kärtner, Phys. Rev. Lett., in press (2003)
60. K. Postava, H. Sueki, M. Aoyama, T. Yamaguchi, Ch. Ino, Y. Igasaki, and M. Horie, J. Appl. Phys. **87**, 7820 (2000)
61. G. Farkas and C. Tóth, Phys. Lett. A **168**, 447 (1992)
62. S. E. Harris, J. J. Macklin, and T. W. Hänsch, Opt. Commun. **100**, 487 (1993)
63. N. A. Papadogiannis, B. Witzel, C. Kalpouzos, and D. Charalambidis, Phys. Rev. Lett. **83**, 4289 (1999)
64. P. M. Paul, E. S. Toma, P. Breger, G. Mullot, F. Augé, Ph. Balcou, H. G. Muller, and P. Agostini, Science **292**, 1689 (2001)
65. A. de Bohan, P. Antoine, D. B. Milosević, and B. Piraux, Phys. Rev. Lett. **81**, 1837 (1998)
66. G. Tempea, M. Geissler, and T. Brabec, J. Opt. Soc. Am. B **16**, 669 (1999)
67. A. Baltuska, Th. Udem, M. Ulberacker, M. Hentschel, E. Goulielmakis, Ch. Gohle, R. Holzwarth, V. S. Yakovlev, A. Scrinzi, T. W. Hänsch, and F. Krausz, Nature **421**, 611 (2003)
68. M. Drescher, M. Hentschel, R. Kienberger, G. Tempea, C. Spielmann, G. A. Reider, P. B. Corkum, and F. Krausz, Science **291**, 1923 (2001)
69. M. Hentschel, R. Kienberger, Ch. Spielmann, G. A. Reider, N. Milosevic, T. Brabec, P. Corkum, U. Heinzmann, M. Drescher, and F. Krausz, Nature **414**, 509 (2001)
70. J. Itatani, F. Quéré, G. L. Yudin, M. Yu. Ivanov, F. Krausz, and P. B. Corkum, Phys. Rev. Lett. **88**, 173903 (2002) and the references cited therein
71. T. Brabec and F. Krausz, Rev. Mod. Phys. **72**, 545 (2000)
72. T. Tritschler O. D. Mücke and M. Wegener, Phys. Rev. A, submitted (2003)
73. Z. Chang, A. Rundquist, H. Wang, M. M. Murnane, and H. C. Kapteyn, Phys. Rev. Lett. **79**, 2967 (1997)
74. A. Paul, R. A. Bartels, R. Tobey, H. Green, S. Weiman, I. P. Christov, M. M. Murnane, H. C. Kapteyn, and S. Backus, Nature **421**, 51 (2003)
75. L. S. Brown and T. W. B. Kibble, Phys. Rev. A **133**, 705 (1964)

76. J. E. Gunn and J. P. Ostriker, The Astrophysical Journal **165**, 523 (1971)
77. E. Esarey, S. K. Ride and P. Sprangle, Phys. Rev. E **48**, 3003 (1991)
78. S. Chen, A. Maksimchuk, and D. Umstadter, Nature **396**, 653 (1998)
79. F. He, Y. Lau, D. P. Umstadter, and T. Strickler, Physics of Plasmas **9**, 4325 (2002)
80. K. W. D. Ledingham, I. Spencer, T. McCanny, R. P. Singhal, M. I. K. Santala, E. Clark, I. Watts, F. N. Beng, M. Zepf, K. Krushelnick, M. Tatarakis, A. E. Dangor, P. A. Norreys, R. Allott, D. Neelly, R. J. Clark, A. C. Machacek, J. S. Wark, A. J. Cresswell, D. C. W. Sanderson, and J. Magill, Phys. Rev. Lett. **84**, 899 (2000)
81. T. E. Cowan, A. W. Hunt, T. W. Phillips, S. C. Wilks, M. D. Perry, C. Brown, W. Fountain, S. Hatchett, J. Johnson, M. H. Key, T. Parnell, D. M. Pennington, R. A. Snavely, and Y. Takahashi, Phys. Rev. Lett. **84**, 903 (2000)
82. K. Boyer, T. S. Luk, and C. K. Rhodes, Phys. Rev. Lett. **60**, 557 (1988)
83. T. Tajima and G. Mourou, Phys. Rev. Special Topics **5**, 031301 (2002)
84. G. Mourou and D. Umstadter, Scientific American, May 2002, 63 (2002)
85. H. K. Avetissian, A. K. Avetissian, G. F. Mkrtchian, and Kh. V. Sedrakian, Phys. Rev. E **66**, 016502 (2002)
86. S. W. Hawking, Nature **248**, 30 (1974)
87. W. G. Unruh, Phys. Rev. D **14**, 870 (1976)
88. E. Yablonovitch, Phys. Rev. Lett. **62**, 1742 (1989)
89. P. Chen and T. Tajima, Phys. Rev. Lett. **83**, 256 (1999)
90. K. L. Shlager and J. B. Schneider, IEEE Antennas and Propagation Magazine **37**, 39 (1995)
91. K. S. Yee, IEEE Trans. Antennas Propagat. **14**, 302 (1966)
92. J. P. Berenger, Journal of Comp. Phys. **114**, 185 (1994)
93. E. L. Lindman, Journal of Comp. Phys. **18**, 66 (1975)

10. Solutions of Exercises

Solution of Exercise 1.1:

In a "dark" room held at room temperature, one still has the unavoidable black-body radiation. The corresponding intensity I of electromagnetic radiation is given by Planck's law

$$I = \int_0^\infty \frac{h}{c_0^2} \frac{f^3}{e^{hf/k_B T} - 1} \, df = \frac{h}{c_0^2} \left(\frac{k_B T}{h} \right)^4 \frac{\pi^4}{15} .$$

For $T = 300\,\text{K}$ one gets $I = 0.7 \times 10^{-2}\,\text{W/cm}^2$. Only a very (!) small fraction of that intensity, however, is visible light (photon energies hf in the interval $[1.5\,\text{eV}, 3.0\,\text{eV}]$). To roughly estimate that fraction, we remember that most of the light stems from the long-wavelength end of the visible spectrum, let us say from the interval $[1.5\,\text{eV}, 1.6\,\text{eV}]$. The center frequency of that interval is $f = 3.7 \times 10^{14}\,\text{Hz}$, its width is $df = 2.4 \times 10^{13}\,\text{Hz}$. This leads to an estimated intensity of visible light at $T = 300\,\text{K}$ of $\underline{I = 10^{-23}\,\text{W/cm}^2}$. This value corresponds to a flux of about one photon through your finger tip in six hours time – it does not get any darker at room temperature.

Note that it takes 22 orders of magnitude to get from the intensity of a "dark" room to that of the sun on the earth. We will go another 29 orders of magnitude upwards in intensity at the end of this article.

Solution of Exercise 2.1.:

Under these conditions, the intensity $I \propto \sqrt{\epsilon}\,\tilde{E}_0^2$ (see Eq. (16)) is the same in air and in the dielectric. In air $\epsilon = 1$, in the dielectric $\epsilon = 10.9$. $\Rightarrow \tilde{E}_0$ in the dielectric is given by $4 \times 10^9\,\text{V/m}/\sqrt{\sqrt{10.9}} = 2.2 \times 10^9\,\text{V/m}$.

Solution of Exercise 2.2.:

There is no unique answer to this unprecise question ... The most stringent way to interpret the question is that all frequency components of the pulse must lie in the visible, i.e., roughly in the photon energy interval $[1.5\,\text{eV}, 3.0\,\text{eV}]$ – one octave. Let us assume that all frequency components have the same amplitude and the same phase, which leads to the shortest pulse. This "box-spectrum" with a FWHM of $\hbar\delta\omega = 1.5\,\text{eV}$, $\Leftrightarrow \delta\omega = 2.28\,/\text{fs}$, leads to $\text{sinc}^2(t)$–pulses of $\delta t = 2\pi \times 0.8859/\delta\omega = 2.44\,\text{fs}$ in duration (for the duration-bandwidth product $\delta\omega\,\delta t$ see footnote in section 2.3.). With a carrier frequency of $\hbar\omega_0 = 2.25\,\text{eV}$, $\Leftrightarrow 2\pi/\omega_0 = 1.84\,\text{fs}$ period of light, these pulses contain 1.3 optical cycles. The electric field of these pulses with $\delta\omega/\omega_0 = 2/3$ is very nearly similar to that depicted in Figs. 3(b) and (c), where $\delta\omega/\omega_0 = 0.60$ has been chosen.

By the way: The number of locked modes is given by the width of the spectrum devided by the distance between adjacent modes, i.e., by the ratio $\delta\omega/\Delta\omega$. With $\Delta\omega = 2\pi \times 100\,\text{MHz}$ from example 2.2. and $\delta\omega = 2.28\,/\text{fs}$ this leads to 3.6×10^6 locked modes.

Solution of Exercise 3.1.:

The relaxation for p_{vc} translates into a relaxation of u and v via Eq. (58) according to $\dot{u} = -u/T_2$ and $\dot{v} = -v/T_2$. Thus, the complete optical Bloch equations including dephasing read

$$\begin{aligned}
\dot{u} &= +\Omega\, v - \frac{u}{T_2} \\
\dot{v} &= -\Omega\, u - 2\,\Omega_{\text{R}} w - \frac{v}{T_2} \\
\dot{w} &= +2\,\Omega_{\text{R}} v\,.
\end{aligned}$$

In the stationary limit we have $\dot{w} = 0$. From the third line it follows that $v = 0$, hence $\dot{v} = 0$. Inserting $v = 0$ into the first line we furthermore get $u = 0$. From the second line with $\dot{v} = u = v = 0$ we finally get $w = f_{\text{c}} - f_{\text{v}} = 0$, i.e., with $f_{\text{c}} + f_{\text{v}} = 1$, the occupation of the excited state and of the ground state are both 50%, the system is transparent. Note that the only point at which the strong dephasing assumption has entered is the existence of a stationary limit. According to the equations of motion for u and v, this limit is reached on a timescale comparable to T_2.

If the inversion additionally experiences relaxation towards the ground state, i.e., towards $w = -1$, according to $\dot{w} = -(w+1)/T_1$, with the occupation relaxation time or longitudinal relaxation time T_1, the steady-state inversion is generally smaller then zero, i.e., $-1 \leq w \leq 0$. In the limit $T_1 \to 0$, we get $w = -1$ ($\Leftrightarrow f_{\text{c}} = 0$ and $f_{\text{v}} = 1$) from the third line of the optical Bloch equations, i.e., from $\dot{w} = 0 = +2\,\Omega_{\text{R}} v \;-\; (w+1)/T_1 \;\approx\; -\,(w+1)/T_1$.

Solution of Exercise 3.2.:

a) The dipole moment of one oscillator is $(d_{\text{cv}}\, u)$ (see Eqs. (48) and (58)) or $(e\, x)$. The peak value of $u = 1$ is reached for a Rabi oscillation with $\tilde{E}_0 = 4 \times 10^9\,\text{V/m}$ in GaAs (see Fig. 10), hence, for e.g. $d_{\text{cv}} = 0.5\,\text{e}\,\text{nm}$ (GaAs) we obtain $\underline{x_0 = 0.5\,\text{nm} \approx a}$ $\propto (\tilde{E}_0)^0$. (Actually, for even much larger fields with $\Omega_{\text{R}} \gg \Omega$, u approaches zero, as the

Bloch vector rotates in the vw-plane only.)

b) According to Eq. (65), we have

$$x_0 = \frac{e\tilde{E}_0}{m_e \omega_0^2}$$

$\propto (\tilde{E}_0)^{+1}$. For ZnO parameters ($m_e = 0.24 \times m_0$, $\hbar\omega_0 = 1.5\,\text{eV}$) we obtain $\underline{x_0 = 0.56\,\text{nm} \approx a}$.

c) Starting from Eq. (63) within the effective mass approximation and with the initial conditions $x(0) = \dot{x}(0) = 0$, the electron displacement is $x(t) = \frac{1}{2}\frac{e}{m_e}\tilde{E}_0\,t^2$. The peak displacement $x = x_0$ is reached at half the Bloch oscillation period, i.e., at $t = \pi/\Omega_{\text{Bloch}}$, where the velocity $\dot{x}$ changes sign. Thus, we have

$$x_0 = \frac{1}{2}\frac{e}{m_e}\tilde{E}_0\left(\frac{\pi}{\Omega_{\text{Bloch}}}\right)^2 = \frac{\pi^2\hbar}{2m_e a}\frac{1}{\Omega_{\text{Bloch}}}$$

$\propto (\tilde{E}_0)^{-1}$. For ZnO parameters ($m_e = 0.24 \times m_0$, $a = 0.5\,\text{nm}$) this leads to $\underline{x_0 = 1.5\,\text{nm} \approx 3\,a}$.

To conclude the result of exercise 3.2.: Even for laser intensities around $I = 2 \times 10^{12}\,\text{W/cm}^2$, $\Leftrightarrow$ $\tilde{E}_0 = 4 \times 10^9\,\text{V/m}$ (see example 2.1.), the classical crystal electron displacements stay on the order of one lattice constant a.

Solution of Exercise 5.1.:

In order to obtain a non-vanishing *"THG in disguise of SHG"* contribution, the low-frequency end of the THG spectrum needs to overlap with spectrometer frequency $2\omega_0$. For sinc^2-pulses and within the $\chi^{(3)}$-limit, the width of the THG spectrum with carrier frequency $3\omega_0$ is three times that of the fundamental spectrum, i.e., it is given by $3\,\delta\omega$. This leads to the condition $2\omega_0 = 3\omega_0 - 3\,\delta\omega/2 \Leftrightarrow \delta\omega/\omega_0 = 2/3$, equivalent to a spectral width of one octave (see solution of exercise 2.2.). With the duration-bandwidth product of $\delta\omega\,\delta t = 2\pi \times 0.8859$ for an unchirped sinc^2-pulse (see footnote in section 2.3. or solution of exercise 2.2.) and with $\hbar\omega_0 = 1.5\,\text{eV}$, this corresponds to a maximum pulse width of $t_{\text{FWHM}} = \delta t = 3.6\,\text{fs}$. Note, however, that the *"THG in disguise of SHG"* signal would still be arbitrarily small at this point. Thus, the 5 fs pulses discussed in section 5.2. can only lead to significant *"THG in disguise of SHG"* deep inside the non-perturbative regime.

11. Important Symbols and Constants

a, lattice constant of a crystal
a_{e}^{0}, peak electron acceleration
α, absorption coefficient
$\vec{B}(\vec{r},t)$, $\vec{B}$–field or magnetic field
$\tilde{B}_0$, peak of magnetic field envelope
c, medium speed of light
c_0, vacuum speed of light with $c_0 = 2.998 \times 10^8\,\mathrm{m/s}$
$\chi = \chi^{(1)}$, linear optical susceptibility
$\chi^{(N\neq 1)}$, nonlinear optical susceptibilities
d_{cv}, dipole matrix element for valence band to conduction band transitions
$\vec{D}(\vec{r},t)$, $\vec{D}$–field
$\delta\omega\,\delta t$, duration-bandwidth product with, e.g., $\delta\omega\,\delta t = 2\pi \times 0.8859$ for $\mathrm{sinc}^2(t)$-pulses
$\Delta\omega$, mode spacing in a mode-locked laser oscillator
$\Delta\phi$, pulse-to-pulse phase slip of the light field
e, elementary charge with $\mathrm{e} = 1.6021 \times 10^{-19}$ As
$\vec{E}(\vec{r},t)$, electric field
$\tilde{E}(t)$, electric field envelope
$\tilde{E}_0$, peak of electric field envelope
E_{g}, semiconductor band gap energy
$\langle E_{\mathrm{kin}}\rangle$, ponderomotive energy
ϵ, material dielectric function
ϵ_{b}, background dielectric constant
ϵ_0, vacuum dielectric constant with $\epsilon_0 = 8.8542 \times 10^{-12}\,\mathrm{AsV^{-1}m^{-1}}$
$\mathcal{E}$, normalized electric field strength (relativistic regime)
f_{e}, f_{h}, electron and hole occupation numbers, respectively
$f_\phi = 1/t_\phi$, carrier-envelope offset (CEO) frequency
$f_{\mathrm{r}} = 1/t_{\mathrm{r}}$, repetition frequency
$\vec{F}$, force vector
$\hbar$, Planck's constant h divided by 2π, with $\hbar = 0.658\,\mathrm{eV\,fs}$ or $\hbar = 1.054 \times 10^{-34}$ Js
$\vec{H}(\vec{r},t)$, $\vec{H}$–field
$\mathcal{H}$, Hamiltonian
I, light intensity
I_0, reference intensity
$\vec{j}$, electrical current density
$\vec{k}$, electron wave vector
k_{B}, Boltzmann's constant with $k_{\mathrm{B}} = 1.3804 \times 10^{-23}\,\mathrm{J/K}$
$\vec{K}$, wave vector of light
l, sample thickness
L, length of the laser resonator
λ, wavelength of light
m_0, free electron (rest) mass with $m_0 = 9.1091 \times 10^{-31}$ kg
m_{e}, effective electron mass or relativistic electron mass, respectively
m_{h}, effective hole mass

$\vec{M}(\vec{r},t)$, magnetization
μ_0, vacuum permeability with $\mu_0 = 4\pi \times 10^{-7}\,\mathrm{VsA^{-1}m^{-1}}$
n, refractive index
n_2, nonlinear refractive index
n_{eh}, electron hole density
$\mathcal{O}$, arbitrary quantum mechanical observable
ω, spectrometer frequency
ω_0, carrier frequency of laser pulse
ω_{c}, cyclotron frequency
Ω, optical (interband) transition frequency
Ω_{Bloch}, Bloch frequency
Ω_{R}, Rabi frequency
$\tilde{\Omega}_{\mathrm{R}}$, envelope Rabi frequency
p, electron momentum
p_{vc}, transition amplitude for valence band to conduction band transitions
$\vec{P}(\vec{r},t)$, optical polarization
ϕ, carrier-envelope offset (CEO) phase
ρ, electrical charge density
$\vec{S}(\vec{r},t)$, Poynting-vector
S_{RF}, spectral radio frequency power density
t, time
$t_{\mathrm{FWHM}} = \delta t$, full width at half maximum of the temporal intensity profile of a laser pulse
τ, time delay of a Michelson interferometer
T, temperature
T_1, longitudinal relaxation time
T_2, transverse relaxation time
Θ, pulse area
$\tilde{\Theta}$, envelope pulse area
$(u, v, w)^{\mathrm{T}}$, Bloch vector
$U(t)$, photomultiplier voltage
$\vec{v}$, velocity vector
w, inversion
x, classical displacement
x_0, peak classical displacement

12. Appendices

12.1. APPENDIX: 1D FINITE-DIFFERENCE TIME-DOMAIN ALGORITHM

As an experimentalist, one is often deeply impressed by numerical solutions of the Maxwell equations. It is the aim of this brief appendix to show that – at least in one dimension – exact numerical solutions of the nonlinear Maxwell equations are actually rather simple. A more complete overview on numerical solutions of the Maxwell equations based on finite-difference time-domain (FDTD) algorithms can, e.g., be found in Ref. [90].

For a plane electromagnetic wave propagating along the z-direction with the electric field $\vec{E}$ polarized along x, the magnetic field $\vec{B}$ is directed along y. Under these conditions, Maxwell's equations (4) in S.I. units reduce to the two coupled first-order partial differential equations

$$\frac{\partial E(z,t)}{\partial z} = -\frac{\partial B(z,t)}{\partial t} \tag{112}$$

$$\frac{\partial H(z,t)}{\partial z} = -\frac{\partial D(z,t)}{\partial t}. \tag{113}$$

In a non-magnetic medium we furthermore have

$$B(z,t) = \mu_0 \, H(z,t) \tag{114}$$

and

$$D(z,t) = \varepsilon_0 \, E(z,t) + P(z,t)\,, \tag{115}$$

with the medium polarization $P(z,t)$.

In order to implement this *initial value problem* on a computer, we discretize space and time according to

$$E_{M,N} = E(M\,\Delta z, N\,\Delta t)\,, \tag{116}$$

$$H_{M,N} = H(M\,\Delta z, N\,\Delta t)\,, \tag{117}$$

and P correspondingly. M and N are integers, the step sizes Δz and Δt have to be sufficiently small. Their actual choice will be discussed below. The central idea of Ref. [91] is to displace the positions of the electric and magnetic field by $\Delta z/2$ in space and $\Delta t/2$ in time, respectively (see Fig. 45). This leads to a simple iterative scheme: Replacing the partial derivatives in Eq. (112) by finite fractions immediately delivers

$$H_{M+\frac{1}{2},N+\frac{1}{2}} = H_{M+\frac{1}{2},N-\frac{1}{2}} - \frac{\Delta t}{\mu_0 \Delta z}\left(E_{M+1,N} - E_{M,N}\right). \tag{118}$$

At some point in time, the initial value of the electric as well as magnetic field must be known at all positions z. If the magnetic field on the RHS of Eq. (118) is, e.g., known at time $t = (N - \frac{1}{2})\Delta t$ and the electric field is known at time $t = N\Delta t$, the magnetic field at time $t = (N + \frac{1}{2})\Delta t$ at all coordinates $z = (M + \frac{1}{2})\Delta z$ can directly be calculated from Eq. (118). Accordingly, Eq. (113) leads to

$$E_{M,N+1} = E_{M,N} - \frac{\Delta t}{\epsilon_0 \Delta z}\left(H_{M+\frac{1}{2},N+\frac{1}{2}} - H_{M-\frac{1}{2},N+\frac{1}{2}}\right) - \frac{1}{\epsilon_0}\left(P_{M,N+1} - P_{M,N}\right). \tag{119}$$

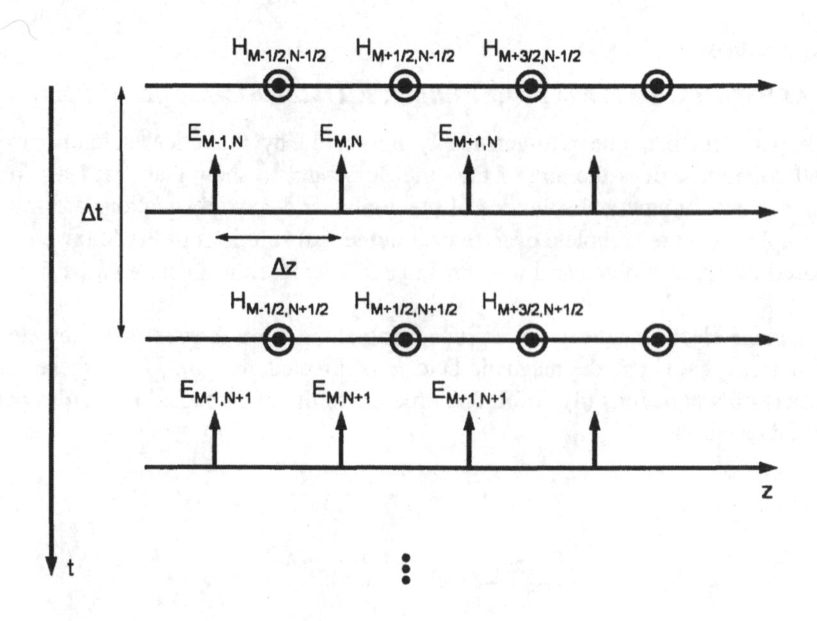

Figure 45: Illustration of the one-dimensional finite-difference time-domain (FDTD) discretization and iteration scheme. The spatial step size is Δz, the temporal step size Δt, M and N are integers. The electric field $\vec{E}$ is parallel to the x-direction, the magnetic field $\vec{H}$ parallel to y, and the wave is propagating along z.

On the RHS of Eq. (119), the magnetic field at time $t = (N + \frac{1}{2})\Delta t$ and the polarization and electric field at time $t = N\Delta t$ are known at this point of the calculation. $E_{M,N+1}$ and $P_{M,N+1}$ have to be calculated for all coordinates $z = M\Delta z$. In vacuum, $P = 0$ holds and $E_{M,N+1}$ can again be directly calculated. The same holds for a linear optical material with $P = \epsilon_0 \chi E$, in which case ϵ_0 has to be replaced by $\epsilon_0 \to \epsilon\,\epsilon_0$ and P disappears. In nonlinear optics, P is generally a nonlinear functional of E. If, for example, we have a $\chi^{(2)}$-medium with $P = \epsilon_0 \chi^{(2)} E^2$, we obtain $P_{M,N+1} = \epsilon_0 \chi^{(2)} E^2_{M,N+1}$, and Eq. (119) is a quadratic equation in $E_{M,N+1}$ allowing to determine the electric field at time $t = (N+1)\Delta t$ for all coordinates $z = M\Delta z$. This then allows to calculate the magnetic field at time $t = (N + \frac{3}{2})\Delta t$ via Eq. (118), etc., which completes the iterative scheme. The knowledge of the electric field $\vec{E}$ and the magnetic field $\vec{H}$ delivers the Poynting vector $|\vec{S}| = |\vec{E} \times \vec{H}| = EH$, hence also the light intensity I (see section 2.2.).

Typically, the step size Δz should be chosen smaller than one tenths of the smallest medium wavelength. Δt should be smaller than $\Delta z / v^{\max}_{\text{phase}}$, where $v^{\max}_{\text{phase}}$ is the largest (anticipated) phase velocity of the problem.

In practice, one has to be careful not to obtain artifacts from the spatial boundaries of the simulated region. The artificial reflections from these boundaries can, in principle, be delayed to very long times by making the simulated region sufficiently large. This "brute force" approach can, however, be rather CPU-time consuming. A faster and more elegant approach is to suppress these reflections by so-called absorbing boundary conditions [92] or by a projection operator technique [93]. The latter has been used for the calculations presented in this article. It also allows for injection of the optical pulses from one side.

12.2. APPENDIX: "THE SQUARE-WAVE APPROXIMATION"

Let us continue our discussion on high harmonics from two-level systems of section 6.1. and derive simple analytical expressions which help us to obtain a better understanding of the behavior of the two-level system in the regime of extreme nonlinear optics, especially of the peaks at positions ω in the optical spectrum corresponding to peaks at the spectral position of even harmonics of the carrier frequency of light ω_0.

The Bloch equations (59) describe rotations of the Bloch vector on the Bloch sphere. Within the regime of extreme nonlinear optics, the behavior becomes "enriched" by the fact that one of the rotation frequencies, namely $2\Omega_R(t)$, itself oscillates with the carrier-frequency of light and periodically changes sign. This oscillation is sinusoidal, yet, one might ask whether it is really so important that it is sinusoidal. Having in mind what we have said about the "frozen-wave approximation" in section 6.1., it is straightforward to extend that result to piecewise constant electric fields $E(t)$ or Rabi frequencies $\Omega_R(t)$, respectively [72]. This leads us to investigating the "square-wave approximation" in which we approximate the Rabi frequency for constant envelope via

$$\Omega_R(t) = \Omega_R \cos(\omega_0 t + \phi) \rightarrow \frac{2}{\pi}\, \Omega_R\, \mathrm{sign}(\cos(\omega_0 t + \phi)) \tag{120}$$

where the signum function is defined as $\mathrm{sign}(x) = +1$ for $x > 0$, $\mathrm{sign}(x) = -1$ for $x < 0$ and $\mathrm{sign}(x) = 0$ for $x = 0$. The prefactor $2/\pi$ ensures that the average Rabi frequency within half an optical cycle is the same for the "square-wave approximation" and the exact problem. In that half of the optical cycle where the Rabi frequency is positive (negative), the Bloch vector rotates via the Matrix $\mathcal{M}_+$ ($\mathcal{M}_-$), where $\mathcal{M}_\pm$ results from $\mathcal{M}$ by replacing $\Omega_R \rightarrow \pm\,(2/\pi)\,\Omega_R$ in eqs. (91) and (92). For more than half an optical cycle, the dynamics of the Bloch vector is described by

$$\begin{pmatrix} u(t) \\ v(t) \\ w(t) \end{pmatrix} = \mathcal{M}_{\mathrm{tot}}(t) \begin{pmatrix} u(0) \\ v(0) \\ w(0) \end{pmatrix}, \tag{121}$$

where the total matrix $\mathcal{M}_{\mathrm{tot}}$ is a product of simple analytical (3×3) rotation matrices: For times t after the optical pulse with integer number of cycles of light N, where $\Omega_R(t) = 0$, we have

$$\mathcal{M}_{\mathrm{tot}}(t) = \mathcal{M}_0\left(t - N\frac{2\pi}{\omega_0}\right)\left(\mathcal{M}_-\left(\frac{\pi}{\omega_0}\right)\mathcal{M}_+\left(\frac{\pi}{\omega_0}\right)\right)^N, \tag{122}$$

where $\mathcal{M}_0$ results from $\mathcal{M}$ by replacing $\Omega_R \rightarrow 0$ in eqs. (91) and (92). $\mathcal{M}_0$ describes nothing but a rotation in the uv-plane with the transition frequency Ω and can be simplified to

$$\mathcal{M}_0(t) = \begin{pmatrix} \cos(\Omega t) & +\sin(\Omega t) & 0 \\ -\sin(\Omega t) & \cos(\Omega t) & 0 \\ 0 & 0 & 1 \end{pmatrix}. \tag{123}$$

Within the optical pulse, we obtain for times t with $\Omega_R(t) > 0$

$$\mathcal{M}_{tot}(t) = \mathcal{M}_+\left(t - N_t\,\frac{2\pi}{\omega_0}\right)\left(\mathcal{M}_-\left(\frac{\pi}{\omega_0}\right)\mathcal{M}_+\left(\frac{\pi}{\omega_0}\right)\right)^{N_t}, \tag{124}$$

and for times t with $\Omega_R(t) < 0$

$$\mathcal{M}_{tot}(t) = \mathcal{M}_-\left(t - \left[N_t + \frac{1}{2}\right]\frac{2\pi}{\omega_0}\right)\mathcal{M}_+\left(\frac{\pi}{\omega_0}\right)\left(\mathcal{M}_-\left(\frac{\pi}{\omega_0}\right)\mathcal{M}_+\left(\frac{\pi}{\omega_0}\right)\right)^{N_t}. \tag{125}$$

Here we have introduced the integer number of cycles N_t completed up to time t, which is given by

$$N_t = \mathrm{int}\left(\frac{\omega_0 t}{2\pi}\right). \tag{126}$$

The value of the integer function $\mathrm{int}(x)$ is given by the largest integer $\leq x$.

We first test the "square-wave approximation" by depicting its solutions in Fig. 46. Parameters are identical to those of the exact numerical calculations in Fig. 39, which allows for a direct comparison. The overall qualitative agreement is amazing, especially for the (a) parts. There, $\Omega/\omega_0 = 1$ (resonant excitation), which is nothing but the generalization of Rabi flopping and Mollow triplets. For instance, the periodically occuring constrictions of the repelling Mollow sidebands at even integers ω/ω_0 versus Ω_R/ω_0 with period $\pi/2$ (see discussion in section 6.1.) are very nicely reproduced. For off-resonant excitation ($\Omega/\omega_0 = 5$) in (b), the square-wave approximation is less convincing. This aspect can be understood intuitively. For resonant excitation ($\Omega/\omega_0 = 1$), the transition frequency resonantly enhances those frequency components of the square-wave with frequency ω_0. Thus, the artificial higher harmonics of the square-wave at frequencies $3\omega_0$, $5\omega_0$, ... are relatively suppressed. Clearly, the "square-wave approximation" does not properly recover the limit of linear optics, in the sense that $u(t)$ is not sinusoidal in that limit (as it should be), equivalent to higher harmonics of the carrier frequency ω_0 in the Fourier domain. Thus, the lower RHS of Figs. 46(a) and (b) (which is dark in Figs. 39(a) and (b)) is an obvious artifact of the "square-wave approximation". This artifact is unimportant because we are rather interested in the regime of extreme nonlinear optics.

The simplest cases of commensurability of the frequencies ω_0, Ω, and Ω_R within the "square-wave approximation" are given by

$$\Omega_{eff}\,\frac{\pi}{\omega_0} = M\,2\pi\,, \tag{127}$$

with integer M, for which we have

$$\mathcal{M}_+\left(\frac{\pi}{\omega_0}\right) = \mathcal{M}_-\left(\frac{\pi}{\omega_0}\right) = \begin{pmatrix} 1 & 0 & 0 \\ 0 & 1 & 0 \\ 0 & 0 & 1 \end{pmatrix}. \tag{128}$$

Under these conditions, an integer number of Rabi flops is completed after half an optical cycle π/ω_0. Inserting

$$\Omega_{eff} = \sqrt{4\left(\frac{2}{\pi}\right)^2 \Omega_R^2 + \Omega^2} \tag{129}$$

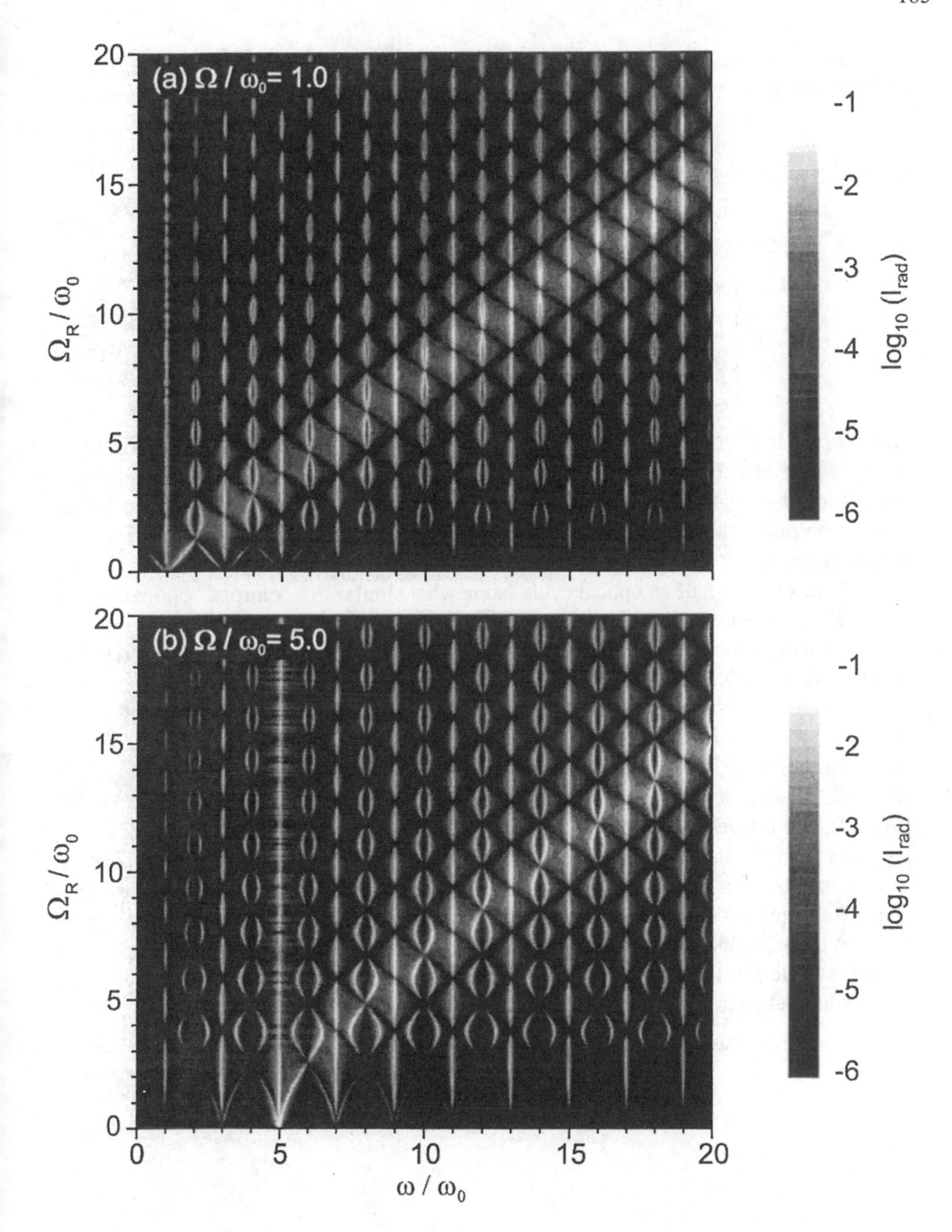

Figure 46: As Fig. 39, but based on the analytical solution of the two-level system Bloch equations within the "square-wave approximation". (a) $\Omega/\omega_0 = 1$, (b) $\Omega/\omega_0 = 5$. Taken from Ref. [72].

into Eq. (127), we get that **commensurability occurs for specific Rabi frequencies** given by the relation

$$\boxed{\frac{\Omega_{\mathrm{R}}}{\omega_0} = \frac{\pi}{2}\sqrt{M^2 - \frac{1}{4}\left(\frac{\Omega}{\omega_0}\right)^2}}\,, \tag{130}$$

with $M = 1, 2, 3, \ldots$ For these Rabi frequencies, **peaks at even integers**

$$\boxed{\frac{\omega}{\omega_0} = \frac{\Omega_{\mathrm{eff}}}{\omega_0} = \sqrt{\frac{16}{\pi^2}\left(\frac{\Omega_{\mathrm{R}}}{\omega_0}\right)^2 + \left(\frac{\Omega}{\omega_0}\right)^2} = 2\,M} \tag{131}$$

are observed in the optical spectrum – apart from the less interesting peaks at odd integers ω/ω_0, which also occur in traditional nonlinear optics. These peaks at even integers ω/ω_0 form the bright band in Fig. 46, whereas the other even integers ω/ω_0 are absent in the spectrum. This bright band also occurs in the exact numerical solutions (Fig. 39). There, in contrast to the "square-wave approximation", the instantaneous Rabi frequency $\Omega_{\mathrm{R}}(t)$ varies within half an optical cycle (somewhat similar to a "chirped" optical pulse), which also introduces peaks at other even integers ω/ω_0. *This altogether shows that the constrictions formed by the crossing Mollow triplets in Fig. 39 can be interpreted as points of commensurability of the carrier frequency of light ω_0, the transition frequency Ω, and the peak Rabi frequency Ω_{R}. Here an integer number of Rabi flops is completed after half an optical cycle and, thus, peaks at even integers ω/ω_0 occur in the optical spectrum.* For, e.g., $M = 1$ and $\Omega/\omega_0 = 1$ in Eq. (130), we get $\Omega_{\mathrm{R}}/\omega_0 = \sqrt{3}\,\pi/4 \approx 1.36$ (Fig. 46), which roughly agrees with $\Omega_{\mathrm{R}}/\omega_0 \approx 1$ in the exact numerical calculations (Fig. 39). For integers $M \gg \Omega/\omega_0$ we get $\Omega_{\mathrm{R}}/\omega_0 = M\,\pi/2$. This period of $\pi/2$ is also precisely found in the exact numerical calculations (Fig. 39). For large Rabi frequencies, commensurability is easily achieved and these "even harmonics" become the rule rather than the exception - despite the fact that the two-level system has inversion symmetry. In between these points of commensurability, it takes some optical cycles to again approach the initial state. In the Fourier domain, this obviously corresponds to nearby sidebands around those even integers ω/ω_0 (see Figs. 39 and 46).

5. CAROTENOID EXCITED STATES-PHOTOPHYSICS, ULTRAFAST DYNAMICS AND PHOTOSYNTHETIC FUNCTIONS

Tomáš Polívka and Villy Sundström

Department of Chemical Physics, Lund University,
Box 124, SE-221 00 Lund, Sweden

1. Introduction

Carotenoids are, along with chlorophylls, the most abundant pigments found in nature. They are present in most organisms including humans, but can be synthesized only by plants and microorganisms. While they are perhaps best known for their bright colors, they have well-documented multiple functions in nature: they serve as light-harvesting pigments in almost all photosynthetic organisms covering a region of the visible spectrum not accessible by (bacterio)chlorophylls ((B)Chl) and they protect against excessive light by quenching both singlet and triplet states of (B)Chls.[1-5] Outside photosynthesis, they are known as efficient quenchers of dangerous singlet oxygen and various reactive radicals by intercepting the chain of oxidative reactions.[6] There is accumulating evidence that this antioxidative function is a key mechanism of protection against various diseases including cancer, atherosclerosis and macular degeneration in humans.[7,8] Yet, knowledge of the detailed molecular mechanism of such actions is so far very limited.

Most of the experimental and theoretical effort aiming for deeper knowledge of carotenoid excited state dynamics performed in recent years was driven by ability of carotenoids to transfer energy of sunlight absorbed by their singlet excited states to (B)Chls, thereby actively participating in the light-harvesting process. The first study providing a clear evidence of the light-harvesting function of carotenoids was performed more than 60 years ago.[9] It represented a milestone that promoted further interest in studies of carotenoid excited states. Due to the absence of spectroscopic methods allowing to follow excited state dynamics of carotenoids, the time period until the late eighties was characterized by studies relying mostly on measurements of fluorescence excitation spectra. Experiments carried out during this period, which are reviewed for example by Govindjee,[10] established the involvement of carotenoid excited states in energy transfer between carotenoids and (B)Chls, and demonstrated that quantum yields of carotenoid-(B)Chl transfer are not far from unity in some cases. Along with the efforts to understand details behind the light-harvesting function of carotenoids, a number of studies of carotenoids, and their close relatives polyenes, in solution was performed. In 1972, an important breakthrough in the field of carotenoid photophysics was reported; it was demonstrated that the absorbing state of longer polyenes is not the lowest excited state, but that an additional, 'dark' state is located between the ground state and the absorbing state.[11,12] This discovery radically affected further research on carotenoid photophysics, as a major part of the experimental and theoretical studies was directed towards localization of this dark state and elucidation of its role in light harvesting and photoprotection. Further theoretical investigations of polyenes showed that other dark states occur in the vicinity of the absorbing state,[13] and these additional dark states were only recently tackled by experimental methods.[14,15] The studies on

B. Di Bartolo and O. Forte (eds.), Frontiers of Optical Spectroscopy, 187-219.

polyenes were vitally important for understanding of carotenoid photophysics and the reader is referred to reviews by Orlandi[16] or Hudson[17] for a more detailed account.
The availability of spectroscopic methods enabling to explore dynamics of excited states on (sub)picosecond time scales became in the late eighties another turning point in carotenoid photophysics. Since the pioneering experiment performed by Wasielewski and Kispert who were the first to measure the S_1 lifetime of β-carotene,[18] a large number of successive experiments, often supported by theoretical investigations, attempted to disentangle the complicated picture of carotenoid excited state dynamics in both solution and various light-harvesting complexes[2,3,19]. In this chapter we will focus on experimental approaches to locate the dark S_1 state of carotenoids and to reveal its role in light-harvesting processes. We will start by discussing properties of carotenoid excited states in solution, as the knowledge of relaxation pathways in isolated carotenoids is a necessary prerequisite to understand carotenoid dynamics in complex systems. Then we deal with carotenoids in light-harvesting systems with the aim of understanding mechanisms and pathways of carotenoid-B(Chl) energy transfer.

2. Excited state structure of carotenoid molecules

The diversity of carotenoid functions is unmatched by any other class of natural pigments. The functional variety is directly related to their unique spectroscopic properties resulting from the structure of the carotenoid molecule (Figure 1). The central pattern repeated in all carotenoids is a backbone consisting of alternating single and double carbon bonds that form a conjugated π-electron system responsible for most of the spectroscopic properties of carotenoids. Until recently, most of our knowledge of properties of carotenoid excited states was largely based on spectroscopic studies of polyenes that belong to the same idealized C_{2h} point symmetry group as carotenoids. In terms of C_{2h} symmetry labels, the two low-lying singlet excited states denoted $2^1A_g^-$ and $1^1B_u^+$, are responsible for most of the spectroscopic properties of carotenoids. Since the ground state of carotenoids is of A_g^- symmetry, the one-photon transition between the ground state and the $2^1A_g^-$ state is symmetry forbidden, while that between the ground state and the $1^1B_u^+$ state is allowed. In 1972, experiments on the polyene diphenyloctatetraene,[12] supported by calculations[11] established that the lowest excited state in this polyene is located below the strongly allowed $1^1B_u^+$ state, making the lowest energy transition symmetry forbidden. This discovery aroused a number of experimental and theoretical studies of polyenes, which established that the forbidenness of the lowest energy transition is a common feature of all polyenes having conjugation length $N>3$.[13] This 'reverse' ordering of the excited states is also a central feature of all carotenoids occurring in nature, since most of them have a conjugation length between 7-13 (Figure 1).
While the $1^1B_u^+$ state, in terms of molecular orbitals, is described by a simple HOMO→LUMO transition, the theoretical explanation of the fact that the $2^1A_g^-$ state lies below the $1^1B_u^+$ state requires involvement of highly excited configurations that are able to take into account electron-electron interactions.[13,20-22] To describe properly the properties of the $2^1A_g^-$ state, at least doubly excited configurations must be used, because this state represents a mixture of singly and doubly excited configurations. In terms of molecular orbitals, the singly excited configurations correspond to (HOMO-1)→(LUMO) and (HOMO)→(LUMO+1) transitions, while the most important doubly excited configuration is the (HOMO, HOMO)→(LUMO, LUMO) transition.[16] For polyenes, involving doubly excited configurations it is usually enough to push the $2^1A_g^-$ state below the $1^1B_u^+$, but quantitative agreement between experimental and theoretical $2^1A_g^-$ energies required even higher configuration interaction.[13,20-22] The doubly-excited character of the $2^1A_g^-$ state can be rationalized as involving two 1^3B_u triplet excitations that are coupled to an overall singlet state, which explains the

fact that the energy of the $2^1A_g^-$ state is about twice the energy of the lowest triplet state.[13] Besides these two states of ionic (B_u^+) and covalent (A_g^-) nature, other covalent states formed by various combinations of triplet excitations coupled to a singlet state exist in the excited state manifold of polyenes and carotenoids. From a spectroscopic point of view, the most interesting ones are the $1^1B_u^-$ and $3^1A_g^-$ states that are formed as a combination of 1^3B_u and 1^3A_g, and 1^3A_g and 1^3A_g triplet states, respectively,[13] because the energies of these states approach the energy of the absorbing $1^1B_u^+$ state as the conjugation length increases to N~11.[13] The detailed descriptions of the electronic properties of the excited states of polyenes can be found elsewhere.[16,17,23]
Throughout this chapter, we use the following notation to describe excited states of carotenoids. In the symmetry notation, we will omit the spin index for singlet states. For description of excited states, we will follow the conventional notation used throughout literature about carotenoid excited states to date, thus the absorbing $1B_u^+$ state will be called S_2 state and the dark $2A_g^-$ state S_1 state. When other states are discussed, they will be designated by their symmetry label.

2.1 The S_2 state of carotenoids

The strong absorption of carotenoids (extinction coefficients of the order of 10^5) in the blue-green spectral region is caused by the strongly allowed S_0-S_2 transition.[24] The S_0-S_2 transition of carotenoids usually exhibits a characteristic three-peak structure corresponding to the lowest three vibrational levels of the S_2 state (Figure 3). The energy gap between vibrational peaks of ~1350 cm^{-1} results from the combination of two symmetric vibrational modes with energies of ~1150 cm^{-1} (C–C stretch) and ~1600 cm^{-1} (C=C stretch). The resolution of vibrational peaks in the absorption spectrum is an important spectroscopic measure as it reflects certain structural properties of a carotenoid molecule. The vibrational structure is well resolved for linear carotenoids such as lycopene or spheroidene, while a clear loss of resolution of vibrational bands is observed for carotenoids having conjugation extended to various end groups. A typical example is β-carotene, for which the conjugation extends to the terminal β-ionylidene rings.

2.1.1 The S_2 lifetime

While the radiative lifetime of the S_2 state as calculated by the Strickler-Berg equation[25] is in the range of nanoseconds, measured quantum yields of the S_2 emission of the order of 10^{-5} showed that internal conversion on the subpicosecond time scale governs the S_2 dynamics.[26,27,28] Thus, information about the dynamics of the S_2 state relies solely on ultrafast time-resolved techniques, particularly on fluorescence up-conversion and transient absorption data. Application of these techniques to a number of carotenoids employed in photosynthetic light harvesting established that the S_2 lifetime is in the range of 100-300 fs, [28,29-31] and dependent on both conjugation length and solvent parameters. A detailed fluorescence up-conversion study of the effect of solvent properties on the S_2 dynamics of β-carotene[28] demonstrated that increasing polarizability of solvent decreased the S_2 lifetime of β-carotene from 180 fs in *n*-hexane to 130 fs in CS_2 due to the polarizability shift of the S_2 state, making the S_2-S_1 gap smaller and thus facilitating the internal conversion. For all solvents, the S_2 decay was successfully fitted by a monoexponential decay with no rise component, signaling that no dynamical Stokes shift is present. Similar results were obtained with excitation of β-carotene into higher vibrational levels[31] and were further confirmed also for the linear carotenoid neurosporene.[30] The effect was explained in terms of sub-100 fs vibrational relaxation in the S_2 state. While the solvent effect on the S_2 lifetime was successfully explained using the S_2-S_1 energy gap, the effect of conjugation length and carotenoid structure on the S_2 lifetime is more difficult to explain. For a number of different carotenoids of various conjugation lengths it was observed that the S_2

lifetime, contrary to expectations based on the energy gap law, decreases with increasing conjugation length.[29-31,32,33] A consistent explanation of these observations is still missing, but it has been suggested that a higher density of accepting vibrational modes facilitates the S_2-S_1 internal conversion for longer carotenoids and that states located between the S_2 and S_1 states may affect the internal conversion relaxation process.[14,15,34-37,38]

2.2 The S_1 State of carotenoids

As described in the previous section, after being promoted to the S_2 state a carotenoid molecule undergoes fast relaxation on a time scale 100-300 fs (depending on environment and structure of the carotenoid) to the lowest singlet excited state S_1, implying that properties of the S_1 state are crucial for understanding photophysics of carotenoids. Because of the forbidden nature of the S_0-S_1 transition, knowledge about energetics and dynamics of this key excited state was very limited. Since the theoretical prediction in 1972 of the $2A_g^-$ state being the lowest excited state in longer polyenes and carotenoids,[11] it took more than 20 years until the energy of this state could be determined experimentally for longer carotenoids.

2.2.1 Energy of the S_1 state

The first attempt to localize the S_1 state experimentally was carried on in 1977 by Thrash et al. who analyzed the Raman excitation profile of β-carotene and located its S_1 energy above 17000 cm^{-1}.[39] However, this rather high value was questioned by subsequent studies of polyenes[40] and carotenoids[41] that utilized S_1 emission of shorter polyenes/carotenoids. On the basis of a theoretical analysis of excited states of polyenes,[13] one can deduce that the S_1 energy of carotenoids decreases with conjugation length, and this decrease is slightly steeper than for the S_2 state, making the S_2-S_1 gap larger for longer carotenoids. This is the reason for the characteristic switch between S_2 and S_1 emission observed for conjugation lengths around 8. Thus, carotenoids with $N \leq 8$ exhibit appreciable S_1 emission with a maximum corresponding to the 0-2 vibrational band of the S_1-S_0 transition, signaling a substantial displacement of the S_0 and S_1 potential surfaces.[42,43] Consequently, spectral analysis of the S_1 emission spectra enabled to determine the 0-0 origin of the S_1-S_0 transition. Both Cosgrove[40] and DeCoster[41] used knowledge of the spectral origin of the S_1-S_0 transition for shorter polyenes/carotenoids to extrapolate the S_1 energy of longer ones. They showed that for a carotenoid having 11 C=C bonds such as β-carotene, the S_1 energy must lie substantially lower than the 17000 cm^{-1} proposed by Thrash et al,[39] and on the basis of the extrapolation they concluded that the S_1 state of β-carotene is located between 13000-14000 cm^{-1}.[40,41] With knowledge of S_1 lifetimes of longer carotenoids (see below), the extrapolation method was further improved by application of the energy gap law for radiationless transitions,[44] which led to refinement of the β-carotene S_1 energy to 14100 cm^{-1}.[45] In addition, the energy gap law was used to estimate S_1 energies of a number of biologically important carotenoids such as spheroidene, zeaxanthin, violaxanthin, antheraxanthin canthaxanthin, diatoxanthin and diadinoxanthin.[45-47] Nevertheless, experimental verification of these extrapolations was necessary and quest for a reliable experimental method that would enable the determination of the S_1 energy became one of the dominating streams in carotenoid photophysics in recent years.

Some properties of the S_1 state can be monitored via its characteristic, strong S_1-S_N excited state absorption occurring in the visible spectral range which spectral profile is shown in Figure 3. Due to symmetry reasons, the final S_N state must be of B_u^+ symmetry to account for such a strong transition. Consequently, it must be present also in the ground state absorption spectrum. Inspection of the UV part of the absorption spectra of carotenoids reveals rather broad absorption bands below 300 nm[48,49] that were assigned to states of B_u^+ symmetry.[48] Then, knowing the energies of the S_1-S_N and S_0-S_N absorption bands from transient absorption and steady-state experiments, it would be possible to calculate the energy of the S_1 state. However, attempts to determine the S_1 energy by this method failed due to two reasons. First, higher B_u^+ states in the absorption spectrum are rather structureless. Therefore, assessment the 0-0 origin of the S_0-S_N transition suffers from a significant error. Second, the S_1-S_N excited state absorption (ESA) band overlaps with ground state bleaching, thus the S_1-S_N ESA band observed in experiment does not correspond to the spectral profile of the S_1-S_N transition as the bleaching cuts its higher-energy part (Figure 3). Nevertheless, the shape and position of the S_1-S_N band is, similarly to the ground state absorption spectrum, a fingerprint of a carotenoid molecule. As shown in Figure 3, dependence of the S_1-S_N maximum on conjugation length always follows that of the S_0-S_2 transition, but the shift of the S_1-S_N band is usually larger. Even though no clear vibrational structure of the S_1-S_N band was observed so far, the vibrational structure of the absorption spectrum is imprinted into the shape of the S_1-S_N ESA band. Linear carotenoids with highly resolved vibrational peaks in their ground state absorption spectrum have a sharp, narrow S_1-S_N ESA band, while significant broadening of the S_1-S_N ESA is observed for carotenoids with less-structured absorption spectrum (Figure 3). Thus, although S_1-S_N ESA can give information about certain properties of the S_1 state (especially about lifetime as described below), different techniques are needed to determine the S_1 energy.

Until the late nineties, experimental studies to locate the S_1 energy relied on progress in fluorescence detection techniques that enabled detection of weak S_1 emission even for carotenoids with N>9. Due to very low quantum yields of the S_1 emission, on the order of 10^{-5}-10^{-7},[26,27] and due to substantial overlap with the dominating S_2 emission for longer carotenoids, these measurements required extremely precise detection to analyze the spectral profile of the S_1 emission. The first room temperature fluorescence spectrum of β-carotene reported by Bondarev[50] displayed a weak, distinct band that was assigned to S_1 emission, with its 0-0 spectral origin being located at 13200±300 cm^{-1}. This value is significantly below the values predicted by extrapolations,[45] but, as shown later, this inconsistency was mainly due to rather poor signal/noise ratio that did not allow resolution of the vibronic structure of the S_1 emission and, consequently, separation of vibronic bands from overlapping contributions of S_2 and S_1 emissions. Nevertheless, the important conclusion of this experiment was that the S_1 energy is much less sensitive to solvent polarizability than the S_2 state: while S_2 emission shifts significantly going from *n*-hexane to CS_2, the weak S_1 emission is only little affected.[50] Later, more accurate measurements of β-carotene S_1 emission revealed vibronic structure of the S_1 emission band[51] that enabled a more detailed analysis of the vibronic transitions. On the basis of the fact that the maximum of the S_1 emission corresponds to the 0-2 vibrational band, these authors concluded that the 0-0 origin of the S_1-S_0 transition must be completely hidden under the red tail of the more intense S_2 emission band and the low S_1 energy reported by Bondarev et al.[50] was concluded to be one vibronic band (~1300 cm^{-1}) too low due to their wrong assignment of vibronic bands. Spectral analysis performed by Andersson et al.[51] positioned the S_1 energy of β-carotene in CS_2 at 14200±500 cm^{-1}, which, despite rather high error, was in good agreement with the energy gap law extrapolations. A slightly higher value of 14500±150 cm^{-1} was reported on the basis of analysis of S_1 emission of β-carotene in *n*-hexane at 170 K.[52]

Motivated by the key importance of the S_1 energy for understanding energy transfer pathways, the S_1 energy of several other carotenoids playing role in photosynthetic light harvesting was also located by measurements of weak S_1 emission. Generally, the obtained S_1 energies are in good agreement with the energy gap law predictions and they also fit to extrapolations based on S_1 energies of shorter carotenoids, measured by fluorescence spectroscopy.[45] The S_1 energies of various carotenoids are summarized in Table 1.

Raman spectroscopy is also a useful tool to obtain S_1 energies of carotenoid molecules. The ground state Raman spectrum of carotenoids in the frequency region 1000-1800 cm^{-1} is dominated by two Raman lines at ~1500 cm^{-1} and ~1150 cm^{-1} (exact frequencies of these lines depend on carotenoid), which correspond to symmetric stretching C=C and C-C vibrations of the conjugated backbone.[53] Under the conditions of resonance Raman when the excitation light is in resonance with an electronic transition of the studied molecule, the intensities of the Raman lines are greatly enhanced. Thus, tuning the detection to the frequency of either C=C or C-C stretching modes and scanning the excitation light over a broad spectral range, a resonance Raman profile can be obtained. The high sensitivity of Raman line intensities to the resonance conditions enables detection of even dark states with very low transition dipole moments. S_1 energies of several carotenoids have been measured with this method, generally in good agreement with other methods.

Our own work lead to an approach employing femtosecond time-resolved spectroscopy for locating carotenoid S_1 states.[49] The method is based on the fact that the S_1-S_2 transition is symmetry allowed and thus corresponds to a strong, easily detected transition. A femtosecond excitation pulse populates the lowest vibrational band of the well-characterized S_2 state to avoid contribution from vibrational relaxation within this state. The excited molecule then relaxes to the S_1 state on the time scale of 50-300 fs. By scanning the wavelength of the probe pulse within the time window dictated by the lifetime of the S_1 state, the S_1-S_2 resonance can be mapped. Given the known spectral profile of the S_2 state and lifetime of the S_1 state, the S_1-S_2 transition must 1) reflect the spacing between the S_2 vibrational bands; 2) decay with the S_1 lifetime. If these two conditions are fulfilled, the energies of the vibrational bands of the S_1-S_2 transition can be determined. The precise location of the carotenoid S_1 state can then be obtained, since its 0-0 spectral origin can be calculated from the spectral origins of S_0-S_2 and S_1-S_2 transitions.[49] Although this approach is quite straightforward, its application was complicated by the necessity to have tunable femtosecond pulses in the 1-2 μm spectral range where the S_1-S_2 transition is expected. Along with availability of sources generating such pulses in the late nineties, this method became a powerful tool to study the properties of the S_1 state of carotenoids. Typical S_1-S_2 profiles of two carotenoids with different conjugation lengths are shown in Figure 4.

The first application of this method was used to locate the S_1 energies of two carotenoids, violaxanthin and zeaxanthin, that participate in the xanthophyll cycle of higher plants.[49] Transient absorption spectra recorded for both carotenoids in the spectral region 900-1600 nm resembled well the characteristic three-peak structure of the S_2 state, with vibrational spacing matching perfectly that observed in the ground state spectrum (see Figure 4 for the S_1-S_2 spectrum of violaxanthin). In addition, the whole near-infrared transient absorption spectrum decayed with a time constant of 8.8 ps for zeaxanthin and 25 ps for violaxanthin, thus agreeing with the known S_1 lifetimes of these carotenoids (see Table 2). Consequently, it was concluded that the recorded transient absorption spectrum indeed reflected the spectral profile of the S_1-S_2 transition and the well-resolved vibrational bands allowed for the location of the 0-0 origin of the S_1-S_2 transition, peaking at 7010 cm^{-1} (zeaxanthin) and 6950 cm^{-1} (violaxanthin). Then, with the knowledge of the S_1-S_2 and S_0-S_2 transition energies, the position of the S_1 level of both carotenoids was determined by a simple subtraction. To improve the precision, both the 0-0 and 0-1 transitions of the steady state and transient absorption spectra were used, leading to an S_1 energy of 14030 ± 90 cm^{-1} for zeaxanthin and 14470 ± 90 cm^{-1} for violaxanthin.[49] Although these values fall into the range of energies expected for carotenoids with these conjugation lengths, using an experimental approach exploiting the allowed S_1-S_2 transition (while the methods used before relied exclusively on detection of weak signals originating from the forbidden S_1-S_0 transition) revealed a number of interesting new features that ignited lively discussions about the S_1 state of carotenoids.

First, the S_1-S_2 spectra of violaxanthin (N=9) and zeaxanthin (N=11) are almost the same, indicating that the S_1-S_2 energy gap is only little affected by the conjugation length, and both the S_1 and S_2 states are shifted by about the same energy with increased conjugation. In addition, the difference between the S_1 energies of these two carotenoids is significantly smaller than predicted by extrapolations using the energy gap law; while the S_1 energy of zeaxanthin is close to the expected one, for violaxanthin having a conjugation length 9 the extrapolations predicts an S_1 energy above 15000 cm^{-1},[54] substantially higher than the 14470 cm^{-1} extracted from the S_1-S_2 spectra. Thus, this new approach demonstrated that extrapolations from the energy gap law are not straightforward when carotenoids with different structures are compared. This conclusion was later confirmed by detecting S_1 emissions of violaxanthin and zeaxanthin.[55]

Since recording of S_1-S_2 spectra proved to be a very useful technique to establish S_1 energies of carotenoids, subsequent studies of various carotenoids followed. A more detailed study, including temperature dependence, was performed on spheroidene (N=10), and revealed additional subtleties of the S_1-S_2 spectra.[56] The analysis of the S_1-S_2 spectrum supported by simulations proved that the observed bands correspond to three vibrational levels of the S_2 state, with the 0-0 transition being the strongest one (Figure 4), suggesting only a small shift between the S_1 and S_2 potential surfaces. Interestingly, the energy of the 0-0 spectral origin of the S_1-S_2 transition at 7300 cm^{-1} put the spheroidene S_1 state at 13400±90 cm^{-1}, about 800 cm^{-1} lower than the energy obtained from fluorescence and resonance Raman measurements. This intriguing result was explained as a result of different carotenoid configurations occurring in the S_1 state. Thus, it was proposed that the crucial difference leading to different results when different techniques are used lies in the fact that while the S_1-S_2 technique probes an allowed transition, fluorescence and resonance Raman monitors a forbidden transition. With a statistical distribution of conformers in the S_1 state, fluorescence and resonance Raman techniques detects predominantly the conformations deviating from the idealized C_{2h} symmetry, since these have the S_1-S_0 transition least forbidden. An exactly opposite situation holds for the allowed S_1-S_2 transition, because the least distorted conformations give the dominant contribution to the S_1-S_2 spectrum. The difference in S_1 energies obtained by the different techniques is consequently a result of that each technique probes a different subset of S_1 state conformations.[56] This implies that the spheroidene S_1 energy of 13400 cm^{-1} obtained from the S_1-S_2 spectra corresponds to the S_1 energy of the *all-trans* configuration, while the higher energy of 14200 cm^{-1} was assigned to conformations deviating from the ideal C_{2h} symmetry. This hypothesis is further supported by the fact that the 0-0 band of the S_1-S_2 transition exhibited a clear asymmetry, having a shoulder on the low energy side. If the S_1 energy is calculated on the basis of this shoulder, a nearly perfect match with the energies determined from S_1 fluorescence and resonance Raman was found. Accordingly, it was concluded that the low energy shoulder in the S_1-S_2 spectra is due to the distorted S_1 conformations.[56]

To date, S_1-S_2 spectra of violaxanthin,[49] zeaxanthin,[49] spheroidene,[56] lycopene,[57] rhodopin glucoside,[58] spirilloxanthin,[59] peridinin,[60,61] spheroidenone,[62] siphonaxanthin[62] and fucoxanthin[62] have been reported. The resulting S_1 energies, together with those obtained by other techniques are summarized in Table 1. Except for rhodopin glucoside, for which emission or resonance Raman data are not available so far, the S_1 energies extracted from S_1-S_2 spectra are systematically lower than those determined by fluorescence and/or resonance Raman spectroscopy, with a difference depending on carotenoid structure. The difference is almost negligible for the carbonyl carotenoids peridinin, fucoxanthin and siphonaxanthin, but the S_1 state properties of these carotenoids are markedly different from those without a carbonyl group[60,61]. The difference found for the linear carotenoids spheroidene and lycopene (800 cm^{-1}) is about twice that revealed for violaxanthin (400 cm^{-1}) and zeaxanthin (500 cm^{-1}). The latter have a more complicated structure leading to a more restricted ability to form conformers than for their linear counterparts. The 'conformational' hypothesis is further supported by measurements on the long (N=13), linear carotenoid spirilloxanthin. For this carotenoid, the 0-0 band of the S_1-S_2 transition is significantly broader, signaling a much wider distribution of S_1 conformers. In addition, spirilloxanthin is known to isomerize spontaneously on a time scale of minutes.[63] Thus ground state conformers are inevitably present in measurements of both S_1-S_2 spectra and fluorescence. Therefore, the actual difference between these two techniques is slightly smaller than for other linear carotenoids (Table 1).

2.2.2 The S_1 Lifetime of Carotenoids

Although not directly useful for determination of the S_1 energy, measurements of the S_1-S_N ESA signals have been widely used to obtain S_1 lifetimes. The first work in the late eighties by Wasielewski and Kispert gave the S_1 lifetimes of β-carotene (N=11, 8.4 ps), canthaxanthin (N=13, 5.2 ps) and β-apo-8' carotenal (N=10, 25.4 ps).[18] This pioneering work was followed by a number of studies on various carotenoids that established not only the S_1 lifetimes, but also their dependence on conjugation length, carotenoid structure and environment (Table 2 and Figure 5). A systematic dependence of the S_1 lifetime on conjugation length was demonstrated for a few series of carotenoid analogs. For example, measurements on spheroidene analogs yielded lifetimes of 400 ps (7), 85 ps (8), 25 ps (9), 8.7 ps (10), 3.9 ps (11), 2.7 ps (12) and 1.1 ps (13).[64] A similar dependence was found for a β-carotene series, although the actual lifetimes were slightly different: 282 ps (7), 96 ps (8), 52 ps (9) and 8.1 ps (11).[65] Apart from the conjugation lengths 7-13 that are characteristic of carotenoids occurring in natural systems, a few studies of carotenoids with both shorter and longer conjugation lengths added more experimental data to the conjugation length dependence of the S_1 lifetimes. A S_1 lifetime of 2.7 ns was found for a very short β-carotene analog with N=5,[65] and an even longer S_1 lifetime was estimated for the naturally occurring *cis*-phytofluene (N=5) on the basis of the quantum yield of the S_1 emission.[66] On the longer conjugation side, S_1 lifetimes of 2.5, 1.1 and 0.5 ps were obtained for β-carotene analogs with 13,[67] 15 and 19[68] conjugated C=C bonds, respectively, confirming the decrease of the S_1 lifetime with conjugation length as was successfully explained by the energy gap law.[45,64]

The effect of carotenoid structure on S_1 lifetimes can be evaluated from a number of studies measuring S_1 lifetimes of various carotenoids. In fact, observed changes of the S_1 lifetime can be in most cases rationalized as a deviation of carotenoid structure from the ideal polyene C_{2h} symmetry, causing a decrease of the effective conjugation length. As an example, one can consider carotenoids having N=11. For the linear unsubstituted carotenoid lycopene, the S_1 lifetime is 4 ps.[69,70,71] For spheroidenone, a linear carotenoid containing 10 C=C bonds and one C=O group, the S_1 lifetime is prolonged to 6 ps,[62,72] because the conjugated carbonyl group that is in *s-cis* position relative to the C=C conjugated backbone makes the effective conjugation length slightly shorter than for the full 11 C=C conjugation in lycopene. Further prolongation of the S_1 lifetime to 9 ps is noticed for β-carotene,[71,73] in which two C=C bonds are located in the terminal β-ionylidene rings leading to even shorter effective conjugation.

In contrast, various substituents that are not in contact with the conjugated backbone have no effect on S_1 lifetime (see Table 2). Rhodopin glucoside (N=11), possessing a glucoside ring at the end of the linear conjugated chain, has the same S_1 lifetime of 4 ps as its non-substituted counterpart lycopene. Similarly, violaxanthin (N=9), with its terminal rings containing epoxy groups decoupled from the conjugated backbone, has a lifetime of 24 ps, the same as observed for the unsubstituted N=9 carotenoid neurosporene.

While there is in fact no information about temperature dependence of the S_2 lifetime, a few studies of the temperature effect on the S_1 lifetime were performed. Single-photon counting measurements of the S_1 emission of mini-9-β-carotene (N=9) revealed a significant increase of the S_1 lifetime from

60 ps at 293 K in toluene to 130 ps at 77 K in 3-methylpentane.[51] A comparable increase of the S_1 lifetime was observed for a shorter mini-8-β-carotene (100 ps at 293 K and 280 ps at 77 K).[51] A markedly smaller effect was reported for spheroidene, yielding an S_1 lifetime of 8 ps at room temperature and 9.5 ps at 186 K in *n*-hexane.[56] Studies of several carotenoids in various solvents have shown that for carotenoids without a conjugated carbonyl group there is almost no solvent effect on the S_1 lifetimes. However, this rule is dramatically violated for carbonyl carotenoids, having unique S_1 state properties.[60,74]

2.3 Other Excited States of Carotenoids

Calculations on polyenes predicted additional states besides the S_1 and S_2 states. From extrapolations of state energies to longer polyenes it became apparent that two other excited states of $1B_u^-$ and $3A_g^-$ symmetry are of interest in carotenoids, since especially the $1B_u^-$ state approaches the S_2 state for N~9 and it should be located in between the S_1 and S_2 states for N>10.[13] As a result, for most of naturally occurring carotenoids this state can be involved in the excited state dynamics. The $3A_g^-$ state was predicted to be in the vicinity of the S_2 state for N~13,[13] and thus above the S_2 state for most of carotenoids, but it could affect the dynamics of the longest carotenoids such as spirilloxanthin. Since transition from the ground state to both $3A_g^-$ and $1B_u^-$ states is forbidden (for $1B_u^-$ it is both one- and two-photon forbidden),[13] location of these states suffered from the same problems as of the S_1 state. However, for the $3A_g^-$ and $1B_u^-$ states the experimental difficulties are further enhanced by the fact that these states, if populated, will relax quickly to the S_1 state, limiting their lifetime to the hundred femtosecond time scale. Resonance Raman and time resolved experiments have given some information about these states. Also other excited states, S* and $S^\#$, have been suggested and we refer to our recent review[75] for a more detailed discussion of these states.

3. Excited states of Carotenoids in Pigment Protein complexes

Most of the biological functions of carotenoids are carried out when carotenoids are bound to specific proteins. Although a number of carotenoid-binding proteins are known to date, excited state properties of carotenoids were studied in detail only in photosynthetic light-harvesting complexes. There are two main reasons for the lack of knowledge about carotenoid excited states in other carotenoid-binding proteins. First, structures of a few light-harvesting complexes are known in great detail; the conformation of a carotenoid and its interaction with the protein environment deduced from the structure facilitates the studies of excited state properties. Second, in light-harvesting complexes, carotenoids serve as antenna pigments, making their excited states directly involved in the light-harvesting process, a vital function in all photosynthetic organisms. For other carotenoid-binding proteins, the relation between excited states properties and biological function of carotenoids is not always apparent, although for example carotenoid-binding proteins found in the human eye are potential candidates for involvement of carotenoid excited states in photoprotection. However, even for carotenoid-binding proteins that are not involved in light-triggered reactions, studies of carotenoid excited states can be helpful for investigating the interactions between carotenoids and protein environment, which can be of key importance for understanding the molecular functions of carotenoids in various biological tissues.

3.1 Carotenoids in Light Harvesting Complexes of Photosynthetic Purple Bacteria

Properties of excited states of carotenoids in pigment-protein complexes are best understood in light-harvesting complexes of purple bacteria, because the detailed structural knowledge available for these proteins provides an ideal platform for experimental and theoretical investigations of energy transfer processes between carotenoids and BChl. To date, structures of two light-harvesting complexes were resolved in great detail. In 1995, the LH2 complex from *Rhodopseudomonas* (*Rps.*) *acidophila* containing the carotenoid rhodopin glucoside was determined with 2.5 Å resolution,[76] and the 2.4 Å structure of the LH2 complex from *Rhodospirillum* (*Rs.*) *molischianum* containing lycopene followed shortly after.[77] The structure, depicted in Figure 6, revealed the BChl-a and carotenoid molecules arranged in a ring with its axis perpendicular to the membrane plane. The elementary building blocks of LH2 complexes are αβ-polypeptide pairs that bind two strongly coupled BChl-a molecules absorbing at around 850 nm (B850), one BChl-a molecule having an absorption band at 800 nm (B800), and one carotenoid molecule spanning the membrane. The carotenoid is in close contact with both the B800 and B850 molecules with the closest distances 5.4-9.3 Å to the central Mg atoms of the neighboring BChl-a molecules.[76] The main difference between the two LH2 structures, apart from accommodating different carotenoids, is that while the LH2 of *Rs. acidophila* exhibits nine-fold symmetry, LH2 of *Rs. molischianum* contains only eight elementary building blocks. A hydrogen bond between rhodopin glucoside and the B800 BChl was found by quantum chemical calculations on LH2 of *Rps. acidophila.*[78] The pigment arrangement of LH2 allows carotenoids to serve as efficient donors in the light-harvesting process, but it was also discovered that they are good probes of processes involving BChl-a molecules, because they are very sensitive to electric fields. The carotenoid band shifts observed in LH2 complexes were interpreted as resulting from the local electric field associated with excitation of nearby BChl-a molecules,[79] and this explanation was also supported by calculations.[80] Linear dichroism measurements showed that the transition dipole moment of rhodopin glucoside in LH2 is 9.1° off axis from the π-electron conjugated chain,[81,82] and it was concluded that this deviation is not due to perturbations by the protein environment but it is rather a consequence of the single-double bond alternation in carotenoids. Moreover, intermolecular π-π stacking interactions between lycopene and the surrounding aromatic amino acid residues were found by means of quantum chemical calculations on the LH2 structure of *Rs. molischianum.*[83] These interactions were suggested to be a molecular mechanism for binding of carotenoids in LH2, and as shown later for other complexes, the role of π-π stacking interactions in carotenoid binding may be a general feature of photosynthetic proteins.[84]
Importantly, a refinement of the structure of LH2 from *Rps. acidophila* to 2.0 Å was reported recently, revealing a second carotenoid molecule in the elementary building block that was not resolved in the initial structure.[85] The second rhodopin glucoside is located at the periphery of the LH2 complex and it adopts a *cis*-configuration. Nevertheless, it was suggested that the observed *cis*-form could result from the preparation procedure and the actual *in vivo* configuration is likely to be *all-trans.*[85] Earlier chemical analyses determined a carotenoid/BChl ratio of 1:2 in a few LH2 complexes.[86,87] It suggests that this situation is not unique for *Rps. acidophila*, but it is likely that also other LH2 complexes possess the second carotenoid. It is, however, worth noting that spectral reconstitution resulted in a ratio of ~1:3 for *Rps. acidophila* and *Rs. molischianum*,[88] indicating that the second carotenoid on the periphery of LH2 might be removed during preparation in some cases. Since all studies of carotenoid-BChl energy transfer in LH2 complexes were so far interpreted in terms of only one carotenoid in the elementary building block (with the exception of ref. 89), this finding will certainly influence our understanding of carotenoid function in LH2 complexes.
Besides the two structures described above, structural information with lower resolution is also available for other complexes. Cryo-electron microscopy studies of the LH2 complex from *Rb. sphaeroides* revealed nine-fold symmetry of this complex and suggested that the structure is similar to the one from *Rps. acidophila.*[90] Similarly, an 8.5 Å resolution electron microscopy projection map

of LH1 from *Rhodospirillum* (*Rs.*) *rubrum*[91] revealed a ring consisting of 16 subunits, each of which accommodates two BChl-a molecules and the carotenoid spirilloxanthin.

That carotenoids can transfer energy to BChls was known long before the X-ray structures were resolved. On the basis of fluorescence excitation spectra, carotenoid-BChl energy transfer efficiencies of 80-100% were reported for *Rb. sphaeroides* containing spheroidene (N=10),[92-94] while lower values between 35-70% were obtained for *Rps. acidophila* containing rhodopin glucoside (N=11).[94-96] The carotenoid-BChl energy transfer yield drops to ~30% for LH1 of *Rs. rubrum* containing spirilloxanthin (N=13).[97] Less than 25% efficiency was obtained for *Rps. palustris* accommodating rhodovibrin (N=12) as the dominating carotenoid.[94]

Initial suggestions regarding mechanisms and pathways of carotenoid-BChl energy transfer in LH2 and LH1 complexes proposed energy transfer via the S_1 state, because of extremely fast deactivation of the S_2 state. The forbidden nature of the S_1 state led to a suggestion that the Dexter electron exchange mechanism is active in this process.[98-100] However, the first time-resolved experiments performed in the early nineties showed that following excitation of carotenoids, the B850 Q_y state was populated in less than 200 fs, signaling that both the S_1 and S_2 states may be efficient donors in the carotenoid-BChl energy transfer process.[101,102] Calculations of carotenoid-BChl energy transfer for neurosporene also led to the conclusion that both the S_1 and S_2 states can be active in energy transfer.[103] When LH2 structures to atomic resolution became available,[76,77] a number of experimental and theoretical investigations of energy transfer mechanisms and pathways between carotenoids and BChls confirmed the initial proposal that both the S_1 and S_2 states are involved in the energy transfer processes. It was also concluded that the precise pathways and directions of energy flow are governed mainly by the conjugation length of the involved carotenoid. In addition, new experimental approaches and improved methods of data analysis recently revealed new energy transfer pathways that contribute to the overall carotenoid-BChl energy transfer in light-harvesting complexes of purple bacteria. A schematic representation of energy transfer pathways within the LH2 complex is depicted in Figure 7. Below we will give a detailed account of these processes.

3.1.1 Energy transfer via the S_2 state

Absorption spectra of carotenoids in LH2 and LH1 complexes resemble well those obtained for carotenoids in solution, except for a red shift of ~1000 cm^{-1} caused by interaction with the protein environment (Figure 6). The vibrational sub-structure of the carotenoid S_2 state in LH2 is usually very similar to that obtained in solution, although for some carotenoids the protein environment represents a confinement of the carotenoid structure, leading to a better resolution of vibrational bands of the S_2 state in protein.

The first experimental data on energy transfer via the S_2 state was obtained by fluorescence up-conversion. S_2 emission decays of spheroidene in LH1 and LH2 complexes of *Rb. sphaeroides* yielded time constants of 55 fs and 80 fs, respectively.[29] Comparing these results with the markedly longer decays in solution (~150-250 fs), it was concluded that energy transfer via the S_2 state takes place with time constants of 90 fs for LH1 and 170 fs for LH2 (corresponding efficiencies of 65% and 47%, respectively), demonstrating that energy transfer can successfully compete with S_2-S_1 internal conversion. On the basis of spectral overlap, it was concluded that energy transfer via the S_2 state occurs to the Q_x state of Bchl.[29] Similar results were reported for the B800-B820 LH2 complex from *Rps. acidophila*, for which analysis of up-conversion kinetics supported by calculations yielded an S_2-Q_x energy transfer rate of (120-150 fs)$^{-1}$. The S_2-Q_x channel was concluded to be the dominating one, accounting for 60% of the total energy transfer efficiency.[104] Later, up-conversion experiments were carried out on the wild type B800-B850 LH2 complex of *Rps. acidophila* and a complex lacking the B800 BChls. Upon combining the results for these two LH2 complexes,

Macpherson et al. addressed the question of energy acceptors in the carotenoid-BChl energy transfer.[105] An observed overall efficiency of S_2-mediated energy transfer of 51% confirmed the results obtained for the B800-B820 complex,[104] and it was in addition shown that 20% of the S_2 population transfers energy to B800 BChl, while the rest (31%) goes to B850.
The fast and efficient S_2 energy transfer pathway was also observed by transient absorption measurements. A study of the LH2 complex from *Chromatium purpuratum* accommodating the carotenoid okenone (N=11+C=O) revealed S_2-mediated energy transfer characterized by a time constant of 50-100 fs.[89] In addition, it was shown that shaping of excitation pulses could control the ratio between S_2-S_1 internal conversion and energy transfer. For LH2 from *Rps. acidophila*, an increase, by a factor 1.4, of S_2-S_1 internal conversion was achieved by optimizing the envelope and phase of the excitation pulses,[106] demonstrating that the actual S_2 energy transfer efficiency depends also on other parameters. Higher efficiencies of 60-70% for the S_2-mediated energy transfer were reported for the LH2 complex from *Rb. sphaeroides* G1C that has neurosporene (N=9) as the main carotenoid.[107] Comparing the results on LH2 complexes containing carotenoids with N=9-13, there is a certain trend, suggesting higher efficiency of the S_2 energy transfer for shorter carotenoids. This trend is supported by experiments on LH2 of the carotenoidless mutant *Rb. sphaeroides* R26 with incorporated spheroidene analogs of different conjugation lengths. With the help of S_1-S_2 internal conversion rates measured in solution, overall carotenoid-BChl energy transfer efficiencies, and S_1-mediated energy transfer efficiencies, the efficiencies of S_2 energy transfer were calculated. Although the LH2s with spheroidene analogs having N=8,9 have slightly less efficient S_2 energy transfer than LH2 with spheroidene (N=10), the decrease of S_2 efficiency from 50% (spheroidene) to 12% for the analog having N=13, supported the observed trend of decreasing S_2-mediated energy transfer efficiency with increasing conjugation length.[108] Although less efficient for very long carotenoids, the S_2 energy transfer route operates in all purple bacterial antenna complexes studied so far with efficiencies in the range 30-70%. In a number of cases it represents the dominant energy transfer pathway between carotenoid and BChl.
The effective competition of S_2 energy transfer with the S_2-S_1 internal conversion was also rationalized theoretically. The contribution of the electron-exchange (Dexter) energy transfer mechanism[109] to the S_2 energy transfer is negligible and the Förster-type mechanism dominates.[104,104,110] However, contrary to the dipole-dipole interaction used in the Förster formula,[111] full Coulombic interaction is usually used to calculate couplings between carotenoids and BChl. Even without knowledge of detailed structural information, Nagae et al.[103] calculated the Coulombic couplings between the S_0-S_2 transition of neurosporene and the Q_x transition of BChl-a, for a few hypothetical configurations. They concluded that S_2 energy transfer via this route can be faster than 100 fs, provided the proper orientation of donor and acceptor is realized.[103] Determination of the LH2 structure provided detailed information about the mutual orientation of pigments within LH2, allowing more precise calculations of couplings between carotenoids and BChl-a. Krueger et al. applied an advanced method, the so-called transition density cube method, to calculate full Coulombic couplings between pigments in both B800-B820[104] and B800-B850[112] LH2 complexes from *Rps. acidophila*. This method replaces the vector description of the transition dipole moments by three-dimensional transition density volumes, which is expected to give a more accurate account of the interaction between molecules.[112] The couplings of the S_0-S_2 transition of rhodopin glucoside with all possible transitions of neighbouring BChl-a were calculated for the B800-B820 complex, yielding couplings larger than 100 cm^{-1} for β-B820 Q_x, B800 Q_y and also for the B800 Q_y transition of BChl-a located in the neighboring building block. However, due to the small value of the spectral overlap integral for the Q_y states, only the S_2-B820 Q_x yielded an appreciable energy transfer rate of $(240\ fs)^{-1}$.[104] Similar results were obtained for the B800-B850 complex. Although the actual couplings were slightly different from those calculated for the B800-B820 complex, the S_2-B850 Q_x

channel represented the dominating route, characterized by an energy transfer rate of $(135\ fs)^{-1}$.[112] The results of these calculations are in very good agreement with the observed depopulation rates of the S_2 state, but absence of significant coupling of the S_2 state to the B800 Q_x contradicts the experimental observation that a substantial part of the S_2 route in *Rps. acidophila* leads toward B800.[105] On the other hand, similar calculations using full Coulombic couplings performed on the basis of the LH2 structure of *Rs. molischianum* containing the carotenoid lycopene, yielded appreciable couplings of the S_0-S_2 transition with both the B850 and B800 BChls, resulting in S_2 energy transfer times of 200 and 250 fs for the B850 and B800 acceptors, respectively.[110] Essentially the same results were obtained by calculations of the lycopene-BChl couplings by means of the collective electronic oscillators algorithm.[113] An energy transfer rate of $(120\ fs)^{-1}$ was calculated for the S_2-Q_x channel, proposing an S_2 depopulation time of 69 fs for lycopene in LH2. Thus, while the calculations reproduce very well the observed rates of S_2-mediated energy transfer and confirm the BChl-a Q_x transitions being the energy acceptors in the S_2-mediated energy transfer, there are still contradictions regarding the branching ratio between B850 and B800 acceptors.

3.1.2 Energy transfer via the S_1 state

Efficient energy transfer via the S_1 state requires the S_1 energies of carotenoids to be higher than those of the acceptor states. Until recently, no information about S_1 energies in LH2 and related light-harvesting complexes was available, mainly due to the fact that the most used techniques to locate S_1 energies of carotenoids, detection of S_1 fluorescence and measurements of resonance Raman profiles, were not applicable to such complex systems. However, since it was known that S_1 energies are usually insensitive to solvent properties (except carbonyl carotenoids that are not typical for purple bacterial light-harvesting complexes), it was reasonable to assume that the S_1 energies in LH2 and LH1 complexes should be very close to those determined for carotenoids in solution. Under this assumption, S_1 energies of carotenoids with $N \leq 10$ were estimated to be high enough to transfer energy to the Q_y states of both B800 and B850 BChls. The first experimental evidence of this fact was provided by investigation of LH2 complexes from the *Rb. sphaeroides* R26 mutant with incorporated spheroidene analogs of different conjugation lengths. The efficiencies of S_1-mediated energy transfer were calculated from S_1 lifetimes of carotenoids in solution and in LH2, obtained from S_1-S_N ESA decays. While for conjugation lengths $N \geq 11$ the S_1 energy transfer was undetectable, a quenching of the S_1 state due to energy transfer was observed for shorter carotenoids.[108] This result was explained in terms of spectral overlap between hypothetical S_1 fluorescence and the B850 absorption band, which became small for longer carotenoids. Another direct experimental verification of the S_1-mediated energy transfer in the LH2 complex from *Rb. sphaeroides* was obtained by means of two-photon fluorescence excitation.[114] Forbidden for a one-photon transition, the S_1 state can be excited directly via a two-photon transition.[115] By measuring B850 emission after two-photon excitation of the S_1 state (achieved by exciting in the 1200-1500 nm spectral range), Krueger et al. demonstrated active participation of the spheroidene S_1 state in energy transfer. In addition, the two-photon excitation spectrum enabled determination of the spectral profile of the S_0-S_1 transition, placing the 0-0 energy of spheroidene in LH2 at 13900 cm^{-1},[114] confirming the similarity of the S_1 energies in LH2 and in solution. It is worth noting, however, that the 0-0 band of the S_0-S_1 transition was unreasonably weak, having about 40 times less intensity than the 0-1 band. Although no clear explanation of this phenomenon was offered, later experiments suggested that it might be due to the fact that the two-photon absorption spectrum reflects the energy transfer efficiency that does not necessarily mirror the Franck-Condon intensities of the vibrational bands. Indeed, a possibility of energy transfer from higher vibrational levels of the S_1 state of spheroidene was demonstrated later.[116,117]

Zhang et al. used three LH2 complexes, accommodating different carotenoids, to investigate the dependence of S_1-mediated energy transfer efficiency on conjugation length of the carotenoid.[118] On the basis of measured S_1 lifetimes of neurosporene (N=9), spheroidene (N=10) and lycopene (N=11) in solution and LH2, a dramatic decrease of S_1 energy transfer efficiency was observed; it dropped from 94% (neurosporene) to 82 % (spheroidene) and further to less than 30 % (lycopene), as the conjugation length increased from 9 to 11. Similar results were obtained for spheroidene and rhodopin glucoside in LH2 complexes from *Rb. sphaeroides* and *Rps. acidophila* by means of measurements of S_1-S_N kinetics after two-photon excitation of the S_1 state.119 These authors confirmed the ~1.8 ps S_1 lifetime of spheroidene in LH2, but for rhodopin glucoside a longer S_1 lifetime (6-7 ps) than in solution was observed, suggesting that there is no energy transfer via the S_1 state for rhodopin glucoside.[119] A few subsequent studies confirmed this trend and established that for carotenoids with N=11 the S_1 state is too low to achieve efficient energy transfer via the S_1 state, although the long S_1 lifetime of rhodopin glucoside observed by Walla et al.[119] was not confirmed by other authors. Instead, an S_1 lifetime of 3.7 ps was measured for rhodopin glucoside in LH2, while values of 4.1-4.9 ps were recorded in solution, showing that the S_1 energy transfer efficiency is around 5% in LH2.[105] The same S_1 lifetime of 3.7 ps for rhodopin glucoside in LH2 was measured recently,[120] while a slightly longer value (4 ps) was determined by Polívka et al.[58] A study of a *Rb. sphaeroides* mutant synthesizing the longer lycopene instead of spheroidene present in wild type LH2, proved that the efficiency of S_1 energy transfer is fully determined by the conjugation length of the carotenoid, while the protein environment plays only a negligible role. When lycopene is present in the complex, the efficiency of S_1-mediated energy transfer drops dramatically to about 20%, while more than 80% is achieved in the wild type.[57] Further evidence that carotenoids with N=11 are on the edge of capability to transfer energy via the S_1 state was provided by studies of LH1 complexes from *Rs. rubrum* containing spirilloxanthin (N=13). Within experimental error, no difference between S_1 lifetimes in solution and in the LH1 complex was found.[121,122] The same result was obtained for LH2 complexes from *Rb. sphaeroides* R26 reconstituted with spirilloxanthin,[117] suggesting that no S_1-mediated energy transfer is possible for spirilloxanthin because the S_1 energy is too low for efficient overlap with BChl Q_y states.

The significance of spectral overlap for the S_1-mediated energy transfer was recently confirmed by direct measurements of the S_1 energies of spheroidene and rhodopin glucoside in LH2 complexes by recording the S_1-S_2 spectra (Figure 8). S_1 energies of 13400±100 cm^{-1} and 12550±150 cm^{-1} were determined for spheroidene and rhodopin glucoside, respectively,[58] showing that the difference of 850 cm^{-1} makes a dramatic change in the energy transfer pathways in these two LH2 complexes. In the case of spheroidene, the S_1 energy extracted from the S_1-S_2 spectra is the same as that determined by this method in solution, while for rhodopin glucoside the S_1 energy is 250 cm^{-1} lower than that in solution.[58] This difference was explained by confinement of rhodopin glucoside in the LH2 structure, which narrows the distribution of conformers presented in solution.

Together with the spectral overlap, calculations of the couplings between the S_0-S_1 transition and Q_y transitions of both B800 and B850 demonstrated that the observed S_1 transfer rates can be explained in terms of the same energy transfer mechanism as for the S_2 route. Although the very small transition dipole moment of the S_0-S_1 transition led initially to a suggestion that the Dexter mechanism had to be invoked,[98-100] more detailed calculations later showed that the Dexter contribution is negligible and that higher-order Coulombic and polarization interactions dominate the S_1-mediated energy transfer.[103,110,112,113,123] In a few works, the Coulombic couplings for the S_1 state were obtained by scaling down the couplings calculated for the S_2 state, using an estimate that the transition dipole moment of the S_0-S_1 transition is about 4-6% of the S_0-S_2 transition. Although the results of this approach were promising, the calculated rates could not reproduce the measured energy transfer rates.[110,118] To resolve this problem, an increase of S_0-S_1 Coulombic coupling via intensity

borrowing from the allowed S_2 state due to S_2-S_1 mixing was proposed.[110,118] Under the assumption that the degree of mixing is inversely proportional to the square of the S_1-S_2 energy gap, Zhang et al. calculated S_1 energy transfer rates for LH2 from *Rs. molischianum* that were reasonably close to the measured values.[118] Further improvement of the agreement between experiment and theory was obtained by calculations of Coulombic couplings by means of time-dependent density functional theory (TTDFT).[124,125] This method, which allows *ab initio* calculations of the S_1 couplings with Q_y states of BChl, confirmed that small mixing between S_2 and S_1 states plays an important role in the Coulombic coupling. S_1 energy transfer rates for LH2 complexes from three different species of purple bacteria were calculated using the Coulombic couplings obtained from TDDFT, and the obtained values (9 ps)$^{-1}$ (*Rs. molischianum*), (1.2 ps)$^{-1}$ (*Rb. sphaeroides*) and (3 ps)$^{-1}$ (*Rb. sphaeroides* G1C)[124] are very close to the experimental ones of (12 ps)$^{-1}$, (2.4 ps)$^{-1}$ and (1.4 ps)$^{-1}$.[118] For LH2 from *Rps. acidophila*, an S_1 energy transfer time >25 ps was calculated,[124] also in a good agreement with the very low efficiency obtained experimentally.[105,120]
Regarding energy acceptors in the S_1-mediated energy transfer route, both B800 and B850 Bchls are capable of accepting energy from carotenoids, but the S_1-B800 channel seems to be the dominating one. In experiments employing incorporation of spheroidene into LH2 from the carotenoidless *Rb. sphaeroides* R26 mutant lacking B800 BChl, the efficiency of energy transfer via the S_1 pathway reached only 35%.[117] This is significantly less than ~80% observed for the LH2 containing both B800 and B850,[57,58,118,119,126] signaling that the main pathway involves B800 as an acceptor. The same conclusion was reached for LH2 from *Rps. acidophila*. Although the S_1 efficiency is only 4-5% in the wild type complex,[105120] removing the B800 BChls led to a complete absence of S_1 energy transfer.[105]
Similarly to the S_2-mediated energy transfer, the experimentally observed branching ratios are not fully consistent with calculations. Whereas for *Rb. sphaeroides* the branching ratio matches that obtained from TDDFT calculations, yielding a ratio of ~1:5 (B850:B800), for *Rps. acidophila* the calculated branching ratio is 3:2,[119] in obvious contradiction with experiments.[105]
In addition to the S_1-mediated energy transfer occurring from the thermalized S_1 state discussed above, a fractional energy transfer channel from the vibrationally hot S_1 state was also proposed. This type of energy transfer was first revealed in the LHCII complex from higher plants,[127] and later proved to be active also in the LH2 complex of *Rb. sphaeroides*. Using global analysis of their data, Papagiannakis et al. observed the decay of a species associated spectrum corresponding to the vibrationally hot S_1 state, suggesting the presence of an energy transfer channel via the hot S_1 state.[116] The contribution of this pathway to the total energy transfer efficiency was only 5%. Interestingly, no such pathway was found when spheroidene was incorporated into the carotenoidless LH2 *Rb. sphaeroides* R26 mutant lacking B800. This was interpreted in terms of B800 being the only acceptor for this energy transfer pathway.[117] Similarly, the energy transfer pathway via the hot S_1 state is absent in the LH2 complex from *Rps. acidophila*,[120] the LH1 complex from *Rs. rubrum*,[121] and LH2 complex from *Rb. sphaeroides* R26 reconstituted with spirilloxanthin,[117] demonstrating that this channel is of minor importance for overall energy transfer in light-harvesting complexes from purple bacteria (Figure 7). Also the S* state mentioned above has been suggested to be active in energy transfer from carotenoids to bacteriochlorophyll. For a discussion of this topic we refer to our recent review.[75]

3.2 Carotenoids in pigment proteins of green plants

The antenna complexes associated with Photosystem II (PSII) in green plants consist of a number of proteins belonging to the Lhcb gene family.[128] A higher structural complexity of these proteins, which usually bind more than one type of carotenoid in addition to Chl-a and Chl-b molecules,[129] results in a more complicated network of possible energy transfer and internal conversion pathways.

Nonetheless, the energy transfer between carotenoids and Chls was extensively studied in the recent few years, because in addition to the light-harvesting function, singlet excited states of carotenoids in these complexes were also suggested to play a photoprotective role, in which the 'back' energy transfer from Chl to carotenoid was the central issue. Yet, the understanding of energy transfer pathways between carotenoids and Chl molecules in light-harvesting complexes of green plants is poorer than for their purple bacterial counterparts.

The most abundant protein of the Lhcb family is LHCII that is located at the periphery of PSII, while minor light-harvesting proteins called CP29, CP26 and CP24 are located close to the core of the PSII supercomplex.[130] Structural knowledge of the Lhcb proteins is based on the crystal structure of LHCII whose structure was determined with 3.4 Å resolution.[131] The structure shows that the native form of LHCII has a trimeric organization, and each of the monomeric subunits (Figure 9) binds 12 Chl molecules and 2 carotenoids. The 12 Chls usually occur in a ratio of 7 Chl-a and 5 Chl-b, but slight variation has been observed.[132] Since the structural data did not allow differentiation of Chl-a and Chl-b, the exact assignment of Chls in the LHCII structure is still matter of debate. Contrary to purple bacteria, in which the carotenoid content is species-dependent and more than 100 different carotenoids were detected in various species,[133] only five different carotenoids are accumulated in light-harvesting proteins of green plants: β-carotene, lutein, violaxanthin, neoxanthin and zeaxanthin.[134] The two carotenoids in the LHCII structure were assigned to luteins, but later biochemical studies showed that while the L1 carotenoid site is exclusively occupied by lutein, the L2 site also binds violaxanthin though with a lutein/violaxanthin affinity of 80/20.[135] Further biochemical analysis revealed two additional carotenoid-binding sites present in LHCII: the N1 site, specific for neoxanthin,[136] and a violaxanthin selective site V1.[137] The V1 site is easily removed during detergent treatment and seems to be occupied preferentially in plants grown at high light conditions.[137] The violaxanthin molecule occupying this site, however, is not able to transfer energy to Chl.[138] Although the sequence homology between LHCII and the other proteins of the *Lhc* family, particularly CP29, CP26 and CP24 complexes, suggests that these proteins are not structurally much different, each of the proteins has a unique carotenoid composition. It seems that all proteins except LHCII have only two carotenoid binding sites.[139] While the L1 site is selective for lutein in all complexes, the L2 site can accommodate lutein or violaxanthin in all complexes. In CP26 and CP29 the L2 site even shows affinity to neoxanthin.[139,140] Under certain conditions the L2 site can also be occupied by zeaxanthin and this was suggested to trigger a structural change of the protein leading to quenching of Chl fluorescence,[139,141] which is the origin of the photoprotective mechanism called non-photochemical quenching (NPQ). This mechanism protects against damage caused by excessive photon flux by dissipating the energy thermally in PSII.[142,143] It involves a cascade of processes triggered by high light conditions resulting in enzymatic conversion of violaxanthin to zeaxanthin by means of the xanthophyll cycle. Although a fundamental understanding of NPQ is still lacking, it is clear that one of the crucial components is a zeaxanthin-induced quenching of Chl-a excited state, as presence of this carotenoid reduces markedly the Chl-a fluorescence lifetime.[141,144-146] Since a direct quenching by the S_1 state of zeaxanthin was suggested as one of the possible mechanisms,[46] studies of carotenoid S_1 dynamics in LHCII and related complexes became an important part of investigations of NPQ mechanisms.

Most of the studies of carotenoid-Chl energy transfer were performed on the LHCII and CP29 proteins. While LHCII is the only one whose structure is known to fine detail, the CP29 is favorable for studies of carotenoid-Chl energy transfer as it contains only 2 Chl-b molecules. This minimizes the overlap between the carotenoid S_2 state and the Chl-b Soret band peaking at around 475 nm (Figure 9). To overcome problems with the presence of a few different carotenoids with overlapping absorption spectra preventing selective excitation of only one carotenoid species, recombinant proteins that allow reconstitution with a single carotenoid species[135] were also used.[145-148] A number

of studies established that the overall carotenoid-Chl energy transfer efficiency in the Lhcb family of proteins is in the range 70-90 %,[5,132,145,147-152] and that the majority of the energy transfer occurs via the S_2 state (for scheme of energy transfer pathways see Figure 10).
It can be estimated that the efficiency of energy transfer from carotenoid S_1 states to chlorophylls is very low. Thus, knowing the energy of the lowest Q_y transition of Chl-a in LHCII-type complexes of ~14600 cm^{-1} and S_1 energies of carotenoids estimated either from measurements in solution or the energy gap law (Table 1), a simple comparison with LH2 complexes of purple bacteria suggests that the S_1-mediated pathway should be inefficient in the LHCII-type complexes. As demonstrated for LH2, the S_1 state should be at least 800 cm^{-1} above the acceptor state to achieve a sufficient overlap between the hypothetical S_1 emission and absorption of the acceptor. While such an arrangement often occurs in LH2, for LHCII-type complexes it implies that the S_1 energy of the carotenoid should be at least 15400 cm^{-1}, which is higher than expected for carotenoids occurring in these complexes (see Table 1). These expectations of inefficient energy transfer from carotenoid S_1 states have been verified by experiment[75]. However, it has been found that because of the relatively slow vibrational relaxation[70,73,] in the carotenoid S_1 state, energy transfer to chlorophyll can occur from higher S_1 vibrational states. Comparing the two-photon excitation spectra of LHCII with those of lutein and β-carotene in solution, it was concluded that the energy transfer must occur from the vibrationally hot S_1 state with a rate of $(250\ fs)^{-1}$ and an efficiency of ~50%.[119] Energy transfer from the hot S_1 state was recently confirmed by transient absorption measurements of the CP29 complex,[148] but the observed transfer rate is slower, $(700\ fs)^{-1}$, leading to modest efficiency of only ~10% (see Figure 10 for summary).
Above, we concluded that energy transfer from the thermalized carotenoid S_1 state is inefficient, since for the carotenoids present in LHCII-type of complexes the S_1 state has a lower energy than the Chl-a Q_y state. This opens up the possibility of back energy transfer from Chl-a to the S_1 state of carotenoids. This energy transfer route was first suggested by Frank et al. who proposed the so called 'molecular gear shift mechanism' as a mechanism for NPQ.[46] In the initial proposal based on S_1 energies estimated from the energy gap law, violaxanthin was supposed to have its S_1 energy above the Q_y transition of Chl-a serving as energy acceptor, while zeaxanthin had an S_1 energy sufficiently low to act as quencher of excited Chl-a. Then, exchange of violaxanthin to zeaxanthin by means of the xanthophyll cycle would enable increase of quencher concentration, explaining the mechanism of NPQ. Later, direct measurements of the S_1 energies in both solution[49,55] and LHCII[146] challenged this mechanism, since the S_1 energies of both violaxanthin and zeaxanthin were shown to be below the Chl-a Q_y state. Nevertheless, although the initially proposed 'molecular gear shift' would not work, the location of the zeaxanthin S_1 state below the Chl-a Q_y transition still allows direct quenching. To detect directly the possible Chl-a quenching by the zeaxanthin S_1 state represents a nontrivial task, because the fastest component of the Chl-a lifetime observed in the quenched state has a time constant of ~500 ps.[141,144,146] If this component were due to energy transfer from Chl-a to zeaxanthin, the S_1 state of zeaxanthin would have essentially zero population due to its 10 ps lifetime, making it almost impossible to detect in a transient absorption experiment. On the other hand, such conditions would make the zeaxanthin S_1 state a very effective sink for Chl-a excitations. Therefore, although experimental data could not unambiguously assign the mechanism of quenching, a direct quenching by means of singlet energy transfer was considered as a possibility.[139,146] However, contrary to the previous suggestions,[46] the key role of zeaxanthin was not due to the intrinsic photophysical properties of zeaxanthin but rather due to its ability to promote a conformational change allowing such a transfer.[139,141,153]

Direct involvement of the zeaxanthin S_1 state in NPQ was also shown in a transient absorption study of intact thylakoid membranes of *Arabidopsis thaliana*.[154] After excitation of either Chl-b or Chl-a by ~100 fs pulses at 664 and 683 nm, respectively, an excited state absorption in the spectral region 530-

570 nm was observed directly after excitation. On the basis of the spectral profile reconstructed from a few kinetic traces measured across this region, having decays characterized by a time constant of ~10 ps, it was concluded that this signal corresponded to the S_1-S_N spectrum of zeaxanthin. Importantly, using transgenic *A. thaliana* plants, the intensity of the S_1-S_N signal was correlated to the presence of the PsbS protein that is known to be an essential part of NPQ,[155] thus confirming the relation between population of the zeaxanthin S_1 state and NPQ. Although the quenching mechanism could not be disclosed from the experimental data, either energy transfer from Chl to the zeaxanthin S_1 state or formation of a zeaxanthin-Chl heterodimer was suggested.[154] It is worth noting that this phenomenon can represent a different quenching mechanism than decrease of the Chl-a lifetime observed earlier. While the S_1-S_N signal was observed only for intact thylakoid membranes, the shortening of the Chl-a lifetime was evidently detected even in isolated monomeric LHCII-type complexes.[141,145,146] Similar zeaxanthin-induced quenching was also observed in artificial zeaxanthin-pheophorbide dyads where pyropheophorbide emission lifetime was markedly shortened when covalently linked to zeaxanthin[156,157] or in a mixture of carotenoids and Chl-a embedded in DMPC liposomes.[158] Moreover, the S_1-S_N spectrum of zeaxanthin observed for *A. thaliana*[154] is apparently blue-shifted from that recorded for LHCII reconstituted with zeaxanthin.[146] Therefore (since the detection of S_1-S_N spectrum of zeaxanthin after Chl excitation is connected with presence of PsbS protein), zeaxanthin detected in these experiments might be located outside the LHCII-type complexes, somehow interacting with the PsbS protein. One possibility could be the PsbS-zeaxanthin complex discovered recently.[159]

It must be noted that besides the Chl-zeaxanthin energy transfer and formation of Chl-zeaxanthin heterodimer, other mechanisms that do not involve the S_1 state of zeaxanthin were also proposed. A quenching of Chl singlet state by electron transfer from carotenoids to Chl was suggested on the basis of observations on artificial carotenoporphyrin dyads.[160-162] Another hypothesis proposed that the essential function of zeaxanthin is to promote changes in structure and/or organization of the light-harvesting complexes that eventually lead to quenching conditions. One such mechanism involves a structural change facilitating a formation of Chl-a excitonic pairs, which than play a role of quenchers.[143,163] A long-range change involving aggregation and/or desaggregation of antenna complexes was also proposed.[164,165-167] In these processes, zeaxanthin is not directly involved in quenching but acts as a trigger of the particular process. Very recently, 'activation' of zeaxanthin, achieved by binding zeaxanthin to the PsbS protein was proposed to be an important mechanism to trigger energy dissipation in plants.[159] Which of these mechanisms play the most important role in NPQ is not known yet. The experimental data collected recently did not lead to general consensus. Nevertheless, given the complexity of the NPQ process, it is quite possible that all of these mechanisms have a certain place, and all together lead to the effective protection against excess light in higher plants.

Acknowledgements.

We wish to thank Tõnu Pullerits, Donatas Zigmantas and Helena Hörvin Billsten from our department for many fruitful discussions and important contributions to the work surveyed in this chapter. We also thank Roberto Bassi for providing us with Figure 9 and Tõnu Pullerits for making Figure 6. The research at Lund University is supported by the Swedish Research Council, the Knut and Alice Wallenberg Foundation, and the Crafoord Foundation. T. P thanks the Swedish Energy Agency for financial support.

FIGURES and TABLES

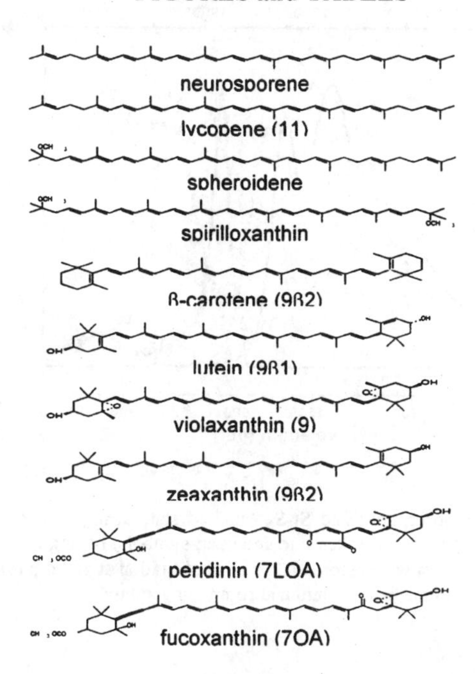

Figure 1. Molecular structures of carotenoids frequently used for studies of excited state dynamics. Conjugation length is denoted in parentheses as follows: N - number of conjugated C=C bonds in the linear conjugated backbone; β*n* – conjugation is extended to *n* C=C bonds located at a terminal ring; O – conjugation extended to a carbonyl group; LO – conjugation extended to a carbonyl group located at a lactone ring; A – conjugation extended to an allene moiety.

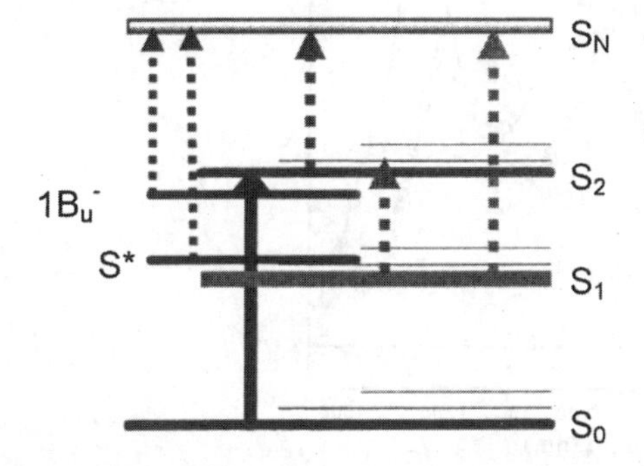

Figure 2. Simplified energy level scheme of a carotenoid molecule including transitions corresponding to transient signals occurring after excitation (blue arrow). The S_N state in this scheme represents only a symbolic final state for S_1-S_N (green), S_2-S_N (purple), $1B_u$-S_N (black) and S*-S_N (black) transitions. In reality, the final states of these transitions must be of different symmetry and therefore the S_N state in the scheme consists actually of four different states.

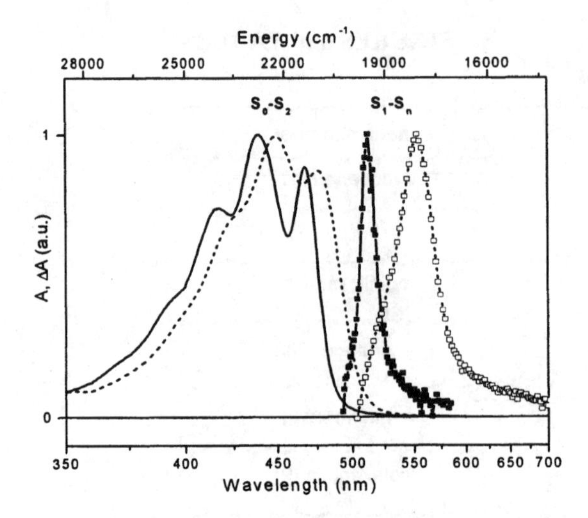

Figure 3. The S_0-S_2 (ground state absorption) and S_1-S_N (excited state absorption) transitions of violaxanthin (solid line and full symbols) and zeaxanthin (broken line and open symbols) having conjugation lengths of 9 and 11 (9β2), respectively. The S_1-S_N spectra were recorded 3 ps after excitation at 480 nm (violaxanthin) and 490 nm (zeaxanthin). All data were obtained in methanol solution at room temperature.

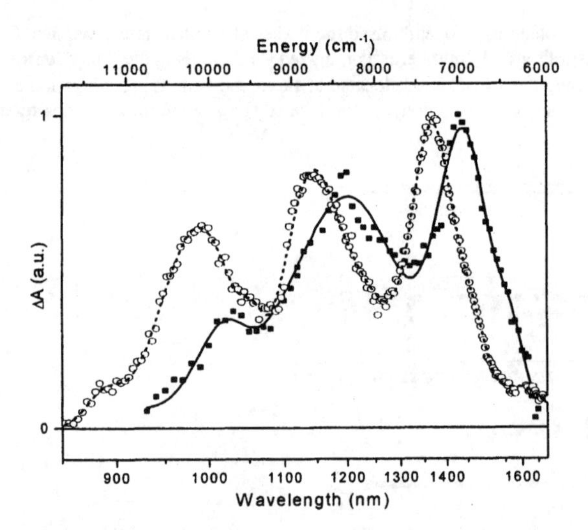

Figure 4. Transient absorption spectra of spheroidene in *n*-hexane (open circles) and violaxanthin in methanol (full squares) in the spectral region 850-1750 nm representing the spectral profile of the S_1-S_2 transition. The spectra were recorded 3 ps following excitation at 490 nm (spheroidene) and 480 nm (violaxanthin). The solid and broken lines are results of fitting the spectra to a sum of Gaussian profiles.

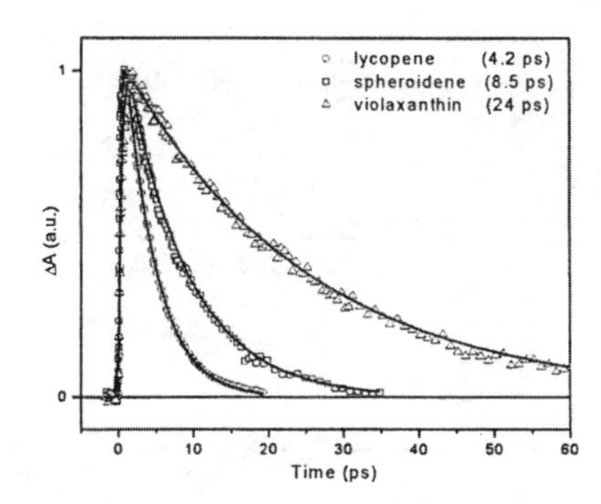

Figure 5. Kinetics representing decay of the S_1 state recorded at the maximum of the S_1-S_N transition of lycopene (N=11) in *n*-hexane (circles), spheroidene (10) in *n*-hexane (squares) and violaxanthin (9) in methanol (triangles). Solid lines represent monoexponential fits of the decays yielding S_1 lifetimes of 4.2 ps (lycopene), 8.5 ps (spheroidene) and 24 ps (violaxanthin). Excitation wavelengths were 480 nm (violaxanthin), 490 nm (spheroidene) and 510 nm (lycopene).

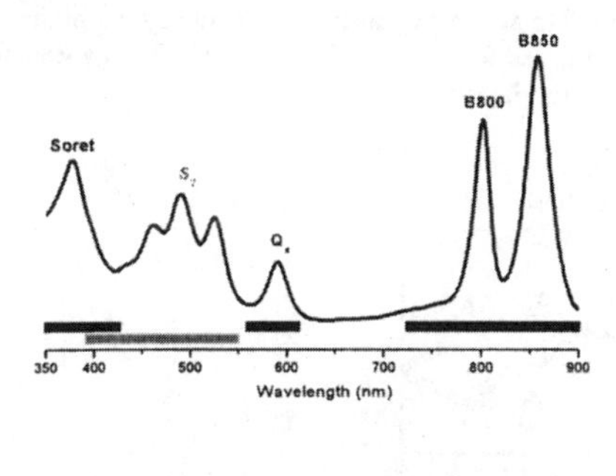

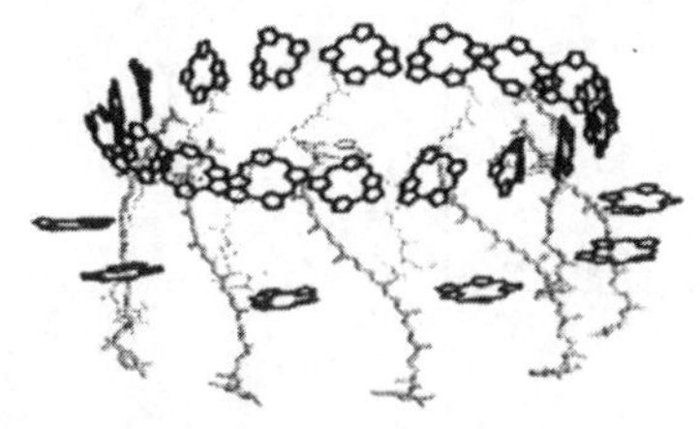

Figure 6. Bottom: arrangement of carotenoid and BChl-a molecules within the LH2 antenna complex of *Rps. acidophila*. The B800 BChl-a molecules are horizontally oriented, and the tightly coupled B850 BChl-a molecules are vertically oriented. The carotenoids are snaking between the BChl molecules. Top: absorption spectrum of the LH2 complex from *Rps. acidophila*. The horizontal bars show spectral regions of BChl-a (upper bars) and carotenoid (lower) absorption bands.

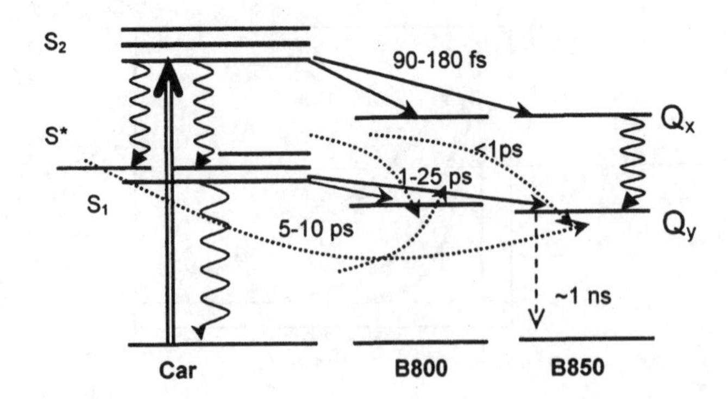

Figure 7. Schematic representation of energy levels and energy transfer pathways between carotenoids and BChl-a in the LH2 complex. Double arrow represents excitation of the S_2 state of a carotenoid. Wavy arrows denote intramolecular relaxation processes, while the dashed arrow represents the long-lived BChl-a fluorescence. Solid arrows represent the dominating energy transfer channels involving the S_2 and S_1 states, although the S_1 channel can be completely suppressed in some LH2 species. The dotted lines represent minor energy transfer channels that usually contributes only fractionally to the total energy transfer: the pathway via higher vibrational levels of the S_1 state and the pathway via the S* state observed for some LH2 complexes. Energy transfer pathways are labeled by the corresponding time constant. See text for details.

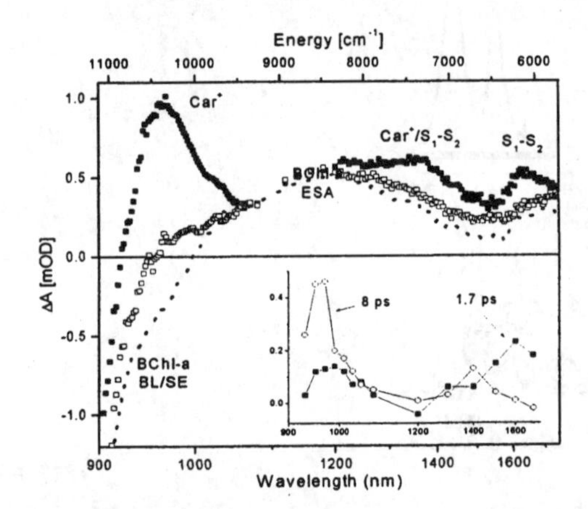

Figure 8. Near-infrared transient absorption spectra of LH2 complexes from *Rb. sphaeroides* recorded at 2 ps (full squares) and 20 ps (open squares). Excitation wavelength is 515 nm. The transient absorption spectrum measured after excitation of the B850 band is shown for comparison (dotted line). The B850 band was excited at 850 nm and the transient spectrum after 850 nm excitation was normalized to have the same magnitude of B850 bleaching as the transient spectra excited in the carotenoid region. Inset shows spectral profiles of the amplitudes of the 1.7 ps (full squares) and 8 ps (open circles) decay components as extracted from multiexponential global fitting of the kinetics, demonstrating different origin of the 960 nm band (carotenoid radical) and 1600 nm band (S_1-S_2 ESA).

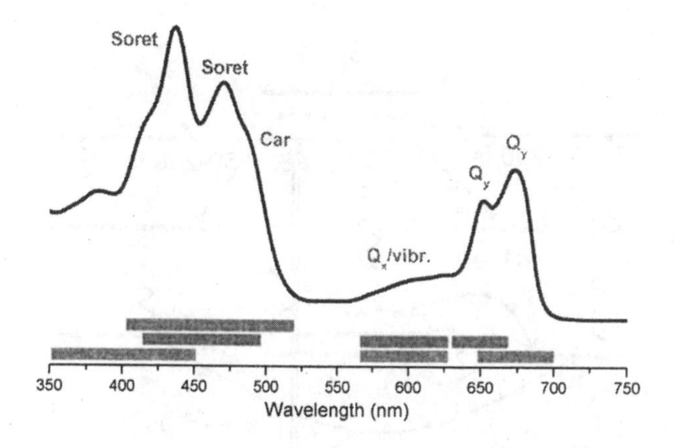

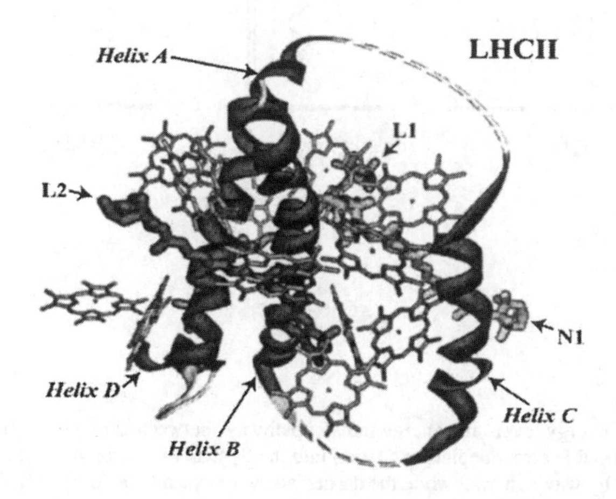

Figure 9. Bottom: Model structure of the LHCII antenna protein based on crystalographic data and in vitro reconstitution and mutational analysis. Two luteins (L1) and (L2) were revealed in the crystal structure, while neoxanthin (N1) was identified by biochemical studies. Chlorophyll molecules are depicted as tetrapyrole rings. Top: Absorption spectrum of the LHCII complex.

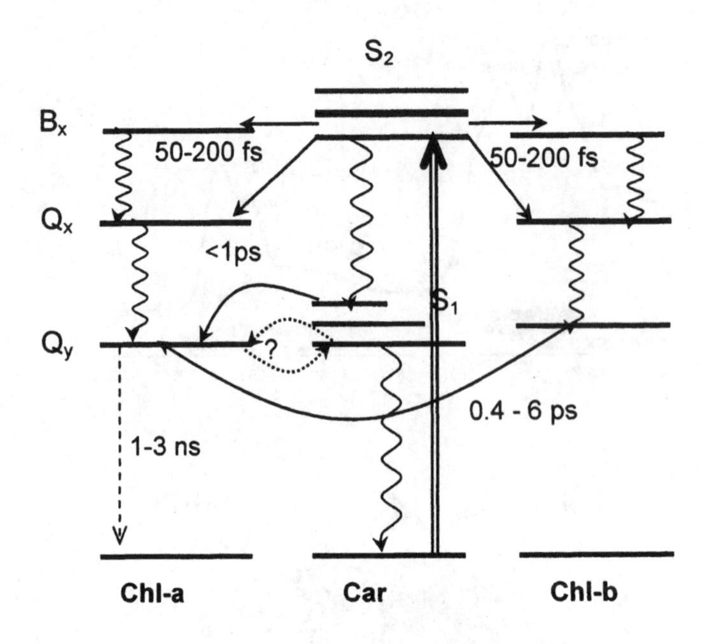

Figure 10. Scheme of energy levels and energy transfer pathways between carotenoids and Chl molecules in the LHCII complex. The double arrow depicts excitation into the S_2 state of a carotenoid. Intramolecular relaxation processes are denoted by wavy arrows, while the dashed arrow represents the long-lived Chl-*a* fluorescence. Solid arrows represent the energy transfer channels confirmed by time-resolved studies. The dotted lines represent possible energy transfer channels involving energy transfer between thermalized carotenoid S_1 state and Chl-a, and back energy transfer from Chl-a to the S_1 state of a carotenoid. The known energy transfer pathways are labeled by the corresponding time constant. See text for detail

Table 1. Energies of the S_2 and S_1 States of Carotenoids.[a]

Carotenoid[b]	N[c]	S_2 energy	S_1 energy			
			Fluorescence	**Raman**[d]	**S_1-S_2**	**2-photon**
Spirilloxanthin	13	19000	11900	11780	11560±200	
anhydrorhodovibrin	12	19400	12500	12195		
Lycopene	11	19900	13300	13200 12920	12500±150	
rhodopin glucoside	11	20000			12800±200 12550±150[e]	
Spheroidenone	10O	19500			13200±100	
Spheroidene	10	20700	14200[f]	14200	13400±90[g]	13900±150[e]
β-carotene	9β2	20840	14200±500 14500[h]	14670 14500		~14500
Zeaxanthin	9β2	21010	14550±90 14610±40[i]		14030±90 13850±200[j]	
Lutein	9β1	21200			14050±300[j]	<15100 ~15300[k]
Violaxanthin	9	21230	14880±90 15580±60[i]			
nonaene*	9	21510	15120±220			
Neurosporene	9	21300	15300			
m9-β-carotene*	7β2	22500		15750		
Siphonaxanthin	8O	20870	~16000		16610±200	
Fucoxanthin	7OA	20960	~16000		16520±200	
Peridinin	7LOA	20620	16200 * 16700 *		16100±200	18500-18850 * ~16200 *
octaene*	8	22420	16840±170			
heptaene*	7	23530	18160±40			

a) The values correspond to room temperature measurements unless stated otherwise. Solution measurements were performed in *n*-hexane except the following cases: violaxanthin, zeaxanthin, lutein and rhodopin glucoside in methanol; fucoxanthin and siphonaxanthin in CS_2; lutein in octanol; β-carotene in octane.
b) Carotenoids marked by asterisk are synthesized molecules that do not occur naturally.
c) Conjugation length. See caption of Figure 1 for explanation.
d) All resonance Raman measurements were carried out with crystalline carotenoid microcrystals deposited on KBr discs.
e) In LH2 complex.
f) At both room temperature and 170 K.
g) In both LH2 complex and *n*-hexane solution. The same value was obtained also in solution at 185 K.
h) At 170 K.
i) At 77 K in EPA glass.
j) In LHCII complexes reconstituted with a single carotenoid species.
k) In native LHCII complexes containing lutein, violaxanthin and neoxanthin with stoichiometry approximately 1.8:1:0.2

*) Slightly different values obtained in different experiments

Table 2. S_1 Lifetimes of Carotenoids without Conjugated Carbonyl Group.[a]

Carotenoid[b]	N[c]	τ_{S1} (ps)
dodecapreno-β-carotene*	17β2	0.5
decapreno-β-carotene*	13β2	1.1
tetradehydrospheroidene*	13	1.1
spirilloxanthin	13	1.4
anhydrorhodovibrin	12	2.2
didehydrospheroidene*	12	2.7
didehydrospheroidene*	11	3.9
Lycopene	11	4-4.7
Rhodopin glucoside	11	4.2-4.8
15,15-*cis*-spheroidene	10	7
β-carotene	9β2	9-11
zeaxanthin	9β2	9
spheroidene	10	8-9.5
diatoxanthin	9Aβ	13.3
Lutein	9β1	14
antheraxanthin	9β1	14.4
Nonaene*	9	18
neurosporene	9	21.2
diadinoxanthin	9A	22.8
violaxanthin	9	24
dihydrospheroidene*	9	25.4
neoxanthin	8A	35
m9-β-carotene*	7β2	52
octaene*	8	68
tetrahydrospheroidene*	8	85
m8-β-carotene*	6β2	96
m7-β-carotene*	5β2	282
Heptaene*	7	290
tetrahydrospheroidene*	7	407
m5-β-carotene*	3β2	2700

a) all the values refers to measurements at room temperature
b) carotenoids marked by asterisk are synthesized carotenoids that do not occur naturally
c) conjugation length. See caption of Figure 1 for description.

References

(1) Frank H. A.; Cogdell R. J. In *Carotenoids in Photosynthesis*; Young, A. J., Britton, G., Eds.; Chapman and Hall, 1993, p. 252.
(2) Frank H. A.; Cogdell R. J. *Photochem. Photobiol.* **1996**, *63*, 257.
(3) Koyama, Y.; Kuki, M.; Andersson, P.O.; Gillbro, T. *Photochem. Photobiol.* **1996**, *63*, 243.
(4) Ritz, T.; Damjanović. A.; Schulten, K.; Zhang, J.-P.; Koyama, Y. *Photosynth. Res.* **2000**, *66*, 125.
(5) van Amerongen, H.; van Grondelle, R. *J. Phys. Chem. B* **2001**, *105*, 604.
(6) Edge, R.; Truscott, T. G. In *Photochemistry of Carotenoids*; Frank, H. A., Young, A. J., Britton, G. Cogdell, R. J., Eds.; Dordrecht, The Netherlands, 1999; p. 223.
(7) Landrum, J. T.; Bone, R. *Arch. Bioch. Biophys.* **2001**, *385*, 21.
(8) Nishino, H. *J. Cell. Biochem.* **1997**, *27*, 86.
(9) Dutton, H. J.; Manning, W. M.; Duggar, B. M. *J. Phys. Chem* **1943**, *47*, 308.
(10) Govindjee, In *Photochemistry of Carotenoids*; Frank, H. A., Young, A. J., Britton, G. Cogdell, R. J., Eds.; Dordrecht, The Netherlands, 1999; p. 1.
(11) Schulten, K.; Karplus, M. *Chem. Phys. Lett.* **1972**, *14*, 305.
(12) Hudson, B. S.; Kohler, B. E. *Chem. Phys. Lett.* **1972**, *14*, 299.
(13) Tavan, P.; Schulten, K. *Phys. Rev. B* **1987**, *36*, 4337.
(14) Sashima, T.; Koyama, Y.; Yamada, T.; Hashimoto, H. *J. Phys. Chem. B* **2000**, *104*, 5011.
(15) Furuichi, K.; Sashima, T.; Koyama, Y. *Chem. Phys. Lett.* **2002**, *356*, 547.
(16) Orlandi, G.; Zerbetto, F.; Zgierski, M. Z. *Chem. Rev.* **1991**, *91*, 867.
(17) Hudson, B. S.; Kohler, B. E.; Schulten, K. In *Excited States*, Ed. Lim, E. C.; Academic Press, New York, 1982; Vol. 6, p.1.
(18) Wasielewski, M. R.; Kispert, L. D. *Chem. Phys. Lett.* **1986**, *128*, 238.
(19) Christensen, R. In *Photochemistry of Carotenoids*; Frank, H. A., Young, A. J., Britton, G. Cogdell, R. J., Eds.; Dordrecht, The Netherlands, 1999; p. 137.
(20) Schulten, K.; Ohmine, I.; Karplus, M. *J. Chem. Phys.* **1976**, *64*, 4422.
(21) Tavan, P.; Schulten, K. *J. Chem. Phys.* **1979**, *70*, 5407.
(22) Tavan, P.; Schulten, K. *J. Chem. Phys.* **1986**, *85*, 6602.
(23) Fuss, W.; Haas, Y.; Zilberg, S. *Chem. Phys.* **2000**, *259*, 273.
(24) Britton, G. In *Carotenoids*; Britton, G., Liaaen-Jensen, S. and Pfander, H, Eds.; Birkhäuser Verlag, Basel, 1995; p.13.
(25) Strickler, S. J.; Berg, R. A. J. *J. Chem. Phys.* **1962**, *37*, 814.
(26) Frank, H. A.; Desamero, R. Z. B.; Chynwat, V.; Gebhard, R.; van der Hoef, I.; Jansen, F. J.; Lugtenburg, J.; Gosztola, D.; Wasielewski, M. R. *J. Phys. Chem. A* **1997**, *101*, 149.
(27) Frank, H. A.; Josue, J. S.; Bautista, J. A.; van der Hoef, I.; Jansen, F. J.; Lugtenburg, J.; Wiederrecht, G.; Christensen, R. L. *J. Phys. Chem B*, **2002**, *106*, 2083.
(28) Macpherson, A. N.; Gillbro, T. *J. Phys. Chem. A* **1998**, *102*, 5049.
(29) Ricci, M.; Bradforth, S. E.; Jimenez, R.; Fleming, G. R. *Chem. Phys. Lett.* **1996**, *259*, 381.
(30) Akimoto, S.; Yamazaki, I.; Takaichi, S.; Mimuro, M. *Chem. Phys. Lett.* **1999**, *313*, 63.
(31) Akimoto, S.; Yamazaki, I.; Sakawa, T.; Mimuro, M. *J. Phys. Chem. A* **2002**, *106*, 2237.
(32) Macpherson, A. N.; Arellano, J., B.; Fraser, N. J.; Cogdell, R. J.; Gillbro, T. *Biophys. J.* **2001**, *80*, 923.
(33) Mimuro, M.; Akimoto, S.; Takaichi, S.; Yamazaki, I. *J. Am. Chem. Soc.* **1997**, *119*, 1452.

(34) Sashima, T.; Nagae, H.; Kuki, M.; Koyama, Y. *Chem. Phys. Lett.* **1999**, *299*, 187.
(35) Zhang, J.-P.; Inaba, T.; Watanabe, Y.; Koyama, Y. *Chem. Phys. Lett.* **2000**, *332*, 351.

(36) Rondonuwu, F. S.; Watanabe, Y.; Zhang, J.-P.; Furuichi, K.; Koyama, Y. *Chem. Phys. Lett.* **2002**, *357*, 376.
(37) Fujii, R.; Inaba, T.; Watanabe, Y.; Koyama, Y.; Zhang, J. P. *Chem. Phys. Lett.* **2003**, *369*, 165.
(38) Rondonuwu, F. S.; Watanabe, Y.; Fujii, R.; Koyama, Y. *Chem. Phys. Lett.* **2003**, *376*, 292.
(39) Thrash, R. J.; Fang, H. L. B.; Leroi, G. E. *J. Chem. Phys.* **1977**, *67*, 5930.
(40) Cosgrove, S. A.; Guite,, M. A.; Burnell, T. B.; Christensen, R. L. *J. Phys. Chem.* **1990**, *94*, 8118.
(41) deCoster, B.; Christensen, R. L.; Gebhard, R.; Lugtenburg, J.; Farhoosh, R.; Frank, H. A. *Biochim. Biophys. Acta* **1992**, *1102*, 107.
(42) Christensen, R. L.; Goyette, M.; Gallagher, L.; Duncan, J.; DeCoster, B.; Lugtenburg, J.; Jansen, F. J.; van der Hoef, I. *J. Phys. Chem. A* **1999**, *103*, 2399

(43) Frank, H. A.; Josue, J. S.; Bautista, J. A.; van der Hoef, I.; Jansen, F. J.; Lugtenburg, J.; Wiederrecht, G.; Christensen, R. L. *J. Phys. Chem B*, **2002**, *106*, 2083.

(44) Engelman, R.; Jortner, J. *J. Mol. Phys.* **1970**, *18*, 145.
(45) Chynwat, V.; Frank, H. A. *Chem. Phys.* **1995**, *194*, 237.
(46) Frank, H. A.; Cua, A.; Chynwat, V.; Young, A.; Gosztola, D.; Wasielewski, M. R. *Photosynth. Res.* **1994**, *41*, 389.
(47) Frank, H. A.; Cua, A.; Chynwat, V.; Young, A.; Gosztola, D.; Wasielewski, M. R. *Biochim. Biophys. Acta* **1996**, *1277*, 243.
(48) Shima, S.; Ilagan, R. P.; Gillespie, N.; Sommer, B. J.; Hiller, R. G.; Sharples. F. P.; Frank, H. A.; Birge, R. R. *J. Phys. Chem. A* **2003**, *107*, 8052.
(49) Polívka, T.; Herek, J. L.; Zigmantas, D.; Åkerlund, H.-E.; Sundström, V. *Proc. Natl. Acad. Sci. U.S.A.* **1999**, *96*, 4914.
(50) Bondarev, S. L.; Knyukshto, V. N. *Chem. Phys. Lett.* **1994**, *225*, 346.
(51) Andersson, P.-O.; Bachilo, S. M.; Chen, R.-L.; Gillbro, T.; *J. Phys. Chem.* **1995**, *99*, 16199.
(52) Onaka, K.; Fujii, R.; Nagae, H.; Kuki, M.; Koyama, Y.; Watanabe, Y. *Chem. Phys. Lett.* **1999**, *315*, 75.
(53) Robert, B. In *Photochemistry of Carotenoids*; Frank, H. A., Young, A. J., Britton, G. Cogdell, R. J., Eds.; Dordrecht, The Netherlands, 1999; p. 189.
(54) Frank, H. A.; Cua, A.; Chynwat, V.; Young, A.; Gosztola, D.; Wasielewski, M. R. *Photos. Res.* **1994**, *41*, 389.
(55) Frank H. A.; Bautista J. A.; Josue J. S.; Young A. J. *Biochemistry* **2000**, *39*, 2831.

(56) Polívka, T.; Zigmantas, D.; Frank, H. A.; Bautista, J. A.; Herek, J. L.; Koyama, Y.; Fujii, R.; Sundström, V. *J. Phys. Chem. B* **2001**, *105*, 1072.

(57) Billsten, H. H.; Herek, J. L.; Garcia-Asua, G.; Hashøj, L.; Polívka, T.; Hunter, C. N.; Sundström, V. *Biochemistry* **2002**, *41*, 4127.
(58) Polívka, T., Zigmantas, D., Herek, J. L., He, Z., Pascher, T., Pullerits, T., Cogdell, R. J., Frank, H. A., Sundström, V. *J. Phys. Chem. B* **2002**, *106*, 11016.
(59) Papagiannakis, E.; Stokkum, I. H. M.; van Grondelle, R.; Niederman, R. A.; Zigmantas, D.; Sundström, V.; Polívka, T. *J. Phys. Chem. B* **2003**, *107*, 11216.
(60) Zigmantas, D.; Polívka, T.; Hiller, R. G.; Yartsev, A.; Sundström, V. *J. Phys. Chem. A* **2001**, *105*, 10296.
(61) Zigmantas, D.; Hiller, R. G.; Yartsev, A.; Sundström, V.; Polívka, T. *J. Phys. Chem B.* **2003**, *107*, 5339

(62) Zigmantas, D.; Hiller, R. G.; Sharples, F.P.; Frank, H. A.; Sundström, V.; Polívka, T. *Phys. Chem. Chem. Phys.*, submitted.

(63) Lindal, T.-R.; Liaaen-Jensen, S. *Acta Chem. Scand.* **1997**, *51*, 1128.
(64) Frank, H. A.; Desamero, R. Z. B.; Chynwat, V.; Gebhard, R.; van der Hoef, I.; Jansen, F. J.; Lugtenburg, J.; Gosztola, D.; Wasielewski, M. R. *J. Phys. Chem. A* **1997**, *101*, 149.
(65) Andersson, P.-O.; Bachilo, S. M.; Chen, R.-L.; Gillbro, T.; *J. Phys. Chem.* **1995**, *99*, 16199.
(66) Andersson, P.-O.; Takaichi, S.; Cogdell, R. J.; Gillbro, T. *Photochem. Photobiol.* **2001**, *74*, 549.
(67) Yoshizawa, M.; Aoki, H.; Ue, M.; Hashimoto, H. *Phys. Rev. B* **2003**, *67*, 174302.

(68) Andersson, P.-O.; Gillbro, T. *J. Chem. Phys.* **1995**, *103*, 2509.
(69) Fujii, R.; Inaba, T.; Watanabe, Y.; Koyama, Y.; Zhang, J. P. *Chem. Phys. Lett.* **2003**, *369*, 165.
(70) Zhang, J.-P.; Chen, C.-H.; Koyama, Y.; Nagae, H. *J. Phys. Chem. B* **1998**, *102*, 1632.

(71) Billsten, H. H.; Zigmantas, D.; Sundström, V.; Polívka, T. *Chem. Phys. Lett.* **2002**, *355*, 465.
(72) Frank, H. A.; Bautista, J. A.; Josue, J.; Pendon, Z.; Hiller, R. G.; Sharples, F. P.; Gosztola, D.; Wasielewski, M. R. *J. Phys. Chem. B* **2000**, *104*, 4569.
(73) de Weerd, F. L.; van Stokkum, I. H. M.; van Grondelle, R. *Chem. Phys. Let.* **2002**, *354*, 38.
(74) Frank, H. A.; Bautista, J. A.; Josue, J.; Pendon, Z.; Hiller, R. G.; Sharples, F. P.; Gosztola, D.; Wasielewski, M. R. *J. Phys. Chem. B* **2000**, *104*, 4569.
(75) Polivka, T; Sundström, V. Chem. Rev. **2004**, 104.

(76) McDermott, G.; Prince, S. M.; Freer, A. A.; Hawthornwaite-Lawless, A. M.; Papiz, M. Z.; Cogdell, R. J.; Isaacs, N. W. *Nature* **1995**, *374*, 517.
(77) Koepke, J.; Hu, X.; Muenke, C.; Schulten, K.; Michel, H. *Structure* **1996**, *4*, 581.
(78) He, Z.; Sundström, V; Pullerits, T. *FEBS Lett.* **2001**, *496*, 36.
(79) Herek, J. L.; Polívka, T.; Pullerits, T.; Fowler, G. J. S.; Hunter, C. N.; Sundström, V. *Biochemistry* **1998**, *37*, 7057.
(80) He, Z.; Sundström, V.; Pullerits, T. *Chem. Phys. Lett.* **2001**, *334*, 159.
(81) Dolan, P. M.; Miller, D.; Cogdell, R. J.; Birge, R. R.; Frank, H. A. *J. Phys. Chem. B* **2001**, *105*, 12134.

(82) Georgakopoulou, S.; Cogdell, R. J.; van Grondelle, R.; van Amerongen, H. *J. Phys. Chem. B* **2003**, *107*, 655.
(83) Wang, Y.; Hu, X. *J. Am. Chem. Soc.* **2002**, *124*, 8445.
(84) Mao, L.; Wang, Y.; Hu, X. *J. Phys. Chem. B.* **2003**, *107*, 3963.
(85) Papiz, M. Z.; Prince, S. M.; Howard, T.; Cogdell, R. J.; Isacs, N. W. *J. Mol. Biol.* **2003**, *326*, 1523.
(86) Germeroth, L.; Lottspeich, F.; Robert, B.; Michel, H. *Biochemistry* **1993**, *32*, 5615.
(87) Evans, M. B.; Cogdell, R. J.; Britton, G. *Biochim. Biophys. Acta* **1988**, *935*, 292.
(88) Arellano, J. B.; Bangar Raju, B.; Naqvi, K. R.; Gillbro, T. *Photochem. Photobiol.* **1998** 68, 84.
(89) Andersson, P. O.; Cogdell, R. J.; Gillbro, T. *Chem. Phys.* **1996**, *210*, 195.
(90) Walz, T.; Jamieson, S. J.; Bowers, C.M.; Bullough, P. A.; Hunter, C. N. *J. Mol. Biol.* **1998**, *282*, 833.
(91) Karrasch, S.; Bullough, P. A.; Ghosh, R. *EMBO J.* **1995**, *14*, 631.

(92) Cogdell, R. J.; Hipkins, M. F.; MacDonald, W; Truscott, T. G. *Biochim. Biophys. Acta* **1981**, *634*, 191.
(93) van Grondelle, R.; Kramer, H.; Rijgersberg, C. *Biochim. Biophys. Acta* **1982**, *682*, 208.
(94) Angerhofer, A.; Bornhäuser, F.; Gall, A.; Cogdell, R. J. *Chem. Phys.* **1995**, *194*, 259.
(95) Angerhofer, A.; Cogdell, R. J.; Hipkins, M. *Biochim. Biophys. Acta* **1986**, *848*, 333.
(96) Chadwick, B.; Zhang, C.; Cogdell, R. J.; Frank, H. *Biochim. Biophys. Acta* **1987**, *893*, 444.
(97) Rademaker, H.; Hoff, A. J.; van Grondelle, R.; Duysens, L. N. M. *Biochim. Biophys. Acta* **1980**, *592*, 240.
(98) Naqvi, K. R. *Photochem. Photobiol.* **1980**, *31*, 523.
(99) van Grondelle, R. *Biochim. Biophys. Acta* **1985**, *811*, 147.
(100) Cogdell, R. J.; Frank, H. A. *Biochim. Biophys. Acta* **1987**, *895*, 63.
(101) Shreve, A. P.; Trautman, J. K.; Frank, H. A.; Owens, T. G.; Albrecht, A. C. *Biochim. Biophys. Acta* **1991**, *1058*, 280
(102) Trautman, J. K.; Shreve, A. P.; Violette, C. A.; Frank, H. A.; Owens, T. G.; Albrecht, A. C. *Proc. Natl. Acad. Sci. U.S.A.* **1990**, *87*, 215.
(103) Nagae, H.; Kakitani, T.; Katoh, T.; Mimuro, M. *J. Chem. Phys.* **1993**, *98*, 8012.
(104) Krueger, B. P.; Scholes, G. D.; Jimenez, R.; Fleming, G. R. *J. Phys. Chem. B* **1998**, *102*, 2284.
(105) Macpherson, A. N.; Arellano, J., B.; Fraser, N. J.; Cogdell, R. J.; Gillbro, T. *Biophys. J.* **2001**, *80*, 923.
(106) Herek, J. L.: Wohlleben, W.; Cogdell, R. J.; Zeidler, D.; Motzkus, M. *Nature* **2002**, *417*, 533.
(107) Zhang, J.-P.; Inaba, T.; Watanabe, Y.; Koyama, Y. *Chem. Phys. Lett.* **2001**, *340*, 484.
(108) Desamero, R. Z. B.; Chynwat, V.; van der Hoef, I.; Jansen, F. J.; Lugtenburg, J.; Gosztola, D.; Wasielewski, M. R.; Cua, A.; Bocian, D. F.; Frank, H. A. *J. Phys. Chem. B* **1998**, *102*, 8151.
(109) Dexter, D. L. *J. Chem. Phys.* **1953**, *21*, 836.
(110) Damjanović, A.; Ritz, T.; Schulten, K. *Phys. Rev. E* **1999**, *59*, 3293.
(111) Förster, T. In *Modern Quantum Chemistry*; Sinanoglu, O., Ed.; Academic Press, New York, 1965; p. 93.
(112) Krueger, B. P.; Scholes, G. D.; Fleming, G. R.; *J. Phys. Chem. B* **1998**, *102*, 5378.
(113) Tretiak, S.; Middleton, C.; Chernyak, V.; Mukamel, S. *J. Phys. Chem. B* **2000**, *104*, 9540.
(114) Krueger, B. P.; Yom, J.; Walla, P. J.; Fleming, G. R. *Chem. Phys. Lett.* **1999**, *310*, 57.
(115) Shreve, A. P.; Trautman, J. K.; Owens, T. G.; Albrecht, A. *Chem. Phys. Lett.* **1990**, *170*, 51.
(116) Papagiannakis, E.; Kennis, J. T. M.; van Stokkum, I. H. M.; Cogdell, R. J.; van Grondelle, R. *Proc. Natl. Acad. Sci. U.S.A.* **2002**, *99*, 6017.
(117) Papagiannakis, E.; Das, S. K.; Gall, A.; Stokkum, I. H. M.; Robert, B.; van Grondelle, R.; Frank, H. A.; Kennis, J. T. M. *J. Phys. Chem. B* **2003**, *107*, 5642.
(118) Zhang, J.-P.; Fujii, R.; Qian, P.; Inaba, T.; Mizoguchi, T.; Koyama, Y.; Onaka, K.; Watanabe, Y. *J. Phys. Chem. B* **2000**, *104*, 3683.
(119) Walla, P. J.; Linden, P. A.; Hsu, C.-P.; Scholes, G. D.; Fleming, G. R. *Proc. Natl. Acad. Sci. U. S. A.* **2000**, *97*, 10808.

(120) Wohlleben, W.; Buckup, T.; Herek, J. L.; Cogdell, R. J.; Motzkus, M. *Biophys. J.* **2003**, *85*, 442.
(121) Gradinaru, C. C.; Kennis, J. T. M.; Papagiannakis, E.; van Stokkum, I. H. M.; Cogdell, R. J.; Fleming, G. R.; Niederman, R. A.; van Grondelle, R. *Proc. Natl. Acad. Sci. U.S.A* **2001**, *98*, 2364.
(122) Okamoto, H.; Ogura, M.; Nakabayashi, T.; Tasumi, M. *Chem. Phys.* **1998**, *236*, 309.
(123) Scholes, G. D.; Harcourt, R. D.; Fleming, G. R. *J. Phys. Chem. B* **1997**, *101*, 7302.
(124) Hsu, C.-P.; Walla, P. J.; Head-Gordon, M.; Fleming, G. R. *J. Phys Chem. B*, **2001**, *105*, 11016.

(125) Walla, P. J.; Linden, P. A.; Hsu, C.-P.; Scholes, G. D.; Fleming, G. R. *Proc. Natl. Acad. Sci. U. S. A.* **2000**, *97*, 10808.

(126) Lindal, T.-R.; Liaaen-Jensen, S. *Acta Chem. Scand.* **1997**, *51*, 1128.
(127) Walla, P. J.; Linden, P. A.; Ohta, K.; Fleming, G. R. *J. Phys. Chem. A* **2002**, *106*, 1909.
(128) Green, B. R.; Pichersky, E.; Kloppstech, K. *Trends Biochem. Sci.* **1991**, *16*, 181.
(129) Bassi, R.; Pineau,B.; Dainese, P.; Marquardt, J. *Eur. J. Biochem.* **1993**, *212*, 297.
(130) Boekema, E. J.; van Roon, H.; Breemen, J. F. L.; Dekker, J. P. *Eur. J. Biochem.* **1999**, *266*, 444.
(131) Kühlbrandt, W.; Wang, D. N.; Fujiyoshi, Y. *Nature* **1994**, *367*, 614.
(132) Das, S. K.; Frank, H. A. *Biochemistry* **2002**, *41*, 13087.
(133) Takaichi, S. In *Photochemistry of Carotenoids*; Frank, H. A., Young, A. J., Britton, G. Cogdell, R. J., Eds.; Dodrecht, The Netherlands, 1999; p. 39.
(134) Dellapenna, R. In *Photochemistry of Carotenoids*; Frank, H. A., Young, A. J., Britton, G. Cogdell, R. J., Eds.; Dordrecht, The Netherlands, 1999; p. 21.
(135) Croce, R.; Weiss, S.; Bassi, R. *J. Biol. Chem.* **1999**, *274*, 29613.
(136) Croce, R.; Remelli, R.; Varotto, C.; Breton, J.; Bassi, R. *FEBS Lett.* **1999**, *456*, 1.
(137) Ruban, A. V.; Lee, P. J.; Wentworth, M.; Young, A. J.; Horton, P. *J. Biol. Chem.* **1999** *274*, 10458.
(138) Caffari, S.; Croce, R.; Breton, J.; Bassi, R. *J. Biol. Chem.* **2001**, *276*, 35924.
(139) Bassi, R.; Caffari, S. *Photosynth. Res.* **2000**, *64*, 243.

(140) Bassi, R.; Croce, R.; Cugini, D.; Sandonna, D. *Proc. Natl. Acad. Sci. U.S.A.* **1999**, *96*, 10056.
(141) Crimi, M.; Dorra, D.; Bösinger, C. S.; Giuffra, E.; Holzwarth, A. R.; Bassi, R. *Eur. J. Bioch.* **2001**, *268*, 260.
(142) Demmig-Adams, B.; Adams, W. W. *Trends Plant Sci.* **1996**, *1*, 21.
(143) Horton, P.; Ruban, A. V.; Walters, R. G. *Annu. Rev. Plant Physiol. Plant Mol. Biol.* **1996**, *47*, 655.
(144) Richter, M.; Goss, R.; Wagner, B.; Holzwarth, A. R. *Biochemistry* **1999**, *38*, 12718.
(145) Frank, H. A.; Das, S. K.; Bautista, J. A.; Bruce, D.; Vasilev, S.; Crimi, M.; Croce, R.; Bassi, R. *Biochemistry* **2001**, *40*, 1220.
(146) Polívka, T.; Zigmantas, D.; Sundström, V.; Formaggio, E.; Cinque, G.; Bassi, R. *Biochemistry* **2002**, *41*, 439.
(147) Croce, R.; Müller, M. G.; Bassi, R.; Holzwarth, A. R. *Biophys. J.* **2001**, *80*, 901.
(148) Croce, R.; Müller, M. G.; Caffari, S.; Bassi, R.; Holzwarth, A.; *Biophys. J.* **2003**, *84*, 2517.
(149) Walla, P. J.; Yom, J.; Krueger, B. P.; Fleming, G. (2000) *J. Phys. Chem. B* **2000**, *104*, 4806.
(150) Connelly, J. P.; Müller, M. G.; Bassi, R.; Croce, R.; Holzwarth, A. *Biochemistry* **1997**, *36*, 281.
(151) Gradinaru, C. C.; van Stokkum, I. H. M.; Pascal, A. A.; van Grondelle, R.; van Amerongen, H. *J. Phys. Chem. B* **2000**, *104*, 9330.
(152) Peterman, E. J. G.; Gradinaru, C.; Calkoen, F.; Borst, J. C.; van Grondelle, R.; van Amerongen, H. *Biochemistry* **1997**, *36*, 12208.
(153) Moya, I., Silvestri, M., Cinque, G. and Bassi R. *Biochemistry* **2001**, *40*, 12552.
(154) Ma, Y.-Z.; Holt, N. E.; Li, X.-P.; Niyogi, K.; Fleming, G. R. *Proc. Natl. Acad. Sci. U.S.A.* **2003**, *100*, 4377.
(155) Li, X.-P.; Björkman, O.; Shih, C.; Grossman, A. R.; Rosenquist, M.; Jansson, S.; Niyogi, K. K. *Nature* **2000**, *403*, 391.
(156) Shinoda, S.; Osuka, A.; Nishimura, Y.; Yamazaki, I. *Chem. Lett.* **1995**, 1139.

(157) Shinoda, S.; Tsukube, H.; Nishimura, Y.; Yamazaki, I.; Osuka, A. *J. Org. Chem.* **1999**, *64*, 3757.
(158) He, Z.; Kispert, L. D.; Metzger, R. M.; Gosztola, D.; Wasielewski, M. R. *J. Phys. Chem. B* **2000**, *104*, 6302.
(159) Aspinall-O'Dea, M.; Wentworth, M.; Pascal, A.; Robert, B.; Ruban, A.; Horton, P. *Proc. Natl. Acad. Sci. U.S.A.* **2002**, *99*, 16331.
(160) Hermant, R. M.; Lidell, P. A.; Lin, S.; Alden, R. G.; Kang, H. K.; Moore, A. L.; Moore, T. A.; Gust, D. *J. Am. Chem. Soc.* **1993**, *115*, 2080.
(161) Cardozo, S. L.; Nicodem, D. E.; Moore, T. A.; Moore, A. L.; Gust, D. *J. Braz. Chem. Soc.* **1996**, *7*, 19.
(162) Fungo, F.; Otero, L.; Durantini, E.; Thompson, W. J.; Silber, J. J.; Moore, T. A.; Moore, A. L.; Gust, D.; Sereno, L. *Phys. Chem. Chem. Phys.* **2003**, *5*, 469.
(163) Crofts, A. R. and Yerkes, C. T. (1994) *FEBS Lett. 352*, 265-270.
(164) Naqvi, K. R; Melø, T. B.; Raju, B. B.; Jávorfi, T.; Simidijev, I.; Garab, G. *Spectrochim. Acta A* **1997**, *53*, 2659

(165) Young, A. J.; Phillip, D.; Ruban, A. V.; Horton, P.; Frank, H. A. *Pure Appl. Chem.* **1997**, *69*, 2125.

(166) Grudziński, W.; Krupa, Z.; Garstka, M.; Maksymiec, W.; Swartz, T. E.; Gruszecki, W. I. *Biochim. Biophys. Acta* **2002**, *1554*, 108.
(167) Garab, G.; Cseh, Z.; Kovács, L.; Rajagopal, S.; Várkonyi, Z.; Wentworth, M.; Mustárdy, L.; Dér, A.; Ruban, A. V.; Papp, E.; Holzenburg, A. Horton, P. *Biochemistry* **2002**, *41*, 15121.

6. SPECTROSCOPY OF QUANTUM WELLS AND SUPERLATTICES

CLAUS KLINGSHIRN
Institut für Angewandte Physik
Universität Karlsruhe and
Center for Functional Nanostructures (CFN)
Wolfgang Gaede Str. 1
76131 Karlsruhe
Germany

Abstract: After a short introduction to the electronic properties of solids, especially of semiconductors, we present examples for interband and intersubband spectroscopy, preferentially on CdS/ZnSe structures. After a further introduction to phonons, their properties in CdS/ZnSe superlattices will be reviewed.

1. Prolog

The author presents two contributions to this school. The first one on spectroscopy of quantum wells and superlatttices can in most parts not be considered as "investigating extreme physical conditions" in contrast to the second one, though it covers the spectral range from a few meV to a few eV and concentrates on new materials of reduced dimensionality.

The motivation of treating this topic can be illustrated in various ways e.g. by the following anecdote: In August 1845 Friedrich Wilhelm IV, King of Prussia visited the University of Bonn. (The Rheinland was at that time a province of Prussia). At the University he visited the observatory and the astronomer Prof. Friedrich Wilhelm Argelander who was rather famous at that time. Jovially the king asked "Na, Argelander, was gibt's Neues am Himmel? (Well, Argelander, what is new in the sky?) and he obtained the answer: "Kennen Ihre Majestät schon das Alte?" (Does his majesty already know the old?).

Other ways to say it are: "one should never underestimate the pleasure of the audience to hear things which they already know", "repetitio est mater studiorum" and according to the director of this school we are all students.

Finally and much simpler, the material outlined here will be used by the author and other lectures of the school in the following. The second contribution "Excitonic Bose Einstein Condensation versus "Electron-Hole Plasma Formation" reaches indeed the frontiers of optical spectroscopy and treats extreme conditions. As we shall see, excitonic Bose condensation has been predicted 40 years ago but is still an unsolved problem.

B. Di Bartolo and O. Forte (eds.), Frontiers of Optical Spectroscopy, 221-250.

2. Introduction to Electronic Properties

In this section we give a short and basic introduction to some of the electronic properties of solids in general and especially to semiconductors. Then we present in chapter 3 a classification of quantum wells and superlattices. Details can be found in many textbooks on solid state physics like [1] or semiconductor physics [2-4], from which the Figs. in chapters 2, 3 and 6 are taken, generally with modifications.

In Fig. 1 we explain how a band structure, i.e. a reduced zone scheme (b) develops starting either from the atomic orbitals of the constituent atoms (d) or from free electrons (a). If we bring the atoms closer together, the discrete atomic orbitals below the ionisation continuum shown schematically in Fig. 1d, start to overlap and to split (Fig. 1c) in a similar way as the frequency of n identical harmonic oscillators splits into n close laying frequencies under weak coupling.

The number n is of the order of 10^{23} cm^{-3} for a solid. This means one atomic orbital splits into about 10^{23} sublevels per cm^3. Since the band width is typically of the order of one to ten eV, these levels can be considered for any practical purpose as a continuum. If the atoms are arranged periodically in space, it can be shown, that (quasi-) momentum $\hbar\mathbf{k}$ is a good quantum number allowing the transition from Fig. 1c to Fig. 1b.

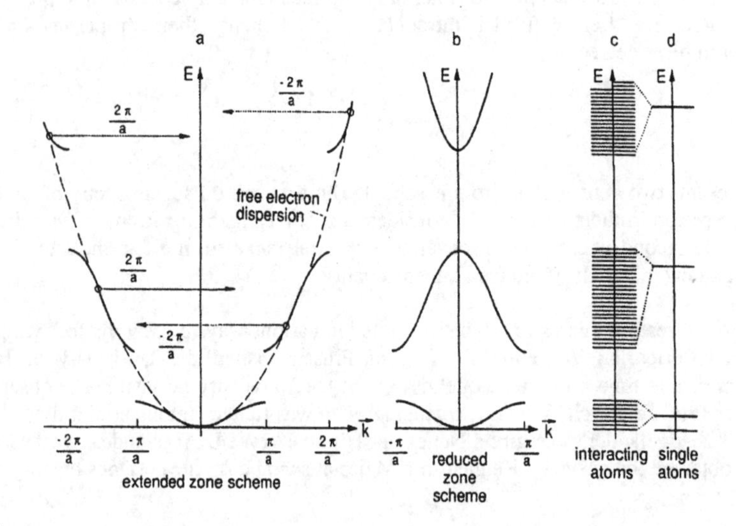

Figure 1: The appearance of an electronic band structure in the reduced zone scheme (b) starting from free electrons in a weak periodic potential (a) or from the atomic orbitals of the constituent atoms (d, c). Schematic! From [4].

Free electrons on the other hand have a non relativistic parabolic dispersion relation

$$E(\mathbf{k}) = \hbar^2\, \mathbf{k}^2 \Big/ 2m_0 = \frac{\hbar^2\left(k_x^2 + k_y^2 + k_2^2\right)}{2m_0} \tag{1}$$

where **k** is the wave-vector and m_0 the free electron rest mass. See the dashed line in Fig. 1a.

If we introduce a weak periodic potential, the plane waves of the free electrons are scattered off this potential. For a general **k**, the scattered waves cancel essentially. There are however special wave vectors, for which all back-scattered waves interfere constructively resulting together with the incident wave in a standing wave. This standing wave can have its antinodes at the position of the mimina or of the maxima of the weak periodic potential resulting for the same **k** vector in two different potential energies, while the kinetic energy is the same in both cases. This fact explaines the appearance of the gaps at theses specific **k** vectors. See the solid lines in Fig. 1a. Theses specific **k** vectors are the boarders of the Brillouin-zones given for a simple cubic lattice by positive and negative integer multiples of π/a. Furthermore it can be shown that the dispersion relation of every crystal is periodic in **k**-space with integer multiples of the vectors of the reciprocal lattice. These reciprocal lattice vectors are given again for an idealized simple cubic lattice by

$$g_i = n\frac{2\pi}{a}; \quad n = \pm 1, \pm 2, \ldots; \quad i = x, y, z \tag{2}$$

resulting in

$$E(\mathbf{k}) = E(\mathbf{k} + \mathbf{g}) \tag{3}$$

This fact allows to shift all "outer" banches of the dispersion relation into the first Brillouin zone extending from $-\pi/a$ to $+\pi/a$ as shown in Fig. 1a by the horizontal arrows, resulting in the so-called reduced zone scheme of Fig. 1b. In other words, we can understand the development of bands from discrete atomic levels through the overlap of the atomic orbitals and the appearance of gaps by the action of a periodic potential on the continous parabolic dispersion of free electrons.

The next step is to populate the bands with the available electrons according to Fermin-Dirac statistics. If we end up at T = 0K with a partly filled band we have a metal. If we have at T = 0 completely filled, so-called valence bands and completely empty so-called conduction bands, we have a semiconductor or an insulator. If the energy gap E_g between the energetically highest valence band (for chemists HOMO = highest occupied molecular orbital, if we consider a crystals as a huge molecule) and the lowest empty conduction band (LUMO = lowest unoccupied molecular orbital) falls in the range

$$0 < E_g \leq 4\text{eV} \qquad \Rightarrow \text{semiconductor} \tag{4a}$$

Actually diamond with $E_g \approx 5.5$eV is considered as one of the semiconductors with the largest gap.

We have usually insulators for E_g above 4eV

$$4\text{eV} \le E_g \qquad \Rightarrow \text{insulator} \tag{4b}$$

and so-called semimetals for vanishing gap i.e. for a situation where conduction and valence band touch

$$0 \approx E_g \qquad \Rightarrow \text{semimetal} \tag{4c}$$

If one looks up band structures of real semiconductors (or other crystalline solids) e.g. in [5] one notices that they are much more complex than Fig. 1b. This is mainly a consequence of the fact, that we have to fold back the dispersion relation in three dimensions. To illustrate this statement, we show in Fig. 2 the result of the back folding of the free electron dispersion for a diamond lattice into the first Brillouin zone for two different directions in **k** space, and for the potential of a Si crystal which crystallizes in the diamond structure, neglecting spin but including some more directions in **k**-space. The Γ-point is always the center of the first Brillouin zone i.e. **k** = 0.

Close to their extrema, the various bands tend to be parabolic. This allows to treat the states essentially as free particles, however with a modified so-called effective mass m given by

$$\frac{1}{m} = \frac{1}{\hbar^2} \frac{\partial^2 E}{\partial k^2} \tag{5a}$$

or more generally in tensor notation

$$\left(\frac{1}{m}\right)_{ij} = \frac{1}{\hbar^2} \frac{\partial^2 E}{\partial k_i \partial k_j} \tag{5b}$$

This means the curvature and thus the effective mass may be anisotropic i.e. it depends on the direction in **k**-space.

For a chemists approach to band-structures see e.g. the contribution by C. Ronda to this book. In the contribution by C. Ronda another intuitive introduction to the band structure has been given starting from a chemists point of view on chemical bonding. It has been shown e.g. that the character of the states changes from the top to the bottom of a band, which is also true in real semiconductors.

A real semiconductor like Si tends, however, to be more complex. As shown in Fig. 3 there is a transition from simple s and p orbitals to sp^3 hybridization with decreasing distance. The fact that there are two identical atoms in the primitive unit cell results in a backfolding of the bands. Actually the top of the valence band has p character and the bottom of the conduction band is of s-type. Furthermore the three dimensional densities of states shows for parabolic bands close to the extrema the well know square-root dependence of the density of states on energy in three dimensional systems.

In the periodic table (table 1) we show, where the most widely investigated and technologically relevant semiconductors can be found. In column IV we find diamond (C), silicon (Si) and germanium (Ge). Graphite is a semimetal, tin a metal except for the modification grey tin, which is also a semimetal. Lead (Pb) is finally only a metal. There is obviously a tendency of decreasing band gap in a column with increasing atomic number:

$$E_g^c \approx 5.5eV,\ E_g^{Si} \approx 1.1eV;\ E_g^{Ge} \approx 0.7eV\ E_g^{grey\,Sn} \approx 0eV$$

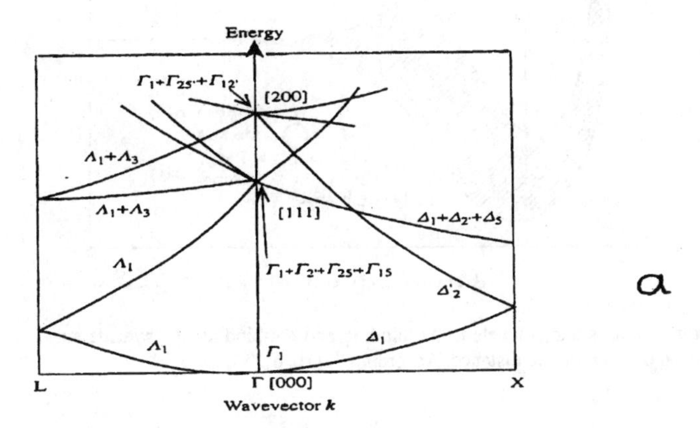

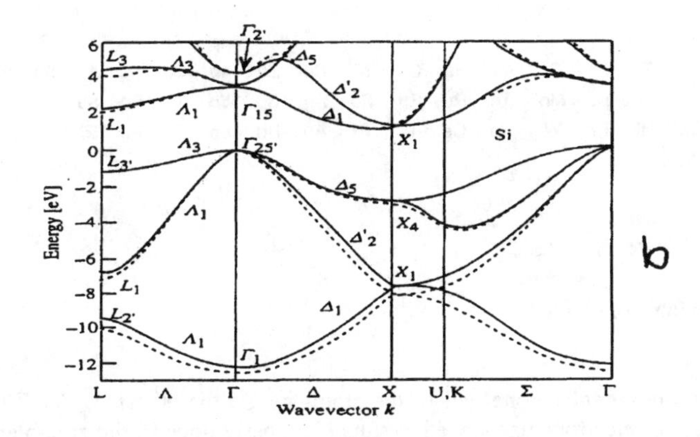

Figure 2: Folding back of the free electron dispersion of eq. (1) into the first Brillouin zone for a diamond structure and vanishing periodic potential (a) and for a weak periodic potential. Solid and dashed lines are two different theoretical approximations (b). According to [3].

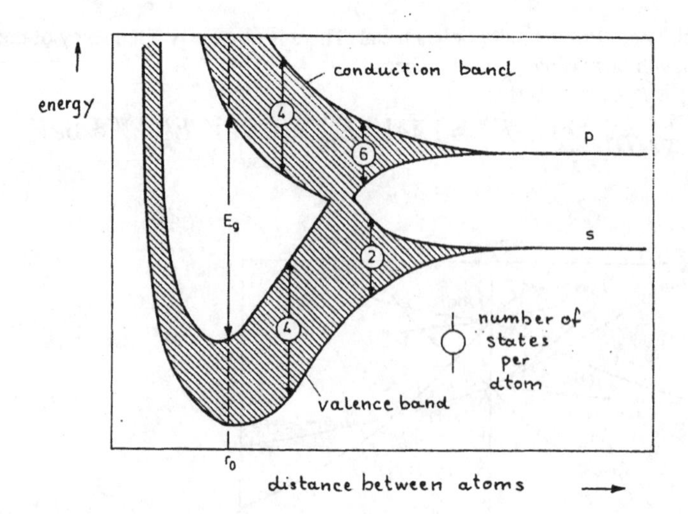

Figure 3: Transition from atomic s and p levels to the binding and antibinding sp^3 hydride states in a covalent semiconductor as a function of inter atomic distance. According to [1b, 6].

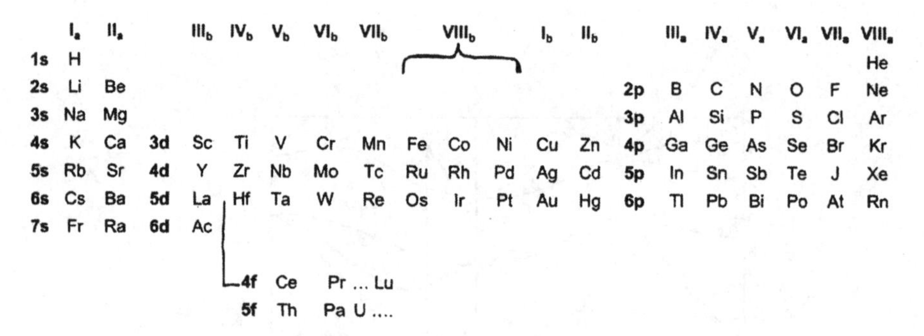

	I_a	II_a		III_b	IV_b	V_b	VI_b	VII_b	VIII_b			I_b	II_b		III_a	IV_a	V_a	VI_a	VII_a	VIII_a
1s	H																			He
2s	Li	Be												**2p**	B	C	N	O	F	Ne
3s	Na	Mg												**3p**	Al	Si	P	S	Cl	Ar
4s	K	Ca	**3d**	Sc	Ti	V	Cr	Mn	Fe	Co	Ni	Cu	Zn	**4p**	Ga	Ge	As	Se	Br	Kr
5s	Rb	Sr	**4d**	Y	Zr	Nb	Mo	Tc	Ru	Rh	Pd	Ag	Cd	**5p**	In	Sn	Sb	Te	J	Xe
6s	Cs	Ba	**5d**	La	Hf	Ta	W	Re	Os	Ir	Pt	Au	Hg	**6p**	Tl	Pb	Bi	Po	At	Rn
7s	Fr	Ra	**6d**	Ac																
					4f	Ce	Pr ... Lu													
					5f	Th	Pa U													

Table 1: The periodic table of elements

In a Ge crystal we can now replace one half of the atoms by Ga the others by As. This procedure leaves the total number of electrons unchanged, results if properly done in the zinc-blende structure and leads to the so called III-V semiconductors. These are all compounds of B, Al, Ga and In with N, P, As or Sb. Repeating this procedure again, leads to the II-VI semiconductors i.e. the compounds of Zn, Cd and Hg with O, S, Se or Te. Another repetition results finally in the I-VII compounds of Cu or Ag with Cl, Br, and J. The chemical binding which is purely covalent for the group IV semiconductors gets increasingly and finally dominantly ionic when going to the III-V, II-V and I-VII materials.

In Fig. 4 we show some more details of the bandstructures of various semiconductors. The bands show deviations from parabolicity for larger k-vectors especially pronounced e.g. in InSb. A semiconductor is said to have a direct gap, or to be a direct semiconductor, if the band extrema occur at the same points of the Brillouin zone (generally at the □-point). Examples are GaAs, InP, GaN most of the II-VI compounds or the Cu-halides. If the extrema occur at different points in k-space, the material is indirect like Si (conduction band minimum in the □-direction close to the X-point, the Fig. 4 shows the anisotropic surfaces of constant energy) or Ge (conduction band minima at the L-point). Some III-V materials are also indirect like GaP, AlAs or the Ag-halides. If spin is included the valence-band

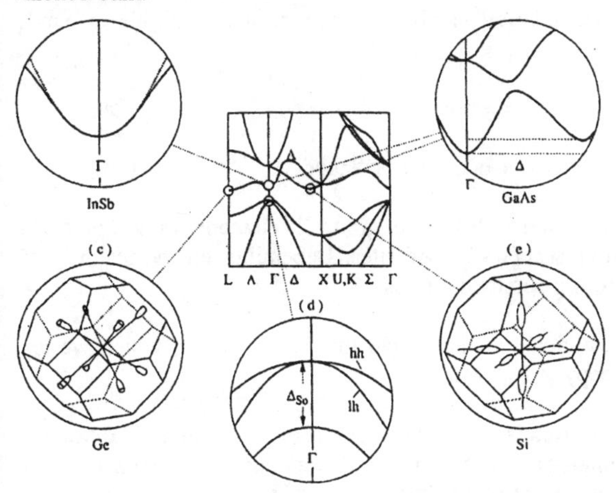

Figure 4: Details of the band structures of various semiconductors (a). According to [2].

maximum (which occurs in all of the above mentioned semiconductors at the □-point) shows a rather complex behaviour. There are two valence bands, split by spin-orbit interaction $□_{so}$. The $□_8$ band is generally the upper one. It is four fold degenerate at $\mathbf{k} = 0$ and splits for $\mathbf{k} \neq 0$ into two twofold degenerate bands with different curvature, which are consequently labelled heavy hole (hh) and light hole (lh) band.

The concept of holes or defect electrons is based on the fact, that it is easier to describe the behaviour of the few unoccupied states in an otherwise filled valenceband than the many electrons in this band. The holes have positive charge and momentum and spin opposite to those of the removed electron.

The next question is now, how can we create electrons and / or holes in semiconductors? The first possibility which may come to mind is thermal excitation of carriers across the gap. Since E_g is in the range of eV and $k_BT = 25$meV at room temperature, the Boltzmann term

$$\exp\left\{-E_g / k_B T\right\} \tag{6}$$

is extremely small. If this would be the only possibility to create electrons and / or holes, the investigation of semiconductors would be a merely academic problem without any technical relevance.

The overwhelming technical relevance of semiconductors and their application in all types of diodes, bipolar- and field effect transistors, thyristors etc. comes from the fact that semiconductors can be doped, some of them even both n- and p-type.

Doping means to introduce intentionally atoms which form levels just below the conduction band (donors) or just above the valence band (acceptors) which exchange carriers with these bands according to

$$D^0 \text{<->} D^+ + e_{CB} \quad \Rightarrow \text{donor} \tag{7a}$$

$$e_{VB} + A^0 \text{<->} A^- \text{ or } A^0 \text{<->} A^- + h \quad \Rightarrow \text{acceptor} \tag{7b}$$

Group V elements act e.g. as donors on lattice sites in Si or Ge, group III elements as acceptors. Si can act as donor on Ga sites in GaAs and as acceptor on As sites. Depending on the majority of carriers one speaks of n- or p-doping or of n- or p-semiconductors.

A third possibility to create carriers is via optical excitation and this is the one in which we are obviously interested in a school on spectroscopy.

In optical inter-band excitation (Fig. 5a) an electron is lifted or excited from the valence band into the conduction band under absorption (or annihilation) of a photon of sufficient energy. Simultaneously a hole is created. This means, that all optical interband excitation (recombination) processes create (annihilate) two particles, namely an electron and a hole. This is however not the whole story. Electron and hole attract each other via their Coulomb potential, resulting in a Hydrogen- or positronium atom like series of states (the so-called excitons) below the gap. See Fig. 5b. The ionisation continuum coincides with the band gap. The excitonic Rydberg energy Ry* is given in simpliest (Wannier exciton) approximation by

$$Ry^* = 13.6eV \frac{1}{\varepsilon^2} \cdot \frac{\mu}{m_0} \quad with\, \mu = \frac{m_e m_h}{m_e + m_h} \tag{8}$$

The excitonic Bohr radius a_B^* is given by

$$a_B^* = a_B \cdot \varepsilon \, m_0 / \mu \tag{9}$$

Typical values for semiconductors are

$$4meV \le Ry^* \le 200meV \tag{10a}$$

and

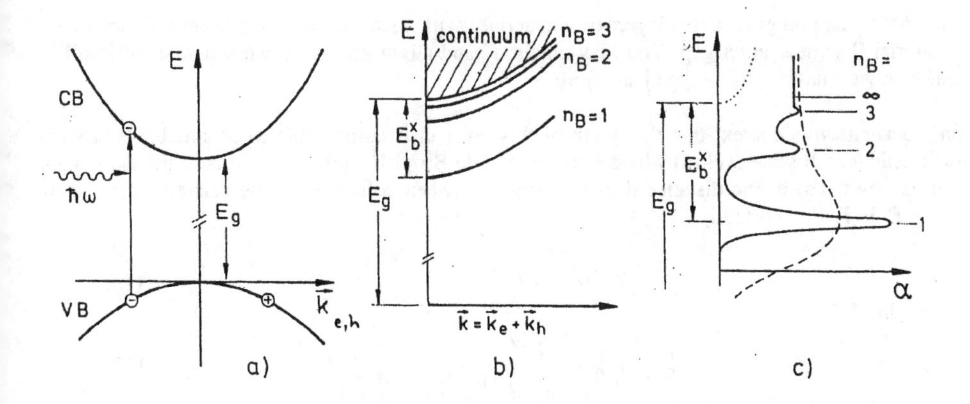

Figure 5: Optical interband excitation (a) the dispersion of excitons in a direct gap semiconductor (b) and the absorption spectrum for a dipole-allowed band-to-band transition (c) (schematic).

$$0.5nm \leq a_B^* \leq 200nm \tag{10b}$$

The excitons can be directly created by absorption of light quanta of an energy given by $E_g - Ry\frac{1}{n_B^2}$. For a direct gap semiconductor with a dipole allowed band-to-band transition this results at low temperature in a hydrogen like series of exciton absorption peaks followed by an almost constant absorption in the ionization continuum (Fig. 5c), which is significantly enhanced compared to the square-root absorption spectrum expected under the same conditions but neglecting the e-h Coulomb interaction band (Sommerfeld enhancement). See the dotted line in Fig. 5c.

Many examples of such spectra and of the corresponding resonances in the reflection spectra and a discussion of corrections to this simple model or of the situation for indirect gap materials are found in [2-5].

3. Quantum Wells and Superlattices

By advanced epitaxial growth techniques like molecular beam epitaxy (MBE), metal-organic chemical vapour deposition (MOCVD) also known as metal organic vapour phase epitaxy (MOVPE) and their various variants, it is possible to grow semiconductor structures almost layer by layer on an atomic scale.

Imagine now, that you grow a thin layers of a material I with a smaller band gap between thick layers of a material II with a larger gap. You may obtain a band alignment as shown schematically in Fig. 6a. Later on we shall call this a type I structure.

"Thin" means in this context that the width of this layer d_I is comparable to or smaller than some characteristic length scale l_c (11a) where l_c can be the de Brogie length λ_B which is in the Boltzmann limit given by (11b) or the distance d of free flight between collisions of the carriers (11c) or the excitonic Bohr Radius a_B (11d)

$$d_A \leq l_c \tag{11a}$$

$$l_c = \lambda_B = h \Big/ \sqrt{3k_B T m} \tag{11b}$$

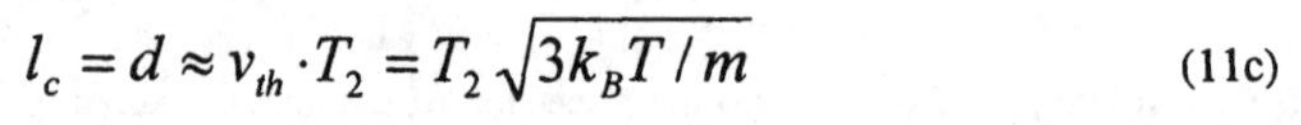

$$l_c = d \approx v_{th} \cdot T_2 = T_2 \sqrt{3k_B T / m} \tag{11c}$$

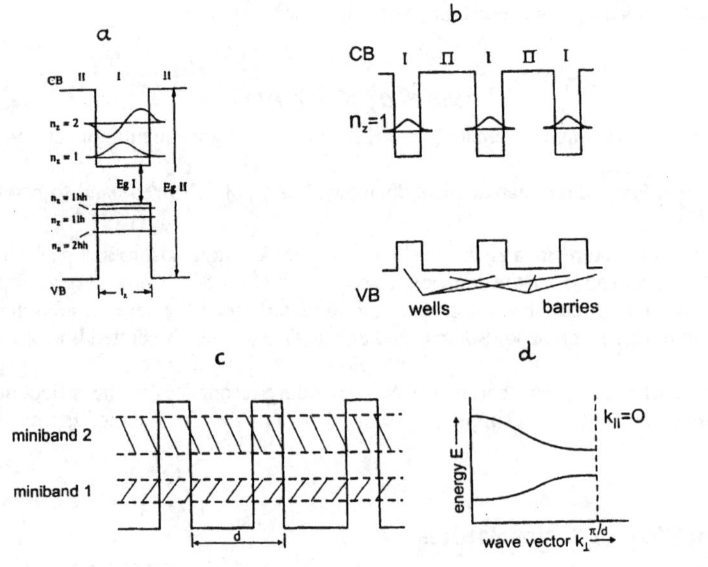

Figure 6: A single quantum well (SQW) of type I (a) a multiple quantum well (b) and a superlattice (only the conduction band is shown) (c) and the mini-band dispersion (d).

with v_{th} = thermal velocity and T_2 time between collisions or phase relaxation time which is also temperature dependent.

$$l_c = a_B^* = a_B^H \cdot \varepsilon \cdot {}^{m_0}\!/\!{}_{\mu} \tag{11d}$$

Then the electrons and holes are confined in growth (or z-) direction to material I which is called a quantum well, while they can move still as free particles in x and y directions. There is a finite number n_{zi} of such quantized states with quantization energies E_i since the barrier height is finite. We show in Fig. 6a the first two quantized electron and hh states and the first lh state.

The dispersion relation of such quasi two dimensional states is given for simple isotropic and parabolic bands with effective mass m by

$$E(\mathbf{k}_{\parallel}) = E_i + \frac{\hbar^2}{2m}\left(k_x^2 + k_y^2\right) \tag{12}$$

For the valence band shown e.g. in Fig. 4 the in-plane dispersion relation is more complex as shown e.g. in [2-5]. Especially the heavy hole and the light hole states have different quantization energies E_{ihh} and E_{ilh}, respectively.

For optical interband transitions the selection rules are in the simplest case

$$n_{zi}^{CB} - n_{zi}^{VB} = 0 \tag{13a}$$

and for intersubband transitions

$$n_{zi}^{CB} - n_{zi}^{CB} = \pm 1 \tag{13b}$$

The interband transitions are again modified by electron-hole pair Coulomb interaction resulting in hh and lh exciton states for every interband transition. We shall see examples below. The excitons in such quantum wells have higher binding energies and oscillator strength compared to those of the parent bulk quantum well material. For details see e.g. [2-5, 7].

Typical values of l_z are around 10nm. With absorption coefficients of the order of $10^4 cm^{-1}$ not much signal is obtained in transmission spectroscopy of a single quantum well. Therefore one grows often multiple quantum well samples (Fig. 6b) for which the barrier is so thick that the wave functions of the quantized layers do not overlap.

If the barrier layers are on the other hand so thin, that the wave functions overlap, this overlap leads again to the formation of now so-called mini-bands in the mini Brillouin zone extending from

$$-\frac{\pi}{d} \le k_z = \frac{\pi}{d}, \quad d = d_I + d_{II} \tag{14}$$

as shown in Fig. 5d. In this case the carriers are delocalized again in z-direction, however with an effective mass depending on the overlap i.e. on the barrier thickness.

If one grows a material with a smaller band gap between barriers with larger gap, an alignment of the band edges as in Fig. 6 or 7a is not necessary. The so-called type I band alignment occurs e.g. for $GaAs/Al_{1-y}Ga_yAs$; $In_{1-y_1}Ga_{y_1}N/Ga_{1-y_2}Al_{y_2}N$ or $Cd_{1-y}Zn_ySe/ZnSe$ where we gave the well material first. If electrons and holes are confined in the two different materials one has a type II staggered band alignment (Fig. 7b) with examples like *CdS/ZnSe* or *InP/GaAs*. If the conduction band of one material overlaps with the valence band of the other one speaks about type II misaligned (Fig. 7c) and the combination of a semiconductor with a semimetal is finally called type III (Fig. 7d).

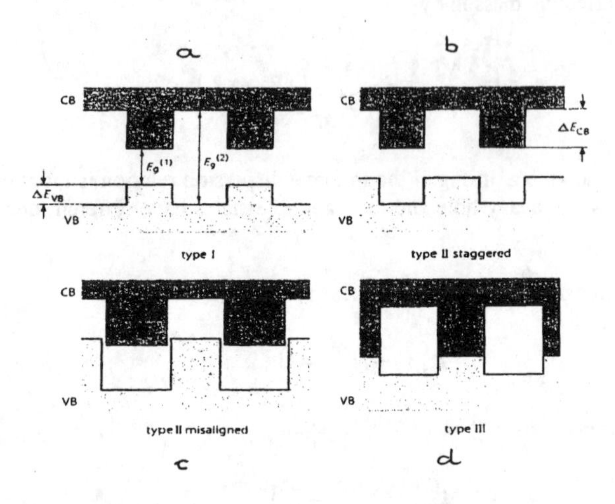

Figure 7: Various bandalignments: type I (a), type II staggered (b), type II misaligned (c), type III (d).

Various layers grow most easily on top of each other if the lattice constants and the coordination of atoms e.g. tetrahedral are equal and if the type of binding (covalent or ionic) is similar. In this sense the combination of $GaAs/Al_{1-y}Ga_yAs$ is almost ideal and possibly the most widely investigated material combination.

The growth of an ionic material on a covalent one or vice versa is difficult to impossible even if the lattice constants are similar (e.g. ZnS on Si). The growth of materials with similar type of binding but different lattice constant leads to strained layers like Si/Ge, InAs/GaAs, CdSe/CdS, CdS/ZnSe or CdSe/ZnSe.

If the strained layer grows pseudomorphic to the substrate it accumulates with increasing thickness more and more elastic deformation energy which will be released above a critical thickness either by (highly unwanted) formation of misfit dislocations or by the formation of growth islands. See [4, 7].

4. Interband Spectroscopy

In this section we show a few examples of interband spectroscopy.

Fig. 8 shows absorption spectra of two different MQW samples. The heavy hole and light hole excitons with main quantum number $n_B = 1$ can be seen for the transitions between the first, second and partly third electron- and hole subbands according to the selection rule (13a).

The higher states of the Hydrogen series merge generally with the ionization continuum.

At low temperatures, the absorption and emission spectra are usually inhomogeneously broadened, caused by alloy disorder in well and / or barrier and by fluctuations of the well width and interdiffusion at the interfaces, which is more important for narrower wells due to the l_z^{-2} dependence of the quantization energies E_i. In good $GaAs/Al_{1-y}Ga_yAs$ MQW samples the widths of absorption and luminescence bands is the order of or below 1meV (Fig. 9a) larger fluctuations especially of the well width lead to wider bands and a considerable Stockes Shift between absorption and emission maxima, as shown in Fig. 9b.

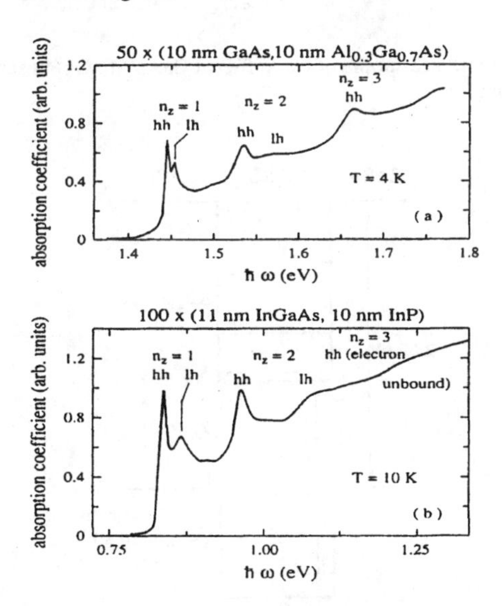

Figure 8: Absorption spectra of $GaAs/Al_{1-y}Ga_yAs$ and of $In_{1-y}Ga_yAs/InP$ MQW samples. According to [4].

We give now some results for the type II system CdS/ZnSe summarized in [8, 9]. Fig. 10 shows the band alignment with some data and the four quantized electron levels in the CdS layer. The numbers given in Fig. 10 show that the energy of the first quantized exciton can be tuned over a large spectral range from 2.0eV almost up to the band gap or more precisely the exciton energy of ZnSe at 2.8eV for vanishing CdS layer thickness.

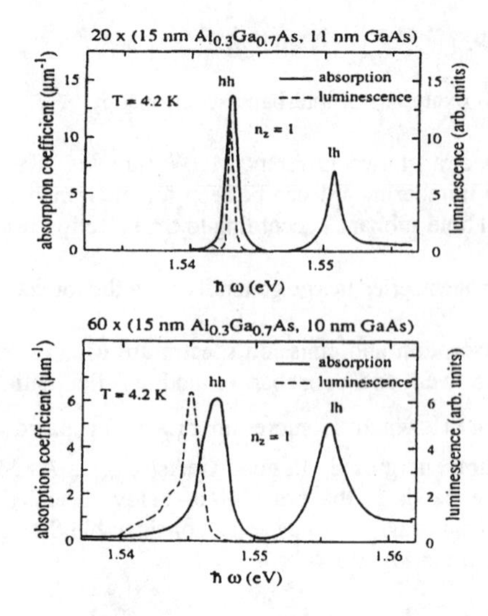

Figure 9: Absorption and luminescence spectra of two different GaAs/$Al_{1-y}Ga_yAs$ samples with different degrees of well width fluctuations. From [4].

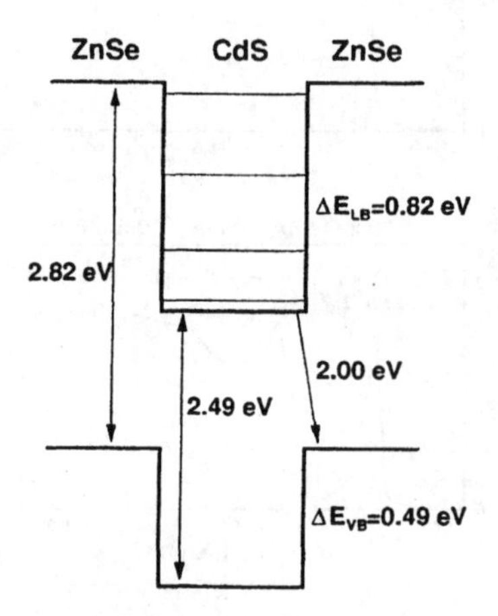

Figure 10: Band alignment of CdS/ZnSe. From [8-10].

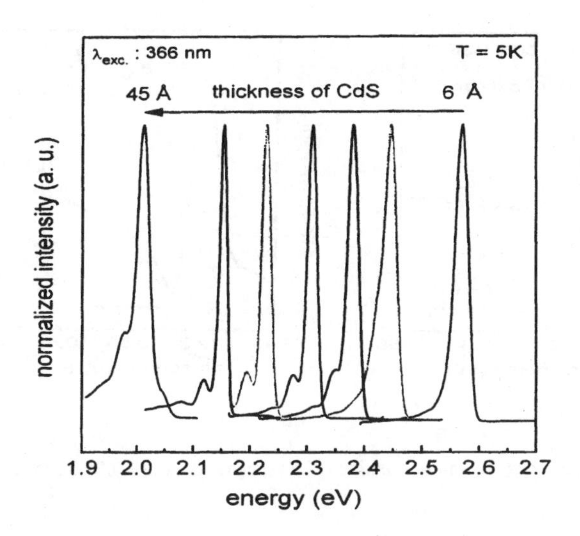

Figure 11: Luminescence spectra of coherently strained CdS/ZnSe quantum well samples with various CdS layer thickness. From [9, 10].

In Fig. 11 we show low temperature luminescence spectra for various CdS layer thicknesses, from 4.5nm down to 0.6nm which cover almost the whole above mentioned range. For details see [8-10].

In Fig. 12 we show a difference between a CdS/ZnSe superlattice and a multiple quantum well sample. The photo-luminescence spectra (PL) are similar. In contrast the photoluminescence excitation spectra, which are similar to the absorption spectra, are different. For the MQW the hole resides in the ZnSe, the electron in the CdS and the spatial overlap is restricted to the interface region. Consequently the exciton binding energy is small and is completely masked by the inhomogenous broadening. In the SL, the electron and holes are delocalized. They overlap more and consequently the exciton binding energy is larger and a pronounced exciton resonance appears in the PLE spectrum. The different overlap of the electron and hole wave functions manifests itself also in different lifetimes. For a MQW sample with $d_{ZnSe} \gg d_{CdS}$ the luminescence decay time is after pulsed excitation in the range of 20 to 30ns. It decreases both with decreasing d_{CdS}, because the electron wave function tunnels more into the ZnSe, and with decreasing d_{ZnSe} because the hole wave function becomes more delocalized. Consequently the luminescence decay time can reach the (sub-)ns regime for short period CdS/ZnSe SL's [8, 9, 11].

A phenomenon, that is observed very frequently in the spectroscopy of semiconductor systems with disorder like in bulk alloy crystals (e.g. [12, 13] and in QW or SL is a so-called S-shape behaviour (e.g. [14, 15, 16] for the CdS/ZnSe system) of the luminescence maximum as a function of temperature. Actually it is an n shaped behaviour but the first name has been introduced and is difficult to change, though it is wrong.

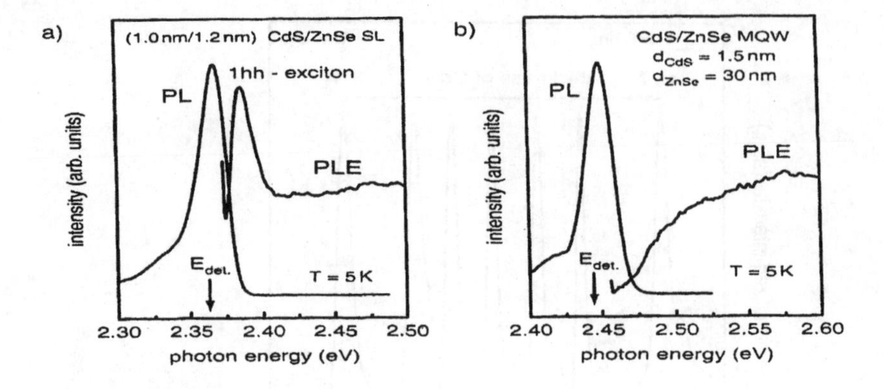

Figure 12: Photoluminescence (PL) and ~excitation (PLE) spectra of a CdS/ZnSe SL (a) and a MQW sample (b). From [8-10].

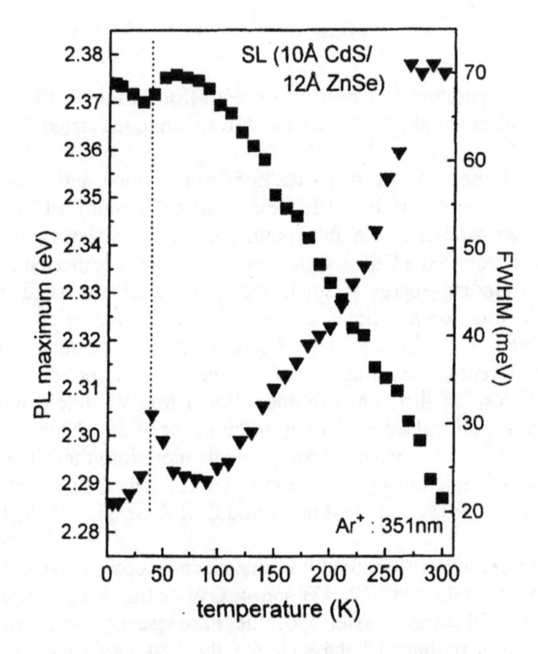

Figure 13: The temperature dependence of the emission maximum of a (1nm CdS/1.2nm ZnSe) superlattice (full squares) and of the half width of the emission band (full triangles) under excitation with the 351nm line (3.53eV) of an Ar^+ laser. According to [8, 9, 14-16].

The phenomenon is shown in Fig. 13. The emission maximum shifts first to the red with increasing temperature, shows then a distinct blue shift, accompanied with a maximum in the spectral half-width and then approaches the temperature dependence of the band gap resulting again in a red shift.

The general interpretation is always in terms of disorder produced localized tail states. For details see e.g. [8, 9, 14-16] and references therein. At low temperature the localized states, including metastable ones, are populated under above band gap excitation randomly, resulting in an inhomogeneously broadened emission band. With increasing temperature there are first thermally activated hopping and tunnelling processes of the electron-hole pairs (excitons) or of single carriers to deeper localized states resulting in a red shift. With further increasing temperature, thermal activation becomes possible to higher localized and extended states which have a much higher density of states explaining the blue shift. The excitons reach then a thermal distribution and the emission follows essentially the temperature dependence of the gap resulting again in a red shift.

This model has been treated in various approximations [14-16] and can be verified by the following experiment from [8, 14-16].

If one increases the spatial resolution of the inhomogenously broadened emission band at low temperature, it decays into individual, narrow lines originating from various localization sites (Fig. 14a).

The global or average maximum shifts with increasing temperature in the way shown in Fig. 13 and 14b, c, while the spectral position of the emission from individual lines follows from the very beginning the temperature dependence of the gap as seen in Fig. 14b, c.

As a last example we shown an effect, which can be observed in CdS/ZnSe SQW. See [9, 17, 18]. In such a structure, the hole should be in principle completely delocalized in ZnSe and the electron confined in the CdS layer. A small attraction results from the e-h Coulomb interaction. This effect is enhanced since both CdS and ZnSe are slightly n-type, even without intentional doping. The capture of electrons in the CdS well results in a negative space charge in the CdS layer, in a curvature of the bottom of the well and in an electric field F in the barrier which drives the holes towards the interface. See the insert of Fig. 15a.

An increasing field F results in a increasing localization and quantization of the hole at the interface and a decreasing CdS well width results in an increasing tuneling of the electron wave function into the barrier, as seen in Fig. 15a where the calculated and measured average lifetimes are given as a function of the CdS well width with the field strength F as a parameter. The experimental data obtained under weak cw-excitation agree for small well width with close to zero field calculations and for larger well width with data for 0.5×10^4 V/cm, since wider wells tend to accumulate more carriers from the ZnSe.

With increasing cw pump power more and more electrons are accumulated in the well resulting in an increasing confining potential for the holes and consequently in an overall blue shift of the emission maximum as shown in Fig. 15b. The self-consistent calculation takes the density dependent electron- and hole envelope functions and their decreasing lifetimes into account.

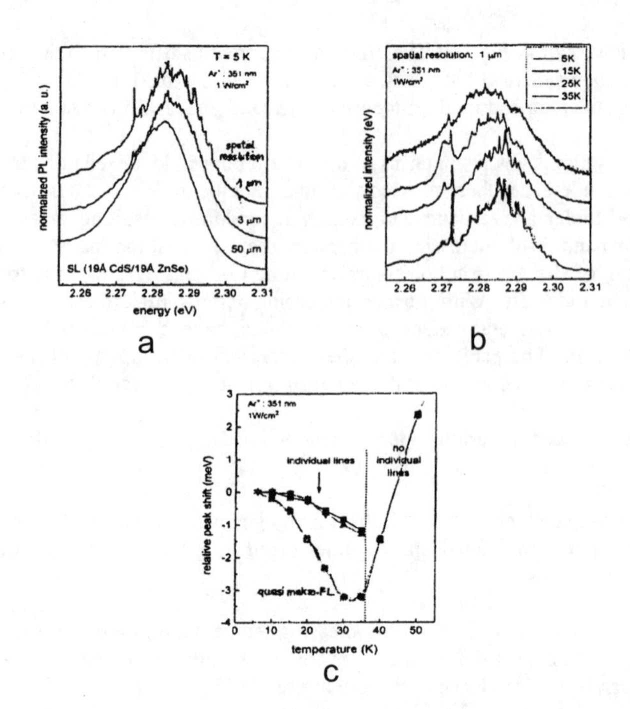

Figure 14: Micro-photoluminescence spectra of a (1.9nm CdS, 1.9nm ZnSe) SL for T = 5K and various detection spot diameters (a) spectra for 1μm and various temperatures (b) and the temperature dependence of the global maximum and that of individual lines (c). From [6, 14 – 17].

5. Intersubband Transitions

If there are some carriers in a well it is also possible to investigate transitions between the various quantized levels in one band, the so-called intersubband transitions. Many references for this topic can be found in [21-24]. We concentrate here again an CdS/ZnSe MQW samples.

If the band itself is simple and parabolic, originating e.g. from s-type atomic orbitals, as is the case for the conduction band of II-VI compounds, the selection rules are also simple. See eq. (13b). Furthermore it can be seen, that the dipole moment of a transition from the $n_z = 1$ to the $n_z = 2$ level is oriented normal to the quantum well (see Fig. 6a) and this means that only that component of the electric field of the incoming IR radiation couples to this transition.

A consequence is, that the transition is forbidden for normal incidence of the light beam. Oblique incidence does not help much, because the beam propagates in the sample almost normal to the well because of Snellius law.

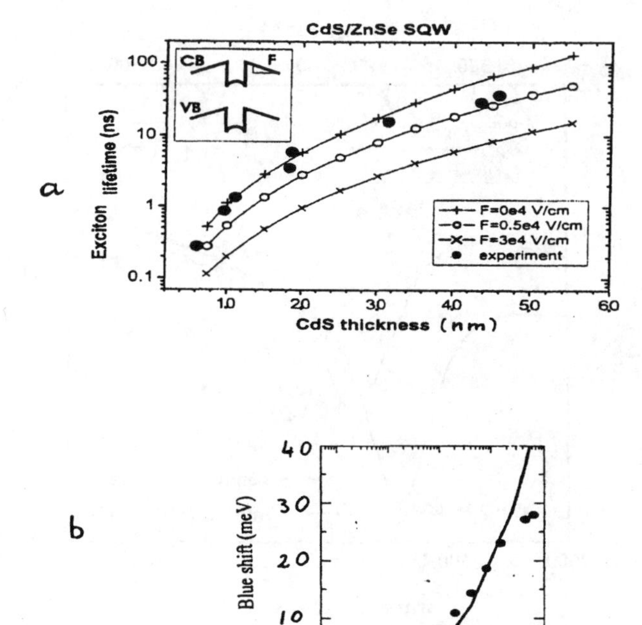

Figure 15: The lifetime of excitons in a CdS/ZnSe SQW as a function of the well width with the electric field F defined in the inset as a parameter (a) and experimental and calculated data for the excitation induced blue-shift of the emission maximum (b) from [9, 18].
For polarized light the emission spectra depend on the properties of the interface see [8, 19] and references therein for other material combinations like ZnSe/BeTe [20].

A better solution is (apart from a grating coupler) a geometry as shown in the l.h. insert of Fig. 16. A beam polarized perpendicular to the plane of incidence (s-pol) does not couple to the transition and can be used as a reference beam in the Fourier spectrometer while the orthogonal one (p-polarisation) has a reasonable field component normal to the well. The carriers in the CdS well can be created by doping. Here Cl-donor doping in the well has been used which has the advantage over modulation-doping in the ZnSe barrier that almost no space charges are built up, however it increases the scattering rate of the carriers from ionized donors. The ratio of the transmission spectra in the allowed and forbidden polarisations T_P/T_s for samples with various well widths and a carrier density around 3 □ 10^{12}cm^{-2} show clearly the transition $n_z = 1 \rightarrow n_z = 2$. The transition shifts to the blue with decreasing well width as expected and in agreement with theory [21-24].

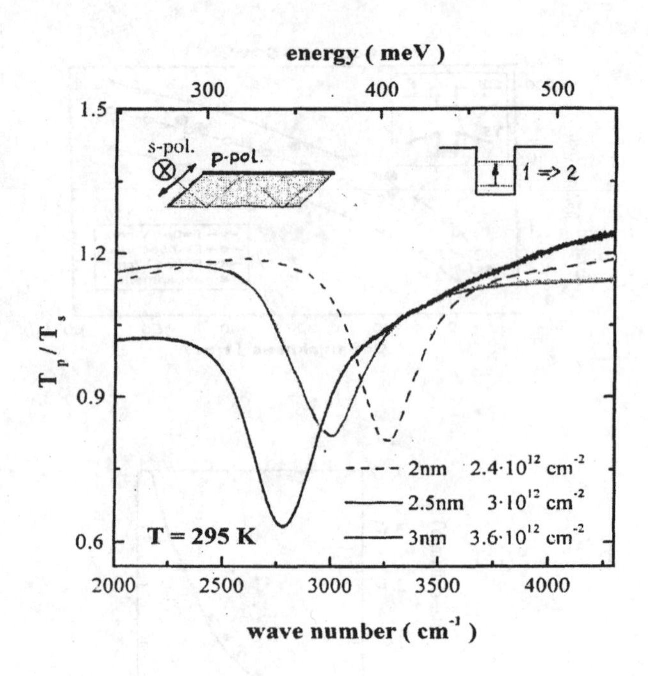

Figure 16: Intersubband transitions of CdS:Cl/ZnSe MQW samples with various CdS well thicknesses. From [21–24].

In Fig. 17a we show a similar normalized transmission spectrum as in Fig. 16, however for a MQW sample with much lower doping level of 6 ▯ $10^{10}cm^{-2}$ resulting in a much smaller signal.

If this sample is additionally excited by a cw Ar^{+} laser one increases the electron density in the CdS well resulting in a measurable signal now in a plot $T_{P\,mit}$ / $T_{P\,ohne}$ i.e. the transmission spectrum in p-polrization with additional pump beam normalized by the one with the same polarization but without pump.

As shown in [21, 24] one can observe with an interband pump beam also intersubband transitions in the valence band. Further examples can be found in [21, 24] for the interminiband transitions in CdS/ZnSe SLs, which have a characteristic lineshape with a tail to higher energies resulting from miniband formation and the corresponding modification of the density of states [3, 4, 7].

6. Phonons in Bulk Semiconductors

Before we come to phonons in superlattices, we give here again a short and very elementary introduction to phonons in bulk, crystalline solids.

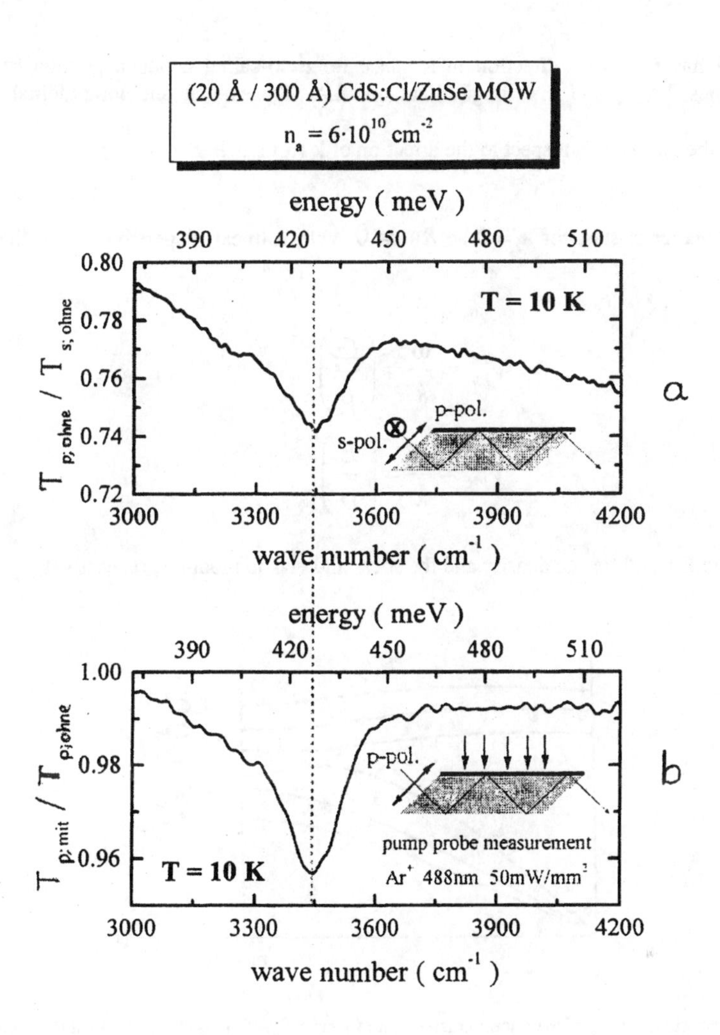

Figure 17: The transmission spectrum as in Figure 16 for a much lower doped sample (a) and the transmission change induced by optical interband pumping with the 488nm line (2.54eV) of an Ar^+ laser. From [21–24].

In Fig. 18 we show schematically the phonon dispersion in the first Brillouin zone of a crystal with two different atoms in the primitive unit cell, compare to Fig. 1b for the electronic dispersion.

Every crystal has for every direction in **k** space three so-called acoustic phonon branches, two transversal ones $TA_{1,\,2}$ and one longitudinal one LA. Transversal and longitudinal refer to the elongation of the atoms with respect to the direction of **k** (for small $k << \frac{\pi}{a}$).

The acoustic branches start for **k** = 0 at $\hbar\omega = 0$ with a linear dispersion giving the velocity of sound

$$v_{Ph}^{TA} = \frac{\omega}{k} = \left(\frac{G}{\rho}\right)^{1/2} \qquad (15a)$$

$$v_{Ph}^{LA} = \frac{\omega}{k} = \left(\frac{E}{\rho}\right)^{1/2} \qquad (15b)$$

where ρ, G and E stand for the density and the shear and elastic modulus, respectively.

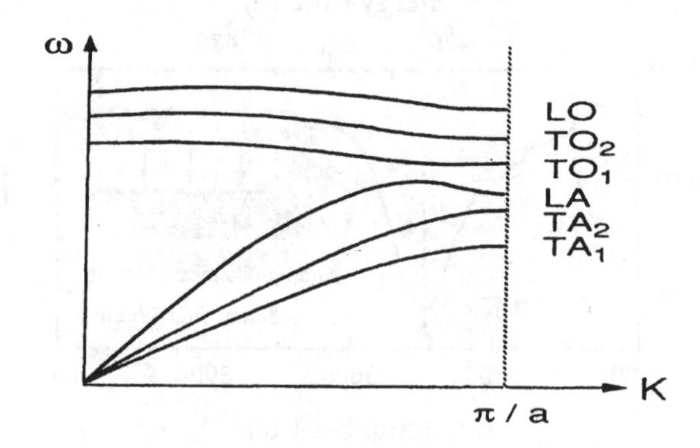

Figure 18: Schematic drawing of the phonon dispersion of a crystal with two atoms per primitive unit cell.

Since E ≥ G one has always $v^{TA} \leq v^{LA}$.The two TA branches may be degenerate or not, depending on the symmetry of the crystal and on the orientation of **k** relative to the crystal axes.

For increasing **k,** i.e. decreasing phonon wavelength, the acoustic phonon branches deviate from a linear dispersion e.g. in the way shown in Fig. 18.

If the crystal has more than one atom in the primitive unit cell, e.g. s atoms, then there are 3s – 3 optic branches above the acoustic ones. If the atoms carry an electric charge, some of these branches may couple to the electromagnetic radiation field resulting in a finite transverse longitudinal splitting at k = 0.

In crystals with inversion symmetry, optical phonons are either IR- or Raman active. Without inversion symmetry they may be both.

For more details on phonons in bulk samples see e.g. [1 – 5].

7. Phonons in Superlattices

In a superlattice one has, as discussed already in chapter 3 an artifical periodicity in growth direction leading to a mini-Brillouin-zone for the **k**-vectors in this direction (14).

Two different things can now happen with the phonon dispersion in this direction:
- backfolding, if the corresponding phonon branches in materials A and B (or I and II) overlap energetically
- confinement, if they do not overlap.

The acoustic branches necessarily overlap, since their dispersions start always at the origin We discuss therefore backfolding for acoustic phonons and with respect to the selection rules for the LA branches.

One defines in the so-called Rytov model [8, 25-29] an average longitudinal velocity of sound V_{LA}^{eff} by

$$V_{LA}^{eff} = \frac{d}{d_A / V_A + d_B / V_B} \tag{16}$$

Then the dispersion relation of the (linear part) of the LA-branch is backfolded into the first mini-Brillouin zone as shown in Fig. 19. The selection rules for Raman scattering say that in a backward configuration only the backfolded LA phonons can be observed for parallel polarisation of the incident and scattered beam and that small, but finite wave vector **q** results from **k**-conservation in the scattering process

$$q = \frac{4\pi n(\omega_i)}{\lambda_i} = 2k_i(\omega_i) \tag{17}$$

where n_i and λ_i are the refractive index and vacuum wavelength of the incident and backscattered light beams. Furthermore one expects, that the intensities decrease rapidly with increasing order m of the backfolded branch and that the intensities for the even numbers are smaller than those for the odd numbers and vanish even for $d_A = d_B$.

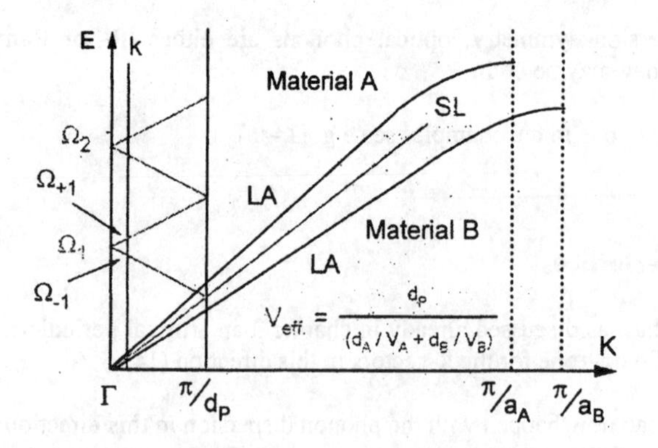

Figure 19: **Backfolding of the lower, linear part of the LA phonons in the first mini-Brillouin. From [8, 27].**

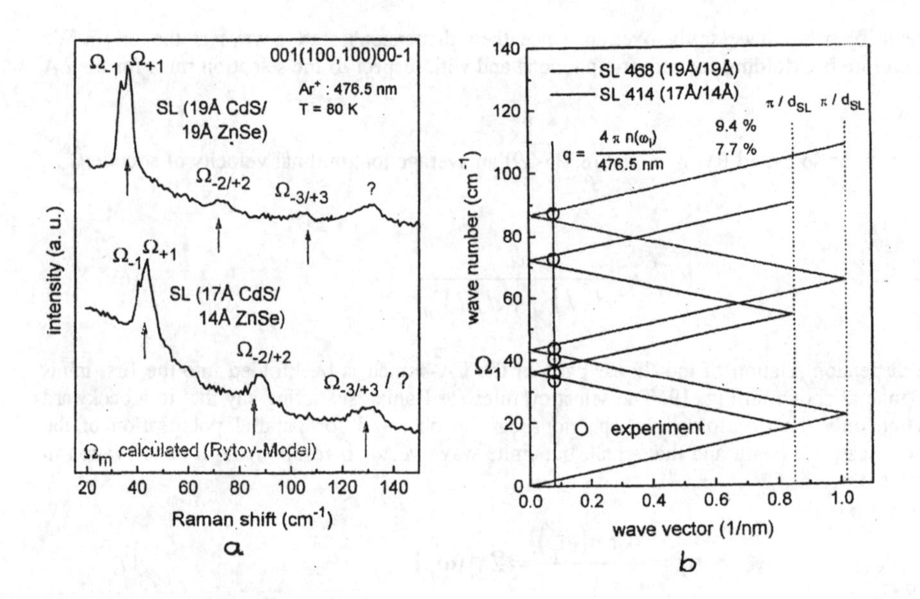

Figure 20: **Raman spectra of two different CdS/ZnSe SL in the spectral range of backfolded longitudinal acoustic phonons (FLAPS). From [8, 26, 27].**

In Fig. 20 we show the Raman spectra for two different SL and find all expectations verified. For Ω_1 one can even see the expected splitting caused by the finite q in (17). The calculated energies Ω_1 are indicated in Fig. 20a by arrows and the experimental data in Fig. 20b by circles.

The peak labelled by "?" coincides for one SL with the third backfolded LA phonon but does not shift with the SL period. Its origin is not completely clear, it could be connected with an electronic Raman process e.g. at donors, which are unintentionally present in many II-VI crystals (see chapter 5).

If the branches of the phonons in materials A and B do not overlap energetically, a phonon excited in material A does not find any resonant partner in material B (and vice versa). Consequently the excitation remains confined in material A with nodes at the interface or a very rapidly decaying tail. Due to the confinement, only discrete wavelength λ_n and vectors $\mathbf{k}_n$ in growth direction can occur given by

$$k_n = {n\pi}/{d_A}; \quad n\frac{\lambda_n}{2} = d_A \tag{18}$$

Since the dispersion of the optical phonons decreases generally with increasing **k** as shown in Fig. 18 the corresponding phonon frequencies are ordered

$$\omega_{n+1} < \omega_n \tag{19}$$

In Fig. 21 we show the calculated longitudinal displacement patterns of the first two confined LO phonon modes in CdS and ZnSe in a superlattice, the composition of which is indicated in the first line labelled IF.

The selection rules indicate that the odd (even) numbers of the confined phonon modes are observed for crossed (parallel) polarizations of incident and scattered beams in a backward Raman scattering experiment [8, 27].

Fig. 22 shows a Raman spectrum in the range of the optical phonons of CdS and ZnSe for the two above mentioned polarisations. There is a small peak from CdS TO modes. The dominant structures come as expected from the LO modes. The above given polarisation selection rules are especially obvious for the ZnSe modes, while they are obscured by the mode IF in the CdS region. This mode will be discussed later.

The confined modes of superlattices with various ZnSe layer thicknesses allow to reproduce the bulk LO phonon dispersion over a wide fraction of the first Brillouin zone as shown in Fig. 23. The bulk dispersion had to be shifted by 4cm^{-1} to lower energies due to the strain in the SL, otherwise the agreement between experiment and theory is excellent.

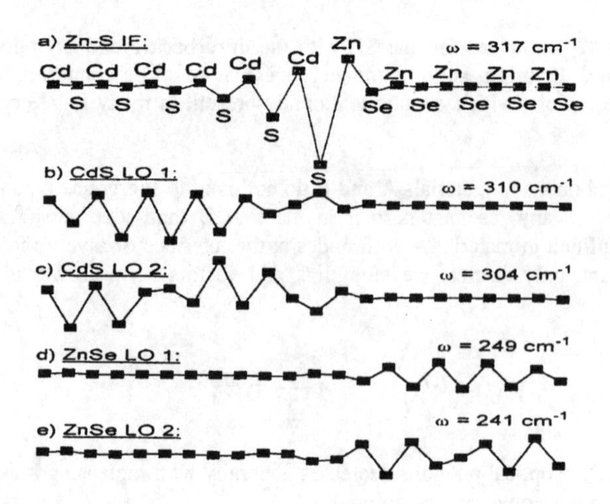

Figure 21: Calculated displacement patterns of CdS and ZnSe LO phonons in a SL. From [8, 26, 27].

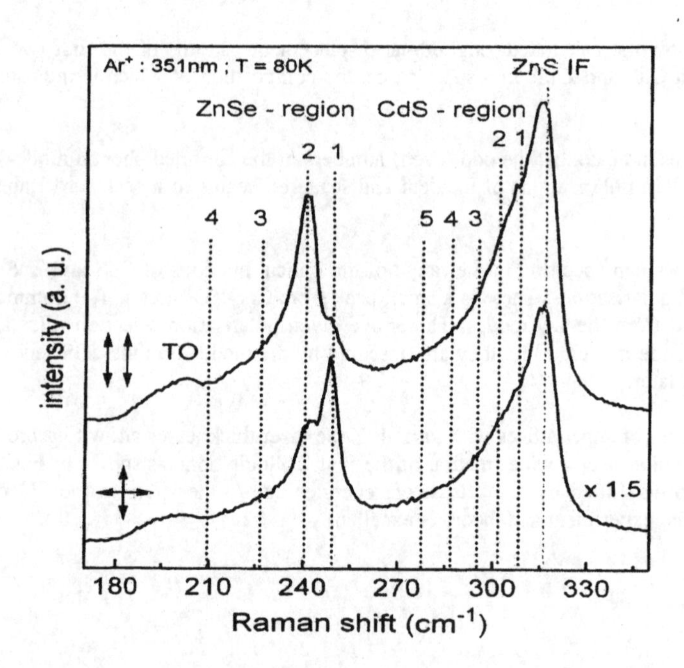

Figure 22: Raman spectra of a (1.7nm CdS/1.4nm ZnSe) SL in backward scattering configuration. From [8, 26, 27].

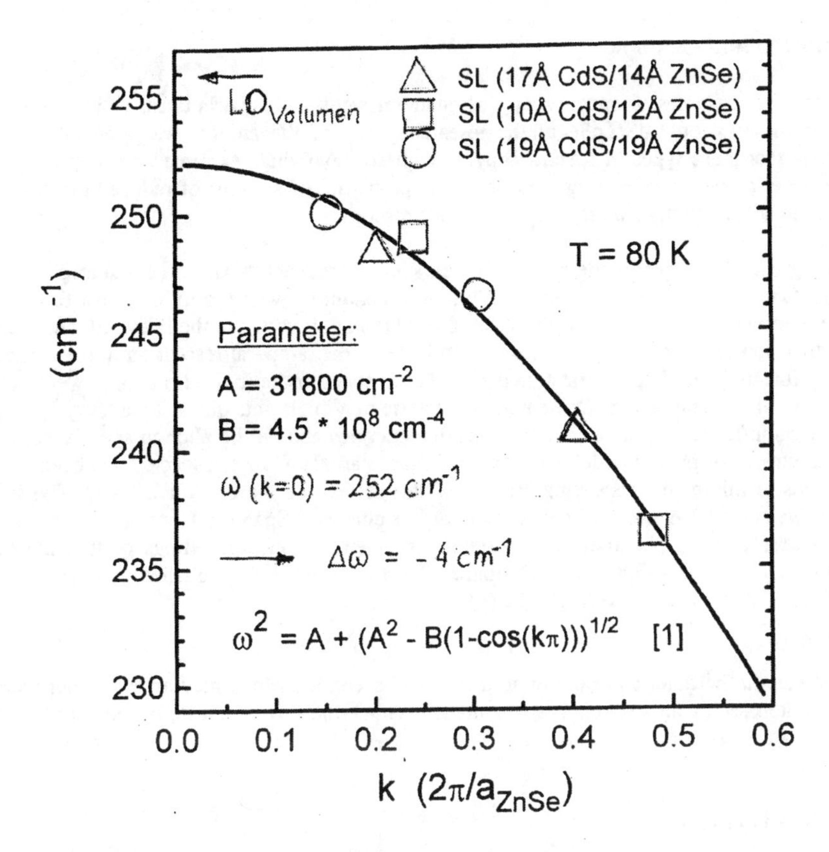

Figure 23: The dispersion of LO phonons in ZnSe compared to data from the first two confined modes in CdS/ZnSe SL. From [8, 26, 27].

The peak labelled IF in Fig. 22 is due to a type of interface modes, which are especially pronounced in superlattices without common anion nor cation as CdS/ZnSe. At the interface a ZnS bond can occur. Since Zn and S are the lighter ones out of the two cat- and anions, this mode is situated energetically above the highest CdS mode and can thus be seen as a distinct structure. Indeed one sees in Fig. 21, that the displacement pattern has a maximum at the interface in contrast to the confined CdS and ZnSe modes and decays (exponentially) on both sides the faster the further away the corresponding bulk modes are. The CdSe interface mode not shown here couples in contrast to the ZnSe modes and does not show up as individual structure [8, 26, 27].

More examples for the properties of phonons in quantum wells are compiled e.g. in [7, 28, 29] and the references given therein.

8. Conclusion and Outlook

With these few examples of spectroscopy of quantum wells and superlattices, which were mainly selected from work on CdS/ZnSe of the research group of the author, we hope that we could demonstrate that these types of spectroscopy, though not involving "extreme" conditions, contain a lot of beautiful physics and exciting insights in the properties of systems of reduced dimensionality, thus reaching at least partly frontiers of optical spectroscopy.

Once the systems have been understood, one can start to approach "extreme conditions" in various directions. One can increase the spatial resolution to conditions where only one or a few localized states are observed, one can enhance the spectral resolution to investigate the linewidth and a possible spectral diffusion of these states. Is it possible to increase the temporal resolution in the range of the dephasing time or of the free polarization decay of such states or the (pulsed) pump power to come to high density effects and lasing. One could apply extremely high uniaxial or hydrostatic pressure to investigate its influence on the electronic transition energies and on the phonon spectra, resulting in data for electronic or phononic deformation potentials. Various of these aspects have been treated in contributions to this or to preceeding schools of this series like the one on Ultrafast Dynamics of Quantum Systems in 1997 or on Spectroscopy of Systems with Spatially Confined Structures 2001. Another access to such phenomena in literature is given by the proceedings of the international workshops on "Nonlinear Optics and Excitation Kinetics" (NOEKS) published in phys. stat. sol. b **146**, **159** (1), **173** (1), **188** (1), **206** (1), **221** (1), **238** (1).

We investigate in the other contribution to this school scenarios which may appear under increasing excitation intensity in an exciton system, namely either their Bose-Einstein condensation or the transition to an electron-hole plasma.

Acknowledgements:

The author thanks the director of the school, Prof. Dr. B. Di Bartolo for the invitation to come and to lecture in Erice, the Deutsche Forschungsgemeinschaft for financial support of his various research activities and his coworkers and colleagues for their fruitful work and cooperation. Concerning especially the CdS/ZnSe activities the thanks are due to Drs. M. Grün, M. Dremel, S. Petillon and E. Kurtz for the growth of the samples, Drs. A. Dinger, M. Schmidt and dipl. phys. H. Priller for the interband spectroscopy, Prof. Dr. H. Kalt and his PhD candidate Benédicté DalDon for the spatially resolved data, Drs. M. Göppert and again A. Dinger, dipl. phys. R. Becker and Ch. Meier together with Prof. Dr. J. Geurts (Würzburg) and his coworkers Drs. V. Wagner, J. Liang and Prof. Dr. V. Burlakov (at that time in Troitsk) for the Raman and intersubband measurements.

References:

1. a) Kittel, Ch. (1986), *Introduction to Solid State Physics 6th edn.*, Wiley, New York.
   b) Ibach, H., Lüth, H. (1991), *Festkörperphysik 3rd edn*, Springer, Berlin, Heidelberg.
   c) Madelung, O. (1981), *Introduction to Solid State Theory*, Springer Series in Solid-State Sciences 2, 2nd edn.
2. Madelung, O. (1970), *Grundlagen der Halbleiterphysik*, Heidelberger Taschenbücher.
3. Yu, Y.P. and Cardona, M. (1996), *Fundamentals of Semiconductors*, Springer, Berlin.
4. Klingshirn, C. (1997), *Semiconductor Optics*, 2nd printing, Springer, Berlin.
5. Rössler, U., ed. (2001) *Landolt-Börnstein*, New Series, Group III, Vol. **41**, Springer, Berlin.
6. Shockley, W. (1950), *Electrons and Holes in Semiconductors*, Van Nostrand, New York.
7. Klingshirn, C., ed. (2001), *Landolt-Börnstein*, New Series, Group III, Vol. **34** C1, Springer, Berlin.
8. Dinger, A. (2000), Dissertation, Karlsruhe, *Exzitonische und phononische Eigenschaften von CdS/ZnSe-Quantenheterostrukturen*, ISBN: 3-8265-7876-7.
9. Schmidt, M. (2002), Dissertation, Karlsruhe, *Optische Anisotropie und Lumineszenzdynamik in CdS/ZnSe Quantentrögen*, ISBN: 3-936231-58-3.
10. Dinger, A., Petillon, S., Grün, M., Hetterich, M. and Klingshirn, C. (1999), Conduction Band Offset of the CdS/ZnSe Heterostructure, *Semiconductor Science and Technology* **14**, 595.
11. Schmidt, M., Priller, H., Dal Don, B., Dremel, M., Grün, M., Kalt, H. and Klingshirn, C. (2001), High Excitation Effects and Relaxation Dynamics in CdS/ZnSe Single Quantum Wells, *phys. stat. sol. (b)* **229**, 643.
12. Kozanecki, A., Werner, Z., Rzewuski, H., Loferski, J.J. (1978), A Study of Excitonic Spectra in CdS_xSe_{1-x} Crystals, *phys. stat. sol. (b)* **89**, 313.
13. Swoboda, H.-E., Majumder, F.A., Renner, R., Weber, Ch., Sence, M., Jie, Lu, Noll, G., Göbel, E.O., Vaitkus, J. and Klingshirn, C. (1988), Nonlinear Optical Proberties of the System $CdS_{1-x}Se_x$, *phys. stat. sol. b* **150**, 749.
14. Dinger, A., Baldauf, M., Petillon, S., Hepting, A., Lüerßen, D., Grün, M., Kalt, H. and Klingshirn, C. (2000), Exciton Localization in Cubic CdS/ZnSe Type II Quantum Well Structures, *Journ. of Crystal Growth* **214/215**, 660.
15. Tarasenko, S.A., Kiselev, A.A., Ivchenko, E.L., Dinger, A., Baldauf, M., Klingshirn, C. and Kalt, H. (2000), Non-Monotonous Temperature Dependence of the Spectral Maximum of Photoluminescence in CdS/ZnSe Superlattices, Proc. 8th Int. Symp. on Nanostructures: Physics and Technology, St. Petersburg, Russia, 2000, Ioffe Institute p. 157-160.
16. Reznitsky, A., Klochikhin, A., Permogorov, S., Sedova, I., Sorokin, S., Ivanov, S., DalDon, B., Kurtz, E., Kalt, H. and Klingshirn, C. (2003), Metastable Exciton States in ZnCdSe Quantum Wells with Nanoislands, 11th International Symposium on Nanostructures: Physics and Technology, St. Petersburg, Russia, June 23-28, 2003, p. 206.
17. Schmidt, M., Priller, H., Dal Don, B., Dremel, M., Grün, M., Kalt, H. and Klingshirn, C. (2001), High Excitation Effects and Relaxation Dynamics in CdS/ZnSe Single Quantum Wells, *phys. stat. sol. (b)* **229**, 643.
18. Priller, H., Schmidt, M., Dremel, M. Grün, M. Dal Don, B., Toropov, A. Ivchenko, E.L., Kalt, H. and Klingshirn, C., Density Dependent Luminescence Properties of CdS/UnSe Single Quantum Wells, Proc. Intern. Conf. II-VI Compounds, Niagara Falls, Sept 2003, in press.
19. Schmidt, M., Grün, M., Petillon, S., Kurtz, E. and Klingshirn, C. (2000), Polarized Luminescence in CdS/ZnSe Quantum Well Structures, *Appl. Phys. Lett.* **77**, 85.
20. Platonov, A.V., et al. (1999), Giant Electro-optical Anisotropy in Type-II Heterostructures, *Phys. Rev. Lett.* **83**, 3546.
    Yakovlev, D.R., et al. (2000), Orientation of chemical bonds at type-II hetero interfaces probed by polarized optical spectroscopy,, *Phys. Rev. B* **61**, 2421.
21. Göppert, M. (2000), Dissertation, Karlsruhe, Infrarotspektroskopie an CdS/ZnSe-Heterostrukturen und am Volumenhalbleiter Cu_2O, ISBN: 3-8265-8154-7.
22. Göppert, M., Becker, R., Petillon, S., Grün, M., Maier, C., Dinger, A. and Klingshirn, C. (2000), Investigation of Mid-Infrared Intersubband Transitions in CdS:Cl/ZnSe Qantum Wells, *Journ. of Crystal Growth* **214/215**, 625.

23. Göppert, M., Becker, R., Petillon, S., Grün, M., Maier, C., Dinger, A. and Klingshirn, C. (2000), Intersubband and Interminiband Transitions in CdS/ZnSe Heterostructures, *Physica E* **7**, 89.
24. Göppert, M., Grün, M., Maier, C., Petillon, S., Becker, R., Dinger, A. Storzum, A., Jörger, M. and Klingshirn, C. (2002), Intersubband and Interminiband Spectroscopy of Doped and Undoped CdS/ZnSe Multiple Quantum Wells and Superlattices, *Phys. Rev. B* **65**, 115334.
25. Dinger, A., Hetterich, M., Göppert, M., Grün, M., Klingshirn, C., Weise, B., Liang, J., Wagner, V. and Geurts, J. (1999), Growth of CdS/ZnS Strained Layer Superlattices on GaAs(001) by Molecular-Beam Epitaxy with Special Reference to their Structural Properties and Lattice Dynamics, *Journal of Crystal Growth* **200**, 391.
26. Dinger, A., Becker, R., Göppert, M., Petillon, S., Grün, M., Klingshirn, C., Liang, J., Wagner, V. and Geurts, J. (2000), Lattice Dynamical Properties of Cubic CdS/ZnSe Strained-Layer Superlattices, Japan, *Journ. of Crystal Growth* **214/215**, 676.
27. Dinger, A., Göppert, Becker, R., M., Grün, M., Petillon, S., Klingshirn, C., Liang, J., Wagner, V. and Geurts, J. (2001), Lattice Dynamics of CdS/ZnSe Strained Layer Superlattices Studied by Raman Scattering, *Phys. Rev. B* **64**, 245310-1.
28. Göppert, M., Hetterich, M., Dinger, A., O'Donnell, K.P. and Klingshirn, C. (1998), Infrared Spectroscopy of Confined Optical and Folded Acoustical Phonons in Strained CdSe/CdS Superlattices, *Phys. Rev. B* **57**, 13068.
29. Dinger, A., Becker, R., Göppert, M., Petillon, S., Hetterich, M., Grün, M., Klingshirn, C., Lian, J., Weise, B., Wagner, V. and Geurts, J. (1999), Raman- and FTIR Investigation of CdS Based Strained Layer Superlattices, *phys. stat. sol. (b)* **215**, 413 (1999)

7. LASERS FOR FRONTIER SPECTROSCOPY

GIUSEPPE BALDACCHINI
ENEA- Centro Ricerche Frascati
UTS Tecnologie Fisiche Avanzate
C.P. 65 - 00044 Frascati, Roma, Italy

Abstract

The first laser has been invented in 1960 by using the red light from a ruby crystal, and since then the laser field exploded almost exponentially, and thousands of different materials, in the state of solids, liquids, vapors, gases, plasmas, and elementary particles have lased up to now from less than 1 Å to more than 1 mm. Many of them have been used with outstanding results both in basic science, and in industrial and commercial applications, by changing for ever the same lifestyle of humankind. As far as spectroscopy is concerned, the laser light has started an unprecedented revolution because of its unique properties as monochromaticity, coherence, power, brightness and short-pulse regime, unrivaled by any other natural and artificial light source. Spectroscopy applications increased qualitatively and quantitatively with the laser sources themselves, and they are still proceeding in parallel with the moving of the laser field towards new territories. Apart the opening up of new regions of the electromagnetic spectrum, like the terahertz gap, and the outstanding increase of the output power which is giving rise to completely new spectroscopic effects, the improvement of laser sources and auxiliary equipments is producing a growth of traditional laser spectroscopy with superior resolution and sensitivity. Moreover, spectroscopic techniques and laser light contributed to the development of new chemical and physical processes which have been used to fabricate photonic materials with new spectroscopic properties enriching the laser field itself, in a virtuous cycle spectroscopy→laser→material and back to spectroscopy with no end in sight.

1. Introduction

A *laser* emission was produced for the first time in 1960 in the red region of the spectrum from a rod of crystalline ruby [1], but its story began much earlier with the observation of natural optical phenomena, which have been always at the center of curiosity of mankind since ancient times.

As a matter of fact, the origins of this fantastic invention which has revolutionized the science of light date back to the dawn of the human civilization. Indeed, luminous phenomena, not connected with the well known burning fire or lightning, have been recorded in ancient times both in the East and in the West. There are clear references in the classic literature at light emitted by animated bodies, like bacteria in the sea, and inanimate bodies, like fabulous precious stones, and it seems certain that in Japan and China luminous paint was known more than 1000 year ago [2], but only in Europe with the Renaissance do the records assume a more credible character [3]. Well known was the stone, a ruby or more probably a diamond, in the ring of Catherine of Aragon, ~ 1530, which

B. Di Bartolo and O. Forte (eds.), Frontiers of Optical Spectroscopy, 251-288.

luminesced at night, or a diamond described by Benvenuto Cellini in 1568 which shone after exposure to light. So interesting were considered at that time such unusual phenomena that Conrad Gesner wrote the first known book on luminescence with the rather long title *A short treatise on rare and marvelous plants that are called lunar because they shine at night and incidentally on other things which shine in darkness,* and published it in Zurich in 1555.

However, the greatest discovery in the entire history of inorganic luminescence occurred in Bologna where Vincenzo Casciarolo, a cobbler by profession but an alchemist at heart, used to bake minerals of various extraction in the hope of finding a way to transform them in solid gold. He never obtained gold, but one day of 1603 a stone, after being heated, started to shine in the dark and became famous as the *petra luminifera bononiensis*. The material, probably containing barium sulphate, immediately aroused the greatest interest among the scholars of the time, among them also Galileo Galilei who understood that the emitted light was due to internal properties of the material itself, the first artificial phosphor.

Other discoveries of phosphor materials followed afterwards, but no real breakthroughs were made until the introduction of the *prism* spectroscope at the beginning of the nineteenth century. It is true that the dispersive power of prisms was experimented by Newton and was even known before as an aesthetic effect in some glass items fabricated in Venice, but only later on, with the introduction of the *slit* in 1802 by W.H. Wollaston, it was utilized systematically in laboratory instruments. Anyway, in 1852 G.G. Stokes noticed that the light emitted by the phosphors is always degraded, in the sense that its wavelength is longer than that of the exciting light. Because he was working with fluorspar, calcium fluoride, he also decided to call *fluorescence* the light emitted, while only in 1888 the more general and still used term *luminescence* was introduced by E. Wiedemann. The Stokes law, or Stokes shift as it is better known, represented the first accurate scientific description of the luminescent phenomena, which remains substantially valid nowadays.

However, a true understanding of the luminescence was only possible at the beginning of the twentieth century with the discovery of *quantum mechanics*, where all physical systems are reduced, as far as the energy is concerned, to a series of discrete energy levels which can interact with the electromagnetic radiation of frequency resonant with the energy difference of the previous levels. Soon, it was also found that the radiation could be *absorbed or emitted spontaneously*, the latter process explaining at last the hidden nature of the luminescent phenomena, which for centuries aroused the fantasy of many scholars and baffled their efforts of comprehension [3]. Moreover, it was found later on that the radiation, besides being emitted spontaneously, could also be induced to be emitted by the same radiation in a process called *stimulated emission*.

This last process was very difficult to be observed experimentally and only later on in 1954, well after the spectacular scientific advancements during World War II, it was put to work in the microwave region of the spectrum with the realization of the first *maser*, acronym from *microwave amplification by stimulated emission radiation*, working with ammonia gas [4].

2. The rise of lasers

2.1 DISCOVERY OF LASERS

After the realization of the first maser, theoretical and experimental efforts were devoted to the extension of the electromagnetic (e.m.) radiation amplification at higher frequencies, i.e. in the optical range and beyond. But, it was immediately recognized that such frequency jump posed new problems, the toughest ones concerning the stimulated emission and the optical cavity. Indeed, it was well known that spontaneous emission grows with the third power of the frequency, and so in the optical region the stimulated emission has to compete with it more toughly than in the microwave region. Moreover, optical cavities allowed many resonance modes and their technology was not developed, contrary to the monomode microwave cavities well known for their application in radar devices.

Anyway, without going into too much details which is beyond the purpose of this work, within a few years the previous problems were brilliantly solved, and atomic, molecular, and semiconductor materials were proposed as active materials [5,6,7]. However, the first *optical maser* did not work with any such materials but with a ruby crystal [1]. It is worthwhile to note that the inventor, T.H. Maiman, knew very well this kind of crystal having worked in the maser field with it, and he was aware of its optical spectroscopy with the well known intense red light emission at 694 nm.

Anyway, before describing the subsequent developments of this new important optical field, it is worth remarking that, although the invention of the first optical maser is attributed to Maiman and the paternity of the field in general to C.H. Townes, N.G. Basov, and A.M. Prokhorov, who obtained the Nobel Prize in 1964 "for fundamental work in the field of quantum electronics, which has led to the construction of oscillators and amplifiers based on the maser-laser principle", the first recorded idea of such device has been attributed to Gordon Gould. Indeed, in November 13, 1957, he uses for the first time the world *laser* for *light amplification by stimulated emission of radiation,* and in 1988 his various patents were issued by the USA Patent Office, so closing a legal battle which lasted for more than 20 years [8].

Having established the priorities of the laser invention, it is necessary to say that many other people were working in the laser field with different material systems and technical approaches. Indeed, later on in 1960 a He-Ne laser emitted at 1.15 μm [9]. Only two years later this system lased at the classical 632.8 nm red line [10].

In 1961 the Nd^{3+} ion in $CaWO_4$ produced laser emission at 1.06 μm [11], but only in 1964 the well known Nd:YAG system was tested positively [12].

Soon after in 1962, laser emission at 904 nm was obtained at the junction of a p-n semiconductor, a gallium arsenide diode [13,14,15].

The laser boom continued with increasing momentum to such an effect that by the end of 1963 gases and vapors produced more than 200 different wavelengths from the visible to the middle infrared, various crystals with impurities displayed 40 different wavelengths from the visible to the infrared and, besides 6 different semiconductor diode lasers, also several glasses, plastics and liquids did show laser emission [5]. It is evident that it would not be possible to give justice to the whole story of this frantic and amazing endeavor in the present work, which has so been limited only to the major breakthroughs.

Laser emissions in the visible region of the e.m. spectrum from argon ions [16] and at $\lambda \sim 10$ μm from CO_2 [17] were found in 1964.

The color center laser was demonstrated in 1965 at a wavelength of 2.7 μm [18] and, although it was practically developed only ten years later [19] for a complex series of human and scientific events [20], it represented and still represents today a model case for the much wider class of vibronic lasers which are an important part of the family of tunable solid state lasers.

In 1965 the first laser emission driven by chemical reactions was obtained in hydrogen chloride [21], while tunable laser emission was also demonstrated in a liquid containing an organic dye, the first dye laser [22].

During the seventies the discovery of completely new types of lasers slowed down considerably, and only a few of them are worthy of being mentioned hereafter.

In 1970 the first laser in the violet and ultraviolet region of the spectrum was a liquid Xe dimer emitting at 1760 Å [23], while the more reliable excimers appeared only in 1974 starting with an emission at 2828 Å from XeBr [24].

Elementary particles produced laser emission in 1977, when the first free-electron laser worked at 3.4 μm [25]. The active material consisted of a beam of electrons, and an "optical cavity" was still used to capture the e.m. radiation emitted by the oscillating electrons moving in a periodic magnetic field.

X-ray emission was observed in 1981 by using highly ionized carbon [26] and a plasma produced by nuclear explosions [27]. The x-ray laser can hardly be compared with the previously described types of lasers as far as the material, the excitation, and the cavity are concerned. Indeed, very often the term superradiance instead of laser emission is utilized to describe it.

Finally, in 1982 the Ti:sapphire lased for the first time [28], and now with its continuous emission in the interval 650-1000 nm is one of the most successful tunable solid state laser, although not the most efficient one.

2.2 ENERGY LEVELS OF LASER EMISSION

Apart x-ray and free electron lasers, all the other lasers mentioned in the previous section are based on radiative transitions among atomic, molecular, and band energy levels, and the vast majority of them can be well described by only *four different energy schemes*.

Figure 1a shows the energy levels of the Cr^{3+} ion in a sapphire (Al_2O_3) crystal, i.e. a ruby, involved in the functioning of the first laser emission. For that the ruby is excited by light, originally by a flash lamp but any other equally intense optical source would suffice, from the ground $^4A_2$ state to the higher $^4F_2$ and $^4F_1$ bands, from where the electron excitation decays during a few ps nonradiatively in the doublet ^2E level which eventually emits light at 694 nm with a lifetime of 3 ms. This optical cycle has, however, the drawback that the lower emission level is also the ground state of the Cr^{3+} ion, and so it is very difficult to obtain a continuous inversion population between the excited ^2E and ground $^4A_2$ state, which is one of the prerequisite for having light amplification. Indeed, in this case the condition $U\tau \gg 1$ must be fulfilled, where U is the pumping rate out of the ground state, and it was a fortunate coincidence in ruby that the lifetime of the ^2E emitting level is long enough so that practical pumping is possible. Anyway, this laser is not very efficient, and usually it works only in pulsed regime. The previous situation is much more clear in Fig.1b where the essential *three level energy scheme* has been sorted out from the many ruby levels. It is immediately evident that only a very powerful pumping source is able to populate level 3 more than level 1, being level 2 only transitory for the excited electron.

Contrary to the previous three-level energy scheme of ruby, Fig.2a shows the case of the Nd^{3+} ion in YAG crystal (Yttrium Aluminum Garnet, $Y_3Al_5O_{12}$), where the population inversion condition, $U\tau>0$, is less demanding, being satisfied for any pumping rate $U\neq0$ out of the ground state. Indeed, because nonradiative transitions last a few ps, the state $^4I_{11/2}$ is never appreciably populated during the excitation, and so the inversion with the state $^4F_{3/2}$, which is involved in the laser emission, is secured for any intensity pumping. The present situation is commonly known as a *four level energy scheme*, as in Fig.2b where level 1 is the ground state, 2 the excited state, 3 the relaxed excited state, and 4 the unrelaxed ground state. The adjective *relaxed* is used for level 3 because it is reached after a nonradiative transition where the energy is delivered to the lattice vibrations, i.e. phonons, and this last process is usually known as *relaxation*.

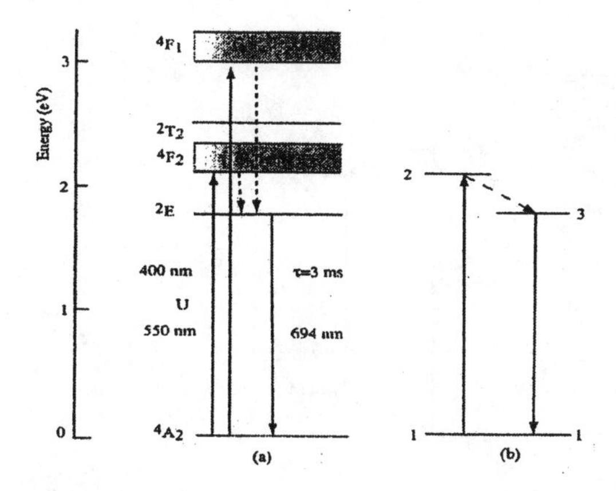

Figure 1. Energy levels of Cr^{3+} in sapphire (Al_2O_3), and radiative (continuous lines) and nonradiative (dashed lines) transitions connected with the laser emission at 694 nm, the so called R_1 line, (a), and the three essential energy levels (b), where level 3 is known as the relaxed excited state.

rather wide energy bands, which is the usual case for point defects, i.e. color centers, atomic and molecular ions, impurities, etc., in insulating crystals. This new class of solid state materials is often called *vibronic* because of the coupling between the electronic excitation and the lattice vibrations, and their complex optical properties can be described by means of the *configurational coordinate diagram*, as shown in Fig.3. Here the diagram is limited to only two states and the conduction band for reasons of simplicity, but in general it can contain any number of states. Anyway, the vibronic state is represented by a parabola which describes the energy of the electron as a function of the coordinate Q, in practice the distance of the neighbor ions of the host crystal from the defect. The two parabolas representing the ground and the excited state have the minima at different coordinates, Q_e-Q_g is a measure of the electron-lattice coupling, and the horizontal lines inside the parabolas represent the vibration levels of the lattice, in this case only one vibrational mode is shown. An optical excitation produces a vertical transition, A for absorption, from the minimum of the ground state, GS, to the excited state in a highly excited vibrational level. The vibronic system relaxes in a matter of a few ps to the minimum of the excited state, which is known as relaxed excited state, RES. The excitation can remain in this peculiar state for a relatively long time, from ms to ns, before returning

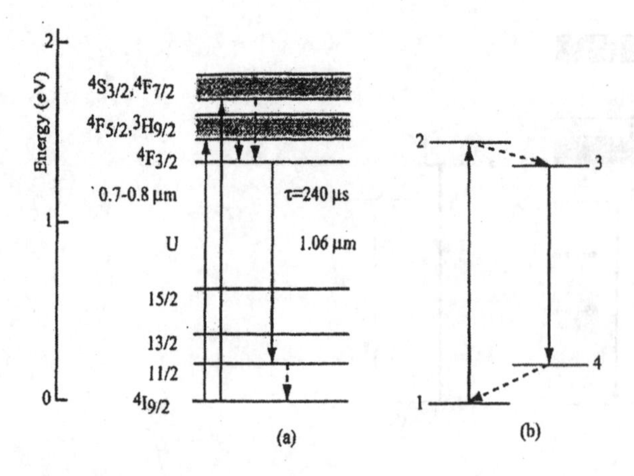

Figure 2. Energy levels of Nd^{3+} in YAG (Yttrium Aluminum Garnet, $Y_3Al_5O_{12}$), and radiative (continuous lines) and nonradiative (dashed lines) transitions connected with the laser emission at 1.06 μm (a), and the four essential energy levels (b), where level 3 is known as the relaxed excited state.

In the previous two notable cases, the energy levels connected with the laser emission are atomic-like, but by looking carefully at higher energies in Figs.1 and 2 it is easily observed that the levels are to the ground state parabola with a vertical transition accompanied very often by light emission. Soon afterwards, a further lattice relaxation completes the optical cycle to the GS. However, the emission which is also called ordinary luminescence, OL, is not the only radiative process which takes place after excitation. Indeed, little beyond the obvious Rayleigh line, a structured emission reveals the presence of Raman scattering, RS, which contains detailed information on the lattice modes coupled to the electronic transitions. Still beyond that and up to OL, there is a very weak emission tail known as hot luminescence, HL, which originates from decaying processes during the relaxation. Moreover, in the case the temperature is high enough, a thermal excitation process can subtract a fraction of excitations from the RES to the conduction band in this case, but to any other upper state in general, so decreasing the optical efficiency of the optical cycle. Anyway, RS and HL are usually a few order of magnitude less intense than OL, which together with the absorption A remain the most conspicuous optical effects during the optical cycle. From the laser point of view the system is equivalent to a four level energy scheme, with the difference that both the absorption and the emission are now broad bands as a consequence of the interaction with the lattice. In conclusion, all solid state materials can be described as in Fig.3, where the electron-lattice coupling can be weak in the rare-earth ions, strong in transition metal ions, and very strong in color centers, with the consequence of narrow emissions, as in Figs.1 and 2, or very broad emission bands, as in Ti:Sa [28] and color centers [29]. In these last two cases it is possible to have efficient laser emissions tunable within the broad emission bands.

In Figs.1, 2, and 3 the electron excitation is localized at the site of the defect center, while the same excitation is completely delocalized in the semiconductor material, where the energy bands refer to the whole sample under study. This is a consequence of the electrons moving all over the semiconductor crystal, while in the defect centers they belong to each defect. Figure 4a describes the energy status of an intrinsic semiconductor crystal at T=0 K, where the valence band is full while the conduction band above the energy gap is empty of electrons. Some electrons of the valence band can be excited radiatively in the conduction band as in Fig.4b, and the same excited electrons can recombine with the holes in the valence band with the emission of light, which may have laser character in presence of an optical cavity around the crystal and enough pumping power. This is the operation principle of a semiconductor laser, but hardly an efficient way to have laser emission which is instead obtained by electrical current excitation as in Figs.4c and 4d. In a junction p-n, the bands in the two differently doped crystals assume at equilibrium different energy values in order to avoid any current flow as in Fig.4c. However, in case a forward voltage
is applied the levels of the two Fermi seas are different, see Fig.4d, and a current is promoted with the *combination of electrons and holes in the active region of the junction.* Radiation is emitted with high efficiency approximately at the energy of the gap, and so it is possible to have laser emissions in a wide region of the spectrum, by choosing among the many existing semiconductor materials with different gaps.

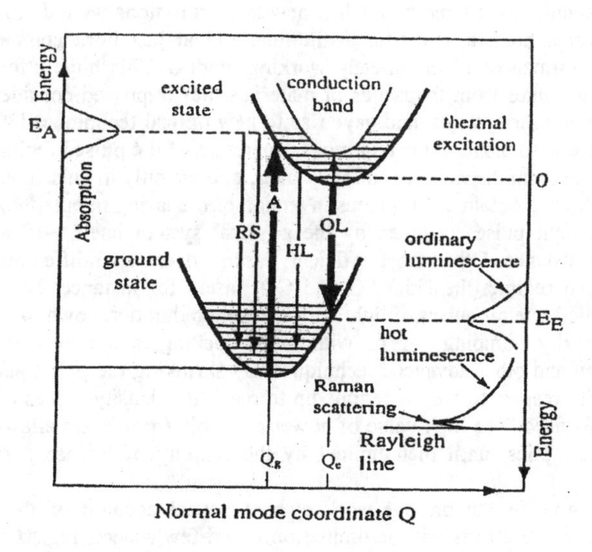

Figure 3. Configuration coordinate diagram illustrating the absorption and emission phenomena of vibronic materials, i.e. impurities and point defects in insulating crystals. See text for details.

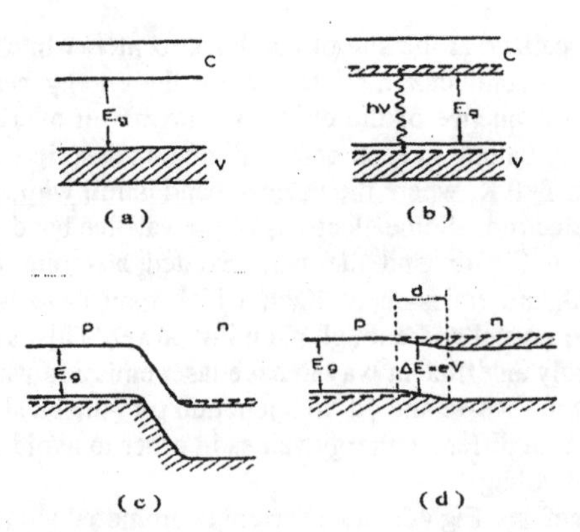

Figure 4. Energy bands and operation principles of semiconductor lasers in a normal material, (a) and (b), and in a junction, (c) and (d). See text for details.

2.3 LASER PANORAMA AND OPTICAL TECHNOLOGIES IN THE NINETIES

Going back to laser development, at the same time when new laser emissions were discovered almost daily, studies were performed both at scientific institutions and at just born commercial industries, to improve also the performances of the already working models. This huge effort has produced thousands of new emission lines from thousands of materials, has improved considerably the efficiency of the already existing laser systems, and gave rise to new optical techniques [30,31]. Among the latter ones, it is worthwhile to mention the enormous progresses of the pulsed regime. As it is well known, a few lasers, among them the excimer and ruby ones, operate only in pulsed regime, mainly because of inherent difficulties to obtain a continuous inversion of the lasing population. As a consequence, the duration of the light pulses is given by the physical system itself, ~10 ns for excimer and ~ 1 ns for ruby, for instance. Other more efficient lasers operate indifferently in continuous wave (cw) and pulsed regime, the Nd:YAG and CO_2 lasers for instance. However, almost all the lasers can be compelled to emit pulses of light with controlled duration down to ~10 ns with Q switching, ~1 ns with cavity dumping, ~1 ps with mode locking, and a few fs with compression by negative dispersion and other advanced techniques. By shrinking the pulse, also the peak power has been increased with respect to the cw regime up to reach the density values of 10^{15} W/cm^2 and more, when properly focused. The high value of power available from lasers allowed an incredible expansion of non-linear optics, until then limited by the scarcity of intense pumping sources.

It would be impossible to give in the present work even a partial account of the laser developments up to a decade ago, so the efforts will be limited only to a few aspects regarding the successful and parallel stories of two families of laser materials, semiconductors and solid state crystals, which have been and are still utilized extensively in spectroscopy.

As reported previously, semiconductor diode lasers, SDL, were introduced in the laser race quite early, but for more than a decade they remained more an interesting scientific curiosity than a tool of any practical use. Indeed, they were not reliable, the efficiency was very low, most of them needed low temperatures to operate, and they could hardly survive to a few temperature cycles.

However, in due time the techniques of handling semiconductor materials improved so much, also as a by-product of the modern integrated electronic industry, that it was possible to fabricate reliable diode lasers, most of them slightly tunable, emitting from the visible region of the spectrum up to about 30 μm [32,33]. Above ~2.5 μm they work only at low temperatures, 4-200 K, and the power decreases with increasing wavelength falling well below 1 mW for lead salt diodes above 10 μm. Below ~2.5 μm they work also at room temperature and the maximum power of about 100 mW is emitted in the near infrared region, although also the new discovered blue SDLs are improving very fast [34]. The TDLs have found many scientific and technological application niches for their high emission efficiency, easiness of handling and, in particular, in spectroscopy for their tunability from the visible up to the infrared region of the spectrum. However, because of their poor spatial and spectral characteristics, the high frequency noise, the easy-facet damage by power output, which does not allow a regime of short and powerful pulses, they could not compete for specialized applications with other well established laser sources, like solid state lasers, SSL.

The latter ones, following vast and detailed studies of luminescence in old and new solids [35,36], are based on thousands of laser active crystals in the range 0.3-5 μm. However, the big majority of laser emissions did not receive any attention beyond the experimental demonstration, and only a few of them were developed up to the point to be widely spread among the scientific and technical communities. The latter lasers include the ruby, the Nd:YAG, and the Ti:sapphire [28]. A large amount of work has been devoted to find new reliable hosts especially for high power regimes, but up to now only the crystal $SLiSrAlF_6$ (LiSAF) has shown some promising characteristics when doped with Cr^{3+} [37]. Very important among SSLs are the vibronic lasers based on point defects in dielectric crystals, because of their broad and continuous tunability from about 0.7 to 4 μm. The power output depends strongly on the pumping power and schemes, dimensions of the crystals, cooling etc., but it can easily reach a few W in cw and GW in pulsed regime, while they operate mostly at room temperature with the exception of color center lasers above 1.5 μm which require liquid nitrogen temperature [38].

Contrary to SDLs, all SSLs possess very good spatial and spectral characteristics which mark them as unique laser sources for a lot of applications which require well defined optical modes, beam shape, single frequency operation, smooth pulse shape, etc.. However, all SSLs are marred by an inherent weakness which is related to their low overall efficiency from plug to beam energy, which seldom is getting over 0.5%. This low figure is due in part to intrinsic factors like the relaxation processes, shown in Figs.1, 2 and 3 for ruby, Nd:YAG, and vibronic systems in general, which are unavoidable, and mostly to the poor overlapping between the exciting spectrum of the pumping optical source, usually an arc lamp, and the absorption spectrum of the active optical material. Figure 5 show the absorption spectrum of Nd^{3+} in YAG and, superimposed, the typical emission spectrum of a xenon lamp usually used to pump SSLs. No more than 5% of the luminous energy of the lamp, which apart a few strong lines in the infrared emits like a black body radiation source, is transferred to the Nd^{3+} ions. When the active ions are excited by a laser line, like in Ti:sapphire or in color centers, both pumped with optical efficiencies up to 50% by an argon ion laser, the overall efficiency is still very low, because the efficiency of the argon laser itself is hardly higher than 0.1 percent. This unfortunate situation changed abruptly with the availability of SDLs, whose emission lines can match exactly the absorption transitions of SSL. In particular the GaAlAs diode can pump with an optical efficiency of 50% the $^4F_{5/2}$-$^4I_{9/2}$ transition of the Nd:YAG at ~809 nm, see Fig.6, by producing the 1.06 μm laser radiation with an overall efficiency of ~20%! This new pumping technology has been applied to obtain for instance a diode pumped SSL with superior qualities as compactness, beam quality and frequency stability but a modest power output of ~30 mW, called monolithic nonplanar ring oscillator [39], and another SSL with near-diffraction limited beam quality at a level of power output of 147 W, called disk laser [40]. In conclusion, this new possibility of merging the good beam

qualities of SSLs and the peculiar pumping capacities of the SDLs opened completely new perspectives for the development of SSLs and SDLs altogether.
Anyway, at the beginning of the nineties more than thousands of laser emissions at different wavelengths and bands in practically every form of matter, solid, liquid, gas, plasma, and also elementary particles have been discovered and utilized [41]. All these emissions have been approximately reported in Fig.7 classified accordingly to the different states of matter as a function of wavelength. A general look at it is really impressive because, in practice, laser devices fill the e.m. spectrum from soft x-ray to the far-infrared and beyond. However, it should be reminded that there is not usually a continuous wavelength tuning inside each family, and on this respect gas and solid state lasers represents two extreme cases. Gas lasers emit at discrete wavelengths, like excimers in the UV and CO_2 in the IR, and only when high pressure is used, as in the case of CO_2, it is possible to get a limited tunability. Instead, solid state lasers, with a few notable exceptions, like ruby and Nd-YAG, include the vibronic family almost continuously tunable from 0.7 to 4.0 μm. The average power, not reported in Fig.7, varies from μW of lead salt diode lasers to kW of carbon dioxide laser, with peak powers up to TW for pulsed devices with duration as short as a few tens fs. However, these achievements should not surprise anybody, because they are a logical consequence of the quantum nature of the matter and light and, practically, any kind of material could become a potential candidate for laser emission, once the required conditions for laser action are satisfied, which in time has proved easier than it would have been expected four decades ago.

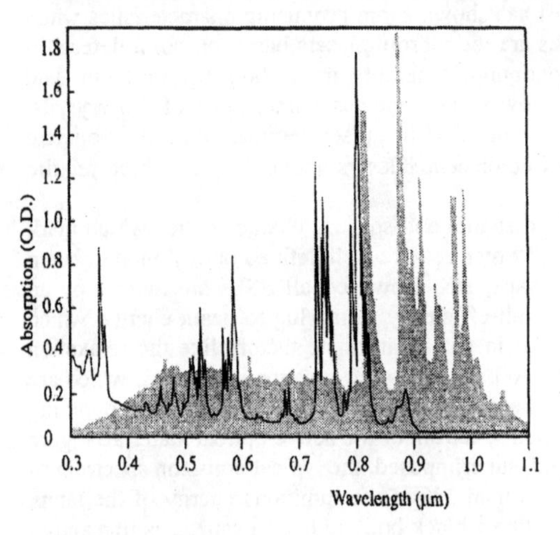

Fig.5 - Absorption spectrum at room temperature of Nd^{3+} ions in YAG, full line, and the typical emission spectrum of a xenon lamp, gray area.

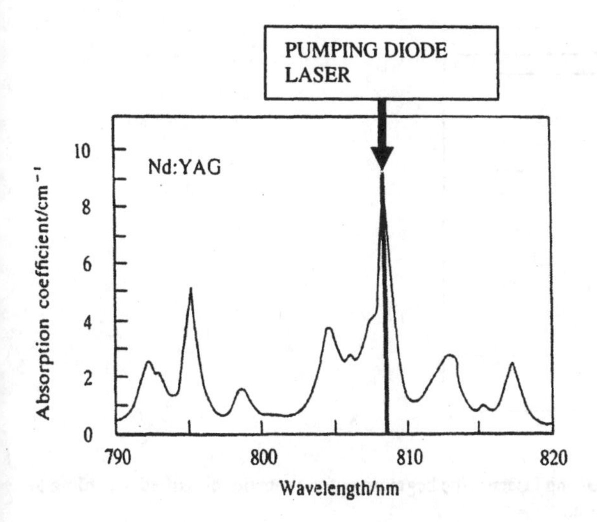

Fig.6 - Absorption spectrum at room temperature of Nd^{3+} ions in YAG expanded around 810 nm, thin full line, and the emission line of a GaAlAs semiconductor diode laser, thick full line.

3. Spectroscopy with lasers

3.1 PROPERTIES OF LASERS RELATED TO SPECTROSCOPY

Since the realization of the first lasers, both the discoverers and experts alike realized the enormous potentialities of this new tool. Indeed, although those new light sources did not display better efficiencies with respect to the old classical sources of lights, they possessed unique characteristics which were and are still unparalleled by any other light source. In particular, laser radiation is *coherent* and *monochromatic*, carries *high power* with *high brightness*, and can be *pulsed* very easily, all properties that have been used up to now in hundredths of applications, among them alignment, automation and robotics, bar readers, communication, displays, holography, isotope enrichment, laser fusion, material treatment, medicine, metrology, nonlinear optics, optical storage, plasma diagnostics, photochemistry, printing, ranger finder, remote sensing, spectroscopy, surgery, ultrafast processes, weapons.

As far as spectroscopy is concerned, all five previous properties of lasers perform their part, although monochromaticity is the main actor. Figure 8 show the absorption spectrum of water vapor measured with one of the first available TDL [42]. It is easily observed that the resolution power of the new system is more than 10^5, a figure at least two order of magnitude bigger with respect to the standard spectrophotometers using prisms or gratings. This high resolution not only allowed for the first time the separation of nearby absorption lines, as the triplet in Fig. 8 which could not have been resolved with classical optical systems, but it also opened the possibility to study the lineshape of the single lines, which bears the imprint of the atomic and molecular interactions, a possibility limited to very few cases before the invention of laser sources, and mostly dreamed by the experts of spectroscopy until that time.

Moreover, the lasers did not limit themselves to increase the resolution, to which later on the improved laser sources still added several orders of magnitude with respect to the already notable one

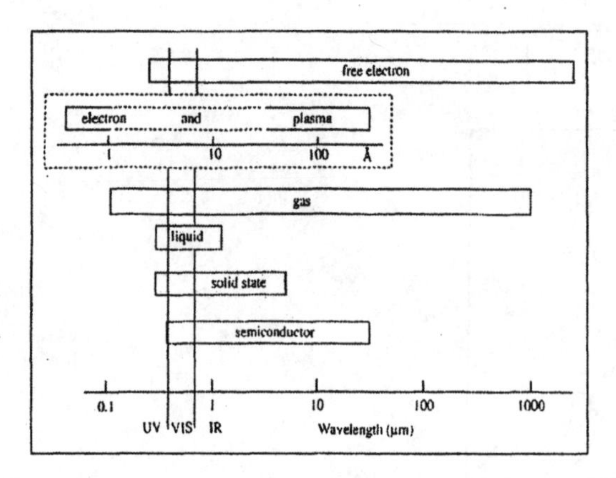

Figure 7. Approximate spectral ranges of the existing lasers at the beginning of the nineties classified according to the state of matter of the active material.

of Fig.8, but also the sensitivity was very much increased for the intrinsic functioning of the laser itself. Figure 9 shows the intensity of a color center laser emission over its continuously tuning band, which contains several absorption features of atmospheric water and carbon dioxide [43]. The rather strong intensity of the molecular absorption is due to intracavity amplification effects in only a few cm atmospheric path, which in a classical optical system would have amounted to an extremely weak absorption barely observable. In this case no care has been paid to select a single mode and so the resolution, although larger than in a classical spectrophotometer, is worst than that reported in Fig.8, but the intensity of the absorption is at least two order of magnitude bigger than that reported in the same Fig.8, as it results by taking into account the very different optical paths in the two experiments.

The previous two examples indicate that laser sources have provided the field of spectroscopy with much more than the simple addition of their exceptional characteristics. Indeed, they have opened such a Pandora box that it would be impossible to describe it summarily in this limited work, for which it is referred mostly to specialized literature [44,45]. So, in the following the matter will be limited only to the description of a few spectroscopic laser techniques which, however, well represent the great revolution aroused by the laser invention in the then rather old and dozing spectroscopic field. However, this brief description in not limited to spectroscopy alone, but it is also propaedeutical to a better understanding of the same laser sources.

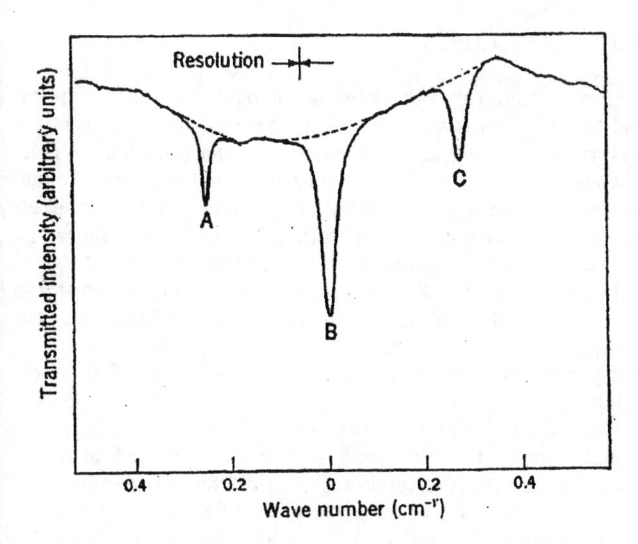

Figure 8. Absorption lines in the ν_2 band of water vapor at 1880 cm^{-1} measured over a 7.4 m air path with a tunable diode laser [42].

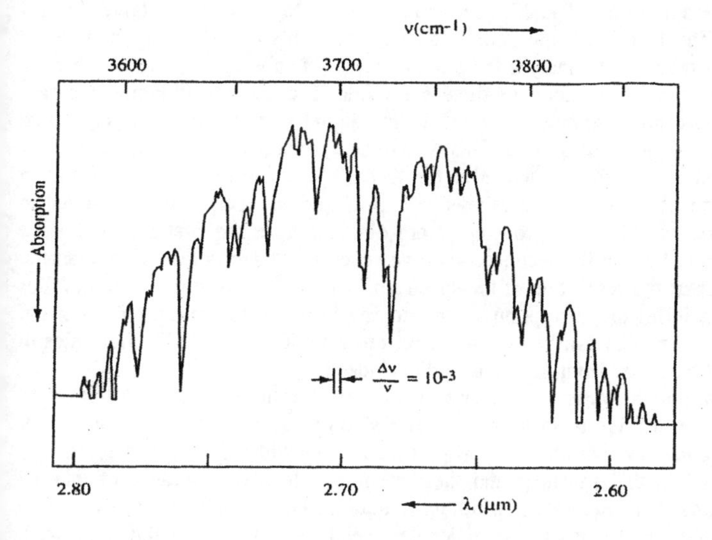

Figure 9. The emission intensity of the KCl:Li $F_A(II)$ color center laser is tuned in a spectral range rich of absorption lines from water vapor and carbon dioxide, contained in small traces in the atmosphere [43]. See text for details.

3.2 LASERS AND SPECTROSCOPIC TECHNIQUES

The most simple and diffuse spectroscopic technique is the *linear absorption* of the optical beam through the sample, solid, liquid or gas, which consists essentially in measuring the intensity of the beam before and after having traversed the sample as a function of wavelength. Figure 8 is an example of such measurements with a laser beam. It is evident that the spectrum is very much rough, although its novelties in comparison with the spectra performed by using classical light sources are outstanding as recalled in the previous section. However, the final form of the laser spectrum can be improved, together with the signal to noise ratio, by using special electronic techniques.

Figure 10 shows four spectra of ammonia around 781 cm^{-1} at room temperature taken by mean of a TDL spectrometer with an optical path 24 m long, and on each spectrum there are four absorption lines not yet identified then, but hundred times less intense than other well known ammonia lines [46]. The top signal shows the four absorption lines on the slope of the TDL mode, and it has been obtained by averaging 100 times the signal from the detector while the spectral interval was swept at 100 Hz. The second signal from the top has been obtained by modulating the intensity of the laser beam at 400 Hz and analyzing the signal from the detector with a lock-in system and 1s constant time, while the frequency of the TDL is swept through the spectral interval slowly, let say 10 m. The two spectra are almost equal, which is the consequence of the two apparently different averaging methods, which instead are substantially similar in principle; in practice there are advantages of one method on the other one only when there are instabilities of temperature, laser beam intensity, etc., i.e. when the parameters of the whole system, spectrometer and sample, change during the relatively long averaging time. The third spectrum from the top has been obtained by fast modulating the frequency of the TDL while the same frequency is swept slowly through the spectral interval, and by using a lock-in system to analyze the output signal. The absorption lines assume the typical derivative shape and the initial slope of the laser mode is canceled, allowing a bigger amplification of the small absorption signals. As a result, the signal to noise ratio is very much improved with respect to the two previous spectra. This way to analyze signals is usually called *first derivative lock-in method*, and it can also be extended to the *second derivative lock-in method*, and so on, up to arrive to n^{th} subsequent derivative analysis with, in principle, an ever improving signal to noise ratio, but in practice no substantial improvements are obtained after the first few derivative steps. Anyway, Fig.10 shows at bottom the second derivative spectrum which is less noisy than the previous one, as expected. Apart the resolution of the spectrum, which does not improve in the four methods, the increasing intensity of the absorption signal moving from top to bottom indicates that this simple measuring ammonia system can be used to detect down to 500 ppb of NH_3 molecules in the atmosphere, a very useful result for environmental studies, indeed.

Going back to Fig.8, although rough and distorted it displays the three absorption lines with well defined widths, which are not due to instrumental effects. As it is well known, an atomic or molecular gas possesses *Lorentzian absorption lines* with a natural linewidth determined essentially by the energy levels involved in the transition and their lifetimes. However, because of thermal motion, the *Doppler effect* adds to the natural linewidth up to obtain a *Gaussian absorption line* with a larger width, which depends on the atomic or molecular weight and the temperature as well. Moreover, when the gas density increases the collisions add still more to the linewidth, transforming the absorption lines in *Voigtian curves* witch tend to become Lorentzian curves at high pressures, again as at the beginning of this story but with a much larger linewidth. Lasers are particularly good at studying all these complex steps, because of their spectral purity and tunability.

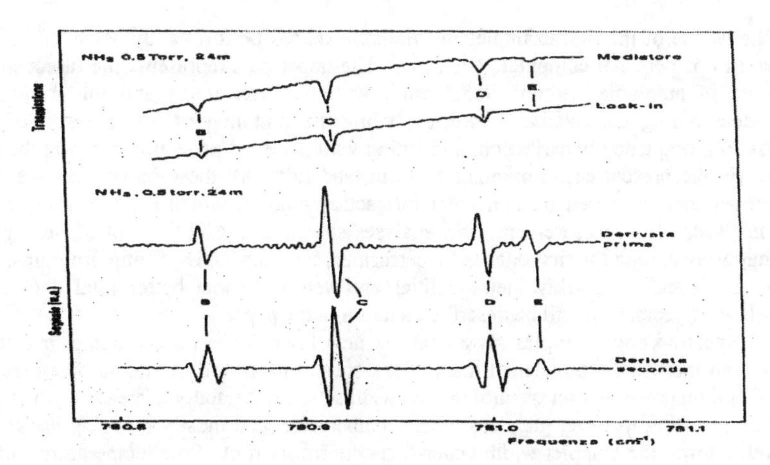

Figure 10. Absorption spectra of ammonia vapors in air obtained by using a TDL and a multipass cell at room temperature [46]. See text for details.

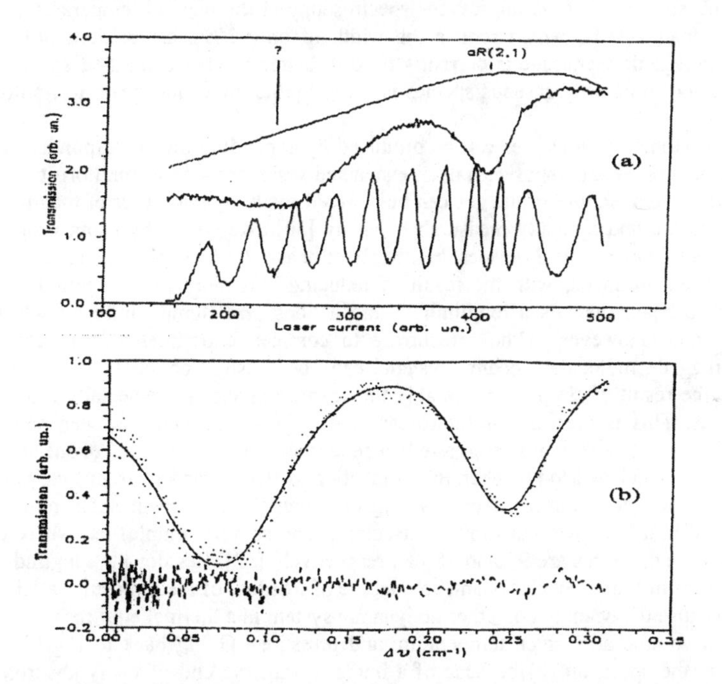

Figure 11. Absorption line of the aR(2,1) transition of $^{15}NH_3$ at 988.396 cm^{-1}, pressure 71 Torr and path length 1 m, and of an unassigned transition at 988.200 cm^{-1} which belongs to the more abundant species $^{14}NH_3$ (a). Normalized absorption profile of the two above transitions (b), the solid line represents the best fit of the experimental points, and the bottom line shows the residuals [47].

Figure 11 shows one of the first examples of *lineshape studies* performed by using TDLs and the calculus capacities of personal computers (PC) [47]. The upper part represents the direct linear absorption spectrum of ammonia around 988.3 cm^{-1}, with the laser mode and the fringes of interferences for determining the relative frequency, while the bottom part is the result of the intensity normalization, frequency linearization, and fitting with a special program, which at the end delivers the shape (in the present case Lorentzian), width, and shift. All these parameters are then compared with proper theories based on molecular interaction models, which are so submitted to severe experimental tests. The measurements and analyses shown in Fig.11 are not of very good quality because they were among the first ones to be performed, but both TDLs, fitting programs, and PCs increased so much and so quickly their qualities and versatility that better final data were obtained in the following years, as it will be described later on in the paper.

The previous spectroscopy examples show that the new laser devices are capable to resolve structures smaller than the usually broadened lines of the spectra measured in atomic or molecular gases. But such resolution power is useless until the linewidths are substantially decreased, which can be accomplished by resorting to lower gas density and temperature, and mostly by using *molecular beams*. As it is well known, the Doppler width depends on the square root of the temperature and so, by decreasing the temperature of the gas in especially built cells, it is possible to reveal more fine details in the spectra. However, this method is severely limited by the low vapor pressure of most of the atomic and molecular species of some interest for spectroscopy at the desired temperatures, so that temperatures lower than 200 K, which reduce the width of only 20%, are seldom utilized. Anyway, although this cryogenic technique is currently used to better resolve congested spectra, a much more efficient method of cooling gaseous species is given by the molecular beam technology [48,49].

A speedy stream of atoms or molecules can be produced by expanding the correspondent gas (also solid materials can be heated and vaporized) in an evacuated vessel through a small orifice, and in the expansion process the temperature of the gas can be lowered easily by two order of magnitude by maintaining a still sizeable and optically measurable density [50]. Moreover, by using properly designed skimmers and mobile mechanical devices, highly directional and monoenergetic atomic and molecular beams have been produced, with the result of reducing still more the temperature, by keeping in mind that in a beam there is a longitudinal and a transversal temperature, which are usually different each other. However, without recurring to complex experimental apparatus, a straightforward and rugged molecular beam system can be easily coupled to a laser spectrophotometer with the result of simplifying notably the absorption spectra, especially those of rather complex molecules. This is the case of fluoro-carbon-12, CF_2Cl_2, which has been studied intensively in the recent past because it was suggested to reduce the total amount of ozone in the atmosphere [51], a matter still debated today when the production of fluoro-carbon (freon) gases has been halted in most industrialized countries. Anyway, Fig.12 shows the spectra of such molecule taken in a normal cell and in a free (without limiting structures, and so very simple) jet, where the rotational and vibrational temperatures are 90 and 180 K, respectively [52]. Besides the longitudinal and transversal temperatures of the molecular beam, any degree of freedom of the molecules attains a different temperature, because the beam is not a thermodynamic system in a thermal equilibrium, as it is the case of a gas in a volume at a given temperature and pressure. Going back to Fig.12, the spectrum (a) differs from the spectrum (b) because of a bump toward the end of many absorption lines, and this feature represents the starting point of a Q branch whose lines have been numbered as reported in the figure. Only the cooling in the free jet allowed the precise determination of the head of the branch, which is completely blurred out in the spectrum at room temperature in the cell.

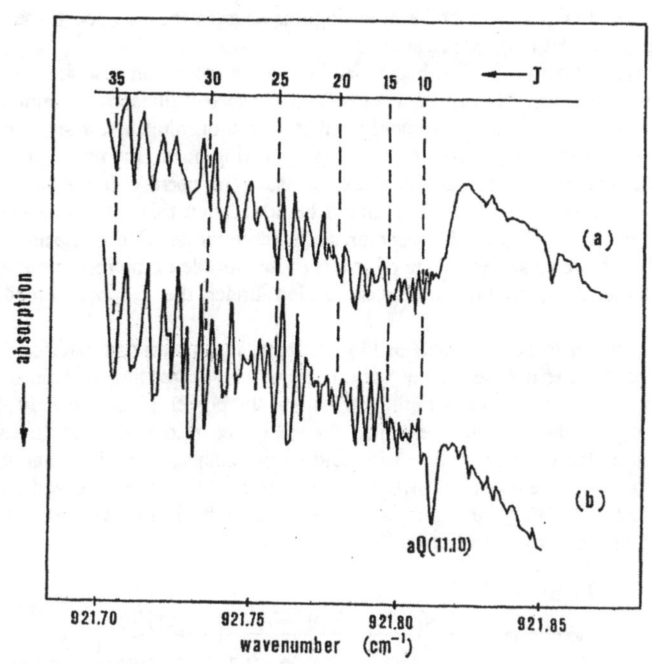

Figure 12. Diode laser spectrum of CF_2Cl_2 in a free jet (a) and in a cell at room temperature (b). The ammonia line aQ(11,10) has been used to calibrate the frequency scale [52].

However sophisticated a molecular beam can be considered, the spectroscopic technique is still the same as in the previous linear case with the cell, where the optical radiation is produced by a laser and revealed by a detector after having been absorbed by the material under study. The same linear absorption method holds also when the absorbed energy is not revealed by an optical detector, as a difference between the input and the output intensity, but rather by the effects produced by the absorbed energy on the gas itself. In one case the absorption is revealed through its heating locally the gas which generates sound waves picked up by a microphone, *photo-acoustic spectroscopy*. In another case, the energy of the absorbed radiation unbalances the equilibrium threshold of an electrical discharge which is so started on the onset of absorption, *opto-galvanic spectroscopy*. Both experimental methods were well know much before the invention of the laser. For instance, the photo-acoustic effect has been discovered by Edison, and the opto-galvanic effect was experienced in the early gas-discharge experiments, but both of them were revitalized by the laser sources which possess the high resolution and the high intensity they needed to be usefully utilized in spectroscopy [53,54].

Especially important in the two previous cases was the intensity of the laser beam, which increases the sensitivity of the techniques up to be competitive with other ones, but without changing the character of the fundamental linear absorption. On the contrary, the intensity of the laser sources can also be used to open the vast realm of *non linear optical phenomena*, which have originated new forms of *spectroscopy* like *saturation*, *two-photon*, and *Raman*. Really these effects were known before the invention of the laser, but only with these new singular sources they ceased to be mere scientific curiosities, however important in basic research.

If a monochromatic laser beam, the pump, is absorbed by an atomic or molecular transition line, it can happens at certain values of the intensity that the energy is not absorbed linearly any more, and the lineshape of the transition develops what is commonly called a hole-burning, i.e. a sharp dip in the much wider absorption curve. If at the same time a counter-propagating beam, the probe, identical to the previous one but much weaker, is switched on, it can measure the narrow dip with a much higher resolution than the linear absorption method. It should be added that this saturation method needs some kind of tunability on the side of the laser or some modulation of the spectroscopic structure under observation, but it allows an increasing of orders of magnitude in the resolution of the spectral features, like the hyperfine structures otherwise buried under the Doppler broadened absorption lines.

The linear absorption is well known to be the result of the first term in an expansion series of the matter-light interaction, where the coefficients of the various terms are decreasing in values with increasing order, but at the same time increasing with the intensity to the power of the order. In short, the first term is proportional to the intensity, linear response, the second term increases as the square of the intensity, the third term as the cube of the intensity, and so on. Hence, it is clear that at low intensity levels only linear effects can be observed, while the high order terms are observed only at high levels of intensity, and are called in general nonlinear effects. Clearly, in this peculiar context the laser has changed the field of optics as never before [55,56].

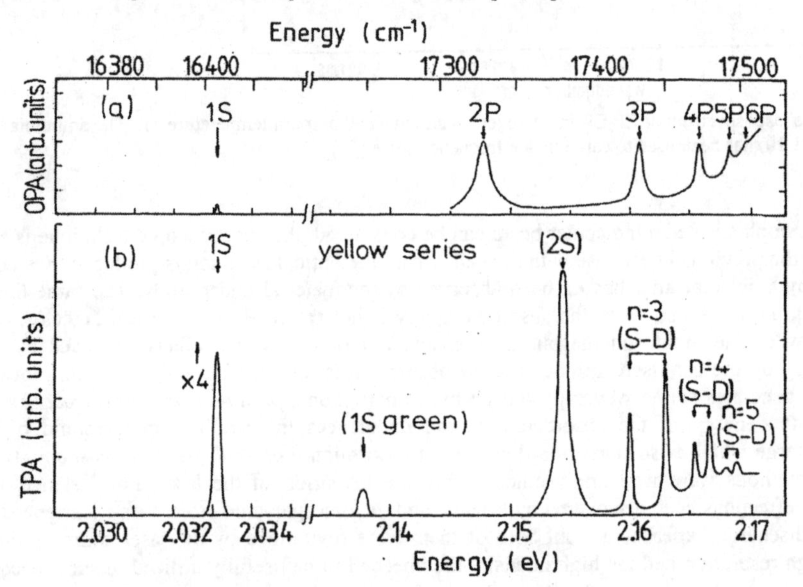

Figure 13. One- (a) and two-photon (b) absorption spectra of excitons in Cu_2O measured at liquid helium temperature, by using laser sources to excite the exciton gas in the semiconductor and to reveal the spectroscopic structure. Let note the different energy positions of the various peaks in the two series [57].

One of these nonlinear effects is the two-photon absorption, which has been discovered in 1931 with intense classical light sources, and where a transition is excited by two photons having each half the energy of the transition itself, and the absorption probability is proportional to the square of the intensity. To be noted that the same optical radiation would not be absorbed at all at low intensity. But, why this non linear spectroscopy, which requires powerful optical sources and also other technical tricks, is so worthy to be pursued in the first place? The answer to this apparently silly question is given by Fig.13, which reports the absorption spectrum of the excitons in a Cu_2O crystal measured by one and two photons spectroscopy [57]. The excitons, which are couples of bound electrons and holes like an atomic system, are generated in the energy gap of the semiconductor by a high power laser, a Raman-shifted Nd-YAG laser at 0.6496 eV, and their quasi-atomic spectrum measured by a tunable dye laser pumped by a krypton laser, a rather complex experimental apparatus. Anyway, Fig.13 show in (a) and (b) two different spectra which correspond to the normal linear spectrum and the two-photon spectrum, respectively. This apparently strange result is a consequence of the energy levels and transition rules of the exciton transitions in the crystal. Indeed, like an atomic system, the exciton possesses several energy levels above the ground state, which can be reached by light absorption according to the transition rules and the parity of the levels. Without going in more details, it suffices here to say that certain levels are allowed and other ones are forbidden to one photon absorption, while in general the contrary case holds for two photon absorption. This means that one photon absorption reaches different levels with respect to two photon absorption, with the result that the two spectra look differently as in Fig.13. In this case the two series of absorption peaks are complementary to each other, and it is clear that the two-photon absorption technique has opened a new spectroscopic dimension for the optical transitions. The previous case of two-photon absorption in a solid material applies just the same in gaseous or liquid materials, clearly with different experimental approaches as required by the circumstances.

Another prominent non linear effect concerns the interaction between the light beam and matter, where the incident photons undergo inelastic collisions with atoms, molecules, and crystals. A small fraction of photons emerges with slight lower or higher energy with respect to before the collision, having left or taken the energy difference to or from the electronic, vibrational, rotational energy levels of the illuminated material. This effect is called Raman from the name of the Indian Scholar who discovered it in 1927 and, in practice, consists in the emerging light beam retaining the imprint of the material itself. As an example for crystalline matter, Fig.14 shows the Raman scattering on silicon, and it consists of a very intense central peak at the same frequency as the laser source, due to Rayleigh elastic scattering, and of two almost symmetrical spectra, called Stokes and anti-Stokes, which contain in the present case some lattice vibrations of the crystal, i.e. surface phonons, R, and acoustic transversal, T, and longitudinal, L, phonons [58]. With the Raman spectroscopy it is possible to measure the frequencies of the various vibrational modes and, eventually, the variations due to the presence of impurities or defects which are easily detected, while for gases the information regards the rotational, vibrational, and electronic structure with very high resolution. However, by looking more carefully at Fig.14, it is easily realized that the new emerging frequencies have very week intensities with respect to the scattered original light, which is a consequence of their high order generation process, and the separation from the central line is very small. Both previous observations indicate that only laser sources could have given such resolved spectra with their high intensity and high spectral purity. However, it is worth noticing that the experimental results in Fig.14 are reported as scattered points denoting the presence of consistent noise due to the still primitive laser source used in the experiment, an argon ion laser at 514.5 nm. The improvements of lasers in the following years have very much improved the Raman spectroscopy, as will be shown later on in the paper.

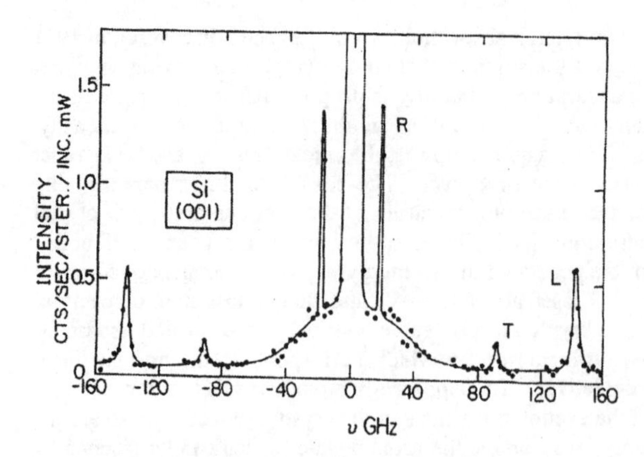

Figure 14. Raman (also called Brillouin for small energy differences, as in this case) scattering spectrum of Si at room temperature excited at 514.5nm by an argon ion laser. Because of the relatively high temperature used, both Stokes (right) and anti-Stokes (left) spectra have been measured, and as expected they are symmetrical with respect to the center energy. The peaks correspond to surface phonons (R), acoustic transversal (T), and longitudinal (L) phonons, i.e. part of the vibrational spectrum of silicon lattice.

4. Advancements of laser and spectroscopy

4.1 PRESENT PANORAMA OF LASERS AND MARKET

At the end of the nineties the laser panorama did not change dramatically from what depicted in Fig.7, but new improved models and optical materials have been introduced, poorly covered gaps were better filled, and a few spectral regions have been added up to obtain the new situation reported in Fig.15 [59].

SDLs extended from VIS to the blue region and below with the introduction of the nitride semiconductor materials [60], and above 30 μm up to 3 mm, a region called far infrared or terahertz still not easily accessible for lacking of well developed technology.

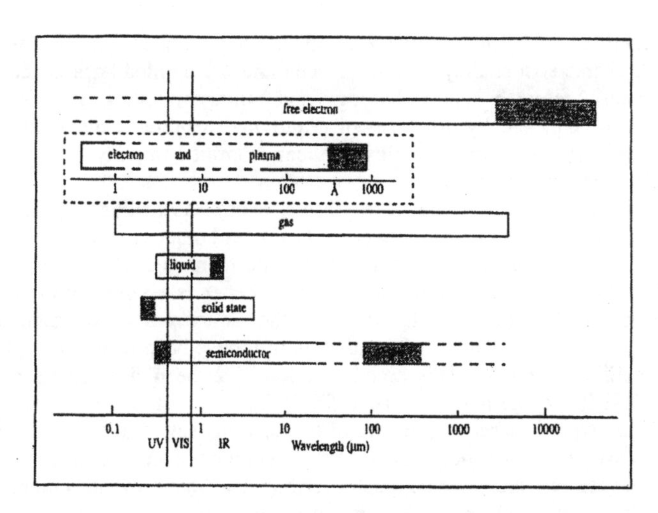

Figure 15. Approximate spectral ranges of laser sources around the year 2000 classified according to the state of matter of the active material. The gray areas show the laser emissions added between 1990 and 2000 with respect to Fig.7, while the dashed lines refer to regions under development.

However, germanium intervalence band lasers are working up to 300 µm, and a new approach called terahertz time-domain spectroscopy is opening the region above 300 µm [61].

SSLs moved in the UV down to about 0.2 µm, but essentially stabilized themselves with more reliable materials and technologies, similarly to dye (liquid in the figure) lasers which also extended a little more in the IR.

Highly ionized plasmas extended beyond 400 Å up to almost 1000 Å, and free electron lasers moved well above 1 mm, and are moving below the UV region of the spectrum.

The average power of the laser sources remained more or less the same, from µW for lead salt lasers to kW of carbon dioxide lasers, but the time domain moved with pulsed devices having durations as short as a few fs or less [62] and peak powers in the TW range.

At moment laser technology has reached high levels of sophistication but, while improvements or new lasers were and are still expected for a few types or families, for other ones there have not been marked advancements lately. Among the latter ones, excimer, helium-neon, helium-cadmium, ion, dye, copper vapor, color center, ruby, carbon dioxide, far-infrared gas, and other ones, which cannot be listed here for sake of space, have indeed reached a high degree of maturity, with some exceptions for the excimer lasers which recently have been subjected to increased interests for better reliability and higher beam qualities [63]. A discussion apart deserve free electron and x-ray lasers. They both look exceptional optical tools, the first ones for their broad tunability and power [64], and the second ones for their short wavelength emissions [65]. Lately, both of them went through new and interesting developments. Free electron lasers have been demonstrated to be able to emit efficiently in the very ultraviolet and soft x-rays region [66] with a promising brightness ten time bigger with respect to any existing x-ray source. Soft x-rays lasers, which have been realized with highly ionized plasmas produced by high-power laser on solid targets, have been obtained more recently also with plasma confined in narrow capillary channels [67] with unique spatial and temporal properties. For these characteristics they both have been and are applied with success in spectroscopy, bio-medicine, material research, nano-technology, and military weapons. But, although they went, as discussed above, through lengthy processes of improving theoretical knowledge and experimental systems, they still remain very complex devices confined mainly in research laboratories, which is especially true for free electron lasers.

At this points, leaving aside free electron and x-ray lasers, exotic lasers like inversionless lasers, nuclear pumped laser, two-photon lasers, micro-spherical lasers, and other ones which are still developed and used only in specialized laboratories, it is worthwhile to have a close look at the commercial side of the laser field in order to gain useful insights for a better assessment of the whole matter [68]. Figure 16a displays the global market of lasers during the last ten years. The lasers have been divided in two groups, SDLs and all the other ones, which in turn have been further on splitted among the most important families in Fig.6b. This second addition has been deemed necessary because of the fast surging of SDLs on the market since 1996. Indeed, until 1995 one graph was enough to describe the sales of all types of lasers, having all of them similar sale figures, but later on the sales of SDLs, which are mostly used in optical storage and telecommunication applications and only after the 10th place in basic research, increased so much that they represent nowadays more than 70% of all sales. Conversely, the other lasers are used, in the order, in material processing, medical therapeutic and at the 3rd place in basic research, well known not to be adequately supported, which explains to some extent why SDLs, used in chatting and entertaining, are sold much more. Anyway, as far as the total sales are concerned, apart the small decreasing in 1993 due mainly to a general world recession, they have shown up to 2000 an unprecedented growth, especially for SDLs, that was projected in 2001 to 11.5 Billion$, a value about 30% more with respect to 2000 and well outside the scale of the figure. Contrary to the expectations, in 2001 the assessed total market went down by almost 40% to a value of 5.6 Billion$, a gasping debacle which raised many doubts on the long established methods used for the annual predictions, incredibly wrong this time.

Without entering in a detailed discussion of this unusual event, it is worth observing that it has been originated mainly by the loss of the ten year momentum of the information technology growth, which crumbled also because of the beginning of the end of the "new economy". Since about 90% of the applications of SDLs refer to telecommunication and optical storage, it is clear why the largest decreasing of sales occurred to SDLs alone. Moreover, the crisis of the information technology hit also the material-processing industries which reflected, for instance, on the excimer lasers forming the basis of any lithography equipment. Indeed, the excimer laser market has been the second hard hit, as shown in Fig.6b. However, the other kinds of lasers have kept in 2001 more or less the same commercial figures of 1999, with actually an increase for SSLs. In 2002 there has been a still consistent decreasing of sales for SDLs and a general stagnation for the all the other lasers, but the forecasts for 2003 indicate a small but significant increase of the whole market, well at least as in 1999. Anyway, by discarding the abrupt increase and decreasing of sales during 1999-2001, due mostly to the unrealistic expectations of the telecommunication industry, the constant growth of the laser market is quite evident in Figs.16a and 16b since 1995, especially for SDLs and SSLs. Now, it is well known that a lot of R&D went toward these two kinds of laser sources, because scientific and technological novelties are very much expected from them. So, market, investments, and exciting scientific progress are still moving together in some conspicuous laser families, which have not yet reached a stage of technical maturity.

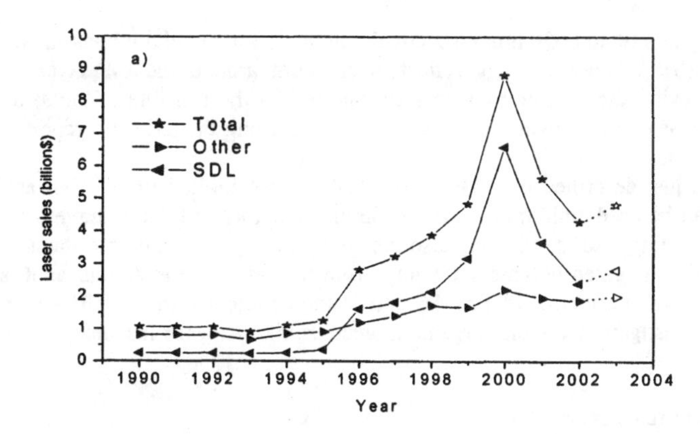

Figure 16. Worldwide commercial laser sales in the last ten years from data reported in Laser Focus World. All types of lasers divided between SDLs and all the other ones are reported in (a), while only four types of lasers among the other ones with appreciable figures have been reported in (b).

4.2 NEW KINDS OF LASERS AND OPTICAL MATERIALS

As it has been discussed in the previous section, the most steadily scientific and technological growth belongs to SDLs and SSLs, which is also a consequence of the recent symbiosis of these two families of lasers. Indeed, while both of them have their well separated fields of application, SSLs pumped by SDLs have demonstrated more simplicity, power and efficiency with respect to the same lasers pumped by lamps. So, diode-pumped SSLs have opened new perspectives both for high and low power laser systems [69]. On this respect, the biggest diode-pumped laser system is being realized at the National Ignition Facility at Livermore, USA, to deliver 1.8 MJ in 1 ns [70], while the much more tiny diode-pumped microlaser can easily produce peaks powers of 80 kW during 300 ps or 100 mW of cw output [71,72].

However small a microlaser can be, about 10 mm of cavity and active material, microchip lasers can be smaller by an order of magnitude, and still be efficient and even more stable. They are monolithic miniaturized devices displaying single longitudinal and transverse mode operation, and low threshold. Figure 17 shows the schematic drawing of one of them which, with a pump power of 150 mW at 980 nm, delivers 25 mW at 1535 nm in a TEM_{00} mode with M^2=1.06 (M^2 is usually referred to as the beam quality, being M^2=1 for a gaussian beam) [72].

The dimensions of the previous microchip lasers cannot be reduced any further for technological problems arising from the energy dissipation and the thickness of the active crystals. But, by renouncing to power it is possible to circumvent the previous limitations and to reduce by more than two order of magnitude the longitudinal dimension of these laser devices. Indeed, by using different techniques for producing thin films, a certain number of active materials can be used between layers of dielectric compounds functioning as Bragg reflectors. Among them it is worth mentioning films of LiF colored with low energy beams of electrons. Figure 18 shows one such sample, where the vertical dimension is of the order of a few μm by excluding the silica substrate [73]. The LiF film has been colored by 6 keV electrons which produce color centers, and among them

F_3^+ and F_2 point defects emitting at 530 and 650 nm, respectively, upon 458 nm excitation, as shown by the dashed curve in Fig.18(b). When the Bragg reflectors are built around the LiF layer by completing a symmetric microcavity, the emission changes as reported by the full line of the same Fig.18(b), which together with a severe narrowing of the cone emission along the axis, not reported here, implies a laser type operation.

Laser devices as the ones just described or slightly modified are not limited to LiF, but any other suitable active material can be used amid microcavities. Semiconductors as InGaAs have been used recently with new and interesting results regarding also the ability to engineer photonic states in coupled cavities [74]. Also solid conjugated polymers are very promising as active materials, as it is the case of PPPV blended in a passive matrix of PMMA and lasing efficiently in the blue-green region [75]. It is still open for investigation whether organic molecules, like Alq_3 for instance, which has found widespread application

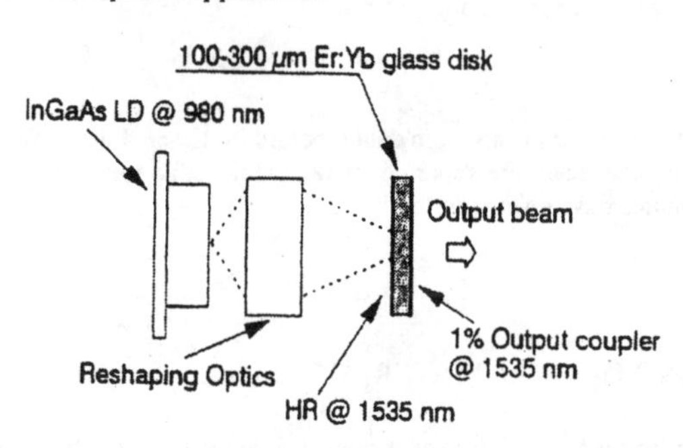

Figure 17. An Er-Yb:glass microchip laser showing from right to left the tiny crystal+cavity, an adapting optics, and the pumping TDL [72]. The whole laser can be very tiny, of the order of a few mm, by using well known optoelectronic techniques of miniaturization.

in light emitting devices for its very promising optical properties [76], can be usefully utilized as active material for photonic structures, although a laser emission has been demonstrated recently [77]. However, the energy of the 500 ps output pulse was limited to a few nJ, mainly a consequence of the low conductivity of the material itself, which is a problem common to all organic materials.

Anyway, the biggest revolution in both types of lasers, i.e. SDLs and SSLs, happened because new optical active materials have been discovered and nanostructure technology is introducing new optical structures.

Among the former ones, by using layers of AlGaN instead of GaN, transparent ultraviolet LED (Light Emitting Diode) structures have been fabricated with the result of moving the blue SDLs still towards the high energy spectrum [78]. These new diodes emit 1 mW at 350 nm, a substantial improvement on the previous situation. Among the latter ones there are the photonic crystals (PC) in one, two and three dimensions, also called photonic band gaps (PBG) because of the analogy of photons in PGBs with electrons in metals and semiconductors [79]. These new dielectric structures display strong photon suppressing properties, i.e. only photons with given wavelengths and polarization states can propagate in PGBs [80,81,82], and so they promote laser action [83]. One recent example of new structures and known materials is given by a two-dimensional PBG coupled to a

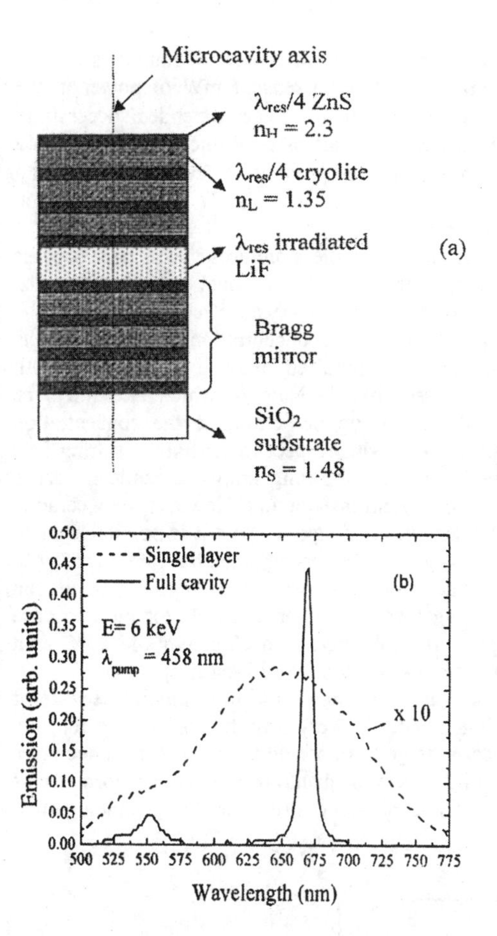

Figure 18. Schematic representation of a colored LiF based microcavity, showing the various Bragg reflector layers and the active LiF film (a). Room temperature photoluminescence spectra of the colored LiF film in a single layer and microcavity configuration (b). By curtesy of R.M. Montereali [73].

SDL which deliver 2.6 mW at 1.53 μm in cw operation at room temperature in a highly pure spectral mode, see Fig.19 [84].

By using the sophisticated technology of molecular beam epitaxy [85], as in the previous case which is however limited to a few layers, it has been possible to fabricate sandwiches of hundreds of nanometric-sized layers of different semiconducting materials, which mimic in the energy domain a series of quantum levels in cascade where the electrons can fall by moving under a driving electrical field, and so producing photons in the infrared region of the spectrum [86]. This singular SDL, called quantum cascade laser (QCL), is not bound to the energy-gap value of the materials utilized, like the classical SDLs, but rather to their thickness and spatial sequence, because of subsequent and appropriately different quantum confinement effects. In practice, peak powers of hundreds of mW have been obtained, and cw regimes maintained up to 175°C with tens of mW of power in the wavelength region from 3.4 to 17 μm [87]. Lately, their functioning has been extended successfully in the far infrared spectral region, and Fig.20 shows the emission intensity of one such laser at 4.45 THz (67.4 μm) [88]. The 1.25 mm long and 180 μm wide device produces a few mW at 8 K in a single mode, and improvements are pursued in order to raise at least above 77 K, liquid nitrogen temperature, the uncomfortably low operating temperature.

The ability to manipulate the laser-active materials and/or the structures which may support them at microscopic level, is also delivering new opportunities with the novel ceramic materials. Indeed, very often the performances of SSLs are limited by the active crystal itself which cannot be doped above a certain value, due to limiting segregation factor and concentration quenching effects on the emission [89], and cannot be as big and shaped as required because of crystal growth limitations. Usually, crystals are substituted by doped glasses, like the Nd:glass which takes the place of the much more expensive Nd:YAG crystal. But, glasses do not eliminate the concentration quenching effects, and their physical properties are inferior with respect to crystals, resulting in a lower efficiency of the final laser system. On the other side, the long known ceramic materials usually possess a sizeable light scattering, which cannot sustain laser action. However, new ceramic materials have lately been fabricated with very little scattering of light, so that it is possible now to use them as efficient active materials. They are not as good as the best crystals, but rather similar to the average-quality crystals. However, they can be fabricated with the highly desired ion concentration, in various shapes and sizes, and also together with other materials for multipurpose functions, at low prices, and still possessing an optical-optical conversion efficiency of ≈25%, as it has been recently demonstrated in polycrystalline 1% Nd:YAG ceramic rod laser [90].

The previous new ceramic materials differ from the well known old ones mainly because of smaller grain size and innovative pressing technologies, and the very tiny dimensions are for sure determining their interesting optical properties. In case the grain dimensions are further reduced to a few nm, then new quantum phenomena appear as in the case of quantum dots, known formerly as colloids. These nanoparticles can be coupled to periodic photonic structures for enhancing their optical properties, as it happens to CdSe

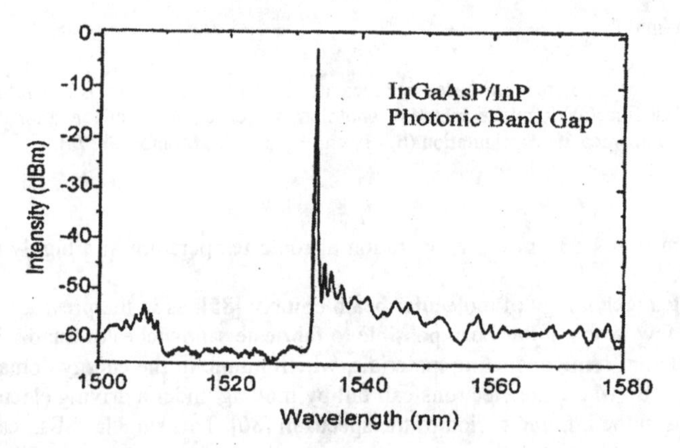

Figure 19. Laser emission from a photonic crystal coupled to a semiconductor diode laser based on InGaAsP and InP layers grown by molecular beam epitaxy. The spectrum shows a side mode suppression greater than 42 dB [84].

quantum dots, whose light emission increases by almost a factor 3 when embedded in a half-wavelength one-dimensional cavity, i.e. two distributed Bragg reflectors, [91].

In the material field, also lasers and spectroscopy are playing an important role. Indeed, small structures can be easily realized by merging together optical processes and chemical properties, as in the case shown in Fig.21. The solid structures have been fabricated with a modelocked Ti:sapphire laser and photoinitiator resins by means of multiphoton spectroscopy [92]. In particular the resin contains the two-photon absorbing molecule Lucirin-TPOL, and the short pulse of the focused laser promotes the photopolimerization of the solution which, after washing out the still liquid part, results in microscopic features smaller than the diffraction limit of the optical system, as the four interconnected towers in Fig.21. It is evident as this new chemical and physical technology can be broadly utilized for the realization of photonic materials like PBGs.

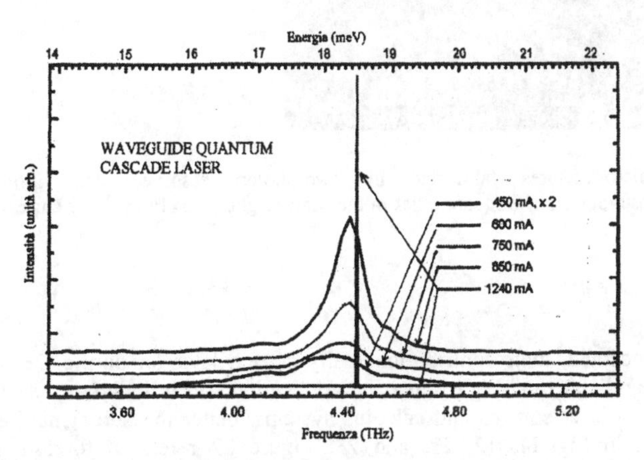

Figure 20. Emission intensity of a waveguide quantum cascade laser operating at 8 K in the THz region. The intensity grows nonlinearly with the excitation current up to reach the laser threshold at 880 mA. The laser emission at 1240 mA, which has been reduced by several orders of magnitude in intensity with respect to the other curves, possesses a side mode suppression of 20 dB [88].

4.3 LATEST DEVELOPMENTS OF LASER SPECTROSCOPY

In section 3.2 several examples of laser spectroscopy have been described in some detail, and the advantages of the laser sources over the classical optical ones for pursuing spectroscopy have been widely discussed. It was very clear from the beginning that the laser sources were going to open a new golden age for spectroscopy, not only by improving already known techniques but also by paving the way for new ones, with the adding of the new calculus possibilities offered by PCs. These expectations have been fulfilled in the following years to such an extent that *spectroscopy is commonly referred to as before and after the sixties*, so sharp has been the divide between the two periods of time.

Figure 21. Three-dimensional microstructures fabricated by two-photon absorbed near infrared photopolimerization by using 100 fs laser pulses. The square relief in the center right is 15x15 μm^2. By curtesy of T. Baldacchini [92].

As far as TDLs are concerned, their contribution really started later on during the late seventies when they became more reliable, and Fig.8 shows an application still in its infancy at that time [42]. Later on, with improving laser sources and calculus systems, better measurements have been performed, as those shown in Figs.11, 12, 22, and 23. Figure 22 refers to foreign gas broadenings and shifts of an ammonia transition line at room temperature. The two important molecular parameters have been measured at the same time with great precision as never before, and the different effects of N_2, O_2, H_2, Ar, He on the lineshape of the line aQ(9,9) are clearly observed [93]. At the same time the potentialities of the SDLs to detect small traces of molecular gases were fully developed, and Fig.23 gives an example of their versatility [94]. It represents one hour of continuous measurements of methane content in the atmosphere of Moscow while moving along its streets, and a big increase of methane was detected nearby a natural gas refilling station. The single peak structure was due both to the movement of the van and to the wind blowing from different directions, and so changing the amount of methane in the measuring device.

The photo-acoustic spectroscopy has been so refined lately that it offers now the possibility to monitor gases at ppbv levels, as in the case of methane produced by a single cockroach insect (Periplaneta Americana) [95]. These studies are important not only for agricultural purposes but also for environmental effects, if one keeps in mind that 25% of world methane production is originated by insects, and that methane is 20 time more effective than carbon dioxide as a green house effect gas. Also opto-galvanic spectroscopy has been so much improved to be used in precision spectroscopic measurements of uranium isotopes [96], which is a matter of utmost concern for industrial and environments purposes.

A very high sensitivity has been also reached by the Raman effect, which previously was used mainly to study the inner structure of matter, but lately also the impurities in the same matter. Figure 24 shows the Raman spectrum obtained by pumping with the 476.5 nm line of an Ar^+ laser a natural diamond where, between the Brillouin components on the left and the only Raman component

on the right, a single weak peak Δ' indicates the presence of boron impurities [97]. It is worth observing the practically inexistence of noise in the spectrum, especially in comparison with the old results displayed in Fig.14. This situation is going to improve still further, because the Raman efficiency depends on the fourth power of the carrier frequency and on a superlinear power of the carrier intensity. And frequencies of SDLs and SSLs are moving towards the violet region, their intensity is increasing, and the intensity is much more stable than an ion gas laser! Besides basic research, Raman spectroscopy is also widely applied in various field as medicine, the case illustrated in Fig.25. It represents the Raman response of 1-mm spot size on the retina of a human volunteer when excited with the 488 nm line of an Ar^+ laser [98]. The three peaks reveal the presence of carotenoid molecules which are normally contained in human retina. Their absence or an amount smaller than normal seems to be related to age
macular degeneration, an invalidating diseases witch can be probably prevented with an early detection by using the previous simple Raman analysis.

High laser intensities are also required to detect gases dispersed in non transparent media where the new technique GASMAT (*gas in scattering media absorption spectroscopy*) can be applied. Indeed, spectroscopic analysis of the weak scattered light can still be performed if the laser beam is intense enough, and with success as in the case of oxygen in wood [99].

Not so is the case, with reference to high intensity, of a new technique where the absorption of gases in a cell is measured by the time it takes for a laser signal to decay while bouncing between two highly reflective mirrors [100]. Here it is not the intensity of the laser beam which plays a mayor role, but the highly sophisticated requirements for the mirror optical properties and optical elements in general. Indeed, the sensitivity of the method which is called *cavity ring-down spectroscopy* is proportional to the number of passages trough the cell, and so to the degree of reflectivity of the mirrors. However, such measurements would have been impossible without lasers. In principle this method isequivalent to the well known linear absorption in a multipass cell, but in practice it is the result of a sophisticated technique which requires also short pulsed lasers.

Different is the case of the *correlation spectroscopy*, where the laser can play an important role in improving the sensitivity of the system but it is not essential. Indeed, the method consists in comparing the experimental spectrum with other ones already known, and from the correlation it is possible to measure the amount of the unknown gas. However, the spectra of many gases should be at disposal in order to perform the comparison in real time, which is not an easy task. So, it has been found recently that a computer can generate diffractive optical elements that synthesize the spectra of various gases, solving at the roots the problem [101]. It is worth noticing that this method does not need to possess necessarily a high resolution to have a high sensitivity, because the comparison of the spectra occurs in a large spectral portion where there are many spectroscopic features that should correspond each other both in the exact spectral position and in intensity, contrary to what happens to slightly tunable SDLs which look at single and isolated resonances, as in Figs.8 and 10.

Up to now, pulsed lasers have been described as functional to some types of classic spectroscopic methods, as saturation, two-photon, cavity ring-down, lifetime, etc., but in reality new

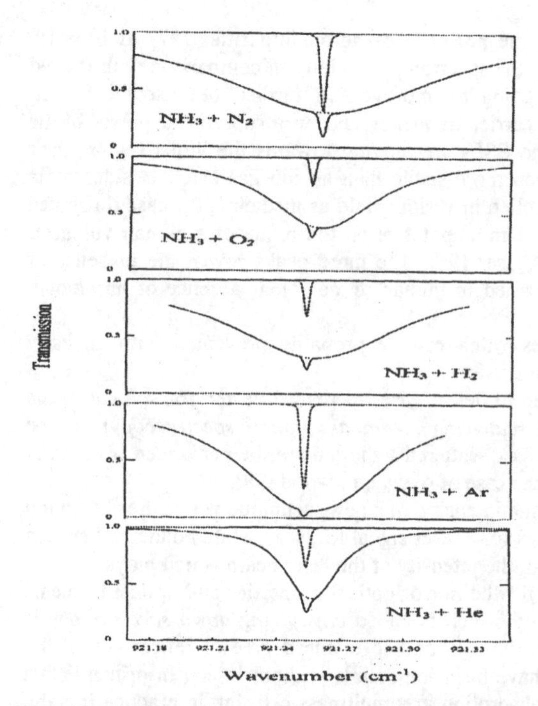

Figure 22. Lineshape of the aQ(9,9) ammonia transition induced by collisions with N_2, O_2, H_2, Ar, and He moving top-down. The spectra are measured in a cell 19 cm long, at 301 K, and at a foreign-gas pressure of about 460 Torr with a few Torr of ammonia. The sharp peak in each spectrum is given by a reference cell of NH_3 at low pressure, and it is used to measure the shift of the broadened line [93].

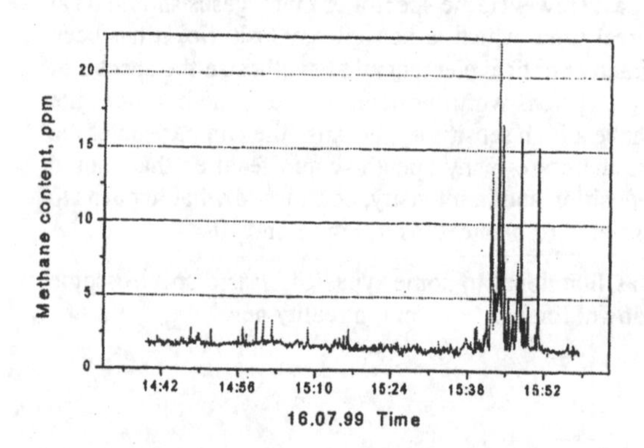

Figure 23. Methane content in the atmosphere of Moscow as measured by a mobile TDL instrument on 16 July 1999, during one hour interval. The intense peaks on the right are originated by methane leaking from a gas refilling station when the van car was passing nearby [94].

fields of investigation concerning the same quantum nature of matter are opened up by short pulse excitation. When matter is excited by a light pulse, it is possible in particular circumstances to generate an excitation at identical quantum levels witch is described by a wave packet. This process is usually called a *coherent excitation*, i.e. a quantum state of superimposed quantum waves with the same intensity and phase. This coherent status changes in time because of population decay by spontaneous emission, i.e. the intensity of the single components of the wave packet is decreasing with life-time T_1, longitudinal relaxation time, which is measured usually in the normal incoherent spectroscopy. The same status changes in time because of phase-perturbing interactions, i.e. the phases of the single components go astray, with the result of the intensity of the wave packet decreasing with decay-time T_2, transverse relaxation time, which can be measured in coherent spectroscopy. Generally, $T_2<T_1$, and the transverse relaxation time varies in different materials as in the following:

- 10-100 fs in atomic collisions,
- 50-300 fs in molecular relaxation in dyes,
- 100-1000 fs in molecular vibrations,
- 100 fs-10 ps in chemical reactions,
- $\approx$ 1 ps in interactions in liquids,
- 1-10 ps in point defect in solids,
- $\approx$ 40 ps in semiconductor quantum dots.

Although coherent spectroscopy has been introduced before the invention of lasers, limited however to a few fortunate and accidental cases, the adding of ultra fast light pulses allowed for the first time the direct measurement of wave packets coherently excited, their motion on the potential energy surfaces, and their ultimate fate [102].

However, the values given above for T_2 pose severe restrictions on the pulse duration of the exciting and probing laser, which should be much shorter than T_2 itself. Indeed, the transverse relaxation time is only the envelope of quantum oscillations with periodicity order of magnitude shorter. In the case of color centers in KBr, while T_2=2.6 ps the oscillation period is T=0.4 ps which can be measured only with pulse duration of less than 40 fs [102], as it is clearly shown in Fig.26. In this case, it is worth reminding that the radiative lifetime of the relaxed excited state is T_1=1.1 μs, although at room temperature, as in this experiment, the real lifetime is shortened by five order of magnitude by thermal excitation effects. Coherent optical control of the quantum state has been also attained on single quantum dots, where T_2=40 ps but the measured luminescence oscillations are in the ps regime or less [104]. With analogous techniques it has also been possible to follow the molecular dynamics after laser excitation, which in $[Ru(bpy)_3]^{2+}$ has shown a coupling to the surrounding solvent and a different one that follows an energy potential internal to the molecule, with different time scales ranging from 50 fs to 1 ps [105]. It is evident how the previous examples and considerations stimulate the quest for shorter and shorter light pulses in all frequency domains of the electromagnetic spectrum.

A last mention deserves the opening up to spectroscopy of a new region loosely defined in the frequency range from 0.1 to 10 THz (1 THz=10^{12} Hz) or wavelength range from 3 mm to 30 μm. In the literature it is often referred to as *terahertz (THz) gap*, because until recently it was not an easily accessible region for lacking of well developed technologies both for sources and detectors. Indeed, the microwave region, well served by electronic devices, is limited to a few mm, while only a few optical devices operate efficiently above 30 μm. It is true, however, that some free electron lasers (FEL) emit intense radiation in this region, but these complex laser sources operate only in specialized laboratory, and possess light beams with peculiar time-energy structures which strongly limit their utilization. Lately, semiconductor laser devices (germanium intervalence band) and new technologies (terahertz time-domain) are closing the gap from the low frequency side, while quantum cascade lasers (QCL) are entering into the gap from the high frequency side with promising results [88,106].

The opening up of this last frontier of electro-optical technology and spectroscopy is expected to be very much rewarding [107]. Indeed, apart the basic research interests, applications are awaiting, for instance, in fields like spectroscopy of light molecules, spectroscopy of doped semiconductors, characterization of high temperature superconductors, optical pump and THz probe of materials, THz wave imaging, genetic analysis and cancer detection, THz bio sensors. Terahertz technology and spectroscopy have grown greatly in the last decade, and they are going to be more and more pervasive with the improvements of the related technologies, which comprise also laser sources.

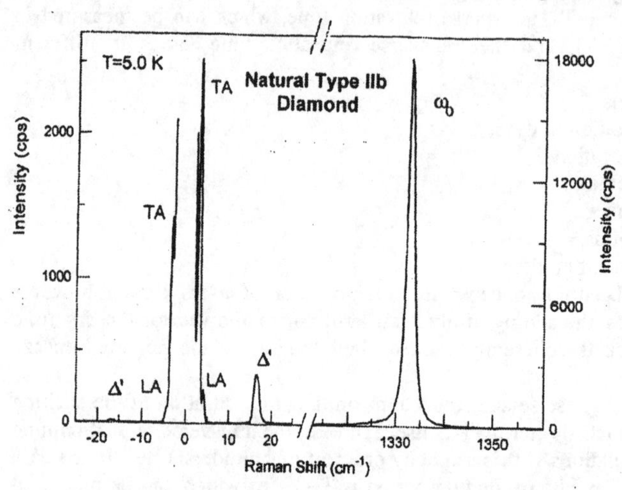

Figure 24. Raman spectrum of a diamond containing boron impurities obtained at 5 K by pumping with the 476.5 nm line of an Ar^{+} laser. The intense resonances belong to the various vibrations of the diamond lattice, while the peak Δ' is due to boron [97].

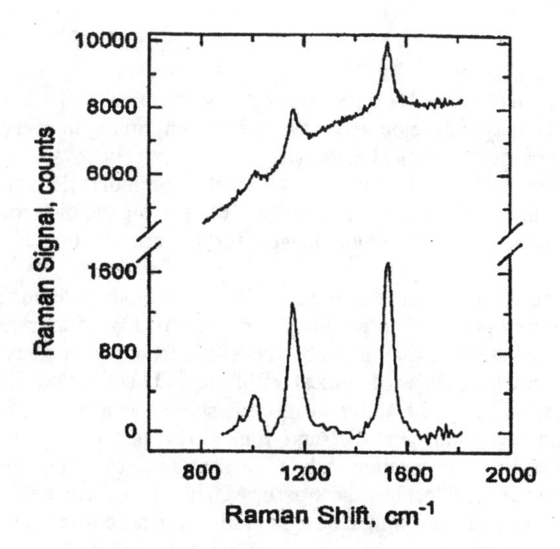

Figure 25. Raman spectrum of a spot in the human retina excited by the 488 nm line of an Ar^+ laser. The top trace is the original spectrum before the subtraction of fluorescence background [98].

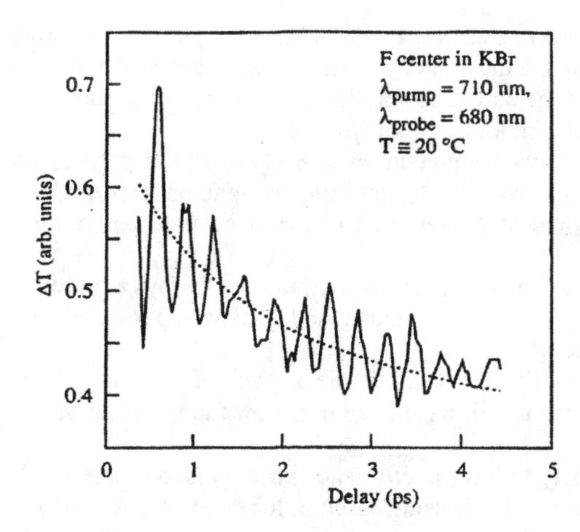

Figure 26. Transmission changes ΔT of the probe pulse as a function of delay time between pump and probe pulses on the absorption band of F centers in KBr at room temperature. The exponential dashed line represents the relaxation decay toward the RES, τ=2.6 ps, while the much faster modulation derives from the coherent oscillations of the electronic wave packet in the vibronic potential [102].

5. Conclusions

At moment laser technology has reached a high level of sophistication, and while only a few new lasers have been devised, the mayor development have been in improving in every respect the old laser types and in introducing new active optical materials.

On the second category belongs the dendrimer laser, where the dye molecule is nestled in the branches of a large polymer dendrimer molecule, in this way partially solving the old problems of quenching effects and heat dissipation, a very common plague for all organic optically active materials [108].

On the first category belong the most successful SSLs, Yb:YAG diode side-pumped and diode end-pumped [109], which are both projected to produce more than 10 kW of average output power [110]. Among them, also the disk laser configuration has benefited from higher dopant levels and improved optical schemes, by reaching 70% of optical efficiency [111]. In the disk laser, invented at the University of Stuttgart in 1992, the active medium is shaped as a thin disk with its plane surface acting as heat sink and cavity mirror at the same time [112]. In this way no thermal lensing effects develop at the expenses of absorption, which is increased by using multi-pass technologies, well known in atomic and molecular spectroscopy. This is an example of how spectroscopy can help laser developments, which can then be again utilized in spectroscopy.

This virtuous circle where spectroscopy played a central role in the discovery and development of laser sources, and conversely lasers contributed to the improvement of spectroscopy, which in turn allowed the realization of new laser devices and to improve the already existing ones, is a symbiosis as old as the laser itself. Still today, laser and spectroscopy are so strongly intertwined that new exchanges of basic and technological knowledge are producing outstanding results. Among them a few are listed as it follows:

- in the *disk laser*, as just described, the active element should be as thin as possible in order to dissipate easily the heath produced by the pumping. As a consequence the optical absorption is very much reduced, and so the output power. The problem has been solved by using optical *multipass techniques* developed long ago for *spectroscopy*,
- the *two-photon* absorption is a well known method in *spectroscopy*, and is particularly efficient when *lasers* are used as optical sources. Today, the two-photon absorption is utilized to fabricate complex microscopic structures, that can also be used as *photonic materials* for new laser emissions,
- *powerful lasers* produce *soft x-ray* bursts when focused on any material. X rays are used in *photolithography* to fabricate *photonic circuits* for active and passive optical devices, having smaller dimensions and higher resolutions,
- *luminous patterns* are efficiently and easily produced on the surface of *LiF crystals* by using soft x-rays as above. The emissions in LiF have been used since long ago in pulsed *tunable solid state lasers*,
- *Two-photon* polymerisation is used firstly to built micro-nanometric tools, and afterwards two-photon radiation moves the same tools, by using laser induced forces, to perform useful tasks in the nano-world.

The last sentence is not a fruit of fantasy but it has been implemented recently, where the gradient force of the radiation pressure exerted by the focused laser beam is the driving motor of the micro mechanism, just fabricated with the chemical- laser technique [113]. This fruitful relationship, which is a consequence of the quantum mechanical nature of the matter, is not going to abate but, on the contrary, new and surprising results will be obtained with the increasing capability of manoeuvring the matter at atomic level.

The pulsed regime of lasers, which is of paramount importance for laser applications where short time domains and high peak powers are required, as we have seen in the last section, has also witnessed some improvements.

The ultrashort pulse record of 4.5 fs from a Ti:sapphire laser in combination with other techniques [114] has been further lowered at 3.9 fs more recently [115], which can hardly be improved because of the limits set by the laser field oscillation cycle in the visible spectral range. However, such limits are removed by moving to extreme UV and x-ray sources, where a pulse duration of 0.650 fs has been obtained at 90 eV by resorting to nonlinear processes [116]. On this respect, also x-ray table-top lasers [117], and FELs generating high-brightness hard x-rays down to less than 2.5 Å [118] possess very interesting short pulse properties, and they will allow new experiments in spectral regions up to now of difficult access.

Besides the well known application fields of short pulses, a new spectroscopy is rapidly developing in nuclear matter. Indeed, photo-nuclear reactions can be induced by the actual lasers which reach an intensity of 10^{21} Wcm^{-2}, and there are plans to built lasers 100 time more intense [119]. High energy x-rays and 100 MeV electrons have been already produced by firing powerful lasers on solid targets, as gold for instance, and much more will be accomplished in the near future, as the production of positron and electron couples, a new frontier for spectroscopy.

Going back to the main purpose of this work, it is a common opinion among the experts of the laser field that further achievements should be expected from technological advancements in general and in material science in particular. Indeed, the fabrication of micro lasers, photonic band gap lasers, and quantum cascade lasers is the result of our ability to assemble atoms and molecules in new artificial structures. This complex game is actually at its infancy, so that others promising results should be obtained especially from the development of new materials and nanotechnology. As far as spectroscopy is concerned, the accessible spectrum has been consistently increased in the last decade. In particular, the fast development of laser sources and auxiliary equipments, above all computers, prompted a wide diffusion of laser spectroscopy with superior resolutions and sensitivity. Moreover, excitations in gas, liquid and solid materials are being investigated systematically, especially by using time resolved spectroscopy in the fs regime and, recently, also below. As a consequence, new luminescent materials with greater efficiency may appear soon, especially in frontier regions like THz and x-rays.

By looking again at the business side of the laser field, it has been often taken as a rule that an increasing market has always produced bigger investments and, as a consequence, exciting scientific progresses. This relationships between market and laser development has been apparently shaken by the great fall during 2001-2002, see section 4.1, but at a second sight it resulted more an industrial debacle than a scientific one. Indeed, the commercial crisis has been originated by the crumbling of the information technology which hit severely SDLs, to a less extent excimer lasers, and not at all the other families of lasers. In particular SSLs show positive trends especially in connection with SDL pumping and better technologies. So, the investments did not decrease in the laser fields still under development, and the previous parallelism between bigger sales and better technologies remains a good omen for future trends of R & D in lasers in general.

In conclusion, the laser field looks well alive, and new surprises are awaiting ahead with significant results also in basic and applied spectroscopy.

Acknowledgments

The author is indebted with Tommaso Baldacchini and Rosa Maria Montereali for providing unpublished scientific material.

REFERENCES

[1] T.H. Maiman, Nature, **187**, 493-494 (1960).
[2] S. Nakajima and M. Tamatani, *History of Phosphor Technology and Industry*, in Phosphor Handbook, S. Shionoya and W.M. Yen, eds., CRC Press, Boca Raton, 1998, pags. 821-857.
[3] E. Newton Harvey, *A History of Luminescence, From the Earliest Times Until 1900*, The America Philosophical Society, Philadelphia, 1957.
[4] J.P. Gordon, H.J. Zeiger, and C.H. Townes, Phys. Rev. **95**, 282-284 (1954) .
[5] M. Brotherton, *Masers and Lasers*, McGraw-Hill Book Company, New York, 1964.
[6] Bela A. Lengyel, Am. J. Phys. **34**, 903-913 (1966).
[7] M. Bertolotti, *Masers & Lasers*, An Historical Approach, Adam Hilger Ltd, Bristol, U.K., 1983.
[8] N. Taylor, *The Inventor, the Nobel Laureate, and the Thirty Year Patent War*, Simon & Schuster, Inc., New York, 2000.
[9] A. Javan, W.R. Bennet, and D.R. Herriot, Phys. Rev. Lett. **6**, 106-110 (1961).
[10] A.D. White and J.D. Ridgen, Proc. IRE, **50**, 1697 (1962).
[11] L.F. Johnson and K Nassau, Proc. IRE, **49**, 1704-1706 (1961).
[12] E. Geusic, H.M. Marcos, and L.G. Van Uitert, Appl. Phys. Lett. **4**, 182-184 (1964).
[13] R.N. Hall, G.E. Fenner, J.D. Kingsley, T.J. Soltys, and R.O. Carlson, Phys. Rev. Lett. **9**, 366-368 (1962).
[14] M.I. Nathan, W.P. Dumke, G.Burns, F.H. Dill and G. Lasher, Appl. Phys. Lett. **1**, 62-64 (1962).
[15] T.M. Quist, R.H. Rediker, R.J. Keys, W.E. Krag, B. Lax, A.L. Mcwhorter, and H.J. Zeiger, Appl. Phys. Lett. **1**, 91-92 (1962).
[16] W.B. Bridges, Appl. Phys. Lett. **4**, 128-130 (1964).
[17] C.K.N. Patel, Phys. Rev. **136A**, 1187-1193 (1964).
[18] B. Fritz and E. Menke, Sol. Stat. Commun. **3**, 61-63 (1965).
[19] L.F. Mollenauer and D.H. Olson, Appl. Phys. Lett. **24**, 386-388 (1974).
[20] G. Baldacchini, in *Spectroscopy and Dynamics of Collective Excitations in Solids*, B. Di Bartolo, ed., Plenum Press, New York, 1997, p. 495-517.
[21] J.V.V. Kaspar and G.C. Pimentel, Phys. Rev. Lett. **14**, 352-354 (1965).
[22] P.P. Sorokin and J.R. Lankard, IBM Journal of R&D, **10**, 162- (1966).
[23] N.G. Basov, V.A. Danilychev, Yu. M. Popov, and D.D. Khodkevich, JETP Lett. **12**, 329-331 (1970).
[24] S.K. Searles and G.A. Hart, Appl. Phys. Lett. **27**, 234-245 (1975).
[25] D.A.G. Deacon, L.R. Elias, J.M.J. Madey, G.J. Ramian, H.A. Schwettnan, and T.I. Smith, Phys. Rev. Lett. **38**, 892-894 (1977).
[26] D. Jacobi, G.J. Pert, S.A. Ramsden, L.D. Shorrock, and G.J. Tallents, Opt. Commun. **37**, 193-196 (1981).
[27] C.A. Robinson, Jr, Aviation Week and Space Technology, Feb. 23, 25-27 (1981).
[28] P.F. Moulton, Optics News, **8**, 9, Nov-Dec 1982.
[29] *Physics of Color Centers*, W.B. Fowler, ed., Academic Press, New York, 1968.
[30] O. Svelto, *Principles of Lasers*, Plenum Press, New York, 1989.
[31] J. Hecht, *The Laser Guidebook*, TAB Books, Blue Ridge Summit, 1992.
[32] H. Kressel and J.K. Butler, *Semiconductor Lasers and Heterojunction LEDs*, Academic Press, New York, 1977.
[33] *Semiconductor Lasers, Past, Present and Future*, G.P. Agrawal, ed., AIP Press, Woodbury, N.Y., 1995.
[34] S. Nakamura, T. Mukai, and M. Senoh, Appl. Phys. Lett. **64**, 1687-1689 (1994).
[35] G. Blasse and B.C. Grabmaier, *Luminescent Materials*, Springer-Verlag, Berlin, 1994.
[36] A.A. Kaminskii, Phys. Stat. Sol. **A148**, 9-79 (1995).
[37] S.A. Payne, L.L. Chase, L. K. Smith, W.L. Kway, and H. W. Newkirk, J. Appl. Phys. **66**, 1051-1056 (1989).
[38] W. Gellermann, J. Phys. Chem. Solids, **52**, 249 (1991)
[39] T.K. Kane, A.C. Nilsson, and R.L. Byer, Opt. Lett. **12**, 175-177 (1987).
[40] U. Brauch, A. Giesen, M. Karszewski, C. Stewen, and A. Voss, Opt. Lett. **20**, 713-715 (1995).
[41] M.J. Weber, *Handbook of Laser Science and Technology*, CRC Press, Boca Raton, 1982.
[42] F.A. Blum, K.W. Nill, P.L. Kelley, A.R. Kalawa, and T.C. Harman, Science, **177**, 694-695 (27 August 1972).
[43] G. Baldacchini, G.P. Gallerano, D. Censi, M. Tonelli, P. Violino, U.M. Grassano, M. Meucci, and A. Scacco, Revue Phys. Appl. **18**, 301-306 (1983).

[44] *Laser Spectroscopy of Atoms and Molecules*, H. Walther, ed., Springer Verlag, Berlin, 1976.
[45] W. Demtroeder, *Laser Spectroscopy*, Springer-Verlag, Berlin, 1996.
[46] G. Baldacchini, A. Bellatreccia, and L. Nencini, ENEA Internal Report RT/TIB/89/29, 1989.
[47] G. Baldacchini and J. Baltussen, EUROTRAC Newletter, No.9, 23-25 (1992).
[48] W. Gerlach and C. Stern, Ann. Physik, **74**, 673 (1924).
[49] I.I. Rabi, J. R. Zacharias, S. Millman, and P. Kusch, Phys. Rev. **53**, 318 (1938).
[50] *Molecular Beams and Low Density Gasdynamics*, P.P. Wegener, ed., Marcel Dekker, Inc., New York,1974.
[51] M.J. Molina and F.S. Rowland, Nature, **249**, 810 (1974).
[52] G. Baldacchini, S. Marchetti, and V. Montelatici, Appl. Phys. **B29**, 269-272 (1982).
[53] V.P. Zharov and V.S. Letokhov, *Laser Optoacoustic Spectroscopy,* Springer Ser. Opt. Sci., Vol.37, Springer, Berlin, 1986.
[54] *Optogalvanic Spectroscopy*, R.S. Steward and J.E. Lawler, eds., Hilger, London, 1991.
[55] P.N. Butcher and D. Cotter, *The Elements of Nonlinear Optics*, Cambridge University Press, Cambridge, 1990.
[56] *Non Linear Optics in Signal Processing*, R.W. Eason and A. Miller, eds., Chapman & Hall, London, 1993.
[57] D. Frohlich, R. Kenklies, and Ch. Uihlein, Phys. Rev. Letters, **43**, 1260-63, (1979).
[58] J.R. Sandercock, Solid State Commun. **26**, 547-551 (1978)
[59] M.J. Weber, *Handbook of Laser Wavelengths*, CRC Press, Boca Raton, 1999.
[60] S. Nakamura, *GaN and related luminescent materials*, in Phosphor Handbook, S. Shionoya and W.M. Yen, ed s., CRC Press, Boca Raton, 1999, pags. 293-306.
[61] D.M. Mittleman, M. Gupta, R. Neelamani, R. G. Baraniuk, J. V. Rudd, and M. Koch, Appl. Phys. **B68**, 1085-1094 (1999).
[62] K.J. Schafer and K.C. Kulander, Phys, Rev. Lett. **78**, 683-641 (1997).
[63] T. Letardi, A Baldesi, S. Bollanti, F, Bonfigli, P. Di Lazzaro, F. Flora, G. Giordano, D. Murra, G. Schina, and C. E. Zheng, in *High-Power Laser in Manifacturing*, X. Chen, T. Fujioka, and A. Matsunawa, eds., SPIE vol. 3888, 2000, pags. 587-597.
[64] H.P. Freund and T.M. Autonsen, *Principles of Free Electron Laser*, Chapman and Hall, London, 1996.
[65] *X-ray Lasers*, S. Svanberg and C-G. Wahlström, eds., IOP, Bristol, 1996.
[66] I. Flegel and J. Rossbach, Europhys. News, Nov-Dec 2000, pags. 12-14..
[67] J.J. Rocca, F.G. Tomasel, M.C. Marconi, V.N. Shlyaptsev, J.L.A. Chilla, S.T. Szapiro, and G. Giudice, *Discharge-pumped soft x-ray laser in neon-like argon*, Phys. Plasmas, **2**, 2547 (1995).
[68] The data on the laser Market have been taken by the specialized Journal Laser Focus World, January and, more recently, February issues, of the various years.
[69] L. Marshall, *Diode-pumped lasers begin to fulfill promises*, Laser focus World, June 1998, pag.63.
[70] W. Hardin, *World's largest laser shots photonics to a new level*, Photonics Spectra, May 1998, pag. 104.
[71] P. Guillot, *Microlasers: Short Pulses Increase Applications*, Photonics Spectra, February 1998, pags. 143-146.
[72] P. Laporta, S. Taccheo, S. Longhi, O. Svelto, and C. Svelto, Opt. Mat. **11**, 269-288 (1999).
[73] A. Belarouci, F. Menchini, H. Rigneault, B. Jacquier, R.M. Montereali, F. Somma, P. Moretti, and M. Cathelinaud, Opt. Mat. **16**, 63-67 (2001).
[74] M.S. Skolnick et al., J. Luminesc. **87-89**, 25-29 (2000).
[75] G. Wegmann, H. Giessen, A. Greiner, and R.F. Mahrt, Phys. Rev. **B57**, R4218-R4220 (1998).
[76] G. Baldacchini, S. Gagliardi, R.M. Montereali, and A. Pace, Philosophical Magazine, **B82**, 669-680 (2002).
[77] D. Schneider et al. Appl. Phys. **B77**, 399-402 (2003).
[78] T. Nishida and N. Kobayashi, Appl. Phys. Lett. **82**, 1-3 (2003).
[79] E. Yablonovitch, J. Opt. Soc. Am. **B10**, 283-295 (1993).
[80] M.J. Bloemer and M.S. Scalora, Appl. Phys. Lett. **72**, 1676 (1998).
[81] J.C. Knight, J.Broeng, T.A. Birks, and P.St. Russel, Science, **282**, 1476 (1998).
[82] J. Schilling, F, Muller, S. Matthias, R.B. Wehrspohn, and U. Gosele, Appl. Phys. Lett. **78**, 1180 (2001)
[83] K. Inoue, M. Sasada, J. Kawamata, K. Sakoda, and J.W. Hans, Jpn. J. Appl. Phys. **38**, L157 (1999).
[84] T.D. Happ, M. Kamp, and A. Forchel, Appl. Phys. Lett. **82**, 4-6 (2003).
[85] A.Y. Cho, J. Cryst. Growth, **111**, 1-13 (1991).

[86] J. Faist, F. Capasso, D.L. Sivco, C. Sirtori, A.L. Hutchinson, and A.Y. Cho, Science, **264**, 533-556 (1994).
[87] F. Capasso, C. Gmachl, D.L. Sivco, and A.Y. Cho, Physics World, June 1999, pags. 27-33.
[88] A. Tredicucci, Il Nuovo Saggiatore, **18** no 3-4, 26-33 (2002).
[89] F. Auzel, F. Bonfigli, S. Gagliardi, and G. Baldacchini, J. Lumin. **94-95**, 293-297 (2001).
[90] Murai, K. Takaichi J. Lu, T., T. Uematsu, K. Misawa, M. Prabhu, J. Xu, K. Ueda, H. Yagi, T. Yanagitani, A.A. Kaminskii, and A. Kudryashov, Appl. Phys. Lett. **78**, 3586 (2001).
[91] C.B. Poitras, M. Lipson, H.Du, M.A.Hahn, and T.D. Krauss, Appl. Phys. Lett. **82**, 1-3 (2003).
[92] T. Baldacchini, R.A. Farrer, J. Moser, J.T. Fourkas, and M.J. Naughton, Synthetic Metals, **135-136,** 11-12 (2003).
[93] G. Baldacchini, F. D'Amato, G. Buffa, O. Tarrini, M. De Rosa, and F. Pelagalli, *Tunable Diode Laser Measurement of Self and Foreign Broadening and Shift Versus Temperature of Seven Ammonia Transitions of the ν_2 Band.* Report ENEA RT/INN/97/29, 1997, and J. Quant. Spectrosc. Rad. Transfer, **68**, 625-633 (2001).
[94] G. Baldacchini and F. D'Amato, *Innovative Trace Gas Analyser System Based on Tunable Diode Lasers.* Final Report, Contract nr. INTAS-93-1230-ext, December 1999. Dr. A. Nadezhdinskii and A. Berezin of the General Physics Institute in Moscow have been the Russian Coordinators of the Project.
[95] F.G.C. Bijnen, F.J.M. Harren, J.H.P. Hackstein, and J. Reuss, Appl. Opt. **35**, 5357-5368 (1996).
[96] C. M. Barshick, R. W. Shaw, J. P. Young, and J. M. Ramsey, Anal. Chem. **67**, 3814 (1995).
[97] H. Kim, R. Vogelgesang, A.K. Ramdas, S. Rodriguez, M. Grimsditch, and T.R. Anthony, Phys. Rev. **B57**, 15315 (1998).
[98] W. Gellermann, I. Ermakov, M.R. Ermakova, and R. W. McClane, J. Opt. Soc. Am. **A19**, 1172-1181 (2002).
[99] M. Sjoholm, G. Somesfalean, J. Alnis, S. Andersson-Engels, and S. Svanberg, Opt. Lett. **26**, 16-18 (2001).
[100] S. Cheskis, I. Derzy, V.A. Lozovsky, A Kachanov, and D. Romanini, Appl. Phys. **B66**, 377-381 (1998).
[101] M.B. Sinclair, M.A. Butler, A.J. Ricco, and S.D. Senturia, Appl. Opt. **36**, 3342-3348 (1997).
[102] B. Garraway, S. Stenholm, and K-A. Suominen, *Adventures in Wave Packet Land*, Physics World, April 1993, pags.46-51.
[103] N. Nisoli, S. De Silvestri, O. Svelto, R. Scholz, R. Fanciulli, V. Pellegrini, F. Beltram, and F. Bassani, Phys. Rev. Lett. 77, 3463-3466 (1996).
[104] N.H. Bonadeo, J. Erland, D. Gammon, D. Park, D.S. Katzer, and D.G. Steel, Science, **282**, 1473-1476 (1998).
[105] A.T. Yeh, C.V. Shank, and J.K. McCusker, Science, **289**, 935-938 (2000).
[106] R. Kohler, A. Tredicucci, F. Beltram, H.E. Beere, E.H. Linfield, A.G. Davies, D.A. Ritchie, R.C. Iotti, and F. Rossi, Nature,**417**, 156-159 (2002).
[107] B. Ferguson and Xi-C. Zhang, Nature Materials, **1**, 26-33 (2002).
[108] S. Yokoyama, A. Otomo, and S. Mashiko, Appl. Phys. Lett. **80**, 7 (2002).
[109] D.W. Hughes and J.R.M. Barr, J. Phys. D: Appl. Phys. **25**, 563-586 (1992).
[110] D.S. Sumida, A.A. Betin, H. Bruesselbch, R. Byren, S. Matthews, R. Reeder, and M.S.Mangir, Laser Focus World, June 1999, pags. 63-70.
[111] U. Brinkmann, Laser Focus World, June 1999, pags. 31-33.
[112] U. Brauch, A. Giesen, M. Karszewski, C. Stewen, and A. Voss, Opt. Lett. **20**, 713-715 (1995).
[113] S. Maruo, K. Ikuta, and H. Korogi, Appl. Phys. Lett. **82**, 133-135 (2003).
[114] M. Nisoli, S. De Silvestri, O. Svelto, R. Szipocs, K. Ferencz, Ch. Spielmann, S. Sartania, and F. Krausz, Opt. Lett. **22**, 522-524 (1997).
[115] T. Kobayashi, value obtained in 2001, private communication (2002)
[116] M. Hentschel et al., Nature, **414**, 509 (2002).
[117] G. Tomassetti, A. Ritucci. A. Reale, L. Palladino, L. Reale, S.V. Kukhlevsky. F.Flora, L. Mezi, J. Kaiser, A. Faenov, and T. Pikuz, Eur. Phys. J. **D19**, 73-77 (2002).
[118] S.W. Milton et al., Science, **292**, 2037 (2001).
[119] K.W.D. Ledingham, R.P. Singhal, P. McKenna, and I. Spencer, Europhysics News, July/August 2002, pags.120-124.

8. COHERENT SPECTROSCOPY OF STRATIFIED SEMICONDUCTOR MICRO- AND NANOSTRUCTURES

V.G. LYSSENKO
Institute of Microelectronics Technology,
Chernogolovka, Moscow district, 142432 Russia

1. Introduction [1]

The enhancement or inhibition of the interaction between light and matter is of great interest, for the fundamental and applications reasons. In perfect direct-gap bulk semiconductors, absorption at low temperatures is dominated by relatively strong and narrow excitonic transitions. At room temperature weakly bounded excitonic states are ionised, and absorption band extends over the whole conduction and valence bands. Therefore, there has been a lot of attempts to develop semiconductor structures, which would have sharper optical features due to higher exciton binding energy or redistribution of oscillator strengths due to band edge density of states reconstruction. For example, impurity or localised defects, disturbing periodicity of crystal lattice, can form sharp exciton-impurity transitions near band edge of bulk semiconductors due to their well-defined, atom-like energy levels.

Another way to obtain sharp optical features of semiconductor structures is through electronic band-gap engineering by the use of quantum-confined structures: quantum wells, wires or dots. The improved light-matter interaction relies on the reducing of one degree of freedom, yielding more spectrally concentrated optical features: absorption, reflection or gain.

Besides the quantum-confined scheme of low dimensional structures, sharp optical features may be obtained through photon-mode selection. Yablonovitch noticed [2] that analogously to band structure of electronic states in periodic lattice of semiconductors, light waves propagation in periodic structures also give rise to formation of allowed and forbidden photonic states. He suggested using periodic distribution of optical properties to create a photonic ban gap structures, which relies on multiple reflections from periodically distributed scatters. Placing inside a photonic ban gap structure an active material with energies within the photonic bandgap gives complete suppression of light-matter interaction: no photon mode can interact with the active material. Deviation from periodicity in photonic ban gap structure can localise light and alter transmission and emission properties of semiconductor material within photonic ban gap structure, analogously to formation of impurity states in periodical lattice of semiconductor. Then, a

B. Di Bartolo and O. Forte (eds.), Frontiers of Optical Spectroscopy, 289-335.

localised photon state build up, similar to localised electronic states of chemical impurities or defects in otherwise perfect solids.

Microcavity (MC) is one-dimensional realisation of photonic ban gap structure, consisting of two parallel high-reflective mirrors and "optical impurity" – active semiconductor layer - between them. When dimension of the "impurity layer" in periodical structure is equal to the quarter- or half-wavelength of light, one or few resonant modes are enhanced by multiple constructive interference. All other modes are suppressed by multiple destructive interferences. All electron-hole pairs not matching a photon cavity mode either recombine nonradiatively or scatter to the photon-matched electronic quantum states through energy and momentum relaxation.

2. Maxwell's equations [3]

Time evolution and spatial distribution of electric $\boldsymbol{E}$ and magnetic $\boldsymbol{H}$ field amplitudes in media with dielectric constant ε, permeability $\mu \sim 1$, and conductivity σ obeys set of Maxwell's equations [2]. From Maxwell's equations follows "wave" equation

$$\nabla^2 \boldsymbol{E} = \frac{\mu\varepsilon}{c^2}\frac{\partial^2 \boldsymbol{E}}{\partial t^2}, \tag{2.1}$$

describing propagating of electromagnetic wave with electric field $\boldsymbol{E}$ in the nonconducting ($\sigma = 0$) isotropic and nondispersive medium. Semiconductor can be characterized by index of refraction n, the ratio of speed of light in a vacuum c and its speed in a medium v

$$n = \frac{c}{v} = \sqrt{\varepsilon\mu} \cong \sqrt{\varepsilon} \tag{2.2}$$

In simplest case of monochromatic electromagnetic wave it is periodic function of time $E\sim\cos(\omega t+\phi)$, therefore $E_t'' = -\omega^2 E$, and spatial distribution of electric field E along z-axis is described by equation

$$\Delta \boldsymbol{E} + \frac{\omega^2}{c^2}\boldsymbol{E} = 0. \tag{2.3}$$

Solutions of equations (2.1) and (2.3) are plane waves, oscillating with frequency ω and propagating in space with wavevector $k = \omega n / c$

$$E = \mathrm{Re}[E_0 e^{-i\omega(t-z/c)}] = \mathrm{Re}[E_0 e^{i(\bar{k}\cdot\bar{r}-\omega t)}]. \tag{2.4}$$

For conductive nondispersive medium, characterized by conductivity $\sigma \neq 0$, from Maxwell's equations follows "telegraph" equation

$$\nabla^2 \vec{E} = \mu\sigma \frac{\partial \vec{E}}{\partial t} + \mu\varepsilon \frac{\partial^2 E}{\partial t^2}, \tag{2.5}$$

which contains a damping term $\partial\vec{E}/\partial t$. In spite of different forms of equations (2.1) and (2.5), second equation can be reduced to first one by introducing "complex" dielectric constant

$$\hat{\varepsilon} = \varepsilon - i\frac{\sigma}{\omega}.$$

For monochromatic wave with frequency ω spatial distribution of the electric field in absorbing medium is described by Helmholz equation

$$\Delta\vec{E} + \omega^2\mu(\varepsilon - i\frac{\sigma}{\omega})\vec{E} = 0. \tag{2.6}$$

It is identical to equation (2.3) derived for nonconducting media if the "real" index of refraction n is replaced by "complex" index of refraction

$$\hat{n} = \sqrt{\hat{\varepsilon}} = n - ik = n(1 - i\kappa) \tag{2.7}$$

In hetero- or stratified structure index of refraction changes abruptly across interfaces. In the absence of surface current and surface charge at the interfaces the tangential components of the electric and magnetic vectors must be continuous across the interface. From above mentioned boundary conditions follows, that the angle of incidence θ_i, angle of reflection θ_r, and angle of transmission θ_t must be related by (Fig. 3.1)

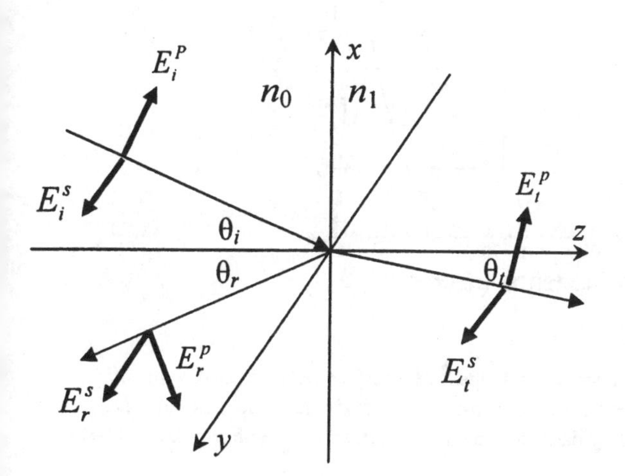

Figure 2.1. Amplitudes of TE(s)- and TM(p) –polarised incident (i), reflected (r) and transmitted (t) waves near interface $z = 0$ between two semi-infinite media with refractive indices n_0 and n_1.

$$\theta_i = \theta_r$$
$$n_1 \sin\theta_i = n_2 \sin\theta_t \tag{2.8}$$

independently of polarization, where n_1 and n_2 are indices of refraction of the first and second medium, respectively.

3. Transmission and Reflectivity [3]

Consider first spatial distribution of electromagnetic wave amplitudes near interface between two nonabsorbing semi-infinite media with refraction indices n_0 and n_1. A plane wave incident on the surface $z = 0$, the plane of incidence being the plane $x0z$, the angle of incidence θ_i and the angle of refraction θ_t (Fig. 3.1). Since the incident wave with electric field of amplitude E_i is a plane wave, the solutions for the transmitted E_t and reflected E_r waves will be also plane waves. Tangential components of electric fields should be continuous across the boundary. Hence we have for complex Fresnel transmission $t_{01}^{s,p} = E_t^{s,p} / E_i^{s,p}$ and Fresnel reflection

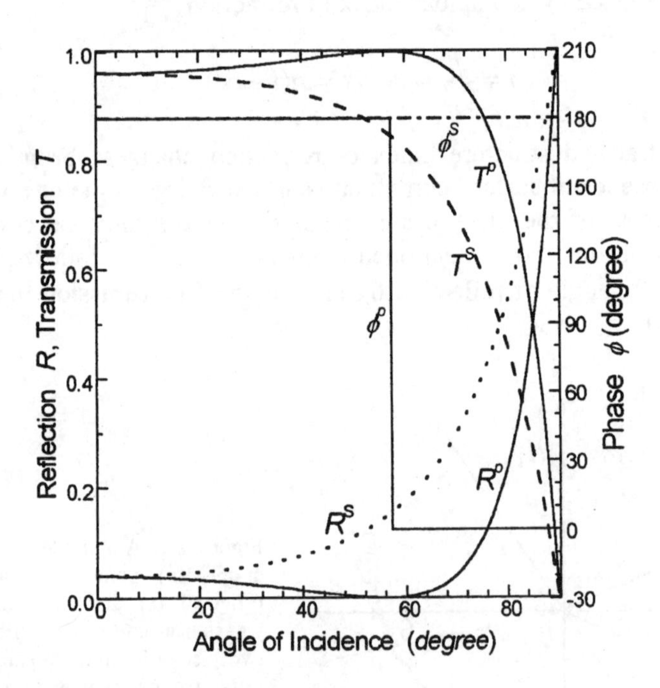

Figure.3.1 Reflectivity $\Re = |r_{01}|^2$ and transmission $T = |t_{01}|^2$ of interface between two semi-infinite media with refractive indices n_0=1 and n_1=1.5 for *s*- (dashed or dotted lines) and *p*-(solid lines) polarized light. Phases ϕ_{01} are presented by dash-dotted and solid lines for *s*- and *p*-polarized light.

$r_{01}^{s,p} = E_r^{s,p} / E_i^{s,p}$ coefficients for s- and p-polarized plane waves

$$r_{01}^{p} = \frac{E_r^p}{E_i^p} = \frac{n_0\cos\theta_1 - n_1\cos\theta_0}{n_0\cos\theta_1 + n_1\cos\theta_0} = \frac{\tan(\theta_1-\theta_0)}{\tan(\theta_1+\theta_0)}$$

$$t_{01}^{p} = \frac{E_t^p}{E_i^p} = \frac{2n_0\cos\theta_0}{n_0\cos\theta_1 + n_1\cos\theta_0} = \frac{2\sin\theta_1\cos\theta_0}{\sin(\theta_1+\theta_0)\cos(\theta_1-\theta_0)} \tag{3.1}$$

$$r_{01}^{s} = \frac{E_r^s}{E_i^s} = \frac{n_0\cos\theta_0 - n_1\cos\theta_1}{n_0\cos\theta_0 + n_1\cos\theta_1} = \frac{\sin(\theta_1-\theta_0)}{\sin(\theta_1+\theta_0)}$$

$$t_{01}^{s} = \frac{E_t^s}{E_i^s} = \frac{2n_0\cos\theta_0}{n_0\cos\theta_0 + n_1\cos\theta_1} = \frac{2\sin\theta_1\cos\theta_0}{\sin(\theta_1+\theta_0)} \tag{3.2}$$

Reflectance $\boldsymbol{R}_{01}^{s,p} = |r^{sp}|^2$ and transmittance $\boldsymbol{T}_{01}^{s,p} = |t^{s,p}|^2\, n_1\cos\theta_1 / n_0\cos\theta_0$ of the interface between two media with refractive indices n_0 and n_1, defined as the ratios of reflected or transmitted energy to incident energy, are presented at Fig.3.2 as a function of incident angle for s- or p-polarized light. Phases $\phi_{01}^{s,p}$ of reflected or transmitted s- and p-polarised light are presented by dash-dotted and solid lines.

From equations (3.1) and (3.2) one can see that $t_{01}^{p} = 1 + r_{01}^{p}$, $t_{01}^{s} = 1 + r_{01}^{s}$, so that, for the case $n_0 > n_1$, transmitted amplitudes exceed incident ones.

4. Multiple-beam interference [3]

Consider a plane-parallel transparent dielectric plate of refractive index n_1, bounded on either side by semi-infinite nonabsorbing media of refractive indices n_0 and n_2. Plane wave of monochromatic light is incident upon the plate at angle θ_0 and is subdivided into reflected and transmitted parts. Such division occurs at both interfaces each time the beam crosses them (Fig. 4.1). Therefore the amplitudes of the transmitted and reflected beams are obtained by summing the multiple reflected and transmitted amplitudes, accounting change of their phases. The expression given below will be valid for either direction of the polarisation provided that r and t are given the appropriate values from equations (3.1)-(3.2).

After a travel of the plate with thickness h_1, wave with vacuum wavelength λ_0 changes phase by

$$\delta_1 = \frac{2\pi}{\lambda_0} n_1 h_1 \cos\theta_1 \tag{4.1}$$

The complex reflected amplitude r is thus given by

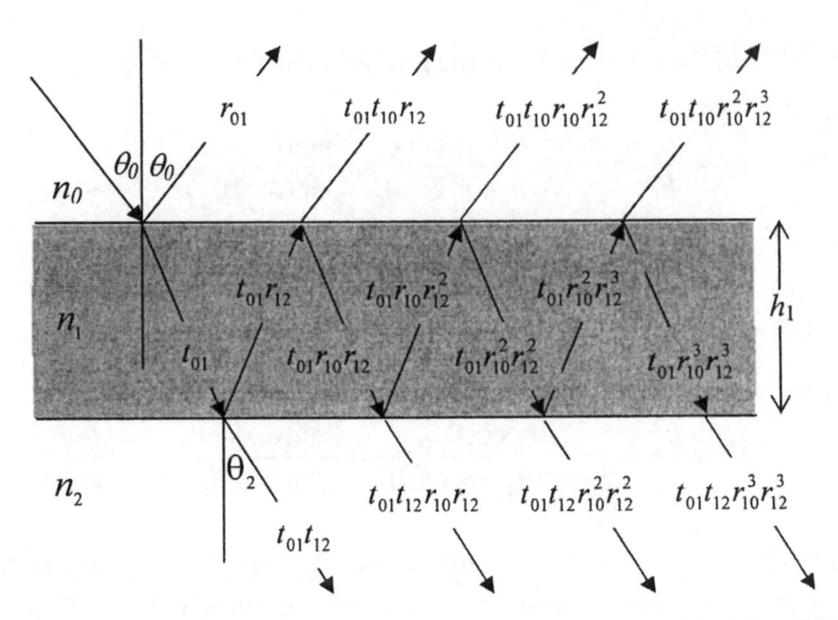

Figure 4.1. Amplitudes of transmitted and reflected plane waves in plane-parallel media of thickness h and refractive index n_1, surrounded by semi-infinite media of refractive indices n_0 and n_2.

$$\begin{aligned} r &= r_{01} + t_{01}t_{10}r_{12}\exp(-i2\delta_1) + t_{01}t_{01}r_{10}r_{12}^2\exp(-i4\delta_1) + \ldots \\ &= r_{01} + \frac{t_{01}t_{01}r_{12}\exp(-i2\delta_1)}{1 + r_{10}r_{12}\exp(-i2\delta_1)} \end{aligned} , \tag{4.2}$$

where r_{ab} (= $-r_{ba}$) and t_{ab} (=2- t_{ba}) are the reflection and transmission coefficients for waves propagating from medium a to medium b.

Analogously, the amplitudes of the transmitted light is:

$$\begin{aligned} t &= t_{01}t_{12}\exp(-i\delta_1) + t_{01}t_{12}r_{01}r_{12}\exp(-i3\delta_1) + t_{01}t_{12}r_{01}^2r_{12}^2\exp(-i5\delta_1)^{-i5\delta} + \ldots \\ &= \frac{t_{01}t_{12}\exp(-i\delta_1)}{1 + r_{01}r_{12}\exp(-i2\delta_1)} \end{aligned} \tag{4.3}$$

For nonabsorbing media reflectance $\Re = |r_{01}|^2 = |-r_{10}|^2$ and transmittance $\mathrm{T} = t_{01} \cdot t_{10}$ of the interface between media with refractive index n_0 and n_1 satisfy energy conservation condition $\mathrm{T} + \Re = 1$, so that $t_{01}t_{10} = 1 - r_{01}^2$ and

$$r = \frac{r_{01} + r_{12}\exp(-i2\delta_1)}{1 + r_{01}r_{12}\exp(-i2\delta_1)} \tag{4.4}$$

In case of identical surrounding media with equal refractive indices $n_0=n_2$ transmittance T and reflectance R of layer "1"sandwiched between two identical media "0"can be expressed in terms of interface reflectance $\Re = |r_{01}|^2$

$$R_{010} = \frac{2(1-\cos 2\delta_1)\Re}{1+\Re^2 - 2\Re\cos 2\delta_1} = \frac{4\Re\sin^2\delta_1}{(1-\Re)^2 + 4\Re\sin^2\delta_1} = \frac{F\sin^2\delta_1}{1+F\sin^2\delta_1}, \tag{4.5}$$

$$T_{010} = \frac{(1-\Re)^2}{1+\Re^2 - 2\Re\cos 2\delta_1} = \frac{(1-\Re)^2}{(1-\Re)^2 + 4\sin^2\delta_1} = \frac{1}{1+F\sin^2\delta_1}, \tag{4.6}$$

where the parameter F is defined as

$$F = \frac{4\Re}{(1-\Re)^2}. \tag{4.7}$$

As shown in Fig.4.2, the intensity of the transmissivity minima falls, and the maxima become sharper, when $\Re$ is increased. When $\Re \to 1$, F becomes large. Then the transmitted light consists of narrow fringes separated by deep minima.

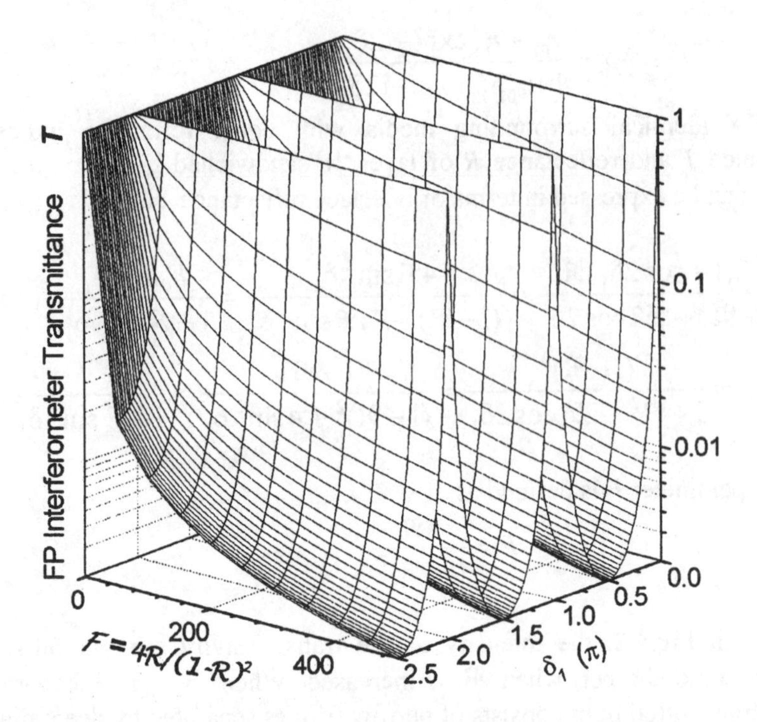

Figure 4.2. Transmittance of non-absorbing Fabry-Pérot Interferometer as a function of phase $\delta_1 = 2\pi n_1 h_1 \cos\theta_1 / \lambda_0$ and $F = 4\Re/(1-\Re)^2$.

The FWHM Δ of the fringes is determined from relation

$$\frac{1}{1+F\sin^2\dfrac{\Delta}{4}} = \frac{1}{2}, \qquad \text{therefore} \quad \Delta = \frac{4}{\sqrt{F}}. \tag{4.8}$$

5. Refraction and reflection at the surface of an absorptive medium

Most of derived above formulae can be extended on absorptive media by replacing the real refractive index n by a complex term $\hat{n} = n - ik$. Since $\hat{n}$ is complex, θ_t in accordance with the refraction low

$$\sin\theta_t = \frac{\sin\theta_i}{\hat{n}} \tag{5.1}$$

is also complex, therefore no longer has significance of an angle of refraction. The

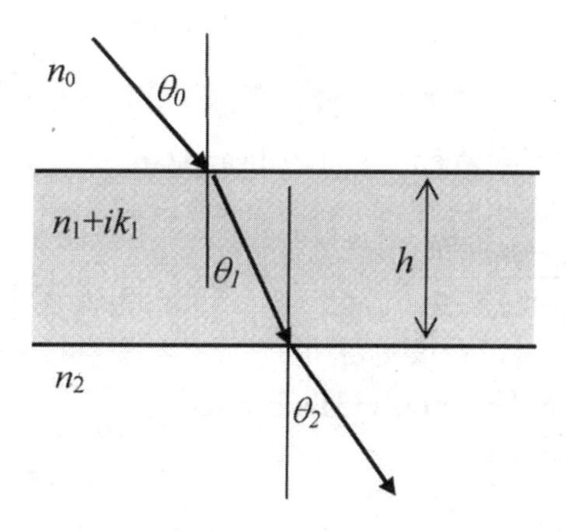

Figure 5.1. Absorptive layer of complex refractive index n_1-ik_1 surrounded by two dielectrics with real indices n_0 and n_2.

Fresnel coefficients and expression for r and t then also become complex. The phase changes on reflection are no longer necessarily 0 or π. Moreover, the reflectivities and phase changes on the two sides of the film are different when the bounding media are of different refractive index.

In simplest case, when the media on the two surfaces of the layer are identical, (4.6) and (4.6) hold provided we interpret $\Re$ as the reflectivity for internal reflection $r_{10} = r_{12}$, and replace δ_1 by

$$\delta' = \frac{2\pi}{\lambda_0} n_1 h_1 \cos\theta_1 + \varphi,$$

where φ is phase change on internal reflection. If A is the fraction of the absorbed light,

$$T = \left(1 - \frac{A}{1-R}\right)^2 \frac{1}{1 + F\sin^2\delta'}.$$

Absorption diminishes the intensity of the transmitted light, and phase φ is equivalent to an increase of thickness of the layer by $\varphi\lambda_0/2\pi n_1\cos\theta_1$.

If the first medium is a dielectric, the reflected wave has a real phase factor. The amplitude components E_i^s, E_i^p of the incident wave and corresponding components E_r^s and E_r^p of the reflected wave are related by (3.1, 3.2). Since θ_1 is now complex, so are the ratios $r_{01}^s = E_r^s / E_i^s$ and $r_{01}^p = E_r^p / E_i^p$, i.e. characteristic phase changes occur on reflection; thus incident linearly polarized light will in general become elliptically polarized on reflection at the absorptive surface. Let ϕ_s and ϕ_p be the phase change, and ρ^S and ρ^P the absolute values of the reflection coefficients, i.e.

$$r^S = \rho_{01}^S e^{i\phi^S}, \quad r^P = \rho_{01}^P e^{i\phi^P}. \tag{5.2}$$

For normal incidence ($\theta_I = 0$), the distinction between r^S and r^P disappears, therefore replacing n by $\hat{n}$ in (23), we get

$$r_{01}^S = r_{01}^P = \frac{n_0 - n_1 + ik_1}{n_0 + n_1 - ik_1}, \tag{5.3}$$

which gives, for reflectance $\boldsymbol{R} = r \cdot r^*$ of the absorptive media surface

$$\boldsymbol{R}^S = \boldsymbol{R}^P = \frac{(n_0 - n_1)^2 + k_1^2}{(n_0 + n_1)^2 + k_1^2}. \tag{5.4}$$

Figure 5.1. Absorptive layer of complex refractive index n_1-ik_1 surrounded by two dielectrics with real indices n_0 and n_2.

Consider now a plane-parallel absorbing film situated between two dielectric media (Fig.5.1). It is convenient to set [3]

$$\hat{n}_1 \cos\theta_1 = u_1 + iv_1, \tag{5.5}$$

Using the Fresnel law, the real u_1 and v_1 can be expressed in terms of the angle of incidence θ_0 and refractive indices n_0, n_1 and k_1 of the first two media.

$$u_1^2 - v_1^2 = n_1^2 - k_1^2 - n_0^2 \sin^2\theta_0, \tag{5.6}$$

$$u_1 v_1 = n_1 k_1. \tag{5.7}$$

In the case of electric vector perpendicular to the plane of incidence (TE or *s*-wave) we have, replacing in (3.2) $n_1 \cos\theta_1$ by $u_1 + iv_1$:

$$r_{01}^S = \rho_{01}^S \exp(i\phi_{01}^S) = \frac{n_0 \cos\theta_0 - (u_1 + iv_1)}{n_0 \cos\theta_0 + (u_1 + iv_1)}. \tag{5.8}$$

From (5.8) we have equations for amplitude and phase of reflected beam

$$\rho_{01}^S = \sqrt{\frac{(n_0 \cos\theta_0 - u_1)^2 + v_1^2}{(n_0 \cos\theta_0 + u_1)^2 + v_1^2}}, \qquad \tan\phi_{01}^S = \frac{2v_1 n_0 \cos\theta_0}{u_1^2 + v_1^2 - n_0^2 \cos\theta_0}. \tag{5.9}$$

Calculated intensities $|\rho_{12}^S|^2$ and $|\rho_{12}^P|^2$, phases ϕ^S and ϕ^P of reflected from silver surface *s*- and *p*-polarized light are presented at Fig. 5.2.

From (3.2) follow equations for amplitude τ_{01}^S and phase χ_{01}^S of transmission $t_{01}^S = \tau_{01}^S \exp(i\chi_{01}^S)$ at the first "01" interface

$$\tau_{01}^S = \frac{2n_0 \cos\theta_0}{\sqrt{(n_0 \cos\theta_0 + u_1)^2 + v_1^2}}, \qquad \tan\chi_{01}^S = -\frac{v_1}{n_0 \cos\theta_0 + u_1}. \tag{5.10}$$

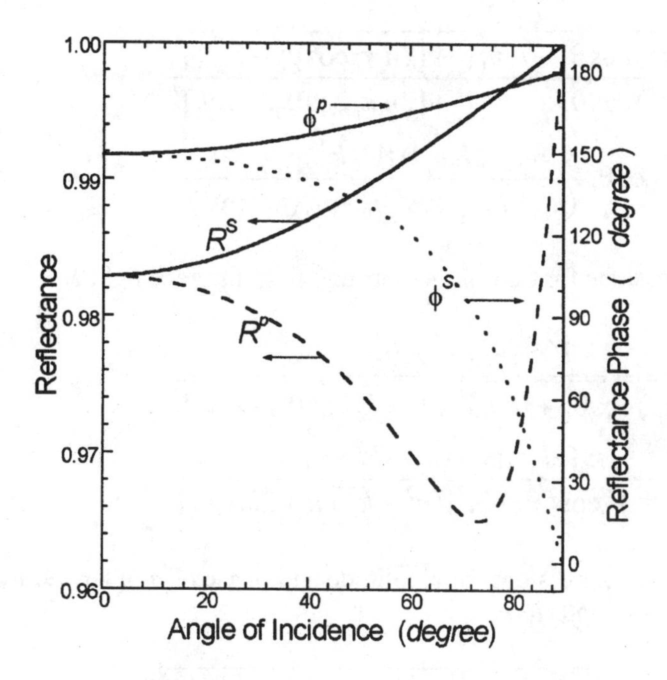

Figure 5.2. Intensities $R_{01}=|r_{01}|^2=\rho_{12}^2$ and phases ϕ_{01} of reflected from silver surface *s*- and *p*-polarized light. Real and imaginary refractive indices of Ag at 550 nm are n_1=0.06, k_1=3.6.

Analogously we obtain expressions for reflection and transmission at the second interface:

$$\rho_{12}^S = \sqrt{\frac{(n_2\cos\theta_2 - u_1)^2 + v_1^2}{(n_2\cos\theta_2 + u_1)^2 + v_1^2}}, \qquad \tan\phi_{12}^S = \frac{2v_1 n_2 \cos\theta_2}{n_1^2 + v_1^2 - n_2^2\cos^2\theta_2}, \quad (5.11)$$

$$\tau_{12}^S = \sqrt{\frac{4(u_1^2 + v_1^2)}{(n_2\cos\theta_2 + u_1)^2 + v_1^2}}, \qquad \tan\chi_{12}^S = \frac{v_1 n_2 \cos\theta_2}{u_1^2 + v_1^2 + u_1 n_2 \cos\theta_2}. \quad (5.12)$$

The angle θ_2 is determined from θ_0 by means of the formula

$$n_2 \sin\theta_2 = n_0 \sin\theta_0 . \quad (5.13)$$

Expression for amplitudes and phases of *p*-polarized (TM-wave) reflected plane waves could be deduced from (3.1). Amplitude and phase of the reflected from first interface *p*-polarized light are

$$\rho_{01}^{p}=\sqrt{\frac{[(n_1^2-k_1^2)\cos\theta_0-n_0u_1]^2+[2n_1k_1\cos\theta_0-n_0v_1]^2}{[(n_1^2-k_1^2)\cos\theta_0+n_0u_1]^2+[2n_1k_1\cos\theta_0+n_0v_1]^2}},$$
$$\tan\phi_{01}^{p}=2n_0n_1^2\cos\theta_0\frac{2k_1u_1-(1-k_1^2)v_1}{(n_1^2+k_1^2)^2\cos^2\theta_0-n_0^2(u_1^2+v_1^2)}. \quad (5.14)$$

Corresponding expressions for transmitted through first interface light are

$$\tau_{01}^{p}=\frac{2(n_1^2+k_1^2)\cos\theta_0}{\sqrt{[(n_1^2-k_1^2)\cos\theta_0+n_0u_1]^2+[2n_1k_1\cos\theta_0+n_0v_1]^2}}, \quad (5.15)$$

$$\tan\chi_{01}^{p}=\frac{n_0[2k_1u_1n_1-(n_1^2-k_1^2)v_1]}{(n_1^2+k_1^2)^2\cos^2\theta_0+n_0^2[(n_1^2-k_1^2)u_1+2k_1v_1n_1]}. \quad (5.16)$$

Analogously we obtain expressions for amplitude and phase of *p*-polarized light, reflected from the second interface:

$$\rho_{12}^{p}=\sqrt{\frac{[(n_1^2-k_1^2)\cos\theta_2-n_2u_1]^2+[2n_1k_1\cos\theta_2-n_2v_1]^2}{[(n_1^2-k_1^2)\cos\theta_2+n_2u_1]^2+[2n_1k_1\cos\theta_2+n_2v_1]^2}}, \quad (5.17)$$

$$\tan\phi_{12}^{p}=\frac{2n_2\cos\theta_2[2k_1u_1n_1-(n_1^2-k_1^2)v_1]}{(n_1^2+k_1^2)^2\cos^2\theta_2-n_2^2(u_1^2+v_1^2)}, \quad (5.18)$$

From the knowledge of the quantities ρ_{01}, ϕ_{01}, etc., the complex reflection coefficient of the absorbing film may be evaluated. It is useful to set

$$\eta=\frac{2\pi}{\lambda_0}h_1, \quad (5.19)$$

so that

$$\beta=\frac{2\pi}{\lambda_0}\hat{n}_1h_1\cos\theta_1=(u_1+iv_1)\eta. \quad (5.20)$$

The equation (4.4) now becomes

$$r=\rho\exp(i\delta_r)=\frac{\rho_{01}e^{i\phi_{01}}+\rho_{12}e^{-2v_1\eta}e^{i(\phi_{12}+2u_1\eta)}}{1+\rho_{01}\rho_{12}e^{-2v_1\eta}e^{i(\phi_{01}+\phi_{12}+2u_1\eta)}}. \quad (5.21)$$

From (5.2) we obtain the expressions for the reflectivity $R=|\rho|^2$ and for the phase change δ_r on reflection:

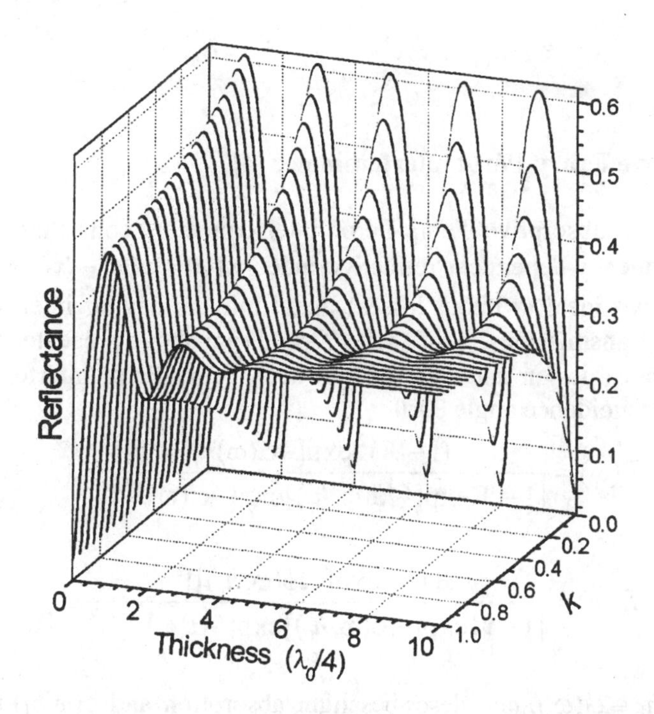

Figure 5.3. The reflectivity of absorbing layer with real part of refractive indices n_1=3.5 as a function of its imaginary part k_1 and optical thickness h, surrounded by semi-infinite substrate of n_2=1.5 and air with n_0=1.

$$R = |r|^2 = \rho^2 = \frac{\rho_{01}^2 + 2\rho_{01}\rho_{12}e^{-2v_1\eta}\cos(\phi_{12} - \phi_{01} + 2u_1\eta) + \rho_{12}^2 e^{-4v_1\eta}}{1 + 2\rho_{01}\rho_{12}e^{-2v_1\eta}\cos(\phi_{01} + \phi_{12} + 2u_1\eta) + \rho_{01}^2\rho_{12}^2 e^{-4v_1\eta}}, \quad (5.22)$$

$$\tan\delta_r = \frac{\rho_{12}(1 - \rho_{01}^2)\sin(2u_1\eta + \phi_{12}) + \rho_{01}(e^{2v_1\eta} - \rho_{12}^2 e^{-2v_1\eta})\sin\phi_{01}}{\rho_{12}(1 + \rho_{01}^2)\cos(2u_1\eta + \phi_{12}) + \rho_{01}(e^{2v_1\eta} + \rho_{12}^2 e^{-2v_1\eta})\cos\phi_{01}}, \quad (5.23)$$

where phase change on reflection δ_r is referred to the interface 0-1. Formulae (22,23) are valid for s- and p-polarized waves. In the former case one must substitute for ρ and ϕ the value given by (6) and (9), in the latter case those given by (13) and (16).

Fig. 5.3 illustrates the dependence of the reflectance on the thickness h_1 of the absorbing layer and imaginary part of refractive index k_1. For a nonabsorbing film R are periodic functions of the film thickness h_1, with a period of one wavelength. Absorbing is seen to reduce the amplitude of the successive maxima.

For a thick film, we have from (5.22) and (5.23)

$$R \cong \rho_{01}^2, \qquad \delta_r \cong \phi_{01}. \tag{5.24}$$

6. Absorptive Fabry-Pérot interferometer [4]

Considers now absorptive Fabry-Pérot interferometer, consisting of layer with complex frequency-dependent refractive index $\hat{n}(\omega) = n(\omega) - ik(\omega)$ and thickness h_1 between two ideal (without losses) metal mirrors having intensity reflectivity $\Re = |r_{01}|^2$ and transmission $T = t_{01}t_{10}$ ($\Re + T = 1$). From above presented formulae (4.3) and (5.22) transmission and reflection of such Fabry-Pérot interferometer are for perpendicular incidence angle $\theta_0 = 0$

$$T(\omega) = \frac{(1-\Re)^2 \exp[-\alpha(\omega)h_1]}{|1-\Re\exp[-i\phi(\omega,h_1)]\exp[-\alpha(\omega)h_1]|^2}, \tag{6.1}$$

$$R(\omega) = \frac{\Re\,|1-\exp[-i\phi(\omega,h_1)]|^2}{|1-\Re\exp[-i\phi(\omega,h_1)]\exp[-\alpha(\omega)h_1]|^2}, \tag{6.2}$$

where $\alpha(\omega)h_1 \equiv 2k(\omega)h_1\omega/c$ describes light absorption and $\phi(\omega,h_1) = 2n(\omega)h_1\omega/c$ is the phase shift caused by one optical round trip of the light between mirrors separated by a distance h_1, c – is the vacuum speed of light. Consider first the empty cavity case, for which $n(\omega)$ and $\alpha(\omega)$ are described by the frequency-independent background index of refraction n_B and absorption α_B, respectively. In the case of resonance the reflection reaches minima and transmission reaches maxima when the phase shift is an integral multiple of 2π. For a small energy deviation $\hbar\delta\omega$ from the cavity resonance phase $\phi = 2m\pi + 2\,\delta\omega\, n_B h_1/\hbar c$, where $m = \pm1, \pm2, \ldots$ The transmission can be expressed using expansion of phase-dependent exponent in (7): $\exp(i\phi) = 1 + i2\delta\omega n_B h_1/\hbar c + \ldots$,

$$T(\omega) = \frac{(1-\Re)^2 \exp(-\alpha_B h_1)}{|(1-\Re e^{-\alpha_B h}) - i2\Re e^{-\alpha_B h_1}\delta\omega n_B h_1/\hbar c|^2}. \tag{6.3}$$

Therefore the Fabry-Pérot transmission has a Lorentzian form, $T(\omega) \propto |\delta\omega - i\Delta|^{-2}$ with the energetic HWHM

$$\Delta = \frac{\hbar c}{2n_B h}\frac{(1-\Re e^{-\alpha_B h})}{\sqrt{\Re e^{-\alpha_B h}}}. \tag{6.4}$$

7. Basic physics of microcavities [5]

The simplest semiconductor microcavities consist of two Distributed Bragg

Reflectors (DBRs) on either side of a cavity region, as shown schematically in Figure 7.1. The DBRs are multiple repeats of alternating high- and low-index layers, each with thickness $\lambda_0/4n_j$, where n_j is refractive index of layer j material. This gives a broad-band high-reflectivity region centred on λ_0, called the stop-band, with oscillating side-lobes on either side. In the stop band, the mirror reflectivity is given by [3, 6]

$$R = 1 - 4\frac{n_0}{n_c}\left(\frac{n_L}{n_H}\right)^{2N} \tag{7.1}$$

where n_L, n_H , n_c and n_0 are the refractive indices of the low- and high-index layers, the cavity material and the external medium respectively, and N is the number of pairs of mirror layers. The cavity between the two mirrors has thickness L_c chosen to be an integer multiple of $\lambda/2n$ in the medium. The semiconductor microcavity is very similar to the above-considered Fabry–Pérot structure with ideal metal mirrors, therefore most of the standard Fabry–Pérot results for the reflectance, transmittance etc apply.

The simple physical picture of MC is that in order to form a Fabry-Pérot resonance the round-trip phase shift has to be an integer multiple of 2π. Further insight into normal-mode coupling can be gained from a graphical solution of the resonance condition

$$\phi(\omega, L_c) \equiv 4\pi n(\omega)L_c/\lambda_0 = 2\pi m, \qquad m = \pm 1, \pm 2,$$

which allows to find resonance frequencies of MC and give further insight into MC property. This condition can be presented as

$$n(\lambda) = \frac{m}{2L_c}\lambda_0. \tag{7.2}$$

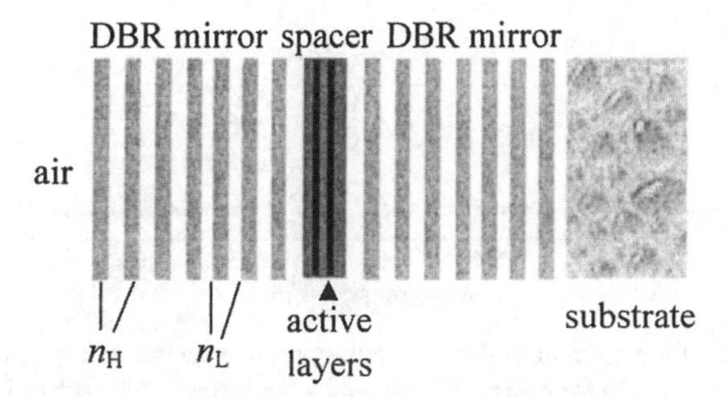

Figure 7.1. Schematic of a planar MC consisting of two DBRs, spacer layer of thickness L_c with two active layers in the cavity-field antinode. DBR mirrors consist of several layers with Low and High refractive indices n_L and n_H and thicknesses $\lambda_0/4n$.

For an empty cavity, the left side of Eq. (7.2) is just a frequency-independent constant n_c (horizontal dotted line at Fig. 7.2.b) and the right side is a straight (short

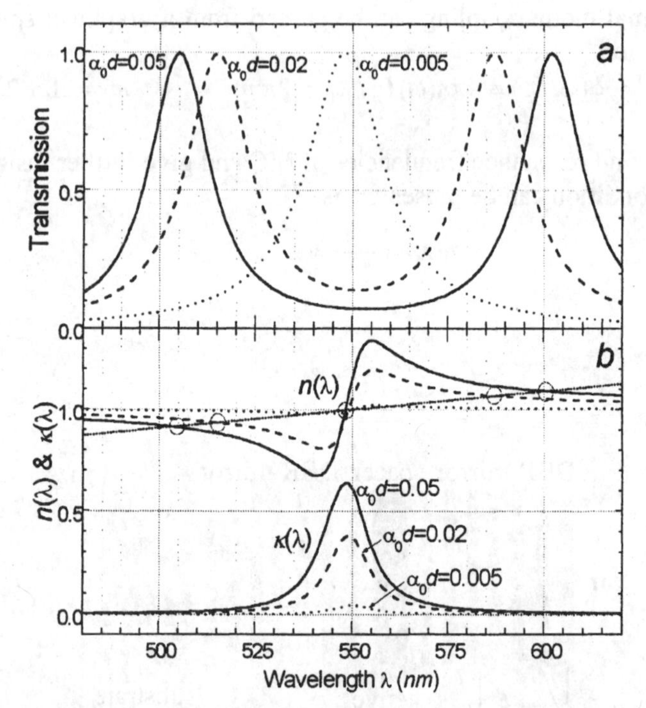

Figure 7.2. (a) Spectra of the normalised MC transmission. (b) Spectral dependence of the refractive index $n(\lambda)$ and extinction $\kappa(\lambda)$ for three different optical densities of MC active layer $\alpha_0 d$=0.005 (dotted); 0.02 (dashed) and 0.05 (solid) lines. Crossings of $n(\lambda)$ with straight dotted line $\lambda/2L_c$ (sircels) determine spectral positions of the MC transmission maxima.

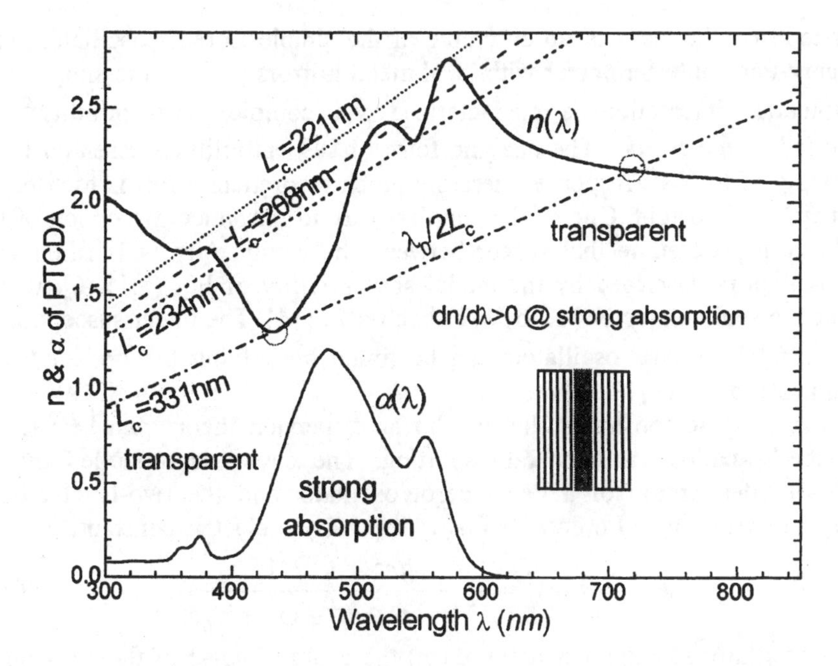

Figure 7.3. Spectra of absorption coefficient $\alpha(\lambda)$ (lower solid curve) and refractive index $n_{\mathrm{PTCDA}}(\lambda)$ (upper solid curve) of organic semiconductor PTCDA and straight lines $\lambda/2L_c$ (dashed or dotted lines) for different thicknesses L_c of MC spacer. Intersections of $n_{\mathrm{PTCDA}}(\lambda)$ with these lines determine resonant maxima of MC transmission.

- dotted) line, shown in Fig. 7.2 (b). The single intersection of these two straight lines gives the wavelength of the single longitudinal mode of the cavity. For a many-oscillator cavity the left-hand side of Eq. (7.2) varies rapidly in the vicinity of the resonance; if the transition is sufficiently narrow and strong, three intersections result, as depicted in Fig. 7.2. The very high absorption at the central solution results in very low transmission there; the other two solutions give the wavelengths of the two transmission peaks.

For arbitrary absorption lineshapes, one can use Kramers-Kronig relations to obtain $n(\lambda)$ and use this graphical method to find the MC peak positions as a function of detuning determined by L_c. In Fig. 7.3 we have plotted the rather complicated absorption spectrum of organic crystal PTCDA, reconstructed from Kramers-Kronig relation spectrum of refractive index $n_{\mathrm{PTCDA}}(\lambda)$, and straight lines $\lambda/2L_c$ for different thicknesses L_c of microcavity spacer. Intersections of these straight lines with $n_{\mathrm{PTCDA}}(\lambda)$ give spectral positions of MC transmission maxima.

7.1. LORENTZ OSCILLATOR

Up to now spectral dependence and nature of dielectric function and complex refractive index was not specified. The many aspects and ideas of MC physics can be understood on the basis of an ensemble of the simple Lorentz oscillators in a planar Fabry-Pérot interferometer with ideal metal mirrors [15]. In the simple case the light-matter interaction is characterized via complex susceptibility $\hat{\varepsilon}$ or refractive index $\hat{n} = n - ik$. The analytic form of susceptibility is essential to an understanding of the MC response, therefore much theoretical effort is focused on the its microscopic origin. One of the simplest way to introduce dispersion of the susceptibility is to assume that spacer between the cavity mirrors is filled with Lorentz oscillators described by the model susceptibility of Eq. (8.1) [6]; it was first applied to semiconductor MC by Weisbuch *et al* [14]. The linear susceptibility $\hat{\varepsilon}$ of a system of Lorentz oscillators can be found by solving the optical Bloch equations in steady state [17,18].

Zhu *et al* [15] demonstrated that the linear dispersion theory can be used to describe the so-called vacuum Rabi splitting. The cavity is modelled by the standard Airy description of a Fabry-Pérot oscillator and the two-level atomic system of thickness d by a Lorentz oscillator dispersive dielectric constant:

$$\varepsilon(\omega) = \hat{n}(\omega)^2 = \varepsilon_B + \frac{Nfe^2}{m}\frac{1}{\omega_0^2 - \omega^2 - i\gamma_0\omega}, \tag{7.3}$$

where f is oscillator strength per atom, e (m) the charge (mass) of the electron, N the oscillator density, ω_0 the resonance frequency and γ the oscillator linewidth. The frequency-dependent absorption coefficient $\alpha(\omega)$ and refractive index $n(\omega)$ of the Lorentz oscillators are given

$$n(\omega) = n_B - \alpha_0 \frac{c}{2\pi\omega}\frac{(\omega - \omega_x)\gamma_0}{4(\omega - \omega_0)^2 + \gamma_0^2}, \quad \alpha = \alpha_0 \frac{\gamma_0^2}{4(\omega - \omega_0)^2 + \gamma_0^2} \tag{7.4}$$

where $\alpha_0 = 2Nfe^2 / mc\gamma$ is absorption coefficient in the line center $\omega = \omega_0$.

The transmission $T(\omega)$, reflectivity $R(\omega)$ of the cavity + atom system is given by equations (6.1, 6.2), where now

$$\phi(\omega, L_c) = \frac{4\pi}{c}\{(\omega - \omega_c)L_c + d[n(\omega) - n_B]\omega\} \tag{7.5}$$

is the phase experienced by the field upon completion of a round-trip through the cavity , αd is the single–pass absorption. For an empty cavity with sufficiently large finesse F (4.7), the empty-cavity resonance width is given by $\gamma_c = c/2L_cF$. Dispersion of linear absorption and refraction alter the transmission $T(\omega)$ through cavity with Lorentz oscillators: now it may exhibit a structure which is completely different from that found in empty-cavity case. The location and even the number of transmission extrema may change.

Consider first simplest case of low absorption ($\alpha_0 d << 1$), high cavity finesse ($\Re \cong 1$), cavity resonance ($\omega_0 \approx \omega_c$), small detuning [$(\omega-\omega_0) << L_c/c$], and comparable Lorentz oscillator and cavity resonance widths ($\gamma_c \approx \gamma_0$).

The zeros of $\phi(\omega,L_c)$ determine the peak positions in MC transmission spectra, and the slope $\partial\phi/\partial\omega$ at the zeros provides a measure of the resonance widths. In Fig. 7.2 (b), the dispersions of refractive index and absorption coefficient (7.4) are plotted for several optical densities of MC active layer. At $\alpha_0 \rightarrow 0$, $n(\omega) \equiv n_B$, therefore expression for $n(\lambda)$ contains a single nonzero term corresponding to a horizontal straight line [dotted line in Fig. 7.2 (b)], indicating that the cavity has a single peak at $\omega = \omega_0$. For $\alpha_0 > 0$, a second dispersion term contributes to $n(\lambda)$ as well. At low α_0, the dispersion changes the slope of $n(\lambda)$ at $\omega \approx \omega_0$ thereby broadening the cavity transmission resonance. At higher α_0, $n(\lambda)$ is so distorted by the dispersive term that it actually passes through zero three times. One zero, as in the empty-cavity case, occurs at $\omega = \omega_0$, and two new zeros, located symmetrically about $\omega = \omega_0$, appear [Fig. 7.2(a)]. The absorptive part of oscillator response (7.4) destroys the central transmission peak and slightly shifts the remaining two peaks away from the zeros of the $\phi(\omega,L_c)$ [see Fig. 7.2(a)].

Consider now the coupling between the cavity mode and excitonic states in active layers grown within the cavity. In order to obtain a significant interaction between the two, the exciton energy is chosen to be close to resonance with the cavity mode. The coupling is determined by the exciton oscillator strength and the amplitude of the cavity field at the active layer position. It is characterized by energy, the vacuum Rabi splitting, which, for active layer placed close to the electric field antinodes, is given by [6]

$$h\Omega_i \approx 2\hbar\sqrt{\frac{2\Gamma_0 c N_{ex}}{n_c L_{eff}}} \tag{7.6}$$

where N_{ex} is the exciton density in active layers in the cavity , $\hbar\Gamma_0$ is the radiative width of a free exciton and can be expressed in terms of the exciton oscillator strength per unit area [7, 13], f_{ex}, as

$$\hbar\Gamma_0 = \frac{\pi}{n_c}\frac{e^2}{4\pi\varepsilon_0}\frac{\hbar}{m_e c} f_{ex} \tag{7.7}$$

In the strong coupling regime, where the vacuum Rabi splitting is greater than the widths of the cavity and exciton modes, this corresponds to a measurable splitting in the optical spectrum when the two modes anticross. Figure 7.4 shows a calculated anticrossing obtained at tuning of cavity resonance ω_c through the exciton resonance ω_x by changing cavity thickness L_c. Since only the cavity mode couples directly to external photons, away from resonance the excitonic mode becomes predominantly weak.

A more detailed analysis reveals that the two peaks in the cavity transmission are approximately Lorentzian in shape, occur at the frequencies $\Delta = \omega - \omega_0 = \pm\,\Omega/2$, where

$$\Omega = \sqrt{\Omega_{\max}^2 - \frac{\gamma_0^2 + \gamma_c^2}{4}} \text{ with } \Omega_{\max} = \sqrt{\frac{\alpha_0 d \gamma_c c}{2\pi L_c}} \propto \sqrt{\frac{N_{ex} f_{ex}}{L_c}}, \tag{7.8}$$

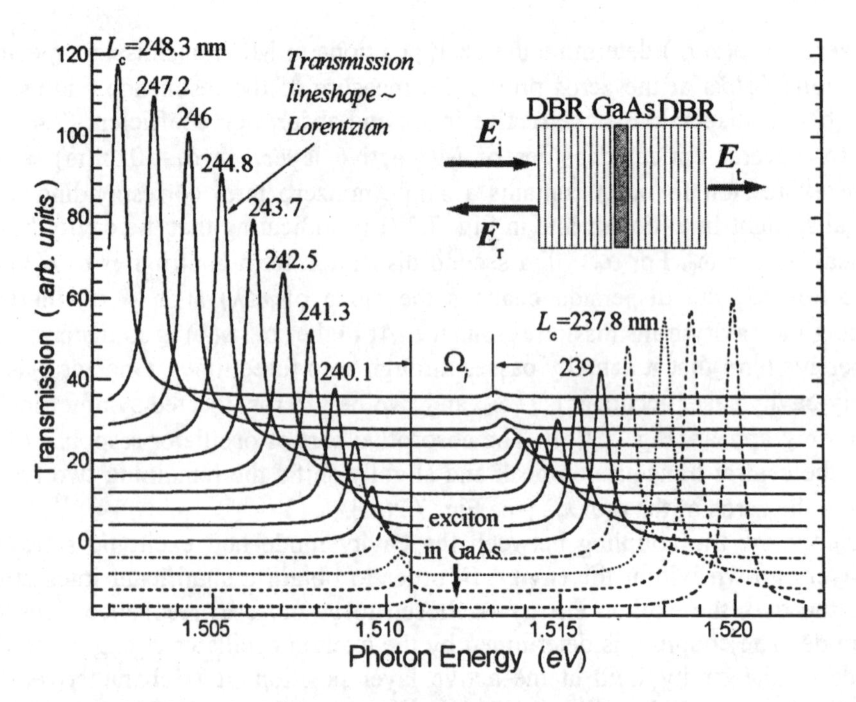

Figure 7.4. Dependence of calculated by transfer matrix model transmission spectra of MC with GaAs active layer on cavity thickness L_c. Due to strong excitonic absorption and refractive index dispersion cavity resonance splits into 3 transmission lines. Central line at 1.512 eV invisible due to strong absorption. Intensities of other two lines are dependent on detuning from excitonic

have an intensity

$$I = \frac{T^2 \gamma_c^2}{(1-\Re)^2 (\gamma_c + \gamma_0)^2} \tag{7.9}$$

and a FWHM of $\gamma_c' = (\gamma_c + \gamma_0)/2$, that is the average of the uncoupled oscillator and cavity widths. The maximum splitting is only a function of total oscillator strength and the cavity size, but not of the finesse. The splitting will be resolved when $\Omega >> (\gamma_c + \gamma_0)/2$.

The properties of complicated MC structures consisting of multilayer DBRs and cavity with dispersive refractive index $n_c(\omega)$ are easily calculated using transfer matrix theory [3] to propagate the electromagnetic field through the structure. The calculated photon field distribution for the MC structure, resembling presented at Fig. 7.1, at the cavity mode wavelength $\lambda_c = 2L_c/m$, is shown in Figure 7.5 and shows enhancement of the electric field amplitude at the centre of the cavity by a factor of ~10, relative to the external field.

Values for the vacuum Rabi splitting, $\hbar\Omega_i$, can also be obtained from the transfer matrix model by including the excitons directly in the calculation, as a contribution to the dielectric constant in the active layer material [15]. This is an

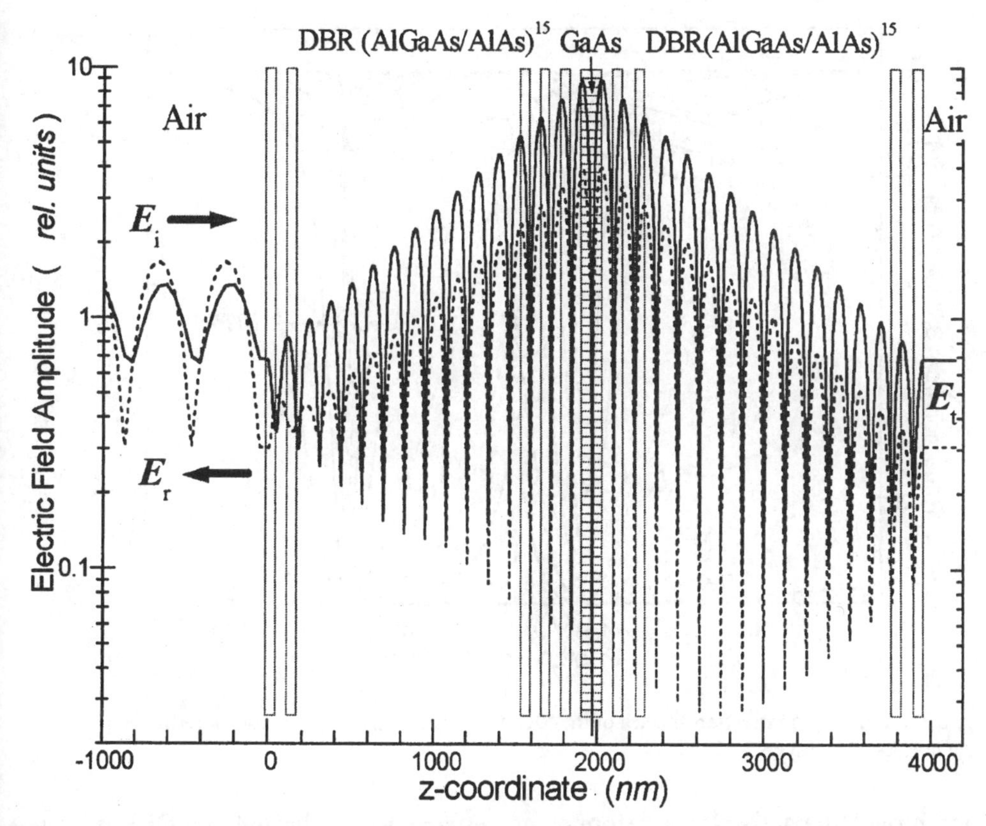

Figure 7.5. Calculated photon field distribution in GaAs cavity with 15-layers GaAlAs/AlAs DBRs at resonant (solid line) and slightly off-resonant (dashed line) wavelengths. The ~10 times field enhancement in the GaAs cavity region is visible. Amplitudes of intracavity and transmitted fields are maximal at resonance wavelength $\lambda_c=2L_c/m$.

appropriate treatment since, in the linear regime, the exciton can be treated as a classical oscillator, which appears as a dispersive term in the dielectric constant. A more accurate treatment of the exciton contribution to the dielectric constant, taking into account the polariton dispersion in the active material, has been given by Pau *et al* [8].

A reflectivity spectra calculated from the transfer matrix model for an MC are presented in Figure 7.6 for different numbers N of DBR's layers. These spectra exhibit many of the important features discussed so far. Two coupled-mode dips of nearly equal intensity are observed arising from the coupled exciton–cavity modes. These dips are superimposed on the high-reflectivity (~100%) stop-band of the DBR mirrors. Inspection of figure 7.6 shows that for high-finesse structures FWHM of the reflectivity dips decrease up to ~1 meV. Splitting between two lines are nearly independent on reflection of DBRs, as already pointed out in Eq. (7.8). High excitonic absorption in GaAs active layer results in addition structure at excitonic resonance $\hbar\omega_x=1.512$ eV. The high finesse leads to many passes of the

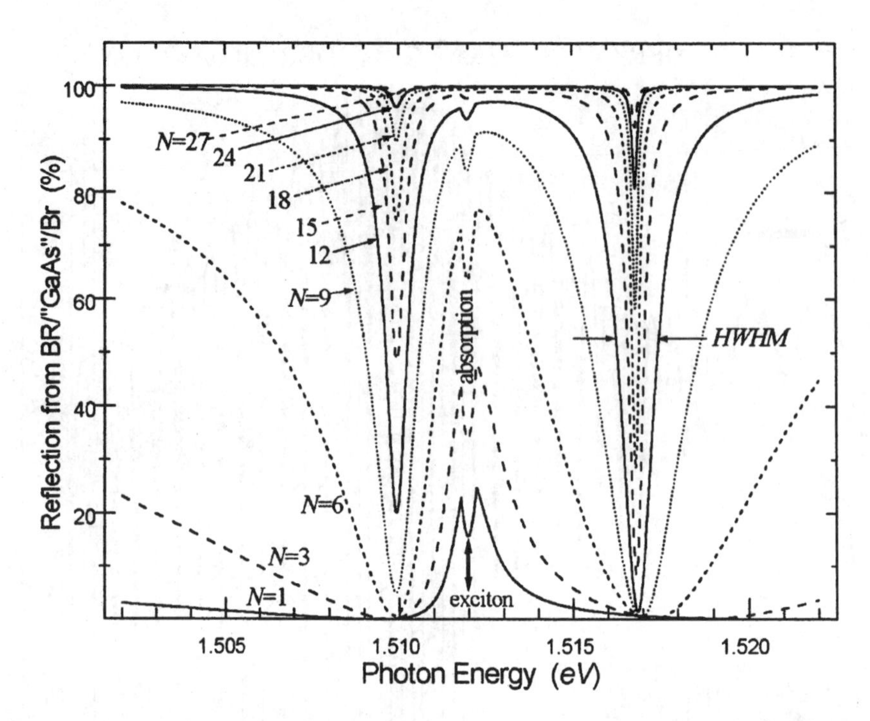

Figure 7.6. Calculated reflection spectra of the MC with GaAs active layer between DBRs consisting of N GaAlAs/AlAs layers.

light through the active layer before photon escape occurs through the DBR mirrors and thus results in the high levels of absorption which occur on resonance.

7.2. CAVITY POLARITONS

The metal mirrors of the Fabry- Pérot interferometer and DBRs of the MC force the axial wavevector k_z in the medium to be $2\pi/L_c$. A planar cavity provides no confinement perpendicular to the z-axis, so the photon has an in-plane dispersion. Therefore the cavity photon energy is approximately

$$E = \frac{hc}{n_c} k = \frac{hc}{n_c} \sqrt{\left(\frac{2\pi}{L_c}\right)^2 + \boldsymbol{k}_{\parallel}^2} \tag{7.10}$$

This dispersion is parabolic for small $k_{\parallel}$, and so it can be described by a cavity photon effective mass $m_{phot} = hn_c/cL_c$. This very small mass is typically $\sim 10^{-5} m_e$ [10]. Such dispersions can be measured directly in angle tuning experiments: moving away from normal incidence in a reflectivity measurement introduces an in-plane component to the photon wavevector [3]. In-plane wavenumbers up to $k_{\parallel} \sim 10^7$ m^{-1} can be probed in this way.

Experiments involving non-normal incidence can also be modelled by including an appropriate in-plane wavevector for the field. These calculations give the energy of the cavity mode, in terms of the structural parameters, and also its homogeneous broadening due to tunnelling through the barriers. A small amount of absorption has to be added in the mirrors and active layer for the calculation to obtain good agreement with the experimental cavity mode widths.

An even simpler model than the transfer matrix simulations can be obtained by treating the cavity and photon modes as coupled oscillators, with coupling matrix element $\hbar\Omega_i/2$. This captures most of the essential physics of the cavity polaritons and has the advantage of being simple enough to solve analytically. The various parameters, such as the vacuum Rabi splitting, the cavity and photon mode widths, are treated phenomenologically, to be obtained either by separate calculation or from fitting to experimental data.

The simplest situation which can be treated like this is a single QW in a microcavity. The coupling between the exciton and cavity oscillators is described by a 2x2 matrix Hamiltonian:

$$H = \begin{pmatrix} \hbar\omega_x & \hbar\Omega_i/2 \\ \hbar\Omega_i/2 & \hbar\omega_c \end{pmatrix} \tag{7.11}$$

Here, $\hbar\omega_x$ and $\hbar\omega_c$ are the energies of the exciton and cavity modes respectively and $\hbar\Omega_i$ is the vacuum Rabi splitting. This Hamiltonian is easily diagonalized, to give eigenvalues

$$\hbar\omega_\pm = \frac{\hbar\omega_x + \hbar\omega_c}{2} \pm \sqrt{(\hbar\Omega_i)^2 \pm (\hbar\omega_x - \hbar\omega_c)^2}\,. \tag{7.12}$$

The coupled oscillator model can readily be used to calculate microcavity in-plane dispersions [3]. To do this, the uncoupled exciton and cavity photon energies are made k dependent according to their dispersions. The polariton dispersion is then obtained by solving the coupled oscillator problem for each value of k.

If accurate values of $\hbar\Omega_i$ are to be extracted from experimental data, care is required to account properly for the broadening. This is most obviously done by adding an imaginary part to the exciton and photon energies, corresponding to a homogeneous broadening of the oscillator. For the cavity mode, this is probably appropriate, as the linewidth is believed to be mainly homogeneous, originating from the tunnelling decay of the photon through the mirrors. For the exciton, by contrast, the dominant broadening mechanism is inhomogeneous, because of disorder. The inhomogeneous linewidth of an active layer exciton is typically a few meV, while the homogeneous width is at most a small fraction of an meV. Hence it is not really appropriate to treat the exciton linewidth as due to homogeneous broadening. However, this is frequently done, because a better treatment is considerably more difficult. The reason why some sort of broadening needs to be included is that the separation of the spectral features is reduced by the broadening, so the measured splitting at resonance is less than $\hbar\Omega_i$ [15]. Moreover, the modifications to the splitting in absorption, transmission and reflectivity are different [18].

As shown in [6-8], the penetration of the cavity field into the DBR means that, in some of the formulae, the Fabry–Pérot cavity length L_c has to be replaced by a significantly larger effective length:

$$L_{eff} = L_c + L_{DBR} \tag{7.13}$$

where L_{DBR} is the penetration length into the DBRs and is given by [6]

$$L_{DBR} = \frac{\lambda}{2n_c} \frac{n_L n_H}{n_H - n_L} \tag{7.14}$$

where λ is the wavelength of light in the cavity. For typical GaAs/AlAs structures $L_{DBR} \sim (3\text{–}4)L_c$. As a result of the field penetration into the DBRs, the cavity mode frequency for a microcavity is given by $\omega_m = (L_c\omega_c + L_{DBR}\omega_s)/L_{eff}$ [9], where ω_c is the Fabry–Pérot frequency defined by the length of the cavity and ω_s is the frequency of the centre of the DBR stop band. For $\omega_c \neq \omega_s$, as may arise from imperfectly controlled growth, the observed Fabry– Pérot frequency is no longer equal to ω_c, and is in fact more sensitive to ω_s than to ω_c, since L_{DBR} is significantly greater than L_c. Because of the fact that the mirrors have finite transmission probability, the cavity mode has finite width Δ_c (full width at half-maximum), given by, for $R \to 1$ [6],

$$\Delta_c = \frac{c(1-R)}{n_c L_{eff}} \tag{7.15}$$

This width can be considered as homogeneous lifetime broadening of the confined cavity mode, brought about by the decay through the mirrors. A typical width of ~1 meV corresponds to a cavity mode lifetime of ~4 ps. The smallest reported "empty" cavity width of 0.13 meV [10], gives a finesse, defined by mode separation divided by width, of about 6000.

8. Angle-dependent properties [5]

The cavity photon modes of a MC have strong in-plane dispersion, which is not quantized. Following from equation (7.10) the energy of a photon with quantized wavevector $k_z = 2\pi/L_c$, and in-plane wavevector $\boldsymbol{k}_\parallel$ in the medium, is given by

$$E(\boldsymbol{k}_\parallel) = E_0\sqrt{1 + \left(\frac{\hbar c \boldsymbol{k}_\parallel}{E_0 n_{eff}}\right)^2} \tag{8.1}$$

where n_{eff} is the effective refractive index of the structure [9] and $E_0 = hc/n_{eff} L_c$ is the photon energy for $\boldsymbol{k}_\parallel = 0$. Equation (8.1) corresponds to strong in-plane dispersion which can be characterized by a very small in-plane mass of $\sim 10^{-5} m_0$. Each $\boldsymbol{k}_\parallel$ in-plane photon mode couples only with exciton states of the same $\boldsymbol{k}_\parallel$ to satisfy the requirement of wavevector conservation. The resulting coupled modes, the cavity polaritons, also show strong in-plane dispersion, but with marked perturbation in the region of strong interaction between the two modes.

$\boldsymbol{k}_{\parallel}$ is related to the external angle of incidence by

$$\boldsymbol{k}_{\parallel} = \frac{E(k)}{\hbar c}\sin\theta_i \tag{8.2}$$

and as a result a particular $\boldsymbol{k}_{\parallel}$ mode can be selected simply by varying the external angle of incidence of θ_i. Elimination of $\boldsymbol{k}_{\parallel}$ in equations (8.1) and (8.2) leads to the following expression for the energy of the Fabry-Pérot mode as a function of θ_i:

$$E(\theta) = \frac{E_0 n_{eff}}{\sqrt{n_{eff}^2 - \sin^2\theta_i}} \tag{8.3}$$

Tuning of the exciton–cavity interaction can thus be achieved simply by varying θ_i. Compared with other tuning techniques such as cavity thickness, application of external electric or magnetic field, angle tuning has the particular advantage that the exciton states are independent of angle to a very good approximation. As a result the exciton–cavity interaction potential is also independent of angle, leading to a very straightforward way to study polariton-tuning phenomena. Furthermore, the polariton dispersion can be investigated directly by measuring optical spectra as a function of the external angle θ_i.

One of the notable features of the angle dependence is that there is a splitting between the TE and TM modes for $\theta_i > 0$. This arises from the slightly different phase shifts and penetrations of the optical modes into the Bragg mirrors L_{DBR} for the two polarizations. Its magnitude is determined by the energy difference between the cavity mode and the centre of the DBR stop bands, as shown by analytical calculations in [9]. Additional results of the work of Baxter *et al* [19] include the observation of a marked polarization dependence of the dip intensities and linewidths [20], interaction of the cavity mode with active layer excited state transitions and broadening of the cavity mode as it becomes degenerate with exciton continuum states. The polarization dependence of the intensities at finite angle was found to be in agreement with the predictions of the transfer matrix simulations and was shown to arise physically from the differing degree of matching between upper and lower mirror reflectivities in TE and TM polarizations.

8.1. ANGLE AND POLARISATION DEPENDENCE OF OFF-RESONANTLY EXCITED EMISSION FROM ORGANIC MC.

We have investigated stationary and picosecond time-resolved emission from "organic" microcavity, consisting of 120 nm layer of spin-coated J-aggregate in PVA matrix (n_c=1.54) surrounded by silver mirror and DBR, made of 9 couples of

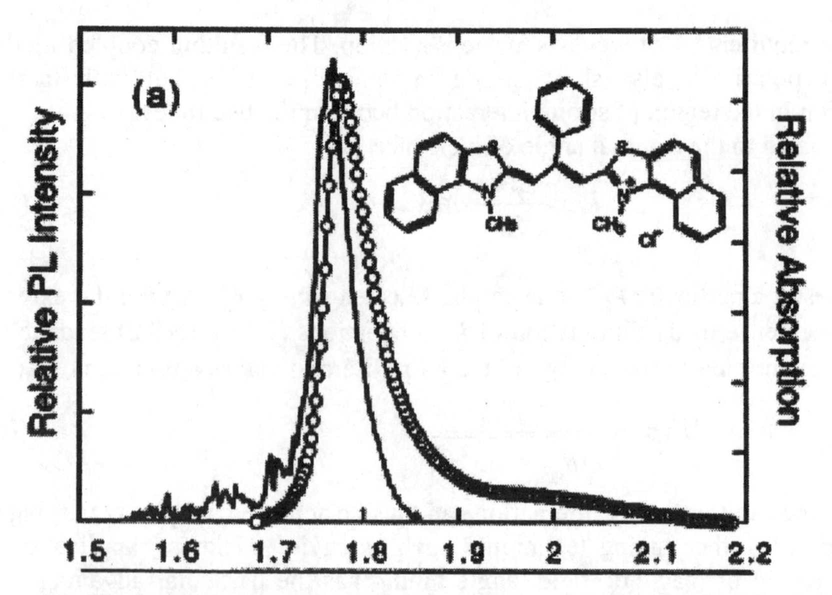

Figure 8.1. The absorption (open circles) and photoluminescence emission (solid line) spectra of a thin film of cyanine dye *J*- aggregates dispersed in a PVA matrix. The inset shows the chemical structure of the specific cyanine dye studied here.

$\lambda/4n_i$ layers of SiO_2 (n_L =1.45) and Si_xN_y (n_H =1.54). Absorption and emission spectra of the J-aggregate layer, measured at room temperature outside of microcavity, are presented at figure 8.1 by open circles and solid lines. They consist of single main peak at approximately 1.78 eV, FWHM ~ 25-50 meV, and order-of-magnitude weaker short- or long-wavelength tails.

At near resonance 633 nm excitation by HeNe laser emission from investigated MCs always consist of two lines: independent on observation angle free exciton line in spectral range 665-695 nm with FWHM ~50 nm and polariton emission, which spectral position was dependent not only on MC thickness L_c, but also on angle of observation θ_o. At $\theta_o \approx 45^\circ$ two emission lines approach each other, producing due to anticrossing two slightly overlapping exciton-polariton lines.

At off-resonance 442 nm CW excitation by HeCd laser in addition to above-mentioned exciton and polariton lines several high-energy lines have been observed in spectral range 480-600 nm of transmission lobes of DBR mirror. Hence at HeCd excitation emission consists of 3 types of lines: (see, for example, Figure 8.4 for observation angle 45°)

1) Single line with FWHM ~15-25 meV, centered at approximately 1.78 eV due to emission of J-aggregate (exciton). Intensity, width and polarization ratio of this line are dependent on observation angle θ, while spectral position is angle-independent.

2) Single line due to emission from "normal cavity mode", situated in stop-band of DBR mirror, with all parameters (including spectral position) strongly dependent on angle of observation θ. At angle~45° spectral position of this lines

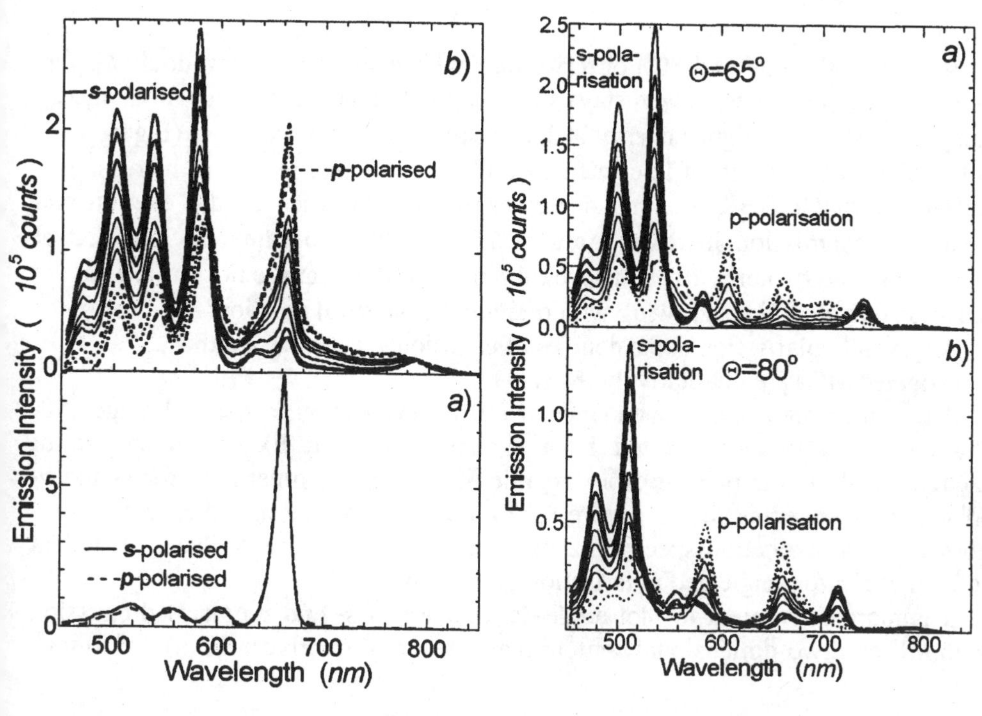

Figure 8.2 Room - temperature emission spectra of MC with J-aggregate in PVA for two angles of observation θ: *a*) θ=0; *b*)θ=45° and different polarisation angles α. Thick solid lines are for α = 0°(s-polarisation), 10 and 20°. Broken lines present emission at p-polarisation (α=90°), α=80° and 70°. Emissions at intermediate polarisation angles a presented by thin lines.

Figure 8.3. Emission spectra of the MC #14 with J-aggregate in PVA for θ=45° (*a*) and 80^0 (*b*). Emission spectra for s-polarisation and small polarisation angle α=10and 30° are presented by thick solid lines, for p-polarisation and large α= 80 and 70° are presented by broken lines, for intermediate polarisation angels – by thin solid lines.

crosses (or anti-crosses) above-mentioned J-aggregate transition.

3) Several emissions bands due to high-energy lobes of DBR mirror above its stop-band. Their positions and polarization properties are strongly dependent on angle of observation.

At observation angle perpendicular to MC surface (see Fig. 8.2 a) excitonic lines dominated in emission spectra at all polarisation angles α. Spectral positions, shapes and intensities of emission lines are independent on polarisation angle.

At θ_o=45° (Fig.8.2 b), due to anticrossing between exciton and cavity mode transitions they produce doublet 640-665 nm, better visible in *s*-polarization. *s*-polarised emission due to high-energy transmission maxima of DBR are much stronger then exciton (665 nm) and low-energy DBR transmission maximum (785 nm) emissions. Changing of polarisation angle from 0° (*s*-polarisation) to 90° (*p*-polarisation) results in decreasing of high-energy line intensities and simultaneous increase of exciton and cavity polariton emission. Moreover, at these changes high-energy lines shift to lower energies while lowest line (785 nm) shifts to higher energy.

At higher (θ_o = 65 and 80°) observation angles exciton-polariton doublet splits into two separate lines, lower stays on free exciton position while high-energy component shifts to higher energy at increasing observation angle α (Figure 8.3). Polarisation dependence of line intensities is the same like at 45°: at increasing α emission through DBR maxima decreases while intensities of the exciton and cavity mode emission increase. At α = 65 and 80°, it became obvious that at increasing α high-energy lines in spectral range 500-600 nm are not shifted but are substituted by new lines growing with α at another spectral position.

Angle and polarisation dependencies of emission spectra for another investigated MCs are qualitatively the same.

Angle and polarisation dependence of DBR transmission in spectral range 450-650 nm can also be extracted from excitation spectra. Excitation spectra of excitonic emission at 665 nm from organic MC at room temperature for *s*- and *p*-polarisations are presented on Figure 8.4 for different excitation angles. As follows from Fig. 8.4, excitation spectra are modulated with period and phase dependent both on excitation angle and polarisation of excitation light.

It is more convenient to plot angle-dependence of *s*- and *p*-polarised emission from MC as a two-dimensional contour map, presented on Figure 8.5 a) and b) for

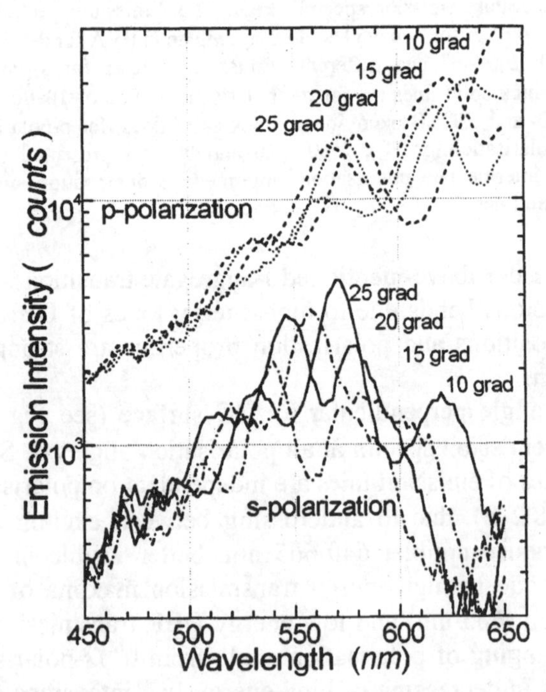

Figure 8.4. Excitation spectra of excitonic luminescence at 665 nm from organic MC at room temperature for four different angles of excitation of *s*-(lower group of lines) and *p*-(upper group of lines) polarised light.

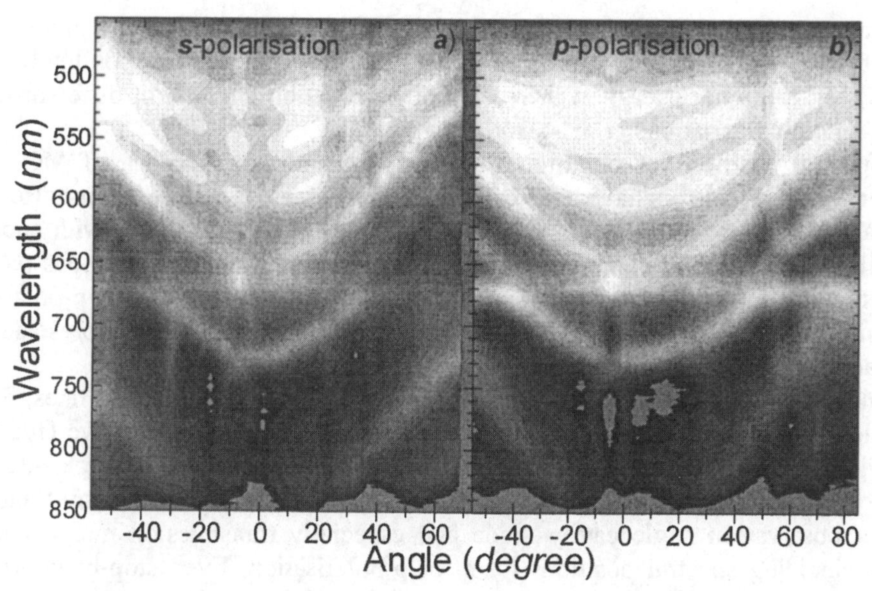

Figure 8.5. Contour plot of experimentally measured emission spectra of organic MC for different observation angles θ_o for *s*- (left panel) and *p*-(right panel) polarisations.

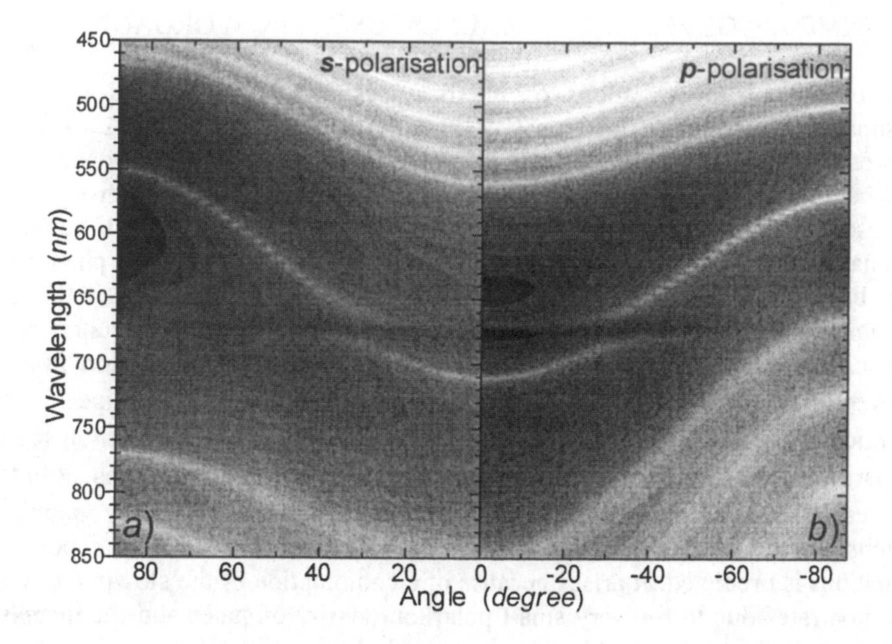

Figure 8.6. Contour plot of the calculated transmission of the MC with Lorentz oscillator medium between silver and DBR mirrors as a function of wavelength and angle of observation for *s*- (left panel) and *p*-(right panel) polarisations.

s- and *p*-polarised emission. Anticrossings of exciton and cavity modes are observed at ~30-35° and 40-45° for *s*- and *p*-polarisations correspondingly. All lines shift to higher energy at increasing of observation angle with different rates for s- and p-polarisations.

Spectral positions of polariton states and transmission maxima of MC with silver and DBR mirrors have been calculated using transfer matrix method. Results of calculations are presented at Figure 8.6. Single Lorentz oscillator with energy, oscillator strength and damping coinciding with these reported in previous papers has simulated J-aggregate layer. Spectral positions of allowed s- and p-polarised transitions in MC with J-aggregate as a function of external angle are presented at left and right panels correspondingly.

Calculation explains different spectral positions of exciton-polariton lines, high- and low-energy transmission maxima of MC with Lorentzian-type active layer. At small observation angle spectra for both polarisations are similar. At angle ~30-40° cavity mode anticrosses excitonic transition, what results in polariton splitting. At larger observation angle cavity mode and especially transmission maxima have non-coinciding spectral positions for s- or p-polarisation. Even stop-band widths are different for different polarisations. Calculations qualitatively and partly quantitatively coincide with experimental observation and allow determining contribution of active layer and each of mirrors to observed wavelength and angle dependencies.

8.2. TIME-RESOLVED PHOTOLUMINESCENCE FROM ORGANIC MC.

Most of optical investigations of microcavities have been carried out using transmission or reflection spectroscopy. Photoluminescence (PL) or PL excitation spectroscopy can also yield important information about microcavity properties.
Influence the exciton or carrier relaxation and recombination on PL from MC at non-resonant excitation has been investigated theoretically in [21-23]. Influence of the small polariton density of states close to $\boldsymbol{k} = 0$ on PL relaxation phenomena nave been also discussed in [24]. It follows from these works that the recombination dynamics is mostly influenced by the uncoupled exciton states radiating into the leaky modes of the cavity, arising for large exciton in-plane wavevectors $\boldsymbol{k}$ [23]. The decay of the exciton population is determined by the dynamics of the large $\boldsymbol{k}$ mostly exciton-like states of the lower polariton branch with large $\boldsymbol{k}$. Luminescence from the upper and lower polariton branches close to $\boldsymbol{k} = 0$ results due to phonon scattering from this large-$\boldsymbol{k}$, thermally populated excitons. In the radiative region close to $\boldsymbol{k} = 0$ a bottleneck in the polariton populations is predicted. It arises because of a combination of the slowing down of relaxation rates due to the very small polariton density of states and the increased radiative recombination rates due to mixing with the cavity modes.

Experimental investigation of CW PL phenomena is contained in the work [25].

These experiments were carried out in the $3\lambda/2$ cavity with non-resonant excitation at 115 K, where the thermal energy is of the order of the vacuum Rabi splitting. Simultaneous measurements were made from the front of the sample, where PL from the strong coupling states is observed and from the edge where only PL of uncoupled excitonic states is seen. In investigated microcavity, PL spectra is described very well by the absorption spectrum multiplied by a Boltzmann distribution, therefore the PL spectrum arises from fully thermalized polariton distributions at a given detuning. Another observation of this work is that, moving away from resonance, the photon-like branch of the polariton was found to be more luminescent than the exciton-like mode, the photon-like branch reaching a maximum intensity at medium detuning. This behaviour is shown to be a consequence of the high absorption that occurs in high-finesse microcavities containing strongly absorbing active layer. The total absorption is not maximum on resonance, since on resonance the intracavity single-pass absorption is sufficiently high that the photon field in the cavity does not build up as strongly as arises for medium detuning; on resonance the cavity is only relatively weakly coupled to the outside world. The maximum total absorption, and hence maximum PL, is instead shown to arise when a photon, in a single round trip, has the same probability of being absorbed as of escaping from the cavity, i.e. when the round trip absorption is 50% and transmission equals internal absorption (impedance matching condition). In investigated microcavities this condition is not fulfilled on resonance, but instead arises 5–10 meV from resonance. In detuning the cavity away from the exciton, the intracavity single round trip absorption drops but approaches that needed for the above impedance matching. Thus the total cavity absorption increases, and so does the PL intensity.

PL experiments at temperatures of 5 K and 30 K and resonant excitation of cavity polaritons have been carried out in [23]. For temperatures above 30 K the PL spectra were found to be thermalized with only a weak dependence on excitation energy. On the other hand, at 5.2 K the PL spectra were very sensitive to excitation energy, indicating that the PL does not arise from a thermalized distribution. For thermalization to occur the exciton fraction of the polaritons must interact effectively with acoustic phonons. According to [26] polariton lifetime is ~7 ps, whereas the acoustic phonon scattering times at 5 K and 30 K are 26 ps and 4.2 ps, respectively. Therefore at $T \geq 30$ K polaritons are expected to thermalize before emission by fast scattering, whereas at $T \sim 5$ K the radiative lifetime is faster than the acoustic phonon scattering time and emission occurs before thermalization is complete.

What is surprising – distribution of relative intensities of three types of emission lines at Fig. 8.2 and 8.3 are inverse to presented at Fig. 8.1: low-energy emission of “cavity mode” is weakest while high-energy emission, transmitted through DBR maxima above stop-band nave highest intensities at room temperature. Therefore question arise: is it due to non-equilibrium distribution of excitations (hot luminescence), or that is result of strong spatial inhomogeneity of J-aggregate and PVA matrix?

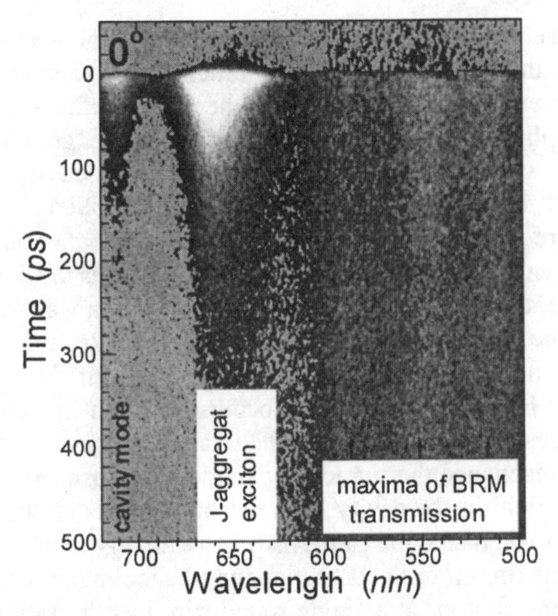

Figure 8.7. Gray-scale representation of the measured by streak-camera room-temperature emission decay of organic MC at picosecond 435 nm excitation for observation angle $\theta_0=0^0$. Emission spectra consist of free-exciton emission at ~660 nm, cavity mode emission at ~710 nm and 3 high-energy bands at DBR.

Gray-scale plot of microcavity emission kinetics at pulsed laser excitation (pulse duration ~2 picoseconds, average power 5 mW, repetition rate 80 MHz, excitation wavelength 435 nm), recorded by streak-camera with time resolution ~ 5 ps, presented at Figure 8.7 for observation angle 0°. As at CW excitation, time-resolved luminescence consists of exciton, "cavity mode" and emission through DBR lobes. Relations between their intensities differ from recorded at CW excitation: excitonic line has strongest initial amplitude, emission of cavity mode is comparable or weaker than excitonic emission, while high-energy emission lines have weak initial amplitudes. Rise-times ≤ 5 ps for all 3 different types of lines are approximately the same, no visible delay in emission is detected. Decays of exciton and cavity mode are two-exponential/ while decays of all high-energy emission bands are monoexponential. Initial decay time of cavity mode at 0° is ~15-18 ps, which change after ~200 ps to decay time ~120-400 ps. Exciton start to decay with decay time ~42-48 ps, after ~200ps it also change decay tame to 120-400 ps. High-energy part of emission decay always with same decay time 570 ±60 ps. We attribute these long-lived luminescence bands to emission of PVA matrix, which at large delays ≥ 0.5 ns partially transfer their excitations to low-energy excitonic states, therefore at large delays all lines decays synchronously with approximately the same decay times ~0.6 ns.

At observation angle 30° (Figure 8.9) cavity mode and high-energy emission band are shifted to higher energy. Cavity mode approaches to and nearly merges with excitonic emission line. Low-energy part of common emission band, which

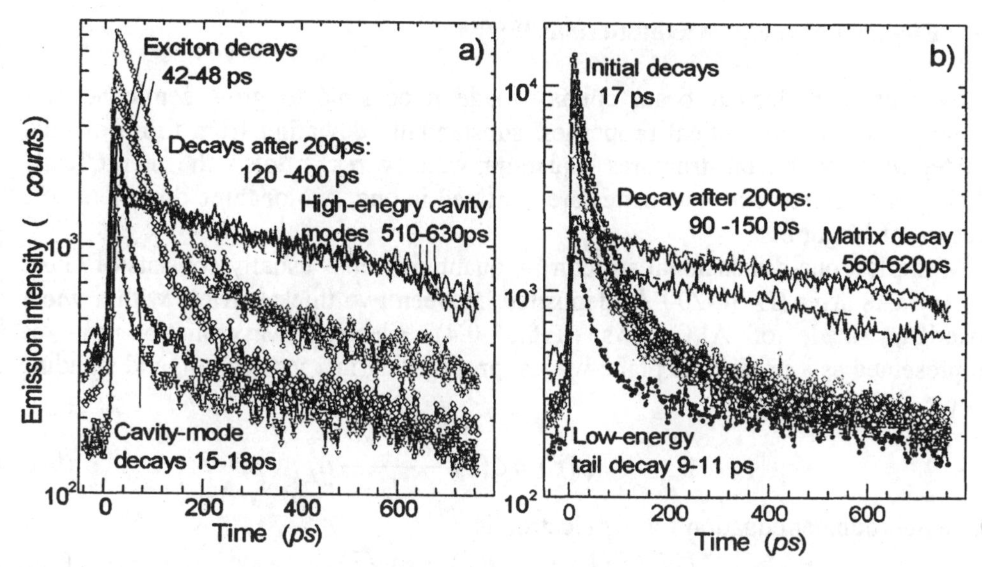

Figure 8.8. Picosecond kinetics of three different types of the organic MC emission (solid lines and symbols) and their exponential fits (broken lines) at observation angle θ=0°. Solid lines – emission of high-energy bands. Lowest three-angles and squares – emission of cavity-mode state. Other symbols – emission of the middle and high-energy parts of the exciton line.

Figure 8.9. Same as Fig.8.8 for observation angle θ = 30° Solid lines – emission of high-energy bands. Circles - emission of the low-energy tail of the exciton-polariton doublet. Other symbols – emission of the middle and high-energy parts of the exciton-polariton doublet.

has photon-like nature, decay with initial time ~9-11 ps, then gradually change decay time to 90-150 ps. Central and high-energy parts of exciton-photon band decay initially with time 17 ps, after ~0.2 ns decay rate characterized by 100-150 ps. High-energy emission bands (which we attribute to matrix emission) decay monoexponentially with decay time 560-620 ps, that is like at angle 0°.

Analysing two last figures, we can come to conclusion, that after ps laser pulse all states are "immediately" excited, then decay in accordance with their lifetimes. Lifetime of cavity mode is mainly radiative lifetime of the order 9-11 ps. Probably, radiative lifetime of exciton is approximately 20 ps, therefore they decay with this time in direction 30°. In perpendicular to MC plane direction radiative decay is suppressed, therefore observed at 0° excitonic decay is ~2.5 times longer. In other words, emitted in perpendicular direction light is reflected by cavity mirrors and reabsorbed, forming exciton. Such mutual conversions of excitons into photons and back results in longer decay time. At large delays part of conserved in PVA matrix energy relax into lower excitonic or cavity mode states, therefore at large delays all states decay with same time ~0.6 ns.

Our results confirm suppression of excitonic spontaneous emission in tuned off-resonance microcavity and explain abnormal intensity distribution at CW excitation by substantial differences in emission decay times of matrix, excitonic and cavity mode states.

9. Electron envelope wavefunctions $\Psi(z)$

Advances in molecular beam epitaxy made it possible to grow semiconductor nanostructures with optical properties, substantially deviating from those of bulk materials. In such nanostructures – quantum wells, wires or dots – the energetically low-lying electron and hole states are confined in one, two or three directions to a region of length L.

Simplest one-dimensional structure – quantum well – usually consists of GaAs or InGaAs layer (L~30-200 Å) deposited between two thick barriers with a wider bandgap, made of $Al_xGa_{1-x}As$ (x~0.05-0.4). Electron wavefunction can be represented as a product of plane-waves, propagating in x,y direction, and standing wave, depending on z

$$\psi(\vec{r}) = \zeta(z)\frac{e^{i(k_x x + k_y y)}}{L}u_L . \tag{9.1}$$

The Schrödinger equation for the electron is

$$[H_e(\vec{r}) + V(z)]\psi(\vec{r}) = E\psi(\vec{r}) . \tag{9.2}$$

For x-y plane there is no quantum confinement, therefore the electrons can move freely in plane QW with momentum $\boldsymbol{k}_\perp$. We assume that the electron energy is

$$E_e(\vec{k}) = E_g + \frac{\hbar^2(k_\perp^2 + k_z^2)}{2m_e}, \tag{9.3}$$

where m_e is the effective mass of electron. Replacing $\hbar k_z \to i\hbar\partial/\partial z$ we find the equation for standing-wave envelope $\zeta(z)$:

$$\left[-\frac{\hbar^2}{2m_e}\frac{\partial^2}{\partial z^2} + V(z)\right]\zeta(z) = E_z(z) . \tag{9.4}$$

Confinement potential V(z) has form

$$V(z) = \begin{Bmatrix} 0 & for\,|z| < L/2 \\ V_c & for\,|z| > L/2 \end{Bmatrix} . \tag{9.5}$$

In quantum well at $|z|<L/2$ solutions are $\zeta(z) = A\sin(k_z z) + B\cos(k_z z)$ and

$$k_z^2 = 2m_e\frac{E_z}{\hbar^2} . \tag{9.6}$$

Outside of the QW solutions are

$$\zeta(z) = C_\pm e^{\pm\kappa z} \text{ with } \kappa^2 = 2m_B\frac{V - E_z}{\hbar^2} . \tag{9.7}$$

Wavefunctions and their derivatives should matches at the interfaces $\pm\, z/2$, therefore solutions are

$$\zeta_{even}(z) = \begin{cases} B\cos k_z z & |z| < L/2 \\ Ce^{-\kappa z} & z > L/2 \\ Ce^{\kappa z} & z < -L/2 \end{cases} \quad \text{with} \quad \sqrt{E_z}\tan\left(\sqrt{m_e \frac{E_z}{2\hbar^2}}L\right) = \sqrt{V - E_z}$$

$$\zeta_{odd}(z) = \begin{cases} A\sin k_z z & |z| < L/2 \\ Ce^{-\kappa z} & z > L/2 \\ -Ce^{\kappa z} & z < -L/2 \end{cases} \quad \text{with} \quad \sqrt{E_z}\cot\left(\sqrt{m_e \frac{E_z}{2\hbar^2}}L\right) = \sqrt{V - E_z}$$

The number of bound states in the QW depends on the depth of the QW V. If $V > 0$, there is always at least one bound state. If more than one bound state exists, the symmetry between the successive higher states alternates, until one reaches the highest bound state.

The task to experimentally demonstrate the spatial extent of a wave function in sign and amplitude has been attracting considerable experimental attention in several fields of physics. While experimental techniques were developed to measure the quantum state of an atom or molecule [27, 28], and to control wave packets and reconstruct their constituents in amplitude and relative phase [29, 30], solids, at first glance, seem not to be a very promising playground for the investigation of the coherent temporal evolution of wave functions. In contrast to discrete atomic energy states, they feature continuous energy bands leading to complicated dynamics. Besides, the corresponding dephasing times in a solid are orders of magnitude smaller, which leaves only a narrow time window for observation.

Despite these unfavourable features, semiconductors offer one unique advantage. Namely, the opportunity to grow heterostructures, which gives freedom to design virtually any model potential system for the holes and electrons, has spurred interest and led to a variety of beautiful experiments demonstrating the wave nature of the carriers [31–34].

Possibility of the reconstruction of a Wannier– Stark state in amplitude and sign by resolving the spatial origin of the emitting polarisations by a transient multiple-grating experiment is demonstrated below.

9.1. DESCRIPTION OF INVESTIGATED SUPERLATTICE.

Semiconductor superlattices (SLs) are artificial analogy of crystals where the periodic modulation of conduction and valence bands in one spatial direction (usually chosen as the z direction) leads to periodic potentials for both electrons and holes [35]. With no electric field applied, minibands form from the overlapping confined quantum-well (QW) states, resulting in wave functions extended over the whole structure, similar to Bloch states in bulk material. With an electric field F applied, the wave functions localise spatially, and the energy spectrum forms the Wannier–Stark ladder [36], a set of equidistant levels $E_n(F) = E(0) + meFd$, where m

is the integer ladder index, e is the elementary charge and d is the length of the elementary cell in the direction of growth.

In our case, this periodic potential sequence was slightly modified: the structure used comprises 15 strongly coupled quantum wells with their well width monotonically increasing from 22 to 36 monolayers (MLs) of GaAs (1ML ≈2.8Å), respectively, separated by barriers of 10 MLs of $Ga_{0.3}Al_{0.7}As$. The external electric field was applied via a semitransparent Schottky contact and an ohmic contact. For experiments in transmission geometry the n-doped substrate was removed by chemical wet etching.
Increasing the well width by one monolayer results in an energetic shift of the 1s exciton interband resonance of about 1.7 meV, enabling selective interband optical excitation of one particular direct transition within one well. This is in contrast to a strictly periodic superlattice, where always all quantum wells are excited. If the line width of the interband transition is below 1.7 meV, information about the spatial position of a wave function is gained by its spectral position. In electric field-dependent PL and differential reflection experiments we derived a linewidth of 0.8–1.2 meV (FWHM). From the observed non-diagonal transitions in the differential reflection experiments we estimated a built-in electric field of 5 – 7 kV/cm.

In cross-correlation experiments in optical interferometry, the amplitudes of the optical "test" wave packet constituents are gained by correlating them with a known "probe" laser pulse and de-convoluting the second-order spectrally resolved signal. Analogously, we correlate a "test" Wannier–Stark state (arbitrarily chosen to be centred in one particular well) with a set of spectrally (and thus spatially) close "probe" states, hereby increasingly overlapping "test" and "probe" states, while recording the quantum interferences of the generated wave packet.

In a partially degenerate four-wave-mixing (FWM) [37] geometry, a first laser pulse with wave vector $\boldsymbol{k}_1$, after passing through a pulse shaper [38], which provides one or two spectrally narrow (0.3 meV FWHM) lines, each resonantly exciting the 1s exciton interband transition in one particular well only, composes ensembles of one or two wave functions, i.e. both the "test" and "probe" states. Each exciton ensemble is centred at a different spatial location and gives rise to coherent first-order polarisations $\boldsymbol{P}_j^{(1)}$. The optical electric field of the delayed, spectrally broad co-linearly polarised 100-fs pulse propagating in direction $\boldsymbol{k}_2$, together with the polarisations $\boldsymbol{P}_j^{(1)}$, induced by pulse $\boldsymbol{k}_1$, creates a set of gratings, propagating in directions $\pm(\boldsymbol{k}_2\text{-}\boldsymbol{k}_1)$, on which pulse $\boldsymbol{k}_2$ self-diffracts.

The diffracted signal in directions $\pm(2\boldsymbol{k}_2\text{-}\boldsymbol{k}_1)$ is spectrally resolved with an optical multichannel analyser (resolution ~0.02 meV) and recorded with respect to delay time τ. We measured the FWM signal in both reflection and transmission geometry; only data gathered in reflection geometry are presented. The excitation density was $\sim 10^8$ cm^{-2} per well. All experiments were performed at 4.2K.

9.2. EXPERIMENTAL RESULTS.

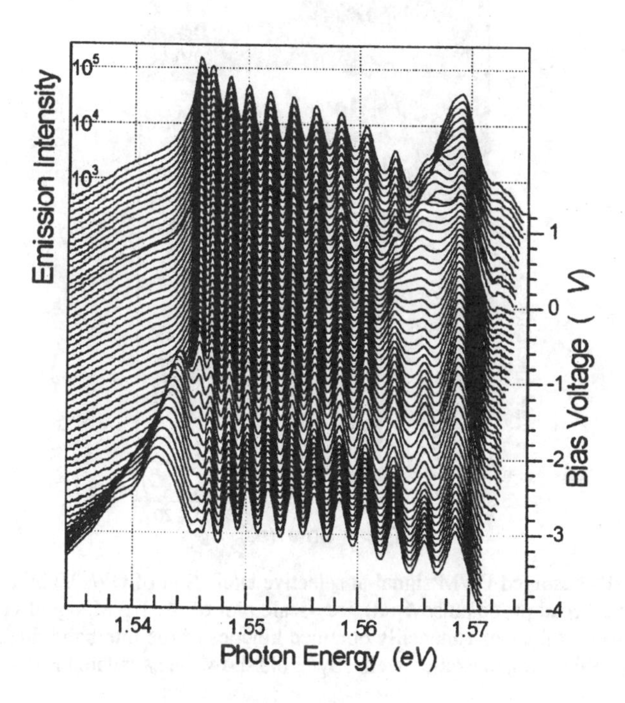

Figure 9.1. Dependence of emission (Logarithm scale) spectra of the graded superlattice emission at CW excitation by HeNe laser and temperature ~10K on bias voltage.

Luminescence of the investigated graded SL at above-band-gap CW HeNe laser excitation for different external electric field from +10kV/cm to –40 kV/cm is presented on Figure 9.1. Spectral position and emission intensities of most intrawell transitions are nearly independent on electric field. Few lowest- and highest-energy transitions are accompanied by broadened field-dependent interband transitions.

Figure 9.2 presents the spectrally resolved FWM signal for increasing delay time τ from one excited quantum well (pulse $\boldsymbol{k}_1$ has a duration of about 2 ps FWHM in the time domain). It shows the following key features of the experiment which are not due to quantum interferences:

(i) the FWM signal from the excited QW reaches its maximum at $\tau_{max} \sim 3.5$ ps and decays exponentially with a decay time $T \sim 4$ ps;

(ii) the weaker FWM signal from neighbouring, unexcited wells is observed to have a rise time of 1 ps only, but decays much faster in a non-monoexponential fashion;

(iii) the FWM signals from remote, non-excited QWs practically coincide with the temporal pulse shape of pulse $\boldsymbol{k}_1$. If the exciting pulse $\boldsymbol{k}_1$ is shaped to

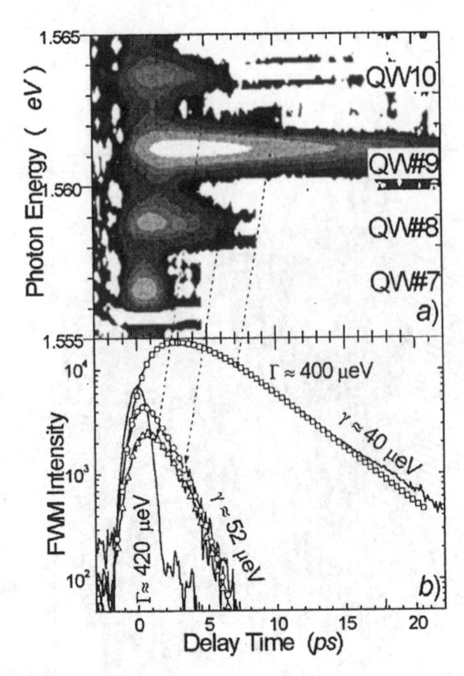

Figure 9.2. Spectrally resolved FWM signal at selective excitation of QW (9) by a spectrally narrow pump pulse $\boldsymbol{k}_1$ and a broad probe pulse $\boldsymbol{k}_2$; **a)** grey-scale plot of the FWM signal versus delay time τ; **b)** solid curves represent the experimentally obtained kinetics of the interband direct transition in the excited well and neighbouring unexcited wells. Symbols show the calculated FWM intensity

consist of two narrow lines resonantly exciting two consecutive quantum wells (the "test" and "probe" wave functions), interferences are observed.

Figure 9.3 shows FWM spectra for three different wave-function combinations, exhibiting the following further features:

(i) the oscillatory modulation of the FWM signal decays with a decay time T_{mod} which is significantly shorter than the FWM signal decay time T;

(ii) the FWM-modulation/signal ratio is much larger in the intermediate, non-excited quantum wells;

(iii) the signal maxima from individual quantum wells have relative phases which do evolve in time.

The spectral laser power in each line was carefully adjusted to yield a maximal signal modulation in the intermittent unexcited well. These spectral weights M_j of the laser later enter the calculation of the macroscopic first-order polarisation. The above features will eventually allow us to reconstruct the wave function.

In Fig. 9.4, experimental FWM data and numerical calculations are compared for the exemplary cases of zero (*a* and *d*), one (*b* and *e*) and two (*c* and *f*) unexcited wells of spatial separation between two transitions excited by pulse $\boldsymbol{k}_1$, which consisted of two narrow lines.

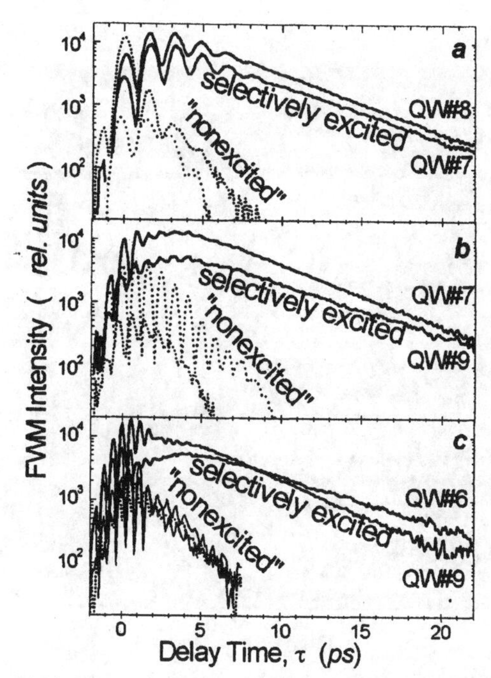

Figure 9.3. Experimental kinetics of spectrally resolved FWM signals from selectively excited (solid lines) and nonexcited (broken or thin lines) with respect to delay time τ. a) two neighbouring wells #8 and #7 are selectively excited by pulse $\boldsymbol{k}_1$ consisting of two narrow lines; b) between excited wells is one intermediate unexcited well; c) between excited wells are two unexcited wells.

9.3. THEORETICAL MODEL

The numerical modelling of the experimental findings is accomplished in three steps. First, the electron wave function $\psi_j(z)$, centred in one particular well j, was calculated by a transfer matrix method, including electron– hole Coulomb interaction. The only parameter used for optimisation, aiming to match the relative FWM intensity and modulation ratio from different wells, was the internal electric field F. A change in F affects the spatial extension and distribution of the electron wave function in different wells.

Figure 9.5 shows calculated electron wave functions excited via a direct interband optical transition from the localized hole in the QW with 29 MLs width. The combined action of Coulomb attraction and internal electric field may result in an asymmetric spatial shape of the wave function. Different regimes can be identified: at -2 kV/cm the electron is localized due to the Coulomb attraction, while having a symmetric shape. The electric field compensates for the asymmetry of the conduction-band potential of the graded superlattice structure. At 2 kV/cm, the electron has overcome the Coulomb binding potential and is delocalised, similarly to field-induced ionisation in atoms. Further increasing the field to 5 kV/cm localises the electron again, with the wave function changing sign in the 31 MLs QW; at 7 kV/cm, the sign change is at the 30ML QW. At yet higher electric

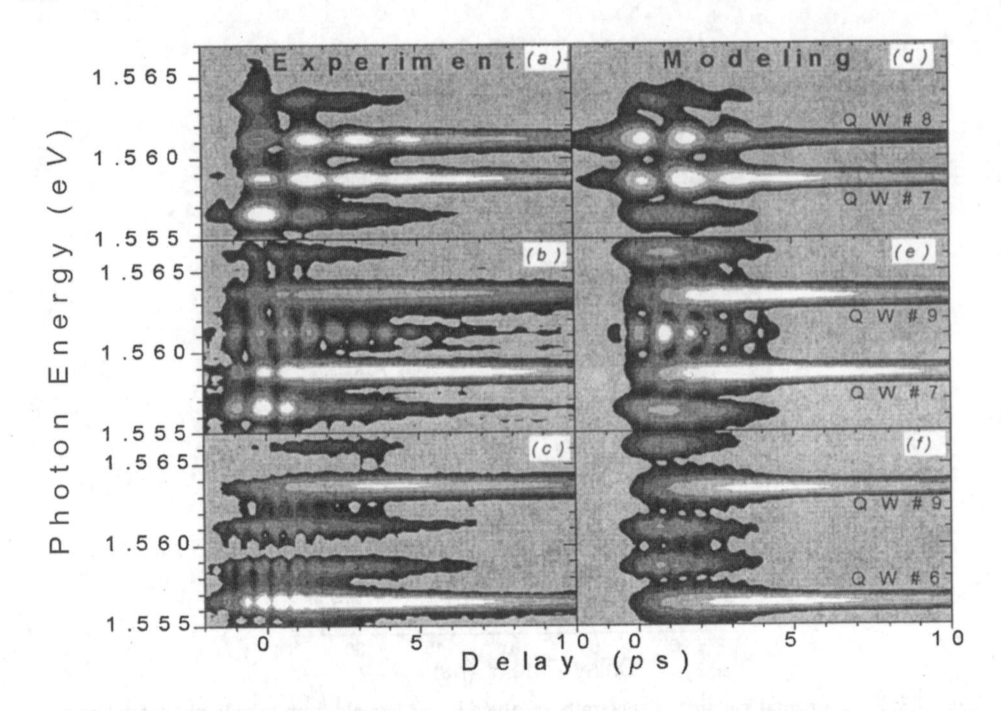

Figure 9.4 Gray-scale representation of experimental (*left*) and modelled (*right*) spectrally resolved FWM signals for selective excitation of two QWs, separated by zero (a, d), one (b, e) and two (c, f) non-excited wells

fields, e.g. 15 kV/cm, the electron localises mainly in the central well [39, 40]. The numerical results shown in Fig. 9.4 were obtained for $F = 5$ kV/cm.

In a second step, we derived spatial weights j_m for a given electron wave function j for each well m close to j, i.e. $m = j, j \pm 1, j \pm 2, j \pm 3$, by integrating over the electron wave function within well m, disregarding the shape of the localised hole wave function in j. In the case where several wave functions j exist at well m, the wave packet $\psi(m, t)$ is summed by adding the weights of the excited electron wave functions, i.e. $\sum_j j_m(t)$ where the weights oscillate in time t with their respective interband transition frequencies.

In the third step, the non-linear signal is calculated. To distinguish between different mechanisms for the induction of a transient non-linear excitonic optical signal, one considers the effect of the optical excitation on the susceptibility (7.3) [41], depending on the oscillator strength f_{ex} of the interband transition in QW_j, being proportional to the square of the dipole transition matrix element μ_j and the square of the exciton relative-motion wave function $V_j(r)$ at $r = 0$. It also depends on is the transition energy $\hbar\omega_j$ and the interband dephasing time γ_j. If one of the quantities in (7.3) is changed during the pulsed excitation, a non-linearity arises. In neglecting a dependence of the exciton wave function on the e–h pair density, i.e. $V_j = const$, a change of f_j is mediated by the Fermi exclusion principle, i.e. phase-space filling (PSF). If, on the other hand, the dephasing rate γ_j is changed during

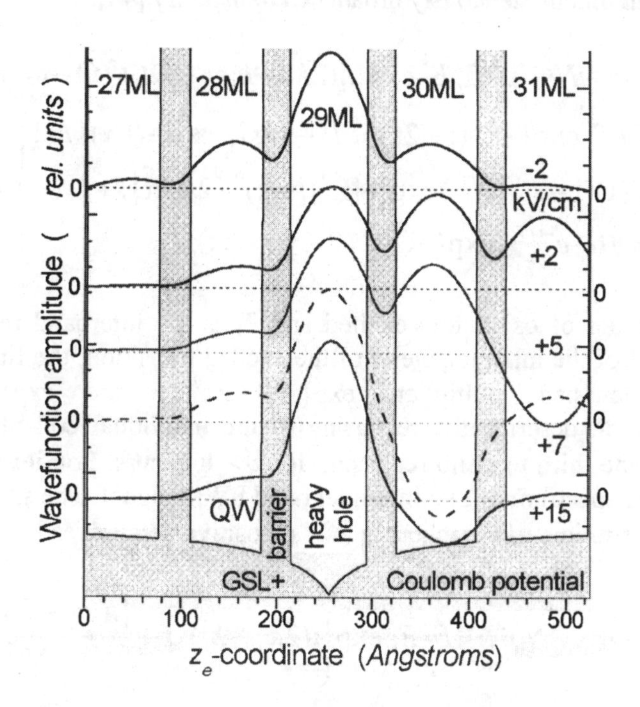

Figure 9.5.The z dependence of the electron envelope wave function in the potential of the graded index superlattice (shown are GaAs QWs with thicknesses from 27 to 31 monolayers, respectively separated by 10 monolayers of Al0.3Ga0.7As), the Coulomb potential of the heavy hole (localised in the 29ML well) and the external electric field F = 15, 7, 5,2 and -2 kV/cm

excitation, the resulting change in dielectric function is labelled excitation-induced dephasing (EID). EID will subsequently lead to a non-linear signal.

For a quantitative understanding of the experimental findings it is necessary to extend the theoretical approach to the third-order response of an inhomogeneously broadened system [42] with both PSF [43] and EID [44-46] as mechanisms responsible for the FWM signal.

We write the optical field of pulses $\boldsymbol{k}_1$ and $\boldsymbol{k}_2$ as

$$E(t) = E_1(t) \exp[\, i(\boldsymbol{k}_1\boldsymbol{r}-\omega_j t)] + E_2(t-\tau) \exp[i(\boldsymbol{k}_2\boldsymbol{r}-\Omega t)]+\text{c.c.} \qquad (9.8).$$

Ω is the centre frequency of the spectrally unshaped laser, E_1 and E_2 are the temporal envelopes of the electric fields of the first and second pulses, respectively, and we introduce an exciton density dependent dephasing rate γ_j of transition j:

$$\gamma_j = \gamma + \sigma n_j\,, \qquad (9.9)$$

where the parameter σ describes the influence of the density n_j on the dephasing rate phenomenologically. It leads to a real-time dependence of the third-order polarisation of an inhomogeneously broadened transition j [44]:

$$P_{inh}^{(3)PSF+EID}(t,\tau) = -iN\hbar^{-3}\mu^4 E_2^2 E_1^* \exp\{i[(2\boldsymbol{k}_2 - \boldsymbol{k}_1)r - \Omega t + (\Omega - \omega_j)(t - 2\tau)]\}$$
$$\times \left\{ \begin{array}{l} \Theta(t-\tau)\Theta(\tau)e^{-\gamma t}\exp\{-\Gamma^2(t-2\tau)^2/2\} + N\sigma T_1 \exp\{-(\Gamma\tau)^2/2\} \\ \times \left[\begin{array}{l} \Theta(t-\tau)\Theta(\tau)(1-e^{-(t-\tau)/T_1})\exp\{-[\Gamma^2(t-\tau)^2/2 - \gamma\tau]\} + \\ \Theta(t)\Theta(-\tau)(1-e^{-t/T_1})\exp\{-\Gamma^2 t^2/2 - \gamma(t-2\tau)\} \end{array} \right] \end{array} \right\} + \text{c.c.} \quad (9.10)$$

N is the number of oscillators excited and T_1 is the interband recombination time. As a result of the inhomogeneity of the excited transition, the first two terms contain the Gaussian multipliers $\exp\{-\Gamma^2(t-2\tau)^2/2\}$ and $\exp\{\{-\Gamma^2(t-\tau)^2/2\}$ respectively, which describe the rephasing of the individual constituents of the polarisation. In the third term no rephasing for $\tau > 0$ occurs. Fourier transforming (4) calculates the spectrum of the time-integrated FWM signal. The spectrum of the PSF term (first term in curly brackets in (4)) at positive delay τ reads:

$$P_{inh}^{(3)PSF}(\omega,\tau) = -i\Theta(\tau)N\mu^4\hbar^{-3}E_2^2E_1^* \exp\left\{ i(2\boldsymbol{k}_2 - \boldsymbol{k}_1)\cdot\boldsymbol{r} + \frac{[i(\omega_j - \omega) + \gamma)]^2}{2\Gamma^2} \right\} e^{-2\gamma\tau}$$

$$\times e^{-i2\tau(\Omega-\omega)} erfc\left\{ \frac{i(\omega_j - \omega) + \gamma}{\Gamma\sqrt{2}} - \frac{\Gamma\tau}{\sqrt{2}} \right\} + \text{c.c.}, \quad (9.11)$$

where *erfc* means the complementary error function. According to (5), the observed FWM signal resulting from PSF is not instantaneous. It has a rise time, which depends on the ratio $\beta = \gamma/\Gamma$. At $\beta \leq 0.1$ the FWM intensity reaches its maximum after a few picoseconds. At larger delay, $\tau > 5/\Gamma$, the FWM intensity decays exponentially, $I_{inh} \propto \exp(-4\gamma\tau)$. The different mechanisms dominating the signal at different delay times also alter the spectrum with respect to τ. During the signal rise time the FWM signal resulting from PSF has a Voigt line shape. For large delays, $\tau > 5/\Gamma$, the line shape becomes Gaussian.

The PSF term (5) suffices to describe several features of the observed FWM signal from quantum wells resonantly excited by pulse $\boldsymbol{k}_1$.

The in z direction extended electron wave function, excited by both pulse $\boldsymbol{k}_1$ and pulse $\boldsymbol{k}_2$ from a hole localised in one well, causes its density to be periodically distributed over neighbouring wells also. Subsequently, the periodic modulation of the dephasing rates affects transitions in neighbouring wells, unexcited but cast over by the extended electron wave function of the excited transition. With the susceptibility (7.3) being inversely proportional to the dephasing rate γ, an exciton density-induced change in the dephasing rate yields a contribution to the non-linear

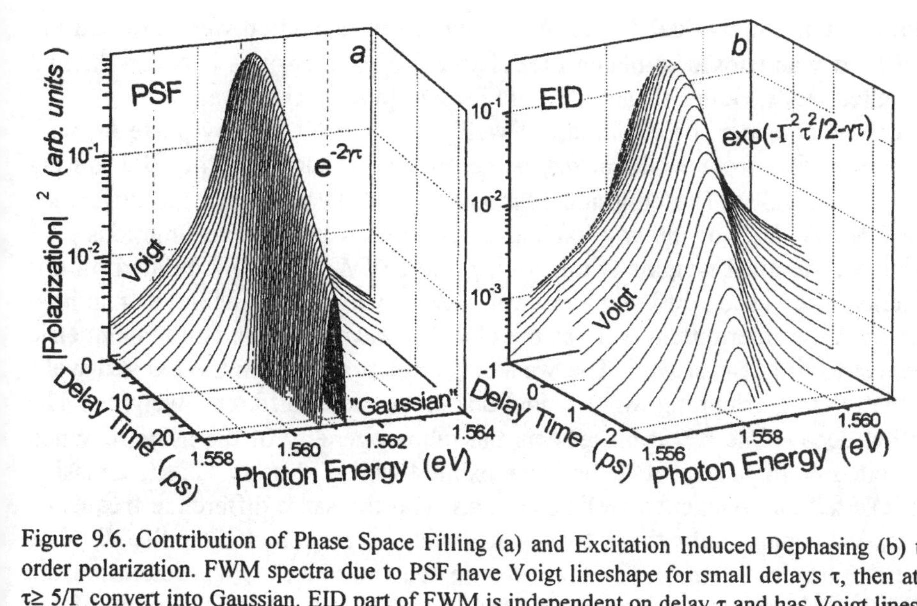

Figure 9.6. Contribution of Phase Space Filling (a) and Excitation Induced Dephasing (b) to third order polarization. FWM spectra due to PSF have Voigt lineshape for small delays τ, then at delays $\tau \geq 5/\Gamma$ convert into Gaussian. EID part of FWM is independent on delay τ and has Voigt lineshape at any delay.

FWM signal. This spatial modulation of the dephasing rate of a transition "unbleached" by pulse $\boldsymbol{k}_1$ is observed from neighbouring wells in which a direct transition is excited. The heterostructure design allows us to explicitly differentiate between FWM signals caused by PSF and EID.

The spectrum of the EID part from (9.10) for positive delay τ (second term in curly brackets) reads

$$P_{inh}^{(3)EID}(\omega,\tau) = \Theta(\tau) i N^2 \mu^4 \hbar^{-3} E_2^2 E_1^* \sigma T_1 \exp\{i(2\boldsymbol{k}_2 - \boldsymbol{k}_1)\cdot \boldsymbol{r} + (\omega + \omega_j - 2\Omega)\tau\} \times$$
$$\exp\left\{-\frac{\Gamma^2\tau^2}{2} - \gamma\tau + \frac{[i(\omega_j - \omega) + \gamma + 1/T_1]^2}{2\Gamma^2}\right\} erfc\left\{\frac{i(\omega_j - \omega) + \gamma + 1/T_1}{\sqrt{2}\Gamma}\right\}. \quad (9.12)$$

The third-order polarisation $P_{inh}^{(3)EID}(\omega,\tau)$ decays for $\tau > 0$ with a Gaussian delay-time dependence, having its maximum shifted to small negative delay $\tau_{max} = -\gamma/\Gamma^2$. The expression for $P_{inh}^{(3)EID}(\omega,\tau)$ for $\tau<0$ is identical, except for a deviation in the decay dependence, i.e. at negative delays $P_{inh}^{(3)EID}(\omega,\tau) \sim \exp\{-\Gamma^2\tau^2/2 + 2\gamma\tau\}$. The spectrum of the EID signal is, in contrast to the PSF term, completely independent of the delay τ. The explanation of the observed experimental features for excitation of one well follows immediately from the distinct mechanisms for the non-linear signal; i.e. τ_{up} and τ_l reflect the inhomogeneous and homogeneous line widths, respectively.

The experiments were modelled using homogeneous inverse line widths of γ = 31–52 μeV and inhomogeneous inverse line widths of Γ = 400–420 μeV (different line widths for the individual QWs). As signals from unexcited wells are entirely due to EID, σ was chosen to match the relative FWM intensities between excited and unexcited wells. Both EID and PSF are present in an excited well.

Let us now consider the excitation of well j -1 and well j +1 by pulse $\boldsymbol{k}_1$, with the FWM emission from well j at $\hbar\omega_j$ being entirely induced by the EID grating (please see Fig. 9.3b). The emission is modulated in delay τ with the difference frequency $\hbar(\omega_{j+1}- \omega_{j-1})$ of the two excited transitions. If both wave functions j +1 and j –1 have identical weights within well j, the FWM is fully modulated. In our experiments, we used different spectral laser weights to achieve maximal modulation in the intermittent well, as the electron wave function is not symmetric with respect to the central well. The weaker modulation ratio in the excited wells shows that the overlapping wave functions have here different weights. The modulation decay time τ_{int} again reflects the inhomogeneity of the system. When two unexcited wells j and j +1 lie between excited wells j -1 and j +2, the emission from all excited and unexcited wells oscillates with the same difference frequency $\hbar(\omega_{j+1}- \omega_{j-1})$, because only these two transitions were excited by pulse $\boldsymbol{k}_1$. The complex amplitudes of the wave functions in the individual QWs determine the exact relative phases of the emissions, and the different rephasing times of the third-order polarisations induced by PSF and EID. With the set of wave functions calculated for an internal electric field of 5 kV/cm and the above-given homogeneous and inhomogeneous line widths, our model yields the observed FWM amplitudes in great detail, and to some extent the complex evolution of the FWM-signal's relative phases.

In summary, by using a graded-index SL, which inherently allows us to spatially resolve the excitation position, we were able to reconstruct the spatial extension and shape of an electron wave function. The consideration of an exciton wave function extended in the growth direction, thereby neglecting the extension of the localised heavy hole, and the distinction of PSF and EID as mechanisms responsible for the FWM signal of our inhomogeneously broadened system, were required to model the experimental results.

10. Acknowledgements

The author would like to thank D. Birkedal, J. Erland, J.M. Hvam, D. Meinhold, B. Rosam, K. Leo, K. Köhler for hospitality, help and useful discussions. Work supported in part by the European Community's Human Potential Programme, Leibniz Prize, RFBR, FTN and contract FIAN #318-03.

11. References

1. Weisbuch, C., Houdre, R., Stanley, R.P. (1995) Microcavities and Semiconductors: The Strong-Coupling Regime, pp. 109-150.
2. Yablonovitch, E. (1993) Photonic bandgap structures, *J. Opt. Soc. Am. B* **10,** 283-305.
3. Born, M. and Wolf, E. (1970) *Principles of Optics*, Pergamon Press, Oxford.
4. Khitrova, G., Gibbs., Jahnke F., Kira M., and Koch, S.W. (1999) Nonlinear optics of normal-mode-coupling semiconductor microcavities, *Rev. Mod. Phys.* **71,** 1591-1639.
5. Skolnick, M.S., Fisher, T.A., and Whittaker, D.M. (1998) Strong coupling phenomena in quantum microcavity structures, *Semicond. Sci. Technol.* **13,** 645–669.
6. Savona, V., Andreani, L. C., Schwendimann, P., and Quattropani, A. (1995) Quantum well excitons in semiconductor microcavities: Unified treatment of weak and strong coupling regimes, *Solid State Commun.* **93,** 733-739.
7. Andreani, L.C., Savona, V., Schwendimann, P., and Quattropani, A. (1994) Polaritons in high reflectivity microcavities: semiclassical and full quantum treatment of optical properties, *Superlattices Microstruct.* **15,** 453-458.
8. Pau, S., Bjork, G., Jacobson, J., Cao, H., and Yamamoto, Y. (1995) Microcavity exciton-polariton splitting in the linear regime, *Phys. Rev. B* **51,** 14 437.
9. Panzarini, G., Andreani, L.C., Armitage, A., Baxter, D., Skolnick, M.S., Roberts, J.S., Kavokin, A.V., Kaliteevski, M.A., Astratov, V.N., and Vladimirova, M.R. (1999) Exciton-light coupling in single and coupled semiconductor microcavities: Polariton dispersion and polarization splitting, *Phys. Rev. B* **59,** 5082-5089.
10. Stanley, R.P., Houdre, R., and Oesterle, U. (1994) *Appl. Phys.Lett.* **65,** 1883.
11. Whittaker, D.M., Kinsler, P., Fisher, T.A., Skolnick, M.S., Armitage, A., Afshar, A.M., and Roberts, J.S. (1996) Motional narrowing in semiconductor microcavities, *Phys.Rev. Lett.* **77,** 4792-4795.
12. Houdre, R., Weisbuch, C., Stanley, R.P., Oesterle, U., Pellandini, P., and Ilegems, M. (1994) Measurement of cavity-polariton dispersion curve from angle-resolved photoluminescence experiments, *Phys. Rev. Lett.* **73,** 2043-2046.
13. Bastard, G. (1988) *Wave Mechanics Applied to Semiconductor Heterostructures* (New York: Halsted)
14. Weisbuch, C., Nishioka, M., Ishikawa, A., and Arakawa, Y. (1992) Observation of the coupled exciton-photon mode splitting in a semiconductor quantum microcavity, *Phys. Rev. Lett.* **69,** 3314-3317.
15. Zhu, Y., Gauthier, D.J., Morin, S.E., Wu, Q., Carmichael, H.J., and Mossberg, I.W. (1990) Vacuum Rabi splitting as a feature of linear-dispersion theory: Analysis and Experimental observations, *Phys. Rev. Lett.* **64,** 2499-2502.
16. Loudon, R. (1973) *The Quantum Theory of Light*, Clarendon, Oxford.
17. Haug, H., and Koch, S.W. (1994) *Quantum Theory of the Optical and Electronic Properties of Semiconductors*, 3rd ed. World Scientific, Singapore.
18. Houdre, R., Stanley, R.P., Oesterle U., Ilegems M., and Weisbuch, C. (1993) Room-temperature exciton-photon Rabi splitting in a semiconductor microcavity, *J. Phys. IV (Paris)* **3,** 51-54
19. Baxter, D., Skolnick, M.S., Armitage, A., Astratov, V.N., Whittaker, D.M., Fisher, T.A., Roberts, J.S., Mowbray, D.J., and Kaliteevski, M.A. (1997) Polarization-dependent phenomena in the reflectivity spectra of semiconductor quantum microcavities, *Phys. Rev.* B **56,** 10032-10035.
20. Kavokin, A.V., and Kaliteevski, M.A .(1995) Excitonic light reflection and absorption in semiconductor microcavities at oblique incidence, *Solid State Commun.* **95,** 859-862.
21. Tassone, F., Piermarocchi, C., Savona, V., and Quattropani, A. (1996) Photoluminescence decay times in strong-coupling semiconductor microcavities, *Phys. Rev.* B **53,** 76 42-7645.
22. Tassone, F., Piermarocchi, C., Savona, V., Quattropani, A., and Schwendimann, P. (1997) Bottleneck effects in the relaxation and photoluminescence of microcavity polaritons, *Phys. Rev. B* **56,** 7554-7563.
23. Savona, V., Tassone, F., Piermarocchi, C., Quattropani, A., and Schwendimann, P. (1996) Theory

of polariton photoluminescence in arbitrary semiconductor microcavity structures, *Phys. Rev. B* **53**, 13051-13062.
24. Pau, S., Bjork, G., Jacobson, J., Cui, H., and Yamomoto, Y. (1995) Stimulated emission of a microcavity dressed exciton and suppression of phonon scattering, *Phys. Rev. B* **51**, 7090-7100.
25. Stanley, R.P., Houdre, R., Weisbuch, C., Oesterle, U., and Ilegems, M. (1996) Cavity-polariton photoluminescence in semiconductor microcavities: Experimental evidence, *Phys. Rev.* B **53**, 10995-11007.
26. Stanley, R. P., Pau, S., Oesterle, U., Houdre, R., and Ilegems, M. (1997) Resonant photoluminescence of semiconductor microcavities: The role of acoustic phonons in polariton relaxation, *Phys. Rev. B* **55**, 4867-4870.
27. Ashburn, J.R. , Cline, R.A., van der Burgt, P.J.M. , Westerveld, W.B., Risley, J.S. (1990) Experimentally determined density matrices for H(n=3) formed in H^+-He collisions from 20 to 100 keV, *Phys. Rev. A* 41, 2407-2410.
28. Leichte, C., Schleich, W.P. , Averbukh, I.S., Shapiro, M. (1998) Quantum State Holography, *Phys. Rev. Lett.* **80**, 1418
29. Weinacht, T.C. , Ahn, J., Bucksbaum, P.H. (1998) Measurement of the Amplitude and Phase of a Sculpted Rydberg Wave Packet, *Phys. Rev. Lett.* **80**, 5508.
30. Chen, X., Yeazell, J.A. Wave-packet reconstruction in a two-electron atom via impulsive isolated core excitation, (1999) *Phys. Rev. A* **60**, 4229-4233.
31. Leo, K., Shah, J., Göbel, E.O., Damen, T.C., Schmitt-Rink, S., Schäfer, W., Köhler, K. (1991) Coherent oscillations of a wave packet in a semiconductor double-quantum-well structure, *Phys. Rev. Lett.* **66**, 201-204.
32. Salis, G., Graf, B., Ensslin, K. Campman, K., Maranowski, K., Gossard, A.C. (1997) Wave Function Spectroscopy in Quantum Wells with Tunable Electron Density, *Phys. Rev. Lett.* **79**, 5106-5109.
33. Dekorsy, T., Kim, A.M.T, Cho, G.C., Hunsche, S., Bakker, H.J., Kurz, H., Chuang, S.L., Köhler, K. (1996) Quantum Coherence of Continuum States in the Valence Band of GaAs Quantum Wells Quantum Coherence of Continuum States in the Valence Band of GaAs Quantum Wells, *Phys. Rev. Lett.* **77**, 3045-3048.
34. Garro, N., Kennedy, S.P., Phillips, R.T., Aichmayr, G., Rössler, U., Vina, L. (2003) Preservation of quantum coherence after exciton-exciton interaction in quantum wells, *Phys. Rev.* B **67**, 121302/1-4.
35. Esaki, L., Tsu, R. (1970) Superlattice and Negative Differential Conductivity in Semiconductors, *IBM J. Res. Dev.* **61**, 61.
36. Wannier, H.G. (1959) *Elements of Solid State Theory*, Cambridge University Press, London.
37. Cundiff, S.T., Koch, M. Knox, W.H., Shah, J., Stolz, W. (1996) Optical Coherence in Semiconductors: Strong Emission Mediated by Nondegenerate Interactions, *Phys. Rev. Lett.* **77**, 1107.
38. Weiner, A.M., Heritage, J.P., Kirschner, E.M. (1988) ,*J. Opt. Soc. Am.* B 5, 1563.
39. Mendez, E.E., Agulló-Rueda, F., Hong, J.M. (1988) Stark Localization in GaAs-GaAlAs Superlattices under an Electric Field, *Phys. Rev. Lett.* 60, 2426.
40. Bastard, G., Bleuse, J., Ferreira, R., Voisin, P. (1989) Wannier-Stark Quantization in Semiconductor Superlattices, *Superlatt. Microstruct.* **6**, 77.
41. Schmitt-Rink, S., Chemla, D.S., Miller, D.A.B. (1985) Theory of transient excitonic optical nonlinearities in semiconductor quantum-well structures, *Phys. Rev. B* **32**, 6601.
42. Erland, J., Pantke, K.-H., Mizeikis, V., Lyssenko, V.G., Hvam, J.M. (1994) Spectrally resolved four-wave mixing in semiconductors: Influence of inhomogeneous broadening, *Phys. Rev. B* **50**, 15 047.
43. Yajima, T., Taira, Y. (1979) Spatial Optical Parametric Coupling of Picosecond Light Pulses and Transverse Relaxation Effect in Resonant Media, *J. Phys. Soc. Jpn.* **47**, 1620.
44. Wang, H., Ferrio, K.B., Steel, D.G., Berman, P.R., Hu, Y.Z., Binder, R., Koch, S.W. (1994) Transient four-wave-mixing line shapes: Effects of excitation-induced dephasing, *Phys. Rev. A* **49**, 1551.

45. Wegener, M., Chemla, D.S., Schmitt-Rink, S., Schäfer, W. (1990) Line shape of time-resolved four-wave mixing, *Phys. Rev.A* **42**, 5675.
46. Sayed, E., Birkedal, D., Lyssenko, V.G., Hvam, J.M. (1997) Continuum contribution to excitonic four-wave mixing due to interaction-induced nonlinearities: A numerical study, *Phys. Rev. B* **55**, 2456.

9. CONSEQUENCES OF EXTREME PHOTON CONFINEMENT IN MICRO-CAVITIES: I. ULTRA-SENSITIVE DEDECTION OF PERTURBATIONS BY BIO-MOLECULES

STEPHEN ARNOLD and MAYUMI NOTO
Microparticle Photophysics Laboratory(MP^{3}L)
Polytechnic University, Brooklyn, N.Y. 11201

FRANK VOLLMER
Center for Studies in Physics and Biology,
Rockefeller University, New York, N.Y. 10021

Abstract

It is becoming possible to confine optical photons within a dielectric microparticle (radius~100μm) for microseconds. This lifetime would allow a free-ranging photon to travel ~300 m in vacuum. In the frequency domain such a mode resonates with a ratio of frequency to line-width Q ~10^9. With such a narrow line-width, the sensitivity to size and refractive index perturbations is extreme. An average size change of less than 1 picometer shifts the resonance line through its complete width. These resonances may be stimulated evanescently by coupling to the guided wave in an optical fiber core. Researchers recently used this fiber-microsphere system to detect hybridization of DNA on a microsphere surface, and found that a single nucleotide polymorphism (SNP, one base mismatch) in a long DNA target could be detected with a signal to noise ratio of 54.

B. Di Bartolo and O. Forte (eds.), Frontiers of Optical Spectroscopy, 337-357.

1. Introduction

In a previous course at this school in June, 2001 one of us (S. A.) taught about the effect which confinement in spherical micro-cavities has on spontaneous and stimulated emission.[1] Although the experimental photon lifetime within a Photonic Atom Mode (Fig.1, a.k.a. Whispering Gallery Mode) in fused silica was of unprecedented length for an optical microstructure (~10^9 oscillations for a 100μm radius sphere), it was never as long as theory would have predicted based on a smooth homogeneous dielectric sphere.

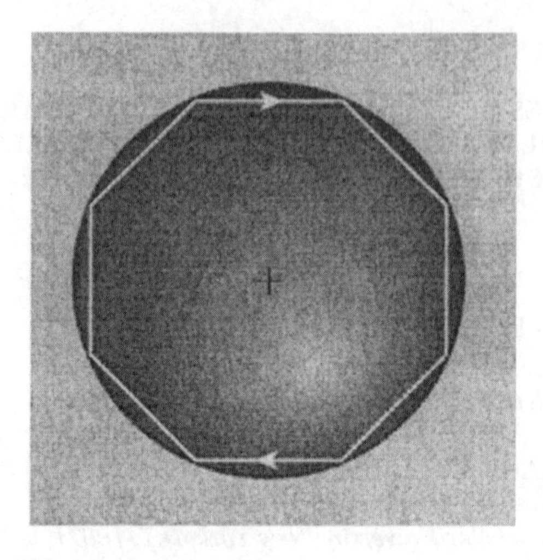

Fig.1 Photonic Atom Mode (a.k.a. Whispering Gallery Mode). Light circumnavigating within a micro-sphere, confined by total internal reflection.

We suspected that the reason was due to sub-nanoscopic perturbations associated with molecular roughness. Opportunity is often built on just such problems. Since such resonant micro-cavities might be particularly sensitive to perturbations, why not cause perturbations by attaching nanoscopic particles to the surface, and measure spectral effects caused by them? Biology offers a multitude of uniform nanoscopic particles in the form of protein, DNA, etc. The result of this work was not only to understand the perturbations,[2,3] but also to generate the world's most sensitive bio-sensor for unlabeled molecules[4,5]. In what follows we will talk about the perturbation of spherical micro-cavities by presenting theoretical ideas first, and then compare the theoretical results with experiment.

2. Simple Considerations

The picture in Fig.1 easily explains photon confinement, but it takes a wave picture to explain spectral discreteness, and to gain a heuristic awareness of the effects of a geometrical perturbation. Fig.2a takes the wave point of view.

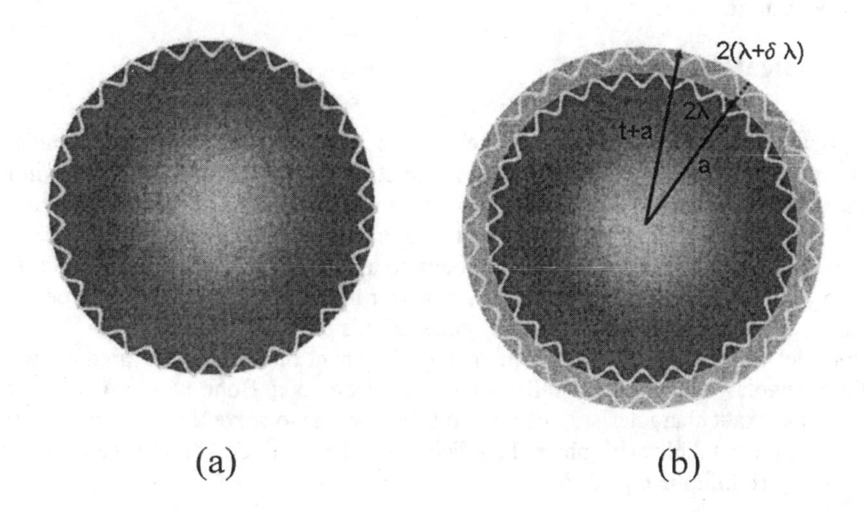

Fig.2 (a) Photonic Atom Mode from a wave point of view; (b) Anticipated wavelength change caused by the addition of a spherically symmetric layer.

One can imagine driving energy into such a mode. To most it should be clear that the only way to drive the mode resonantly is if the wave returns in phase (as shown). In Sec.4 we will outline the means by which this mode can be driven, however at this point we want to characterize our simple mode. There are precisely 30 waves that wrap around the interior circumference in this figure. This will be our mode characteristic or mode number. Suppose now that material of an identical nature and a small thickness t adsorbs on the sphere (Fig.2b). This will cause the wave to circumnavigate a larger circumference, and in order to maintain the same mode number, one might expect the mode wavelength to increase in proportion to the size change. On this basis the fractional increase in wavelength $\Delta\lambda/\lambda$ will be approximately equal to the fractional increase in radius t/a,

$$\frac{\Delta\lambda}{\lambda} \approx \frac{t}{a}. \tag{1}$$

Suppose now that the adsorbing material is 1 nanometer in thickness. For a sphere having a 100 μm radius, the fractional shift in wavelength according to Eqn.1 would be $\sim 10^{-5}$. This is smaller than the resolution of a grating spectrometer, but is a "piece of cake" (i.e. easy) for the microsphere as a spectrometer. The reason is that the resonances of a microsphere are extremely narrow. Resonances with Q's of 10^7

(linewidths of 1 part in 10^7) are considered "broad". But such a "broad" line would shift one hundred times its linewidth for a 1 nm layer. For it to shift just one linewidth requires only a 10 picometer layer (i.e. one tenth the size of a hydrogen atom). So such a small perturbation should be easy to observe.

Our picture thus far is heuristic, and may in fact be incorrect. In Sec. 5 we will see what exact perturbation theory says. However first we will discuss the physical nature of the modes more exactly.

3. Theoretical Approach

As students at this school you likely come from both physical and chemical backgrounds. Extensive theory into electromagnetics is generally not pursued within chemistry departments. However quantum mechanics is common to both. Fortunately an understanding of the electrodynamics of Photonic Atom Modes in a sphere can be gained by reducing the electrodynamic problem to a quantum analog. We will take this approach just as we had in the last school, however here our interest will be to perturb the sphere, and look for particular consequences of this perturbation.

First let us establish the quantum analog. Resonant modes contain photons with quantized angular momentum similar to the electron in a Bohr atom. In order to determine the exact characteristics of these modes one has to solve the Electromagnetic wave equation for a dielectric sphere. In a dielectric microsphere with no excess charge, the governing Helmholtz equation is

$$\nabla^2 \mathbf{E} + k^2 \mathbf{E} = 0. \tag{2}$$

where $\mathbf{E}$ is the electric field, $k = \omega n(r)/c$ is the propagation constant and n(r) is the refractive index. Inspired by the orbiting images displayed in Figs.1 and 2, and the spherical symmetry associated with the problem, we emphasize the importance of the angular momentum by constructing the Laplacian using the angular momentum operator $\hat{L} = -i(\vec{r} \times \nabla)$; $\nabla^2 = \left(\frac{1}{r} \frac{\partial^2 (r)}{\partial r^2} - \frac{\hat{L}^2}{r^2} \right)$. Eqn. 1 becomes

$$\left(\frac{1}{r}\frac{\partial^2(r)}{\partial r^2}-\frac{\hat{L}^2}{r^2}\right)\mathbf{E}+k^2\mathbf{E}=0\,. \tag{3}$$

Considering that the angular momentum operator commutes with its square (i.e. $[\hat{L}^2,\hat{L}]=0$), a solution to Eqn 3 can be written down by examination, $\mathbf{E}=\hat{L}\psi$. This is a so-called TE mode. With this form $\hat{L}$ can be factored through, leaving Eqn. 3 in the form

$$\hat{L}\left[\frac{\partial^2(r\psi)}{\partial r^2}-\frac{\hat{L}^2(r\psi)}{r^2}+k^2(r\psi)\right]=0\,. \tag{4}$$

This is a convenient form since it is clear that the vector Helmholtz equation can be satisfied so long as the scalar equation within the brackets in Eqn. 4 is set to zero;

$$\frac{\partial^2(r\psi)}{\partial r^2}-\frac{\hat{L}^2(r\psi)}{r^2}+k^2(r\psi)\ =0\ . \tag{5a}$$

Eqn. 5a can be re-written in the form of a Schrödinger equation. First we set

$$r\,\psi(\underline{r})=\psi_r(r)\,Y_{\ell,m} \tag{5b}$$

and use $\hat{L}^2 Y_{\ell,m}=\ell(\ell+1)Y_{\ell,m}$, where ℓ is the angular momentum quantum number and $Y_{\ell,m}$ is a Spherical Harmonic function with azimuthal quantum number m. Next we add and subtract $k_0^2\,\psi_r$ from the left in Eqn. 5a. With these modifications Eqn. 4a takes the form

$$\frac{d^2\psi_r}{dr^2}+\{k_0^2-[k_0^2(1-n^2)+\ell(\ell+1)/r^2]\ \}\psi_r=0\,, \tag{6a}$$

which is clearly identified as a Schrödinger-like equation, in which the effective energy is k_0^2 and the effective potential [6]

$$V_{eff}(r;k_0,n,\ell)=k_0^2(1-n^2)+\ell(\ell+1)/r^2. \tag{6b}$$

This potential is an important key for constructing a perturbation theory, and we will return to it often.

Eqn. 6b describes the behavior of light (trapped photon) inside the dielectric microsphere. The first term in the effective potential is negative and leads to dielectric confinement of light, while the second centripetal term is repulsive. The sum of these two terms generates a potential pocket in which photons can be confined. In Fig 3 we have plotted the effective potential for $\ell = 23$, $n = 1.41$, and $k_0 = 10^5$/cm, for a sphere with radius, a = 1.94 μm.

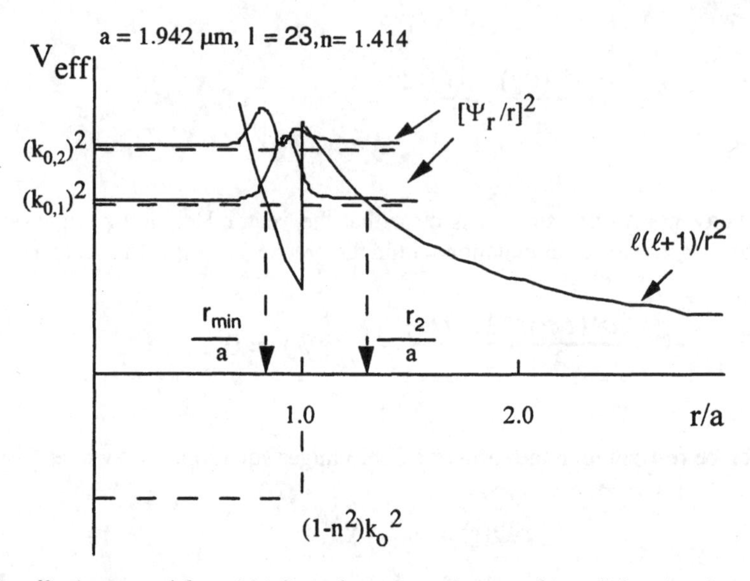

Fig. 3 The effective potential vs. (r/a) for a photonic mode trapped in a dielectric microsphere.

Solutions for ψ^2 at particular values of the effective energy k_0^2 show a substantial buildup of intensity, as seen in Fig.3. These solutions can be stimulated to "resonate" using an external source (i.e. plane wave or evanescent field), and their form conveys a great deal of physical information.

By transforming the independent variable in Eqn 6a from radius r to the product of free space wave vector k_0 times r, we create an convenient "dimensionless length", which at the surface is known as optical size X; $X = k_0 a$. For a given refractive index n, the optical size alone identifies the "dimensionless frequency" of a resonance (i.e. as the size changes, the optical size remains constant). If a particle were to grow in size by a small increment δa of identical material, then the free-space frequency of a given resonance must shrink in the same proportion, i.e. $\delta k_0/k_0 = -\delta a/a$, consistent with Eqn.1.

The first two modes plotted above correspond to the lowest order and next higher order resonance with optical sizes, $(k_{0,1}\ a) = 19.42$ and $(k_{0,2}\ a) = 23.46$. Note that photons in these states are not strictly confined by "classical turning points". The photon can extend into "classically forbidden regions" where the effective energy is lower than the effective potential. Most notably, the probability of locating a photon just outside the particle decays exponentially. This is the region of the so-called "evanescent field" which will be described in more detail in Sec. 4. This field is responsible for the interaction with macromolecules just outside the sphere, and is essential for coupling energy into the sphere from a guided wave.

4. Experimental Insights which grow out of the Fig.3

The evanescence demonstrated in each of the resonant solutions in Fig.3 allows us to construct a means for coupling energy into a photonic atom mode of a microsphere. Once again we can learn from quantum mechanics. The electronic wavefunction of a metal decays exponentially just outside at its surface. When two metals are brought close to each other these exponential tails overlap leading to transfer of electrons between the metals due to tunneling. From a photonic standpoint we can affect the same sort of tunneling by overlapping the evanescent tails between confining dielectric structures. A one-dimensional structure for confinement is an optical fiber. If the core of such a fiber is brought close to a microsphere at the correct frequency then the situation illustrated in Fig.4 can occur.

The diminishing amplitude of the guided wave as it passes the microsphere is a consequence of energy loss in the sphere. Actually the system is an interferometer. Tunneling into the fiber is accompanied by a 90^0 phase shift relative to the wave which proceeds down the fiber. When the light tunnels back into the fiber after many circumnavigations of the microsphere it is subject to an additional 90^0 phase shift. So the wave arriving back from the fiber at resonance is 180^0 out of phase with the wave travelling down the fiber. This leads to a spectral dip in the forward going energy (Fig.5).

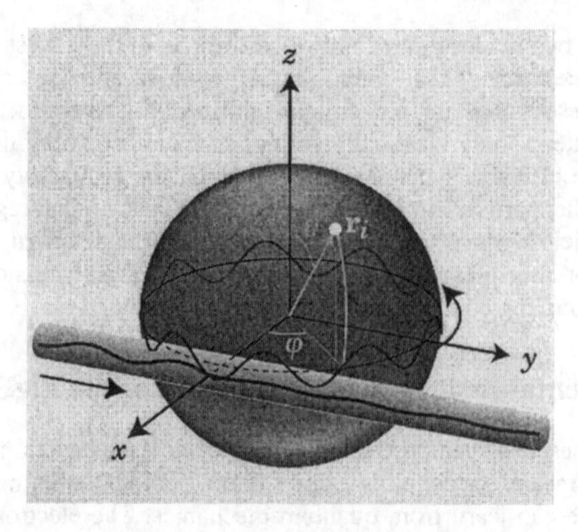

Fig.4 Illustration of coupling of optical energy between a guided wave in an optical fiber and a Photonic Atom mode in a microsphere. The vector $\mathbf{r}_i$ points to a small nano-perturbation.

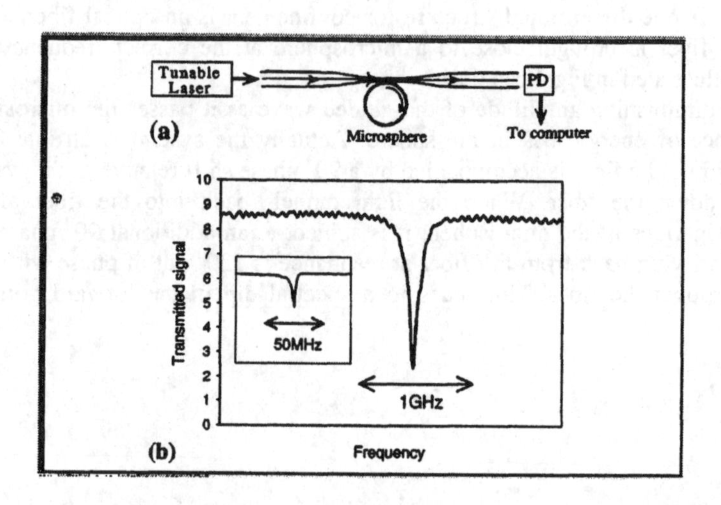

Fig.5 The transmitted light shows a dip corresponding to a Photonic Atom resonance.[7]

A perturbation at the microsphere surface may be expected to lead to a change in the spectral dip (i.e. spectral position or linewidth). The frequency shift of the dip may be expected based on our heuristic model (Eqn.1). This shift will be the basis for transducing the adsorption of bio-molecules. If the surface is conditioned with specific biological recognition elements, the frequency shift can inform us as to the existence of particular bio-molecular interactions (e.g. antibody molecules for antibody-antigen detection, or single stranded DNA for DNA detection through hybridization). Before presenting experiments, we will estimate the shift from perturbation theory.

5. First Order Perturbation Theory: Spherically Symmetric Layer

Before proceeding we remind the reader of our Quantum analog

$$\frac{d^2\psi_r}{dr^2} + \{k_0^2 - [k_0^2(1-n^2) + \ell(\ell+1)/r^2]\}\psi_r = 0 \qquad (6a)$$

This Equation has the usual Schrödinger form, with various analogs listed below;

	Schrödinger	Analog
Energy	E	$E_{eff} = k_0^2$
Potential	V	$V_{eff} = k_0^2(1-n^2) + \ell(\ell+1)/r^2$

Table. 1 Quantum Analogs

Adding a nanoscopic layer to the surface leads to a perturbation in the potential,

$$\delta V_{eff} = \delta[k_0^2(1-n^2)]. \qquad (7)$$

This perturbation will lead to a first order energy shift δE_{eff}. The standard result from quantum mechanics is that the first order perturbation in the energy of a quantum level is

$$\delta E_{eff} = \frac{<\psi_r|\delta V_{eff}|\psi_r>}{<\psi_r|\psi_r>}. \qquad (8)$$

where V_{eff} is the perturbation of the potential. Using our analogs from Table 1 transforms Eqn.8 to

$$\delta(k_0^2) = \frac{< \psi_r | \delta[k_0^2(1-n^2)] | \psi_r >}{< \psi_r | \psi_r >}. \tag{9}$$

The change in the square of the refractive index square is $\delta(n^2) = n_\ell^2 - n_m^2$, where n_ℓ and n_m are the refractive indices of the layer and medium respectively. Furthermore having in mind that the energy k_0^2 will not be constant, we expand Eqn. 9 to;

$$\begin{aligned} \delta(k_0^2)* < \psi_r | \psi_r > = \delta(k_0^2)* < \psi_r | \psi_r > + \\ < \psi_r | -k_0^2(n_\ell^2 - n_m^2) | \psi_r > + < \psi_r | -2k_0 \delta k_0 n_s^2 | \psi_r >, \end{aligned} \tag{10}$$

which reduces to

$$< \psi_r | -k_0^2(n_\ell^2 - n_m^2) | \psi_r > + < \psi_r | -2k_0 \delta k_0 n_s^2 | \psi_r > = 0. \tag{11}$$

Expressing Eqn.11 in terms of integrals;

$$\int_{\text{layer}} -k_0^2(n_\ell^2 - n_m^2) |\psi_r|^2 dv + \int_{\text{all space}} -2k_0 n_s^2 \delta k_0 |\psi_r|^2 dv = 0. \tag{12}$$

With the first integral only taken over a nanoscopic layer on a 100 μm radius sphere, and using continuity for the wave function at the surface

$$\int_{\text{layer}} -k_0^2(n_\ell^2 - n_m^2) |\psi_r|^2 dv \cong -k_0^2(n_\ell^2 - n_m^2) |\psi_r(a)|^2 4\pi a^2 t. \tag{13}$$

In addition with 94% of the square of the wavefunction within the sphere,[8] the second integral can be reasonably taken over just the interior volume.

$$\int_{\text{all space}} -dv |\psi_r(r)|^2 (2k_0)\delta k_0 n_s^2 \cong \int_0^a -dr\, 4\pi r^2 |\psi_r(r)|^2 (2k_0)\delta k_0 n_s^2. \tag{14}$$

Therefore according to Eqn. 12, 13 and 14, the fractional shift in frequency $(\frac{\delta\omega}{\omega} = \frac{\delta k_0}{k_0})$, will be

$$\frac{\delta\omega}{\omega} = \frac{\delta k_0}{k_0} = -\frac{a^2 t |\psi_r(a)|^2 (n_\ell^2 - n_m^2)}{2\, n_s^2 \int dr\, r^2 |\psi_r(r)|^2}. \tag{15}$$

The volume integral in the denominator of Eqn.15 can be related to the surface value of the square of the spherical Bessel function in the limit in which the wavelength is much smaller than the microsphere radius,[9]

$$\int_0^a j_\ell^2(\sqrt{\varepsilon_{rs}}\, k_o r) r^2 dr \approx \frac{a^3}{2} j_\ell^2(\sqrt{\varepsilon_{rs}}\, k_o a) \frac{\varepsilon_{rs} - \varepsilon_{rm}}{\varepsilon_{rs}}. \tag{16}$$

Substituting Eqn.16 in the denominator of Eqn.15 gives a differential shift due to accretion of a layer on the surface, where ε_{rm} is the relative permittivity of the surrounding medium (e.g. water).

$$-\frac{\delta\omega}{\omega} = \frac{\delta\lambda}{\lambda} = \frac{(\varepsilon_{r\ell} - \varepsilon_{rm})}{(\varepsilon_{rs} - \varepsilon_{rm})} \frac{t}{a} = \frac{(n_\ell^2 - n_m^2)}{(n_s^2 - n_m^2)} \frac{t}{a}. \tag{17}$$

In Eqn.17 we have presented the 1st order shift in terms of refractive indices and relative permittivities (i.e. $n^2 = \varepsilon_r$). Eqn.17 agrees with Eqn.1 when $n_\ell = n_s$, as anticipated heuristically, and is consistent with the shift obtained through detailed electromagnetic theory for a layer on a microparticle in vacuum.[10]

One can think of the perturbation problem in terms of molecular properties by imagining the layer to be carved up into a jigsaw pattern. Of course there are few identical shapes that could fit into such a pattern (in a monolayer) without leaving voids. However we will assume that the molecular layer is voidless and return to the question of voids in discussing experimental results in a later section. We will call this model the Voidless Tile Model. It turns out that Eqn.17 is easily transformed by introducing two new parameters: the tile excess polarizability α_{ex} (excluding depolarization effects), and the surface density of tiles σ. By using the constitutive equation $\varepsilon \mathbf{E} = \varepsilon_0 \mathbf{E} + \mathbf{P}$, and applying it to two situations: tile present and tile absent, we arrive at the following expression for the permittivity difference times thickness in the numerator of Eqn.17,

$$(\varepsilon_{r\ell} - \varepsilon_{rm})\, t = \frac{P_\ell - P_w}{\varepsilon_0 E} t = \frac{P_{ex}}{\varepsilon_0 E} t = \frac{p_{ex}\, t}{\varepsilon_0 E\, v_t} = \frac{p_{ex}\, \sigma}{\varepsilon_0 E} = \frac{\alpha_{ex}\, \sigma}{\varepsilon_0} \tag{18}$$

where p_{ex} is the excess dipole moment for a tile, and v_t is the tile volume. Using the expression at furthest right in Eqn.18 we obtain

$$\frac{\delta\lambda}{\lambda} = \frac{\alpha_{ex}\sigma}{\varepsilon_0(n_s^2 - n_m^2)a} \tag{19}$$

This equation is perfectly correct for a voidless layer which is considerably thinner than the evanescent field depth. However, it is approximately true also for a dilute layer so long as it is uniformly distributed and one can neglect local field effects (i.e. depolarization effects which are always present for partially isolated nanoparticles). It turns out that it is a good approximation for a dilute layer of particles considerably smaller in size than the microsphere (< 50 nm on a microsphere 100μm in radius). Under these circumstances the shift will be proportional to the surface concentration as indicated.

Finally we are interested in the possibility for single protein detection. Single protein detection would be possible by looking at steps in the change of $\delta\lambda/\lambda$ with time, and this in turn provides a possible means for separately measuring α_{ex}. Since the light within a WGM circumnavigates the equator ($\theta = \pi/2$) in an orbit which is confined to a thin ring, molecules at polar angles outside the ring cannot influence the mode frequency. The greatest signal comes from molecules which stick at $\theta = \pi/2$. For a TE mode which circulates at the equator $\ell = m$, and the angular intensity is proportional to $|\hat{L}Y_{\ell\ell}|^2$, which for large ℓ is proportional to $|Y_{\ell\ell}|^2$.[11] So the ratio of the frequency shift for a protein at the equator to that averaged over random positions on the surface is enhanced by a factor $EF = 4\pi|Y_{\ell\ell}(\pi/2,\varphi)|^2$. This spatial enhancement EF can be significant. For the average size microparticle anticipated (a ~ 100μm) , $\ell \sim 1000$ and $EF \cong 36$. To obtain the average shift for an individual protein at a random position, we set the surface density in Eqn. 19 to $\sigma = 1/(4\pi R^2)$ with the result $(\delta\lambda/\lambda)_r = \alpha_{ex}/[4\pi\varepsilon_0(n_s^2 - n_m^2)R^3]$. The shift due to a single protein at the equator is $(\delta\lambda/\lambda)_e = EF \times (\delta\lambda/\lambda)_r$, or

$$(\delta\lambda/\lambda)_e = \frac{\alpha_{ex}|Y_{\ell\ell}(\pi/2,\varphi)|^2}{\varepsilon_0(n_s^2 - n_m^2)a^3}. \tag{20}$$

The first test of all of this theory requires attaching biological nanoparticles to the microsphere and tracking dips in resonances.

6. Experimental Setup

To carry out our perturbation experiments it is necessary to construct a situation similar to the concept in Figs.4 and 5. This requires forming a microsphere and an eroded optical fiber, and touching one against the other while measuring the excitation spectrum of the transmitted light through the fiber. The entire system will be immersed in a buffer solution (pH = 7.4) where protein molecules can fold into there natural biological shape. In what follows we will briefly describe the microsphere and fiber fabrication, the sample cell, and our means for taking high resolution spectra of the microsphere.

To fabricate a microsphere one can simply cut a small portion of the single mode fiber (about 10 cm) and then melt the stripped end of the fiber by exposing it to the butane/nitrous oxide microtorch flame (Microflame, Inc.), while rotating the fiber. As the fiber rotates the melting silica will form into a spheroidal shape under the force of surface tension. A spheroid manufactured in this form can have a radius as small as 70 μm. Such a microspheroid is shown in Fig.6b. The excitation fiber is led from the laser source to the detector and held in place while a 1 cm section is eroded using 25% HF acid. The erosion is terminated when this local portion of the fiber reaches 4 μm in diameter. The fiber can be seen in Fig.6b as what appears to be a narrow vertical line in contact with the microspheroid.

The sample cell shown in Fig.6a is used to contain 1 mL of aqueous buffer solution between the two glass slides (top and bottom) by means of surface tension. It is assembled around the the etched portion of the fiber. Two silicone rubbers are placed on each side of the lower glass slide, and the stem of a surface modified microsphere (note: the preparation of the surface will be described) is held between silicone strips on the right side as illustrated in Fig.6a. The opossite side of the cell holds a thermocouple in the same fashion.

The light source has stringent requirements. It must operate at a single frequency with a linewidth smaller than that of a microsphere resonance, and it must be tunable. At first these requirement may seem simple enough until one looks at the specifics. The linewidths of the resonances which are to be interrogated may be as small as 0.00001nm, hardly a width that can resolve with a source consisting of an arc lamp followed by a grating monochromator. For such a source a linewidth below 0.01nm would be challenging to produce. Fortunately the telecommunication community has produced small semiconductor laser with linewidths in line with our needs.[12] However simple heterojunction lasers cannot be tuned without mode hopping, which is unacceptable. To remedy this situation a Bragg grating is imprinted within the laser and resulting device is

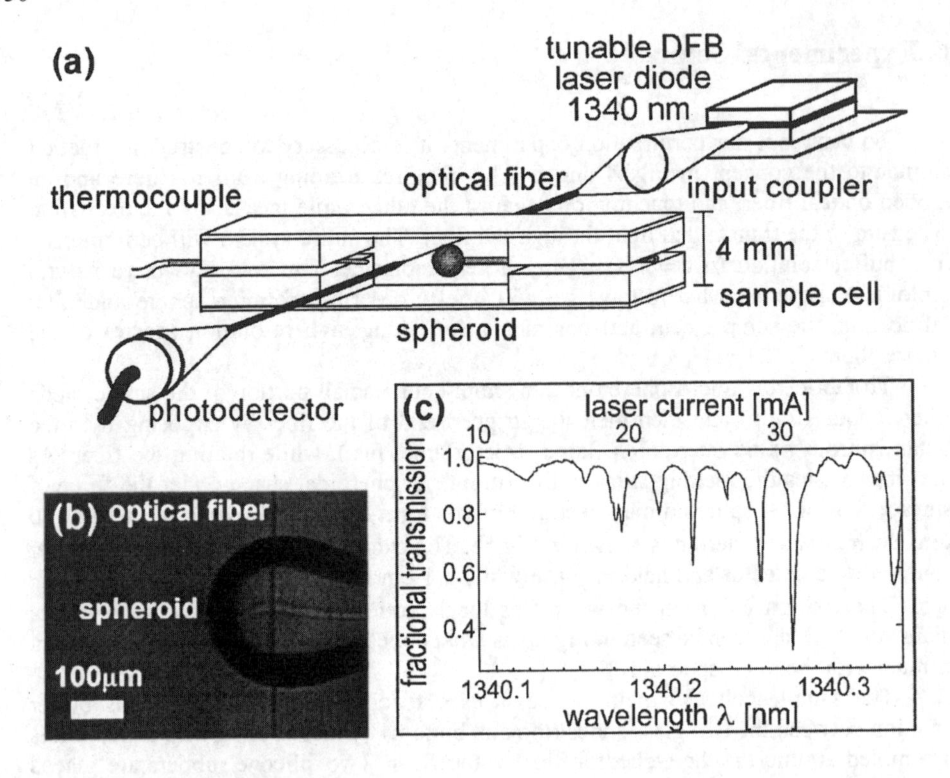

Fig.6 (a) Experimental setup. Optical resonances of a spheroidal glass microparticle are excited by coupling the spheroid evanescently to an eroded single mode optical fiber. (b) Picture of the spheroid coupled to the eroded part of the optical fiber. (c) Resonances dips vs. wavelength.

termed a Distributed Feedback Laser (DFB). The DFB can be easily tuned by changing the drive current or the temperature. For our purposes the temperature was held constant in order to avoid time delays in tuning. The current tuning coefficient, which is ~0.01nm/mA is sufficient to locate a resonance in a slightly spheroidal glass microstructures with a radius of a couple of hundred micrometers. Since our research is highly dependent on precise knowledge of the tuning coefficient, measurements independent from the specifications provided by the manufacturer were needed. These measurements were obtained using a scanning Michelson Interferometer.

Fig. 6c shows a typical spectrum of the microsphere while immersed in the buffer solution. Note that the entire sweep of the laser is only ~0.2 nm, but the resonance lines are considerably narrower ($Q \sim 10^6$). Also note that one of the dips in transmission subtracts 70% from the incident light.

7. Experimental Results

To test our perturbation theory we have chosen to use protein molecules. Typically they are a few nanometers in size. An example of an abundantly available protein is Human Serum Albumin (HSA). In fact, it is the most abundant protein in our blood. The bovine form of the protein (BSA) is available in high purity, and has a known crystal structure[13]. The surface of our microsphere must be treated in order to make the protein "stick".

In our buffer solution BSA carries a negative charge. To promote adsorption to the microsphere the sphere's surface is modified so as to give it a positive charge. To give the microsphere surface a positive charge it is treated with a solution 3-aminopropyltrimethoxysilane. This compound reacts with the surface and self assembles a "rug" of amine groups (NH_2,Fig. 7).

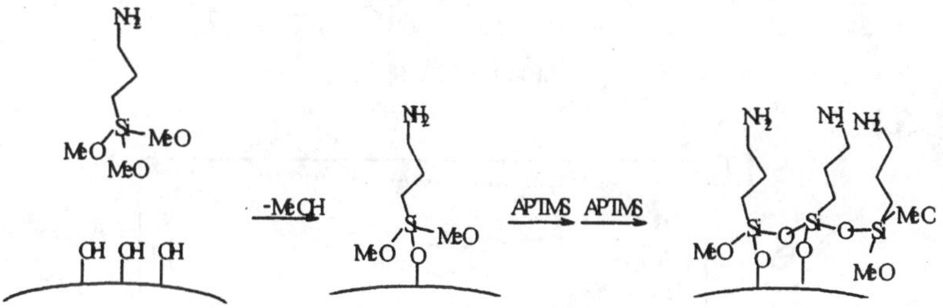

Fig. 7 Chemical action of the silane coupling agent.

The amine groups acquire a positive charge in the buffer solution. So BSA sticks readily to this amine "rug".

BSA is added to the sample cell by injecting a 10 μL of a BSA buffer solution into the sample cell. The expectation for our first experiment was an increase in the wavelength of a resonance. Fig.8a shows that the effect initially went in the opposite direction.

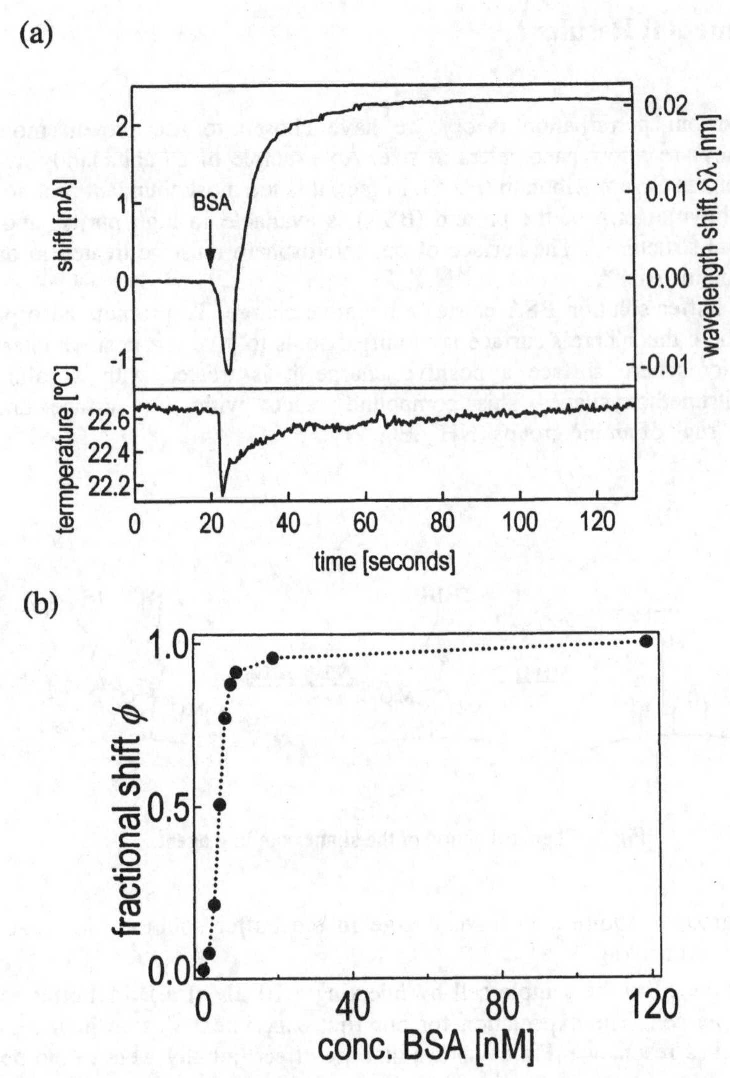

Fig.8 BSA adsorption measurements. (a) shows the shift of the microsphere resonance in real time. (b) adsorption isotherm.

Turns out that the student had retrieved the solution from a cooler so that the overall solution got a bit cooler upon injection, and the microsphere shrunk by a few tenths of a nanometer. This was evidenced by the slight drop in temperature recorded by the thermocouple (lower plot in Fig. 8a). With a little time for equilibrium to be re-established the negative wavelength shift began to recover and the experiment recorded an overall positive shift in wavelength of 0.02 nm. The BSA had adsorbed. The lower plot shows an isotherm for the process. Basically different concentrations of BSA were injected. The threshold for seeing a change was unprecedentedly small for unlabelled adsorption (<1 nM). Although the experiment in Fig.8a was carried out for a solution concentration of 2μM. Fig.8b clearly shows that the adsorption process saturated above ~20nM. The shape reflected here is called a Langmuir isotherm. Where it flattens there is 100% coverage. That does not mean there are no voids, however before dealing with this question we want to show the dependence of the saturation shift on the radius of the microsphere.

Our theoretical results for for the fractional shift in wavelength at saturation (Eqns. 17 and 19) show a distinct inverse size dependence on microsphere radius. Fig.9 displays the measured size dependence.

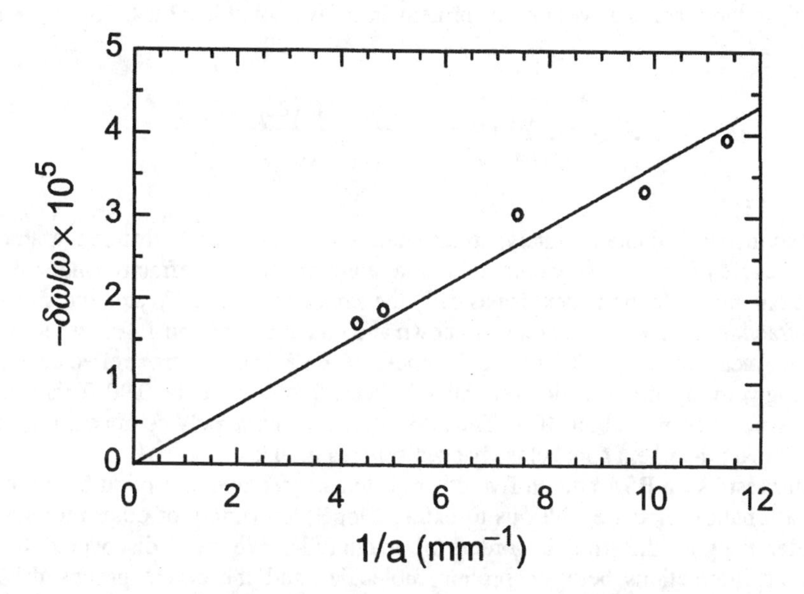

Fig.9 The dependence of the shift in resonance frequency on microsphere curvature.

The theory in this respect is clearly borne out. Furthermore, the slope of the plot (3.6 nm) can be compared directly with theory. From Eqn. 17 the slope is just the fractional shift times the microsphere radius, or more precisely

$$\text{Slope in Fig.9} = -a\frac{\delta\omega}{\omega} = \frac{(\varepsilon_{r\ell} - \varepsilon_{rm})}{(\varepsilon_{rs} - \varepsilon_{rm})}\,t\,. \tag{21}$$

If the permitivity of the layer were the same as that of silica, then the slope would be the thickness of the layer. In fact the permitivity of protein is known to be slightly larger than silica. So the thickness of the layer may be expected to be less than 3.6 nm. Since BSA is known from x-ray crystallography to be a thick heart shaped pancake with a dimension in the plane of ~6 nm and a thickness between 3, and 4 nm our measurement leads us to the conclusion that the largest area of the molecule is against the surface of the silica. This configuration should minimize the coulombic energy between the positively charged amine rug and the negatively charged protein. This conclusion is in agreement with recent neutron reflection experiments which also show that BSA flattens to a 3 nm thickness due to the coulombic tug of the surface.[14]

Based on our existing data and the neutron reflection results we can take our analysis one further step. Since we expect voids to exist, and the wavelength is considerably larger than a molecule, we will express the permitivity of the layer in terms of a mean field approximation.

$$\varepsilon_{r\ell} = f_p \varepsilon_{rp} + (1 - f_p)\varepsilon_{rm} \,. \tag{22}$$

where f_p is the fractional volume of protein in a layer of thickness t. Joining Eqn.22 with Eqn.21,

$$\text{Slope in Fig.9} = -a\frac{\delta\omega}{\omega} = \frac{f_p(\varepsilon_{rp} - \varepsilon_{rm})}{(\varepsilon_{rs} - \varepsilon_{rm})}\, t \,. \tag{23}$$

A protein molecule contains a polypeptide chain which includes 21 different amino acids. Because of this statistical mix of similar components, the refractive index from one protein molecule to the next is basically the same, $n_p \cong 1.50$.[15] Water and silica have refractive indices which are well known and equal to 1.33 and 1.46, respectively. Using our measured slope (3.6 nm), a thickness of BSA from neutron reflectance (3.0 nm) along with the aforemention refractive indices, f_p is found to be 0.90. This means that the void fraction is about 10%. Thus BSA forms an extremely compact monolayer. More compact than hexagonal close packed spheres in 2D.

Our results on BSA and on five other proteins agree with our perturbation model. This perturbation model enables us to extract slopes for variety of other proteins with molecular weights differing by orders of magnitude. We have discovered that the coulombic interactions between protein molecules and the amine groups drive the adsorption of molecules in a particular way, such that proteins deform in a similar aspect ratio (thickness divided by the square root of surface area occupied by protein) upon adsorption.[16] It appears that the Photonic Atom adsorption sensor is an extremely sensitive realtime probe to nanoscopic layers. However, this lesson thus far has only scratched the surface.

The Photonic Atom sensor also acts as a biosensor. It can be used to identify the adsorbing molecule. It is important to note here that the means for doing this is not through conventional spectroscopy. Although chemists and physicists have traditionally learned to identify atoms and molecules by optical spectroscopy (absorption, fluorescents, etc.), there is not light deep in our cells. Yet our cells recognize tens of thousands of different proteins well enough to enable us to function. Much of this recognition is through shape complimentarity aided by sticky physio-chemical interactions. For example in a sensitized person, the allergic reaction to certain toxic proteins on the surface of pollen grain, causes a specific antibody to engulf the invading allergen like a lock covering a key. This highly specific physio-chemical recognition will not occur with other proteins. Another example is the hybridization that occurs between complimentary strands of DNA. If we mimic biology, we would also sense biomolecules through dark interactions. The Photonic Atom Biosensor is ideal for this purpose. Simply by attaching a biorecognition element to its surface, adsorption becomes specific For example. antibody molecules can be attached for the detection of specific antibody-antigen interactions, or single stranded DNA for DNA detection through hybridization. This is not "pie in the sky". The experiments have already been done with the Photonic Atom Biosensor.[4,5]

There is much more to do. Perturbation theory will not simply stop at first order for the "light atom", just as it did not stop at first order for the electronic atom. In addition there is a great deal to be done concerning single molecule detection.

Acknowledgements

Steve Arnold thanks Rino DiBartolo for his kind invitation to lecture at this school and for re-introducing him to the land of Rino's birth.

This was the second time Steve attempted to make a course out of his research. By learning from those who have done this in the past years, he gained a great deal of insight into the technique. Steve is particular indebted to Eric Mazur and Ralph von Baltz for their encouragement.

Without the prodding and encouragement from Ottavio Forte this manuscript would not have been completed on time. Bravo Ottavio.

Finally Mayumi Noto would like to thank the National Science Foundation for her fellowship and S. A. and M. N. would like to thank NASA for supporting work in thermal sensing using Photonic Atoms.

References

1 S. Arnold, Spontaneous Emission within a Photonic Atom: Radiative Decay Rates and Spectroscopy of Levitated Microspheres, in *Spectroscopy of Systems with Spatially Confined Structures*. Ed. by Baldassare DiBartolo (Kluwer Academic Publisher, 2002) pp.465-488

2 S.Arnold, M.Khoshsima, I.Teraoka, S.Holler, F.Vollmer, "Shift of whispering gallery modes in microspheres by protein adsorption," Opt. Lett. **28**, 272(2003).

3 I. Teraoka, S. Arnold, and F. Vollmer , "Perturbation approach to resonance shifts of whispering-gallery modes in a dielectric microsphere as a probe to the surrounding medium," J. Opt.Soc.Am.B **20**, 1937(2003).

4 F. Vollmer, D. Braun, and A. Libchaber, M. Khoshsima, I. Teraoka, and S. Arnold, "Protein detection by optical shift of a resonant microcavity," Appl. Phys. Lett., **80**, 4057(2002).

5 F. Vollmer , S. Arnold, D. Braun , I. Teraoka, A. Libchaber , "DNA detection from the shift of whispering gallery modes in multiple microspheres," Biophysical Journal **85**, 1974(2003).

6 H.M. Nussenzveig, Comments Atomic and. Mol.Phys. **23**, 175(1989).

7 J.C. Knight, G. Cheung, F. Jacques, and T.A. Birks, "Phase-matched excitation of whispering-gallery-mode resonances by a fiber taper," Opt. Lett. 22, 1129(1997).

8 D.Q. Chowhury, S.C. Hill, and M.M. Mazumder, IEEE J. Quantum Electron. **29**, 2553(1993).

9 M. Khoshsima, *Perturbation of Whispering Gallery Modes in Microspheres by Protein Adsorption: Theory and Experiment*, PhD thesis, Polytechnic University, Jan. 2004.

10 L.M. Folan, "Characterization of the accretion of material by a microparticle using resonant ellipsometry", Appl. Optics **31**, 2066(1992).

11. J. D. Jackson, *Classical Electrodynamics*, Wiley, New York (1962), p. 753.

12 M-C Amann, J. Buus, *Tunable Laser Diodes* [Artech House Optoelectronics Library, (1998)]

13. D.C. Carter, X. M. He, S. H. Munson, P.D. Twigg, K. M. Gernet, M. B. Broom, and T.Y. Miller, "Three-dimensional structure of human serum albumin," Science **244**, 1195(1989).

14 T. J. Su, J. R. Lu, R. K. Thomas, and Z. F. Cui. "The effect of pH on the adsorption of bovine serum albumin at the silica/water interface studied by neutron reflection," J. Phys. Chem. B. **103**, 3727(1999).

15 P.A. Cupers, W. Th. Hermens, and H.C. Hernker, "Ellipsometry as a tool to study protein films at a liquid-solid interface", Biochemistry **84**, 56(1978).

16 M. Noto, M. Khoshsima, G. Guan, D. Keng, and S. Arnold, "Molecular mass sensitivity of a whispering gallery mode biosensor," submitted.

10. LUMINESCENCE PROPERTIES OF VERY SMALL SEMICONDUCTOR PARTICLES

CEES RONDA
Philips Research Laboratories

P.O. Box 500145
D-52085 Aachen
Germany
e-mail: cees.ronda@philips.com

Utrecht University, Debye Institute

P.O. Box 80.000
3508 TA Utrecht
the Netherlands
e-mail: c.r.ronda@phys.uu.nl

1. Introduction

In this chapter, we deal with optical properties of quantum dots. These are particles in which electronic properties are depending on the size of the particles. Such quantum dots are very small; generally size effects are expected only for particles smaller than about 10nm.
We will start with explaining the interest in these particles, for didactical reasons. After having gained the attention of the interested reader this way, we will treat the basic quantum mechanical properties of electrons (and holes) in these very small particles. We will describe the properties of a particle in a potential well. In addition, we will briefly touch upon the quantum mechanical description of atoms, taking the simplest ion, the hydrogen atom as example. Then we will treat the properties of electrons in an infinite crystal. Finally we will treat the elementary properties of coupled electrons and holes.
In the next section, we will deal with the quantum mechanical properties of quantum dots, both in the weak and strong confinement limit
In the last section, the manifestation of quantum confinement in optical absorption and luminescence will be elucidated.
This contribution ends with a summary and an outlook.
Some of the material presented here has also been presented in [1], which deals with a very similar topic. In this paper, some recent results are added.

2. Some possible application areas of very small semi-conductor quantum dots

A very prominent feature of quantum dots is the dependence of the optical properties of quantum dots on the dot size: the optical absorption and the emission spectra depend are

B. Di Bartolo and O. Forte (eds.), *Frontiers of Optical Spectroscopy,* 359-393.

material emitting in different colours by solely changing particle size. This property gained much interest, both in industrial and academic research laboratories: dependent on the particle size distribution, relatively narrow emission bands can be observed, enabling to tune the emission bands to spectral regions, optimally adapted to the application in contrast to e.g. line emission generated by rare-earth ions of which the spectral position of the individual emission lines is more or less fixed.

As quantum dots are much smaller than the wavelength of visible light, at least in principle non-scattering emitting layers can be made. This can have interesting applications in f.i. displays, in which scattering induces light loss.

Very small particles also can be used in electroluminescent devices: due to their very small size, both electrons and holes can be injected into the quantum dots without the need of long range electrical conductivity in the nano-crystalline material. Radiative recombination then results in the generation of light.

A schematic example of such a structure is given in fig. 1.

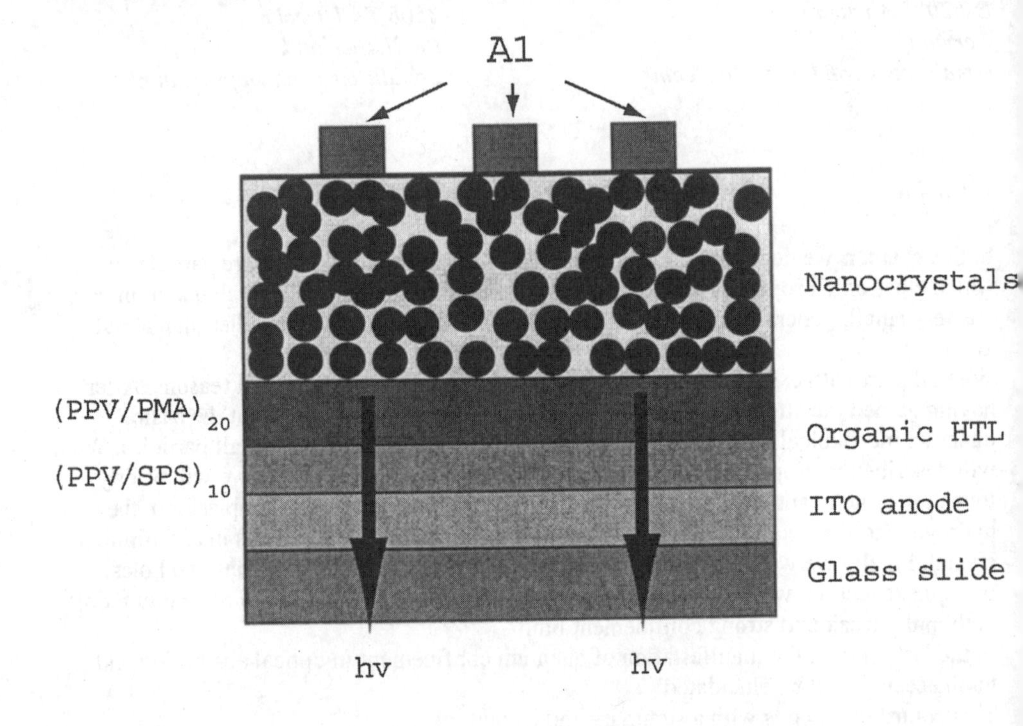

Figure 1: Sketch of an electroluminescent cell based in which emitting nano-crystals are used. The organic layers are applied by layer-by-layer sequential absorption.

Al serves as cathode, ITO as anode and there are two hole-transporting layers, to optimise charge injection into the nano-particles. In this particular structure, the nano-crystal layer itself also serves as electron transporting layer. PPV is the abbreviation for polyphenylenevinylene, PMA means polymethacrylic acid and SPS is short for sulfonated polystyrene. Glass is used as material carrying the electroluminescent structure. The injection efficiency is dependent on the size of the nano-particles as the position of the first excited state and the ground state of the small particles, relative the valence band and conduction band of the electron transport layer varies as a function of the size of the nano-particles.

As the excitation voltages can be very low (typically less than 10 V), the energy efficiency of such electroluminescent cells can be very high.

The normalised electroluminescence and photoluminescence spectra of CdSe nano-particles as a function of the particle size are given in fig. 2. We observe a clear dependence of the emission wavelength on particle size. The emission in the green part of

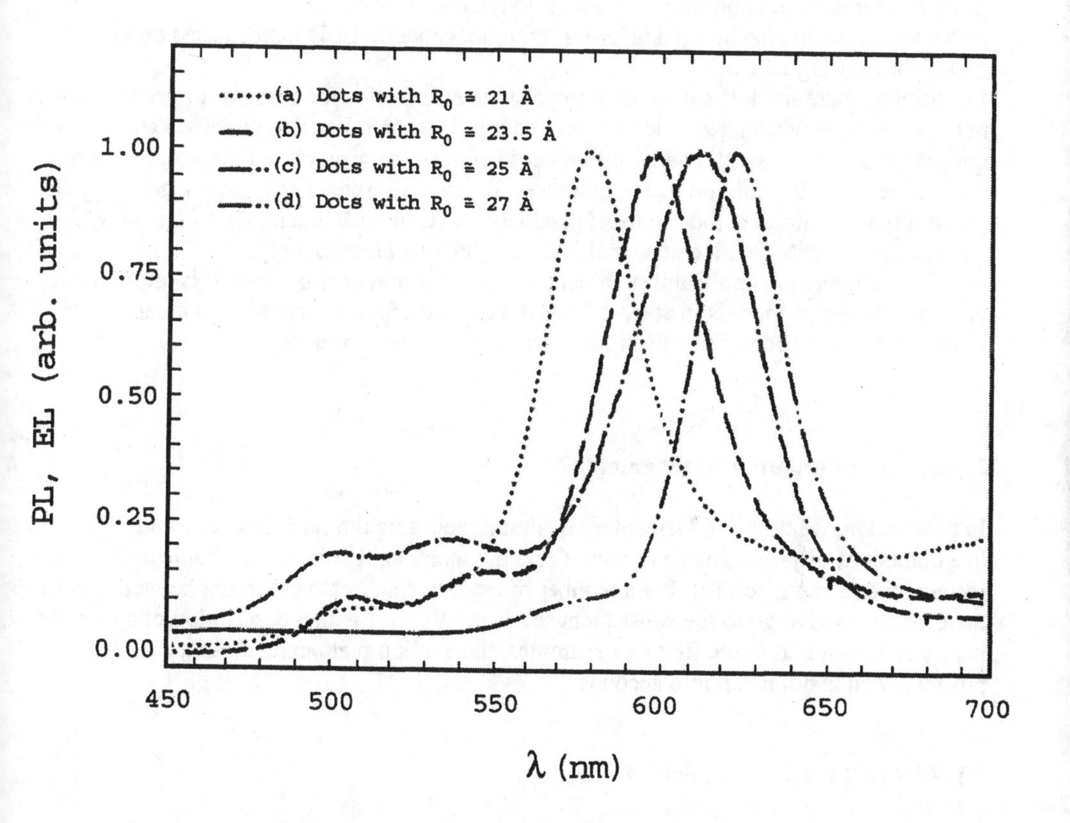

Figure 2: Normalised electroluminescence (and photoluminescence) of CdSe nanoparticles as a function of the particle size.

the spectrum is originating from the hole-transporting layer (PPV in this case). For smaller particles, no EL is observed due to complete quenching of the emission at the surface. The onset value for EL is about 3 – 3.5 V, in principle enabling an energy efficiency larger than 50%. The external quantum efficiency (photons per electron) typically is 0.1%, however.

Apart from the intrinsic efficiency of the quantum dots and the quality of the interface, which determines the voltage needed to generate luminescence, the efficiency of EL devices is also determined by outcoupling efficiencies. In addition, the lifetime of such a device is important for practical applications.
As an alternative, quantum dots can be incorporated in e.g. an organic electroluminescent layer. Energy transfer of excitons in the organic electroluminescent layer to the quantum dots on the one hand likely stabilises the organic material and on the other hand generates an extra degree of freedom to adjust the emission colour.
Nano-particles can also be used in LEDs, f.i. to make white LED lamps based on UV or blue light emitting LEDs.
In addition, quantum dots can be used to generate efficiently IR emission. In general, this is not easy using emission generated by ions, as small lattice relaxation already leads to quenching in case of small energy differences between the ground- and the excited state.
Finally, very small semi-conductor nano-crystals have discrete energy levels, enabling the possibility to manipulate the electrical conductivity (Coulomb blockade). In this way, switches can be made with can be manipulated with one electron only.

To judge the applicability of nano-particles in practical devices, it is important to understand their properties in at least some detail. Therefore, we continue with an elementary description of the electronic properties of quantum dots.

3. Elementary quantum mechanics

In this section, we describe elementary quantum mechanical treatments describing particles in a number of different environments. There are many textbooks on this subject. A recommended one is ref. [2]. For a number of readers, this treatment might be well known; these are advised to go to the subsequent sections. We start with a description of a particle in potential well and in the first two examples, the explicit mathematical shape of the potential well is not taken into account.

3.1. PARTICLE IN A POTENTIAL WELL

For this system, the time independent Schrödinger equation is given by:

$$-(\hbar^2/2m)\partial^2\psi(x)/\partial x^2 + U(x)\psi(x) = E\psi(x) \tag{1}$$

in this equation, m is the mass of the particle, the potential is given by U(x) and E is the energy of the particle with wave function $\psi(x)$.
First we describe the case of a well with width a and with a potential $U(x) = 0$ for $|x| = \leq a/2$ and infinite otherwise. The solutions of equation (1) are even and odd types of expressions (by n, the quantum number, see below):

$$\psi_{odd} = \sqrt{(2/a)} \cos (1/\hbar . \sqrt{(2mE)}x) \quad (2)$$

for odd expressions and

$$\psi_{even} = \sqrt{(2/a)} \sin (1/\hbar . \sqrt{(2mE)}x) \quad (3)$$

for even expressions

These solution are found for $|x| \leq a/2$. Outside this range, $\psi(x) = 0$. The corresponding discrete set of energy levels is given by:

$$E_n = (\pi^2 \hbar^2 / 2ma^2)\, n^2 \quad (4)$$

which means that in this case, the energy levels can be described by only one quantum number n.

The energy separation between two subsequent levels is given by:

$$E_n = \pi^2 \hbar^2 (2n+1) / 2ma^2 \quad (5)$$

Until now, we have calculated values for the kinetic energy of a particle in potential well. As the kinetic energy, momentum p and wave number k are related:

$$E = p^2 / 2m; \qquad p = \hbar k \quad (6)$$

and therefore:

$$p_n = (\pi \hbar / a)n \; ; \quad k_n = (\pi / a)\, n \quad (7)$$

we learn that these quantities also take discrete values.
The wave functions vanish at $x > a$. When a particle exists in the well, the product $\psi\psi^*$ must be nonzero somewhere. This excludes $n = 0$. The minimum energy of a particle is therefore nonzero and given by:

$$E_1 = (\pi^2 \hbar^2 / 2ma^2) \quad (8)$$

Please note that equation (8) can also be derived (apart from a constant) from the Heisenberg uncertainty relation.
For potential wells with walls with a finite height, the nature of the solutions obtained are almost the same, there are, however, a few differences. Above a certain value for the kinetic energy (U_o, corresponding to the height of the well), the states form a continuum, which corresponds to continuous motion. In addition, the probability to find a particle outside the box is larger than zero and the probability increases with increasing n. The number of states inside the well is given by the following expression:

$$a\sqrt{(2mU_o)} > \pi \hbar (n-1) \qquad (9)$$

for n = 1, this condition always holds and therefore, there is at least one state inside the well. The number of states within the well corresponds to the value for n for which equation (9) still holds.
For a particle in a potential well, the dispersion relation (which gives the kinetic energy of the particle as a function of k) consists of points on a parabola, as the energy increases with k^2. The dispersion relation is given by (equations (6)).

$$E = \hbar^2 k^2/2m \qquad (10)$$

Below U_o, again only discrete points are possible, as is the case for any energy level in the case of a particle in a potential well with infinitely high walls. Above U_o, any value of k is possible, and the dispersion curve is a continuous curve, quadratically dependent on k.
In fig. 3, the results obtained are summarized.

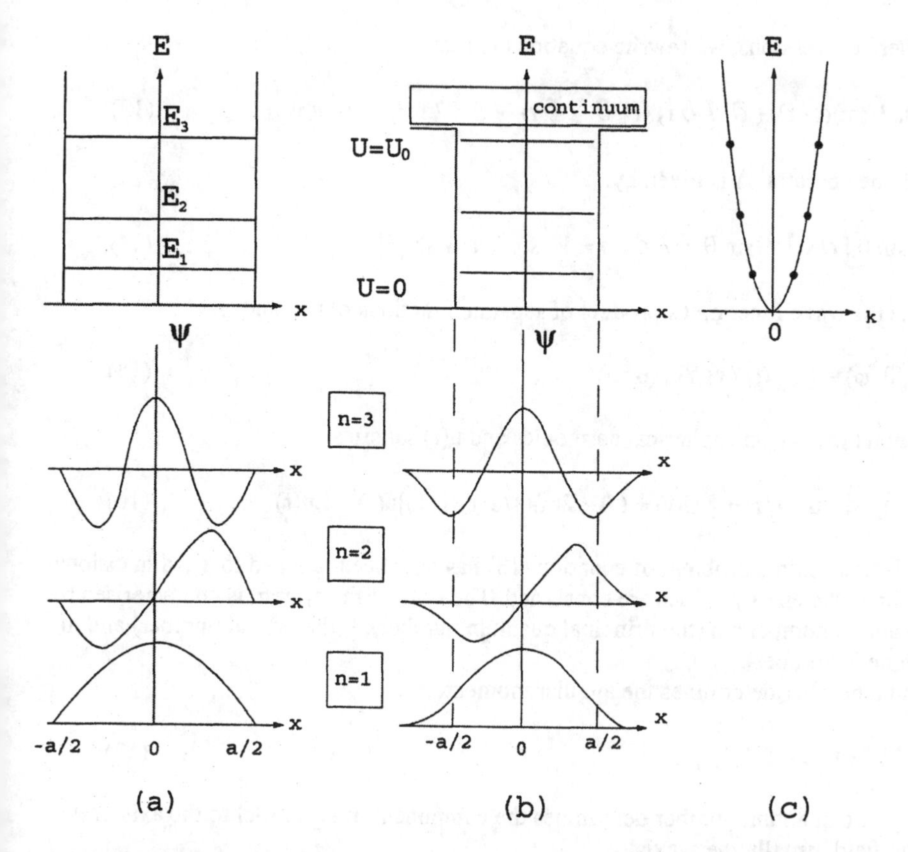

Figure 3: Energy levels and wavefunctions for: a) particle in a well with infinitely high walls; b) particle in a well with finite height; c) gives the dispersion curve for a free particle. The dots on this curves are discrete energy values for a particle in a box and in case of a finite well, above U_0 the dispersion curve for a free particle is obtained.

3.2. PARTICLE IN A SPHERICALLY SYMMETRIC POTENTIAL

In this case, it is convenient to write the Hamiltonian as:

$$H = -(\hbar^2/2m)\nabla^2 + U(r) \tag{11}$$

in which $r = \sqrt{(x^2 + y^2 + z^2)}$. As the system has spherical symmetry, we describe the system in spherical co-ordinates:

$$x = r \sin\theta \cos\varphi;\ y = r \sin\theta \sin\varphi \text{ and } z = r \cos\varphi \tag{12}$$

Using these expressions, we rewrite equation (11) as:

$$H = -(\hbar^2/2m)(1/r^2)(\partial/\partial r)(r^2 \partial/\partial r) - (\hbar^2/2mr^2)\Lambda + U(r) \quad (13)$$

in which the operator Λ is given by:

$$\Lambda = 1/\sin\theta\,[\partial/\partial\theta\,(\sin\theta\,\partial/\partial\theta) + 1/\sin\theta\,\partial^2/\partial\varphi^2] \quad (14)$$

We write the wave function as product of separate functions of r, θ and φ:

$$\psi_{n.l.m}(r, \theta, \varphi) = (u_{n,l}(r)/r)\,Y_{lm}(\theta, \varphi) \quad (15)$$

In equation (15), Y_{lm} are spherical harmonics and u(r) satisfies:

$$-(\hbar^2/2m)\,\partial^2 u/\partial r^2 + [\,U(r) + (\hbar^2/2m)(1/r^2)\,l(l+1)]u(r) = Eu(r) \quad (16)$$

So, the 3-dimensional problem of equation (13) has now been reduced to a 1-dimensional one, as far as the energy values are concerned. The state of the system is characterised by three quantum numbers: n (the principal quantum number), l (the orbital number) and m (the magnetic number).
The orbital number determines the angular momentum L:

$$L^2 = \hbar^2 l(l+1),\ l = 0,1,2,3,.... \quad (17)$$

The magnetic quantum number determines the component of L parallel to the axis of the magnetic field, usually the z-axis:

$$L_z = \hbar m,\ m = 0, \pm 1, \pm 2, ... \pm l \quad (18)$$

The states with different l values are usually denoted as s, p, d, f,..... states. Every state l is (2l+1) fold degenerate, as follows from (18). The parity of the states is given by l: the radial part of the wave function is not sensitive to inversion of r and the spherical functions $Y_{lm}(\theta, \varphi)$ transform as:

$$Y_{lm}(\theta, \varphi) \rightarrow (-1)^l\,Y_{lm}(\theta, \varphi) \quad (19)$$

Taking again a potential well with an infinite barrier, as in the previous section, we obtain for the energy values of this system:

$$E_{n,l} = (\hbar^2/2ma^2)\,\chi^2_{nl} \quad (20)$$

Where χ_{nl} are roots of the spherical Bessel functions with n the number of the root and l the order of the function. For l=0, equation (16) is equal to equation (1).

When the potential well is finite with potential U_o, equation (20) is a good approximation only for:

$$U_o >> (\pi^2 \hbar^2 / 8ma^2) \qquad (21)$$

The right side of this equation simply follows from the Heisenberg uncertainty relation:

$$\Delta p \Delta x \geq \hbar/2 \qquad (22)$$

with $\Delta x = a$, it follows for the energy:

$$\Delta E = \Delta p^2/2m = \hbar^2/8ma^2 \qquad (23)$$

Only for values of U_0 much larger than given by the uncertainty principle, the values of the energy $E_{n,l}$ are determined by the system.
The smallest value for the energy is obtained for the state with l = 0 and n= 1. For this case, the energy $E_{1,0}$ is given by:

$$E_{1,0} = (\pi^2 \hbar^2 / 8ma^2) \qquad (24)$$

For $U_o < (\pi^2 \hbar^2 / 8ma^2)$, no state exists within the well, in contrast to the one-dimensional problem.
Until now, we have derived our equations without knowing exactly the form of the potential. Solutions were nevertheless obtained which were found to depend on the system chosen. For a particle in a one dimensional quantum well, the state of the system can be described by one quantum number only, for a particle in a spherically symmetric potential, three quantum numbers are needed. In the next part we will extend our treatment with a known potential.

3.3. ELECTRON IN A COULOMB POTENTIAL

The Coulomb potential is given by (in all derivations, the quantity $1/(4\pi)$ has been omitted):

$$U(r) = - e^2/r \qquad (25)$$

The equation for the radial part of the wave function can be written as:

$$[\partial^2/\partial r^2 + \varepsilon + 2/\rho - l(l+1)/\rho^2]\, u(\rho) = 0 \qquad (26)$$

In this equation, dimensionless arguments for distance and energy are used:

$$\rho = r/a_o\ ;\ \varepsilon = E/E_o \qquad (27)$$

$a_o = \hbar^2/(m_o e^2) \approx 5.29.10^{-2}$ nm (m_o being the electron mass) and

$E_o = e^2/2a_o \approx 13.6$ eV

Equation (26) has as solution:

$$\varepsilon = -1 / (n_r + l + 1)^2 \equiv -1/n^2 \qquad (28)$$

The number $n = n_r + l + 1$ is the so-called principal quantum number. It has as minimum value 1. n_r determines the number of nodes of the corresponding wave function. For any value of n, n states exists, which differ in l and l runs from 0 to (n-1). In addition, for every l value (2l+1) degenerate states occur, with respect to m = 0, ± 1, ± 2, ... ± l. The total degeneracy is given by:

$$\sum_{l=0}^{n-1} (2l + 1) = n^2 \qquad (29)$$

For n=1 and l=0 (1s-state), the wave function has spherical symmetry with a_o corresponding to the most probable distance (from the centre from which the Coulomb potential originates) where the electron can be found (Bohr radius in atoms). For E > 0, the particle has infinite motion with a continuous spectrum.
So far, we have dealt with one-particle problems. The simplest real quantum mechanical topic that can be treated is that of the hydrogen atom: a particle with a positive charge (proton, mass M_o) and a particle with a negative charge (electron, mass m_o).
The Hamiltonian, describing this system is a two-particle equation and therefore consists of three terms, one term for each particle and a term describing the interaction between the particles:

$$H = -\,[(\hbar^2/2M_o)\nabla_P^2 + (\hbar^2/2m_o)\nabla_e^2 + e^2/|r_P - r_e|] \qquad (30)$$

In this equation, the proton and electron radius vectors are given by r_P and r_e, respectively.

In what follows, we use r for $r_P - r_e$ and R for:

$$R = (m_o r_e + M_o r_p) / (m_o + M_o) \qquad (31)$$

For the masses, we write

$$M = m_o + M_o\ ;\ \mu = m_o\, M_o/(\, m_o + M_o) \qquad (32)$$

We now read equation (30) as:

$$H = -\,[(\hbar^2/2M)\,\nabla_R^{\,2} + (\hbar^2/2\mu)\,\nabla_r^{\,2} + e^2/r] \qquad (33)$$

Equation (33) now describes a Hamiltonian of a free particle with mass M and a Hamiltonian of a particle
with reduced mass μ in the potential $-e^2/r$. The first term is that of the centre of mass motion of the two particle atom which describes continuous motion, the other terms generate internal states, the energies of which we have derived before:

$$E_n = -\,Ry\,/\,n^2 \text{ for } E < 0 \qquad (34)$$

in which $Ry = e^2/2a_B$ and $a_B = \hbar^2/\mu e^2$ (35)

Ry is the Rydberg constant, which corresponds to the ionisation energy of the lowest state and a_B is the Bohr radius of a hydrogen atom. As derived before, the energy difference between neighbouring levels decreases with increasing n and for E>0 the motion of the electron and proton is continuous.
Equation (27) and (35) differ only by μ/m_e. Please note that equation (27) has been derived for a single particle problem. μ/m_e has, for the hydrogen atom, a value of 0.9995, justifying that for the hydrogen atom also the single particle equations are used frequently.
Although elementary, the equations derived and the procedures used will accompany us further. The single particle problem is important for the description of an electron and a hole in nano particles, whereas the two particle equations are important in the description of excitons. Moreover, our treatment of the two particle system has shown that, by using mass renormalisation, using the reduced mass instead of the individual particle masses can be used to treat the problem as a single particle problem, beit at the cost of differentiation between centre of mass translational motion and single particle motion in an effective field.

3.4. PARTICLE IN A PERIODIC POTENTIAL

For a particle in a 1-dimensional periodic potential with period a, the potential energy U satisfies:

$$U(x) = U(x + a) \qquad (36)$$

This implies for the Schrödinger equation:

$$-(\hbar^2/2m)\partial^2\psi(x+a)/\partial x^2 + U(x)\psi(x+a) = E\psi(x+a) \quad (37)$$

which means that the wave function after a translation over a has the same eigen value E. This, in turn, means that the eigen functions may differ in a constant coefficient only:

$$\psi(x+a) = c\psi(x) \quad (38)$$

In view of normalisation:

$$|c| = 1 \quad (39)$$

and therefore:

$$|\psi(x+a)|^2 = |\psi(x)|^2 \quad (40)$$

This means that the probability to find a particle at an interval Δx near x is the same as near to (x + a). This leads to the, to be expected, result that the average spatial distribution of particles possesses the periodic periodicity of the crystal.
After two translations, we obtain:

$$\psi(x + a_{n1} + a_{n2}) = c_{n1}c_{n2}\psi(x) \quad (41)$$

in which we used: $a_n = na$; n = 1, 2, 3,.........

As

$$a_{n1} + a_{n2} = a_{n1+n2}$$

we find:

$$\psi(x + a_{n1} + a_{n2}) = \psi(x + a_{n1+n2}) = c_{n1+n2}\,\psi(x) \quad (42)$$

as $c_{n1}c_{n2} = c_{n1+n2}$,

equation (42) has as solution:

$$c_n = \exp[ika_n] \quad (43)$$

in which k can adopt all values

The resulting wave functions can be written as:

$$\psi(x) = \exp[ikx]\,u_k(x), \text{ with } u_k(x) = u_k(x + a_n) \quad (44)$$

From this equation, we learn that the eigen function of a Hamiltonian with a periodic potential is a plane wave modulated with a period, which is the same as that of the potential. Bloch obtained this famous result for the first time and the respective wave functions are also called 'Bloch functions'.
We now think a bit about the physical meaning of this result. The period of the potential in real space is a. In reciprocal space, the period is given by $2\pi/a$, as we deduce from e.g. equation (43). This means that wave numbers k_1 and k_2 differ by a value $2\pi/na$, with n being any integer number, are equivalent. Taking the minimum energy at $k = 0$, this means that all physically relevant k values are contained in equivalent intervals given by e.g.:

$$-\pi/a < k < \pi/a;\ \pi/a < k < 3\pi/a, \text{ etc.} \tag{45}$$

Each of these intervals contains all non-equivalent k values. Such an interval is called 'Brillouin zone'. The dispersion curve differs from the one of a free particle and has discontinuities at the points:

$$k_n = (\pi/a)\, n, \text{ with } n = \pm 1, \pm 2, \ldots\ldots \pm n \tag{46}$$

as for these values of k, two standing waves exists with different potential energies. As a result, forbidden energy levels occur for which no propagating waves exist, leading to energy gaps, just being the result of the periodic modulation of the potential in the crystal. If one folds all dispersions curves into the first Brillouin zone, see fig. 4, one obtains the reduced zone schemes, in which the allowed and forbidden energy zones can be observed clearly. Please note that k still takes discrete values.

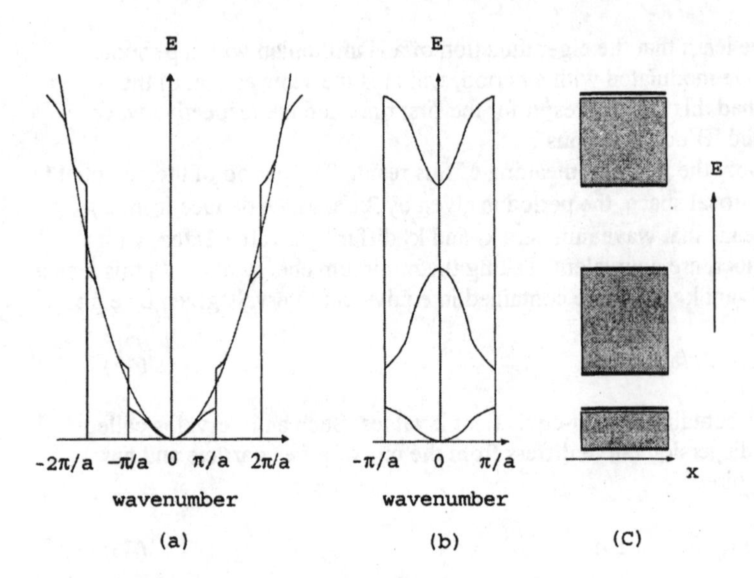

Figure 4: Dispersion curves for particles in a 1-dimensional periodic potential: a) extended zones; b) reduced zone; c) gives resulting energy bands

The quantity

$$p = \hbar k \tag{47}$$

is called quasi momentum. It implies that the quasi momentum is conserved with $2\pi\hbar/a$, which again is a direct consequence of the translational symmetry of the lattice. It differs from momentum due to the conservation law, which finds its origin in the symmetry of the lattice.
For the E(k) relation we now formally write:

$$E = \hbar^2 k^2 / 2m^*(k) \tag{48}$$

Where $m^*(k)$ is usually called effective mass. The effective mass generally is not equal to the rest mass of the particle, due to curvature of the energy bands:

$$E(k) = E_0 + (k-k_o)(dE/dk)|_{k=0} + \tfrac{1}{2}(k-k_o)^2(d^2E/dk^2)|_{k=ko} + \ldots\ldots\ldots\ldots \tag{49}$$

Defining $E_o = 0$ and $k_o = 0$ and looking at the extrema only, which justifies neglecting terms with order higher than k^2, we find:

$$E(k) = \tfrac{1}{2}\, k^2\, (d^2E/dk^2)|_{k=0} \tag{50}$$

This leads to (with equation (48)):

$$1/m^* = 1/\hbar^2(d^2E/dk^2)|_{k=0} \tag{51}$$

This equation strongly resembles the equation for a free particle with mass m as is derived easily from equation (10).

Again formally, Newton's second law can be applied using the effective mass:

$$m^*a = F \tag{52}$$

In the vicinity of a minimum point, the effective mass m* is much smaller than the inertial mass of a particle. In a periodic potential, therefore, a particle can be 'lighter' or 'heavier' than in free space. It can even have a negative effective mass, dependent on the curvature of E(k). The difference in momentum does not vanish, but is transferred to the lattice, adjusting the periodic potential.

4. Electrons in a Crystal

After having treated single particles, we now extend our treatment to a large number of particles in a crystal, i.e. we are treating a system of interacting electrons in a system with translation symmetry.

The Hamiltonian for this system is written as:

$$H = \sum_i \hbar^2/2m\nabla_i^2 - \sum_a \hbar^2/2M\nabla_a^2 + ½\sum_{i\neq j} U_1(r_i - r_j) + ½\sum_{i,a} U_2(r_i - R_a) + \\ + ½\sum_{a\neq b} U_3(R_a - R_b) \tag{53}$$

In this equation, m is the electron mass and M the mass of the nuclei. R is the radius vector of the nuclei and r of the electrons. The last three terms in this equation describe interaction between the electrons, the electrons and the nuclei and between the nuclei, respectively. As the number of particles is in the order of $10^{23}/cm^3$, this equation is not soluble. Therefore a number of approximations is made.
The first approximation is the adiabatic approximation. This approximation relies on the fact that the nuclear mass is much larger than the mass of the electrons. The wave function of the system is then separated into two functions and two Hamiltonians are obtained: one

for the nuclear system and one for the electronic system. We continue with the electron sub system only:

We obtain:

$$H = -\hbar^2/2m \sum_i \nabla_i^2 \psi + \frac{1}{2} \sum_{i \neq j} U_1 (r_i - r_j) \psi + \frac{1}{2} \sum_{i,a} U_2 (r_i - R_a) \psi = E_R \psi \quad (54)$$

In this equation, the nuclear co-ordinates are parameters, not variables.

The second approximation is that core and shell electrons are treated differently. This approximation is justified by the fact that many of the measurable properties of such systems are determined by the valence electrons only (conductivity, optical properties, magnetic properties, etc.). The result is that we now treat ionic cores instead of nuclei.

Using this second approximation, we obtain for the interaction between the electrons:

$$\frac{1}{2} \sum_{i \neq j} U_1 (r_i - r_j) = \frac{1}{2} \sum_{i \neq j} e^2 / |r_i - r_j| \quad (55)$$

If now we take a mean interaction between the electrons and assume a periodic potential, equation (55) reduces to the single-particle equation (equation (11)), which we have already solved. The same kind of solutions is obtained here: we obtain bands, separated by forbidden gaps.
If the bands are partly filled, the material is expected to show metallic conductivity. In e.g. materials with partly filled d-bands, this is not observed. This is due to strong interactions between the d-electrons and here our assumption of a mean interaction is not valid anymore. Systems with completely filled or empty bands are dielectric materials.
The bands which are at least partly filled at T = 0K are called valence bands, the other bands, at this temperature being empty, are called conduction bands. The energy separation between the lowest lying empty band and the highest lying, at least partly, filled band is the energy gap.
When the minimum of the conduction band is observed for the same k value as the maximum of the valence band the material is a direct gap material, otherwise it is indirect. Indirect gap materials might even have a negative value for the energy gap.
The dispersion curve E(k) can be rather complicated in real crystals. By no means, the effective mass can be considered to be constant.
In the next part, we will try to combine the physics derived above with the more intuitive world of chemistry. Both ways of conceptual thinking have there own difficulties and possibilities. Most chemists have a rather localised picture and like to think in real space as chemical reactions take place in real space and are localised. As such, in such of a way of thinking, delocalisation as described by the expressions we just derived and which

physicists usually favour, does not play an important role. So we have to look for a way to combine both worlds. To this end, we first relate the way the bands run, derived from the properties of the wave functions of the atoms. Then we will derive the so-called density of states from the Bloch functions, using which we are able to derive a local picture from the delocalised bands. The interested reader is referred to [3] for a more elaborate treatment of this issue.

If we look at fig.4, we observe that the energy of the band with the lowest energy increases when k increases, starting at k=0. For the second band, the opposite situation is observed. Of course, this result has been derived in the treatment above by first formally deriving the wave functions of the system and then applying the symmetry rules which follow from the periodicity of the system. But: can we also intuitively understand the way the bands run? Yes, there is a way. To this end we look at the bonding character of the delocalised wave function as a function of k.

We start with considering a delocalised wave function built from atomic s-states. For k= 0, we find that all atoms contribute to the bonding in the system, as a clearly visible in fig. 5.

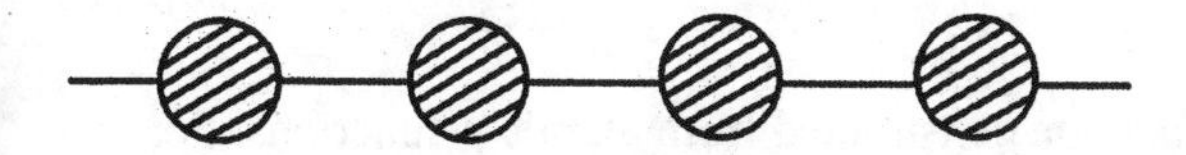

Figure 5: Delocalised wave function constructed from atomic s-orbitals at k= 0

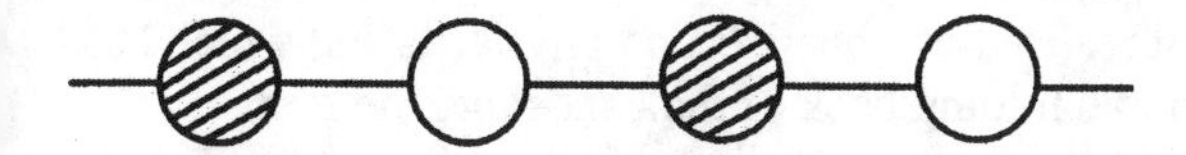

Figure 6: Delocalised wave function constructed from atomic s-orbitals at k= π/a

For k= π/a, we find that the resulting wave function is completely antibonding, as sketched in fig. 6.

So for atomic s-states we expect a dependence of the energy of the Bloch functions as a function of k as given in fig. 7.

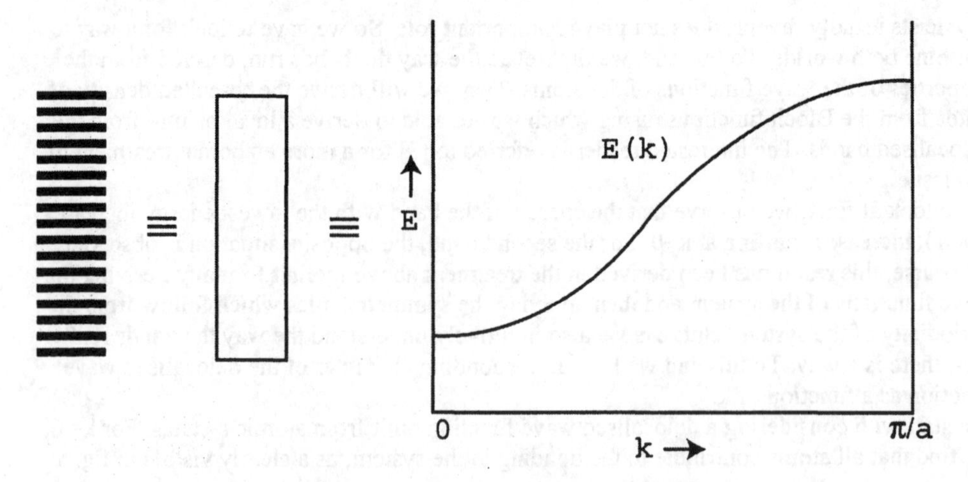

Figure 7: Energy dependence of Bloch functions constructed from atomic s-orbitals as a function of k. The left part of the figure shows the band to be formed from single atomic s-orbitals

Now we consider a delocalised wave function constructed from atomic p-functions. For k=0, we find a complete antibonding wave function, whereas for $k = \pi/a$, we find that the resulting wave function has highest bonding character. So we expect the delocalised wave function to run from highest to lowest energy in increasing k from 0 to π/a, as sketched in fig. 8.

It is exactly this behaviour, which has been derived above. Our comparison between the physicists and the chemist world, however, teaches us more: the bonding character in the bands changes as a function of their energy, and eventually the top of a band can even have antibonding character which is often not realised by chemists and physicists and which has consequences for reactivity. In addition, the folding back of the bands into the first Brillouin zone gives rise to a difference in symmetry behaviour for the different bands as some bands go up and some down as a function of k, which as we now understand reflects the symmetry properties of the atomic orbitals. This should not surprise us, as we learned already in the first section that application of a potential well results in wave functions with a difference in symmetry behaviour.

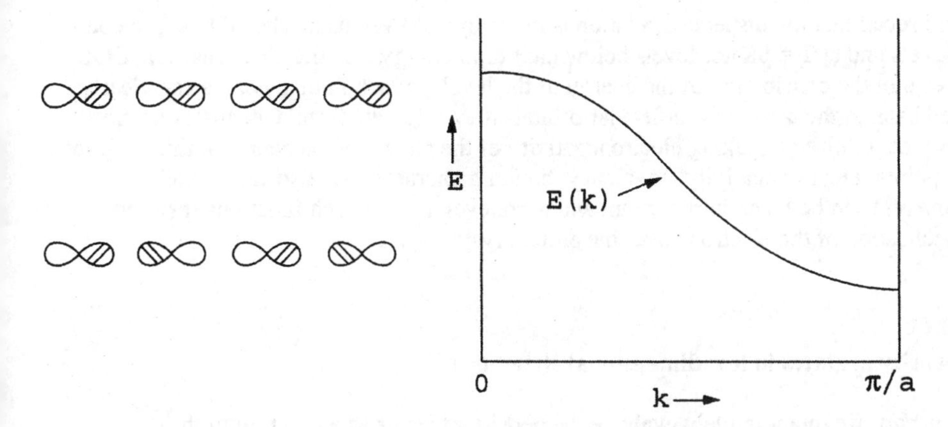

Figure 8: Energy dependence of Bloch functions constructed from atomic p-orbitals as a function of k

The interested reader might try to obtain the energy dependence of the different d-functions on k.

Now we try to obtain a localised picture and try to answer the questions where the electrons are. This implies a transition to real space. The means we use is to calculate the 'density of states' from the dispersion relation, which follows from the Bloch functions. The density of states at the energy E (DOS(E)) is simply the number of states we find in an energy interval dE at E. The DOS(E) is inversely proportional to the slope of E(k) versus k. Graphically, we can understand the transition from the dispersion relation to the DOS as given in fig. 9.

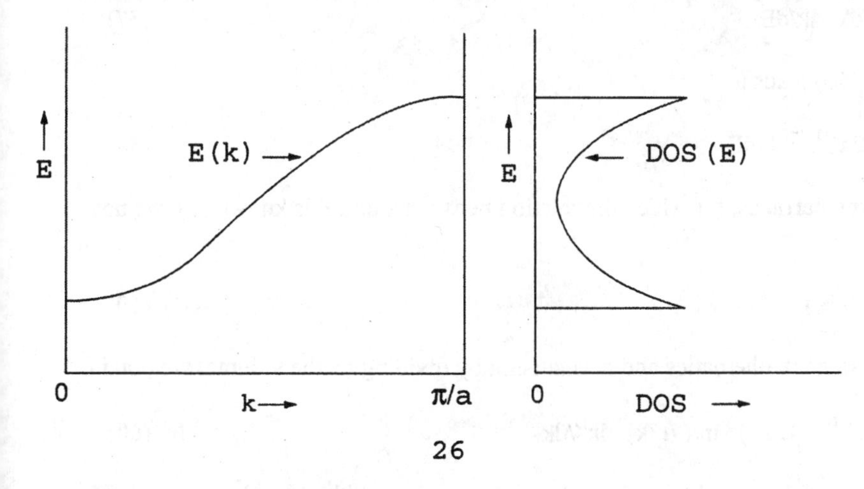

Figure 9: Dispersion curve and corresponding density of states

Please recall that the dispersion relation is given by E(k) versus k. The DOS curve counts the levels and at T = 0K for levels below the Fermi energy also the electrons. The DOS curves plot the distribution of the energy of the levels, which in many cases can clearly traced back to the atomic or molecular orbitals they originate from. Chemists can draw DOS plots intuitively, taking electronegativity of the atoms into account. In this way, local viewpoints (electronegativity, covalency, bonding character but also crystal field arguments) can be taken into account and used, even if the Bloch functions result in delocalisation of the electrons over the entire crystal.

5. Density of states in low dimensional structures

In this part, we quantitatively evaluate the density of states as a function of the dimensionality of the structure. As argued above, the density of states is the number of states present in a certain energy interval. We first derive the number of states in k-space and then calculate the DOS function as a function of E.
It follows from the periodic boundary condition that $k = 2\pi n/L$. n is an integer and L the length of the crystal. This means that n is given by $kL/(2\pi)$. With arbitrary dimensionality (D), this equation reads:

$$n = k^D (L/2\pi)^D \qquad (56)$$

The number of states per energy interval (dn/dE) can be written as:

$$dn/dE = dn/dk.\ dk/dE \qquad (57)$$

Insertion of (56) leads to:

$$dn/dE = (L/2\pi)^D.\ d^Dk/dE \qquad (58)$$

Assuming free particles, for which the relation between k and E is known (6), we now write:

$$dk/dE = m/(\hbar^2 k) \qquad (59)$$

After doing some mathematics and normalising by dividing by the volume, we obtain:

$$DOS(E) = 1/L^D\ .\ (L/2\pi)^D\ m/(\hbar^2 k).\ dk^D/dk \qquad (60)$$

For three dimensions, we obtain (the states are in a sphere with radius k):

$$DOS_3\ (E)\ dE = 1/4\pi^2.(2m/\hbar^2)^{3/2}\ (E - E_g^o)^{1/2}\ dE \qquad (61)$$

Please note that the number of electron states is twice as large.

The results for two dimensions and one dimension are:

$$DOS_2(E)\,dE = 1/\pi^2 .\ m/\hbar^2 dE \tag{62}$$

$$DOS_1(E)dE = 1/(2\sqrt{2}\pi).\ 1/\hbar\ .\ m^{1/2}.\ (E - E_g^o)^{-1/2}dE \tag{63}$$

Finally in a zero dimensional system, the DOS(E) is described by a set of δ- functions.

The results are summarised in fig. 10.

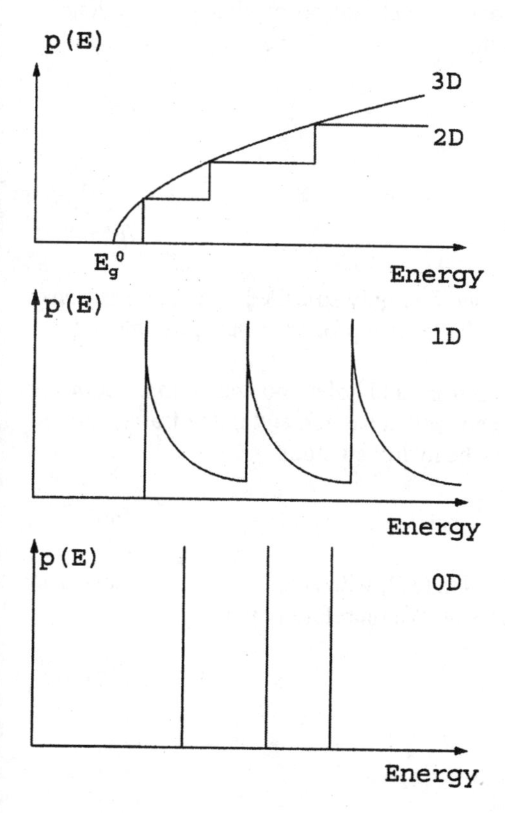

Figure 10: Density of States for 3,2,1 and zero Dimensional Systems

6. Electrons, holes and excitons

Electrons are particles with charge –e, mass m_o and spin ½. In a crystal only the charge and the spin remain the same. As we have learned, in a crystal, the effective electron mass is m_e* and the quasi momentum is $\hbar k_e$. Holes have charge +e, spin ½ , effective mass m_h* and momentum $\hbar k_h$.
We now treat electrons in the valence band and holes in the conduction band as elementary excitations. Using the approach derived above, we obtain an approximate solution in terms of a small number of non-interacting particles. These so-called quasi particles represent excitations of the system, which consists of real particles. The ground state of the system contains no electrons in the conduction band and no holes in the valence band. The first excited state is the state with one electron in the conduction band and one hole in the valence band. Such a transition can e.g. be induced by photon absorption. On applying energy and momentum conservation, we obtain:

$$\hbar\omega = E_g + E_{kin,\,e} + E_{kin,\,h} \tag{64}$$

and

$$\hbar k = \hbar k_e + \hbar k_h \tag{65}$$

The momentum of the photons is very small; therefore only so-called vertical transitions are possible, without a change in momentum. The reverse process is also possible, e.g. leading to luminescence.
We now extend our treatment to interacting electrons and holes and add an interaction term to the Hamiltonian. The resulting quasi particle is called exciton and in the framework of almost free electrons and holes corresponds to the hydrogen atom:

$$H = - [(\hbar^2/2m_e)\nabla_e^2 + (\hbar^2/2m_h)\nabla_h^2] - e^2/\varepsilon\,|r_e - r_h|] \tag{66}$$

Which essentially is the same equation as equation (30), with adaptation of the masses and insertion of the dielectric constant ε of the crystal. We therefore obtain:

$$a_B = \varepsilon\,\hbar^2/(\mu e^2) = \varepsilon\, m_o/\mu\ .\ 0.53\ \text{Å} \tag{67}$$

for the Bohr radius of the electron.

For the reduced mass we obtain:

$$\mu^{-1} = m^{*}{}_{e}^{-1} + m^{*}{}_{h}^{-1} \quad (68)$$

For the Rydberg energy of the system we find:

$$Ry^{*} = e^{2}/(2\varepsilon\, a_{B}) = \mu e^{4}/\,(2\varepsilon^{2}\hbar^{2}) = \mu/m_{o}\,.1/\varepsilon^{2}.\ 13.6\ \mathrm{eV} \quad (69)$$

7. Low dimensional structures

In a semi-conducting material, the wavelength electron and holes, or the Bohr radius of an exciton can be much larger than the lattice constant a_L. On decreasing the dimensions of a structure, we may enter the regime in which at least one of the dimensions is in the same order of magnitude or even smaller than the wavelength of the electron, hole and the Bohr radius of the exciton but still larger than a_L. In such a structure, the elementary excitations we discussed above will be quantum confined. The result is a finite motion in the direction of the confinement axis and infinite in the other direction.
Confinement in one direction results in a quantum well, which we have treated.
Confinement in two directions results in a quantum wire. Confinement in three directions results in an effectively zero-dimensional system: a quantum dot.
In what follows we will treat the electron and hole states in a nano-crystal. We will discuss two limiting cases, the weak confinement regime and the strong confinement regime. In both cases we will assume that we can use the effective mass approximation for the electrons and the holes. We will derive the energy states and the optical spectra. As is the case in other branches of spectroscopy, optical selection rules apply. The selection rules obtained for the nano-crystals will be compared to selection rules known for optical transitions on ions and in conventional solid-state physics.
We will deal with the nano-crystal adopting a three dimensional quantum well with an infinite potential and electrons and holes with an isotropic effective mass.

7.1. THE WEAK CONFINEMENT REGIME

Weak confinement occurs when the radius (a) of the nano-particle is a few times larger than the exciton Bohr radius (a_B). In this case, the exciton centre of mass motion is confined. We can easily derive the energy of the exciton from the results already derived in the previous chapter. The kinetic energy of the exciton is obtained from the dispersion law of an exciton in a crystal in which the kinetic energy of the free exciton is replaced by the solution for a particle in a box:

$$E_{nml} = E_{g} - Ry^{*}/n^{2} + (\hbar^{2}/2Ma^{2})\,\chi^{2}_{ml} \quad (70)$$

The quantum number n describes the internal exciton states, which originate from the Coulomb interaction between electron and hole in the exciton. The two additional numbers m and l describes the states connected to the center of mass motion in the presence of the external barrier. Both sets of numbers have states 1s, 2s, 2p, etc.

The lowest state is given by n = 1, m = 1 and l = 0. Its energy is given by:

$$E_{1S1s} = E_g - Ry^* + \pi^2 \hbar^2/(2Ma^2) \quad (71)$$

which can be written as, using the relations derived above:

$$E_{1S1s} = E_g - Ry^*(1 - (\mu/M)(\pi a_B / a)^2) \quad (72)$$

The last part of the expression gives the blue shift of the first exciton absorption as a function of the particle size. When $a >> a_B$, this shift is small compared to Ry*.
As photons have almost zero angular momentum, optical absorption can only connect states exciton with l=0, as the $\Delta l = \pm 1$ part is already included in the optical transition which connects the p-like valence band states to the s-like conduction band states. This means that the absorption spectrum is given by equation (70) with χ_{m0}, which is πm:

$$E_{nm} = E_g - Ry^*/ n^2 + \pi^2 \hbar^2/(2Ma^2).m^2 \quad (73)$$

7.2. THE STRONG CONFINEMENT REGIME

In the strong confinement regime, the radius (a) of the nano-crystal is much smaller than the exciton Bohr radius (a_B): $a << a_B$, in this case the confinement also has impact on the electron and hole states.
In this situation, the zero-point kinetic energy is much larger than the Ry* value. The electron and the hole do not have bound states corresponding to the hydrogen like exciton. Therefore, in this physical limit, the electron and hole motion may be treated uncorrelated and the Coulomb interaction between electron and hole is ignored. Please note that the Coulomb interaction energy by no means vanishes, its contribution to the ground state energy is even higher than in the bulk crystal. In the strong coupling limit, the Coulomb energy of a free electron-hole pair is unequal to zero, but the zero-point kinetic energy is even much larger.
The energy spectrum of electron and hole are given by, respectively:

$$E_e^{nl} = (\hbar^2/2m_e a^2)\, \chi^2_{nl} \quad (74)$$

and

$$E_h^{nl} = E_g + (\hbar^2/2m_h a^2)\, \chi^2_{nl} \quad (75)$$

Please note that the electron and hole states are described by orbital quantum numbers only (n and l), reflecting the description of an uncorrelated electron and hole. Taking the selection rules into account, in the optical absorption spectra we obtain discrete bands, peaking at:

$$E^{nl} = E_g + (\hbar^2/2\mu a^2)\,\chi^2_{nl} \qquad (76)$$

indicating that only optical transitions are allowed between electron and hole states with the same n and l values. Again the electric-dipole selection rule is obeyed, taking the valence and conduction band character underlying the excitonic transition into account. Here, there is a clear parallel to atoms. In the strong coupling limit, the optical spectrum is determined by the number of atoms (via the size of the quantum dot), a beautiful manifestation of which is given below. An atom has a discrete spectrum, dependent on its atomic number.
As mentioned above, the electron and the hole in the quantum dot show Coulomb interaction. The Hamiltonian describing this system is given by:

$$H = -[(\hbar^2/2m_e)\nabla_e^2 + (\hbar^2/2m_h)\nabla_h^2] - e^2/\varepsilon|r_e - r_h| + U(r) \qquad (77)$$

In which the last term gives the Coulomb interaction between the electron and the hole. Please note the difference with equation (66). As we have seen above, it is this latter potential which does not allow us to use to the mass renormalisation. This problem has been treated by several authors [4-6]. The result for the electron-hole pair in the ground state is:

$$E_{1s1s} = E_g + \pi^2\hbar^2/(2\mu a^2) - 1.786e^2/\varepsilon a \qquad (78)$$

The last term describes the Coulomb interaction between electron and hole. The exciton Rydberg energy is $e^2/2\varepsilon a_B$ and as $a \ll a_B$, this means that the Coulomb energy does not vanish as already stated above. For other optical transitions, the parameter 1.786 has slightly different values. We now continue with a comparison of experimental results with the theory derived above.

8. Quantum confinement in action

In this section, we will describe experimental results of optical investigations concentrating on luminescence. First we will shortly discuss some important techniques for the preparation on the nano-materials. We will then deal with luminescence properties of compound quantum dots and of doped nano-materials.

Quantum confinement studies are done on semi-conductor crystals (in contrast to insulators) in view of the fact the exciton radii are large in semi-conductors (large dielectric constants and small effective mass). Quantum confinement studies are done on semi-conductor crystals typically in the range below 10 nm in size. At larger sizes, no quantum size effects are observed. This can be understood relatively easily: for a dielectric constant $\geq$ 10 and reduced effective masses in the order of 0.01 – 0.1 m_o, the exciton radii are in the order of 1 – 10 nm and their binding energies vary between 0.1 – 10^{-3} eV, as can be estimated with equations (67) and (69).
There are many techniques to prepare quantum dots, and all techniques have their own advantages and disadvantages. Wet chemical precipitation methods are used and in recent years have resulted in relatively narrow particle size distributions. In these methods, the starting materials are dissolved in a liquid, reacted and the resulting nano-particles are not soluble in this liquid. Changing the reaction conditions can vary the size of the particles. In general, the quantum dots have to be protected from agglomerating with each other or from uncontrolled growth. This is achieved by capping, e.g. by application of a coating on top of the quantum dots. Although the particle size distribution obtained can be rather small (+/-15%), nevertheless the emission spectra of such materials still show broad spectral features. Sharp spectral features have been obtained by using quantum dots obtained via epitaxial techniques. There are several different methods; some of them will be discussed:
Growth of islands on a substrate with lattice mismatch (Volmer-Weber)
Growth of a layer, which subsequently forms, islands (Stranski-Krastanow), due to lattice mismatch.
Self organised growth of quantum dots, as found by Nötzel [7]. Using this method, quantum dot structures can be obtained with both a vertical and a lateral ordening in some III-V material systems, also exploiting lattice mismatch. It is an interesting feature of this method that the quantum dots bury themselves in the substrate layer and that one can study the effect of contact between the individual quantum dots.
Laser ablation techniques are also used. Another interesting technique is the incorporation of small particles in zeolites.
Luminescence studies require rather perfect materials with a well-defined surface. In addition, as stated above, the quantum dots should not agglomerate to larger units. This can be achieved by surface treatments with stabilising agents. Interestingly, size selective precipitation can also be used to achieve narrower size distributions of the very small particles, see e.g. [8].
Doped nano-particles, to the best of the authors' knowledge are prepared using wet chemical techniques only.
In general, the optical absorption spectra are richer than the luminescence spectra. In the absorption spectra, absorption features reflecting many excited states are observed. The luminescence spectra discussed here are due to the lowest excited state. Emission from higher excited states is generally absent, due to a rapid relaxation to the lowest excited state. In addition, in many case near energy gap emission is observed and emission with a much larger Stokes Shift, this latter emission generally being due to lattice defects.

8.1. INFRARED ABSORPTION BY NEGATIVELY CHARGED ZNO QUANTUM DOTS

Quantum dots can be charged electrically, e.g. by putting between two electrodes, which are at a different potential. In this way, one or more electrons can be transferred to the nano particles and consequently their spectroscopy can be studied. In case of ZnO, quite a few electrons can be stored on the particles, dependent on the size up to virtually 10 electrons. Optical transitions between conduction band states of the ZnO nano crystals can than be observed, see [9]. As we have derived before, electron states, belonging to the nano particles, of s, p, d, etc. character are expected. The wave functions, belonging to these states are delocalised over the complete quantum dot. Optical transitions between these conduction band like states are expected, obeying the usual parity selection rule, i.e. s-p, p-d optical transitions, etc. are allowed.
In fig. 11, the optical absorption, due to transitions between conduction band like states in ZnO nano particles with a mean diameter of 4.3 nm is given. An experimental complication is that there is not only a size variation but also an occupancy variation (with electrons) in the quantum dots. All these factors haven been taken into account. The spectra obtained can be deconvoluted, and the contribution of each individual peak can be compared to results of tight binding calculations in terms of spectral position and intensity. The agreement obtained is quite good; the interested reader is referred to [9].
Fig. 11a shows the result for quantum dots with a low mean occupation number. This is also reflected in the spectra: the shoulder at the right hand side is assigned to particles with one electron, the left part is due to particles with two electrons. A very small portion of the particles also has one electron in a p state as deduced from the p-d absorption observed. Larger mean sizes allow a higher number of electrons to be stored on the quantum dots, in the example given in fig. 12 up to almost 9 electrons. In this way, the relative contributions of the different transitions to the spectrum can be varied, see fig. 11 and 12. In the larger quantum dots with higher occupation numbers, even f states are involved in the optical absorption process.

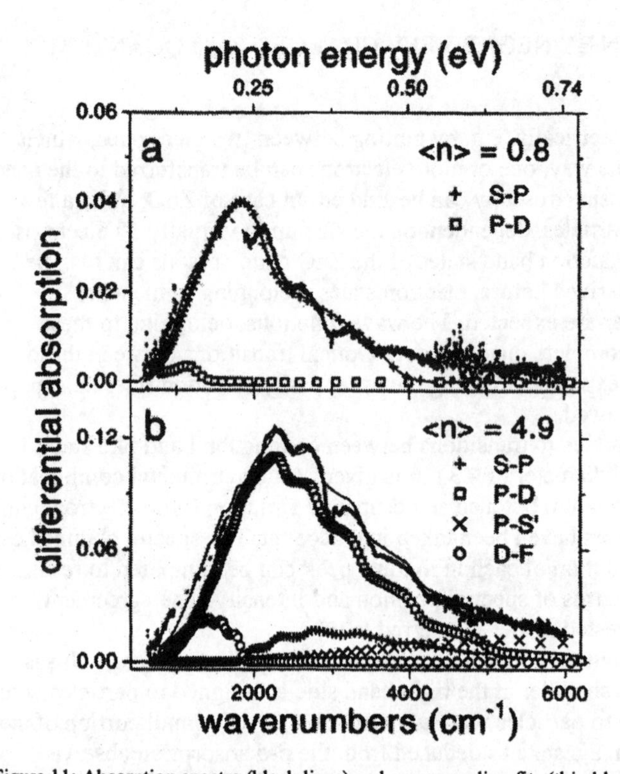

Figure 11: Absorption spectra (black lines) and corresponding fits (thin black lines) for ZnO quantum dot thin films (200 nm) with
mean size 4.3 nm at low and high orbital occupancy. The dip at 3000 cm^{-1} is due to the electrolyte used (propylene carbonate).

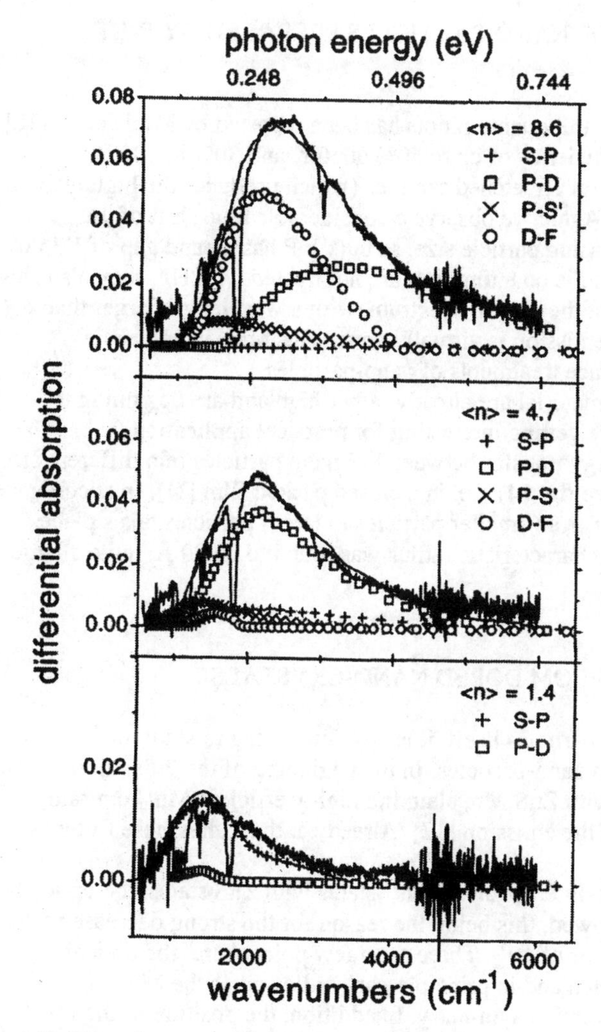

Figure 12: Absorption spectra (black lines) and corresponding fits (thin black lines) of for ZnO quantum dot thin films (200 nm)
with mean size 5.2 nm at low and high orbital occupancy. The sharp spectral features are due to the electrolyte used (propylene
carbonate).

8.2. PHOTOLUMINESCENCE OF NANO-PARTICLES PREPARED BY WET CHEMICAL PRECIPITATION

Highly efficient luminescence for InP quantum dots has been reported by Micic et al. [10] with a relatively high quantum efficiency of up to 60% at 10 K and 30% at 300 K. Absorption- and emission spectra of HF etched samples (which produces the high quantum efficiencies) are given in fig. 13. Again we observe a considerable blue shift of the quantum dot emission with decreasing particle size, as bulk InP has a band gap of 1.35 eV (corresponding to 918 nm). When the quantum dots are not treated with HF, they also show an emission in the deep red part of the optical spectrum, with a wavelength larger than 850 nm., after the HF treatment, this emission is virtually gone. This observation again underlines the importance of surface treatments of nano-particles.
The quantum efficiencies in this material are already rather high and are beginning to approach the range where they are getting interesting for practical applications.
Mimic et al. have also found energy transfer between InP nano particles of a different size when they are in close contact (see fig. 14), i.e. in a closed packed film [11]. In accordance with expectations, energy transfer from smaller particles to larger particles takes place. Using Förster-Dexter theory, the characteristic radius was found to be 90 Å, quite a large value.

8.3. PHOTOLUMINESCENCE FROM DOPED NANO-CRYSTALS

Apart from emission of the nano-particles itself, it is also interesting to study the luminescence properties of doped nano-particles. In the beginning of the 90's, there was a number of publications dealing with ZnS:Mn, claiming highly efficient Mn^{2+} emission and a large decrease in decay time of the emission [12]. Already at that time, these findings were discussed controversially.
Bhargava et al. argued that due to interaction of Mn^{2+} states with ZnS states, the optically forbidden transition becomes allowed, this being the reason for the strong decrease in the emission decay time (about a factor of 10^5). There are a few points here: the optical transition on Mn^{2+} is spin forbidden and in principle parity allowed as the Mn^{2+} ion incorporates on a site without inversion symmetry. In addition, the position of the Mn^{2+} emission band hardly shifts, as compared to bulk ZnS:Mn, which is hardly understandable in view of the claimed strong interaction of Mn^{2+} states with ZnS host lattice states. Finally, Bhargava et al. only performed decay time measurements in the ns range.

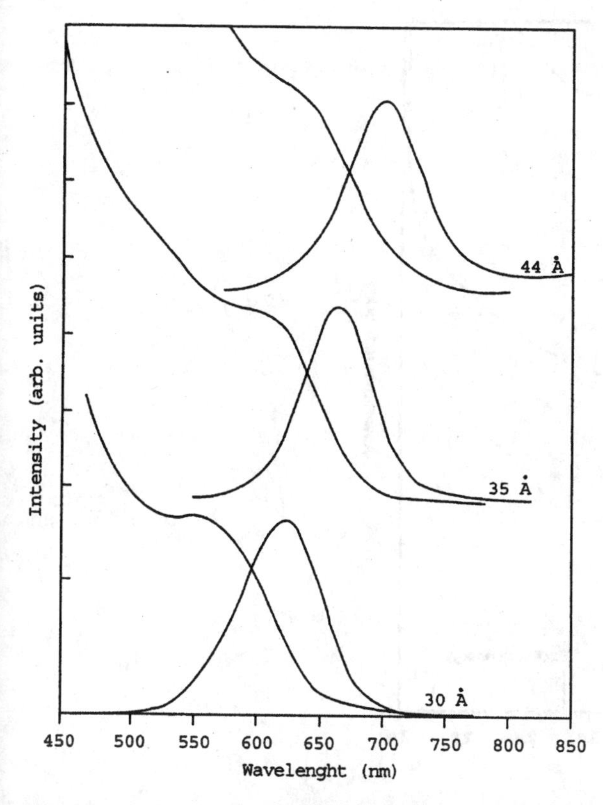

Figure 13: Absorption and emission spectra of InP particles, treated with HF as a function of the particle size. The spectra were recorded at 300K.

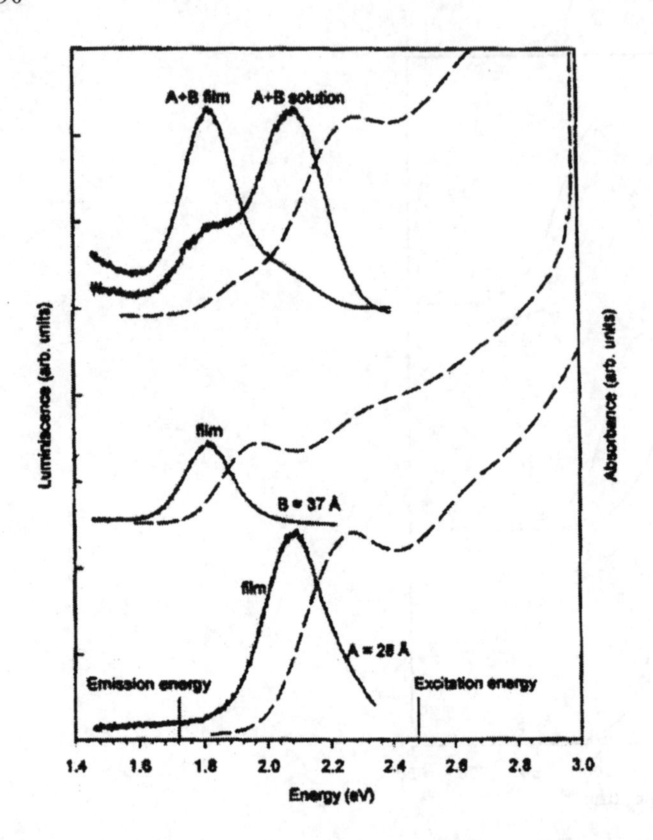

Figure 14: Absorption and emission spectra (excited at 500 nm) of InP quantum dots with
Top: two sizes (2.8 and 3.7 nm) in closed packed films and a mixed solution
Middle: 3.7 nm quantum dots only
Bottom: 2.8 nm quantum dots only

These observations struck the attention of many other researchers and this particular example was even mentioned in a number of textbooks on optical properties on nano-crystals [2] or on luminescence, mentioning this discovery as one of the most important ones in this field in this decade [13,14]. But is it true?
The emission spectrum of ZnS:Mn consists of two bands: one at about 420 nm, which is also observed in undoped ZnS:Mn and one at about 590 nm, which is due to the $^4T_1 \rightarrow {}^6A_1$ transition on Mn^{2+}. Bol et al. [15], have measured the emission spectrum of nanocrystalline ZnS:Mn as a function of the delay time between the laser pulse and the emission These

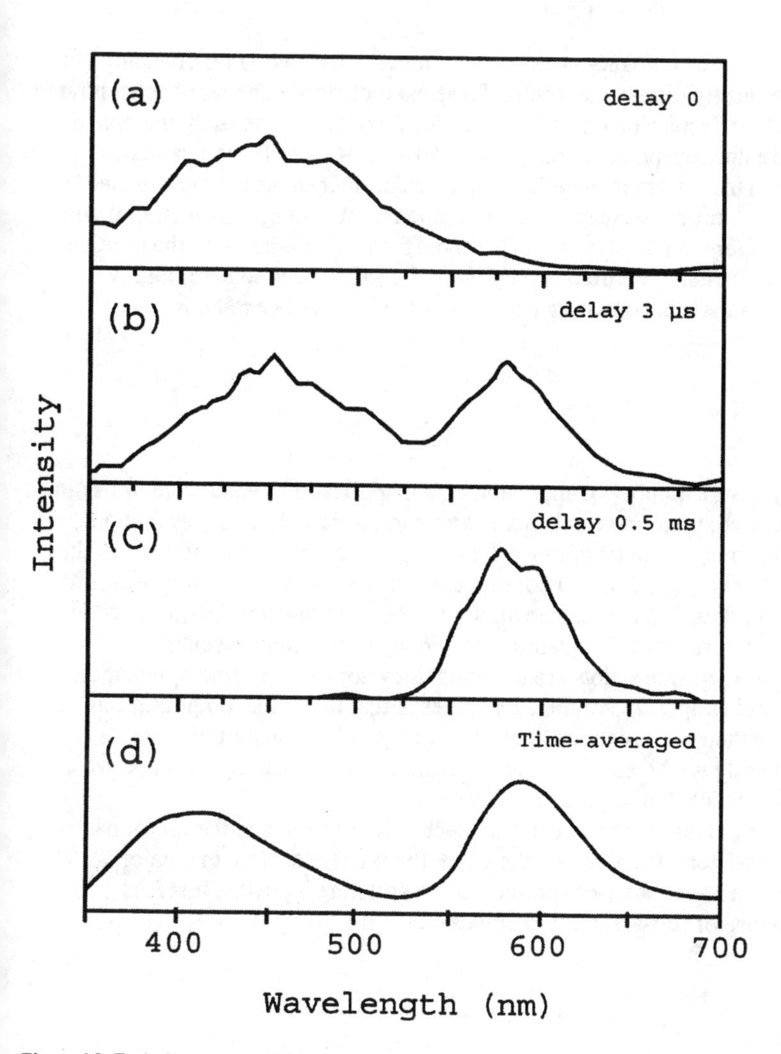

Figure 15: Emission spectra of ZnS:Mn as a function of the delay time and gate width. The delay times and gate widths applied are: a) ~ 0 and 2 μs; b) 3 μs and 200 μs; c) 0.5 and 1 ms and finally d) the time averaged spectrum. All spectra were recorded at 300 K.

observations clearly show that there is no combination of a high luminescent efficiency of Mn^{2+} coupled to a very short decay time (in the ns range) of the emission on this ion. measurement. The results are given in fig. 15. As is clearly visible, the fast emission is the emission centered at 420 nm and the Mn^{2+} emission is still very slow.
The results obtained by Bol and Meijerink imply that the fast emission, with a decay time of 20 ns is not due to emission of Mn^{2+} but due to the emission on the ZnS host, which extends, beit with low intensity, into the region where Mn^{2+} shows its orange emission.

In recent years, also research on nano-particles of commercially used lamp phosphors has been performed. We shortly discuss $LaPO_4$:Ce,Tb as example [16]. The work concentrated on absorption on Ce^{3+} and emission on Tb^{3+}, i.e. on localised states. As such, the optical transitions studied are not dependent on the particle size. However, the non-radiative transitions likely are. This system is therefore of particular interest, as energy transfer is needed to feed the Tb^{3+} emission. Haase et al. succeeded in obtaining nano-material with quantum efficiency larger than 60% (sum of Ce- and Tb emission), beit that the relative Ce^{3+} contribution is larger than in micro crystalline $LaPO_4$:Ce,Tb. Also here, highly efficient luminescent materials, consisting of nano-particles, have been obtained.

9. Outlook

Quantum dots likely have a number of applications. An application treated is their use in luminescent devices. Here major research issues are: narrow particle size distributions, contacting issues, improving of outcoupling efficiency and improving the lifetime of the device. In many cases, energy is lost as a consequence on non-radiative energy losses at the surface of the quantum dots. It has been shown in the literature that application of coatings sometimes result in an increase of the quantum efficiency of the luminescence.
Once efficiently emitting quantum dots are obtained, they are likely to find application as emitters in displays and lamps. One of the advantages is that the same compound can be used to generate more than one colour of light. In addition, when the particle size distribution is sufficiently well controlled, narrow band emissions can be obtained which can be chosen optimally adapted to the human eye.
A second application field might be there use in electronic devices, e.g. the use of nano-crystals in making transistors, f.i. based on the Coulomb blockade. Here the major issues are the contacting of the nano-particles and the way to structure a device. It seems unavoidable to make use of self-assembly methods here.

10. References

1. C.R. Ronda in B. Di Bartolo (ed.), *Spectroscopy of Systems with Spatially Confined Structures*, Kluwer Academic Publishers, Dordrecht, Boston, London, Series II, Mathematics, Physics and Chemistry (**90**), 391
2. S.V. Gaponenko (1998) *Optical Properties of Semi-Conductor Nanocrystals*, Cambridge University Press
3. R. Hoffmann (1988) *Solids and Surfaces, a Chemist's View of Bonding in Extended Structures*, VCH, Weinheim
4. L.E. Brus, (1986) J. Phys. Chem. 90, 2555
5. Y. Kayanuma, (1986) Solid State Comm. 59, 405

6. H.M. Schmidt and H. Weller, (1986), Chem. Phys. Lett. 129, 615
7. R. Nötzel, (1996), Semicond. Sci. Technol. 11, 1365
8. C.B. Murray, D.J. Norris and M.G. Bawendi, (1993) J. Am. Chem. Soc. 115, 8706
9. A. Germeau, A.L. Roest, D. Vanmaekelbergh, G. Allan, C. Delerue and E.A. Meulenkamp, Phys. Rev. Letters 90, 097401, 2003
10. O.I. Micic, J. Sprague, Z. Lu and A.J. Nozik, (1996), Appl. Phys. Lett. 68 (22), 3150
11. O.I. Micic, K.M. Jones, A. Cahill and A. Nozik (1998), J. Phys. Chem. B 102 (49), 9791
12. R.N. Bhargava and D. Gallagher (1994), Phys. Rev. Lett. 72, 416
13. Y. Masumoto (1999) in S. Shionoya and W.M. Yen (eds.), *Phosphor Handbook,* CRC Press, Boca Raton, FL, USA, 78
14. S. Shionoya (1998) in D.R. Vij (eds), *Luminescence of Solids,* Plenum Press, New York, USA, 131
15. A.A. Bol and A. Meijerink (2000), J. of Lum. 87-89, 315
16. D.V. Talapin, A.L. Rogach, A. Kornowski, M. Haase and H. Weller (2001), Nano Letters 1(4), 207

11. AN INTRODUCTION TO THE PHYSICS OF ULTRACOLD ATOMIC GASES

C. J. Pethick
Nordita, Blegdamsvej 17, DK-2100 Copenhagen Ø, Denmark,
H. Smith,
Ørsted Laboratory, H. C. Ørsted Institute,
Universitetsparken 5, DK-2100 Copenhagen Ø, Denmark

It is now eight years since the first experimental realization of Bose–Einstein condensation in a dilute gas, and the study of ultracold atomic gases is well developed. Bose-Einstein condensation has been achieved for a number of different atoms, those involved in the first pioneering experiments, ^{87}Rb, ^{23}Na and ^{7}Li, and in addition ^{85}Rb, ^{1}H, ^{41}K, ^{133}Cs and ^{4}He*, helium atoms in an excited electronic state, the lowest energy electronic spin triplet state. Bose-Einstein condensation has been realized also for an atom with two valence electrons, Yb. Atomic Fermi gases have been cooled to well below the degeneracy temperature, and experiments are underway to detect the transition to the Bardeen-Cooper-Schrieffer state in which atoms are paired in much the same way as in a metallic superconductor. Recently, Bose–Einstein condensation of diatomic molecules of alkali atoms, e.g., ^{6}Li-^{6}Li, has been achieved.

Experimentally the systems are almost ideal. First there are a number of different atoms to choose from, with different atomic properties. Second, the trapping geometry can be varied, thereby enabling one to study, for example, systems in which certain degrees of freedom are frozen out, and the gases behave like lower-dimensional systems. Third, characteristic microscopic length scales are typically of the same order as the wavelength of visible light and since the atoms have strong transitions in this wavelength range, they may be detected optically. Also because of the hyperfine splitting, these methods are sensitive to the specific hyperfine state of an atom. Fourth, a number of atoms have Feshbach resonances, which allow one to tune the atomic scattering length, which determines the strength of interatomic interactions, to

B. Di Bartolo and O. Forte (eds.), Frontiers of Optical Spectroscopy, 395-425.

essentially any desired value. A further advantage is that the systems are tractable theoretically, which has led to a lively interplay between theory and experiment.

Within the compass of these lectures it is impossible to describe more than a small fraction of work that has been carried out in the area. Our philosophy will be to discuss a number of topical subjects in simple physical terms. There are now textbooks on the basic physics of ultracold gases [1, 2], as well as a number of review articles [3, 4], and the reader is referred to these and the original literature for more details.

1 Energy and length scales

To set the scene, we begin with a review of the fundamental energy and length scales for quantum gases. Densities n are typically in the range 10^{12} - 10^{14} cm^{-3}, and therefore spacings r_s between atoms are of order $n^{-1/3} \sim 10^{-4} - 10^{-5}$ cm, that is 1 - 0.1 μm or $\sim 10^4 a_0$, where $a_0 = 4\pi\epsilon_0\hbar^2/m_e e^2$ is the Bohr radius. These estimates are appropriate for alkali gases, while for experiments on hydrogen, the density can be as high as 5×10^{15} cm^{-3}.

Quantum phenomena set in when the de Broglie wavelength $\lambda_{\rm dB}$ of an atom becomes comparable to the distance between particles. For Fermi gases, the Pauli exclusion principle comes into play, and the distribution of atoms tends towards a filled Fermi sea while, for Bose gases, Bose-Einstein condensation sets in. This occurs when

$$\lambda_{\rm dB} = \frac{2\pi\hbar}{p} \sim r_s, \tag{1.1}$$

where p is the particle momentum. The particle energy is therefore of order $p^2/2m \sim \hbar^2 n^{2/3}/m$. This implies that effects of quantum degeneracy become important when particle energies are less than or of the order of

$$E_{\rm deg} \sim \frac{\hbar^2 n^{2/3}}{m}. \tag{1.2}$$

For a thermal gas, this corresponds to a temperature of order

$$T_{\rm deg} \sim \frac{\hbar^2 n^{2/3}}{m k_{\rm B}} \sim \frac{n_{12}^{2/3}}{A}\,\mu{\rm K}, \tag{1.3}$$

where A is the mass number of the atom and n_{12} is the atom number density in units of 10^{12} cm^{-3}. This temperature is of order $0.1 - 1\mu$K under typical experimental conditions. For Fermi gases, this is the magnitude of the Fermi temperature, while for Bose gases it is the temperature at which Bose-Einstein condensation first appears. The atomic unit of energy is $m_e(e^2/4\pi\epsilon_0)^2/\hbar^2 \sim 27$ eV, which corresponds to a temperature of $\sim 3 \times 10^5$ K, so these temperatures are minuscule on the scale of typical atomic energies.

One cannot use ordinary containers for cold atoms, because atoms will be lost by collisions with the walls. Instead one uses magnetic and optical traps. In the case of magnetic traps, one constructs a configuration in which the magnitude of the magnetic field has a local minimum. Atomic magnetic moments are of order the Bohr magneton $\mu_B = e\hbar/(2m_e) \approx 5.8 \times 10^{-5}$ eVT^{-1}, which in temperature units is ~ 0.67 KT^{-1}. In experimental magnetic traps, the scale of magnetic fields is commonly of order 0.1 T, so the depths of traps are of order 0.1 K or less. This indicates the need to cool atoms before they can be contained by magnetic traps. If one denotes the typical spatial scale of the magnetic trap by L, the force constant K for the trap is of order $\mu_B B/L^2$. The angular frequency of atomic motions about the minimum of the magnetic field is therefore of order $(K/m)^{1/2} \sim (\mu_B B/mL^2)^{1/2}$. The spatial scale of magnetic traps is about 0.1 m and, using the estimates above for the magnetic field, one therefore finds frequencies of order 10^2 Hz. Considerably higher frequencies can be achieved by the use of nanostructures on chips. The currents that can be sustained are very much lower, but the characteristic spatial scales of the structures are of order micrometers or less, and consequently the frequencies, which scale as $1/L$, are correspondingly larger.

To construct an optical trap, one uses a focused laser beam. In an electric field, the energy of an atom is changed by the Stark effect by an amount

$$\Delta E = -\frac{1}{2}\alpha(\omega)\overline{\mathcal{E}^2}, \tag{1.4}$$

where $\mathcal{E}$ is the strength of the electric field, α is the polarizability of the atom, ω is the angular frequency of the electric field, and the bar denotes an average over time. It is convenient to introduce the polarizability in atomic units, which is defined as

$$\tilde{\alpha} = \frac{\alpha}{4\pi\epsilon_0 a_0^3}, \tag{1.5}$$

and the energy shift (1.4) is then given by

$$\Delta E = -4\pi a_0^3 \tilde{\alpha}(\omega)\left[\frac{1}{2}\epsilon_0\overline{\mathcal{E}^2}\right]. \tag{1.6}$$

The term in the square brackets is the energy density of the electric field, and therefore the energy shift is equal to the electric field energy in a volume

$4\pi a_0^3\tilde{\alpha}(\omega)$. For an alkali atom at low frequencies, the polarizability in atomic units is of order 300. The large values are due to the existence of low-lying resonance lines in their optical spectra. To give some idea of the size of the magnitude of the polarizability, we recall that for a conducting sphere of radius R the polarizability is $4\pi\epsilon_0 R^3$. Consequently, as far as the polarizability is concerned, an alkali atom behaves like a conducting sphere with a volume $\sim 10^3 a_0^3$, which is enormous compared with the physical size of an atom. If we take as an example a laser beam with a power of 50 mW focused to a minimum diameter d of ~ 50 μm, the energy shift is $\Delta E \approx 3 \times 10^{-31}\tilde{\alpha}(\omega)$ J or, in temperature units, $2 \times 10^{-8}\tilde{\alpha}(\omega)$ K. For frequencies of the laser beam low compared with that of the resonance lines, the depth of the potential is thus of order 10 μK. The frequency of oscillations of an atom transverse to the direction of the laser beam is therefore $\sim (\Delta E/m)^{1/2}/d$, which is typically in the kilohertz range.

The temperatures necessary for producing quantum degenerate gases cannot be achieved by cryogenic means. In all experiments to date, except those on hydrogen, the low temperatures were achieved by laser cooling followed by evaporative cooling. For a discussion of laser cooling, we refer to Ref. [1] and the monograph [5]. The physical idea behind evaporative cooling is to remove atoms having an energy greater than the average, thereby lowering the average energy of the remaining atoms. This has been exploited in traditional low-temperature physics, where cooling by evaporation of molecules or atoms from liquid helium, liquid hydrogen and liquid nitrogen has been one of the standard techniques for more than a century. In the case of trapped gases, the method was first proposed by Hess [8]. Evaporation can be brought about by lowering the depth of the potential well containing the atoms. In magnetic traps, this leads to rates of evaporation which are too slow, so one generally forces the evaporation by applying radio-frequency radiation which changes the state of the atom to one for which the trapping potential is less. In optical traps a steady rate of evaporation is achieved by lowering the depth of the trap with time.

In evaporative cooling, interactions play an important role since, without them, cooling would halt as soon as all atoms originally having an energy sufficient to evaporate had left the trap. Collisions are necessary to repopulate the higher-lying states from which evaporation can occur. We now investigate typical properties of interactions between atoms. As a simple model of an atom, one often imagines it to be a billiard-ball like object with a radius of order a_0. This is not a bad picture for phenomena at relatively high energy, but it is extremely poor for low energy scattering. The reason is the van der Waals interaction, which for large separations r of the two atoms varies as $-\alpha_{\rm vdW} r^{-6}$, where $\alpha_{\rm vdW}$ is a coefficient. While this is small on the scale of typical atomic energies, it is large compared with the kinetic energy of atoms at microkelvin temperatures. A convenient measure of the strength of two-body interactions is the scattering length a, which is the only parameter needed to characterize low-energy scattering. In the limit of zero relative energy, the wave function for the relative motion of a pair of atoms is given by

$$\phi = 1 - \frac{a}{r}, \tag{1.7}$$

where the first term represents the initial wave and the second term represents a spherical scattered wave. For particles that behave as hard spheres, a would be the diameter of the sphere. The scattering length gives the strength of the scattered wave, and it may be determined by solving the Schrödinger equation for the relative motion of two atoms at zero energy. Let us consider the Schrödinger equation for zero energy for a potential of the van der Waals form. The two terms in the equation are the kinetic energy and the van der Waals potential. One may estimate a characteristic length scale r_0 of the zero energy wave function by equating these two contributions. The kinetic energy term is of order $\hbar^2/mr_0^2$ and the magnitude of the van der Waals interaction is of order $\alpha_{\rm vdW}/r_0^6$. Equating these two energies one finds

$$r_0 = \left(\frac{m\alpha_{\rm vdW}}{\hbar^2}\right)^{1/4}. \tag{1.8}$$

It is convenient to express quantities in atomic units. The atomic unit of mass is the electron mass $m_{\rm e}$ and that of length is the Bohr radius $a_0 = 4\pi\epsilon_0\hbar^2/m_{\rm e}e^2$. The van der Waals coefficient has the dimensions of an energy times the sixth power of a length, so the unit is $e^2a_0^5/4\pi\epsilon_0$ and one conventionally writes $\alpha_{\rm vdW} = C_6e^2a_0^5/4\pi\epsilon_0$. It follows from Eq. (1.8) that

$$r_0 = \left(\frac{m}{m_{\rm e}}C_6\right)^{1/4} a_0. \tag{1.9}$$

For alkali atoms in the ground state, the coefficients C_6 are large, and lie in the range 10^3–10^4. The reason for these large values, like the large values of the polarizabilities, can be traced to the existence in these atoms of strong resonance lines in the optical part of the spectrum, as explained in Ref. [1, p.106]. It therefore follows that for alkali atoms, the magnitude of scattering lengths is of order $100a_0$. This result is remarkable, since it indicates that, at low energies, alkali atoms behave as though they had linear sizes two orders of magnitude larger than the size of the atomic core. The actual value of the scattering length depends on properties of the interaction potential at short distances, where it is no longer given by the van der Waals form, and it can have values ranging from $+\infty$ to $-\infty$. However, the estimate (1.9) sets the typical scale of scattering lengths. That elastic cross sections are large, typically of order $4\pi a^2 \sim 10^5 a_0^2$, is important for ensuring that evaporation can take place sufficiently rapidly.

Another relevant energy scale is the spacing of vibrational states of molecules This is given by the typical scale of the kinetic and interaction energies for $r \simeq r_0$, that is $\hbar^2/mr_0^2 = \alpha_{\rm vdW}/r_0^6 = (\hbar^2/m)^{3/2}/\alpha_{\rm vdW}^{1/2}$, which is $(m_e/m)^{3/2}C_6^{-1/2}$ in atomic units. This is typically of order 100 μK in temperature units, and therefore at temperatures of order 1 μK or less, it is generally not important to take into account energy dependence of atomic scattering amplitudes.

The original motivation for investigating dilute atomic gases was to find quantum systems which were tractable theoretically. The criterion for interaction effects to be small is that the magnitude of the scattering length be small compared with the separation between particles:

$$|a| \ll r_s. \tag{1.10}$$

This condition ensures that the most important collision processes in the gas are binary collisions, i. e. when atoms collide, the process may be regarded as a two-body one, unaffected by other atoms. From the estimates given above this condition is well satisfied under most conditions. There are, however, interesting exceptions. By means of Feshbach resonances, which will be discussed in Section 7.2 it is possible to tune scattering lengths to an arbitrary value, and condition (1.10) can be violated. Under such conditions, many-body correlations play a crucial role.

2 Bose–Einstein condensation

The phenomenon of Bose–Einstein condensation in an ideal gas was predicted almost 80 years ago. For a gas in a box, the physics is explained in standard texts [6]. For a general external potential, the Bose-Einstein distribution function is

$$f^0(\epsilon_\nu) = \frac{1}{e^{(\epsilon_\nu-\mu)/kT}-1}, \tag{2.11}$$

where ϵ_ν denotes the energy of the single-particle state for the particular potential under consideration and μ is the chemical potential. The chemical potential is determined as a function of the total number of particles N and T by the condition that the total number of particles be equal to the sum of the occupancies of the individual levels.

In the ground state of the gas of identical bosons, all particles reside in the lowest energy state $\nu = 0$. This implies that the chemical potential, which at zero temperature is the energy required to add a particle to the system, is ϵ_0. At nonzero temperature, the distribution function for all excited states increases. Thus the number of particles in the lowest state must decrease if the total number of particles is held fixed. The important discovery made by Einstein was that for a gas in a three-dimensional box, the occupancy of the lowest state was a nonzero fraction of the total number of particles up to a temperature $\sim T_{\text{deg}}$ [7].

Experiments on dilute gases are generally made in traps whose potential is well approximated by that of an anisotropic harmonic oscillator. One can estimate the Bose–Einstein transition temperature for a trapped gas by assuming that the transition occurs when the central density in the trap is equal to that for Bose–Einstein condensation in a spatially uniform gas, Eq. (1.3). If for simplicity we assume that the trap is isotropic, $V(r) = m\omega_0^2 r^2/2$, with r being the distance from the center of the trap, the spatial size R of the cloud is found from classical arguments using the Boltzmann distribution to be given by

$$k_{\text{B}}T \sim \frac{1}{2}m\omega_0^2 R^2 \tag{2.12}$$

or

$$R \sim \left(\frac{k_{\text{B}}T}{m\omega_0^2}\right)^{1/2}, \tag{2.13}$$

that is of order the thermal velocity of the particle divided by the trap frequency. The central density in the cloud is therefore $\sim N/R^3 \sim N(m\omega_0^2/k_{\text{B}}T_{\text{deg}})^{3/2}$ at the transition temperature $\simeq T_{\text{deg}}$. It thus follows from Eq. (1.3) that the transition temperature is given by

$$T_{\text{deg}} \sim \frac{\hbar\omega_0}{k_{\text{B}}} N^{1/3}. \tag{2.14}$$

In the first experiments, evidence for Bose–Einstein condensation was provided by observations of the properties of clouds when released from an anisotropic trap [9, 10]. Consider first a non-interacting classical gas in a trap which is switched off at time $t = 0$. The position of a particle initially at position $\mathbf{r}_0$ and moving with velocity $\mathbf{v}_0$ is given at a later time t by

$$\mathbf{r}(t) = \mathbf{r}_0 + \mathbf{v}_0 t. \tag{2.15}$$

Let us further assume that the size of the expanded cloud is large compared with the initial size, in which case one may neglect $\mathbf{r}_0$ compared with

$\mathbf{v}_0 t$ in Eq. (2.16). On summing over all particles, one finds for the mean square value of the ith component of the position the expression

$$\overline{r_i^2} \approx \overline{v_{i0}^2} t^2, \tag{2.16}$$

where the bar here denotes an average over all particles. For a classical gas, the mean square velocity of a particle in a given direction is independent of position, and therefore the cloud becomes spherical when it has expanded to much more than its original size. For a quantum mechanical particle in the ground state of an anisotropic harmonic oscillator,

$$V(\mathbf{r}) = \frac{1}{2} m(\omega_1^2 x^2 + \omega_2^2 y^2 + \omega_3^2 z^2), \tag{2.17}$$

where ω_i is the angular frequency of classical motion in the ith direction, the ground state wave function is

$$\phi_0(\mathbf{r}) = \frac{1}{\pi^{3/4}(a_1 a_2 a_3)^{1/2}} e^{-x^2/2a_1^2} e^{-y^2/2a_2^2} e^{-z^2/2a_3^2}, \tag{2.18}$$

where the length

$$a_i = \left(\frac{\hbar}{m\omega_i}\right)^{1/2} \tag{2.19}$$

sets the spatial scale of the wave function. In the directions in which the cloud is tightly confined, the typical particle momenta are larger, by virtue of the Heisenberg uncertainty principle, so on release the cloud will expand more rapidly in the directions in which it was more tightly confined initially. The wave function in the momentum representation is given by

$$\phi_0(\mathbf{p}) = \frac{1}{\pi^{3/4}(c_1 c_2 c_3)^{1/2}} e^{-p_x^2/2c_1^2} e^{-p_y^2/2c_2^2} e^{-p_z^2/2c_3^2}, \tag{2.20}$$

where

$$c_i = \frac{\hbar}{a_i} = \sqrt{m\hbar\omega_i}. \tag{2.21}$$

The mean square velocity in the ith direction is

$$\overline{v_i^2} = \frac{\hbar\omega_i}{4m} = \left(\frac{\hbar}{2ma_i}\right)^2, \tag{2.22}$$

in agreement with the uncertainty principle. When the trap is switched off and the cloud is allowed to expand to a size much greater than its original one, the mean square size of the cloud is thus given by an equation analogous to Eq. (2.16), except that the mean square velocity is given by the quantum mechanical expression (2.22):

$$\overline{r_i^2} \approx \overline{v_{i0}^2} t^2. \tag{2.23}$$

After expansion, the size of the cloud in the ith direction therefore varies as $\omega_i^{1/2}$ and, for an anisotropic trap, the cloud is anisotropic. This is to be contrasted with the isotropic cloud which is expected for a nondegenerate gas after release from an anisotropic trap. The experimental observation of anisotropic expansion provided strong support for the conclusion that a Bose–Einstein condensate had been produced.

3 Interatomic interactions

The properties of clouds are modified significantly by interatomic interactions. An approach which has been extremely successful in treating such effects is that of Gross and Pitaevskii, who adopted a mean-field point of view. The basic idea is to make an ansatz for the many-body wave function which is a product of single-particle wave functions. This is similar to what Hartree did for electrons in atoms. For bosons the problem is simpler than for fermions, since in the fully condensed state, all bosons are in the same single-particle state, $\phi(\mathbf{r})$, and therefore we may write the wave function of the N-particle system as

$$\Psi(\mathbf{r}_1, \mathbf{r}_2, \ldots \mathbf{r}_N) = \prod_{i=1}^{N} \phi(\mathbf{r}_i). \tag{3.24}$$

The single-particle wave function $\phi(\mathbf{r}_i)$ is normalized in the usual way,

$$\int d\mathbf{r} |\phi(\mathbf{r})|^2 = 1. \tag{3.25}$$

This wave function does not contain the correlations produced by the interaction when two atoms are close to each other. These effects are taken into account by using an effective interaction $U_0\delta(\mathbf{r} - \mathbf{r}')$, where $\mathbf{r}$ and $\mathbf{r}'$ are the positions of the two particles. This approach is similar to what is done in the shell model for nucleons in atomic nuclei, where an analogous approximation is made for the wave function, and the nucleon-nucleon interaction, which has a rather complicated structure, is replaced by a simpler effective interaction, for example one of the Skyrme type.

The next task is to relate the effective interaction to the scattering length, which encapsulates all the information required to characterize scattering at low energies. The relation is

$$U_0 = \frac{4\pi\hbar^2 a}{m}. \tag{3.26}$$

To understand this result, let us consider the simpler problem of a single particle of mass M in a spherical box of radius d. The lowest energy s-wave state of a particle is of the form

$$\Phi_0(r) = (C/r)\sin(\pi r/d), \tag{3.27}$$

(the spherical Bessel function j_ℓ for $\ell = 0$), where r is the radial coordinate and C is a normalization constant, which by explicit calculation may be shown to be given by

$$|C|^2 = \frac{1}{2\pi d}. \tag{3.28}$$

The factors in the argument of the sine function in Eq. (3.27) have been chosen so that the wave function vanishes at the surface of the box and is finite for $r \to 0$. The radial wave number for this state is $k = \pi/d$, and the energy of the state is

$$E_0 = \frac{\hbar^2\pi^2}{2Md^2}. \tag{3.29}$$

Next we consider the case when there is a potential acting near the center of the box, and which is characterized by a scattering length a. We shall assume that the range of the potential is small compared with d. In the presence of the potential, the lowest energy s-wave state which extends over the volume of the box has the form $(C'/r)\sin[\pi(r-a)/(d-a)]$ in the region where the potential may be neglected. Here C' is a new normalization constant. The argument of the sine function is such that the wave function vanishes at the surface of the box and at $r = a$, as is required by the condition that the wave function for low energies vary as $1 - a/r$. The wave number of the state is $k = \pi/(d-a)$, and thus for a positive scattering length, the wavelength of the state is reduced from $2d$ to $2(d-a)$. The energy of the state is

$$E = \frac{\hbar^2\pi^2}{2M(d-a)^2}, \tag{3.30}$$

and therefore the energy of the state is changed by an amount

$$\Delta E = E - E_0 = \frac{\hbar^2\pi^2}{2M}\left(\frac{1}{d^2} - \frac{1}{(d-a)^2}\right) \approx \frac{\hbar^2\pi^2 a}{Md^3}. \tag{3.31}$$

If we think about this problem in terms of effective interactions, the shift in the energy of the particle due to the potential should be proportional to the particle density at the origin. For the wave function in the absence of the potential, $|\Phi_0(r=0)|^2 = |C|^2\pi^2/d^2 = \pi/(2d^3)$, and therefore the energy change may be written as

$$\Delta E = \frac{2\pi\hbar^2 a}{M}|\Phi_0(r=0)|^2. \tag{3.32}$$

In general, the change in the energy contains contributions from both the potential energy and also from the kinetic energy. This is illustrated most dramatically for a hard-sphere interaction. The change in the potential energy is then zero, because the wave function always vanishes when the potential is nonzero, and all the energy change is due to changes in the kinetic energy.

Let us now apply this result to two interacting particles, rather than a single particle in an external potential. The arguments above may then be carried out in terms of the relative coordinate of the two particles, by imagining the *relative* motion to be confined within a spherical box. One important difference compared with the case of a single particle in a potential is that in the Schrödinger equation for the relative motion of two particles the reduced mass of the pair of particles enters, rather than just the particle mass. Therefore in the expression for the energy, the mass M must be replaced by reduced mass, which for two particles of mass m is $m/2$. Thus the energy of one particle at point $\mathbf{r}$ due to the presence of a second particle is given by Eq. (3.32), but with M replaced by $m/2$, and the wave function at the origin replaced by the wave function of the second particle at point $\mathbf{r}$:

$$E_{\text{int}} = \frac{4\pi\hbar^2 a}{m}|\phi(\mathbf{r})|^2. \tag{3.33}$$

The average energy due to interaction of two particles is therefore given by multiplying the result (3.32) by the probability of finding the first particle at point $\mathbf{r}$, $|\phi(\mathbf{r})|^2$, and integrating over $\mathbf{r}$:

$$(\Delta E)_{\text{pair}} = \frac{4\pi\hbar^2 a}{m}\int d\mathbf{r}|\phi(\mathbf{r})|^4. \tag{3.34}$$

In the next section we shall apply this result to calculate the energy of a dilute gas.

4 Equilibrium properties of a trapped gas

For a gas so dilute that the typical separation between particles is large compared with the range of the potential and the scattering length, collisions between atoms may be regarded as being essentially binary ones, since the probability of a third atom being close enough to an interacting pair to have a significant effect on the collision is small. Under these circumstances, the energy change due to the interaction may be found by multiplying the energy for a pair of atoms by the number of independent pairs that may be made with N particles, $N(N-1)/2$. Thus

$$E_{\text{int}} = \frac{N(N-1)}{2} U_0 \int d\mathbf{r} |\phi(\mathbf{r})|^4. \tag{4.35}$$

where

$$U_0 = \frac{4\pi\hbar^2 a}{m}. \tag{4.36}$$

The energy shift per particle at point $\mathbf{r}$ is therefore $U_0(N-1)|\phi(\mathbf{r})|^2 = n(\mathbf{r})U_0(1-1/N)$, where $n(\mathbf{r}) = N|\phi(\mathbf{r})|^2$ is the particle density. For a uniform gas, the energy change due to interactions is therefore given by

$$\frac{E_{\text{int}}}{N} \approx \frac{nU_0}{2}, \tag{4.37}$$

when terms of order $1/N$ are neglected.

Interparticle interactions influence the properties of a trapped gas when the shift in energy due to the interaction becomes comparable to or greater than the spacing between single-particle energy levels in the trap, which for an isotropic trap is $\hbar\omega_{\text{osc}}$, where ω_{osc} is the oscillator frequency. Let us estimate the effects of interactions by using perturbation theory, starting from a state in which all particles are in the lowest state of the oscillator. The density distribution in the lowest state of the oscillator has a spatial extent $\sim a_{\text{osc}} = (\hbar/m\omega_{\text{osc}})^{1/2}$, and therefore a typical particle density is $n \sim N/a_{\text{osc}}^3$. The ratio of the interaction energy to the level spacing is therefore

$$\frac{nU_0}{\hbar\omega_{\text{osc}}} \sim \frac{Na}{a_{\text{osc}}}. \tag{4.38}$$

When Na/a_{osc} is small compared with unity, the effects of interparticle interactions may be determined from perturbation theory, while if it is large compared with unity, interactions have a large effect on the trapped cloud.

In experiments, $a_{\rm osc}$ is typically of order one micrometer, while scattering lengths are of order 10^{-2} micrometers. Since typical particle numbers are in the range 10^4 - 10^7, interactions generally have a profound effect.

To understand how large they are, we estimate the radius R of a cloud of interacting bosons in a harmonic trap. We shall employ a variational approach, and shall only make order-of-magnitude calculations. For quantitative details we refer to Refs. [11] and [1, Chapter 6]. The total energy of the cloud may be expressed as the sum of contributions due to the kinetic energy, the potential due to the trap, and the interactions:

$$E = E_{\rm kin} + E_{\rm trap} + E_{\rm int}. \tag{4.39}$$

From the uncertainty principle, the typical momentum of a particle is of order $\hbar/R$, and the kinetic energy $E_{\rm kin}$ for N particles is of order

$$E_{\rm kin} \sim N\frac{\hbar^2}{2mR^2}, \tag{4.40}$$

and the energy due to the trapping potential is given by

$$E_{\rm trap} \sim \frac{N}{2}m\omega_{\rm osc}^2 R^2. \tag{4.41}$$

If particle interactions are neglected, the total energy has a minimum for $R \sim a_{\rm osc} = (\hbar/m\omega_{\rm osc})^{1/2}$, which is consistent with the exact result for the wave function (2.18) for a particle in a a harmonic trap.

Let us now investigate the effects of interactions. The interaction energy may be estimated from Eq. (4.35). Since the magnitude of the wave function is of order $1/R^{3/2}$, as a consequence of the normalization condition, the interaction energy is of order N^2U_0/R^3, which expresses the fact that the energy per particle is of order U_0 times a typical particle density, N/R^3. Let us now calculate the equilibrium radius by minimizing the total energy. For a repulsive interaction, the interaction energy, which decreases with increasing R, will increase the equilibrium radius. When the interaction energy becomes comparable to or greater than the energy of the non-interacting cloud, the kinetic energy, which varies as $1/R^2$ plays little role in determining the equilibrium radius, which may be estimated by minimizing the sum of the interaction energy and the trap energy. By equating the derivative with respect to R of the sum to zero, one finds the condition

$$2E_{\rm trap} = 3E_{\rm int}, \tag{4.42}$$

from which one finds that the equilibrium radius is given by

$$R \sim \left(\frac{Na}{a_{\text{osc}}}\right)^{1/5} a_{\text{osc}}. \tag{4.43}$$

In typical experiments, N is of order $10^4 - 10^6$, a is of order 10^{-8} m, and a_{osc} is of order 10^{-6} m, so Na/a_{osc} is of order $10^2 - 10^4$. Consequently, Bose-Einstein condensed clouds are considerably larger than one would predict for a non-interacting gas.

An attractive effective interaction decreases the size of the cloud, and for sufficiently large and negative Na/a_{osc}, the cloud is no longer stable, and will implode.

To derive a general expression for the equilibrium wave function, it is convenient to adopt a variational approach. The total energy may be written in the form (4.39), and for the wave function (3.24), the kinetic energy is given by

$$E_{\text{kin}} = \frac{N\hbar^2}{2m}\int d\mathbf{r}|\nabla\phi(\mathbf{r})|^2 \tag{4.44}$$

and the energy due to the trap by

$$E_{\text{trap}} = \int d\mathbf{r}V(\mathbf{r})|\phi(\mathbf{r})|^2. \tag{4.45}$$

Combining these results with Eq. (4.35) for the interaction energy, one finds

$$E = N\int d\mathbf{r}\left(\frac{\hbar^2}{2m}|\nabla\phi(\mathbf{r})|^2 + V(\mathbf{r})|\phi(\mathbf{r})|^2 + \frac{(N-1)}{2}U_0|\phi(\mathbf{r})|^4\right). \tag{4.46}$$

For some purposes it is convenient to normalize the wave function in a different way, and introduce the concept of the wave function of the condensed state,

$$\psi(\mathbf{r}) = N^{1/2}\phi(\mathbf{r}). \tag{4.47}$$

The density of particles is then given by

$$n(\mathbf{r}) = |\psi(\mathbf{r})|^2, \tag{4.48}$$

and, with the neglect of terms of order $1/N$, the energy of the system may therefore be written as

$$E(\psi) = \int d\mathbf{r}\left[\frac{\hbar^2}{2m}|\nabla\psi(\mathbf{r})|^2 + V(\mathbf{r})|\psi(\mathbf{r})|^2 + \frac{1}{2}U_0|\psi(\mathbf{r})|^4\right]. \tag{4.49}$$

To find the equilibrium form for ψ, we minimize the energy (4.49) with respect to variations of $\psi(\mathbf{r})$ and $\psi^*(\mathbf{r})$ subject to the condition that the total number of particles

$$N = \int d\mathbf{r} |\psi(\mathbf{r})|^2 \tag{4.50}$$

be constant. The constraint is conveniently taken care of by the method of Lagrange multipliers. One finds

$$-\frac{\hbar^2}{2m}\nabla^2\psi(\mathbf{r}) + V(\mathbf{r})\psi(\mathbf{r}) + U_0|\psi(\mathbf{r})|^2\psi(\mathbf{r}) = \mu\psi(\mathbf{r}), \tag{4.51}$$

where μ, the Lagrange multiplier, is the chemical potential. This equation is the time-independent Gross-Pitaevskii equation. It has the form of a Schrödinger equation in which the potential acting on a particle is the sum of the external potential V and a non-linear term $U_0|\psi(\mathbf{r})|^2$ that takes into account interactions with the other particles. Note that the eigenvalue is the chemical potential, the energy required to add a particle to the system, which for an interacting system is not the same as the average energy per particle.

For a uniform Bose gas, ψ = constant is a solution of the Gross-Pitaevskii equation, with $\mu = U_0|\psi|^2$. In the presence of an external potential, it is generally necessary to solve the equation numerically. However, under many situations of experimental interest, it is a good first approximation to neglect the kinetic energy, in which case the Gross-Pitaevskii equation reduces to

$$[V(\mathbf{r}) + U_0|\psi(\mathbf{r})|^2]\psi(\mathbf{r}) = \mu\psi(\mathbf{r}), \tag{4.52}$$

which is referred to as the Thomas–Fermi approximation since it resembles the Thomas-Fermi approximation in the theory of electrons in atoms and solids. The solution of the Thomas–Fermi equation is

$$n(\mathbf{r}) = |\psi(\mathbf{r})|^2 = (\mu - V(\mathbf{r}))\,/U_0, \tag{4.53}$$

where the right hand side is positive, while $\psi = 0$ elsewhere. The boundary of the cloud is therefore given by

$$V(\mathbf{r}) = \mu. \tag{4.54}$$

For a harmonic trap, the Thomas-Fermi theory predicts a density profile having the form of an inverted parabola, rather than the gaussian form found in the absence of interactions.

5 Dynamics of condensates

A remarkable property of Bose-Einstein condensates is that they can exhibit superfluidity. To understand such phenomena, it is necessary to develop a theory of the dynamics. The natural generalization of equation (4.51) to time-dependent situations is the time-dependent Gross-Pitaevskii equation,

$$-\frac{\hbar^2}{2m}\nabla^2\psi(\mathbf{r},t)+V(\mathbf{r})\psi(\mathbf{r},t)+U_0|\psi(\mathbf{r},t)|^2\psi(\mathbf{r},t)=i\hbar\frac{\partial\psi(\mathbf{r},t)}{\partial t}. \tag{5.55}$$

The physical content of the equation may be brought out by writing $\psi = fe^{i\phi}$. Making this substitution in Eq. (5.55), and dividing by $e^{i\phi}$, one finds by equating the imaginary part of the equation to zero that

$$\frac{\partial(f^2)}{\partial t}=-\frac{\hbar}{m}\boldsymbol{\nabla}\cdot(f^2\boldsymbol{\nabla}\phi). \tag{5.56}$$

This has the form of a continuity equation and it expresses conservation of particle number, since the particle density is given by

$$n=f^2. \tag{5.57}$$

The quantity

$$\mathbf{j}=\frac{\hbar}{m}f^2\boldsymbol{\nabla}\phi \tag{5.58}$$

thus corresponds physically to the particle number current density, and

$$\mathbf{v}=\frac{\mathbf{j}}{n}=\frac{\hbar}{m}\boldsymbol{\nabla}\phi \tag{5.59}$$

is naturally interpreted as the particle velocity. The reason for the continuity equation having exactly the same form as for non-interacting particles,

$$\frac{\partial n}{\partial t}+\boldsymbol{\nabla}\cdot(n\mathbf{v})=0, \tag{5.60}$$

is that the energy of interaction between particles is local and independent of particle momenta. A second equation is obtained by equating to zero the real part of Eq. (5.55) after division by $e^{i\phi}$. This is

$$\hbar\frac{\partial\phi}{\partial t}=-[V(\mathbf{r})+U_0f^2+\frac{1}{2}mv^2-\frac{\hbar^2}{2mf}\nabla^2 f]. \tag{5.61}$$

This is a form of the Josephson relation, which expresses the fact that the phase of the wave function decreases in time at a rate which is the local chemical potential divided by $\hbar$. For a system in an energy eigenstate, the result follows quite simply if one makes use of the fact that the condensate wave function corresponds to a matrix element of the boson annihilation operator, $\hat{\psi}(\mathbf{r})$:

$$\psi(\mathbf{r},t) = \langle N-1|\hat{\psi}(\mathbf{r})|N\rangle. \tag{5.62}$$

Since the eigenstates develop in time as $\exp(-iE_N t/\hbar)$, the matrix element varies as $\exp(-i(E_N - E_{N-1})t/\hbar)$, from which it follows that

$$\frac{\partial\phi}{\partial t} = -\frac{E_N - E_{N-1}}{\hbar} = -\frac{\mu}{\hbar}. \tag{5.63}$$

The latter result follows if the states N and $N-1$ are ground states, since μ is the energy required to add a particle to the ground state. In Eq. (5.61), the first term on the right hand side corresponds to the contribution to μ from the external potential, the second to the interaction energy evaluated for the local density, and the third term to the kinetic energy of the condensate. The final term is due to nonuniformity of the magnitude of the condensate wave function.

To find the equation of motion for the velocity, given by Eq. (5.59), we take the gradient of Eq. (5.61), and the resulting equation is

$$m\frac{\partial\mathbf{v}}{\partial t} = -\boldsymbol{\nabla}(\tilde{\mu} + \frac{1}{2}mv^2), \tag{5.64}$$

where

$$\tilde{\mu} = V + nU_0 - \frac{\hbar^2}{2m\sqrt{n}}\nabla^2\sqrt{n}. \tag{5.65}$$

Apart from the last term in Eq. (5.65), the so-called "quantum pressure term", Eqs. (5.60) and (5.64) have precisely the same form as for the hydrodynamics of a perfect fluid. This remarkable result, which has far-reaching consequences, is due to the fact that the superfluid may be described by two variables, the amplitude and the phase of the condensate wave function (or, equivalently, the particle velocity). In the case of an (ordinary) perfect fluid, which has many degrees of freedom (corresponding classically to the positions and velocities of every particle), a simplified description is possible since it is assumed that collisions are so frequent that the system is in thermodynamic equilibrium locally. The fluid may then be described in terms of the local density, fluid velocity and temperature. The conditions for a

Bose–Einstein condensate are very different, and the reason for the hydrodynamic description being possible is that all particles are in the same quantum state. The hydrodynamic equations for superfluids and ordinary fluids are very similar since they follow from the conservation laws for particle number and momentum.

As an application, we now consider small oscillations of a uniform Bose gas ($V = 0$). In equilibrium, the chemical potential is given by $\mu = nU_0$.

Following the standard procedure employed in, e. g. deriving the equations for sound waves in a classical gas, we linearize the continuity equation (5.60) and the acceleration equation (5.64) by writing $n = n_{\rm eq} + \delta n$ and treating the velocity $\mathbf{v}$ and the changes in the other terms in the equations as small quantities. The result is

$$\frac{\partial \delta n}{\partial t} = -n\boldsymbol{\nabla}\cdot\mathbf{v} \tag{5.66}$$

and

$$m\frac{\partial \mathbf{v}}{\partial t} = -\boldsymbol{\nabla}\delta\tilde{\mu}, \tag{5.67}$$

where $\delta\tilde{\mu}$, which is obtained by linearizing (5.65), is given by

$$\delta\tilde{\mu} = U_0\delta n - \frac{\hbar^2}{4mn}\nabla^2\delta n. \tag{5.68}$$

To simplify notation, we have written n for $n_{\rm eq}$. Taking the time derivative of (5.66) and eliminating the velocity by means of (5.67) results in the equation of motion

$$m\frac{\partial^2 \delta n}{\partial t^2} = \boldsymbol{\nabla}\cdot(n\boldsymbol{\nabla}\delta\tilde{\mu}). \tag{5.69}$$

We look for travelling wave solutions, proportional to $\exp(i\mathbf{q}\cdot\mathbf{r} - i\omega t)$. From Eq. (5.65) the change in $\tilde{\mu}$ is seen to be equal to

$$\delta\tilde{\mu} = \left(U_0 + \frac{\hbar^2 q^2}{4mn}\right)\delta n \tag{5.70}$$

and the equation of motion becomes

$$m\omega^2\delta n = \left(nU_0 q^2 + \frac{\hbar^2 q^4}{4m}\right)\delta n. \tag{5.71}$$

It is convenient to work with the energy of an excitation, ϵ_q, rather than the frequency. Nonvanishing solutions of (5.71) are possible only if the frequency is given by $\omega = \pm\epsilon_q/\hbar$, where

$$\epsilon_q = \sqrt{2nU_0\epsilon_q^0 + (\epsilon_q^0)^2}. \tag{5.72}$$

Here

$$\epsilon_q^0 = \frac{\hbar^2 q^2}{2m} \tag{5.73}$$

is the free-particle energy. This spectrum was first derived by Bogoliubov from microscopic theory in a classic paper [12]. For long wavelengths, the excitations are sound waves with velocity s given by

$$s = \sqrt{(nU_0/m)}. \tag{5.74}$$

This is just what one would expect on the basis of hydrodynamics, since the sound velocity is given by $s = \sqrt{(\partial P/\partial \rho)}$, where P is the pressure and $\rho = nm$ is the mass density [13], and for a uniform Bose gas, the energy density and the pressure are both equal to $U_0 n^2/2$. For short wavelengths ($q \gg ms/\hbar$), the quantum pressure term dominates, and the leading term in the energy is the free-particle energy shifted by an amount nU_0 due to interactions. The length

$$\xi = \hbar/(\sqrt{2}ms) \tag{5.75}$$

(the factor $\sqrt{2}$ is inserted to agree with the usual convention) is referred to as the healing length. On length scales greater than this, mean field effects dominate, while on smaller length scales particles behave more like free particles. The healing length also gives the length scale over which the condensate wave function in a bulk superfluid responds to a localized perturbation, such as the container boundary.

Bogoliubov's calculation resolved one of the great puzzles in the theory of liquid ^{4}He. In 1938, Fritz London suggested that the superfluid properties of liquid ^{4}He were related to the phenomenon of Bose–Einstein condensation [14], and Tisza developed a two-fluid model based on the idea of Bose–Einstein condensation of an ideal Bose gas [15]. Landau objected to Tisza's proposal, his argument being in essence that a weak long-wavelength perturbation of arbitrarily low frequency could remove atoms from the condensed state, and thereby lead to friction. On the basis of rather general considerations, but without invoking explicitly Bose–Einstein condensation, Landau had argued that the elementary excitations of liquid ^{4}He at long wavelengths should be sound waves, and he developed a two-fluid model based on a spectrum of elementary excitations not of the free-particle form [16]. Bogoliubov's calculations showed for the first time how in a microscopic theory, particle interactions could lead to elementary excitations which had a sound-like dispersion relation at long wavelengths. This feature is due to the mean field produced by particle interactions, rather than particle collisions, which make possible a hydrodynamic description of classical gases.

The observation of collective modes of Bose–Einstein condensates in traps is an important tool for investigating condensate properties. The characteristic scale of frequencies may be estimated from results for the uniform gas, since one would expect the fundamental mode of oscillation, corresponding to a purely radial ("breathing") motion to have a period τ roughly equal to the time for a sound wave to travel across the condensate, i. e. $\tau \simeq R/s$. Here s is the sound speed, Eq. (5.74) evaluated at the centre of the cloud, $s \sim (NU_0/mR^3)^{1/2}$, and therefore the frequency of the mode is given in order of magnitude by $\omega \sim (NU_0/mR^5)^{1/2}$, which with the use of Eq. (4.43) is of order the trap frequency ω_0. Repulsive interactions increase the size of the cloud (which varies as $U_0^{1/5}$), but they also increase the sound velocity by the same factor. Many different modes have been observed experimentally, including those corresponding to a motion of the centre of mass of the cloud, quadrupolar modes, and surface modes. A particularly interesting example is the so-called scissors mode in an anisotropic trap. Imagine a gas in an anisotropic trap. If the trap is suddenly rotated by a small angle, a perfect classical fluid would oscillate like a rigid body about the equilibrium position corresponding to the new configuration of the trap. However, as we shall demonstrate in the next section, a condensate cannot rotate like a rigid body and, consequently, a condensate will undergo a motion very different from rigid-body rotation. The observation of this mode provided striking confirmation of fundamental properties of the dynamics of condensates [19].

6 Potential flow and quantized vortices

The fact that the velocity of the condensate is the gradient of the phase of the condensate wave function, Eq. (5.59), leads immediately to the conclusion that

$$\nabla \times \mathbf{v} = \nabla \times \frac{\hbar}{m} \nabla \phi = 0, \tag{6.76}$$

provided the phase does not have a singularity at the point in question. This equation states that the velocity field is irrotational. Consequently, the motions possible for a condensate are a very limited subset of those for a classical fluid. In particular, a condensate cannot rotate like a rigid body: for rotation with angular velocity $\mathbf{\Omega}$, the velocity field is given by $\mathbf{v} = \mathbf{\Omega} \times \mathbf{r}$, from which it follows that $\nabla \times \mathbf{v} = 2\mathbf{\Omega} \neq 0$. The requirement of irrotational flow affects the frequencies of collective modes in condensates, and the good agreement between calculated and observed frequencies provides support for the correctness of the condition that the flow be irrotational.

From the fact that the condensate wave function is single-valued it follows that the change $\Delta\phi$ in the phase of the wave function must be a multiple of 2π, or

$$\Delta\phi = \oint \nabla\phi \cdot d\mathbf{l} = 2\pi\ell, \tag{6.77}$$

where ℓ is an integer. Thus the *circulation* Γ around a closed contour is given by

$$\Gamma = \oint \mathbf{v} \cdot d\mathbf{l} = \frac{\hbar}{m} 2\pi\ell = \ell\frac{h}{m}, \tag{6.78}$$

which shows that it is quantized in units of h/m. The quantization of circulation was first predicted by Onsager in a footnote to a conference paper on classical turbulence, and elaborated by him in a comment following one of the other papers at the same conference [17]. Onsager's profound insight provided the basis for the study of quantized vorticity in both uncharged superfluids and superconductors. The initial impact of Onsager's work was slight, and it was only after its rediscovery by Feynman in the 1950s that systematic investigations of quantized vortex lines began [18, 20].

The simplest example of quantized vorticity is a single vortex line. Consider a bulk condensate in a uniform external potential, and we assume that the wave function is independent of the z coordinate. For the condensate wave function to be single valued, it must vary as $e^{i\ell\varphi}$, where φ is the azimuthal angle and ℓ is an integer. If ρ is the distance from the z axis, it follows from Eq. (6.78) that the velocity is

$$v_\varphi = \ell\frac{h}{2\pi m\rho}. \tag{6.79}$$

The circulation is thus $\ell h/m$ if the contour encloses the axis, and zero otherwise. If $\ell \neq 0$, the condensate wave function must vanish on the axis of the trap, since otherwise the kinetic energy due to the azimuthal motion would diverge. The structure of the flow pattern is thus that of a vortex line. Under most circumstances, vortex lines with $|\ell| \neq 1$ are unstable to decay into a number of vortices with $|\ell| = 1$.

Much work has been done on quantized vortex lines in Bose–Einstein condensed gases, and a review may be found in Ref. [21]. We cannot go into detail here, but will mention some highlights of recent work. Under rotation, a superfluid can mimic solid-body rotation by means of an array of singly-quantized vortex lines in which the number per unit area perpendicular to the rotation axis is equal to $2m\Omega/h$, since this ensures that the average circulation is 2Ω. Experimentally, arrays with some hundreds of vortex lines have been observed, and the lines are arranged on a regular triangular lattice [22]. At low rotation rates, the radius of the vortex core is of order the healing length ξ, Eq. (5.75), but for fast rotation, vortices can be packed together so closely that their cores begin to overlap, and the effective core size is of order the vortex spacing [23]. In this regime, the physics bears similarities to that of the quantum Hall effect [24]. The investigation of properties of rapidly rotating condensate is currently an active field of study. Many novel effects have been seen, including the elastic (Tkachenko) modes of the vortex lattice [25, 26].

7 Other topics

Finally, we mention briefly some other important areas of current research.

7.1 Cold atoms in periodic potentials

Designing and constructing periodic condensed-matter systems with specific electronic properties has been one of the important areas of physics over the past two decades. Similar effects can be achieved for atoms in the periodic potential produced by a standing-wave light field produced by two oppositely-directed travelling waves. If one adds together two light waves propagating in the z direction with electric fields in the x direction given by $\mathcal{E}_0 \cos(kz - \omega t)$ and $\mathcal{E}_0 \cos(-kz - \omega t)$, the total electric field is given by

$$\mathcal{E} = \mathcal{E}_0[\cos(kz - \omega t) + \cos(-kz - \omega t)] = 2\mathcal{E}_0 \cos kz \cos \omega t. \tag{7.80}$$

When averaged over time, the energy shift of an atom in the electric field, given by Eq. (1.4), has the form

$$\overline{\Delta E} = -\alpha(\omega)\mathcal{E}_0^2 \cos^2 kz = -\frac{1}{2}\alpha(\omega)\mathcal{E}_0^2(1 + \cos 2kz). \tag{7.81}$$

The laser beams thus produce a potential acting on the atoms which is sinusoidal in space, hence the name "optical lattice". The period of the lattice is equal to half that of the light. By superimposing two or three pairs of laser beams one can make two- and three-dimensional lattices. The original suggestion of the possibility of making optical lattices dates from 1968 [27] and the first experimental realization of a lattice was made in 1987 [28]. Since then the field has burgeoned, especially after the production of atomic Bose–Einstein condensates.

Many experiments have been performed on such systems, and one recent example is the observation of the transition between a Bose–Einstein condensed state (a superfluid) and one in which atoms are localized on individual sites (an insulator) [29]. This is analogous to the Mott transition for electrons in solids [30]. To set the scene, let us first consider a spatially-uniform, interacting Bose gas at zero temperature. At low temperature, this will display superfluid behavior because the wave function of the condensate is coherent over the whole of the gas. When a three-dimensional optical lattice is applied to the system, the atoms tend to become more and more localized around the lattice sites at which the potential has local minima. Eventually, for sufficiently strong lattice potentials, there will be essentially no transmission of atoms between different potential wells. When that is the case, there is no mechanism for locking together the phases of the wave function in different parts of space, and long-range coherence is lost. The system will behave as an insulator, since application of a weak external field will have no effect because tunneling between wells is absent due to the high potential barriers and the repulsion between atoms on the same site. Note that the insulating state cannot be described within the Gross-Pitaevskii mean-field approach because the number of atoms on each lattice site is essentially fixed,, and there is no phase coherence between different sites. Experimentally, the transition between the two sorts of states is detected by switching the lattice potential off and allowing the gas to expand. When it is a superfluid, there is long-range coherence between atoms on different sites, and one observes an interference pattern determined by the structure of the lattice. In the insulating state, there is no coherence between atoms on different sites, and no interference pattern is seen.

Interparticle interactions can have important effects even when the system is superfluid [31]. One of these is the appearance of swallow-tail features in the band structure. For a single particle in a periodic potential, the wave functions $\phi(\mathbf{r})$ of extended states have the general form

$$\phi(\mathbf{r}) = f(\mathbf{r})e^{i\mathbf{k}\cdot\mathbf{r}}, \tag{7.82}$$

where $f(\mathbf{r})$ has the periodicity as the lattice. This result is referred to as Bloch's theorem, or Floquet's theorem. Also for the Gross-Pitaevskii equation one can find solutions having the usual Bloch form. As shown for a

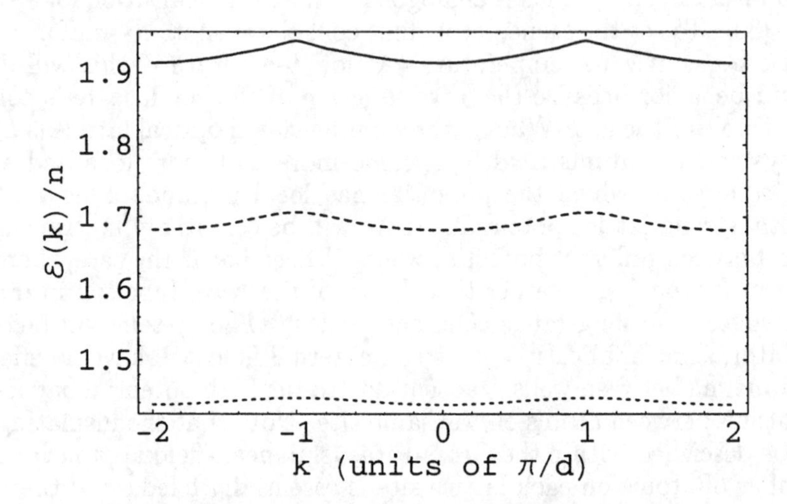

Figure 1: Energy per particle in units of $\hbar\omega$ versus wave number for the lowest band for a Bose-Einstein condensate in a periodic potential. The dotted curve is for $nU_0/V_0 = 0.57$, the dashed one for $nU_0/V_0 = 0.78$ and the full line for $nU_0/V_0 = 0.99$. The quantity $\omega = (2V_0/m)^{1/2}\pi/d$ is the frequency of small oscillations about the minimum of the potential.

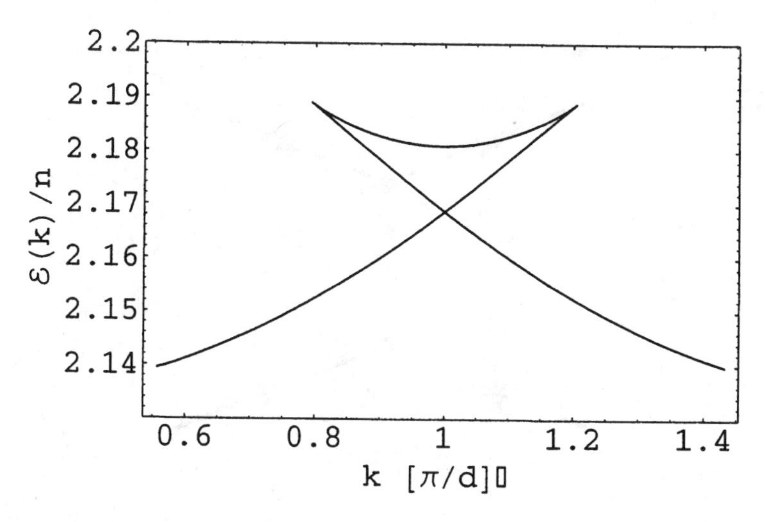

Figure 2: Same as Fig. 1 but for $nU_0/V_0 = 1.2$.

one-dimensional lattice in Fig. 1, with increasing interparticle interaction U_0, the energy spectrum becomes increasingly peaked at $k = \pi/d$ and, remarkably, for $nU_0 > V_0$ the energy per particle becomes triple valued for k close to π/d [32]. These effects may be shown to be related to periodic soliton solutions [34, 33]. Yet another effect related to the interparticle interaction is that the Gross-Pitaevskii equation has solutions which have a periodicity different from that of the lattice. If the particle density is not the same within each lattice cell, the total potential acting on an atom will not have the periodicity of the lattice. For a one-dimensional optical lattice, solutions with periods equal to integer multiples of the spacing of the optical lattice have been found [35]. So far, these new nonlinear effects have not been observed directly in experiment.

7.2 Strong correlations

The original motivation for investigating ultracold atomic gases was to study quantum phenomena and superfluidity under conditions when particle interactions were relatively unimportant, in the sense that the particle spacing r_s is large compared with the magnitude of the scattering length a. In recent years, increasing attention has been paid to situations where correlations are strong, that is $|a| \gtrsim r_s$. The tool employed is what is usually referred to as a

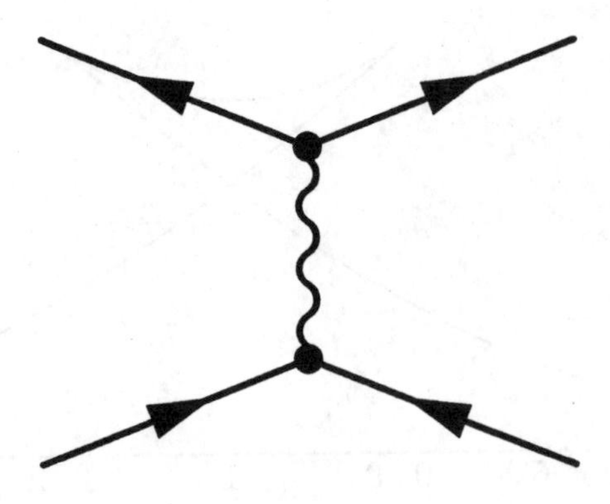

Figure 3: Schematic representation of a process in which two atoms combine to make a molecule, which

Feshbach resonance [36]. The basic idea is that if there is a molecular state of two atoms close to the energy of two low-energy atoms, the process in which atoms are scattered to the molecular state with energy $E_{\rm mol}$, which subsequently decays into atoms again, will shift the energy of the atoms. The process is represented diagramatically in Fig. 2. Quantitatively, the strength of the induced interaction (or the scattering amplitude T for atom-atom scattering) may be estimated by second order perturbation theory, and it is

$$T_{\rm res} = \frac{g^2}{E - E_{\rm mol}}, \tag{7.83}$$

where g is the strength of the matrix element for coupling low-energy atoms to the molecular state, and E is the energy of the two atoms. Thus there is attraction between atoms if the molecular state lies at positive energy and repulsion if the state is at negative energy, i. e. , it is bound with respect to two atoms.

In most situations the magnitudes of induced interactions are limited by the fact that resonances decay, thereby limiting the magnitude of the induced interaction to $\sim g^2/\Gamma$, where Γ is the width of the molecular state. At the low energies of importance in experiments with cold atoms, the density of two-atom states into which the molecule can decay, which varies as $E^{1/2}$ is restricted, and widths can be very small. Consequently, large scattering lengths can be realized and in practice they can be as large as micrometers, two orders of magnitude larger than the typical scale of scattering lengths, Eq. (1.9), when resonances are absent. An important feature of these resonances is that, because the magnetic moment of the molecule is generally not equal to that of two atoms, the energy of the resonance varies with the strength of the magnetic field. For small values of the departure of the magnetic field from the value B_0 at which the molecule has zero energy ($E_{\rm mol} = 0$), one may write

$$E_{\rm mol} = \Delta\mu(B - B_0), \tag{7.84}$$

where $\Delta\mu$ is the difference between the magnetic moment of two atoms and that of the molecule. Consequently, the resonant contribution to the scattering amplitude at zero energy (which is equal to the effective interaction) is given by

$$T_{\rm res}(E = 0) = -\frac{g^2}{\Delta\mu(B - B_0)}. \tag{7.85}$$

We saw earlier that the scattering length is related to the effective interaction by the relation $U_0 = 4\pi\hbar^2 a/m$. In general, there will be background (nonresonant) contributions to the scattering length, and we denote these by $a_{\rm bg}$. Combining these contributions one finds

$$a = a_{\rm bg}\left(1 - \frac{\Delta B}{B - B_0}\right), \tag{7.86}$$

where

$$a_{\rm bg}\Delta B = \frac{mg^2}{4\pi\hbar^2\Delta\mu}. \tag{7.87}$$

Equation (7.86) is the standard phenomenological expression used to fit properties of Feshbach resonances. The relationship between the phenomenologial parameters and the microscopic ones is given by Eq. (7.87). When there is a Feshbach resonance close to zero energy, the scattering length may be tuned to essentially any value, positive or negative.

One application of Feshbach resonances is to produce strongly correlated Bose gases. Theory predicts that when the scattering length becomes large compared with the interparticle spacing, the energy per particle should be of order $(\hbar^2/m)n^{2/3}$ and independent of the scattering length, rather than the low density result $(2\pi\hbar^2/m)na$, Eq. (4.37) [37]. A second application is to create the BCS superfluid state of a mixture of fermions in two different internal (usually different hyperfine) states. This state is similar to that of electrons in conventional superconductors. For strong interactions, transition temperatures are predicted to be as large as $\sim 0.25T_{\rm F}$, where $T_{\rm F}$ is the Fermi temperature [38]. Such temperatures can be realized experimentally, and much effort is being devoted to creating and detecting the superfluid state. A third application is to form cold molecules by sweeping the magnetic field through the resonance value, at which $E_{\rm mol} = 0$. This is the method employed in the recent experiments that realized Bose–Einstein condensation of molecules [39].

8 Concluding remarks

There are many other topics within the field of atomic quantum gases which we have not been able to take up here. One is the properties of low-dimensional systems, which can be created by making traps which are very tight in some directions. Another area important for quantum information is the control of atomic wave functions, including phase information. Cold atomic gases are also being exploited in the development of better frequency standards. The study of cold atoms has brought together workers in diverse subfields of physics, including atomic physics, condensed matter physics, nuclear physics, nonlinear physics, and quantum optics. The field is developing rapidly, and it promises to do so in the foreseeable future.

References

[1] C. J. Pethick and H. Smith, *Bose–Einstein Condensation in Dilute Gases*, Cambridge University Press, Cambridge (2002).

[2] L. P. Pitaevskii and S. Stringari, *Bose–Einstein Condensation*, Oxford University Press, Oxford (2003).

[3] F. Dalfovo, S. Giorgini, L. P. Pitaevskii, and S. Stringari, *Rev. Mod. Phys.* **71**, 463 (1999).

[4] A. J. Leggett, Rev. Mod. Phys. **73**, 307 (2001).

[5] H. Metcalf and P. van der Straten, *Laser Cooling and Trapping*, (Springer, Berlin, 1999).

[6] L. D. Landau and E. M. Lifshitz, *Statistical Physics*, Part 1 (Pergamon, Oxford, 1980), §62.

[7] A. Einstein, Sitzungsberichte der Preussischen Akademie der Wissenschaften, Physikalisch-mathematische Klasse 1924, 261 (1924); 1925, 3 (1925).

[8] H. F. Hess, Phys. Rev. B **34**, 3476 (1986).

[9] M. H. Anderson, J. R. Ensher, M. R. Matthews, C. E. Wieman, and E. A. Cornell, Science **269**, 198 (1995).

[10] K. B. Davis, M.-O. Mewes, M. R. Andrews, N. J. van Druten, D. S. Durfee, D. M. Kurn, and W. Ketterle, Phys. Rev. Lett. **75**, 3969 (1995).

[11] G. Baym and C. J. Pethick, Phys. Rev. Lett. **76**, 6 (1996).

[12] N. N. Bogoliubov, J. Phys. (USSR) **11**, 23 (1947), reprinted in D. Pines, *The Many-Body Problem*, (W. A. Benjamin, New York, 1961), p. 292.

[13] At nonzero temperature, it is important that the derivative should be evaluated for constant entropy per particle. However, for a Bose–Einstein condensate, the entropy is zero, so the restriction on the derivative is unimportant.

[14] F. London, Nature **141**, 643 (1938); Phys. Rev. **54**, 947 (1938).

[15] L. Tisza, Nature **141**, 913 (1938).

[16] L. D. Landau, J. Phys. (U.S.S.R.) **5**, 71 (1941).

[17] L. Onsager, Nuovo Cimento **6**, Suppl. 2, 249 (1949).

[18] R. P. Feynman, in *Progress in Low Temperature Physics*, Vol. 1, ed. C. J. Gorter, (North-Holland, Amsterdam, 1955), Chap. 2.

[19] O. Maragò, S. A. Hopkins, J. Arlt, E. Hodby, G. Hechenblaikner, and C. J. Foot, Phys. Rev. Lett. **84**, 2056 (2000).

[20] The monograph by R. J. Donnelly, *Quantized Vortices in Liquid He II*, (Cambridge University Press, Cambridge, 1991), gives an extensive account of the effects of rotation on superfluid liquid ^{4}He.

[21] A. L. Fetter and A. A. Svidzinsky, J. Phys.; Condens. Matter **13**, 135 (2001).

[22] J. R. Abo-Shaeer, C. Raman, J. M. Vogels, and W. Ketterle, Science **292**, 476 (2001); P. C. Haljan, I. Coddington, P. Engels, and E. A. Cornell, Phys. Rev. Lett. **87**, 210403 (2001); P. Engels, I. Coddington, P. C. Haljan, and E. A. Cornell, Phys. Rev. Lett. **89**, 100403 (2002).

[23] V. Schweikhard, I. Coddington, P. Engels, V. P. Mogendorff, and E. A. Cornell, cond-mat/0308582.

[24] T.-L. Ho, Phys. Rev. Lett. **87**, 060403 (2001).

[25] V.K. Tkachenko, Zh. Eksp. Teor. Fiz. **49**, 1875 (1965) [Sov. Phys. JETP **22**, 1282 (1966)]; Zh. Eksp. Teor. Fiz. **50**, 1573 (1966) [Sov. Phys. JETP **23**, 1049 (1966)].

[26] I. Coddington, P. Engels, V. Schweikhard, and E. A. Cornell, Phys. Rev. Lett. **91**, 100402 (2003).

[27] V. S. Letokhov, JETP Letters **7**, 272 (1968).

[28] C. Salamon, J. Dalibard, A. Aspect, H. Metcalf, and C. Cohen-Tannoudji, Phys. Rev. Lett. **59**, 1659 (1987).

[29] M. Greiner, O. Mandel, T. Esslinger, T. W. Hänsch, and I. Bloch, Nature **415**, 39 (2002).

[30] For theoretical considerations, see M. P. A. Fisher, P. B. Weichman, G. Grinstein, and D. S. Fisher, Phys. Rev. B **40**, 546 (1989); D. Jaksch, C. Bruder, J. I. Cirac, C. W. Gardiner, and P. Zoller, Phys. Rev. Lett. **81**, 3108 (1998).

[31] B. Wu and Q. Niu, New J. Phys. **5**, 104 (2003).

[32] D. Diakonov, L. M. Jensen, C. J. Pethick, and H. Smith, Phys. Rev. **A** **66**, 013604 (2002).

[33] T. Tsuzuki, Journ. Low Temp. Phys. **4**, 441 (1971).

[34] M. Machholm, C. J. Pethick, and H. Smith, Phys. Rev. A **67**, 053613 (2003).

[35] M. Machholm, A. Nicolin, C. J. Pethick, and H. Smith, cond-mat/0307183

[36] They are named after Herman Feshbach, who discussed such effects in the context of nuclear physics.

[37] S. Cowell, H. Heiselberg, I. E. Mazets, J. Morales, V. R. Pandharipande, and C. J. Pethick, Phys. Rev. Lett. **88**, 210403 (2002).

[38] H. Heiselberg, Phys. Rev. B **34**, 3476 (2001); J. Carlson, S. Y. Chang, S. Pieper, and V. R. Pandharipande, Phys. Rev. Lett. **91**, 50401 (2003).

[39] S. Jochim, M. Bartenstein, A. Altmeyer, G. Hendl, S. Riedl, C. Chin, J. H. Denschlag, and R. Grimm Science **302** 2101 (2003); M. W. Zwierlein, C. A. Stan, C. H. Schunck, S. M. F. Raupach, S. Gupta, Z. Hadzibabic, and W. Ketterle, Phys. Rev. Lett. **91**, 250401 (2003); M. Greiner, C. A. Regal, and D. S. Jin, cond-mat/0311172.

12. LASER COOLING AND TRAPPING OF NEUTRAL ATOMS TO ULTRALOW TEMPERATURES

KRISTIAN HELMERSON
Atomic Physics Division, Physics Laboratory
National Institute of Standards and Technology - Gaithersburg, Maryland 20899-8424

1. Introduction

A Bose-Einstein condensate (BEC) of a dilute atomic vapor represents the coldest matter ever produced in a laboratory, or even observed in the universe. The temperature of the atoms, which we usually associate with their translational (kinetic) energy can easily be in the nanoKelvin regime. This is truly an extreme physical condition and the optical techniques developed to create, probe and manipulate Bose-Einstein condensates are at the frontiers of optical spectroscopy.

BEC in a dilute atomic vapor requires sufficiently low densities such that the atoms can be considered as weakly interacting. This in turn necessitates the attainment of extremely low temperatures such that the thermal deBroglie wavelength of the atoms is comparable to the interatomic spacing. To obtain and maintain such extraordinarily low temperatures required the development of cooling and trapping techniques for neutral atoms. Certainly, the techniques developed by the spin polarized Hydrogen community especially evaporative cooling, has been essential in producing Bose-Einstein condensates; however, laser cooling and trapping of neutral atoms has allowed BEC in dilute gases to be observed first with alkalis and has contributed greatly to many of the beautiful experiments with condensates.

These notes are based on three lectures given at the International School of Atomic and Molecular Spectroscopy at the Ettore Majorana Center for Scientific Culture in May of 2003. The first two lectures were a general discussion of the mechanical effects of light and laser cooling and trapping techniques for neutral atoms. The third lecture describes experiments with Bose-Einstein condensed sodium atoms at the National Institute of Standards and Technology (NIST) in Gaithersburg, Maryland, and represents a direct application of many of the wonderful techniques developed for manipulating neutral atoms.

The topics to be covered in the general discussion of the mechanical effects of light, and laser cooling and trapping techniques include:

B. Di Bartolo and O. Forte (eds.), *Frontiers of Optical Spectroscopy*, 427-495.

radiative forces, both spontaneous and dipole; diffraction of atoms by optical standing waves; Doppler cooling and the Doppler cooling limit; deceleration and cooling of atomic beams; optical molasses and the discovery of sub-Doppler laser cooling; new laser cooling mechanisms, including polarization gradient and Sisyphus heating; various topics related to trapping neutral atoms, including magnetic traps, both static and time-averaged orbiting potentials (TOP); laser dipole traps and far-off resonant traps (FORTs); radiation pressure traps and the optical Earnshaw theorem; magneto-optical traps (MOTs); mechanisms limiting the loading of MOTs and techniques to increase the densities, including dark spot MOTs; and finally, the confinement of atoms in optical lattices.

These topics have been selected as representing some of the important developments in the interaction of electromagnetic fields and atoms which have enabled the creation and manipulation of Bose-Einstein condensates. More details on these topics, especially as they are applied to BEC research, can be found in other sets of lecture notes, special journal articles and review articles [1, 2, 3, 4, 5, 6, 7]

In addition, there are other related topics in the manipulation of matter by light; atom interferometry, atom holography, and atom optics in general, which are briefly covered in these notes in the context of NIST's BEC research, but of greater scientific and technological significance. Again, an introduction to these topics as well as detailed discussions can be found in various lecture notes and review articles [8, 9, 10].

The last section of these notes describes experiments with Bose-Einstein condensates. After a brief description of the strategy employed to obtain condensates of sodium atoms in a TOP trap, we discuss the normal incidence and Bragg diffraction of BEC, observation of the motion of BEC in the trap, directional output coupling of atoms from the condensate using stimulated Raman pulses and the demonstration of a quasi-continuous atom laser, as well as an interference experiment to measure the effective coherence length of a BEC. These matter-wave effects with BEC are analogous to effects observed with light waves. When the interactions of the atoms are included, however, effects analogous to those in nonlinear optics such as 4-wave mixing and the generation of solitons can be observed with matter-waves. The notes end with a description of such nonlinear matter-wave optics experiments, which NIST has pioneered.

Before beginning the discussion of the mechanical effect of atom-light interactions, we define for convenience, some frequently used notation and symbols, with frequencies expressed in radians/second. The reader is cautioned that other authors may have used other conventions:

$\delta = \Omega_{laser} - \Omega_{atom}$ is the detuning of the laser frequency from the natural resonant frequency of the atom.

$\Gamma = \tau^{-1}$ is the decay rate of the population in the excited state, the inverse of the natural lifetime, the linewidth of the transition.

$k = 2\pi/\lambda$ is the laser photon wave vector.

k_B is Boltzmann's constant.

Ω is the on-resonance Rabi frequency for a laser field, the precession frequency of the Bloch vector representing the 2-level atom when $\delta = 0$.

$I/I_0 = 2\Omega^2/\Gamma^2$ is the normalized intensity of the laser.

M is the mass of the atom.

$v_{rec} = \hbar k/M$ is the recoil velocity of an atom upon emission or absorption of a single photon.

$E_{rec} = Mv_{rec}^2/2 = \hbar^2 k^2/2M$ is the kinetic energy of an atom having velocity v_{rec}.

2. Radiative forces

Light comes in quantized packets of energy and momentum called photons. The transfer of energy and momentum between the photon and an atom through either coherent or incoherent scattering results in a force exerted on the atom. This is the basis for the mechanical effects of atom-light interactions. Radiative forces or the mechnical effects of light are generally divided into two catagories - the scattering force and the dipole forces.

2.1. THE SCATTERING FORCE

The scattering force, also called the spontaneous force or radiation pressure, is the force exerted on an atom by the incoherent scattering of photons. More precisely it is the force on an atom corresponding to absorption of a photon followed by spontaneous emission. The photon absorbed transfers $\hbar k$ of momentum to the atom. The spontaneous emission of the photon is symmetrically distributed and so the momentum transfer due to spontaneous emission, averaged over many absorption-emission events, is zero. The average force on a two-level atom moving with velocity $\mathbf{v}$ in a plane wave of wave vector $\mathbf{k}$ and detuning δ is

$$\mathbf{F}(\mathbf{v}) = \hbar \mathbf{k} \frac{\Gamma}{2} \frac{I/I_0}{1 + I/I_0 + \left[\frac{2(\delta - \mathbf{k}\cdot\mathbf{v})}{\Gamma}\right]^2}. \tag{1}$$

This force is the maximum scattering rate $\Gamma/2$, times the resonance Lorentzian times the photon momentum, i.e. the rate of absorbing photon momentum. The detuning $\delta - \mathbf{k} \cdot \mathbf{v}$ accounts for the Doppler shift, and the force is large when $|\delta - \mathbf{k} \cdot \mathbf{v}| \leq \Gamma$. The spontaneous force is limited by the rate at which spontaneous emissions can occur. These occur at a rate Γ for excited atoms, whose maximum fractional population is 1/2.

For real, multilevel atoms, the situation can be more complicated. A common occurrence, typical of alkali atoms, is that the ground state is split by the hyperfine interaction into two states separated in frequency by many times the optical linewidth Γ. An atom excited by a laser from one of these hyperfine levels to an optically excited state may decay by spontaneous emission to the other hyperfine level. Transitions from this level are then so far out of resonance that effectively no further absorption occurs and no force is applied to the atom. While various schemes involving selection rules and polarization of the light may be used to avoid this problem of optical pumping, the most straightforward method is to apply a second laser frequency, tuned to resonance between the "wrong" hyperfine state and the optically excited state. This "repumper" keeps the atom out of the wrong ground state and allows the atom to effectively feel the force of the laser acting on the main transition.

Equation 1 is only valid if the force can be meaningfully averaged over many absorption-emission events. If a single event changes the atomic velocity so much that the resonance condition is not satisfied, such an average is not possible. This imposes a validity condition $\hbar k^2/M \ll \Gamma$ on eq. 1. This condition is well satisfied for most atoms of interest in laser cooling. For example, $M\Gamma/\hbar k^2$ is 1200 for cesium laser cooled on its resonance transition at 852 nm and is 200 for sodium cooled at 589 nm.

2.2. The dipole force

The dipole force, also called the gradient force or stimulated force, is the force exerted on an atom due to coherent redistribution of photons. The dipole force can be considered as arising from stimulated Raman events - the absorption and stimulated emission of photons. (Note that the absorption and stimulated emission cannot be thought of as successive and independent events; their correlation is central to the proper understanding of the force.)

We can also understand the dipole force in analogy to a driven, classical oscillator. A harmonically bound charge driven by an oscillating electric field $\mathbf{E}$ has an oscillating dipole moment μ which is in phase with the driving field when driven below resonance and out of phase when driven above resonance. The energy of interaction between the

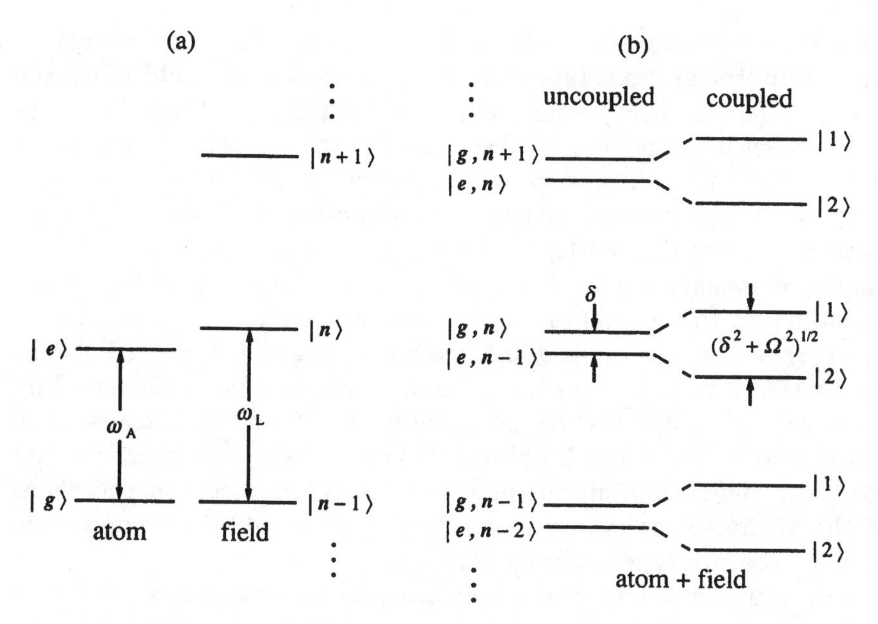

Figure 1. a) Energy levels of a 2-level atom and a laser field in the bare basis, *b*) dressed basis with the atom uncoupled and coupled to the field.

dipole and field is $W = -\mu \cdot \mathbf{E}$. Below resonance the energy is negative and the oscillator will be drawn toward a more intense field, while above resonance it will be drawn to the weaker part of the driving field.

A particularly powerful and intuitive way of describing the dipole force is in the dressed-atom picture. This has been treated in detail by Dalibard and Cohen-Tannoudji [11] and we present the basic idea here. Consider a 2-level atom with ground state $|g\rangle$ and excited state $|e\rangle$. Separately consider a single mode of the radiation field, close to resonance with the atomic transition, having energy levels labeled ...$|n-1\rangle$, $|n\rangle$, $|n+1\rangle$,... according to the number of photons in the mode. The energy levels of these two distinct systems are shown in fig. 1*a* for the case where the photon energy Ω_{L} is greater than the difference in energy between the atomic states, Ω_{A}. This is the "bare" basis in which we consider the atom's energy levels and those of the photon field separately. If we now consider the atom and laser field together as a single system we have the dressed basis (atom dressed by laser photons) of fig. 1*b*. If there is no interaction between the atom and the laser field (as, for example, when the laser field does not spatially overlap the atom's position) the dressed level energies are simply the sum of the atom and field energies. This gives a ladder where each rung is a pair of nearly degenerate energy levels, since an atom in the ground state with n photons in the laser field has nearly the same energy as an atom in the excited state with $n-1$ photons in the field (with the energy difference being equal to the laser detuning). When the atom interacts with the field, the dressed levels (the eigenstates of the full Hamiltonian) become superpositions of ground and excited states and of numbers of photons in the field. By convention, the higher of the two levels in a rung is called $|1\rangle$, and the rungs are labeled by the photon number n associated with the excited state in the uncoupled basis [12]. According to the general rule that interacting energy levels repel each other by an amount depending on the coupling, the spacing between the dressed levels increases from the detuning δ to the "effective Rabi frequency" $\Omega_{eff} = \sqrt{\delta^2 + \Omega^2}$ as the interaction turns on. Because each of the dressed levels has both ground and excited state character, an atom can make spontaneous emission transitions between the rungs of the ladder. These transitions establish an equilibrium population between the two types of levels [11].

Now consider a light field whose intensity varies in space, such as a focused laser beam with a Gaussian intensity profile, or a plane standing wave. The Rabi frequency seen by an atom now varies in space, so the energy of the dressed levels also varies. The dipole force arises from this variation in energy and the relative populations of the dressed levels. Figure 2 illustrates the idea. For laser detuning above resonance ($\delta > 0$) the upper of the two dressed levels is the one that connects to the ground state in the limit of small interaction. This upper level always has the largest population, and its potential tends to repel the atom from the region of most intense light. For $\delta < 0$ it is the lower level that connects to the ground state and that has the higher population; its potential attracts the atom to the higher intensity. The actual dipole force is the average force, weighted by population, for the two potentials. The dipole force is derivable from a potential [11, 13, 14, 15, 16] which can be written as

$$U = \frac{\hbar\delta}{2} log \left[1 + \frac{I/I_0}{1 + (2\delta/\Gamma)^2} \right]. \quad (2)$$

Where the spatial dependence of the potential comes in through the dependence of the intensity I on position.

2.3. Diffraction of atoms by a standing wave

Light can be used to manipulate atoms in a manner analogous to the manipulation of light by material elements such as lens, mirrors and beamsplitters. This field is often referred to as "atom optics," in analogy with conventional optics, and is an active area of research. I will discuss

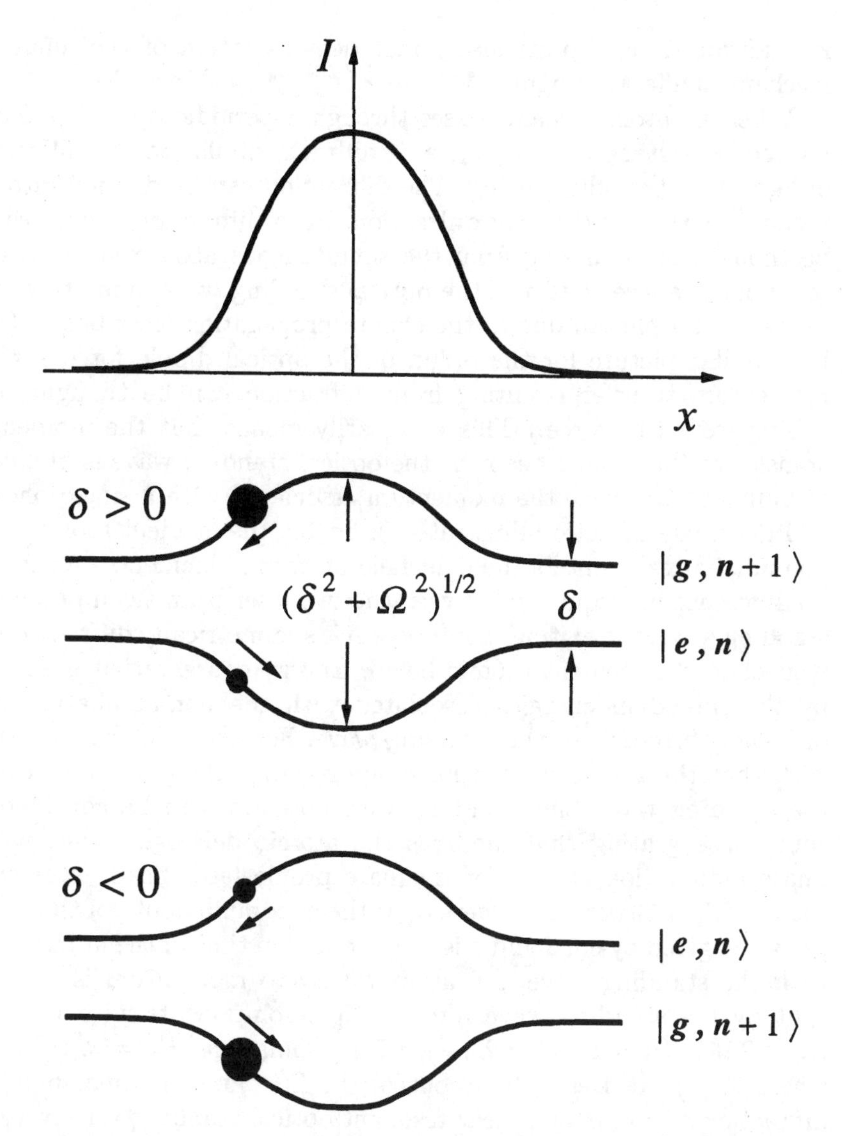

Figure 2. Top: the intensity profile of the laser field. Bottom: the dressed energy levels as a function of position for this intensity profile, for laser detuning above and below resonance. The sizes of the black dots indicate the relative population of the dressed states.

atom or matter-wave optics in more detail in the context of Bose-Einstein condensates, but at this point I would like to describe an effect that illustrates the complementarity between the wave nature of matter and light, namely, the diffraction of atoms by optical standing waves.

In addition, it is a particularly nice demonstration of the quantized mechanical effects of light.

When an atomic beam passes through a periodic optical potential formed by a standing light wave, it diffracts similar to the diffraction of light by a periodic grating. The diffraction can be divided into two regimes, normal and Bragg diffraction. Both diffraction processes can be thought of as arising from the simultaneous absorption of a photon from one laser beam of the optical standing wave, and stimulated emission of a photon due to the counterpropagating laser beam. (This is a similar picture for the origin of the optical dipole force and the momentum transfer resulting from diffraction can be thought of as arising from this force.) This necessarily means that the momentum transfer to the atomic beam by the optical standing wave is quantized in units of $2\hbar k$, twice the momentum associated with a single photon.

In normal diffraction illustrated in fig. 3*a*, the incident atomic beam non-adiabatically enters the light field at normal incidence. As there is no difference in frequency between the two laser beams comprising the standing wave, the exiting atomic beam is symmetrically diffracted with respect to the incident atomic beam. Energy conservation is satisfied by the spread in energies associated with the non-adiabatic "turn-on" and "turn-off" of the standing wave. For short interaction times such that the atoms do not move appreciably along the direction of the standing wave, the standing wave potential can be considered a thin phase grating that modifies the atomic deBroglie wave with a phase modulation, which for a square profile laser beam is given by $\phi(x) = (U_0\tau/\hbar)cos^2(kx)$, where U_0 is the maximum depth of the optical potential given by eq. 2 and τ is the interaction time of the atomic beam with the standing wave. An atom with zero momentum is therefore split by the standing wave into multiple components with momenta $p_n = 2n\hbar k$, $(n = 0, \pm1, \pm2, \ldots)$, with populations $P_n = J_n^2(U_0\tau/2\hbar)$, where $J_n(x)$ are Bessel functions of the first kind. Normal incidence diffraction of atoms by a near resonant optical standing wave was first demonstrated in Pritchard's group at M.I.T. [17] in 1983.

When the atoms enter and exit the monochromatic standing wave adiabatically, energy conservation must be explicitly satisfied in the interaction between the atoms and the light field. In this regime, also known as the Bragg regime, the energy difference of the atom before and after the change of momentum of $2\hbar k$ must come from the photon field. This is typically accomplished by having the atomic beam incident on the standing wave at an angle such that the atoms see a differential Doppler shift between the two counterpropagating laser beams comprising the standing wave. This geometry is shown schematically in fig. 3*b*. Under these conditions, Bragg diffraction can be understood as a

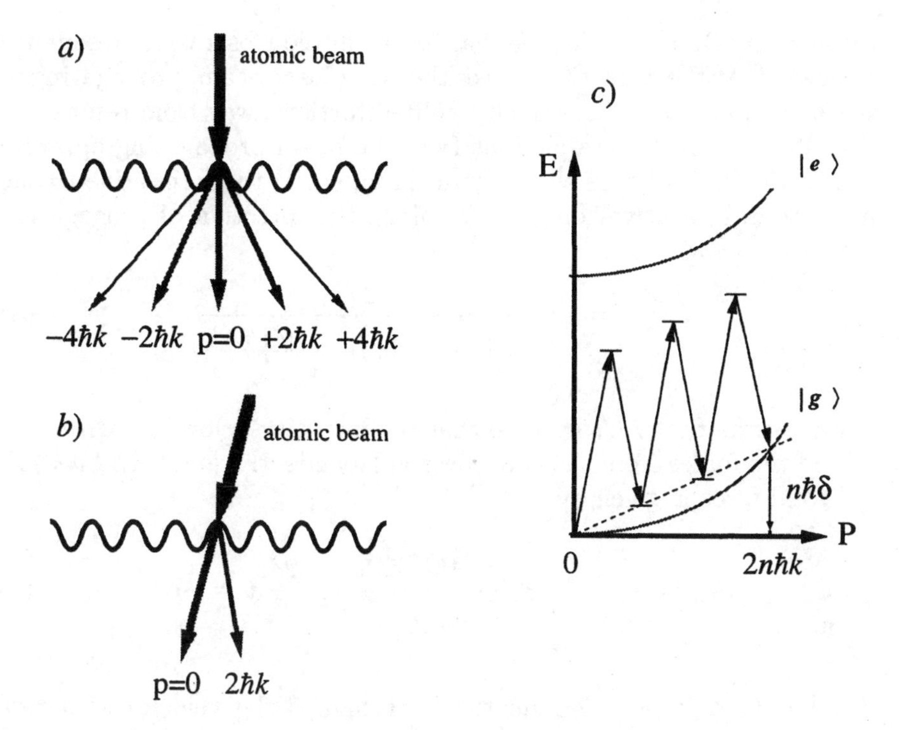

Figure 3. *a*) Normal incidence diffraction *b*) Bragg diffraction *c*) Bragg diffraction as a $2n$-photon Raman transition.

stimulated Raman transition between two momentum states. Figure 3*c* shows nth order Bragg diffraction as a $2n$-photon, stimulated Raman process in which photons are absorbed from one beam and stimulated into the other. Conservation of energy and momentum requires $(n2\hbar k)^2/2M = 2n\hbar kvsin\theta$, where M is the mass of the atom, v is the longitudinal velocity of the atomic beam and θ is the angle of incidence of the atomic beam on the standing wave. Bragg diffraction of atoms by a near resonant optical standing wave was also first demonstrated in Pritchard's group at M.I.T. [18] in 1988.

2.4. Doppler cooling

Doppler cooling results from near resonant radiation pressure acting on an atom and thus it is a particularly instructive example of an application of the scattering force. To see how cooling works we shall consider the one-dimensional motion of an atom in counterpropagating plane waves. If the laser beams are tuned below the atomic resonance, an atom moving in the direction opposite to one of the beams will, because of the Doppler shift, see that beam shifted closer to its resonance frequency. At the same time it sees the other laser beam, propagating in the same direction as its velocity, shifted further away from resonance. The atom absorbs more photons from the beam propagating opposite to its velocity and thus slows down. From eq. 1, the forces due to the positive and negative-going waves, along the direction of progagation are

$$F_{\pm}(v) = \pm \hbar k \frac{\Gamma}{2} \frac{I/I_0}{1 + I/I_0 + \left[\frac{2(\delta \mp kv)}{\Gamma}\right]^2}. \tag{3}$$

If we assume that $I/I_0 \ll 1$, so that the intensity is low enough for us to add the forces from the two waves independently, and that $kv \ll \Gamma$, the total force is given by

$$F_{tot} = F_+ + F_- = \frac{4\hbar k^2 I/I_0}{\left[1 + (2\delta/\Gamma)^2\right]^2} \frac{2\delta}{\Gamma} v = -\alpha v. \tag{4}$$

For $\delta < 0$, α is positive and the force damps the velocity at a rate $\dot{v}/v = -\alpha/M$. The largest damping force is obtained for $\delta \cong -\Gamma/2$.

Atomic motion in a strong standing wave [11, 13, 14, 19] is beyond the scope of this treatment, however a simple approximation for small velocity, ignoring interference of the two beams and stimulated redistribution of photons between the beams [20] shows that the friction coefficient α maximizes at about $\alpha = \hbar k^2/4$ for $I/I_0 = 1$ in each beam and $2\delta/\Gamma = -1$. Under these conditions the velocity damping time $v/\dot{v}$ for sodium, cooled on the 589 nm resonance line, would be 13 μs.

So far we have only considered the average force on an atom in a light field. We should remember that the force arises from discrete photon scatterings, and so must fluctuate about its average. The fluctuations can be thought of as arising from two sources: fluctuations in the number of photons absorbed in a given time and fluctuations in the direction of the spontaneously emitted photons. Both of these effects arise because of the randomness of spontaneous emission.

The fluctuations represent a random walk of the atomic momentum, with each random walk step being of magnitude $\hbar k$. For simplicity we will assume a fictitious one-dimensional situation where photons are emitted as well as absorbed along a single axis. Each scattering event represents two random walk steps, one from the absorption, which could be from either of the counterpropagating beams, and one from the spontaneous emission, which can be in either direction along the axis. The mean square momentum of the atom increases linearly with the number of scattering events (random walk steps), with a rate

$$\frac{d}{dt}\left\langle p^2\right\rangle = 2R\hbar^2k^2 \tag{5}$$

where R is the scattering rate; the factor of 2 comes from the two steps per scattering. We define a momentum diffusion coefficient

$$2D_p = \frac{d}{dt}\left\langle p^2\right\rangle \tag{6}$$

so that D_p/M is the rate of increase in kinetic energy, the heating rate. The damping force decreases the kinetic energy as $\mathbf{F}\cdot\mathbf{v} = -\alpha v^2$, the cooling rate. At equilibrium we set the sum of the heating and cooling rates to zero, finding

$$D_p/\alpha = M\left\langle v^2\right\rangle = k_BT. \tag{7}$$

Here we have replaced v^2 with its mean value and used the equipartition theorem to identify $k_BT/2$ as the mean kinetic energy $M\left\langle v^2\right\rangle/2$ in the single degree of freedom. Using eq. 4 to get α, eqs. 5 and 6 to get D_p, and remembering that the total scattering rate from the two laser beams, each of intensity, I, is

$$R = \Gamma\,\frac{I/Io}{1+(2\delta/\Gamma)^2}, \tag{8}$$

we find, for low intensity and small velocity:

$$k_BT = \frac{\hbar\Gamma}{4}\,\frac{1+(2\delta/\Gamma)^2}{2\delta/\Gamma}. \tag{9}$$

This is the Doppler temperature, and we emphasize that it applies to the fictional one-dimensional case we have constructed. A true 1D experiment, such as cooling an atomic beam along one axis, would produce a lower temperature depending on the distribution of scattered photons in 3D. However, it can be shown [20] that in a symmetrical three dimensional case, the temperature is also given by eq. 9, which is plotted in fig. 4. The temperature minimizes at a detuning of $\delta = -\Gamma/2$ where

$$k_BT_{Dopp} = \frac{\hbar\Gamma}{2}, \tag{10}$$

defines the Doppler cooling limit. To derive the Doppler temperature we assumed that the velocity was small enough that $kv \ll \Gamma$. At the Doppler limit, sodium atoms would have $v_{r.m.s.} = 30$ cm/s, corresponding to a temperature of 240μK and $kv_{r.m.s.} = \Gamma/20$, so the assumption is justified in this case (and most others). We may also ask whether the velocity distribution corresponds to a temperature. That is, is it Maxwell-Boltzmann? It can be shown that for sodium, and similar atoms where the recoil energy $E_{rec} = \hbar^2k^2/2M \ll \hbar\Gamma$, the velocity distribution is indeed very close to being thermal [20]. What happens if the recoil energy is not small? Then we violate the validity condition on eq. 1, and the detuning due to the Doppler shift changes significantly with each emission or absorption. In the limit where the linewidth Γ is small compared to the recoil energy, we feel intuitively that the cooling limit, rather than being related to Γ, as in eq. 10, is related to the recoil energy. Indeed, it can be shown [20, 21] that in this case the lowest temperature attainable (the recoil temperature) is given by

$$\frac{1}{2}k_BT = \frac{\hbar^2k^2}{2M} = E_{rec}, \tag{11}$$

that is, one recoil energy per degree of freedom. While it might seem that this is the ultimate limit of laser cooling, in fact it is possible to break the recoil limit under certain circumstances where the interaction with the light is turned off as the atomic velocity becomes small. One way this may be achieved is by velocity selective coherent population trapping in multilevel atoms [22, 23]. Atoms are optically pumped into a coherent superposition of both internal states and center-of-mass velocities, a superposition that cannot absorb the laser light. Another way is velocity-space optical pumping [24, 25]. Atoms are cooled on a transition with a narrow linewidth, but with a distribution of laser frequencies such that the excitation rate for zero velocity atoms is small or vanishing. This has been accomplished by use the use of pulsed two-photon Raman transitions [26].

The damping force of eq. 4 is similar to the viscous force on an object moving in a fluid. Because of this, the configuration of pairs of counterpropagating laser beams is often called "optical molasses." The viscosity is so high that a velocity corresponding to the Doppler cooling limit is damped out and randomized while the atom travels only a few tens of micrometers, much smaller than the size of the molasses, which is typically a centimeter. Thus the atoms executes a Brownian-like motion with a short mean free path, moving diffusively rather than ballistically [27, 28]. The evidence for this is the long residence time of atoms in optical molasses. Atoms require several seconds to diffuse out of a typical optical molasses [28, 29], whereas, moving ballistically, atoms cooled to the Doppler cooling limit would traverse a region the size of a typical molasses in a few tens of milliseconds.

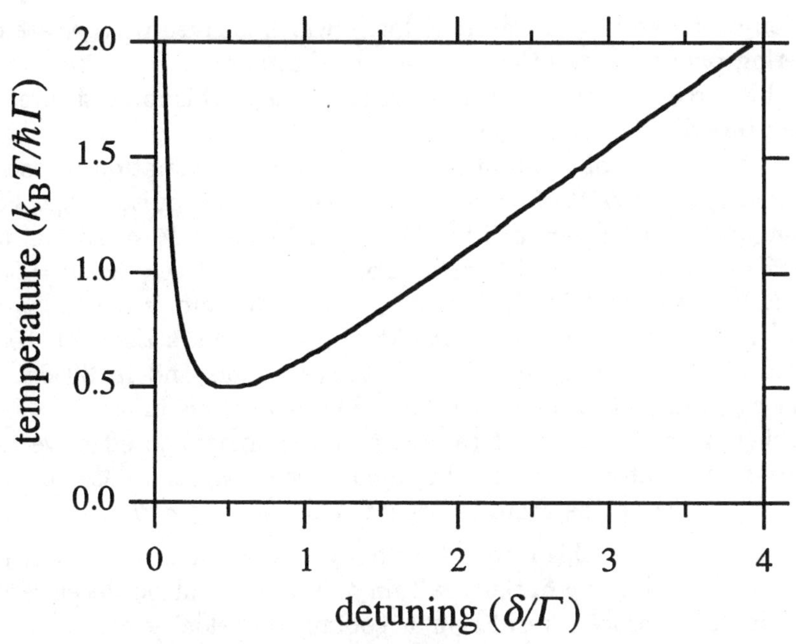

Figure 4. Doppler temperature at low intensity as a function of laser detuning.

3. Deceleration and cooling of an atomic beam

Experimentally, atoms must first be at reasonably slow velocities ($kv \lesssim \Gamma$) before Doppler cooling can be effective. The two general ways this has been accomplished is by laser deceleration of an atomic beam [27, 30] and by collection of slow atoms from a thermal gas [31, 32]. Loading from a thermal gas has the advantage of allowing a more compact and simpler apparatus, while the beam deceleration technique usually allows lower background pressure and faster production of slow atoms.

Deceleration of an atomic beam is usually accomplished by directing a near resonant laser beam so as to oppose the atomic beam. The atoms absorb photons at a rate determined by the intensity of the laser beam, the detuning from resonance and the atoms' velocity. For each photon absorbed, the atomic velocity changes by v_{rec} in the direction of the laser propagation. The spontaneously emitted photons are emitted randomly in a pattern that is symmetric on reflection through the atom, so there is no net average change in the atomic velocity due to these emissions. If the absorption is followed by stimulated emission into the same direction as the incident laser beam (we assume the laser beam is a plane wave), there is no net momentum transfer from the absorption-emission process. Only absorption followed by spontaneous emission contributes to the average force, which is given by the rate of scattering photons times the momentum of a photon. For a two level atom this force is given by eq. 1. At high intensity this force saturates to the value $\hbar k\Gamma/2$.

The acceleration of an atom due to the saturated radiation pressure force is $a_{max} = \hbar k\Gamma/2M = v_{rec}\Gamma/2$, which can be quite large. For sodium with $\lambda = 2\pi/k = 589$ nm, $1/\Gamma = 16$ ns and $M = 23$ a.m.u., $v_{rec} \approx 3$ cm/s and $a_{max} \approx 10^6$ m/s^2. For cesium, with $\lambda = 2\pi/k = 852$ nm, $1/\Gamma = 30$ ns and $M = 133$ a.m.u., $v_{rec} \approx 3.5$ mm/s and $a_{max} \approx 6 \times 10^4$ m/s^2. This acceleration would stop in 50 cm, a thermal, 1000 m/s Na atom scattering ≈33,000 photons in 1 ms and in 80 cm, a thermal, 300 m/s Cs atom scattering ≈84,000 photons in 5 ms.

Implicit in eq. 1 is one of the major impediments to effective deceleration of an atomic beam using a counterpropagating laser beam. The force acting on the atom is large if $|2(\delta - \mathbf{k} \cdot \mathbf{v})| \lesssim \Gamma\sqrt{1 + I/I_0}$. Atoms much outside this resonant-velocity range will experience little deceleration, and atoms intially within this range will be decelerated out of it. This process results in a cooling or velocity compression of a portion of the atomic beam's velocity distribution, and was first observed by Andreev et al. [33]. Atoms initially at the resonant velocity decelerate out of resonance. Other atoms with nearby velocites will also decelerate, those with larger velocities first decelerating into resonance, then to slower velocities out of resonance, while initially slower atoms decelerate to still lower velocities. The atoms will "pile up" at a velocity somewhat lower than the resonant velocity. Both deceleration and cooling occur because a range of velocities around the resonant velocity are compressed into a narrower range at lower velocity. The change in the velocity distribution of an atomic beam with a thermal spread of velocities is illustrated in fig. 5.

The difficulty with the velocity distribution of fig. 5 is that only a small portion of the total velocity distribution has been decelerated by only a small amount. There are a number of possible solutions to this problem, some of which have been discussed in [27]. These include Zeeman tuning [27] where a spatially varying magnetic field compensates the changing Doppler shift as the atoms decelerate so as to keep the atoms near resonance; white-light deceleration [34] where a range of laser frequencies ensures that some light is resonant with the atoms, regardless of their velocity (within the range to be decelerated); diffuse-light deceleration [35] where light impinges on the atoms from all angles so that, with the Doppler shift, some of the light is resonant with each velocity; Stark cooling [36] where a spatially varying electric field is used to Stark shift atoms and keep them near resonance as they Doppler shift

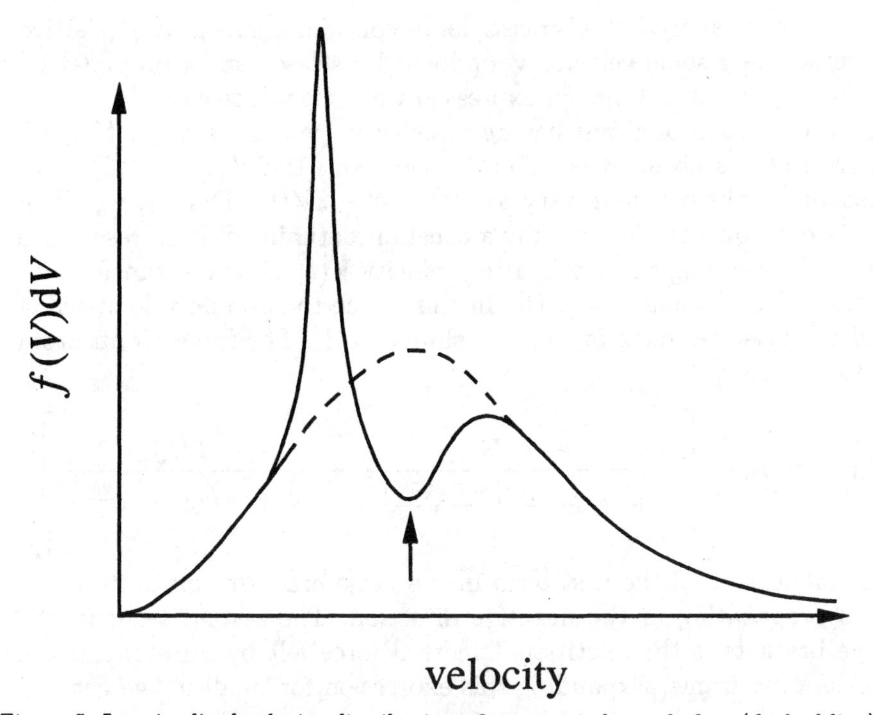

Figure 5. Longitudinal velocity distribution of an atomic beam before (dashed line) and after (full line) interacting with a counterpropagating, fixed-frequency laser. The arrow indicates the velocity resonant with the laser.

due to deceleration; intense standing wave deceleration [37] where the linewidth is sufficiently power broadened to capture a large velocity distribution for laser deceleration; and "chirp cooling" [38] in which the frequency of the laser is swept up, or chirped, in time such that the laser stays in resonance with atoms that have been decelerated, and they continue to absorb photons and decelerate. Zeeman tuning and chirp cooling were among the earliest atomic beam slowing techniques demonstrated and they continue to be the most widely used techniques to date.

3.1. Chirp cooling

In chirp cooling, the frequency of the laser is swept up, or chirped, in time [38]. Because of the chirp, atoms that have been decelerated by the laser stay in resonance, continue to absorb photons, and continue to decelerate. Furthermore, the chirp brings the laser into resonance with additional atoms having lower velocities than the original group around the velocity initially resonant with the laser.

In order to analyze this process, let us consider atoms having positive velocities near some velocity V opposed by a laser beam propagating in the negative direction. We express any atomic velocity as $v = V + v'$. The acceleration of atoms having velocity V ($v' = 0$) is $a = F(V)/M$, where $F(V)$ is given by eq. 1. Therefore, we write $V(t) = V(0) + at$. Also we let the detuning vary as $\delta(t) = \delta' - kV(t)$. That is, we chirp the laser frequency so as to stay a constant detuning δ' from resonance with atoms having the decelerating velocity $V(t)$. Now we transform to a frame decelerating with $V(t)$. In this frame the atomic velocity is v' and the laser detuning is Doppler shifted to δ'. The force on an atom in this frame is

$$F(v') = \hbar k \frac{\Gamma}{2} \left[\frac{-I/I_0}{1 + I/I_0 + \left[\frac{2(\delta' + kv')}{\Gamma}\right]^2} + \frac{I/I_0}{1 + I/I_0 + \left[\frac{2\delta'}{\Gamma}\right]^2} \right] \tag{12}$$

The minus sign of the first term in the large brackets comes from the laser propagating in the negative direction. The second term in the large brackets is the "fictitious" inertial force felt by an atom in the decelerating frame. Expanding this expression for small v', we get

$$F(v') = 2\hbar k^2 \frac{I}{I_0} \frac{(2\delta'/\Gamma)\, v'}{\left[1 + I/I_0 + \left(\frac{2\delta'}{\Gamma}\right)^2\right]^2} \tag{13}$$

The term multiplying v' is minus the friction coefficient α. When $\delta' < 0$, the force opposes the velocity v' and tends to damp all velocities to zero in the decelerating frame, which is $V(t)$ in the laboratory frame. Maximum damping occurs for $I/I_0 = 2$ and $2\delta'/\Gamma = -1$. The final velocity to which the atoms are decelerated is determined in practice by the final frequency to which the laser is chirped. Figure 6 illustrates the results of chirp cooling an atomic beam. All of the atoms in the initial distribution below the velocity resonant with the laser at the beginning of its chirp are decelerated.

The first definitive experiment showing such chirp cooling was in ref. [39], with deceleration to zero velocity first achieved in ref. [40]. The analysis given above is similar to that given in ref. [41]. The robust character of this sort of cooling is evident. Atoms within a range of velocities around $V(t)$ are damped (in velocity) toward $V(t)$. Lower velocities, not initially close to $V(t)$, come within range as the laser chirp brings $V(t)$ into coincidence with them. If the laser intensity changes during the time an atom is being decelerated (because, for example, the laser beam is not collimated), the atoms will continue to decelerate according to the chosen chirp rate, but with a different

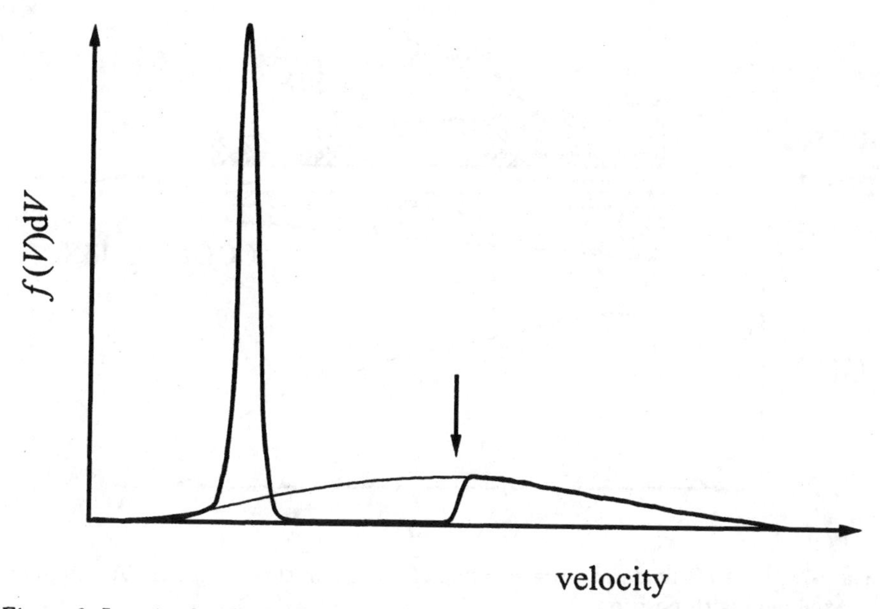

Figure 6. Longitudinal velocity distribution of an atomic beam before (thin line) and after (bold line) deceleration by a chirped laser. The arrow indicates the velocity initially resonant with the laser.

effective detuning δ'. The chosen chirp rate, however, must be consistent with an achievable deceleration with the given I/I_0. That is, the chirp rate must satisfy

$$\dot{\delta} = ka = \frac{\hbar k^2 \Gamma}{2M} \frac{I/I_0}{1 + I/I_0 + \left[\frac{2\delta'}{\Gamma}\right]^2} \tag{14}$$

This means that $\dot{\delta}$ has an allowable upper limit of ka_{max}. We have noted that for the velocities to be damped in the decelerating frame we must have $\delta < 0$ and it is easy to show that the conditions for best damping lead to a deceleration half as large as the maximum.

3.2. Zeeman tuning

In the Zeeman tuning technique, a spatially varying magnetic field is used to keep the frequency of an atom resonant with a counterpropagating laser beam, as the atom slows down by scattering photons from this laser. Because the atoms are slowed down and the spread of the velocity distribution is narrowed, this process is often referred to as "Zeeman cooling." Figure 7 illustrates the general idea of this scheme.

The atomic beam source directs atoms, which have a range of velocities, along the axis (z direction) of a tapered solenoid. This magnet has

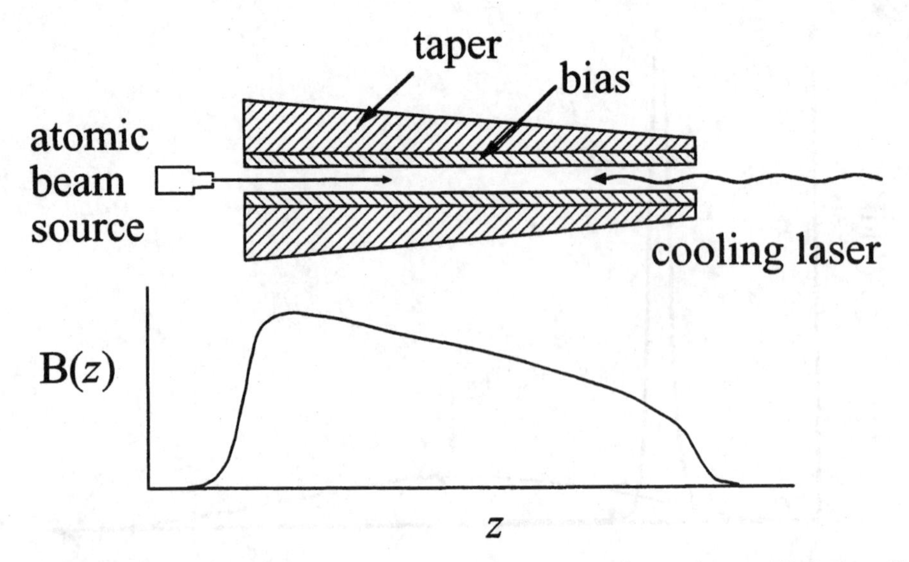

Figure 7. Upper: Schematic representation of a Zeeman slower. Lower: Variation of the axial field with position.

more windings at its entrance end, near the source, so the field is higher at that end. The laser is tuned so that, given the field-induced Zeeman shift and the velocity-induced Dopper shift of the atomic transition frequency, atoms with velocity v_0 are resonant with the laser when they reach the point where the field is maximum. Those atoms then absorb light and begin to slow down. As their velocity changes, their Doppler shift changes, but is compensated by the change in the Zeeman shift as the atoms move to a point where the field is weaker. At this point, atoms with initial velocities slightly lower than v_0 come into resonance and begin to slow down. The process continues with the initially fast atoms decelerating and staying in resonance while initially slower atoms come into resonance and begin to be slowed as they move further down the solenoid. Eventually all the atoms with velocities lower than v_0 are brought to a final velocity that depends on the details of the magnetic field and laser detuning. The magnetic field profile of the tapered solenoid is

$$B(z) = B_0\sqrt{1 - 2az/v_0^2} \tag{15}$$

with $0 \leq 2az \leq v_0^2$. B_0 is the magnetic field producing a Zeeman shift equal to the Doppler shift for atoms with velocity v_0 and $a \leq a_{max}$ is the deceleration rate.

The first experiment on the deceleration of atoms using the Zeeman technique is in ref. [42]. Subsequently, neutral sodium atoms were

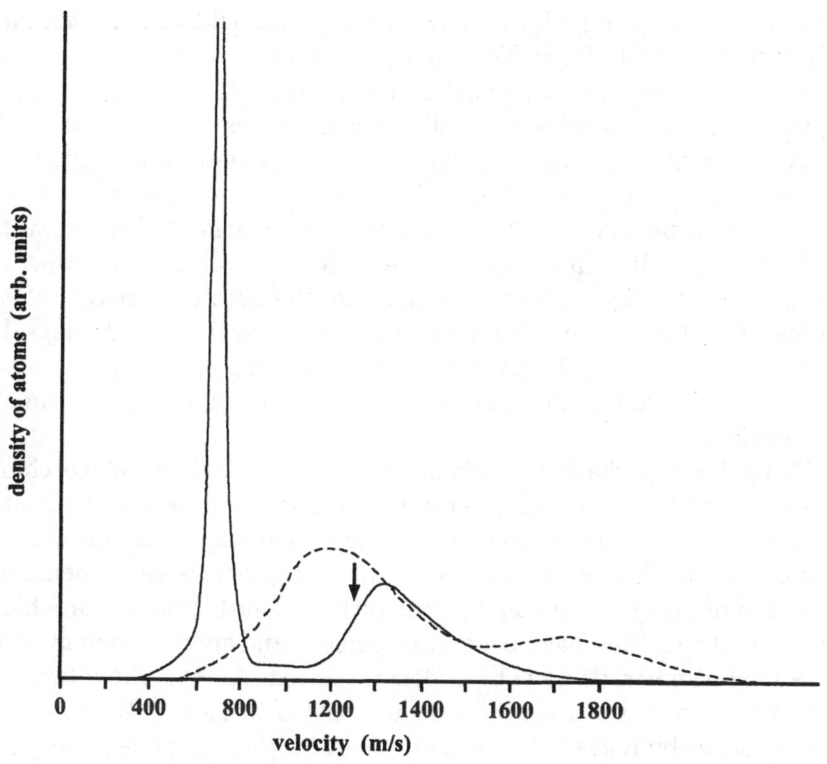

Figure 8. Velocity distribution before (dashed) and after (solid) Zeeman cooling. The arrow indicates the highest velocity resonant with the slowing laser. (The extra bump at 1700 m/s is from $F = 1$ atoms, which are optically pumped into $F = 2$ during the cooling process.)

stopped in ref. [43]. Figure 8 shows the velocity distribution resulting from Zeeman cooling: a large fraction of the initial distribution had been swept down into a narrow final velocity group. The velocity distribution after deceleration was measured in a detection region some distance from the exit end of the solenoid using a a separate detection laser. We were able to determine the velocity distribution in the atomic beam by scanning the frequency of the detection laser and observing the fluorescence from atoms having the correct velocity to be resonant.

4. Traps for neutral atoms

Trapping of atoms usually refers to their confinement by the application of external fields rather than by the use of a material container. In contrast with the trapping of ions by electric and magnetic fields, the trapping forces that can be applied to neutral atoms are relatively weak.

Ions have a charge on which an electromagnetic field can exert a large Coulomb or Lorentz force. Neutral atoms, however, may be acted upon through their permanent magnetic dipole moments or induced electric dipole moments, allowing generally smaller forces to be applied. The strongest traps for neutral atoms have energy depths of only a few kelvin, while ion traps can typically hold room temperature ions, and have trapped particles with energies of a few thousand electron volts.

Another possible difficulty with traps for neutral atoms is that the trapping potentials represent changes in the internal energy of the atoms. That is, the postitions of and, in general, the spacings between energy levels are changed by the trapping fields. This means, in particular, that high accuracy spectroscopy of trapped atoms is problematic.

Neutral atoms have the advantage that the lack of space charge effects means that one can generally trap larger numbers and densities of neutral atoms than of ions. Furthermore, some applications demand that one work with neutral atoms (as for example in Bose condensation of an atomic gas). Fortunately, even rather weak forces are capable of trapping atoms that have been laser cooled, and many different kinds of neutral atom traps have been demonstrated. Among these are:

i) Magneto-static traps, first demonstrated in 1985 [44] rely on the force exerted by a gradient magnetic field on the permanent magnetic dipole moment of an atom such as a ground-state alkali.

ii) Laser dipole traps, first demonstrated in [45] use the dipole force that results from the gradient of the energy of the oscillating dipole moment induced on an atom in an inhomogeneous laser field.

iii) Radiation pressure traps use the scattering force of eq. 1, but are not stable in 3D for two-level atoms. The magneto-optical trap (MOT), using multilevel atoms and an inhomogenous magnetic field, was the first radiation pressure trap to be demonstrated [46].

iv) Magneto-dynamic traps use the micromotion driven by oscillating magnetic field gradients to allow trapping of high-field-seeking states not stably trapped in static magnetic fields. Such a trap, first demonstrated in 1991, is analogous to the radio frequency Paul trap for ions [47].

v) Microwave traps are low-frequency, spontaneous emission free analogs of laser dipole traps. Such a trap was first demonstrated near a magnetic resonance transition [48].

vi) Electrostatic traps, while never demonstrated in 3D, have been proposed [49] for excited atoms.

vii) Gravito-optical traps, which combine optical dipole forces with gravity to produce stable trapping [50], are only one example of hybrid traps that combine different types of forces to achieve trapping of atoms.

viii) TOP traps (for Time-averaged Orbiting Potential) [51], are an important modification of one type of magnetostatic trap. While time dependent, they are not dynamic in the sense of atomic micromotion being essential.

These notes will not treat each of these kinds of traps in detail, nor attempt to give more complete references about them. We shall, however, discuss the radiation pressure, laser dipole and magnetic traps in some more detail below. Since these are the traps that are currently in extensive use in BEC research.

4.1. Dipole force traps

The dipole force discussed in section 21 can be used to trap atoms. A single, focused laser beam, tuned below resonance is the simplest dipole trap and was first proposed by Ashkin in 1978 [52]. As an example, consider sodium atoms interacting on their strongest transition ($I_0 = 6$ mW/cm^2) with a modest power, 10 mW Gaussian laser beam focused to a $1/e^2$ radius of 10 μm. This gives $I/I_0 \cong 10^6$ at the focus. For the detuning maximizing U, we find $U_{max}/k_B \cong 100$ mK. For a 1W beam we find $U_{max}/k_B \cong 1$ K. Such traps can easily confine laser cooled atoms.

The dipole potential is a conservative potential, so it does not have any dissipative mechanism associated with it. It was not until the demonstration of laser cooling in optical molasses [28] that such a trap could be loaded and finally realized [45]. One of the difficulties involved in such trapping is that although the trap is tuned below resonance (the proper sign of the detuning to achieve cooling) the cooling provided by the trapping light does not reduce the thermal energy below the trap depth. Auxiliary cooling [53] is required, but even this is difficult because the inhomogeneous light shifts induced by the trapping laser interfere with the cooling process. Dalibard et al. [54, 55] proposed a solution in which the trapping and cooling are alternated in time, and this procedure was used for the first dipole trap [45].

Another difficulty with a single focus dipole trap is that the radiation pressure force pushes the atoms away from the focus, while the dipole force is attracting them to it. While more complicated, counterpropagating beam geometries can avoid this difficulty [52, 56, 57], one can, at the expense of reduced trap depth, solve the problem by detuning the laser [45]. According to eq. 1, the destabilizing radiation pressure force varies as $1/\delta^2$ (for sufficiently large δ), while the dipole trapping force obtained from the dipole potential given by eq. 2 varies as $1/\delta$. Thus, for large enough detuning, the radiation pressure will be negligible compared to the dipole trapping force.

Large detuning has other advantages, particularly when coupled with multi-level laser cooling. When the detuning is large enough that the trap depth is comparable to the natural linewidth, the thermal energy of laser cooled multilevel atoms can still be considerably less than this depth, which is comparable to the Doppler cooling limit of eq. 10. Furthermore, the optimum detuning for good multilevel laser cooling is at least several times the linewidth, the inhomogeneous light shifts should not much affect the cooling. Such a far-off-resonance-trap (FORT) has been demonstrated to work without the need to alternate cooling and trapping phases [58, 59].

Another advantage [60] of such a FORT is that for sufficiently large detuning the population of the trapped atoms is almost entirely in the ground state. The trap is then nearly free of heating due to spontaneous emission [61] and of many of the collisional perturbations involving excited atoms. FORTs have been used to hold atoms for collision experiments [62] and to trap atoms for evaporative cooling [63].

Variations of the dipole force trap include using the evanescent wave created by total internal reflection of light (detuned blue of resonance) to act as a mirror to reflect atoms [50, 64, 65], crossing two red-detuned laser beams at their foci to achieve a strong gradient in all directions [63] , and crossing sheets of blue-detuned light to form a box bounded by light (with confinement in the vertical direction sometimes provided by gravity) [66, 67].

4.2. Radiation pressure force traps

Radiation force traps use the spontaneous or scattering force to confine atoms. Unlike the dipole force, the force generated by absorption of a photon followed by spontaneous emission does not depend on the gradient of the intensity. Hence larger volume traps can be created using the scattering force compared to the dipole force for a given flux of incident photons. A simple radiation force trap, shown in fig. 9, can be made in 1D using two counterpropagating focused laser beams with separated foci [52]. Midway between the foci the radiation pressure from the two beams is balanced and the net force on an atom is zero. As the atom moves away from this equilibrium point toward one of the foci it is pushed back by the higher intensity near that focus. In contrast to the dipole force trap, radiation pressure trap depths on the order of a kelvin can be achieved with a modest, near saturation intensity, i. e., a few milliwatts per square centimeter [68]. While the trap of fig. 9 produces radiation pressure force trapping along only one axis, the

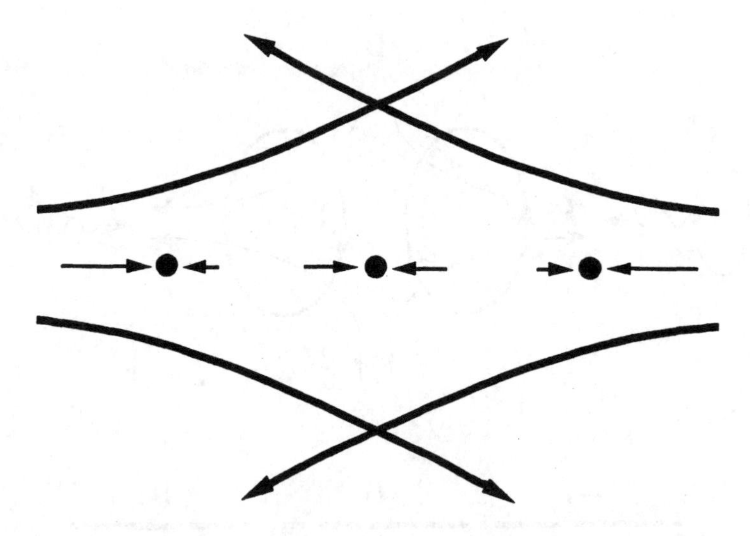

Figure 9. **A one dimensional radiation pressure force trap formed from two counterpropagating, focussed laser beams with separated foci. The length of each of the pairs of arrows indicates the magnitude of the force from the respective laser beams on an atom at the indicated positions.**

dipole force can provide trapping along the two orthogonal axes [52], a configuration first demonstrated in [57].

It is tempting to extend the trap idea of fig. 9 to three dimensions, but it can be shown that such as extension is not possible when the radiation pressure force is proportional to the photon flux from the laser beams. The impossibility of such trapping is related to the fact that the divergence of the Poynting vector is zero, and has been called the optical Earnshaw theorem [69] by analogy with the theorem from electrostatics forbidding stable trapping of a test charge in a charge-free region.

In spite of the optical Earnshaw theorem, which applies to 2-level atoms in a weak, static laser field, it is possible to create a 3D radiation force trap by making use of such features as saturation, multiple levels, optical pumping and Zeeman shifts. The most successful radiation pressure trap that circumvents the Earnshaw theorem is the magneto-optical trap or MOT. Conceived by Dalibard [70] and demonstrated in an MIT-Bell Labs collaboration [46] its principle is illustrated in fig. 10 for 1-D operation and a simple $J = 0 \to J = 1$ transition.

A pair of current-carrying coils with opposing currents creates a quadrupole magnetic field that is zero at the origin and whose vector value is proportional to the displacement from the origin. The simple $J = 0 \to J = 1$ transition gives us a Zeeman shifted transition

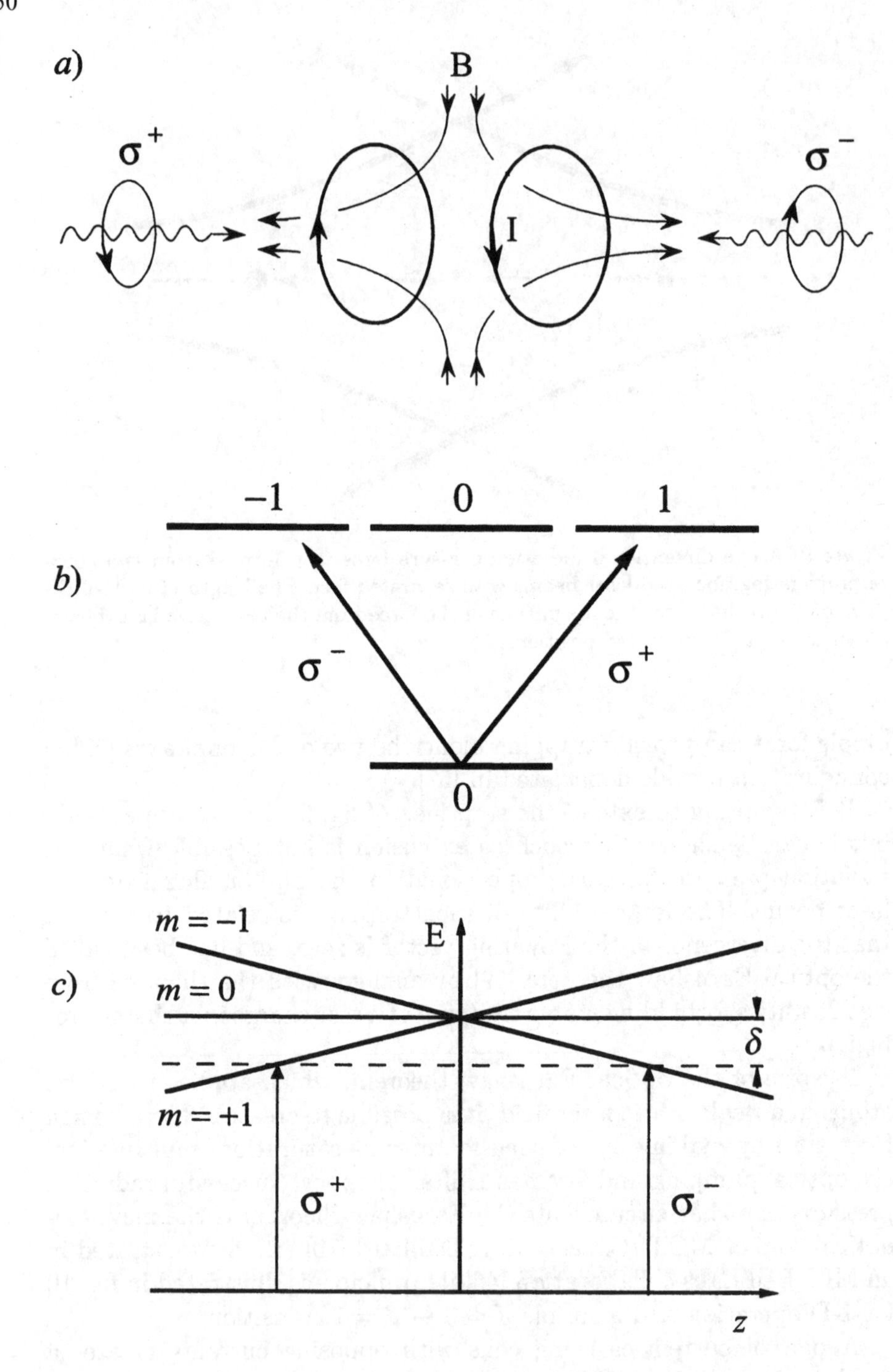

Figure 10. *a*) Magnetic field and laser configuration for a 1-D MOT. *b*) Transition scheme. *c*) Energy levels and transitions in the spatially varying magnetic field. The designation of the m-state is with respect to a space-fixed axis, as is the laser polarization.

frequency with a non-degenerate ground state. (This non degeneracy leads to Doppler rather than sub-Doppler cooling.) Two circularly polarized laser beams with opposite helicity counterpropagate along the coils' axis. The $\sigma^{\pm}$ beam excites atoms to the $m = \pm 1$ excited state respectively. Thus, for a red-detuned laser frequency ($\delta < 0$) atoms displaced in the positive direction will experience a Zeeman shift that brings the $m = -1$ state into resonance with the laser frequency, and the σ^{-} laser beam that excites this state is the one that pushes it back toward the origin. Similarly, an atom displaced in the negative direction is pushed back by the σ^{+} beam. In addition to the restoring force, there is also the usual Doppler-cooling damping force.

The force on an atom with velocity v and position z can be obtained from eqs. 3 and 4 by replacing the effective detuning $\delta \mp kv$ with $\delta \mp (kv + \beta z)$, where βz is the magnitude of the Zeeman frequency shift at position z:

$$F(v,z) \cong \frac{4\hbar k I/I_0}{\left[1+\left(\frac{2\delta}{\Gamma}\right)^2\right]^2}\left(\frac{2\delta}{\Gamma}\right)(kv+\beta z) \tag{16}$$

where in the second expression we are in the limit of low velocity, magnetic field and laser intensity. For negative detuning the above force represents a damped harmonic oscillator. Typical operating conditions for a MOT might be $I/I_0 = 1$, $2\delta/\Gamma = -1$ and $\beta = 10$ MHz/cm. This would lead to an oscillation frequency of about 1 kHz for Na, but with strong overdamping.

While we have considered only the 1D case for a simple transition, the MOT was first demonstrated in 3D, on an atom with a degenerate ground state (Na). The theory in 3D has been worked out in detail for the $J = 0 \rightarrow J = 1$ transition [71] and for transitions allowing sub-Doppler cooling, some insights have been gained by studying the forces in 1D on a moving atom in a magnetic field [72, 73, 74, 75]. Experiments have shown that, with a degenerate ground state, sub-Doppler temperatures are achieved in a MOT [74] along with larger trapping and damping than predicted by the $J = 0 \rightarrow J = 1$ theory.

The MOT has become an important tool in the study of cold atoms. A particularly useful feature is that it can capture atoms from an uncooled, thermal atomic vapor [31, 32]. This often allows considerable simplification of the apparatus compared to one where an atomic beam is first decelerated.

A MOT can concentrate atoms to the point that collisions [76, 77] and radiation pressure exerted by the atoms' fluorescence [78] are factors limiting the density. A major advance in reducing such problems is the "dark spot" MOT [79] in which atoms are, for the most part, optically pumped into a state from which the excitation rate is considerably smaller than in a conventional MOT. In practice this is accomplished, for atoms with ground state hyperfine structure, by pumping the atoms into one of the hyperfine states. Normally laser cooling and trapping of such atoms is performed by applying a separate laser frequency to excite each of the ground hyperfine states to the electronically excited state, ensuring atoms are not pumped into a state from which they cannot be excited. In a dark spot MOT, one of the laser frequencies (the re-pumper) is eliminated from the central region of the trap. Atoms then accumulate in the hyperfine state that would have been excited by the missing frequency. These atoms are rarely excited (only by off-resonant light or indirectly scattered resonant light) so problems due to excited atoms are greatly reduced, even though the atoms are still cooled and trapped. Such techniques have achieved atomic densities near 10^{12} cm^{-3}.

4.3. Magneto-static traps

Both laser dipole and radiation pressure traps involve optical excitation of the atoms being trapped. Although in principle a sufficiently intense laser dipole trap can be tuned far enough off resonance for the excitation rate to be small, it is difficult in practice to make it truly negligible. As a result, optical trapping generally results in some heating of the atoms due to the random nature of spontaneous emission, and cooling is required to keep the atoms in equilibrium. Magnetic traps do not suffer this problem and so can be used to store atoms for long periods of time without the need for additional cooling. On the other hand, such traps only work for atoms having a magnetic dipole moment and only for those states of such atoms whose Zeeman energy increases in increasing magnetic field.

The idea of magnetic trapping was an extension of the magnetic focusing of atomic beams [80] and was discussed by Paul during the 1950s [81]. The first published proposals for magnetic trapping of neutral atoms were in the 1960s [82, 83, 84]. It was not until after the demonstration of laser cooling of neutral atoms that the first magnetic trapping of atoms was achieved [44].

The principle of the magnetic trap can be understood by considering an atom with a Zeeman sub-structure such as that shown in the excited state of fig. 10*c*. (Note that magnetic trapping is generally done on ground state atoms with such Zeeman structure.) Since the energy varies with magnetic field, an atom feels a force in a magnetic field gradient. For states whose energy increases with magnetic field (low-field-seeking states), a trap is formed by the field of the coils in fig. 10*a*.

This quadrupole field is zero on the axis midway between the coils, and its magnitude increases linearly along any line away from this central point. Low-field-seeking atoms experience a restoring force towards and are trapped around this point. (It can be shown [85, 86] that in a current-free region no magnetic field can have a relative maximum in its magnitude, so that high-field-seekers cannot be trapped.) The depth of a magnetic trap is given by the maximum Zeeman shift along the easiest escape path. A magnetic field of 1 mT (10 G) gives a typical shift of 14 MHz, equivalent to 670 μK, so laser cooled atoms are easily trapped by modest fields.

The quadrupole magnetic field was used in the first magnetic trap [44] and is the simplest of magnetic traps. It works for low-field-seekers as long as they do not change their spin orientation with respect to the local field. This means than they must adiabatically follow the changes in the direction of the field as they orbit in the trap. The condition for adiabatic following is that

$$\frac{d\theta}{dt} \ll \Omega_{Zeeman}, \tag{17}$$

where θ is the angle of the magnetic field at the atom's position and Ω_{Zeeman} is the Zeeman frequency separating states of different spin orientation. Since laser cooled atoms move so slowly, they generally experience small field rotation rates, and their motion is usually adiabatic. The quadrupole trap, however, has a point where the magnetic field is zero and the adiabatic condition is impossible to satisfy. Atoms passing sufficiently close to the trap center will fail to follow adiabatically, change their spin orientation (undergo Majorana transitions) and be ejected from the trap [44]. This difficulty can be avoided by a variety of traps that do not have a point of zero magnetic field [87], but such traps are generally not as "stiff" as a quadrupole trap. That is, the restoring force near the center of the trap is not as large as for a quadrupole trap. In practice, non-adiabaticity is usually a problem only for very cold atoms confined very near the center of a quadrupole trap, which is exactly what occurs on the way to BEC.

For achieving Bose-Einstein condensation through runaway evaporative cooling, it is desirable to have a "stiff" trap to maintain a high collision or thermalization rate. For atoms cold enough to Bose condense, the problem of non-adiabatic spin flip transitions at the zero-field point of a quadrupole trap are quite severe [88, 89]. The TOP trap solved this problem by superposing onto the quadrupole field a magnetic field rotating in the plane of symmetry. The rotating field guarantees that the field is not zero at the trap center, and produces a rounding of the sharp-cusp quadrupole trapping potential. This time-averaged orbiting potential (TOP) trap [51], although time dependent, is not a magneto-dynamic trap in the sense that the TOP trap does not rely on micromotion of the atoms. The atoms respond to the time averaged potential at a given point in the trap. A more detailed discussion of the TOP trap can be found in the discussion of the BEC experiment at NIST-Gaithersburg, in section 71 of these lecture notes.

5. Sub-Doppler laser cooling

The early experiments [28, 90] on optical molasses produced a satisfying agreement of the observed temperature and spatial diffusion with the predictions of the theory of Doppler cooling as outlined in the previous section. Some experiments, however, were in disagreement with expectations [91]. Finally, in 1988, careful temperature measurements [92] showed conclusively that atoms in optical molasses were much colder than the Doppler cooling limit. The time-of flight (TOF) method used to measure the temperature in those experiments has become a standard technique and is illustrated in fig. 11. Atoms are collected and cooled in the optical molasses at the intersection of the molasses laser beams. The molasses beams are suddenly extinguished and the released atoms fall toward the probe. As the atoms pass through the probe laser beam they absorb and fluoresce light. This fluoescence is measured, with time resolution, by the detector. The distribution of detected fluorescence in time gives the distribution of times of flight from the molasses to the probe, and the temperature can be determined from that distribution.

5.1. Observation of sub-Doppler temperatures

Figure 12 shows an example of an experimental TOF signal for sodium atoms cooled in optical molasses [20]. The experimental points correspond reasonably well with the predicted signal for a temperature of 25 μK, while 250 μK, about the Doppler cooling limit for Na, is completely inconsistent with the experimental points. Furthermore, the dependence of the temperature on detuning was found to be inconsistent with the theory of Doppler cooling. Figure 13 shows the measured temperature as a function of laser detuning for Na, where the linewidth is 10 MHz. The temperature decreases for detunings larger than $\Gamma/2$ (until the laser frequency approaches resonance with another hyperfine state, about 60 MHz to the red of the chosen resonance). This is in sharp contrast to the prediction of Doppler cooling theory (fig. 13), which has the temperature increasing for detunings greater than $\Gamma/2$.

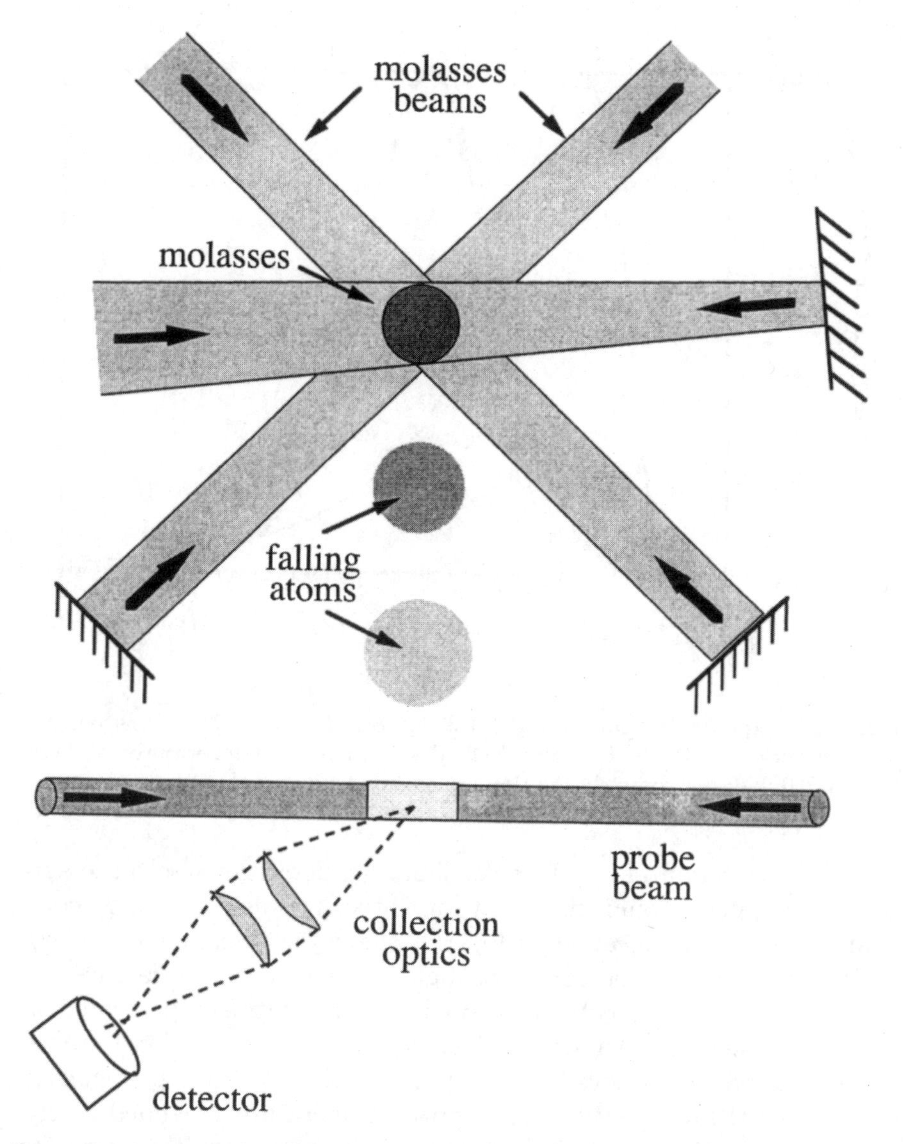

Figure 11. Schematic of the time-of-flight temperature measurement. Atoms are released from the optical molasses and travel ballistically to the probe region. The distribution of arrival times, as measured by the fluorescence seen by the detector, allows one to determine the atomic temperature.

In addition, the temperature was found to be linearly dependent on laser intensity [93], again in contrast to the predictions of Doppler cooling theory, and the temperature was found to depend on magnetic field and laser polarization [92, 93]. These latter facts, particularly, suggest that the magnetic sublevels of the atom play an important role.

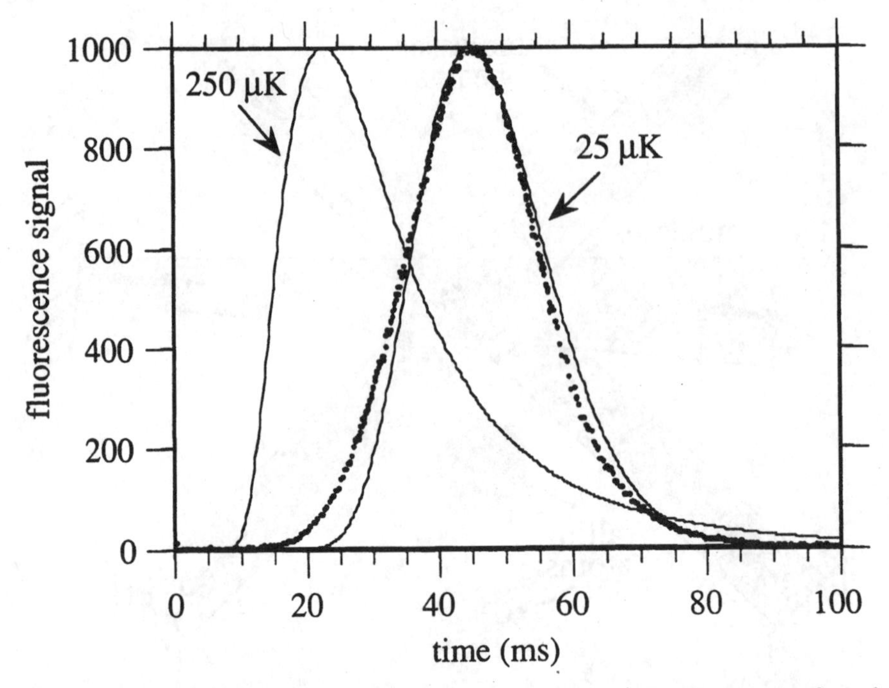

Figure 12. Experimental time-of-flight distribution (points) for Na atoms released from an optical molasses. Expected TOF distributions for temperatures of 25μK and for 250μK (approximately the Doppler cooling limit) are shown.

The observation of sub-Doppler-limit temperatures was quite surprising. Doppler cooling theory at low intensity was simple and compelling. Furthermore, at least for 1D, there was a complete theory, taking into account the effects of high intensity and interference between laser beams that were ignored in the treatment presented in the previous section. The theory had been restricted to 2-level atoms, but it was widely believed that this restriction was not particularly important. At low intensity the Doppler temperature depended on the transition linewidth and the detuning, and these were the same for any of the degenerate Zeeman sublevels in a given alkali hyperfine level. Nevertheless, the magnetic field and polarization dependence of the sub-Doppler-limit temperatures pointed to the importance of the Zeeman sublevels.

5.2. New cooling mechanisms

The explanation for the sub-Doppler temperatures soon came in the form of a new theory of multi-level laser cooling. The key elements of the new theory [72, 94, 95] were optical pumping among the magnetic

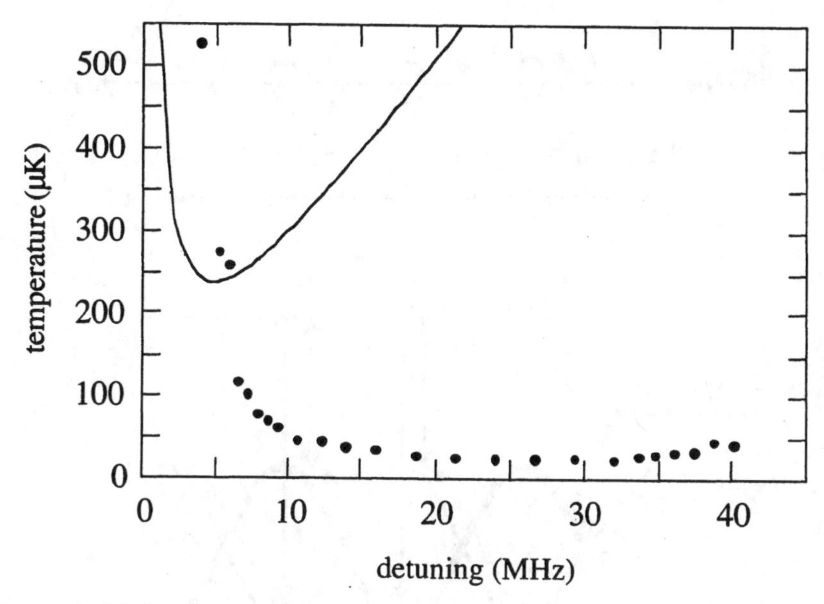

Figure 13. Measured temperature as a function of laser detuning for sodium atoms cooled by a 3D optical molasses. The temperature predicted by Doppler cooling theory is shown by the solid line.

sublevels of the electronic ground level and differential light shifts of the sublevels. While the original theories treated cases where a spatial gradient of the polarization of the optical field was important, it was later demonstrated that such multilevel laser cooling was possible even without polarization gradients [95, 96, 97, 98, 99, 100, 101].

The full theory of multilevel laser cooling is complicated and arose from a deeper understanding of the interaction of light with atoms that developed along side experimental progress on the manipulation of atoms with light. The most important of the multilevel cooling mechanisms is Sisyphus cooling [16, 72, 102, 103]. Semiclassically (when the atom can be considered to be well-localized on the scale of an optical wavelength) the simplified physical picture for Sisyphus cooling can be understood by considering the atom and laser field situation illustrated in fig. 14.

Figure 14*a* shows a 1-D set of counterpropagating beams with equal intensity and orthogonal, linear polarizations. The interference of these beams produces a standing wave whose polarization varies on a subwavelength distance scale. At points in space where the linear polarizations of the two beams are in phase with each other, the resultant polarization is linear, with an axis that bisects the polarization axes of the two individual beams. Where the phases are in quadrature, the

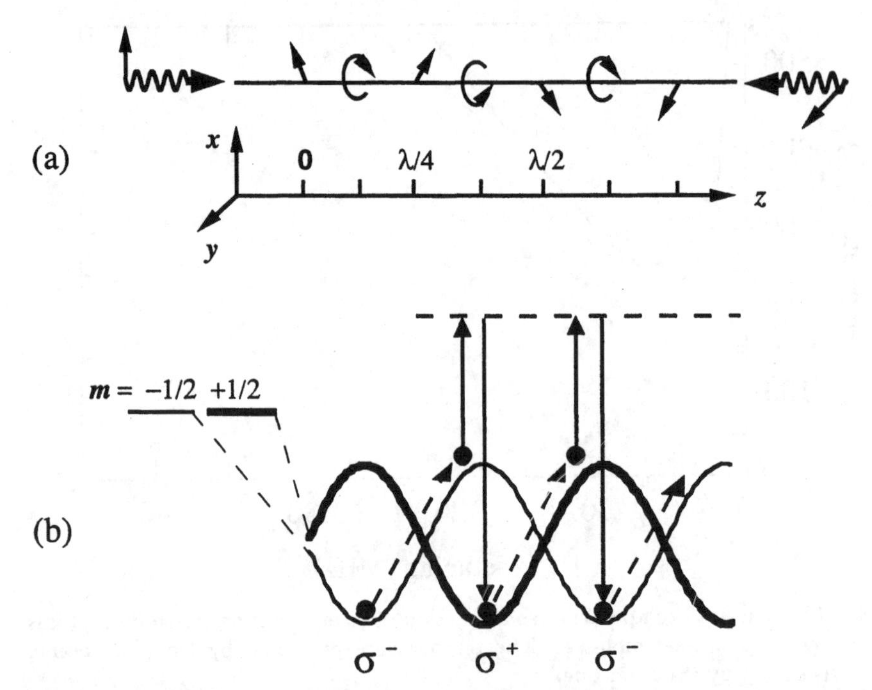

Figure 14. *a*) Interfering, counterpropagating beams having orthogonal, linear polarizations create a polarization gradient. *b*) The different Zeeman sublevels are shifted differently in light fields with different polarizations; optical pumping tends to put atomic population on the lowest energy level, but non-adiabatic motion results in "Sisyphus" cooling.

resultant polarization is circular and at other places the polarization is elliptical. An atom in such a standing wave experiences a fortunate combination of light shifts and optical pumping processes.

Because of the differing Clebsch-Gordon coefficients governing the strength of coupling between the various ground and excited sublevels of the atom, the light shifts of the different sublevels are different, and they change with polarization (and therefore with position). Figure 15*b* shows the sinusoidal variation of the ground-state energy levels (reflecting the varying light shifts or dipole forces) of a hypothetical $J_g = 1/2 \rightarrow J_e = 3/2$ atomic system. Now imagine an atom to be at rest at a place where the polarization is σ^- at $z = \lambda/8$ in fig. 15*a*. As the atom absorbs light with negative angular momentum and radiates back to the ground states, it will eventually be optically pumped into the $m_g = -1/2$ ground state, and simply cycle between this state and the excited $m_e = -3/2$ state. For low enough intensity and large enough detuning we can ignore the time the atom spends in the excited state and consider only the motion of the atom on the ground state potential. In the $m_g = -1/2$ state, the atom is in the lower energy level at $z = \lambda/8$, as shown in fig. 15*b*. As the atom moves, it climbs the potential hill of the $m_g = -1/2$ state, but as it nears the top of the hill at $z = 3\lambda/8$, the polarization of the light becomes σ^+ and the optical pumping process tends to excite the atom in such a way that it decays to the $m_g = +1/2$ state. In the $m_g = +1/2$ state, the atom is now again at the bottom of a hill, and it again must climb, losing kinetic energy, as it moves. The continual climbing of hills recalls the Greek myth of Sisyphus, so this process, by which the atom rapidly slows down while passing through the polarization gradient, is called Sisyphus cooling. In Sisyphus cooling, the radiated photons, in comparison with the absorbed photons, have an excess energy equal to the light shift. While in Doppler cooling, the energy excess comes from the Doppler shift.

In contrast to the case for Doppler cooling (see eq. 4) the friction force is independent of laser intensity, and proportional to detuning at low (but not too low) intensity and low velocity, while the momentum diffusion coefficient is proportional to intensity and independent of detuning. This leads, according to eq. 7, to a temperature that depends linearly on intensity and inversely on detuning. That is, the temperature is proportional to the light shift

$$k_B T \propto \hbar \Delta_{lightshift} = \frac{\hbar \Omega^2}{4\delta} \tag{18}$$

where the expression for the light shift is valid in the limit of low intensity and large detuning.

A less restricted treatment [104] shows that the friction force is not linear in the atomic velocity, nor is the momentum diffusion constant independent of velocity. Nevertheless, the temperature remains approximately linear in the light shift as long as the intensity is sufficiently above a critical intensity. The lower limit to the temperature obtainable by Sisyphus cooling is set by this lowest intensity for which the cooling process works [104, 105]. This is the intensity at which the light shift is comparable to the recoil energy, and it leads to a lower limit for the thermal velocity on the order of a few times the recoil velocity.

These qualitative features of multi-level laser cooling have been confirmed by experiments on atoms cooled in 3D optical molasses [20, 106]. The experimental results showed the linear dependence of the temperature on intensity and light shift for all but the largest intensity at the smallest detuning, outside the limits of validity of the simple results listed above. The constancy of the lowest temperature, once the detuning is large enough, is consistent with its depending only on the recoil energy. The temperature of 2.5 μK obtained for Cs represents an *r.m.s.* velocity of about three recoil velocities, similar to the case for the lowest temperatures observed for Na [20, 97] and Rb [107].

Even more explicit confirmation of the theory of multilevel laser cooling comes from a comparison of one-dimensional experiments with 1D theory. This is discussed below in the section on optical lattices.

6. Optical lattices

An optical lattice is an array of light-shift-induced potential wells created by the interference of a set of intersecting laser beams. A good optical lattice typically cools and confines atoms into low-lying quantum states of center-of-mass motion in the individual wells. That is, most atoms are well localized in some well of the lattice. The idea that atoms experience a periodic potential is already implicit in the image of Sisyphus cooling alluded to in section 52 above. Furthermore, the idea that atoms could be trapped in such periodic light-shift potentials predates even the idea of laser cooling [108]. Atoms were first ordered by such a periodic potential in 1D "channeling" experiments [109, 110] in which the potential of a strong standing wave altered the spatial distribution of an uncooled beam of atoms. The first 3D experiments, which showed localization of atoms in a laser cooled gas [111], observed Dicke narrowing of the spectrum of light radiated by atoms in an optical molasses. The Dicke narrowed spectrum, obtained when laser light scatters from atoms localized to a distance smaller than the wavelength, corresponds to transitions that leave the atom in the same vibrational state of center-of-mass motion in a potential well.

6.1. Localization of atoms in optical lattices

Optical lattices for laser cooled atoms have been constructed in one, two and three dimensions. In each case, the phase relations between the laser beams forming the lattice are important. In a 1D lattice, formed by a pair of counterpropagating laser beams, phase changes simply have the effect of translating the origin of the lattice. In a 2 or 3D lattice formed from 2 or 3 pairs of counterpropagating beams, the situation is more complicated. In order to maintain a lattice that changes only by a translation with changes in phase, one or two relative phases must be fixed and stabilized by interferometric phase comparison [112]. An alternative [113] is to use only three beams in 2D or four beams in 3D. Then one recovers the 1D situation where changes in phase only translate the lattice. Figure 15 shows a four-beam arrangement and the structure of the 3D optical lattice it forms. Atoms are trapped at

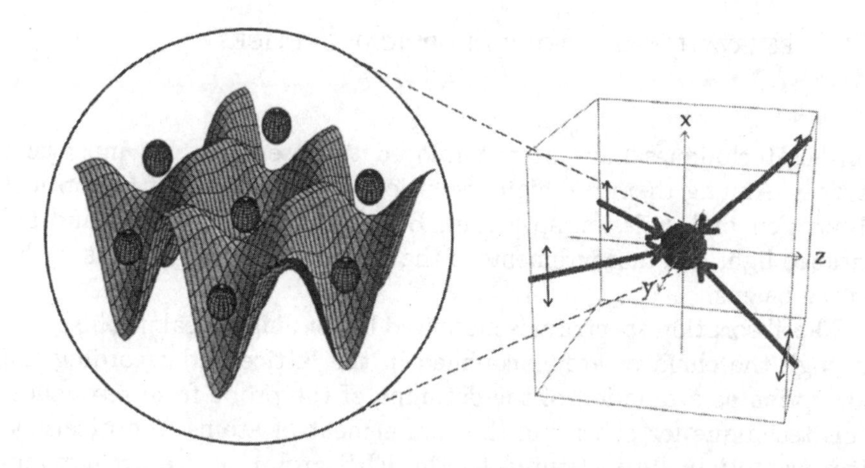

Figure 15. A four-beam arrangement for a 3D optical lattice. The interference of the four beams produces a lattice with centered tetragonal structure. Lattice points, and positions where atoms are trapped correspond to points of pure curcular polarization along the z-axis. For this configuration, the lattice spacing along z is one half of the spacing along x or y.

places where the local polarization of the light field is purely circular, either σ^+ or σ^-, along the symmetry or z-axis.

The optical potential wells of an optical lattice result from the optical dipole force. As such, the detuning from resonance can be chosen to determine the degree to which spontaneous emission is present. Similar to dipole traps, optical lattices can be divided into two classes - near resonant and far off resonant. Strategies for loading and the subsequent localization of atoms in optical lattices depends on which class the lattice falls in. For near resonant lattices, in which substantial spontaneous emission is occuring, atoms are cooled into the lattice by Sisyphus or sub-Doppler cooling mechanisms described in section 51 of these notes. Typically, one starts with a cloud of laser cooled atoms, switches on the lattice beams and subsequently, the atoms cool off and are localized in the potential wells of the optical lattice. For far detuned lattices, where there is little or no spontaneous emission, localization of atoms in the potential wells is accomplished by starting with a cold cloud of atoms and non-adiabatically or rapidly turning on the lattice beams. Those atoms near the bottom of the potential wells at the instant when the lattice beams are turned on will be localized in the wells. The density of atoms trapped in optical lattices is substantially below degeneracy. Typically, only 1/100 of the lattice sites are occupied in a 3D lattice.

6.2. Observation of atoms in optical lattices

Various techniques have been employed to observe atoms in optical lattices. Among them are high resolution measurements of the optical absorption and emission spectrum, Bragg diffraction of near and far detuned light and measurements of the redistribution of photons in the lattice beams.

The absorption spectrum is measured by sending a weak probe beam through the cloud of atoms confined in the lattice and recording the absorption as a function of the detuning of the probe from resonance. This technique for observing the confinement of atoms in optical lattices was originally instituted by the ENS group, and a great many insights into the behavior of atoms in optical lattices have been gained by these experiments. The emission spectrum is obtained by measuring the frequency of the lattice light scattered by the atoms using an optical heterodyne technique. The fluorescent light emitted by the atoms held in the lattice is mixed with a local oscillator derived from the same laser that creates the lattice. This results in a high resolution measurement of the spectrum of emitted light, since any frequency fluctuations due to technical noise in the laser are common mode between the signal and the local oscillator.

Transitions between the quantized vibrational states of atoms in the potential wells of the optical lattice have been observed as sidebands in the optical absorption and emission spectrum. Figure 16 is an emission spectrum of atoms confined in a 3D optical lattice. Quantum mechanically, the sidebands may be thought of as Raman scattered light, with the lower frequency (red) sideband arising from processes beginning on a given vibrational level of a potential well and returning to a higher level, while the blue sideband comes from processes returning to lower vibrational levels. The central peak represents those processes where the vibrational level does not change. The temperature of atoms in optical lattices can be extracted from the relative strengths of the red and blue sidebands. This is possible because the red sideband comes from processes starting on lower vibrational states, and for a thermal distribution these states are more highly populated than the higher states.

The first observations of the sidebands came in 1D experiments in the absorption spectrum [114] and the emission spectrum [115]. Sidebands were subsequently measured in 2 and 3 dimensional optical lattices, and a great many insights into the behavior of atoms in optical lattices have been gained by these experiments. The details of these and other experiments, as well as the theoretical analysis are not covered

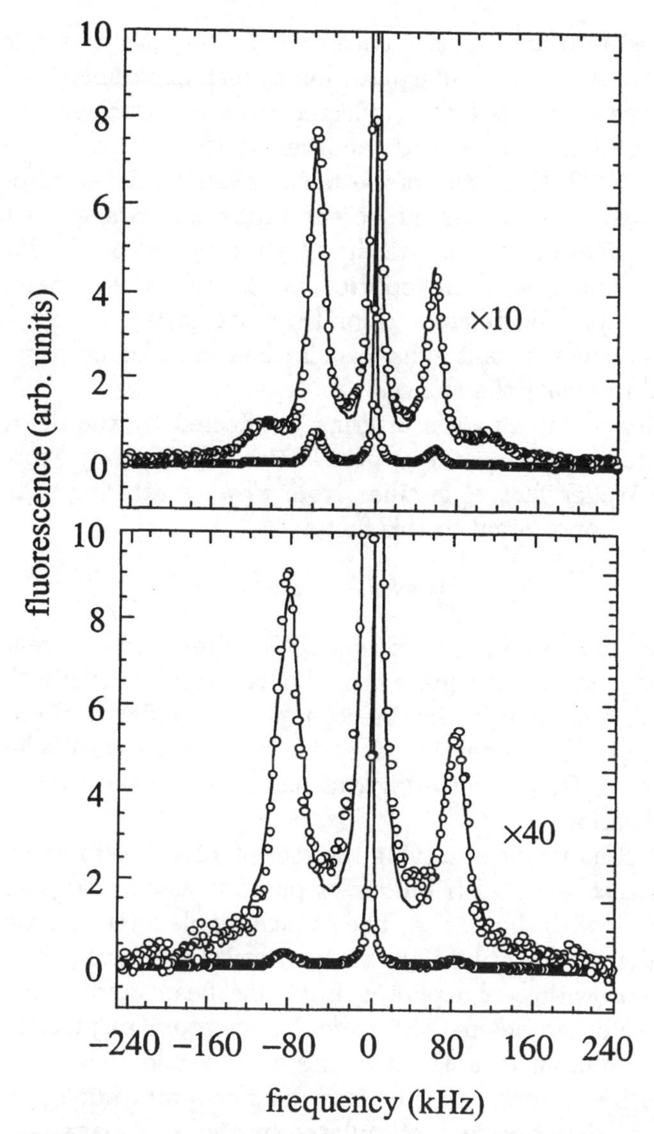

Figure 16. Emission spectra taken at NIST-Gaithersburg for cesium atoms trapped in the lattice of fig. 15. The upper spectrum was from light emitted along the x axis and the lower one from light along the z axis.

in these notes, but can be found in refs [112, 113, 114, 115, 116, 117, 118, 119, 120, 121, 122, 123, 124, 125, 126, 127].

Just as x-rays may be Bragg scattered from 3D solid crystals whose lattice constant is on the order of x-ray wavelengths, light can be Bragg scattered from optical lattices whose lattice constant is on the order of optical wavelengths. While they are closely related, we distinguish Bragg scattering from four wave mixing processes in which a probe beam is directed into an optical lattice and the reflected beam is observed [118, 123]. A common interpretation of such experiments is that one of the lattice beams is Bragg reflected from a grating formed by the interference of the probe with another of the lattice beams. In experiments at NIST [128], the probe beam was introduced after the lattice beams had been turned off, so the probe was Bragg scattered from the arrangement of atoms that had been imposed by the lattice. Similar experiments have been reported in Munich [129]. There, the probe was a completely different color from the lattice laser beams, and made use of the resonant enhancement near a different transition than that used to create the lattice.

The intensity of the Bragg scattering is affected by the degree to which atoms are localized at the optical lattice sites. This is expressed by the Debye-Waller factor, familiar from x-ray scattering. The reflected signal is proportional to this factor

$$F_{D-W} = \exp(-K^2 \Delta x^2) \tag{19}$$

where K is the difference in wavevectors between incident and reflected beams and Δx^2 is the mean square spread of the atomic distribution in a potential well, measured in the direction perpendicular to the Bragg plane. Because of the exponential dependence of the Debye-Waller factor, the amount of Bragg scattering can be extremely sensitive to the degree of localization.

The third technique for observing the motion of atoms in an optical lattice is to detect the redistribution of photons associated with the atoms interacting with the lattice. The optical dipole force experienced by atoms in the lattice potential can be viewed as arising from the simultaneous absorption of a photon from one beam and stimulated emission into another beam. For example, an atom displaced from the potential minimum in a 1D standing wave would tend to simultaneously absorb a photon from the laser beam counterpropagating to the displacement direction and, stimulated by the copropagating laser beam, emit a photon in the same direction: the net result being that the atom recoils toward the potential minimum with twice the single photon momentum transfer from one beam. The photon-redistribution technique detects the change or redistribution of photons amongst the lattice beams as the atoms experience a net force on their center of mass motion due to the optical potentials.

6.3. Dynamics of atoms in optical lattices

Bragg scattering and photon redistribution can be used to study the motion of atoms in optical lattices. In turn, studying the dynamical be-

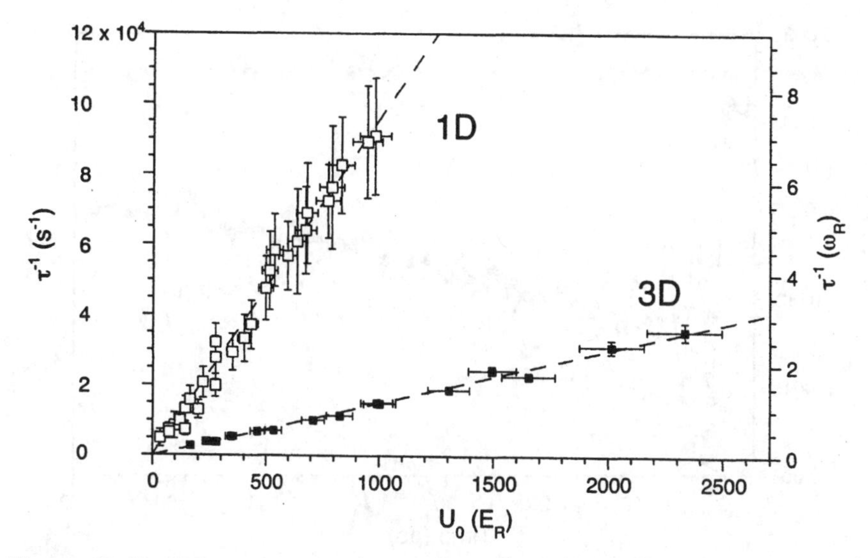

Figure 17. Localization rate τ^{-1} in s^{-1} and in units of the recoil frequency, $w_R = 2\pi \times 2.07$ kHz, versus the potential depth U_0, for fixed detunings.

havior of atoms in optical lattices provides insight into the mechanisms of laser cooling previously infered from steady-state measurements.

We have used the exponential dependence of the Bragg scattering on the localization of the atoms to sensitively measure the spread of the atomic position in the optical potentials. In one such study, we start with a disordered gas of atoms, turn on the optical lattice and observe the Bragg signal after the lattice has been on for varying times. Using the expression for the Debye-Waller factor, we determine the time evolution of the localization from the Bragg signal. We find that for a wide range of parameters the localization decays exponentially to its steady state value [130]. The rate of exponential decay is found to be proportional to the photon scattering rate of atoms trapped at the lattice sites, as shown in fig. 17. In the 1D experiments with Cesium atoms, the localization rate is about 30 times slower than the scattering rate and in 3D it is about 200 times slower. This dramatic difference in the 1D and 3D rates is not understood. Perhaps it is due to the existence of orbits with angular momentum in 3D.

Bragg scattering can also be used to detect changes in the localization caused by deliberate changes in the optical lattice potential. Starting with atoms in steady state, reasonably well localized near the lattice sites, we suddenly increase the intensity of the lattice beams and correspondingly the depth of the potential wells. Figure 18 shows the results of such an experiment in a 1D cesium opitcal lattice [131]. There

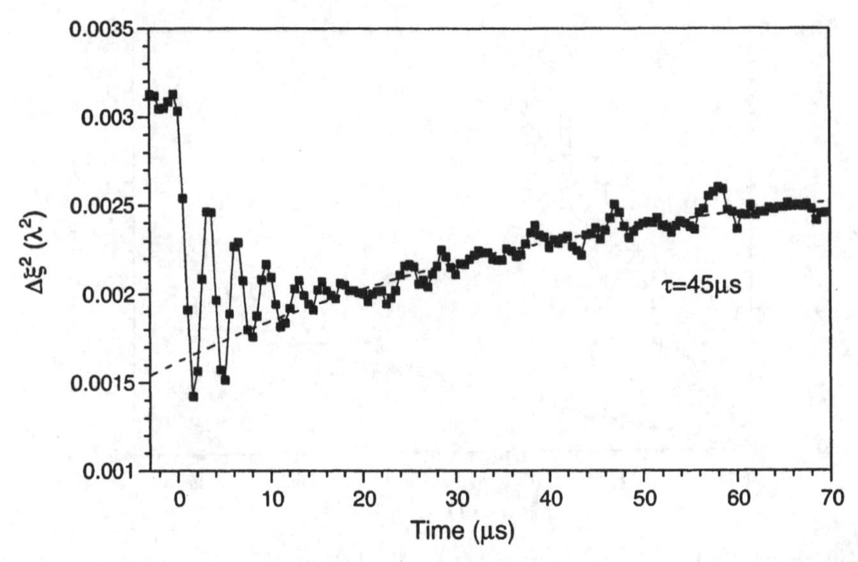

Figure 18. Wave-packet oscillations in a 1D lattice formed by tow counter propagating beams with orthogonal, linear polarization. The ocillations are induced by a sudden increase, at t=0, of the potential depth by a factor of 4.6. The long-term heating is best fitted by an exponential with a time constant of 45 μs (dashed line).

is an initial decrease in the mean square position spread as the atoms have received a kick due to the sudden increase in the lattice potential depth, causing them to accelerate toward the centers of the potential wells. They then oscillate through the centers, producing a breathing motion of the wave-packet at twice the trap vibrational frequency. The oscillations are damped mainly by dephasing due to dispersion of the vibrational frequencies in the anharmonic trapping potential. Over a longer period of time, the average value of the mean squared spread increases, indicating a heating to a higher temperature in the deeper potential well. For the data of fig. 18, this long-term heating is best fitted by an exponential with a time constant of 45 μs, which is much longer than the localization time constant of 6 μs estimated for the deeper potential well [131].

We also observe, using the photon redistribution technique [132], the oscillatory motion of the atomic wave-packets in a 1D optical lattice formed by two, nearly counterpropagating, laser beams with orthogonal, linear polarization. The two beams are directed onto photodiodes and the difference signal is measured; taking the difference signal strongly suppresses noise due to laser intensity fluctuations while doubling the single-beam signal corresponding to the coherent photon transfer. A similar redistributed photon detection technique was recently used for observing the wave-packet motion of atoms also in a 1D

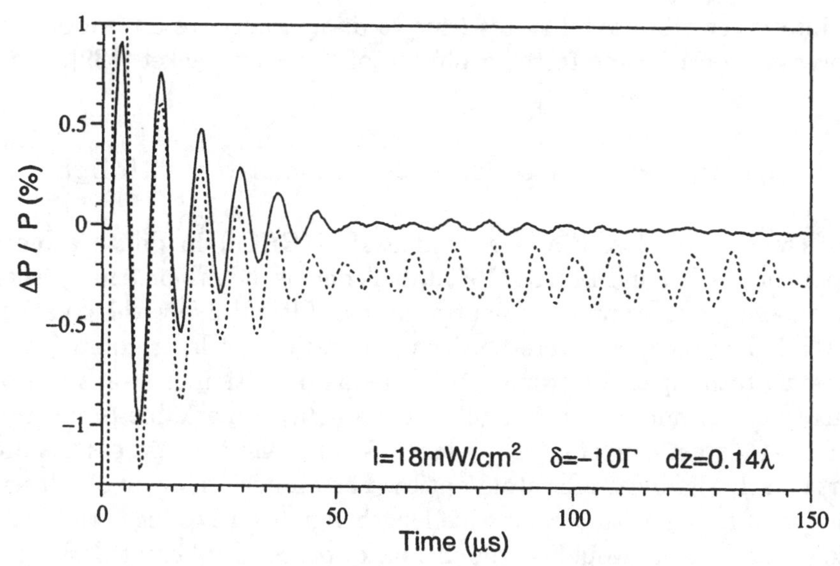

Figure 19. Experimental (solid) and theoretical (dotted) power transfer ratio, $(\Delta P/P)(t)$, between the lattice beams induced by sloshing mode, wave-packet oscillations of atoms in a 1D optical lattice. For clarity, the solid and dotted lines are shifted with respect to each other. The oscillations are initiated by a sudden shift of the lattice at t=0 by the indicated value of dz. The theoretical curve was obtained from QMCWF calculations and is scaled to match the second extremum, because the first is strongly affected by the time constant of the shift.

lattice [133], although in this experiment, the signal was interpreted as arising from recoil-induced resonance.

Starting with the atoms cooled and localized at the bottom of the potential wells, we suddenly shift the potential minima with respect to the center of mass of the atoms by an amount $0 < dz < 0.25\lambda$ using a phase modulator in one of the lattice beams. The atoms then execute a sloshing, oscillatory motion in the displaced lattice. Figure 19 shows a typical experimental result along with the results of quantum Monte-Carlo wavefunction (QMCWF) calculations. The experimental and theoretical curves are in excellent agreement, except for a faster decay in the experimental data. We attribute this to intensity inhomogeneity in the experiment. The initial collapse of the oscillation is due to dephasing, which results from the anharmonicity of the system. The long-time structures evident in fig. 19 represent the first convincing observations of wave-packet revivals in optical lattices. These revivals occurred earlier than we expected based on dispersion due to anharmonicity alone. Instead, our QMCWF theory, which completely treats dispersion and dissipation, shows that anharmonicity, band curvature (resulting from tunneling between adjacent wells) and coherence transfer into lower vibrational states (due to dissipation) are all important processes, contributing to the evolution of the wave-packet [132].

7. Manipulating Bose-Einstein condensates with light

The creation of Bose-Einstein condensates (BECs) in dilute atomic vapors of Rb, Na and Li [134, 135, 136] is one of the major triumphs of laser cooling and trapping of neutral atoms. Alternatively, the creation of BECs has renewed interest in the applications of laser cooling and trapping techniques for atom optics, the manipulation of atoms analogous to the manipulation of light. Of the many remarkable properties of Bose-Einstein condensates, its macroscopic coherence properties are particularly significant for atom optics. The atomic analogue of a laser, a beam of atoms with a coherence length significantly longer than the size of the sample, would be the ultimate source of coherent deBroglie waves. One promising approach is the coherent extraction of atoms from a condensate.

The principal focus of research on Bose-Einstein condensates in the Laser Cooling and Trapping Group of NIST is for atom optics and the realization of the "atom laser," the atomic analogue of the optical laser. Our atom optics experiments are performed on a BEC of sodium atoms produced in a TOP trap. Using optical, stimulated Raman transitions, we have demonstrated normal and Bragg diffraction of condensates and directional output coupling of atoms from a BEC for a quasi continuous-wave (CW) atom laser.

We begin with a brief description of our experimental apparatus and approach for creating a Bose-Einstein condensate. We do this essentially for two reasons: our strategy for achieving BEC is somewhat different than other approaches, and it illustrates an application of many of the techniques of laser cooling and trapping developed over the last 15 years. This is followed by a description of experiments using optical stimulated Raman pulses to manipulate condensates both in and out of traps.

7.1. The tri-axial TOP trap for sodium

Our Bose-Einstein condensates of sodium atoms are produced in a time-averaged orbiting potential or TOP trap [51]. Our TOP trap differs from those in other BEC experiments in two respects. First, all other experiments with TOP traps that have resulted in BEC use rubidium. We are currently the only group making condensates of sodium in a TOP trap. The lighter mass of sodium poses a greater technical challenge in the design of the TOP trap compared to rubidium. Since the oscillation frequency of an atom in a trap is higher for the lighter atom, the frequency of the rotating bias field of the TOP trap must be correspondingly higher so that the atom experiences the time-averaged potential. Typically, sub-Doppler laser cooling can cool a sample of atoms to an energy which is a few times the recoil energy. Since the recoil energy is inversely proportional to the mass of the atom, the TOP trap must also be deeper or stronger to contain the lower mass, laser cooled atoms. Second, the geometry of our TOP trap is different from other TOP traps resulting in a totally anisotropic or tri-axial, time-averaged potential.

A time orbiting potential or TOP trap is a magnetic trap consisting of a quadrupole magnetic field and a constant magnitude rotating bias field. The potential resulting from a superposition of these two magnetic fields, averaged over a rotation period, is harmonic for small displacements. The time-averaged value of the minimum magnetic field in the TOP trap is just the magnitude of the rotating bias field. That is, the rotating bias field has effectively "plugged" the zero field region of the quadrupole field. For a bias field $\mathbf{B}_b = B_0\left(\hat{x}cos(\Omega t) + \hat{y}sin(\Omega t)\right)$,rotating at frequency Ω, with magnitude B_0, in the presence of a general quadrupole field $\mathbf{B}_q = b_q\left(\alpha x\hat{x} + \beta y\hat{y} + \Gamma z\hat{z}\right)$, where the values of α, β and Γ must satisfy Maxwell's equations, the magnitude of the magnetic field averaged over one rotation is,

$$\langle \mathbf{B} \rangle_t = \frac{\Omega}{2\pi} \int_0^{2\pi/\Omega} dt \left| \mathbf{B}_q + \mathbf{B}_b \right| = \frac{2}{\pi} B_{max} \mathrm{E}(m) \tag{20}$$

$$B_{max} = \sqrt{\left(B_0 + b_q r\right)^2 + \left(\Gamma b_q z\right)^2} \tag{21}$$

$$r = \sqrt{(\alpha x)^2 + (\beta y)^2} \tag{22}$$

$$m = 4 B_0 b_q r / B_{max}^2 \tag{23}$$

$\mathrm{E}(m)$ is a complete elliptic integral of the second kind. There is a locus of points where the instantaneous magnetic field is zero, the "circle of death." These points occur where the rotating bias field cancels the quadrupole field, that is, these points corresponding to $z = 0$ and $b_q r = B_0$.

In the standard configuration for the TOP trap fields, such as in the original trap of JILA used to create the first BEC of Rb, the bias field rotates in the symmetry plane of the quadupole field, hence $\alpha = \beta = 1$ and $\Gamma = -2$. For small displacements, the time-averaged potential is given by

$$\langle \mathbf{B} \rangle_t = B_0 + \frac{b_q^2}{4B_0}\left(x^2 + y^2 + 8z^2\right) \tag{24}$$

The spring constant in the radial (x, y) direction is a factor of eight less than the axial direction (z) producing a disk shaped time-averaged potential.

In the NIST-Gaithersburg TOP trap, the bias field rotates in a plane containing the symmetry axis of the quadrupole field. To obtain the time-averaged magnitude of the magnetic field from eqs. 20-23, we make the substitution: $\alpha = 1$, $\beta = -2$, $\Gamma = 1$, $x \to x$, $y \to z$ and $z \to y$. For small displacements, the time-averaged potential is given by

$$\langle \mathbf{B} \rangle_t = B_0 + \frac{b_q^2}{4B_0}\left(x^2 + 2y^2 + 4z^2\right) \tag{25}$$

The spring constants of a magnetic trap given by eq. 25 are in the ratio of 1:2:4 in the x, y and z direction. Such a tri-axial trap is closer to spherical than the JILA TOP trap and better matched for loading from the nearly spherical clouds of laser cooled atoms from the MOT. Unlike the coventional TOP trap of JILA, described by eq. 24, this time-averaged potential is tri-axial, that is, it has no rotational symmetry, which poses an additional challenge for theoretical calculation.

Although the TOP trap has a number of desirable properties, including some independent adjustments of the spring constants in the three principle directions, there are certain limitations associated with trapping a mixed state such the $F = 1$, $m_F = -1$ state of sodium. The most important is the quadratic Zeeman effect which reduces the "effective" magnetic moment of that state.

Equations 24 and 25 show that in order to achieve the stiffest TOP trap for a given radius of the "circle of death" (a radius typically chosen to be larger than the radius of the sample of trapped atoms), the largest possible quadrupole field should be used. If the strength of the quadrupole field is increased then the magnitude of the rotating field must be increased to keep the radius of the "circle of death" constant. The energy of the $F = 1$, $m_F = -1$ state of sodium as a function of magnetic field initially increases linearly (the linear Zeeman effect), but the slope decreases with magnetic field due to the quadratic Zeeman effect. Eventually the energy of this state reaches a maximum around 31.5 mT (315 Gauss) above which it becomes an anti-trapped state. Thus for a fixed radius of the "circle of death," the stiffness of the TOP trap no longer increases linearly with the strength of the quadrupole field, becoming extremely weak for sufficiently large values of the bias field.

7.2. BEC OF SODIUM IN A TOP TRAP

Similar to other BEC experiments, our Bose-Einstein condensate of alkali atoms is produced by evaporative cooling in a magnetic trap loaded with laser cooled and trapped atoms, however, the details of our technique differ from other groups. More specifically, we load a dark spot MOT [79] from an effusive source of sodium at 625 K using Zeeman slowing. The slowing laser beam passes through the trap so that the capture area of the trap subtends a large solid angle of the flux of slow atoms. In order to minimize the effect of the slowing laser on the trapped atoms, we use a hybrid slower geometry [137]. A conventional Zeeman slower is used to slow atoms from about 800 m/s down to about 160 m/s, followed by a short section of reverse slower [138] to slow atoms from about 160 m/s down to a few m/s, which is within the capture velocity of the dark spot MOT. To avoid the loss of atoms in the slowing process due to optical pumping as the atoms pass through the zero field region between the conventional and reverse Zeeman slowing magnets, a second laser frequency is in the slowing beam to pump atoms from the $F = 1$ hyperfine level to the $F = 2$ level where they can continue to be slowed. In addition, a dark spot is placed in the slowing laser beam, creating a shadow in the region of the cloud of atoms in the dark spot MOT [139]. This further reduces the effect of the slowing laser beam on the trapped atoms. Typically, we load more than 10^{10} atoms into the MOT in less than 0.5 seconds.

After loading into the dark spot MOT, the magnetic fields are rapidly switched off and the atoms are further cooled to 200 μK by a brief period of dark molasses (there is a dark spot in the repumper light in the molasses beams), followed by optical pumping of the entire population into the $F = 1$ hyperfine level. The magnetic trap is then rapidly switched on, trapping atoms in the $m_F = -1$ sublevel of the $F = 1$ ground state. Since all three sublevels are present in equal populations at the time when the magnetic trap is turned on, two-thirds of the sample of laser cooled atoms are necessarily lost in the transfer. Experimentally, we find that we are able to trap 4 to 5 $\times 10^9$ sodium atoms in the magnetic trap.

The atoms are confined in the benign environment of a magnetic trap in order to be evaporatively cooled [140, 141]. We have developed two strategies for evaporatively cooling atoms to Bose-Einstein condensation. The first strategy involves evaporatively cooling using rf, the atoms initially trapped in a quadrupole field. This is then followed by rapidly transfering them into the TOP trap and further cooling the sample to condensation again using rf-induced evaporation. The second strategy involves starting with atoms in the TOP trap and evporatively cooling the atoms with the circle of death all the way to condensation. Both strategies produce approximately the same number of final condensate atoms, about 3×10^6, at a BEC transition temperature of 1.2 μK.

The first strategy evolved in response to the quadratic Zeeman effect problem in the $m_F = -1$ state of sodium, discussed in section 71. That is, for the large cloud of atoms initially confined in the TOP trap, the stiffness of the trap could not be increased, while keeping the "circle of death" well outside the cloud, to sufficiently compress the sample for runaway evaporation. Instead, we initially compress and evaporatively cool the sample of atoms trapped in a quadrupole magnetic field. When the sample is sufficiently cold and dense enough, we transfer them to the TOP trap by rapidly switching on the rotating bias field while the quadrupole field is on, after which further evaporative cooling can proceed in the TOP trap. Evaporative cooling in the TOP trap can be achieved by removing the higher energy atoms to untrapped states with either rf-induced transitions or the circle of death. For most experiments we use rf-induced transitions because it allows us greater control and flexibility. By appropriate choice of frequency and power, we can control the final energy of the atoms we are removing as well as the width of cut made into the sample of atoms. In addition, the parameters of the rf can be changed rapidly compared to changing values of magnetic fields for the circle of death. This technique works for a large range of initial numbers of trapped atoms. When the initial number of trapped atoms in the MOT is $\geq 10^9$, we can achieve BEC with atoms in the TOP trap, directly. We initially loose a large number of atoms after transfer into the TOP trap because the circle of death can not be placed sufficiently outside the cloud of atoms. After this initial loss and the cloud of atoms has rethermalized to a lower temperature and higher density, runaway evaporation can be achieve by compressing the sample and removing high energy atoms with the circle of death.

We probe our samples of atoms using absorption imaging technique [134]. The TOP trap is rapidly shut off and after a variable delay, a short laser pulse optically pumps the atoms from $F = 1$ to $F = 2$, after which another short laser pulse resonant with the $3S_{1/2}$, $F = 2$ $\rightarrow 3P_{3/2}$, $F' = 3$ transition is applied to the atoms along the direction of gravity (the x direction). The light absorbed from this laser beam is imaged onto a CCD camera. From this image we extract the transverse spatial dependence of the optical depth along the direction of the probe beam.

7.3. DIFFRACTION OF BEC BY OPTICAL STANDING WAVES

As discussed in section 21 of these lecture notes, atoms can be diffracted by the periodic potential resulting from an optical standing wave. The diffraction produces a coherent splitting of the atomic wave packet and can be thought of as arising from the process of simultaneous absorption of photons from one laser beam and stimulated emission of photons into the other laser beam of the optical standing wave, each process transfering $\pm 2\hbar k$ of momentum to the atoms. There are effectively two regimes of diffraction, normal and Bragg. We have demonstrated diffraction of Bose-Einstein condensates in both regimes.

In the early experiments demonstrating diffraction of atoms by optical standing waves [17, 18], a beam of atoms passed through an optical standing wave and the diffracted beam was detected downstream. In our experiments [142, 143], we start with BEC at a temperature sufficiently below the transition temperature such that no discernable thermal fraction is present. We then adiabatically reduce the strength of the confining potential, which lowers the energy of the condensate by both reducing the mean-field interactions and increasing the size of the condensate wavefunction. We then expose the atoms to a short pulse of the optical standing wave, while they are either still in the TOP trap, or shortly after releasing them from the trap by rapidly shutting off the magnetic fields. Hence the condensates are essentially at rest and we expose them to the optical standing wave temporally. We detect the momentum transferred to the atoms by the diffraction process by taking an absorption image after a sufficient time delay, such that the various atomic wave-packets with different momenta have spatially separated.

Figure 20 is an example of normal diffraction of BEC by a short pulse of a weak and strong optical standing wave. The optical standing wave is applied along the z-axis, 2 ms after the condensate atoms have been released from the adiabatically expanded trap. For a weak pulse there is only a small phase modulation imposed on the cloud of atoms by the optical standing wave, as described in section 22, and only the first diffraction orders with momentum $\pm 2\hbar k$ are observed (fig. 20a). When the pulse intensity is increased by a factor of 5 there is a substantial phase modulation imposed on the released condensate atoms and higher diffraction orders are observed. In fig. 20b, both second and third order diffraction, corresponding to momentum transfer of $\pm 4\hbar k$ and $\pm 6\hbar k$, is clearly evident. The images in fig. 20 were taken 10 ms after the application of the diffraction pulse. The momentum spread of the undiffracted atoms is approximately 0.06 $\hbar k$ and so the diffracted components are clearly resolved.

Figure 20. Diffraction of a BEC by a short pulse, optical standing wave. *a*) For low intensities only first order diffraction into $\pm 2\hbar k$ momentum states is visible. *b*) At higher intensities, higher order diffraction ($\pm 4\hbar k$ and $\pm 6\hbar k$) is observed.

7.4. Diffraction Beyond the Raman-Nath Regime

In the description of normal diffraction by an optical standing wave as a phase modulation of the atomic deBroglie wave, the pulse was considered short enough to be in the Raman-Nath regime, so that the phase modulation was essentially instantaneous. Alternatively, in the picture of diffraction as the simultaneous absorption and stimulated emission of photons, the bandwidth of the standing wave pulse was broad enough such that energy conservation can be satisfied for momentum transfer in both directions. If the laser pulse is left on for a longer time, we will violate the Raman-Nath approximation. The atoms can move along the periodic potential of the standing wave and the phase modulation will spatially average to a constant value. Alternatively, the pulse will have insufficient frequency width to satisfy energy conservation. This regime beyond Raman-Nath leads to periodic focusing and defocusing of the atoms and is relevant for atom lithography.

We have studied the behavior of a BEC in a pulsed optical standing wave beyond the Raman-Nath regime, the details of which can be found in ref. [142]. We observed oscillations in the intensity of the diffracted orders as a function of the laser pulse duration. Our results are in good agreement with a simple model where we project an incoming plane wave state onto a Bloch state basis, accumulate the differential phases due to the different energies of the Bloch bands, and then project back onto momentum eigenstates.

7.5. The Pulsed Talbot Effect

Periodic focusing and defocusing can also be studied within the thin diffraction grating (Raman-Nath) regime. This behavior is known as the Talbot effect. In the optical Talbot effect, coherent light passing through a periodic grating will from an "image" of the grating at a characteristic distance known as the Talbot length. For a phase grating, this "image" corresponds to the initial intensity distribution of the light with the phase distribution of the grating. Unlike light, however, atoms can be initially at rest, and the "reimaging" of the phase grating occurs at integer multiples of the Talbot time. Also, unlike light, atoms can be exposed to a pulsed, phase grating, which leads to a unique manifestation of the Talbot effect. We have demonstrated a new manifestation of the Talbot effect using the diffraction of a BEC by pulsed optical standing waves. In our experiment, the details of which can be found in ref. [144], we start with atoms at rest and apply a short pulse, optical standing wave to diffract the condensate atoms. A second identical diffraction pulse is applied after a variable delay to analyze the temporal evolution of the resulting condensate wavefunction. We observe that the initial phase distribution reimages itself at integer multiples of 10 ms, the Talbot time for our parameters. When the second pulse is applied at odd multiples of half the Talbot time, self imaging of the condensate in momentum space is observed. Intermediate delays produce more complicated momentum-space patterns that are in excellent agreement with theory. The coherent property of the condensate provides signals of very high contrast. In addition, we observe that the dynamics of the short pulse is different from that of a static grating because it has a broad frequency spectrum and hence can add energy to the system. It is the dispersion relation of matter waves, not the path length difference as in the case of static gratings, that results in this new manifestation of the Talbot effect.

7.6. Spatial phase variations across a Bose-Einstein condensate

We have used an unequal arm length interferometer, based on normal diffraction by pulsed optical standing waves, to study the spatial coherence of a BEC, the details of which can be found in ref. [146]. Two optical standing wave pulses of duration 100 ns and separation time δt are applied to the condensate. Each standing-wave phase grating diffracts the condensate, making small "copies" of the condensate displaced in momentum space by twice the momentum of a single photon. As the first copy moves away from the condensate its phase is evolving at $4E_{rec}/h$, where E_{rec} is the single photon recoil energy. (For sodium atoms with an excitation wavelength of 589 nm, E_{rec}/h is 25 kHz.) After the second copy is created at a time δt later, the phase of both copies then evolve at nominally the same rate. The quantum mechanical amplitudes of each copy interfere, and the total number of atoms coupled out of the condensate by the two pulses is measured. The resulting interferogram oscillates at the expected 100 kHz phase evolution of the first copy with respect to the second copy. The decay of the envelope of the interferogram is due to both the spatial overlap of the two copies (since the first copy has moved during δt due to the momentum kick) and on the initial spatial phase variations across the condensate.

When the coherence measurement is made on a condensate held in the trap, we obtain an interferogram whose envelope decays essentially as the spatial overlap of the two coupled out copies. The results are consistent with the trapped condensate having a uniform spatial phase. Hence we have experimentally verified that the trapped BEC, despite being spatially expanded due to the mean-field interaction between the atoms, has a momentum spread that is determined by the Heisenberg uncertainty principle. This result, which we measured in the time domain, was also obtained earlier, from measurements in the frequency domain, by Ketterles group at MIT using Bragg spectroscopy [147]. Alternatively, a released BEC exhibits large phase variations across the condensate as the mean-field interaction is converted into kinetic energy. This is apparent in our measurements where we obtain an interferogram with an envelope that decays much faster than the spatial overlap of the two copies. Our measurements also confirm that the successive, Raman output coupled pulses of atoms in our atom laser are fully coherent.

7.7. Bragg Diffraction of Atoms

In order to Bragg diffract atoms by an optical standing wave, as described in section 22, energy and momentum must be satisfied explicitly for the duration of the pulse. Since diffraction results in the transfer of multiple units of $\pm 2\hbar k$ to the atoms, the final kinetic energy of the atomic ensemble can be different from the initial kinetic energy. For a beam of atoms incident on a fixed standing wave, energy conservation can be satisfied by chosing the angle of incidence such that the energy difference comes from the differential Doppler shift of the two counter-propagating beams of the standing wave, as seen by atoms. In the case where we start with BEC essentially at rest, this differential Doppler shift can be created by moving the standing wave with respect to the atoms.

We create our moving standing wave by having a frequency difference δ between the two counterpropagating waves that make up the standing wave. In the presence of this moving standing wave, an atom intially at rest will simultaneously absorb photons from the higher frequency laser beam and be stimulated to emit photons into lower frequency beam acquiring momentum $\pm 2n\hbar k$ in the process. In order to satisfy energy conservation, the detuning δ must be chosen such that $n\delta = n^2 4E_{rec}/\hbar$, where E_{rec} is the recoil energy.

Figure 21 shows first, second and third order Bragg diffraction of Bose condensed atoms released from the magnetic trap. When the frequency difference between the two lasers is 100 kHz, atoms initially at rest can resonantly absorb a photon from the higher frequency laser beam and be stimulated to emit a photon into the lower frequency beam. The result of this process is a transfer of two units of photon momentum to the atoms, which then travels ballistically with a velocity of 6 cm/s. Similarly, by setting the frequency difference of the lasers to -100 kHz, the momentum transfer to the atoms from the Raman process will be in the opposite direction. Since Bragg diffraction of the atoms can be thought of as a two level system (the initial and final momentum states) coupled by the Raman process, it is possible to transfer all of the atoms to the final momentum state. We have observed first order Bragg diffraction of 100% of the condensate atoms. The amount of transfer was reduced in the images of fig. 22 so that the location of the condensate atoms, initially at rest, could be easily identified. Second and third order Bragg diffraction was observed when the laser detuning was increased to 200 kHz and 300 kHz, respectively. We have observed up to sixth order Bragg diffraction with a momentum transfer of $\approx 12\ \hbar k$ (corresponding to a velocity of 0.35 m/s).

7.8. Raman output coupling: Demonstration of a quasi-CW atom laser

In order to realize an atom laser from BEC, it is necessary to coherently extract the condensed atoms; that is, an atom output coupler is needed. The first demonstration of an output coupler for BEC was reported in 1997 [148], where coherent, rf-induced transitions were used to change the internal state of the atoms from a trapped state to an untrapped one. This method, however, did not allow the direction of the output coupled atoms to be chosen. The extracted atoms fell under the influence of gravity and expanded because of the intrinsic repulsion of the atoms. We have developed a highly directional method to optically couple out a variable fraction of a condensate. We use stimulated Raman transitions between magnetic sublevels to coherently transfer

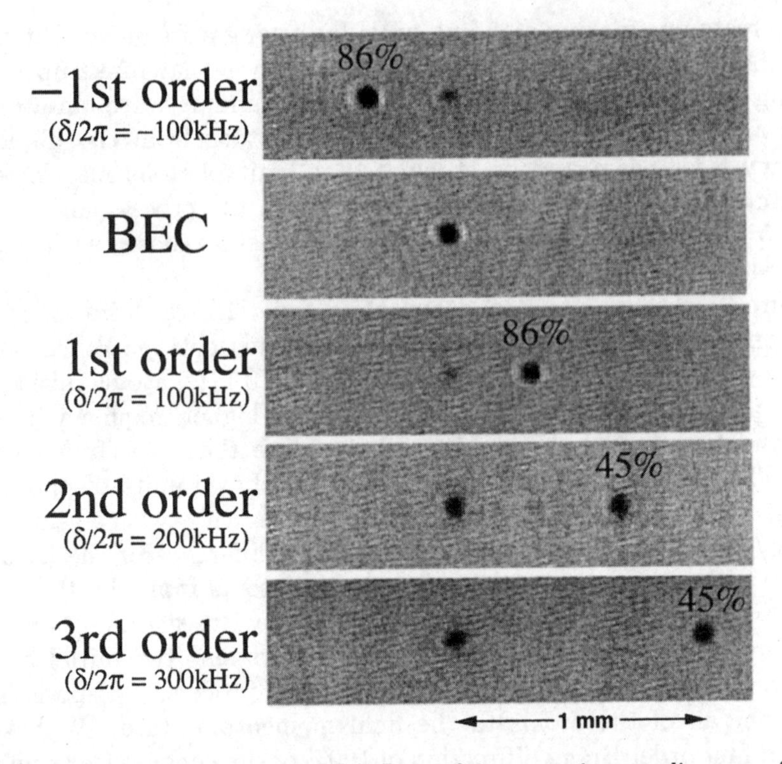

Figure 21. Bragg diffraction of a BEC: By applying a moving standing wave (whose velocity is determined by the frequency difference of the two waves comprising the standing wave) we can Bragg diffract a portion of the condensate into a well-defined momentum state.

trapped condensate atoms to an untrapped state while giving them a momentum kick [145].

The Bragg diffraction of atoms [143] discussed earlier involves a stimulated Raman transition between different momentum states while keeping the atoms in the same magnetic sublevel. If the frequency difference between the lasers includes the additional Zeeman energy between two magnetic sublevels, a simultaneous change in the momentum and internal state of the condensate atoms can be achieved. This is illustrated in fig. 22, where BEC trapped in the $F = 1$, $m_F = -1$ state is transferred to the $F = 1$, $m_F = 0$ state. Two units of photon momentum are transferred in the Raman process, so the cloud of atoms in the $m_F = 0$ state has a velocity of 6 cm/s with respect to those atoms in the $m_F = -1$ state.

We can repeatedly apply the Raman pulses to achieve multiple output coupling of atoms from a BEC. This is shown in fig. 23. In order to avoid changes in the Raman resonance frequency between differ-

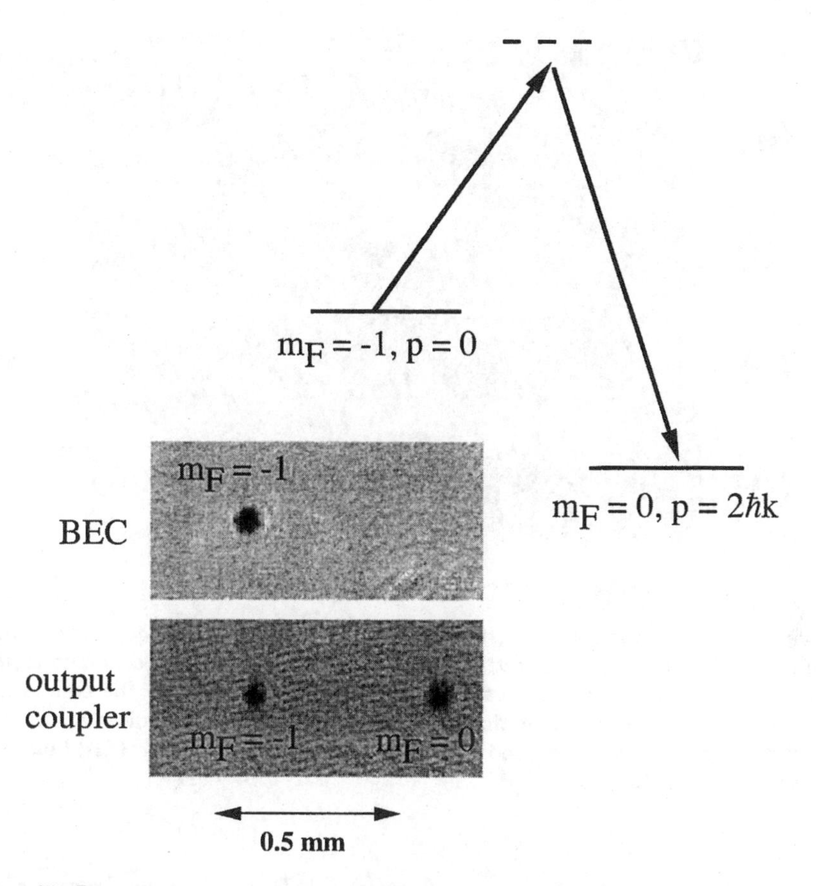

Figure 22. Raman output coupler: A stimulated Raman transition is used to transfer 2 $\hbar k$ of momenta, and change the magnetic sublevel from the trapped $m = -1$ to the untrapped $m = 0$ state. The pictures are absorption images taken after a time-of-flight period.

ent magnetic sublevels we synchronized the application of the Raman pulses to our rotating TOP field. (Our condensate atoms were displaced by gravity away from the zero of the quadrupole field, so that the local magnetic field was modulated by the rotating TOP bias field.) Figure 23*a-c* are optical absorption images of the condensate after one, three and seven Raman pulses respectively. For these images, the TOP trap was held on for a 9 ms window during which time 6 μs Raman pulses were applied at a subharmonic of the rotating TOP bias frequency. The magnetic fields were then extinguished and the atoms were imaged 1.6 ms later. In fig. 23*d*, the TOP trap was held on for a 7 ms window during which time 140 Raman pulses were fired at the 20 kHz frequency of the rotating bias field and the distribution of atoms was imaged 1.6 ms later. The Raman pulse duration was reduced to 1 μs in order to couple

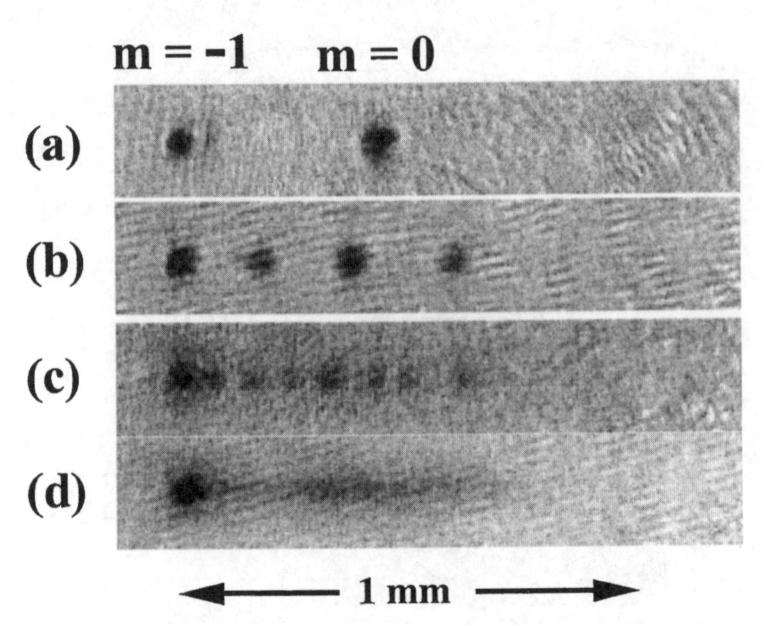

Figure 23. Series of images demonstrating multiple Raman output coupling of atoms from BEC en route to demonstrating a quasi-continuous stream of coherent atoms. In *a-c*), one, three and seven 6 μs Raman pulses were applied to the condensate respectively. *d*) is the result of the application of 1 μs Raman pulses at the full repetition rate of 20 kHz imposed by the frequency of the rotating TOP bias field (140 pulses in 7 ms).

less atoms out of the condensate during each Raman pulse. In the time between two Raman pulses each output coupled wavepacket moves only 2.9 μm. These pulses strongly overlap because this spatial separation of 2.9 μm is much smaller than the $\approx 50\mu$m size of the condensate, therefore the output coupled atoms form a quasi-continuous coherent matter wave.

Our Raman output coupling scheme dramatically reduces the transverse momentum width of the extracted atoms compared to other methods such as rf output coupling [148]. This dramatic reduction occurs because the output coupled atoms have received a substantial momentum kick from the Raman process. If the atoms were simply released from the trap with no momentum transfer, they would undergo a burst of expansion due to the repulsive interactions with the other condensate atoms. In our output coupling scheme, however, this additional expansion energy is primarily channeled into the forward direction. The increase in the transverse momentum width due to the interaction between the atoms is reduced by roughly the ratio of the timescale over which the mean field repulsion acts on the freely expanding condensate, divided by the characteristic time it takes the output coupled atoms to leave the still trapped condensate. In our case, the reduction ratio is about a factor of ≈ 20 which results in a well collimated beam of atoms.

One of the important properties of an optical laser is that the coherence length of the emitted beam of photons is much longer than the size of the cavity. A similar property for an atom laser would be highly desirable. The output coupled beam of our atom laser beam is much longer than the characteristic size of the condensate. Since stimulated Raman transitions are coherent processes, we expect the coherence length of this beam to be much longer than the size of the condensate, but this has yet to measured.

7.9. Nonlinear atom optics with Bose-Einstein condensates

The advent of the laser as an intense, coherent, light source enabled the field of nonlinear optics to flourish. The interaction of light in materials, whose index of refraction depends on the intensity, has led to effects such as multi-wave mixing of optical fields to produce coherent light of a new frequency, and optical solitons, pulses of light that propagate without dispersion. Nonlinear optics now plays an important role in many areas of science and technology. With the experimental realization of Bose-Einstein condensation (many atoms in a single quantum state) and the matter-wave or atom "laser" (atoms coherently extracted from a condensate), we now have an intense source of matter-waves analogous to the source of light from an optical laser. This has led us to the threshold of a new field of physics: nonlinear atom optics [149].

The analogy between nonlinear optics with lasers and nonlinear atom optics with Bose-Einstein condensates can be seen in the similarities between the equations that govern each system. For a condensate of interacting bosons, in a trapping potential V, the macroscopic wave function Ψ satisfies a nonlinear Schrdinger equation [150],

$$i\hbar\frac{\partial\Psi}{\partial t} = \left(-\frac{\hbar^2}{2M}\nabla^2 + V + g\,|\Psi|^2\right)\Psi \tag{26}$$

where M is the atomic mass, g describes the strength of the atom-atom interaction ($g > 0$ for sodium atoms), and $|\Psi|^2$ is proportional to atomic number density.

7.9.1. *Four-wave mixing with matter-waves*

The nonlinear term in Eq.((19) is similar to the third-order susceptibility term, $\chi^{(3)}$, in the wave equation for the electric field describing optical four-wave mixing. We therefore expect that if three coherent

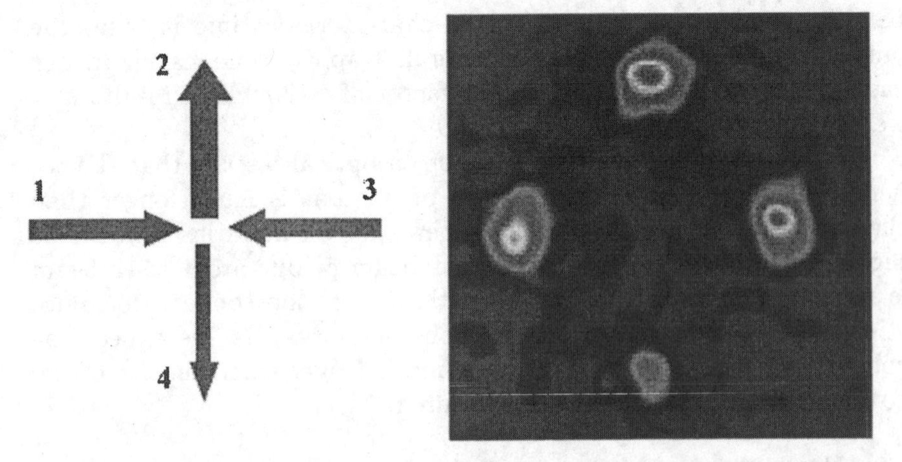

Figure 24. The process of four-wave mixing of matter-waves can always be transformed to a reference frame where the mixing process is degenerate (all of the waves have the same energy; left). The nonlinear term describing the mean-field, s-wave interaction of the atoms is responsible for the four-wave mixing. Atoms from waves 1 and 3 scatter off each other and go off back-to-back. The scattering process can be stimulated by wave 2, so that it is more likely that one of the scattering pairs goes into this wave. By momentum conservation, wave 4 is created. The small cloud of atoms in the image on the right is the fourth wave generated by four-wave mixing of matter-waves.

matter-waves are spatially overlapped with the appropriate momenta, a fourth matter-wave will be produced due to the nonlinear interaction, analogous to optical four-wave mixing. In contrast to optical four-wave mixing, the nonlinearity in matter-wave four-wave mixing comes from atom-atom interactions, described by a mean-field; there is no need for an external nonlinear medium.

Using the atoms from a BEC, we have observed such four-wave mixing of matter-waves. This work is described in detail in ref. [151]. In our four-wave mixing experiment, we used optically induced Bragg diffraction [143] to create three overlapping wavepackets with appropriately chosen momenta. As the three wavepackets spatially separate, a fourth wavepacket, due to the wave-mixing process, is observed (see Fig.24).

The process of four-wave mixing of matter-waves (and also optical waves), can be thought of as Bragg diffraction off of a matter grating. In this picture, two of the matter-waves interfere to form a standing matter-wave grating. The third wave can Bragg diffract off of this grating, giving rise to the fourth wave. An alternative picture of four-wave mixing is in terms of stimulated emission. In this picture it is helpful to view the four-wave mixing process in a reference frame where the process looks like degenerate four-wave mixing; that is, all of the waves have the same energy.

In four-wave mixing, both energy and momentum (corresponding to phase matching) must be conserved. Since atoms, unlike photons, can not be created out of the vacuum we have the additional requirement for matter-waves of particle number conservation. (If we included the rest mass of the atom, particle number conservation is contained in energy conservation.) Given these three conditions, one can show that the only four-wave mixing configurations possible with matter-waves are those that can be viewed in some frame of reference as degenerate four-wave mixing. This is also the geometry of phase conjugation. Fig. 24 shows the four-wave mixing geometry for matter-waves viewed in the degenerate or phase conjugation frame.

In the picture of four-wave mixing as arising from stimulated emission, atoms in waves 1 and 3 can be considered as undergoing an elastic collision. The scattering process results in atoms going off back-to-back in order to conserve momentum, but at some arbitrary angle with respect to the incident direction. (The scattering process is typically s-wave and the outgoing waves can be considered spherical.) In the presence of wave 2, however, this scattering process can be stimulated. There is an enhanced probability that one of the atoms from the collision of waves 1 and 3 will scatter into wave 2. (This probability is enhanced by the atoms in wave 2.) Because of momentum conservation, the enhanced scattering of atoms into wave 2 results in an enhanced number of atoms in wave 4. In this picture, it is obvious that the four-wave mixing process removes atoms from waves 1 and 3 and puts them into waves 2 and 4. This may have some interesting consequences in terms of quantum correlatons between the waves.

7.9.2. *Quantum phase engineering*

A three-dimensional image of an arbitrarily complex object can be constructed by sending light, with sufficient spatial coherence, through the appropriate phase and/or amplitude mask. This is the basic principle behind physical optics, which includes wave phenomena like diffraction and holography. Diffraction can be achieved with a periodic phase and/or amplitude mask; while a more complicated mask is needed to construct a complex holographic image. In each case, the mask modifies the incoming wave and subsequent propagation produces the desired pattern of light. This idea can be readily adapted to atom optics, especially when the "incoming" matter-wave is from a highly coherent source such as a Bose-Einstein condensate.

We have developed a technique to optically imprint complex phase patterns onto a Bose-Einstein condensate in order to create interesting topological states. This technique is analogous to sending a wave through a thin phase mask. The basic idea is to expose the condensate atoms to a short pulse of laser light with a spatially varying intensity pattern. The laser detuning is chosen such that spontaneous emission is negligible. (The phase mask can also serve as an amplitude mask by tuning closer to resonance, so that spontaneous emission is significant.) The pulse duration is sufficiently short such that the atoms do not move an appreciable distance (i.e. the wavelength of light) during the pulse. This is sometimes referred to as the Raman-Nath regime. During the laser pulse, the AC Stark effect or optical dipole potential (see Eq.((2)) shifts the energy of the atoms by $U(r,t)$. Hence the effect on the atomic wavefunction is to "instantaneously" change its phase. This effect can be represented by multiplying the wavefunction by the phase factor $\exp(i\phi(r))$, where $\phi(r) = -\int U(r,t)dt/\hbar$. Since the AC Stark or light shift is proportional to the intensity of the light, any spatial intensity variation in the light field will be written onto the BEC wavefunction as a spatial variation in its phase.

Optically induced phase imprinting is a tool for "quantum phase engineering" of the wavefunction to create a wide variety of states. For example, as discussed in the section on diffraction of the condensate, the application of a short pulse of standing wave light will imprint a sinusoidal phase onto the condensate. The imprinted wavefunction subsequently evolves in momentum states differing by $2\hbar k$. It should be possible to use quantum phase engineering to produce collective states of excitation of the interacting BEC, such as solitons and vortices. The application of a uniform intensity light field to half of the BEC imprints a relative phase difference between the two halves. This phase step is expected to give rise to dark solitons (see following section). Such solitons will propagate with a speed related to the phase difference [152], which can be adjusted by the intensity of the laser pulse.

It should also be possible to produce one or more vortices by applying a laser pulse which has a linearly-varying, azimuthal, intensity dependence [153]. This will produce a topological winding of the BEC phase, which if large enough (i.e. 2π) should produce a vortex. Numerical solutions to a 3-D Gross-Pitaevskii equation [154] show that this is the case; and also show that such a vortex, although unstable because it is created in a non-rotating trap, will live for a sufficient time to be observable. Increasing the phase winding will generate multiple vortices (vortices with more than $\hbar$ of angular momentum are not stable and will immediately split into multiple vortices each with angular momentum $\hbar$). Quantum phase engineering can generate arbitrary phase patterns, and perhaps other interesting quantum states. In this sense, it is a form of atom holography [155]. The technological challenge is mostly one of imaging. Any complicated pattern must be imaged to the size of the BEC, typically of order 50 μm.

7.9.3. *Solitons in a BEC*

Solitons are stable, localized waves that propagate in a nonlinear medium without spreading. They may be either bright or dark, depending on the details of the governing nonlinear wave equation. A bright soliton is a peak in the amplitude while a dark soliton is a notch with a characteristic phase step across it. Equation (26), which describes the weakly interacting, zero-temperature BEC also supports solitons. The solitons propagate without spreading (dispersing) because the nonlinearity balances the dispersion; for Eq.(26) the corresponding terms are the nonlinear interaction $g|\Psi|^2$, and the kinetic energy $-(\hbar^2/2M)\nabla^2$, respectively. Our sodium condensate only supports dark solitons because the atom-atom interactions are repulsive [152, 156] ($g > 0$).

A distinguishing characteristic of a dark soliton is that its velocity is less than the Bogoliubov speed of sound [152, 156] $v_0 = \sqrt{gn/M}$ (n is the unperturbed condensate density) and they travel opposite to the direction of the phase gradient. The soliton speed v_s can be expressed either in terms of the phase step $\delta (0 < \delta \leq \pi)$, or the soliton "depth" n_d, which is the difference between n and the density at the bottom of the notch [152, 156]:

$$v_s/v_0 = cos(\delta/2) = \sqrt{1 - n_d/n} \tag{27}$$

For $\delta = \pi$ the soliton has zero velocity, zero density at its center, a width on the order of the healing length [156], and a discontinuous phase step. As δ decreases the velocity increases, approaching the speed of sound. The solitons are shallower and wider, with a more gradual phase step. Because a soliton has a characteristic phase step, optically imprinting a phase step on the BEC wavefunction should be a way to create a soliton.

7.10. Observation of Solitons in a BEC

We modified the phase distribution of the BEC by employing the technique of quantum phase engineering discussed in an earlier section. The condensate atoms were exposed to a pulsed, off-resonant laser beam, coaxial with the absorption probe beam, with a spatial intensity profile such that only half of the BEC was illuminated. This was accomplished by blocking half of the laser beam with a razor blade and imaging this razor blade onto the condensate. The intensity pattern at the condensate, as observed by our absorption imaging system, had a light to dark (90 % to 0%) transition region of 7 μm. The intensity required to

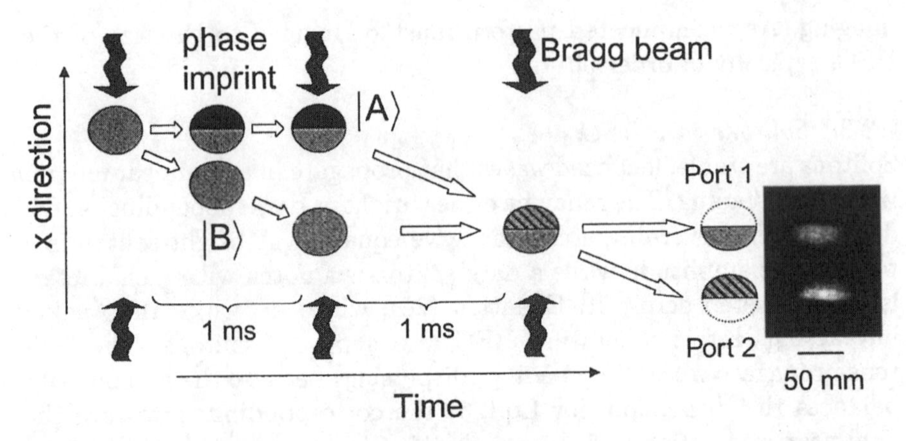

Figure 25. Space-time diagram of the matter-wave interferometer used to measure the spatial phase distribution imposed on the BEC. Three optically induced Bragg diffraction pulses formed the interferometer. Each pulse consisted of two counter-propagating laser beams detuned by about −2 GHz from atomic resonance (so that spontaneous emission is negligible) with their frequencies differing by 100 kHz. The first pulse had a duration of 8 ms and coherently split the condensate into two components $|A\rangle$ and $|B\rangle$ with equal number of atoms. $|A\rangle$ remained at rest and $|B\rangle$ received two photon recoils of momentum. When they were completely separated, we exposed the top half of $|A\rangle$ to a phase imprinting pulse of π, which changed the phase distribution of $|A\rangle$ while $|B\rangle$ served as a phase reference. 1 ms after the first Bragg pulse, a second Bragg pulse of 16 ms duration brought $|B\rangle$ to rest and imparted two photon momenta to $|A\rangle$. When they overlapped again, 1 ms later, a third pulse of 8 ms duration converted their phase differences into density distributions at ports 1 and 2, which appears in the images.

imprint a phase of π was checked with a Mach-Zehnder atom interferometer based on optically induced Bragg diffraction [157, 158] (see Fig. 25). Our Bragg interferometer [159] differs from previous ones in that we can independently manipulate atoms in the two arms (because of their large separation) and can resolve the output ports to reveal the spatial distribution of the condensate phase. When a phase of π was imprinted on one half of the condensate relative to the other half, the two output ports of the interferometer displayed the complementary halves of the condensate.

To observe the creation and propagation of solitons, we measure BEC density distributions with absorption imaging after imprinting a phase step. Figure 26 shows the evolution of the condensate after the top half was phase imprinted with $\phi = 1.5\pi$, a phase for which we observe a single deep soliton (the reason for imprinting a phase step larger than π is discussed below). Immediately after the phase imprint, there is a steep phase gradient across the middle of the condensate such that this portion has a large velocity in the $+x$ direction. This velocity

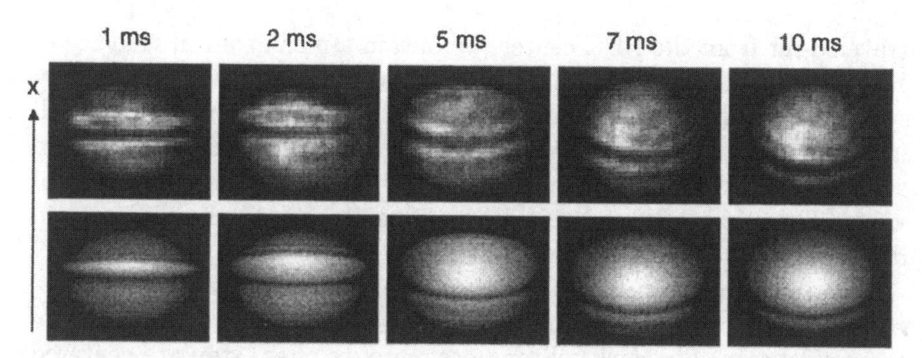

Figure 26. Experimental (upper) and theoretical (lower) images of the integrated BEC density for various times after we imprint a phase of about 1.5π on the top half of the condensate with a $1\mu s$ pulse. The measured number of atoms in the condensate was $1.7(3) \times 10^6$, and this value was used in the calculations. A positive density disturbance moved rapidly in the $+x$ direction and a dark soliton moved oppositely at significantly less than the speed of sound. Because the imaging is destructive, each image shows a different BEC. The width of the images is $70\mu m$.

can be understood as arising from the impulse imparted by the optical dipole force, and results in a positive density disturbance that travels at or above the speed of sound. A dark notch is left behind, which is a soliton moving slowly in the $-x$ direction (opposite to the direction of the applied force).

A striking feature of the images is the curvature of the soliton. This curvature is due to the 3-D geometry of the trapped condensate, and occurs for two reasons. First, the speed of sound v_0 is largest at the trap center where the density is greatest, and decreases towards the condensate edge. Second, as the soliton moves into regions of lower condensate density, we find numerically that the density at its center, $n - n_d$, approaches zero, δ approaches π, and v_s decreases to zero before reaching the edge. This is because the soliton depth n_d rather than its phase offset δ appears to be a conserved quantity in a nonuniform medium.

A clear indication that the notches seen in Fig.26 are solitons, rather than simply sound waves, is their subsonic propagation velocity. To determine this velocity, we measure the distance after propagation between the notch and the position of the imprinted phase step along the x direction. Because the position of our condensate varies randomly from shot-to-shot (presumably due to stray, time-varying fields) we cannot always apply the phase step at the center. A marker for the location of the initial phase step is the intersection of the soliton with the condensate edge, because at this point the soliton has zero velocity. Using images taken 5 ms after the imprint, at which time the soliton has not traveled far from the BEC center, we obtain a mean soliton velocity of 1.8(4) mm/s. This speed is significantly less than the mean Bogoliubov speed of sound v_0= 2.8(1) mm/s. From the propagation of the notch in the numerical solutions (Fig.26, lower images) we obtain a mean soliton velocity, v_s = 1.6 mm/s, in agreement with the experimental value. The experimental uncertainty is mainly due to the difficulty in determining the position of the initial phase step.

From the lower image of Fig.26 at 5 ms, we can extract the theoretical density and phase profile along the x-axis through the center of the condensate. The dark soliton notch and its phase step are centered at $x = -8$ μm. This phase step, $\delta = 0.58\pi$, is less than the imprinted phase of 1.5π. The difference is caused by the mismatch between the phase imprint and the phase and depth of the soliton solution of the nonlinear Schrdinger equation (Eq.(19)): Our imprinting resolution of 7 μm is larger than the soliton width, which is on the order of the healing length (0.7 μm), and we do not control the amplitude of the wave function.

In order to improve our measurement of the soliton velocity, we avoid the uncertainty in the position of the initial phase step by replacing the razor blade mask with a thin slit. This produces a stripe of light with a Gaussian profile (1/e^2 full width $\approx$ 15μm). With this stripe in the center of the condensate, numerical simulations predict the generation of solitons that propagate symmetrically outwards. We select experimental images with solitons symmetrically located about the middle of the condensate, and measure the distance between them. For a small phase imprint of $\phi \approx 0.5\pi$ (at Gaussian maximum), we observe solitons moving at the Bogoliubov speed of sound, within experimental uncertainty. For a larger phase imprint of $\phi \approx 1.5\pi$, we observe much slower soliton propagation, in agreement with numerical simulations. An even larger phase imprint generates many solitons. The results of these experiments on the creation and propagation of solitons can be found in Ref. [159]. Solitons in a BEC have also been observed by a group in Hannover [160].

Much of the experimental work described here was carried out in the Laser Cooling Group led by Bill Phillips in the Atomic Physics Division of the National Institute of Standards and Technology. It is a pleasure to acknowledge the contributions to this work by the other members of the Laser Cooling Group, Steve Rolston and Paul Lett, as well as former member Chris Westbrook. This work would not have been possible without the contributions of many postdocs, students, and visitors to the group.

While not attempting to be a review of all the work on laser cooling of neutral atoms, these notes have drawn extensively, both explicitly and implicitly, on the work of many groups throughout the world that have advanced the art and science of laser cooling. We are indebted to all of those colleagues, both cited and not, who have contributed so much to our understanding of laser cooling and to the pleasure of working in this field.

The BEC experiments were carried out by present and past members of the Laser Cooling Group, Steve Rolston, Jesse Simsarian, Johannes Denschlag, Julien Cubizolles, Ed Hagley, Lu Deng, Mikio Kozuma, Jesse Wen, Yuri Ovchinnikov, Joerg-Helge Muller, Robert Lutwak, Rob Thompson, Aephraim Steinberg, Mike Gatzke, and Gerhard Birkl. These experiments have also benefited greatly from direct interactions with our theoretical colleagues Paul Julienne, Carl Williams, Marek Tippenbach, Yehuda Band, David Feder, Marya Doery, Charles Clark, Mark Edwards and Keith Burnett.

We also gratefully acknowledge the financial support not only of NIST, but of the U. S. Office of Naval Research, NASA and NSF.

References

1. Arimondo E., Phillips W. D. and Strumia F. Editors (1992) *Laser Manipulation of Atoms and Ions, Proc. S.I.F., Course CXVIII* (North Holland, Amsterdam)
2. Dalibard J., Raimond J.-M. and Zinn-Justin J. Editors (1992) *Fundamental Systems in Quantum Optics* (North Holland, Amsterdam)
3. Chu S. and Wieman C. Editors (1989) Feature issue on laser cooling and trapping of atoms, *J. Opt. Soc. Am. B*, **6**, 2020
4. Meystre P. and Stenholm S. Editors (1985) Feature issue on mechanical effects of light, *J. Opt. Soc. Am. B*, **2**, 1706
5. Metcalf H. and van der Straten P. (1994), *Phys. Reports*, **244**, 203
6. Kazantsev A. P., Surdutovich G. I. and Yakovlev V. P. (1990), *Mechanical action of light on atoms* (World Scientific, Singapore)
7. Minogin V. G. and Letokhov V. S. (1987), *Laser light pressure on atoms* (Gordon and Breach, New York)
8. (1992) Special issue on atom optics and interferometry, *Appl. Phys. B*, **54**, 321
9. (1994) Special issue on atom optics and interferometry, *J. de Physique*, **4**, 1877
10. (1996) Special issue on atom optics and interferometry *Quant. Semiclass. Optics*, **8**, 495
11. Dalilbard J. and Cohen-Tannoudji C. (1985) *J. Opt. Soc. Am. B*, **2**, 1707
12. Cohen-Tannoudji C., Dupont-Roc J. and Grynberg G. (1992) *Atom-photon interactions: Basic Processes and Applications* (Wiley, New York)
13. Gordon J. P. and Ashkin A. (1980) *Phys. Rev. A*, **21**, 1606
14. Cook R. J. (1986) *Phys. Rev. A*, **22**, 1078
15. Stenholm S. (1986) *Rev. Mod. Phys.*, **58**, 699
16. Cohen-Tannoudji C. (1992) in Fundamental Systems in Quantum Optics edited by Dalibard J., Raimond J.-M. and Zinn-Justin J. (North Holland, Amsterdam)
17. Moskowitz P. E., Gould P. L., Atlas S. R. and Pritchard D. E. (1983) . Rev. Lett, **51**, 370; Gould P. L., Ruff G. E. and Pritchard D. E. (1986) *Phys. Rev. Lett.*, **56**, 827

18. Martin P. J., Oldaker B. G., Miklich A. H. and Pritchard D. E. (1988) *Phys. Rev. Lett.*, **60**, 515
19. Dalibard J. (1986) *Thse de doctorat d'tat en Sciences Physique, Universit de Paris VI*
20. Lett P. D., Phillips W. D., Rolston S. L., Tanner C. E., Watts R. N. and Westbrook C. I. (1989) *J. Opt. Soc. Am. B*, **6**, 2084
21. Castin Y., Wallis H. and Dalibard J. (1989) *J. Opt. Soc. Am. B*, **6**, 2046
22. Aspect A., Arimondo E., Kaiser R., Vansteenkiste N. and Cohen-Tannoudji C. (1989) *J. Opt. Soc. Am. B*, **6**, 2112
23. Aspect A., Arimondo E., Kaiser R., Vansteenkiste N. and Cohen-Tannoudji C. (1988) *Phys. Rev. Lett.*, **61**, 826
24. Pritchard D. E., Helmerson K., Bagnato V. S., Lafyatis G. P. and Martin A. G. (1987) in *Laser Spectroscopy VIII* edited by W. Persson and S. Svanberg (Springer-Verlag, Berlin), 68
25. Wallis H. and Ertmer W. (1989) *J. Opt. Soc. Am B*, **6**, 2211
26. Kasevich M. and Chu S. (1992) *Phys. Rev. Lett.*, **69**, 1741
27. Phillips W. D., Prodan J. and Metcalf H. (1985) *J. Opt. Soc. Am. B*, **2**, 1751
28. Chu S., Hollberg L., Bjorkholm J., Cable A. and Ashkin A. (1985) *Phys. Rev. Lett.*, **55**, 48
29. Hodapp T. W., Gerz C., Furtlelehner C., Westbrook C. I. and Phillips W. D. (1995) *Appl. Phys. B*, **60**, 135
30. Phillips W. and Metcalf H. (1982) *Phys. Rev. Lett.*, **48**, 596
31. Cable A., Prentiss M. and Bigelow N. P. (1990) *Opt. Lett.*, **15**, 507
32. Monroe C., Swann W., Robinson H. and Wieman C. (1990) *Phys. Rev. Lett.*, **65**, 1571
33. Andreev S., Balykin V., Letokhov V. and Minogin V. (1981) *Pis'ma Zh. Eksp. Teor. Fiz.*, Vol. no. 34, 463 [(1981) *JETP Lett.*, **34**, 442]
34. Hoffnagle J. (1988) *Opt. Lett.*, **13**, 102
35. Ketterle W., Martin A., Joffe M. A. and Pritchard D. E. (1992) *Phys. Rev. Lett.*, **69**, 2483
36. Gaggl R., Windholz L., Umfer C. and Neureiter C. (1994) *Phys. Rev. A*, **49**, 1119
37. Prentiss M. and Cable A. (1989) *Phys. Rev. Lett.*, **62**, 1354
38. Letokhov V. S., Minogin V. G and Pavlik B. D. (1976) *Opt. Commun*, **19**, 72
39. Phillips W. D. and Prodan J. V. (1983) in *Laser-Cooled and Trapped Atoms* edited by Phillips W. D. Natl. Bur. Stand. (U.S.) Spec. Publ., **653**, 137; (1984) *Prog. Quantum Electron.*, **8**, 231; in (1984) *Coherence and Quantum Optics V* edited by Mandel L. and Wolf E. (Plenum, New York), 15; Phillips W., Prodan J. and Metcalf H. (1983) in *Laser Spectroscopy Laser VI* edited by Weber H. and Luthy W. (Springer-Verlag, Berlin), 162
40. Ertmer W., Blatt R., Hall J. and Zhu M. (1985) *Phys. Rev. Lett.*, **54**, 996
41. Salomon C. and Dalibard J. (1988) *C. R. Acad. Sci. Paris*, **306**, 1319
42. Prodan J., Phillips W. and Metcalf H. (1982) *Phys. Rev. Lett.*, **49**, 1149
43. Prodan J., Migdall A., Phillips W. D., So I., Metcalf H. and Dalibard J. (1985) *Phys. Rev. Lett.*, **54**, 992
44. Migdall A., Prodan J., Phillips W., Bergeman T. and Metcalf H. (1985) *Phys. Rev. Lett.*, **54**, 2596
45. Chu S., Bjorkholm J., Ashkin A. and Cable A. (1986) *Phys. Rev. Lett.*, **57**, 314
46. Raab E., Prentiss M., Cable A., Chu S. and Pritchard D. (1987) *Phys. Rev. Lett.*, **59**, 2631
47. Cornell E., Monroe C. and Wieman C. (1991) *Phys. Rev. Lett.*, **67**, 3049

48. Spreeuw R. J. C., Gerz C., Goldner L., Phillips W. D., Rolston S. L., Westbrook C., Reynolds M. W. and Silvera I. F. (1994) *Phys. Rev. Lett.*, **72**, 3162
49. Wing W. (1980) *Phys. Rev. Lett.*, **45**, 631
50. Aminoff C. G., Steane A. M., Bouyer P., Desbiolles P., Dalibard J. and Cohen-Tannoudji C. (1993) *Phys. Rev. Lett.*, **71**, 3083
51. Petrich W., Anderson M. H., Ensher J. R. and Cornell E. A. (1995) *Phys. Rev. Lett.*, **74**, 3352
52. Ashkin A. (1978) *Phys. Rev. Lett.*, **40**, 729
53. Ashkin A. and Gordon J. (1979) *Opt. Lett.*, **4**, 161
54. Dalibard J., Reynaud S. and Cohen-Tannoudji C. (1983) *Opt. Comm.*, **47**, 395
55. Dalibard J., Reynaud S. and Cohen-Tannoudji C. (1984) *J. Phys. B*, **17**, 4577
56. Gould P. L., Lett P. D., Phillips W. D., Julienne P. S., Thorsheim H. R. and Weiner J. (1987) in *Advances in Laser Science III* edited by Tam A., Gole J. and Stwalley W. (American Institute of Physics, New York, N.Y.), 295
57. Gould P. L., Lett P. D., Julienne P. S., Phillips W. D., Thorsheim H. R. and Weiner J. (1988) *Phys. Rev. Lett.*, **60**, 788
58. K. Helmerson (1991) Interdisciplinary Laser Conference (Unpublished, Monterey, CA)
59. Rolston S. L., Gerz C., Helmerson K., Jessen P. S., Lett P. D., Phillips W. D., Spreeuw R. J. and Westbrook C. I. (1992) in *1992 Shanghai International Symposium on Quantum Optics* edited by Yuzhu Wang, Yigui Wang and Zugeng Wang (Proc. SPIE, Shanghai), **1726**, 205
60. Chu S., Bjorkholm J. E., Ashkin A., Gordon J. P. and Hollberg L. W. (1986) *Opt. Lett.*, **11**, 73
61. Miller J. D., Cline R. A. and Heinzen D. J. (1993) *Phys. Rev. A*, **47**, R4567
62. Miller J. D., Cline R. A. and Heinzen D. J. (1993) *Phys. Rev. Lett.*, **71**, 2204
63. Adams C. S., Lee H. J., Davidson N., Kasevich M. and Chu S. (1995) *Phys. Rev. Lett.*, **74**, 3577
64. Cook R. and Hill R. (1982) *Opt. Comm.*, **43**, 258
65. Kasevich M., Weiss D. and Chu S. (1990) *Opt. Lett.*, **15**, 607
66. Davidson N., Lee H. J., Adams C. S., Kasevich M. and Chu S. (1995) *Phys. Rev. Lett.*, **74**, 1311
67. Lee H.-J., Adams C. S., Davidson N., Young B., Weitz M., Kasevich M. and Chu S. (1995) in *Atomic Physics 14* edited by Wineland D., Wieman C. and Smith S. (AIP Press, N.Y.), 258
68. Phillips W. D. (1992) in *Laser Manipulation of Atoms and Ions, Proc. S.I.F., Course CXVIII* edited by Arimondo E., Phillips W. D. and Strumia F. (North Holland, Amsterdam), 289.
69. Ashkin A. and Gordon J. (1983) *Opt. Lett.*, **8**, 511
70. Dalibard J. (1986) personal communication.
71. Walker T. (1994) *Laser Physics*, **4**, 965
72. Dalibard J. and Cohen-Tannoudji C. (1989) *J. Opt. Soc. Am. B*, **6**, 2023
73. Walhout M., Dalibard J., Rolston S. L. and Phillips W. D. (1997) *J. Opt. Soc. Am. B*, **9**, 1997
74. Steane A. and Foot C. (1991) *Europhys. Lett.*, **14**, 231
75. Werner J., Wallis H. and Ertmer W. (1992) *Opt. Comm.*, **94**, 525
76. Prentiss M., Cable A., Bjorkholm J. E., Chu S., Raab E. L. and Pritchard D. E. (1988) *Opt. Lett.*, **13**, 452
77. Sesko D., Walker T., Monroe C., Gallagher A. and Wieman C. (1989) *Phys. Rev. Lett.*, **63**, 961

78. Walker T., Sesko D. and Wieman C. (1990) *Phys. Rev. Lett.*, **64**, 408
79. Ketterle W., Davis K. B., Joffe M. A., Martin A. and Pritchard D. E. (1993) *Phys. Rev. Lett.*, **70**, 2253
80. Ramsey N. F. (1956) *Molecular Beams* Edited by Elliott R. J., Krumhansl J. A., Marshall W. and Wilkinson D. H. (International Series of Monographs on Physics (Oxford University Press, Oxford), 1985
81. Paul W. (1985) personal communication
82. Heer C. V. (1963) *Rev. Sci. Instrum.*, **34**, 532
83. Heer C. V. (1960) in *Quantum Electronics* edited by Townes C. H. (Columbia University Press, New York, N.Y.), 17
84. Vladimirski V. V. (1960) *Zh. Eksp. Teor. Fiz*, **39**, 1062
85. Wing W. (1984) *Prog. Quant. Electr.*, **8**, 181
86. Ketterle W. and Pritchard D. E. (1992) *Appl. Phys. B*, **54**, 403
87. Bergeman T., Erez G. and Metcalf H. J. (1987) *Phys. Rev. A*, **35**, 1535
88. Davis K. B., Mewes M.-O., Joffe M. A., Andrews M. R. and Ketterle W. (1995) *Phys. Rev. Lett.*, **74**, 5202
89. Cornell evaporation ICAP?
90. Sesko D., Fan C. G. and Wieman C. E. (1988) *J. Opt. Soc. Am B*, **5**, 1225
91. Phillips W. D., Gould P. L. and Lett P. D. (1987) in *Laser Spectroscopy VIII* edited by Persson W. and Svanberg S. (Springer, Berlin), 64.
92. Lett P. D., Watts R. N., Westbrook C. I., Phillips W. D., Gould P. L. and Metcalf H. J. (1988) *Phys. Rev. Lett.*, **61**, 169
93. Phillips W. D., Westbrook C. I., Lett P. D., Watts R. N., Gould P. L. and Metcalf H. J. (1989) in *Atomic Physics 11* edited by Haroche S., Gay J. C. and Grynberg G. (World Scientific, Singapore), 633.
94. Dalibard J., Salomon C., Aspect A., Arimondo E., Kaiser R., Vansteenkiste N. and Cohen-Tannoudji C. (1989) in *Atomic Physics 11* edited by Haroche S., Gay J. C. and Grynberg G. (World Scientific, Singapore), 199.
95. Ungar P. J., Weiss D. S., Riis E. and Chu S. (1989) *J. Opt. Soc. Am. B*, **6**, 2058
96. Sheehy B., Shang S.-Q., Watts R., Hatamian S. and Metcalf H. (1989) *J. Opt. Soc. Am. B*, **6**, 2165
97. Weiss D. S., Riis E., Shevy Y., Ungar P. J. and Chu S. (1989) *J. Opt. Soc. Am. B*, **6**, 2072
98. Sheehy B., Shang S.-Q., van der Straten P., Hatamian S. and Metcalf H. (1990) *Phys. Rev. Lett.*, **64**, 858
99. Nienhuis G. (1992) in *Laser Manipulation of Atoms and Ions, Proc. S.I.F., Course CXVIII* edited by Arimondo E., Phillips W. D. and Strumia F. (North Holland, Amsterdam), p 171.
100. Emile O., Kaiser R., Gerz C., Wallis H., Aspect A. and Cohen-Tannoudji C. (1993) *J. Phys. II (Paris)*, **3**, 1709
101. Aspect A., Emile O., Gerz C., Kaiser R., Vansteenkiste N., Wallis H. and Cohen-Tannoudji C. (1992) in *Laser Manipulation of Atoms and Ions, Proc. S.I.F., Course CXVIII* edited by Arimondo E., Phillips W. D. and Strumia F. (North Holland, Amsterdam), 401.
102. Cohen-Tannoudji C. and Phillips W. D. (1990) *Physics Today*, **43**, 33
103. Cohen-Tannoudji C. (1992) in *Laser Manipulation of Atoms and Ions, Proc. S.I.F., Course CXVIII* edited by Arimondo E., Phillips W. D. and Strumia F. (North Holland, Amsterdam), 99.

104. Castin Y., Dalibard J. and Cohen-Tannoudji C. (1991) in *Light Induced Kinetic Effects on Atoms, Ions and Molecules* edited by Moi L., Gozzini S., Gabbanini C., Arimondo E. and Stumia F. (ETS Editrice, Pisa), 5.
105. Castin Y. (1992) Doctoral Dissertation, Ecole Normale Suprieure.
106. Salomon C., Dalibard J., Phillips W. D., Clairon A. and Guellati S. (1990) *Europhys. Lett.*, **12**, 683
107. Gerz C., Hodapp T. W., Jessen , Jones K. M., Phillips W. D., Westbrook C. I. and Mlmer K. (1993) *Europhys. Lett.*, **21**, 661
108. Letokhov V. (1968) *Pis'ma Zh. Eksp. Teor. Fiz.*, **7**, 348. [(1968) *JETP Lett.*, **7**, 272].
109. Salomon C., Dalibard J., Aspect A., Metcalf H. and Cohen-Tannoudji C. (1987) *Phys. Rev. Lett.*, **59**, 1659
110. Prentiss M. G. and Ezekiel S. (1986) *Phys. Rev. Lett.*, **56**, 46
111. Westbrook C. I., Watts R. N., Tanner C. E., Rolston S. L., Phillips W. D., Lett P. D. and Gould P. L. (1990) *Phys. Rev. Lett.*, **65**, 33
112. Hemmerich A., Zimmermann C. and Hansch T. W. (1993) *Europhys. Lett.*, **22**, 89
113. Verkerk P., Meacher D. R., Coates A. B., Courtois J.-Y., Guibal S., Lounis B., Salomon C. and Grynberg G. (1994) *Europhys. Lett.*, **26**, 171
114. Verkerk P., Lounis B., Salomon C., Cohen-Tannoudji' C., Courtois J.-Y. and Grynberg G. (1992) *Phys. Rev. Lett.*, **68**, 3861
115. Jessen P. S., Gerz C., Lett P. D., Phillips W. D., Rolston S. L., Spreeuw R. J. C. and Westbrook C. I. (1992) *Phys. Rev. Lett.*, **69**, 49
116. Courtois J.-Y. and Grynberg G. (1992) *Phys. Rev. A*, **46**, 7060
117. Hemmerich A. and Hansch T. (1993) *Phys. Rev. Lett.*, **70**, 410
118. Grynberg G., Lounis B., Verkerk P., Courtois J.-Y. and Salomon C. (1993) *Phys. Rev. Lett.*, **70**, 2249
119. Hemmerich A. and Hansch T. (1993) *Phys. Rev. A*, **48**, R1753
120. Lounis B., Verkerk P., Courtois J.-Y., Salomon C. and Grynberg G. (1993) *Europhys. Lett.*, **21**, 13
121. Courtois J.-Y., Grynberg G., Lounis B. and Verkerk P. (1994) *Phys. Rev. Lett.*, **72**, 3017
122. Hemmerich A., Zimmermann C. and Hansch T. W. (1994) *Phys. Rev. Lett.*, **72**, 625
123. Hemmerich A., Weidemller M. and Hansch T. (1994) *Europhys. Lett.*, **27**, 427
124. Meacher D. R., Boiron D., Metcalf H., Salomon C. and Grynberg G. (1994) *Phys. Rev. A*, **26**, R1992
125. Hemmerich A., Weidemuller M., Esslinger T., Zimmermann C. and Hansch T. (1995) *Phys. Rev. Lett.*, **75**, 37
126. Jessen P. S. (1993) Ph. D., University of Aarhus
127. Marte P., Dum R., Taieb R., Lett P. and Zoller P. (1993) *Phys. Rev. Lett.*, **71**, 1335
128. Birkl G., Gatzke M., Deutsch I. H., Rolston S. L. and Phillips W. D. (1995) *Phy. Rev. Lett.*, **75**, 2823
129. Weidemuller M., Hemmerich A., Gorlitz A., Esslinger T. and Hansch T. W. (1995) *Phys. Rev. Lett.*, **75**, 4583
130. Raithel G., Birkl G., Kastberg A., Phillips W. D. and Rolston S. L. (1997) *Phys. Rev. Lett.*, **78**, 630
131. Raithel G., Birkl G., Phillips W. D. and Rolston S. L. (1997) *Phys. Rev. Lett.*, **78**, 2928
132. Raithel G., Phillips W. D. and Rolston S. L. (1998) *Phys. Rev. Lett.*, **81**, 3615

133. Kozuma M., Nakagawa K., Jhe W. and Ohtsu M. (1996) *Phys. Rev. Lett.*, **76**, 2428
134. Anderson M. H., Ensher J. R., Matthews M. R., Wieman C. E. and Cornell E. A. (1995) *Science*, **269**, 198
135. Davis K. B., Mewes M.-O., Andrews M. R., van Druten N. J., Durfee D. S., Kurn D. M. and Ketterle W. (1995) *Phys. Rev. Lett.*, **75**, 3969
136. Sackett C. A., Stoof H. T. C. and Hulet R. G. (1998) *Phys. Rev. Lett.*, **80**, 2031; Bradley C. C., Sackett C. A., Tollett J. J. and Hulet R. G. (1995) *Phys. Rev. Lett.*, **75**, 1687
137. Witte A., Kisters T., Riehle F. and Helmcke J. (1992) *J. Opt. Soc. Am. B*, **9**, 1030
138. Barrett T., Dapore-Schwartz S., Ray M. and Lafyatis G. (1991) *Phys. Rev. Lett.*, **67**, 3483
139. Miranda S. G., Muniz S. R., Telles G. D., Marcassa L. G., Helmerson K. and Bagnato V. S. (1999) to be published in Phys. Rev. A, **59**,
140. Hess H. F. (1986) *Phys. Rev. B*, **34**, 3476
141. Masuhara N., Doyle J. M., Sandberg J. C., Kleppner D., Greytak T. J., Hess H. F. and Kochanski G. P. (1988) *Phys. Rev. Lett.*, **61**, 935
142. Ovchinnikov Y., Muller J.-H., Doery M. R., Vrendenbrecht E. J. D., Helmerson K., Rolston S. L. and Phillips W. D. (1999) *Phys. Rev. Lett.*, **83**, 284
143. Kozuma M., Deng L., Hagley E. W., Wen J., Lutwak R., Helmerson K., Rolston S. L. and Phillips W. D. (1999) *Phys. Rev. Lett.*, **82**, 871
144. Deng L., Hagley E. W., Denschlag J., Simsarian J. E., Edwards M. A., Clark C. W., Helmerson K., Rolston S. L. and Phillips W. D. (1999) *Phys. Rev. Lett.*, **83**, 5407
145. Hagley E. W., Deng L., Kozuma M., Wen J., Edwards M. A., Helmerson K., Rolston S. L. and Phillips W. D. (1999) *Science*, **283**, 1706
146. Hagley E. W., Deng L., Kozuma M., Trippenbach M., Band Y. B., Edwards M. A., Doery M., Julienne P. S., Helmerson K., Rolston S. L. and Phillips W. D. (1999) *Phys. Rev. Lett.*, **83**, 3112
147. Stenger J., Inouye S., Chikkatur A. P., Stamper-Kurn D. M., Pritchard D. E. and Ketterle W. (1999) *Phys. Rev. Lett.*, **82**, 4569
148. Mewes M.-O., Andrews M. R., Kurn D. M., Durfee D. S., Townsend C. G. and Ketterle W. (1997) *Phys. Rev. Lett.*, **78**, 582
149. Lens G., Meystre P., and Wright E. W. (1993) *Phys. Rev. Lett.*, **71**, 3271
150. Dalfovo F., Giorgini S., Pitaevskii L. P. (1999) *Rev. Mod. Phys.*, **71**, 463
151. Deng L., Hagley E. W., Wen J., Trippenbach M., Band Y. B., Julienne P. S., Simsarian J. E., Helmerson K., Rolston S. L. and Phillips W. D. (1999) *Nature*, **398**, 218
152. Reinhardt W. P. and Clark C. W. (1997) *J. Phys. B: At. Mol. Opt. Phys.*, **30**, L785
153. Dobrek L., Gajda M., Lewenstein M., Sengstock K., Birkl G. and Ertmer W. (1999) *Phys. Rev. A*, **60**, R3381
154. Feder D. L., Clark C. W. and Schneider B. I. (1996) *Phys. Rev. Lett*, **82**, 4956
155. Fujita J., Morinaga M., Kishimoto T., Yasuda M., Matsui S. and Shimizu F. (1996) *Nature*380, **1996**, 691
156. Jackson A. D., Kavoulakis G. M. and Pethick C. J. (1998) *Phys. Rev. A*, **58**, 2417

157. Torii T., Suzuki Y., Kozuma M., Kuga T., Deng L. and Hagley E. W. (2000) *Phys. Rev. A*, **61**, 041602
158. Giltner D. M., McGowan R. W. and Lee S. A. (1995) *Phys. Rev. Lett.*, **75**, 2638
159. Denschlag J., Simsarian J. E., Feder D. L., Clark C. W., Collins L. A., Cubizolles J., Deng L., Hagley E. W., Helmerson K., Reinhardt W. P., Rolston S. L., Schneider B. I. and Phillips W. D. (2000) *Science*, **287**, 97
160. Burger S., Bongs K., Dettmer S., Ertmer W., Sengstock K., Sanpera A., Shlyapnikov G. V. and Lewenstein M. (1999) *Phys. Rev. Lett.*, **83**, 5198

13. ULTRAFAST STRUCTURAL DYNAMICS IN THE CONDENSED PHASE

MAJED CHERGUI

Ecole Polytechnique Fédérale de Lausanne
Laboratoire de Spectroscopie Ultrarapide de la Phase Condensée
ICMB-BSP, CH-1015 Lausanne-Dorigny
Switzerland

ABSTRACT: We present a review of wave packet dynamics in different systems and stress their relation to information on the ultrafast structural changes. We show how dynamical information is obtained from ultrafast optical pump-probe studies on structural changes in isolated molecular systems, before moving to complex systems, such as pure and doped solids, and biological systems. We end with a presentation of new developments in structural dynamics, which are based on the pump-probe scheme, using as probe ultrashort pulses of electron or X-rays. As detection methods, electron or X-ray diffraction or X-ray absorption spectroscopies are discussed.

1. Introduction:

Observing atoms in the process of chemical reactions, biological functions or vibrations in solids has been the dream of physicists, chemists and biologists for decades until the 1980s. It is then that the implementation of femtosecond laser spectroscopies brought entirely new possibilities for the study of light-induced phenomena in condensed matter physics, in chemistry and in biology. In particular, the ultrafast studies of chemical reactions have given birth to the field of Femtochemistry [1], which was recognized by the Nobel Prize to Ahmed Zewail in 1999.
For the first time, it was possible to conduct observations on time scales that are shorter than single nuclear oscillation periods in solids or molecules. It therefore became possible, in principle, to monitor molecules at various stages of vibrational distortion, recording "stop-action" spectroscopic events corresponding to well-defined molecular geometries far from equilibrium, including stretched and/or bent unstable transient structures. Molecular structures corresponding to such unstable intermediates between reactant and product could be followed in real-time [1,2]. Distorted crystal lattices and other specific out–of-equilibrium structures could be and have been characterized in real-time [3].
To get a feeling why femtosecond time resolution is needed for these "real-time" observations, it is useful to note that an order of magnitude of the speed of atoms in matter is given by the speed of sound: 300 m/s-1000m/s, which translate to 0.3-1.0 Å in 100 fs. Such length scales are precisely the ones one deals with in dynamical processes of molecules, biological systems, and condensed matter.
The key to observing moving structures in real-time is based on the generation of wave packets and on their detection, using suitable light sources for excitation and detection. The link between the optical spectroscopy and the molecular or condensed phase structures is made via the knowledge of potential energy surfaces, which is a prerequisite if one wants to get the structural dynamics. This review deal with the physics of wave packets, first in isolated quantum systems, then we will consider more complex

B. Di Bartolo and O. Forte (eds.), Frontiers of Optical Spectroscopy, 497-519.

systems consisting of an excited centre and a "bath" (such as a doped crystal, a solution or a protein). The last part will deal with the development of new tools that can probe structures in real-time, and are based on the use of ultrashort pulses of electrons and X-rays. While we will often refer to molecular systems in the following, the essence and the conclusions of this review also concern atomic and many-body systems.

2. Historical background: from kinetics to dynamics [4]:

Kinetics is concerned with populations and their evolution over time. In chemistry, one looks at the overall population of educts before a reaction and of products after the reaction, and conclusions can be drawn about the reaction mechanisms. However, there is no better way to identify mechanisms than to observe them in "real-time". The search for an ever greater time resolution to "clock" events in physics, chemistry and biology was a process that took place over decades. In 1889, Svante Arrhenius presented a description of how chemical reactions vary as a function of temperature. His well known formula describes the temperature dependence of the reaction rate k(T):

$$k(T) = A\, e^{-E_a/kT} \tag{1}$$

where k is the Boltzmann constant, T is the temperature (in Kelvin) and E_a is the so-called activation energy, i.e. the height of the barrier up to a hypothetical state called by Arrhenius, the "activated complex". The Arrhenius equation has been, and is still used with success by chemists, physicists and biologists to describe kinetic processes in a large class of media.
In 1931, H. Eyring and M. Polanyi developed the first potential energy surface for the $H_2 + H \longrightarrow H + H_2$ reaction, and Hirschfelder, Eyring and Topley performed the first trajectory calculation with femtosecond steps in 1936 using the then available computing power. This was the birth of a new way of thinking in terms of potential energy surfaces with dynamics occurring through energy valleys and mountains, with the transition state at the saddle point.
Around that time, Eyring, Evans and Polanyi formulated the transition-state theory for chemical reactions, giving an explicit expression for the pre-exponential factor A in the Arrhenius equation by invoking statistical mechanics. The theory gave an analytical expression for the rate constant with a "frequency" for the passage through the transition state. This frequency factor is typically 10^{13} Hertz, the typical value for the frequency of molecular vibrations !
Shortly before these developments, the foundations of quantum mechanics were being laid down. In 1926, Erwin Schrödinger introduced the idea of "wave groups" in order to make a natural connection between quantum and classical descriptions. A year later, Ehrenfest published his famous theorem, outlining the regime for the transition from a quantum to a classical description: the quantum expectation values behave classically in the limit of large quantum numbers. The use of wave groups, now called wave packets in physics remained limited to a few theoretical examples and was only considered for Gedanken experiments. Indeed, it was not possible to synthetise wave packets, as the temporal resolution at that time was of seconds to milliseconds at best. The field had to wait several decades in order to reach the required time resolution.
The long standing efforts to improve the latter were the works of many researchers. G. Porter, R. Norrish introduced the Flash Photolysis technique, which allowed the millisecond time resolution in the second half of the 1940s [5]. By exposing a chemical solution to a heat, a pressure or an electrical shock, M. Eigen achieved the microsecond temporal resolution [6]. The advent of the pulsed nanosecond laser in the mid 1960's [7] and soon after, of the picosecond laser [8] brought about million times better resolution. However, even on the short picosecond time scale, molecular states already reside in eigenstates (see § 3) and there is only one evolution observable, the change of population with time of that state. Hence, with picosecond spectroscopy one is still concerned with kinetics, not dynamics. The advent of femtosecond laser technology, thanks to the works of C. V. Shank and co-workers [9] finally opened the door to the

direct probing of nuclear motion in real-time. We have just entered the attosecond time regime [10], with which it will be possible to probe the dynamics of electrons and of valency.

3. Basic Quantum Mechanics

The realization of these experimental possibilities requires not only adequate time resolution for probing nuclear motion, but also excitation mechanisms suitable for initiating molecular motion in a phase-coherent (i.e., synchronized) manner in order to "observe" the transient structures. This is based on wave packets, since a wave-packet represents a quantum object that is localized in its position coordinate. As an example, for diatomic molecules, the wave packet represents the fact that there is some uncertainty in the separation of the two atoms. Here, we just recall a few points for the description of the dynamical behaviour of quantum systems. The time-dependent Schrödinger equation:

$$H\psi(r,t) = i\hbar\frac{\partial\psi(r,t)}{\partial t} \tag{2}$$

has solutions of the form

$$\psi(r,t) = \varphi(r)e^{-iEt/\hbar} \tag{3}$$

where $\phi(r)$ is the eigenfunction, which is the solution of the time-independent Schrödinger equation and the exponential term represents the time dependence. In general it is rare to find a physical system in just one eigenstate. At t=0, the wave function of the system will be a linear combination of eigenstates given by setting t=0 in eq. 3:

$$\Psi(r,0) = \sum_n a_n\varphi_n(r) \tag{4}$$

Each eigenstate evolves differently, as determined by its energy. In eq. 3, the temporal dependence is the modulation of the phase of the wave function :

$$e^{-iEt/\hbar} = \cos(Et/\hbar) - i\sin(Et/\hbar) \tag{5}$$

which oscillates from 1 to –i to 1 to i ..., at a frequency of $E/\hbar$. Thus the temporal dependence of the wave function is due to the fact that its amplitude oscillates from <0 to >0 at a frequency determined by its energy. As an example, the 1s state of H has an energy E = 0, therefore the phase is a constant. The 2s state has an energy of E= 1.6 10^{-18} J, i.e. the phase oscillates a 2.5 10^{15} Hz (i.e. T= 0.4 fs).
If we now look at the system after a time t, eq. 4 becomes:

$$\Psi(r,t) = \sum_n a_n\varphi_n(r)e^{-iE_nt/\hbar} \tag{6}$$

This is a linear combination of stationary wave functions but is not a solution of the stationary problem. It represents a coherent superposition of states or a wave packet. The a_n coefficients are given by:

$$a_n = \langle\varphi_n|\Psi(t=0)\rangle \tag{7}$$

and are determined by the way the system has been excited. This will also determine the shape of the wave packet. If we consider g(k) to be the weighing of the different eigenfunctions, which is determined by the shape of the exciting pulse (with k being a continuous variable parameter), then eq. 6 becomes an integral over k:

$$\Psi(r,t) = \int_k g(k)\psi_k(r,t)dk \tag{8}$$

Creation of a wave packet supposes the simultaneous excitation of several levels (eq. 6), which are reached due to the width of the exciting pulse. Indeed, if the pulse width is short enough, its spectral width will be broad by virtue of the uncertainty principle, and in this case several levels are excited simultaneously (see figure 1, left). If on the other hand the laser is spectrally sharp as is the case with nsec lasers, then only a single level will be excited and one then looks at populations. The creation of wave packets also imposes a phase relationship between the eigenfunctions composing it (eq. 6), i.e. a coherence. A complete description of the populations and coherences is best given by the density matrix formalism in Liouville space. It is more adequate than a Hilbert space description, which works well only for isolated systems (i.e. in the absence of a dissipative bath). The survival or loss of this coherence can deliver rich information about the system under investigation and its interaction with the environment [11].

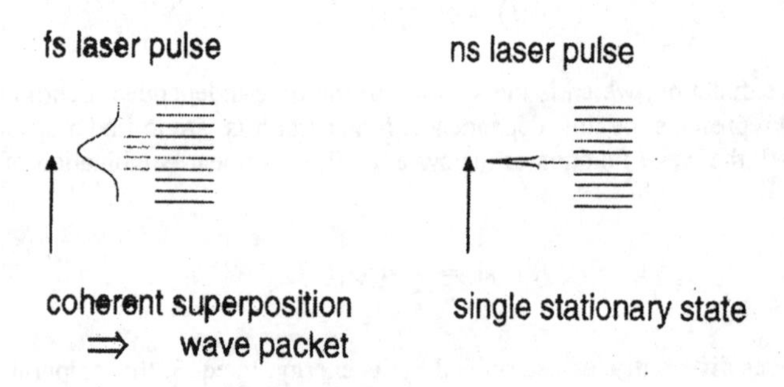

Figure 1. Excitation schemes with an ultrashort, spectrally broad laser pulse (left), and a spectrally narrow but temporally long laser pulse (right)

Wave packet dynamics need not necessarily femtosecond pulses, and it is possible to visualise nuclear motion with psec pulses provided the period of the motion is longer than the pulse duration(see table I). Depending on the physics one wants to investigate, the pulse width of the laser will be selected.

TABLE I: Time-to-width relationships for homogeneous lines

Type of width	Width (cm^{-1})	Time (seconds)
Doppler	10^{-3} - 10^{-1}	$3\ 10^{-8} - 3.\ 10^{-10}$
Rotational	10^{-2} to 10	$3\ 10^{-9} - 3\ 10^{-12}$
Phononic	≤ 100	$\geq 3.\ 10^{-13}$
Vibrational spacings	100 – 4000	$3\ 10^{-13} - 8\ 10^{-15}$
Electronic spacings	≥ 5000	$\leq 6\ 10^{-15}$

. Wave packets dynamics in isolated systems:

he typical approach for recording wave packet dynamics is based on the pump-probe scheme which is lustrated in fig. 2 in the case of a molecular system. Absorption of a first pulse of light (called the pump ulse) brings the molecule from the ground state potential V_1 to the excited state potential V_2. If the pulse is ufficiently short in time (and at least shorter than the oscillation period of the molecule in the excited state $_2$), then a coherent superposition of vibrational levels (a so-called wave packet) is created, which is ocalized in space at the ground state equilibrium configuration at time t=0. The newly created wave packet vill feel the driving force of the excited state potential and will undergo oscillations at a frequency, haracteristic of the vibrational spacings of the excited state potential, if the latter is bound. Key issues here re that: a) the creation of a wave packet implies that all excited molecules in the irradiated volume have e same configuration to within the spatial spread of the wave packet. Thus, the shorter the pump pulse, e sharper defined will the distribution of molecular configurations be. b) The time evolution of the wave acket reflects that of the ensemble of coherently excited molecules, which evolve in phase.

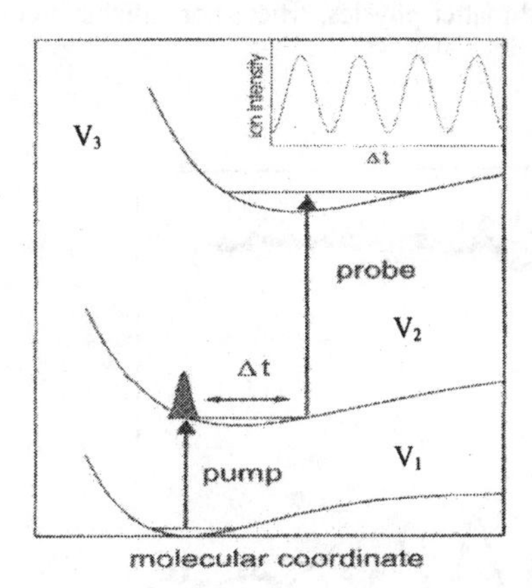

Figure 2: the pump-probe scheme in a molecular system. The pump pulse creates a wave packet on the excited V_2 potential, and the probe pulse samples the dynamics by inducing an absorption from V_2 to V_3. The measured signal can be the absorption of the probe pulse, a fluorescence it induces from V_3 or an ion signal if V_3 is above the ionization limit of the molecule.

he observation of the dynamics of the wave packet is done using a second ultrashort laser pulse (called e probe pulse) that interrogates the ensemble excited by the pump pulse and whose time delay with espect to the latter can be continuously tuned. The probe pulse induces a transition between V_2 and a igher lying state V_3. The choice of the probe wavelength determines the choice of the probed onfiguration as can be seen in fig. 2. Thus changing the probe wavelength allows the probing of different onfigurations. The spatial width of the window in real space, which is opened by the probe pulse, is etermined by its spectral width and by the slope of the potential difference between V_2 and V_3. Each time e wave packet enters the probe window, a signal is detected, which can be the absorption of the probe ulse, a fluorescence signal from V_3 induced by the probe absorption, or an ion signal, if V_3 is an ion otential. The key point here is that the choice of the probe wavelength determines the choice of the

probed configuration. Thus structural dynamics is retrieved from spectroscopic signals thanks to an *a-priori* knowledge of the energetics of the molecular system, i.e. its steady-state spectroscopy.
As an illustration, we look at the decay of the ion-pair states Na^+I^-. The NaI molecule was excited to a covalent state, which crosses the ground ion-pair state. The probe pulse examines the system at a frequency that corresponds either to that of free Na or at a frequency in which the atom absorbs when it is part of the complex. The latter frequency depends on the Na-I distance, so that an absorption (or more often, a laser-induced fluorescence) is obtained each time the molecules return to the corresponding position. Provided the wave packet is not completely spread (or dephased), one obtains a structural and a dynamical information [12]. A typical result is shown in figure 3. The bound Na absorption intensity shows up a series of oscillations that recur with a period of about 1 ps, and decays at a rate at which the molecule dissociates (the system does not dissociate on every outward going swing). The free Na signal also grows in an oscillating manner and shows the same period as before that gives it a chance to dissociate each picosecond or so. Aside from being one of the first examples demonstrating nuclear dynamics, it also showed that a wave packet launched on a surface can cross-over to another surface, through non-adiabatic couplings, without loosing its coherent properties. This is of central importance in chemistry, biology and condensed matter physics, where non-adiabatic couplings are the rule rather than the exception.

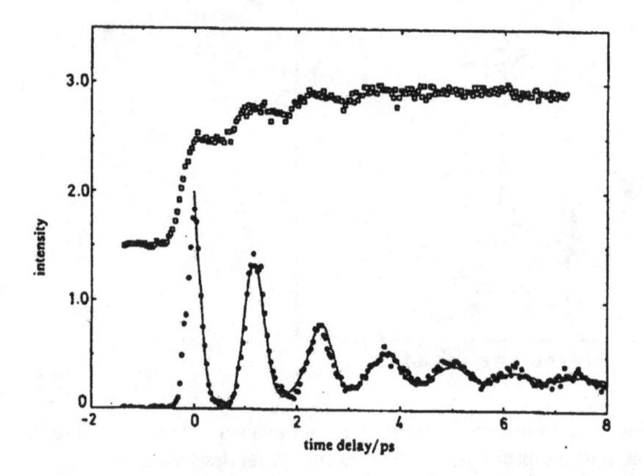

Figure 3: Wave packet dynamics in NaI. The lower trace shows the wave packet dynamics within the electronically excited bound state. The upper trace shows the build up of free Na atoms due to predissciation processes occurring in the excited state and leading to free Na and I atom. After ref. 12.

The pump-probe scheme has been a revolutionary tool to observe motion in real-time of bound states of small diatomic molecules such as I_2, Na_2, and of elementary chemical reactions, such as the dissociation of ICN, HgI_2, to name a few. Its implementation (and variants of it) to the study of systems of even greater complexity has occurred throughout the 1990's, with the study of liquids, solids and biological molecules [2,4,13]. This review is mainly concerned with the latter systems.
To finish this short introduction on wave packets in isolated systems, it is worth mentioning that not only vibrational wave packets have been reported, but also electronic wave packets made up of Rydberg levels of atoms. This implies that the electronic cloud can be localised in time and space, and it was experimentally demonstrated for both radial [14] and angular [15] Rydberg wave packets. We refer the reader to refs [16, 17, 18] for a more complete discussion of wave packets in isolated atomic and molecular systems.

When dealing with many-body systems, several issues arise, which are absent in isolated systems: a) The line widths are broadened due to both inhomogeneous and homogeneous contributions. The inhomogeneous contributions arise from the different interaction each impurity, solute or chromophore is experiencing with its environment. b) Due to fluctuations of the surrounding species atoms, if coherence is imparted to an atomic or molecular entity, does it survive in a solvent, crystal or protein? c) If yes, is there a coherence also imparted to the surroundings? In the case of chemical processes, is coherence of the system transferred or destroyed upon product formation? These are only a few of the questions that arise, bearing in mind that depending on the physical phenomena under study, many others will be asked. In the following, we show a few representative examples of phonon wave packets dynamics in pure and doped solids, and show that vibrational coherence is maintained. We will then discuss the case of biological systems and show the occurrence of vibrational coherences in this case too.

5. Wave packets dynamics in solids

5.1 PHONON WAVE PACKETS IN PURE SOLIDS

Until the beginning of the 1990s, the sole sources of information about phonons in solids were Raman spectroscopy, and in some cases IR spectroscopy. A technique was developed by K. Nelson [3,19,20], which is based on a stimulated Raman process (ISRS or Impulsive Stimulated Raman Scattering). Because of the large spectral width of the ultrashort fsec pump pulse, a stimulated Raman process takes place, which creates phonons (ω_p) by way of two components (ω_{l1} and ω_{l2}) of the excitation pulse (Fig. 4) such that:

$$\omega_p = \omega_{l1} - \omega_{l2} \text{ or } \vec{k}_1 + \vec{k}_2 = \pm\vec{q} \tag{9}$$

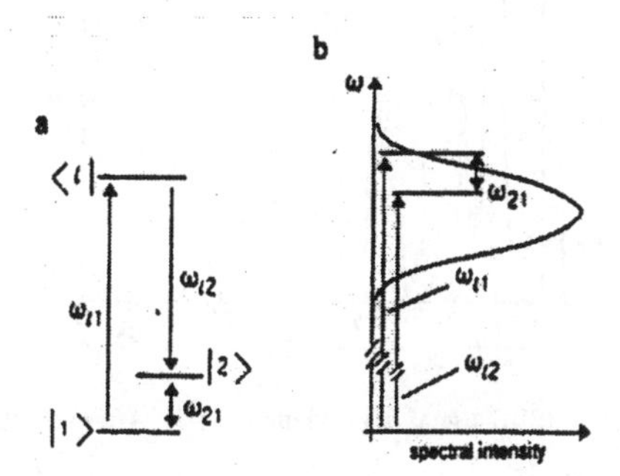

Figure 4: a) Excitation of a Raman transition |1> → |2> via an electronically excited state <l|. b) For a fsec pulse of average carrier frequency ω_l, the two frequency components of the Raman transition are contained within the pulse spectrum. The medium itself selects the frequency pairs suitable to drive the Raman transition.

The excitation pulse has a duration shorter than the phonon period. The fsec excitation excites in phase a large number of oscillators, which are phonons. The ISRS process therefore creates a standing wave pattern, which damps out with time due to dissipation (fig. 5).

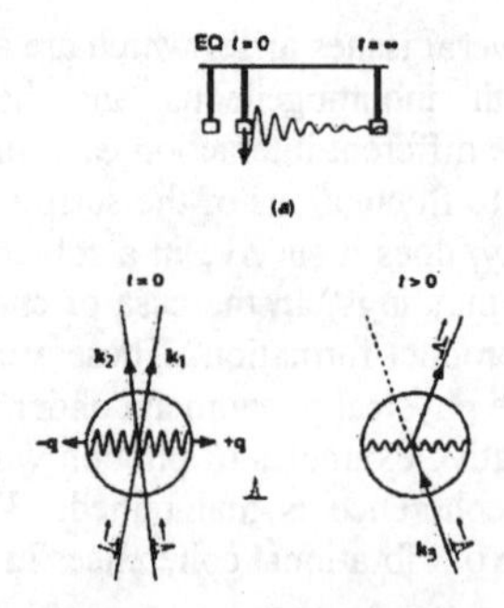

Figure 5:a) Mass-on-spring model for impulsive stimulated Raman scattering (ISRS): at t=0, the laser pulse sets the atom to oscillate coherently, b) Optical geometry for measuring ISRS signal: the ISRS process creates a standing wave pattern that damps out with time. A third pulse scatters on the pattern and is read in another direction due to wave vector conservation. After ref. 20.

A second, probe pulse can record this damping, and its scattering on the sample will reflect the amplitude of the phonon oscillation. Fig. 6 shows the results obtained in the case of a pyrelene crystal at 18 K. The beat pattern is due to 4 Raman modes of the solid. The signal decays with a time constant of ~9 ps. This method is however not very selective because the dispersion curve of optical phonons is almost flat for most solids and all Raman modes satisfy eq. 1. A review on the further developments of this method and on time-resolved vibrational spectroscopy in general is found in ref. 3.

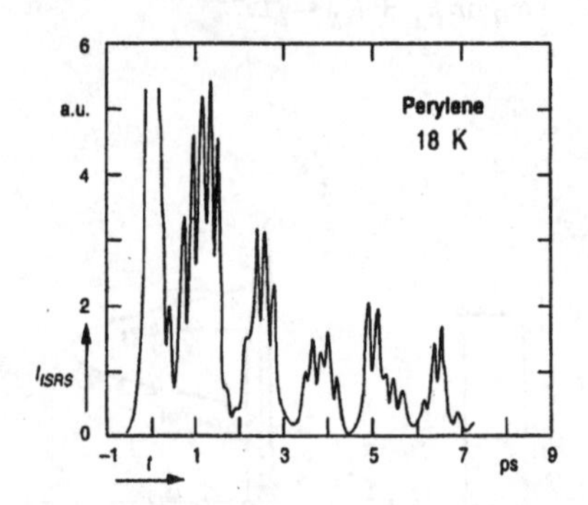

Figure 6: ISRS signal of perylene at 18 K. After ref. 20

In solids with small energy gaps, the phonon vibrations can be observed in reflectivity or transmission, because they directly affect the band gap structure of the solid, thus the dielectric function is affected by the deformation potential. Alternatively, excitation of a small gap solid brings the charge carriers in an excited state and leads to a screening of the lattice potential. This results in an impulsive departure of the ions from their equilibrium position and to a coherent excitation of their oscillations. Fig. 7 shows the relative reflectivity of Ti_2O_3 following excitation by a fsec pulse [21]. A number of examples of coherences in solids are discussed in refs 3 and 22.

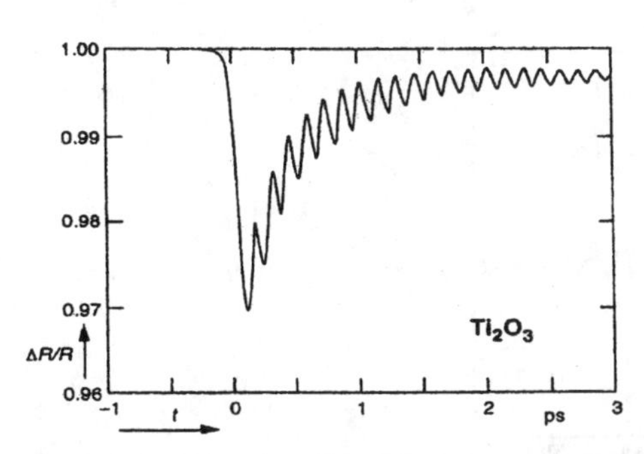

Figure 7: Transient reflectivity changes of Ti_2O_3 after excitation with a fsec laser pulse. After ref. 21.

The above examples concern pure solids where the occurrence of phonon wave packets reflects a collective oscillation of the solid. A complication arises in the case of doped solids as the lattice is perturbed and its long range symmetry changed. The study of doped solids is also interesting as they represent ideal model systems to identify localised vibrations and to clarify the fundamentals of wave packet dynamics in many-body systems. As such, they are a first step towards the understanding of wave packet dynamics in complex systems, such as liquids and biological molecules. In this case, one may ask if vibrational coherence is created in the impurity, how long does it survive, and is there a transfer of coherence to the lattice?

5.2 LOCALISED VIBRATIONAL WAVE PACKETS IN DOPED SOLIDS

The existence of vibrational wave packets of molecular impurities in solids was first demonstrated by Apkarian et al [23,24,25,26] , in the case of solid rare gases at low temperatures, doped with the I_2 molecule. Several groups have also investigated them using classical molecular dynamics simulations [23-27]. The connection between classical trajectories and wave packets has been made by Martens and co-workers [28].

We have studied the dynamics of vibrational wave packets and the energy dissipation mechanisms between a molecular impurity and an Ar crystal in the case of the Hg_2 molecule [27,29]. This system has a very shallow ground state potential and excitation with a UV fsec pulse reaches the deeper D state [29] and launches a wave packet, whose dynamics has been simulated [27]. It is shown in fig. 8 as the average of a swarm of trajectories of the Hg-Hg bond distance as a function of time.

As the system relaxes to the lowest electronic state, the wave packet may dephase as a result of the fluctuations of the medium. One can see the system undergoes several oscillations before complete dephasing occurs.

We have also visualized the motion of the surrounding Ar atoms (fig. 9), and interestingly, one can see that there is a concerted motion between the Hg_2 vibration and the response of the matrix cage in particular those atoms that are in a plane perpendicular to the internuclear axis (compare the oscillation period in fig. 8 and fig. 9a), showing that the coherence of the impurity is transferred to the environment.

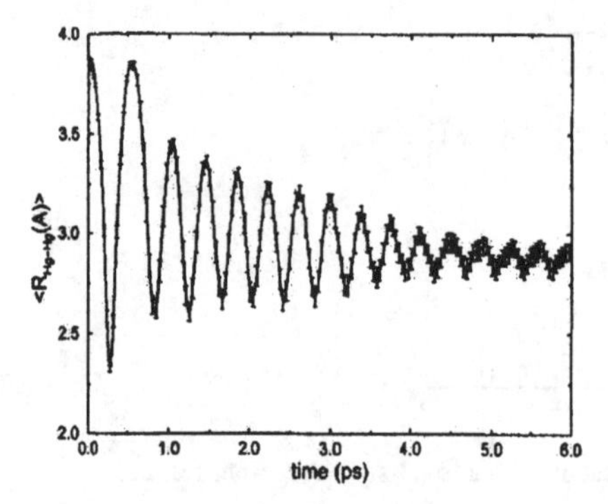

Figure 8: Classical molecular dynamics simulation of the average over all trajectories of the Hg-Hg distance ($R_{Hg\text{-}Hg}$) as a function of time for an excited Hg_2 dimer in solid argon. The error bars result from the average of 100 trajectories. After ref. 27

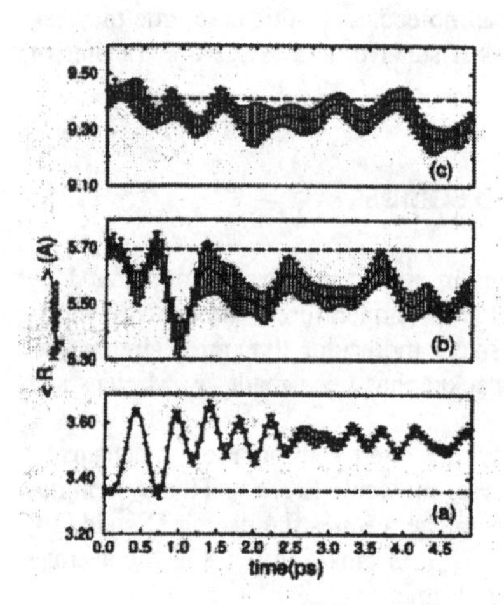

Figure 9: Average over all trajectories of the distance between Ar atoms and the Hg_2 centre-of-mass (COM) for: a) equatorial Ar atoms nearest to the Hg_2 COM in the plane perpendicular to the Hg internuclear axis; b) and c) first and second Ar atoms with respect to the Hg_2 COM in the direction of the internuclear axis, respectively. The error bars result from the average of 100 trajectories. After ref. 27.

To summarize this part:

a. Vibrational coherences of molecular impurities occur in condensed matter, in spite of the presence of a bath. Even more remarkable and of great interest for chemistry is the fact that vibrational coherences have also been observed for solutes in liquid solvents. Examples concern I_2 in liquids [30,31]. Transfer of vibrational coherence resulting from a photochemical reaction in liquids has also been observed in the the

photofragmentation of HgI_2 and I_3^- in liquids, were the departing diatom (HgI, I_2^-, resp.) fragment is ejected with an impulsive (coherent) vibrational excitation [32,33].
b. We also observe transfer of coherence from the impurity to the medium (fig. 9)
c. Therefore dephasing of vibrational motion is not immediate and an oscillator can be out of equilibrium with the solvent. This has important consequences for quantum control of condensed phase chemical reactions.
d. The presence or destruction of coherences depends on a number of parameters, of which the most obvious are the frequency difference between the localised oscillators and the bath oscillators and the coupling between these two types of oscillators. In many of the above cases, the occurrence of vibrational coherences has to do with either a mismatch of oscillator frequencies or a poor coupling between oscillators.

5.3. INTERMOLECULAR VIBRATIONAL WAVE PACKETS IN SOLIDS

In the preceding paragraph, we showed an example where the vibrational coherence imparted to an impurity could be transferred to the surrounding medium. However, photoinduced events in condensed matter start by absorption of a photon, which implies a redistribution of charge in the excited centre that then sets the nuclear dynamics in action. Furthermore, the surrounding species are suddenly subject to another field of forces and will, as a result, also rearrange in order to accommodate themselves to the new situation. This process occurs prior or is concurrent to any internal rearrangements that may take place in the excited centre. The rearrangement of the environment upon excitation of an impurity is a crucial process in many light induced phenomena and is at the origin of radiation-induced defect formation in solids, solvation dynamics in liquids, or dielectric relaxation in biological systems.
We investigated this process in the case of van der Waals (mainly rare gas and hydrogen/deuterium) solids. The advantage of these solids is that they are ordered media, are easily amenable to computer simulations, are not reactive chemically, and are the closest solids to the liquid phase due to their large zero-point fluctuations. They can therefore serve as models for the latter. In order to free ourselves from intramolecular contributions, we choose to excite a molecular impurity (NO) to its lowest electronic state A-3sσ (i.e. principal quantum number n=3, l=0 and its projection on the intramolecular axis λ=0) at its lowest vibrational level v=0. Therefore the molecule can be considered as an atom, as no internal modes are excited (the molecule is blocked in the rare gas matrix, so no rotation is taking place). The fact of exciting the molecule to an s-type Rydberg orbital results in an extended spherically shaped electronic cloud that overlaps the electronic clouds of the surrounding matrix species (see figure 10). This leads to a strong shift to high energies of the absorption bands of the A-state (see fig. 11 for typical examples in solid Ne and solid hydrogens) due to the repulsive interaction between the excited molecule and the surrounding species [34].
The A-state fluorescence is strongly Stokes-shifted (fig. 11) as a result of the medium reorganization, due to the fact that the matrix species are pushed outwards in order to accommodate the extended Rydberg orbital. The steady-state absorption and emission lineshapes of fig. 11 were analysed in detail in a one-dimensional (i.e. assuming one single effective mode) configuration coordinate model in the harmonic approximation (see fig. 12) [34]. This model delivered details of the structural (equilibrium configuration in the excited state) and physical parameters (effective phonon energies, Huang-Rhys factors, etc.) governing the lattice relaxation process. In order to observe the latter in real-time we carried out fsec pump-probe spectroscopy. The principle of which is shown in figure 12.

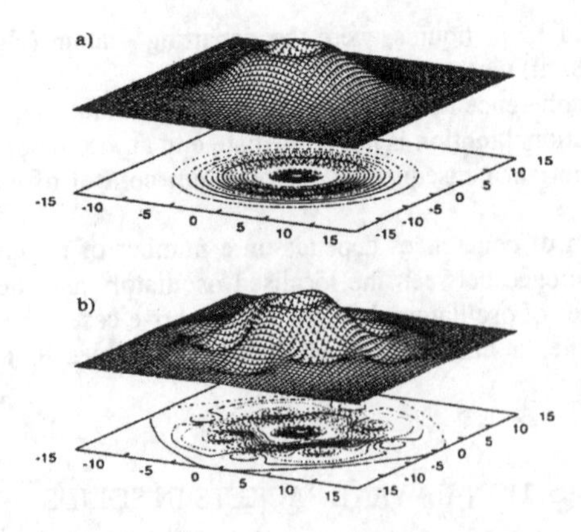

Figure 10: Electron density map of the lowest Rydberg state of NO calculated by a pseudopotential method (ref. 35); a) in the gas phase and, b) in an Argon matrix. The minima in the map are due to avoidance of the Rydberg electron of the Ar atoms, resulting from Pauli repulsion. After ref. 35.

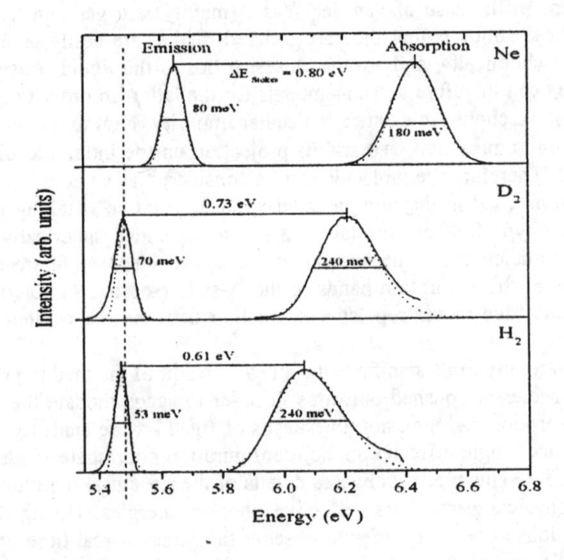

Figure 11: Absorption and emission bands of the lowest Rydberg state of NO in solid Ne and solid hydrogens. After ref. 34.

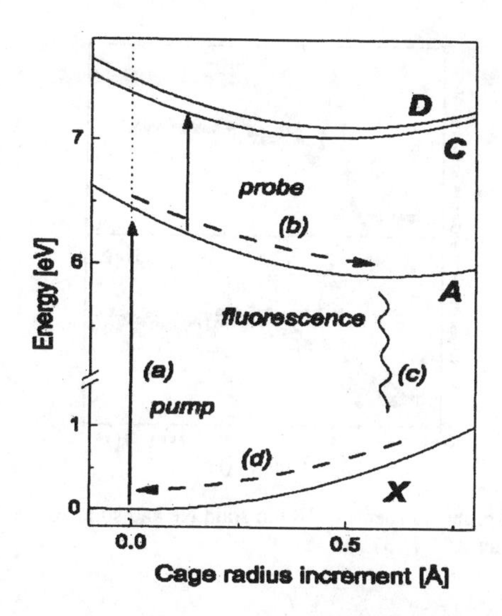

Figure 12: Intermolecular one-dimensional NO-matrix potentials extracted from steady-state spectroscopic studies (see ref. 34) and principle of the pump-probe experiment: a) Excitation process by a fsec pulse, b) Ultrafast relaxation of the environment, c) fluorescence, d) relaxation in the ground state back to the initial configuration. The probe pulse induces absorption of a photon from the excited A-Rydberg state to higher lying ones (C or D). The choice of the probe wavelength determines that of the probed configuration by knowledge of the difference potentials A-C and A-D.

UV pump pulse at around 6 eV, excites the A Rydberg state and triggers the relaxation of the lattice in radial fashion, a second probe pulse whose wavelength can be tuned, samples the dynamics along the A-ate intermolecular potential surface. The choice of the probe wavelength determines the probed nfiguration, so that a complete picture of the structural dynamics can be obtained. Typical results for O in solid Ar and solid Ne are shown in figure 13.

iese transients exhibit a peak at time t=0, followed by an oscillatory pattern and a flat signal at longer nes. The flat signal reflects the fact that the system has fully relaxed at the bottom of the A-state termolecular potential well (fig. 12). The peak at early times is due to the departing wave packets of termolecular modes, which is created by the pump pulse. The interesting oscillatory features reflect the ct that the cage of Ar (Ne) atoms is breathing in a coherent fashion around the excited impurity. terestingly, although the Ne cage is lighter the oscillation period is twice longer than that in solid Ar. is is due to the fact that the Ne lattice is tighter, and the extended Rydberg orbital spans the first 2 ells around the impurity, so that they are simultaneously set in motion, and set simultaneously in otion the next shells, because of the short range nature of the Ne-Ne interaction. The interpretation of r pump-probe data was fully confirmed by classical molecular dynamics simulations [38]. In the case solid hydrogens (not shown here) the pump-probe data shown a one-way coherent process, with no urrence of the cage boundaries [34,36], suggesting that a shock wave is launched in the lattice.
ese results showed that coherence in intermolecular modes of doped solids can be created and served. The remarkable point to stress is that these localized modes occur at frequencies that are within phonon spectrum of the respective solids, yet dephasing requires a finite time to take place.

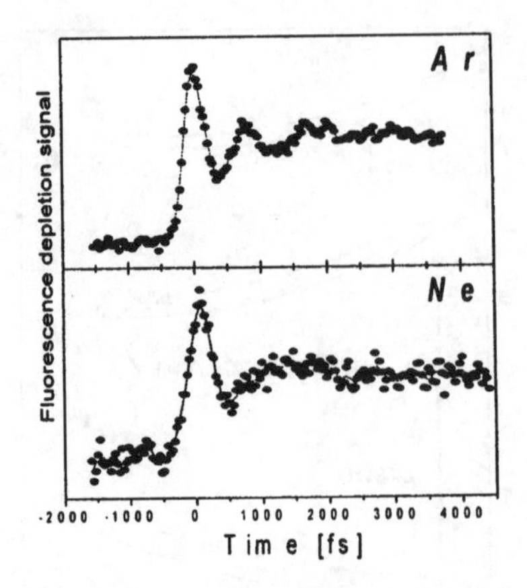

Figure 13: Pump-probe transients of NO in solid Ne and solid Ar, excited to its lowest Rydberg state (after ref. 34, 37 and 38)

6. Wave packet dynamics in biological systems

The lack of an immediate equilibration of an oscillator in a phonon bath is also observed in the case of biological systems. In such systems the observation of vibrational coherences (i.e. motions along specific coordinates) may help establish the link between structural deformation and biological function. Structure plays a crucial role in biology, but the advent of time-dependent structural studies are surely going to unravel many details of the function, as the latter should be seen as a sequence of events over time.
I will only give one example here to demonstrate that despite their complexity, biological systems, in particular proteins, exhibit a wealth of vibrational coherence phenomena. We look at the case of myoglobin [39], as it is one of the most striking examples. The reaction center in such proteins is the so-called heme pocket (fig.14), which consists of a porphyrin plane with an iron atom in its center to which ligands (X=NO, CO, O_2, etc...) bind. This is the origin of the basic respiratory mechanism.

Excitation of the porphyrin near 400 nm (which corresponds to the maximum of its main UV-Visible absorption band, the so-called Soret band) leads to dissociation of the ligand, if it is bound to the Fe atom or simply to excitation of the porphyrin itself, if no ligand is present. In fact, Champion and co-workers observed vibrational coherences in both case. The first case would correspond to the equivalent of a photochemical bond-breaking reaction (i.e. the so-called reaction-driven coherences), while the second case corresponds to (laser) field driven coherences.
Fig. 15 shows the pump-probe transient obtained in the case of the unliganded myoglobin (Mb). The pump was at 433 nm and the probe at 428 nm. A rich pattern of vibrational coherences is observed. Its Fourier transform reveals something like 8 vibrational modes of the protein that are coherently excited by the fsec pulse. It is compared in the lower panel with the resonance Raman spectrum of the same species. The resonance Raman spectrum reflects the modes that are created by the pump pulse and the remarkable resemblance between the two shows that the coherences are an intrinsic feature of the unliganded myoglobin and are produced by the laser field (field-driven coherences).

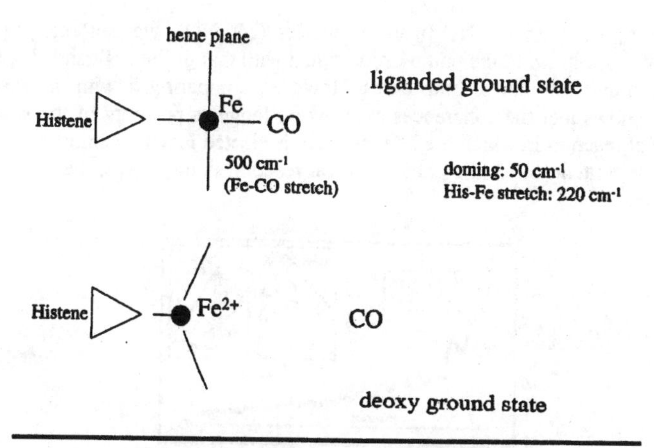

Figure 14: The ligand (CO in this case) binds to the Fe atom of the porphyrin, which is found in the so-called Heme procket. Upon detachment (e.g. by a laser pulse) of the

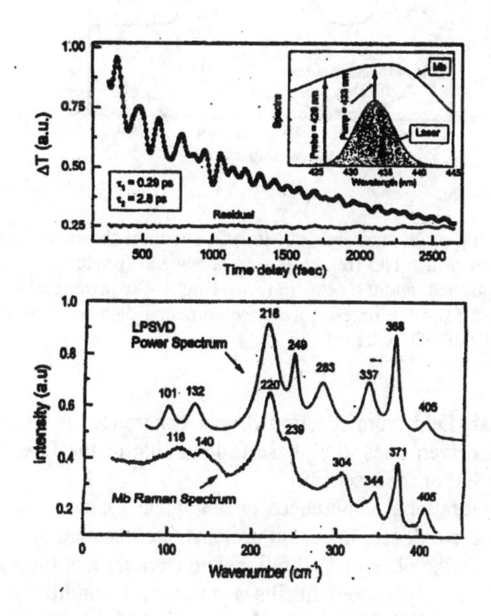

Figure 15: Femtosecond transients from deoxy Mb (upper panel). The inset shows the absorption spectrum of deoxy Mb, the spectrum of the laser pulse (grayed) and the detection wavelength. The lower panel shows the generated power spectrum from the time-resolved transient (upper trace) and the experimental resonance Raman spectrum of deoxy Mb. After ref. 39.

If we repeat the experiment with the NO-liganded species (Mb-NO), we can see (fig. 16) from the comparison of the power spectrum of the pump-probe signal with that of the unliganded species in fig. 15, that the coherences are quite similar in both cases. However, comparing it with the resonance Raman spectrum of Mb-NO, shows that the coherences are now no longer a property of the species, but result from the photochemical reaction in which the NO fragment is ejected leaving behind the porphyrin with a high vibrational energy content. Since the photochemical reaction is impulsive, the observed coherences are now reaction-driven.

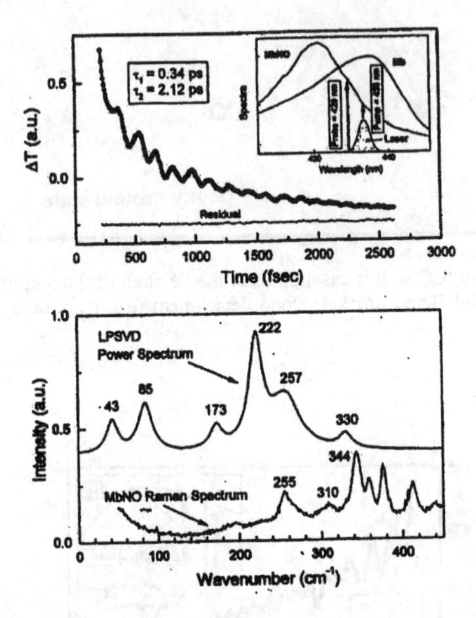

Figure 16: Fsec pump-probe transients of Mb-NO. The inset shows the equilibrium absorption spectrum of Mb-NO (reactant) amd deoxy Mb (product), the spectrum of the laser pulse (grayed) and the detection wavelength. The lower panel compares the power spectrum from the time-resolved experimental data with the resonance Raman specrum of Mb-NO. After ref. 39.

An additional way to distinguish field from reaction driven coherences is to look at the phase of the coherences. In the case of field driven ones, the phase follows that of the laser field. In the other case, there is phase shift compared to that of the laser [39]

Many other instances exist of vibrational coherences in biological systems: in the photosynthetic light harvesting system, in cytochrome c-oxidase, in retinal proteins, etc. (see e.g. ref. 13 for a few examples). However, the relationship between the observed vibrational coherences and the actual biological function is not trivial. Indeed, provided the experiment fulfills a number of conditions (see § 3), and that the system does not dephase them too fast, vibrational coherences will be created and observed. However, whether these reflect the biologically significant nuclear motions remains to be established. One way to circumvent this difficulty is to be able to look at the structure in a more direct way than when using optical spectroscopy.

7. New frontiers of ultrafast structural dynamics

While the relation between probe wavelength and probed configuration is a trivial one in the case of diatomic (see § 4) and some triatomic molecules (provided their steady-state spectroscopy is well established), this relationship becomes ambiguous as the molecular system increases in complexity and size. A further degree of complexity is added to the problem, if we are dealing with condensed phase or biological samples, since intermolecular coordinates come into play. Consequently, experimental approaches are needed, which can overcome the limitations of optical studies for structural determination, while maintaining the high temporal resolution of femtosecond spectroscopy.
Structural techniques such as X-ray diffraction, electron diffraction, X-ray absorption spectroscopy (XAS) and neutron scattering can deliver the answer, provided sufficiently short and intense pulses of X-rays, electrons or neutrons can be produced. While neutrons are a powerful tool for static structural studies, we are not aware of work using them in time-resolved studies. The majority of time-resolved structural studies have been carried out with ultrashort pulses of X-rays or electrons, either using diffraction techniques or X-ray spectroscopies. Conceptually, the methods of ultrafast X-ray and electron diffraction or X-ray absorption are similar to ultrafast optical pump-probe experiments. The principle of a pump-probe X-ray or electron diffraction is shown in figure 18. The essential new ingredient is that the probe is not optical but via an X-ray (or electron) pulse that is diffracted or absorbed (for X-ray absorption only).

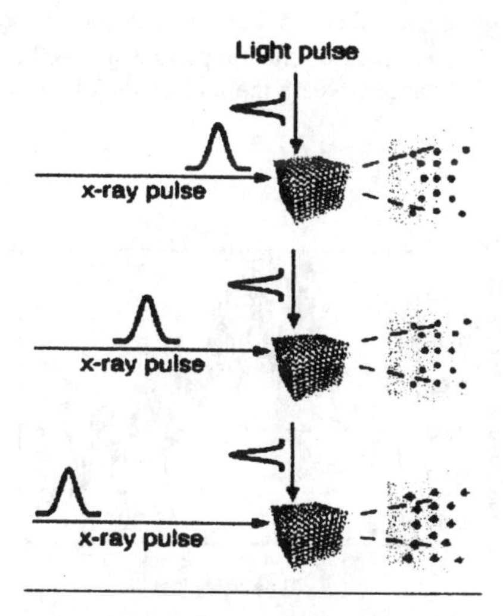

Figure 17: Principle of an ultrafast X-ray diffraction experiment. The optical laser pulse triggers a structural change in the solid, the probe X-ray pulse samples these changes as a function of time delay with respect to the pump pulse. The transient structures by the diffraction spots are visualized on a 2-D detector after the sample. If the solid undergoes coherent phonon oscillations, the diffraction spots will be seen to oscillate, if disorder sets in (e.g. melting), the diffraction spots become blur. The same lay-out is valid for electron diffraction, except that the sample if a gas phase molecule.

An impressive effort over the last 12 years has recently culminated in the demonstration of ultrafast electron diffraction, where transient structures of molecules in the gas phase, including molecules containing low-Z atoms, have been determined with picosecond temporal resolution [40]. The key point to the success of electron diffraction lies in the high scattering cross sections of electrons, which is 5 orders of magnitude larger than that of X-rays. The fact that the energy deposited per scattering event is 2-3 orders of magnitude weaker is certainly attractive in view of destruction due to the probe beam. Extending time-resolved electron diffraction to condensed phases would be a major challenge due to the low penetration depth of electrons in matter, and the technique will most probably have success in the study of surface phenomena.

7. 1 ULTRAFAST X-RAY DIFFRACTION

X-ray techniques, on the other hand, offer the advantage of high penetration depth in matter. Therefore, x-ray techniques using diffraction or absorption offer the advantage to study systems in bulk media, e.g., biological molecules in physiological solutions. Sources of ultrashort X-ray pulses are of two types: a) laser generated plasmas on a metal target, which deliver fsec X-ray pulses at characteristic X-ray emission lines of the target. They are therefore of limited tenability; b) Synchrotrons, which are the most powerful sources of X-rays, with extended tunability, but they deliver 20-100 psec long pulses. In the following, we will show recent examples of applications of these sources in an ultrafast diffraction experiment and an ultrafast x-ray absorption one.
Ultrafast time-resolved diffraction has been used only since the past five to ten years [41,42]. Most studies have dealt with issues of material science, such as the dynamics of acoustic phonons, heating, non-thermal behaviour of materials at or near the melting point, and phase transitions. Here we show a recent example from the group of von der Linde [42]. Coherent lattice vibrations of Bismuth have been probed in a pump-probe configuration, using 4.5 keV femtosecond x-ray pulses from a laser-driven plasma source (see fig. 18). They determined oscillation periods of ~400 fs near the Lindemann stability limit, which corresponds to bond distances, where the long-range order is lost, and melting most likely occurs.

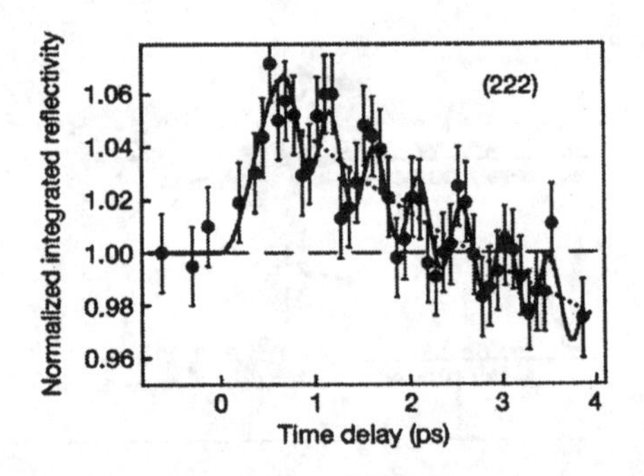

Figure 18: X-ray diffraction efficiency of the (222) reflection as a function of time delay between the optical pump pulse and the X-ray probe pulse. The decrease of the mean value of the X-ray signal (dotted line) is explained by the Debye-Waller effect, and reflects the increasing random component of the atomic motion. After ref. 42.

Aside from allowing an observation of structures in motion, ultrafast x-ray diffraction is interesting in order to extend the range of probed modes in the Brillouin zone, as compared to optical techniques Indeed, any scattering event in a solid (Raman, Brillouin, etc...) satisfies the wave vector conservation given by eq. 9. In the optical domain, the incident and scattered wave vectors lie in the 10^5 cm^{-1} Therefore, the phonon wave vector q, will be of the same order of magnitude. In the X-ray domain, the k's are in the 10^8 cm^{-1} range, so that one samples a larger part of the Brillouin zone. Thus, ultrafast X-ray

diffraction will be very useful for the study of molecular solids, where high frequency modes are common.
Most of the studies so far carried out with ultrafast X-ray diffraction have dealt with solid materials (metals, semi-conductors, or insulators), in which a bulk effect (e.g., coherent phonons, melting, etc.) is induced by the laser and subsequently probed with x-rays [41, 42]. Therefore, nearly all atoms in the interrogated volume participate in the scattering process. For chemistry and biology, one needs structural information on the atomic-scale of time and space during the response of molecular entities to light. Few studies have dealt with chemical or biological systems, and while most of them dealt with bulk samples (e.g., organic solids or protein crystals), none of them concerns the subpicosecond time domain.
In order to obtain full structural information, the largest number of Bragg reflections (see fig. 17) need to be monitored. Upon onset of disorder the spots in the diffractogramme will experience a decrease in intensity. In the above-cited example (fig. 18), only one Bragg reflection was used at a time, which is not sufficient to retrieve a global structure. Two recent exceptions, however, demonstrate that this limitation may be overcome, at least for bulk samples: One is the powder diffraction study of the DMABN molecular crystal at a time resolution of 100 ps, where several Bragg reflections were simultaneously recorded and geometries of different short-lived transients could be extracted [43]. The other is a sequence of 6 snapshots of the Laue pattern of photoexcited MbCO ranging from the subnanosecond to microsecond time scales, showing the movement of photodetached CO from the heme center [44].
Since most of natural and preparative chemistry, and of biology take place in liquid media, a technique that allows visualization of the structural dynamics of dilute species in disordered media is highly needed. Time-resolved X-ray absorption offers such a possibility. A complete review of its capabilities is given in ref. 45. Here we only mention the most important points and a result from our work.

7.2 ULTRAFAST X-RAY ABSORPTION SPECTROSCOPY

X-ray absorption spectroscopy measures the absorption of x-rays as a function of incident x-ray energy E. In such a plot, the x-ray absorption coefficient shows the presence of saw-tooth like features with a sharp ise at discrete energies, called absorption edges. The energy positions of these features are unique to a given absorbing atom. They occur near the ionization energy of inner-shell electrons and contain spectral features due to core-to-unoccupied valence orbital transitions and core-to-continuum transitions. The nomenclature for x-ray absorption features reflects the core orbital, from which the absorption originates. For example, K edges refer to transitions from the innermost $n = 1$ electron orbital, L edges refer to the $n = 2$ absorbing electrons (L_{I} to $2s$, L_{II} to $2p_{1/2}$ and L_{III} to $2p_{3/2}$ orbitals), and M, N, etc., to the corresponding higher lying bound core shells. The transitions are always referred to unoccupied states, i.e., to states with a photoelectron above the Fermi energy (E_F), leaving behind a core hole, and absorption features may appear just below the edge, which correspond to transitions to bound unoccupied levels just below the ionization limit.
Zooming into of these edges one observes a number of fine structures as can be seen in fig. 19 for the case of an iron compound. The region just below, at and just above the edge is called the X-ray Near-Edge Absorption or XANES region. In the region tens to hundreds of eV's above the edge, one finds the so-called the region of Extended X-ray Absorption Fine Structure or EXAFS.

The XANES region contains information both about the electronic and the molecular structure. Indeed, it is sensitive to the oxidation state of the atom of interest. If an electron is removed from it, the ionization potential shifts to higher energies, as more work is needed in order to extract another electron (fig. 19). If an electron is added, the opposite occurs. Thus one can obtain information about the oxidation state of a given atom. This is very important in coordination chemistry and in biology, where biologically active centres undergo changes in the oxidation states, as shown below.
Above the ionization limit the excited electron is often referred to as a photoelectron, and in a solid, depending on its kinetic energy, it can propagate more or less freely through the material. This occurs even in insulators since the excited states are almost always delocalized states (quasi-free states in

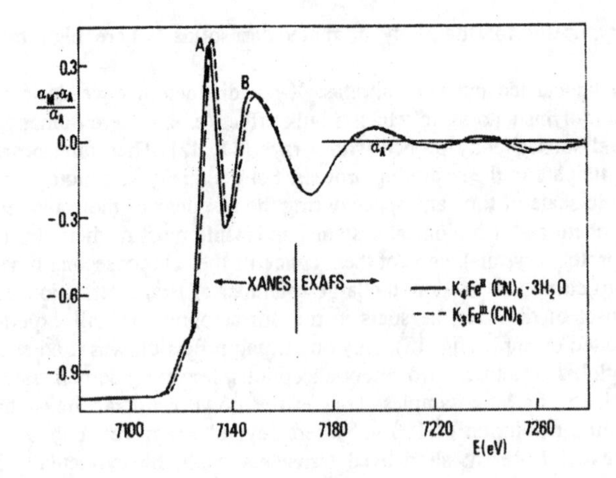

Figure 19: K-edge x-ray absorption spectra of iron in $K_4Fe^{II}(CN)_6$ and $K_3Fe^{III}(CN)_6$ bulk samples. The relative absorption with respect to the high energy background is plotted. After ref. 46.

molecules and conduction band states in solids). Photoelectrons propagate through the solid as a spherical wave with a kinetic energy E_{kin} given by

$$E_{kin} = h\nu - E_B \tag{11}$$

where $h\nu$ is the incident x-ray energy and E_B the binding energy (which may be close to the Fermi energy). The photoelectron wave vector is then defined as

$$k = \frac{2\pi}{\hbar} \cdot \sqrt{2m(h\nu - E_B)} \tag{12}$$

In the EXAFS region, the oscillatory structure is due to the interference between the outgoing photoelectron wave and the wave scattered back at neighboring atoms. It therefore does not exist in the case of the isolated atom. At high photoelectron kinetic energies, the scattering of electrons is such that the only significant contributions to the final state wave function in the vicinity of the absorbing atom comes from paths, in which the electron is scattered only once (single scattering events). The photoelectrons emitted from the excited atom as spherical waves damp out rapidly due to inelastic effects caused by the extended valence orbitals of the nearby-lying atoms. This limits the probed spatial region and, ensures that multiple-scattering effects beyond simple back-scattering can be ignored. The fact that multiple-scattering events can be neglected allows the analysis of the data by a simple Fourier transformation. The absorption coefficient (for the oscillatory part) is written as:

$$\chi(k) = \sum_j S_0^2 N_j \frac{|f_j(k)|}{kR_j^{\,2}} \sin[2kR_j + 2\delta_e + \Phi] \cdot e^{-\frac{2R_j}{\lambda(k)}} \cdot e^{-2\sigma_j^2 k^2} \tag{13}$$

which is the standard EXAFS formula. The structural parameters (for which the subscript j refers to the group of N_j atoms with identical properties, e.g., bond distance and chemical species) are:

a) the interatomic distances R_j,

b) the coordination number (or number of equivalent scatterers) N_j,

c) the temperature-dependent *rms* fluctuation in bond length σ_j, which should also include effects due to structural disorder.

In addition, $f_j(k)=|f_j(k)|e^{i\phi(k)}$ is the backscattering amplitude, δ_e is central-atom partial-wave phase shift of the final state, and $\lambda(k)$ is the energy-dependent photoelectron mean free path (not to be confused with its de Broglie wavelength), and S_0^2 is the overall amplitude factor. Moreover, although the original EXAFS formula referred only to single-scattering contributions from neighboring shells of atoms, the same formula can be generalized to represent the contribution from N_R equivalent multiple-scattering contributions of path length $2R$.

As an example of ultrafast X-ray absorption spectrum, we show the results we obtained on a coordination chemistry compound [47,48]. Aqueous $[Ru^{II}(bpy)_3]^{2+}$ is a model system for Metal-to-Ligand Charge Transfer (MLCT) reactions. Light absorption by $[Ru^{II}(bpy)_3]^{2+}$ results in the formation of a Franck-Condon MLCT singlet excited state, ^{1}(MLCT), which undergoes sub-picosecond intersystem crossing to a long-lived triplet excited state, ^{3}(MLCT), with near-unity quantum yield. The oxidation states of (ground state) Ru^{II} and (photoexcited) Ru^{III} complexes exhibit pronounced differences in their *L*-edge XANES, due to the oxidation shift to higher energies. Removal of the weakest bound electron from the fully occupied $4d_{5/2}(t_{2g})$ level generates a trivalent ruthenium compound, opening up an additional absorption due to the allowed $2p_{3/2} \rightarrow 4d_{5/2}\,(t_{2g})$ (feature *A'*) in addition to an oxidation state induced shift (from B to B'). Thus we observe the appearance of the *A'*/*B'* doublet structure in the trivalent L_{III} XANES, together with the energetic $B \rightarrow B'$ and $C \rightarrow C'$ shift (Fig. 20a), resulting from the change of oxidation state of the Ru central atom and formation of ^{3}(MLCT). Below the Ru L_{III} edge in fig. 20 we also observe the chlorine *K* edge absorption due to the Cl^- counterions of the dissolved $[Ru^{II}(bpy)_3]Cl_2$ sample, which does not contribute to the time-resolved experiment.

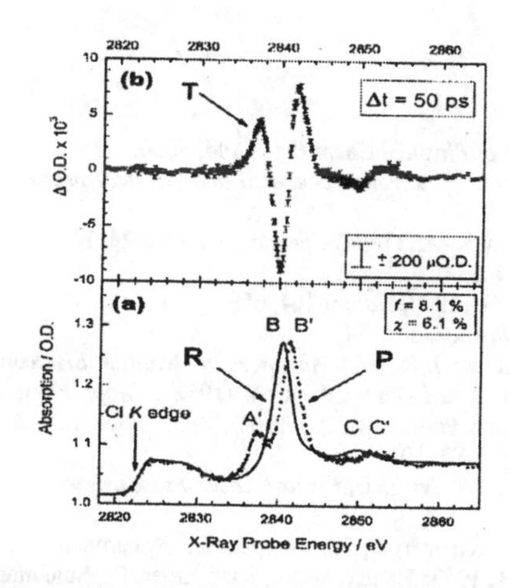

Figure 20: a) Static L_{III}-edge X-ray absorption spectrum of ground state $[Ru^{II}(bpy)_3]$ (trace R) and excited state absorption spectrum (trace P) generated from the transient data cuuve T in b). b) Transient difference X-ray absorption spectrum between the laser excited and unexcited samples at a time delay of 50 ps after the pump laser. After ref. 48.

A crucial aspect of ultrafast pump-probe spectroscopies is to have the ability to scan the time between pump and probe pulses. This was successfully done for the first time between an optical pump pulse and an x-ray probe pulse in our experiment [47,48].

The above example concerns the first evidence for optical pump- X-ray probe spectroscopy in the picosecond time domain. Although it does not concern a structural study *per se*, as it focuses on electronic structure changes, it sets the stage for future ultrafast experiment exploiting the information content of the EXAFS region (see above and fig. 19 and eq. 13).

The development of femtosecond spectroscopy and the ability it has opened to be able to observe transient structures in real-time has revealed details of the photoinduced dynamics in molecules, liquids, solids and biological systems, that were hitherto unthought of. With the advent of ultrafast structural techniques such as electron diffraction, X-ray diffraction and X-ray absorption spectroscopy, we are at the dawn of a new era, where the observation of transient structures will become more direct, and deliver new information and insight.
Going beyond structural dynamics, one area, which we feel has a huge potential, especially with the advent of new and intense source of ultrashort X-rays [45], is the field of non-linear x-ray phenomena, which are the analogue to non-linear optical phenomena. Optical spectroscopy is based on correlation functions of dipoles, x-ray spectroscopy and x-ray scattering are based on the correlation functions of charge densities and of currents, respectively. Extending the formalisms and techniques of non-linear optical methods to the x-ray domain [11], as is being proposed by Mukamel and his co-workers [49] opens a whole new area of Science, where one could look at electron-electron correlations and the flow of charges in a wide class of systems.

ACKNOWLEDGMENTS

I would like to thank the Swiss NSF for its support of our research mentioned in this review, via contracts FN 20-53811.98, 2000-061897.00, 2000-059146.99 and 620-66145.01.

REFERENCES

1. Zewail , A. H. (2000) *Journal of Physical Chemistry A* **104**, 5660.
2. Chergui M. (1996) *Femtochemistry ultrafast chemical and physical processes in molecular systems,* World Scientific: Singapore.
3. Dhar, L. , Rogers, J. A., Nelson, K. A., (1994) *Chemical Reviews* **94**, 157
4. Chergui, M. (2000) *Chimia* **54**, 83
5. Norrish, R. G. W. and Porter, G. (1949) *Nature* **164**, 658
6. Eigen, M. (1954) *Disc. Faraday Soc.* **17**, 194
7. Hellwarth R. W. (1961) in Singer J. R. (ed.), *Advances in Quantum Electronics*, Columbia University Press, New York, p. 334; McClung F. J. and Hellwarth, R. W. (1962) *J. Appl. Phys.* **33**, 828
8. Hargrove L. E., Fork R. L. and Pollack M. A. (1964), *Appl. Phys. Lett.* **5,** 4; McDuff O. P., Harris S. E. (1967), *IEEE J. Quant. Electr.* **QE3,** 101.
9. Shank C. V. (1988) in Kaiser W. (ed.), *Ultrashort Laser Pulses,* Springer-Verlag, Berlin, p.5; *ibid* (1986) Science **233**, 1276
10. Drescher, M., Hentschel, M., Kienberger, R., Tempea, G., Spielmann, C., Reider, G. A., Corkum, P. B.; Krausz, F. (2001) *Science* **291,** 1923; Hentschel, M., Kienberger, R., Spielmann, C., Reider, G. A., Milosevic, N., Brabec, T., Corkum, P., Heinzmann, U., Drescher, M., Krausz, F. (2001) *Nature* **414,** 509.
11. Mukamel, S. (1995) *Principles of nonlinear optical spectroscopy*; Oxford University Press: New York.
12. Rose, T. S., Rosker, M. J. and Zewail, A. H. (1988) **88**, 6672
13. Sundström, V. (1997) Ed. *Femtochemistry and femtobiology ultrafast reaction dynamics at atomic-scale resolution* Imperial College Press: London.; Gaspard, P.(1997) *Chemical reactions and their control on the femtosecond time scale* Wiley, New York.; Douhal, A. (2002) Ed. *Femtochemistry and femtobiology ultrafast dynamics in molecular science*; World Scientific, Singapore.
14. A. Parker *et al* (1986) *Phys. Rev. Letters* **56,** 716; ten Wolde D. *et al*(1988), *Phys. Rev. Letters* 61, 2099 Yeazell A. *et al* (1990), *Phys. Rev. Letters* **64,** 2007.
15. Molander M. *et al* (1986), *J. Phys.* **B19,** L461; Yeazell A. *et al*(1988), *Phys. Rev. Lett.* **60,** 1494; Gaeta B. *et al* (1994), *Phys. Rev. Lett.* **73,** 636.
16. Albert, C. and Zoller, P. (1991) *Physics Reports* **5,** 231.
17. Garraway, B. M.; Suominen, K. A. (1995) *Reports on Progress in Physics* **58**, 365
18. Uzer, T., Ed., (2000) *The Physics and Chemistry of Wavepackets,* John Wiley and Sons Ltd, New York.

19. Wiederrecht, G. P., Dougherty, T. P., Dhar, L., Nelson, K. A., Laird D. E. and Weiner A. M. (1995), *Phys. Rev.* **B51**, 916

20. de Silvestri, S., Fujimoto, J. G., Ippen, E. P., Gamble Jr., E. B., Williams L. R. and Nelson, K. A., (1986) *Chem. Phys. Lett.* **116**, 146

21. Zeiger, H. J., Vidal, J., Cheng, T. K., Ippen, E. P., Dresselhaus G. and Dresselhaus, M. S., *Phys.. Rev.* (1993) **B 45**, 768

22. Mazur E. (1997) in Di Bartolo, B. (Ed.) *Spectroscopy and Dynamics of Collective Excitations in Solids* NATO ASI Series B; Vol. 356, Plenum Press, New York, p. 417

23. Li, Z., Zadoyan, R., Apkarian V.A. and Martens, C.C., (1993) *J. Phys. Chem.* **99,** 7453

24. Apkarian, V. A.; Schwentner, N. (1999) *Chem.Rev.* **99,** 1481 and refs therein

25. Jungwirth, P., Gerber, R. B., (1999) *Chem. Rev.* **99,** 1583

26. Zadoyan, R., .Li, Z., Ashjian, P., Martens C. C. and Apkarian, V. A. (1994) *Chem. Phys. Letters* **218**, 504 ; Zadoyan, R., Li, Z., Martens, C.C. and Apkarian, V.A., (1994) *J. Chem. Phys.* **101,** 6648;

27. Gonzalez, C. R., Fernandez-Alberti, S., Echave J. and Chergui M. (2002) *J. Chem. Phys.* **116,** 3343

28. Borrmann, A.and Martens, C. C., (1995) *J. Chem. Phys.* **102**, 1905; Martens, C. C. (1996) in Chergui M. (Ed.) *Femtochemistry, Ultrafast Chemical and Physical Processes in Molecular Systems*, World Scientific, Singapore.

29. Helbing, J., Chergui, M. and Haydar A. (2000) *J. Chem. Phys.* **113,** 3621.

30. Scherer, N., Jonas, D. M., Fleming, G. R. (1993) *J. Chem. Phys.* **99**, 153

31. Liu, Q., Wan, C. and Zewail, A. H. (1996) *J. Phys. Chem.* **100**, 18666

32. Pugliano, N., Szarka A. Z., Gnanakaran, S., Triechel, M. and Hochstrasser, R. M. (1995) *J. Chem. Phys.* **103,** 6498; Pugliano, N., Szarka A. Z., and Hochstrasser, R. M. (1996) *J. Chem. Phys.* **104,** 5062.

33. Banin U. and Ruhman, S. (1993) *J. Chem.Phys.* **98,** 4391 and (1993) **99,** 9318

34. M. Chergui (2001) *Comptes Rendus de l'Academie des Sciences, Serie IV: Physique Astrophysique* **2** 1453

35. Gross M. and Spiegelmann F. (1998) *Europhys. J.* **D4,** 219

36. Vigliotti, F., Bonacina, L. and Chergui, M. (2002) *J. Chem. Phys.* **116,** 4553

37. Jeannin, C. , Portella-Oberli, M. T., Jimenez, S., Vigliotti, F., Lang B. and Chergui M. (2000) *Chem. Phys. Lett.* **316,** 51

38. Vigliotti, F., Bonacina, L. and Chergui, M. (2003) *Phys. Rev.* **B67,** 115118-1

9. Rosca, F., Kumar, A. T. N. , Ye, X., Sjodin, T. , Deminov A. A. and Champion P. M., (2000) *J. Phys. Chem.* **A104,** 4280

0. Srinivasan, R.; Lobastov, V. A.; Ruan, C. Y.; Zewail, A. H. (2003) *Helvetica Chimica Acta* **86,** 1763, and refs therein

1. Rischel, C., Rousse, A., Uschmann, I., Albouy, P. A., Geindre, J. P., Audebert, P., Gauthier, J. C., Forster, E., Martin, J. L., Antonetti, A. (1997) *Nature* **390,** 490; Rose-Petruck, C., Jimenez, R., Guo, T., Cavalleri, A., Siders, C. W., Raksi, F., Squier, J. A., Walker, B. C., Wilson, K. R., Barty, C. P. J. (1999) *Nature* **398**, 310; Lindenberg, A. M.; Kang, I.; Johnson, S. L., Missalla, T., Heimann, P. A., Chang, Z., Larsson, J., Bucksbaum, P. H.; Kapteyn, H. C., Padmore, H. A.; Lee, R. W., Wark, J. S., Falcone, R. W. (2000) *Phys.Rev.Lett.*, **84,** 111; Cavalleri, A., Toth, C., Siders, C. W., Squier, J. A., Raksi, F., Forget, P., Kieffer, J. C. (2001) *Phys. Rev. Lett.*, **87** art. no.-237401.

2. Sokolowski-Tinten, K., Blome, C., Blums, J., Cavalleri, A., Dietrich, C., Tarasevitch, A., Uschmann, I., Forster, E., Kammler, M., Horn-von-Hoegen, M., von der Linde, D. (2003) *Nature*, **422,** 287

3. Techert, S., Schotte, F., Wulff, M. (2001) *Phys. Rev. Lett.*, **86,** 2030

4. Schotte, F., Lim, M. H., Jackson, T. A., Smirnov, A. V., Soman, J., Olson, J. S., Phillips, G. N., Wulff, M., Anfinrud, P. A. (2003) *Science* **300,** 1944

5. Bressler Ch.and Chergui, M. *Chem. Rev.* (in press)

6. Durham, P. J. (1988) in Prins, R. and Koningsberger, D. C. (Eds) *X-ray absorption principles, applications, techniques of exafs, sexafs and xanes* Wiley: New Cork.

7. Saes, M., Bressler, C., Abela, R., Grolimund, D., Johnson, S. L., Heimann, P. A., Chergui, M. (2003) *Phys. Rev. Lett.*, **90,** art. no.-047403.

. Saes, M. G., W., Kaiser, M., Tarnovsky, A., Bressler, Ch., Chergui, M., Johnson, S. L., Grolimund, D., Abela, R. (2003) *Synchrotron Radiation News*, **16,** 12

. Tanaka, S., Mukamel, S. (2003) *Phys. Rev. A*, **67,** art. no.-033818 and refs therein.

14. LANTHANIDE SERIES SPECTROSCOPY UNDER EXTREME CONDITION

N.P. BARNES
NASA Langley Research Center
Hampton, VA 23681

Abstract

Lasers evolved based on spectroscopic work that preceded them. In turn advanced spectroscopic techniques evolved because of the availability of lasers. Lasers can serve as both pump sources and spectroscopic probe beams while improvements in crystal growth needed for emerging laser technology enable high concentrations of lanthanide series elements. These technology advances can produce extreme effects in laser materials under consideration. Extreme effects include absorption saturation, amplified spontaneous emission, energy transfer, self-quenching, up conversion, and excited state absorption. These effects will be described along with methods of analyzing the data.

1. Introduction

Historically, spectroscopy was performed using incandescent lamps or gaseous discharge lamps on low concentration samples. Although many advances were made using lamps, typically these sources are not very bright. Here brightness is defined as power per unit area and unit solid angle. With a low brightness source, high levels of excitation are very difficult to achieve. Also, spectroscopic samples often contain only a limited amount of lanthanide series atoms. Low concentrations may be a result of crystallographic limitations. For example, the lanthanide series atom should possess the same valance and atomic size as the atom in the crystal that it is replacing. If the former is not true, charge compensation may be required to maintain charge neutrality. If the latter is not true, lattice distortions may result. In either case, all of the lanthanide series atoms may not enjoy the same environment. Different environments will randomly shift spectroscopic features, broaden the features, and complicate the analysis, in general.

Given a sample of a lanthanide series atom in some crystal, the transmission and emission spectra are measured. Typically transmission spectra are measured first so that excitation bands can be identified for the laser material. These spectra may be a function of the polarization. Another measurable parameter is the temporal history of the fluorescence emitted by a particular manifold. Usually the particular manifold under investigation is excited by a short pulse. The fluorescence is recorded as a function of time. Results are often fit to an exponential decay curve.

To be useful for the laser designer, basic spectroscopic data are usually analyzed to obtain more useful parameters. Transmission spectra are corrected for Fresnel losses and then transformed into an absorption coefficient as a function of wavelength. Emission spectra as a function of wavelength are analyzed to produce an emission cross section. Usually, the lifetime is needed to

B. Di Bartolo and O. Forte (eds.), *Frontiers of Optical Spectroscopy*, 521-538.

calibrate the relative emission. Often fluorescence as a function of time can be fit to an exponential function to produce a value for the lifetime. However, in many instances, a simple exponential function is not good fit to the data. More complex temporal histories may indicate processes other than radiative decay. These complex temporal histories often result from extreme conditions.

Extreme spectroscopy can result because of high excitation densities or high concentrations of active atoms or both. High excitation densities can lead to decay curves that deviate from the familiar exponential and nonlinearities in the fluorescence versus the level of excitation. High concentrations can lead to energy transfer effects. That is, energy is exchanged between 2 physically separated atoms without the emission and absorption of photons. There are several distinctive effects that occur under extreme conditions.

Extreme conditions include absorption saturation, amplified spontaneous emission, general energy transfer, self quenching, up conversion, and excited state absorption. With absorption saturation, the spectroscopic probe beam is so intense that a non negligible fraction of the ground state population density is partially depleted. This results in a transmission greater than predicted from the small signal absorption coefficient. Amplified spontaneous emission results when the gain is high enough that a spontaneously emitted photon enjoys a significant gain before it escapes from the excited volume. Energy transfer is the exchange of energy between 2 closely spaced atoms without the need to resort to the emission and subsequent absorption of a photon. Self quenching is an energy transfer process whereby an excited lanthanide series atom shares energy with a similar lanthanide atom. This is usually a deleterious process because it depletes the upper laser manifold population density. Up conversion is often a deleterious process as well because it depletes the upper laser manifold population density. However it accomplishes this by promoting the excited atom to a yet higher manifold where energy is often lost by nonradiative transitions.

2. Absorption Saturation

Absorption saturation can occur if an intense laser pump beam is used. Absorption spectra for lanthanide series atoms usually contain many relatively narrow spectral features. If the wavelength of the pump beam coincides with the peak of the absorption feature, absorption of the pump can be quite strong. Moreover, the laser beam can be confined to a relatively small volume thereby limiting the number of lanthanide series atoms that are illuminated. With an intense pump beam and strong absorption, the excitation can be so high that the number of atoms in the ground manifold is significantly depleted. As depletion increases, transmission of the pump beam increases.

Absorption saturation can be described using an approach originally derived for laser amplifiers. The basic amplifier

$$\frac{dN_2}{dt} = -\frac{dN_1}{dt} = -\frac{N_2}{\tau_2} + \frac{\sigma c}{n}[Z_2 N_2 - Z_1 N_1]N_p \tag{1}$$

$$\frac{dN_p}{dt} + \frac{c}{n}\frac{dN_p}{dz} = \frac{\sigma c}{n}[Z_2 N_2 - Z_1 N_1]N_p \tag{2}$$

where N_1, N_2, and N_p are the population densities of the lower and upper laser manifold and the photon densities, respectively, c is the speed of light, n is the refractive index, σ is the cross section, while Z_1 and Z_2 are Boltzman factors describing the fraction of the manifold population that resides in the lower and upper laser levels. Typically, effective absorption and emission cross sections are used. In this case,

$$\sigma_a = Z_1\sigma. \tag{3}$$

For many lanthanide laser system, essentially all of the active atom population resides either in the upper laser manifold or the ground manifold. If this is approximately true,

$$C_H N_S = N_1 + N_2 \tag{4}$$

Combining these equations,

$$\sigma[Z_2N_2 - Z_1N_1] = \sigma_a\left[\frac{Z_2C_HN_S}{Z_1} - (1+\frac{Z_2}{Z_1})N_1\right] \tag{5}$$

These equations can be solved to yield the transmission of the pump beam through the doped material.

The population density and the transmission through the sample can be obtained by solving the differential equations to yield

$$N = N_0\frac{\exp(\sigma_a N_0 1)}{\left[\exp(\sigma_a N_0 1) - 1 + \exp((\frac{\sigma_a c\gamma_a}{n}) \int_{-\infty}^{t-\frac{nz}{c}} N_p(\tau)d\tau\right]} \tag{6}$$

where the population difference is defined as

$$N = (1+\frac{Z_{2a}}{Z_{1a}})N_1 - (\frac{Z_{2a}}{Z_{1a}})C_AN_S \tag{7}$$

This can be used to determine the transmission of the pump beam through the absorbing crystal. This has been done for 2 different wavelengths in Ho:YAG and the results shown in Figure 1. If the wavelength of the pump beam corresponds to a weak
absorption feature, the transmission is relatively high but does not change significantly with the pump power. If, on the other hand, the wavelength of the pump beam corresponds to a strong absorption feature, the initial transmission is low. However, as the pump beam becomes more powerful, the Ho

atoms available to absorb the pump becomes depleted. This saturation effect allows more transmission. However, saturation may require a significant amount of energy.

3. Amplified Spontaneous Emission

If the gain of the medium is high enough, spontaneously emitted photons will enjoy gain before they escape the pumped volume. Typically lanthanide series lasers are fabricated into long rods, often circular in cross section. These laser rods are pumped so the gain along the length of the laser rod is on the order of exp(3) or about 20. If an atom near the end of the laser rod emits a photon spontaneous in the direction of the other end of the laser rod, it can cause 20 other atoms to also emit a photon. An increase in emitted photons can significantly can increase the rate of loss of active atoms in the excited state and thus decrease the lifetime. Because of total internal reflection, the range of emission angles that enjoy gain along the entire laser rod length can be a significant fraction of 4π.

If there is no amplified spontaneous emission, the rate equations describing the population density can be expressed as

$$\frac{dN_2}{dt} = R_2 - \frac{N_2}{\tau_2} \qquad (8)$$

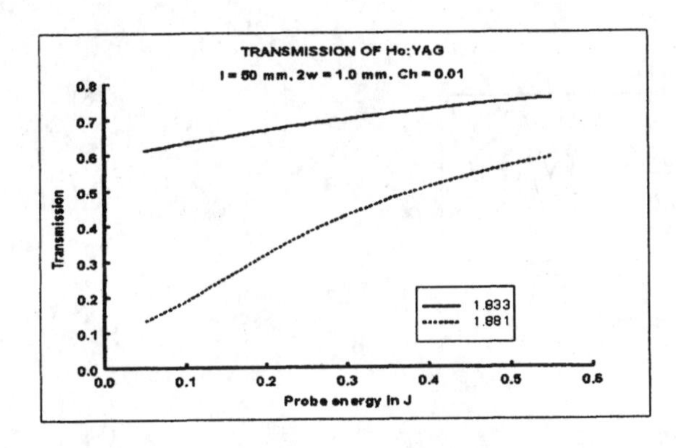

Figure 1. Transmission of Ho:YAG versus probe energy

After the cessation of pumping, R_2, the usual exponential decay of the upper laser manifold population, N_2, with lifetime τ_2 occurs. If amplified spontaneous emission is significant, the differential equation that describes the upper laser manifold population density becomes

$$\frac{dN_2}{dt} = R_2 - \iiint \frac{N_2}{\tau_2} \exp(l\sigma_2 N_2) dV d\Omega d\lambda \tag{9}$$

The integration is over the entire pumped volume, all possible angles of emission, Ω, and emission wavelengths, λ. In this expression, the emission cross section, σ_e, is highly dependent on the wavelength. It has been shown that the integration over wavelength can be approximately independent of the integration over the volume and emission angles [1].

If amplified spontaneous emission is not too high, the 6 fold integration can be approximated as

$$N_2 = N_{20} \frac{\exp(-\frac{t}{\tau_2})}{\left[1 + \sigma_{ea} l_a N_{20} (1 - \exp(-\frac{t}{\tau_2}))\right]} \tag{10}$$

In this expression, σ_{ea} is the average emission cross section and l_a is the average path length that a spontaneously emitted photon travels in the pumped volume. The former can be calculated from spectroscopic data while the latter is often determined from experimental data [1].

Typical fluorescent data as a function of time displays a faster than exponential decay in the influence of amplified spontaneous emission. Data for the fluorescence of Nd:YAG is shown in the Figure 2. Initially the rapid decay can be noted. At longer times, when the gain is lower, the decay becomes very nearly exponential with the characteristic lifetime. A fast initial decay erodes the storage efficiency of high gain lasers.

4. Energy Transfer

Energy transfer, as defined here, is the exchange of a quantum of energy between 2 physically separated atoms. Perhaps the classic case of energy transfer is the Ho:Tm system. A Tm atom residing in the first excited manifold, $^3F_4$, transitions to the ground manifold, $^3H_6$, causing a nearby Ho atom to transition from the ground manifold, $^5I_8$, to the first excited manifold, $^5I_7$. This occurs without the emission and subsequent absorption of a photon. For historical reasons, the atom donating the quantum of energy is labeled as the sensitizer while the atom receiving the quantum of energy is labeled the active atom.

The description of energy transfer was derived in 2 seminal papers 1 by Forster [2] and another by Dexter [3]. The energy transfer parameter, P_{SA}, is given by

$$P_{SA} = \frac{(2\pi)^2}{g_S g_a h} \sum_i \sum_f \iiint \frac{q_e^4}{\kappa R^6} \rho_a(w_a)\rho_s(w_s)$$

$$\left| <\vec{r}_a> \bullet <\vec{r}_s> -3(<\vec{r}_s> \bullet \vec{R})(<\vec{r}_a> \bullet \vec{R}) \right|^2 dE dw_a dw_s \qquad (11)$$

$$= \frac{3h^4c^4Q_a}{64\pi^5 n^4 \tau_s} \int \frac{f_s(E)F_a(E)}{E^4} dE$$

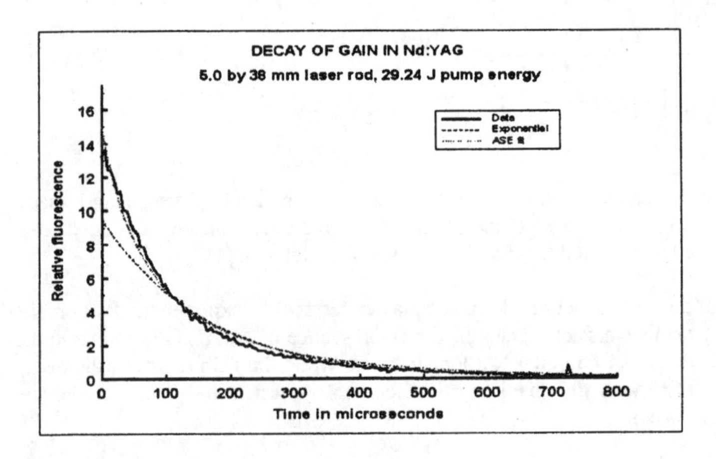

Figure 2. Decay of gain in Nd:YAG versus time

In the above expressions, h is Planck's constant, q_e is the electronic charge, R is the separation between the interacting atoms, $\mathbf{r_a}$ and $\mathbf{r_s}$ are the dipole moments of the interacting atoms, **R** is the vector that connects the interacting atoms, E is the energy, and $\rho_i(w_i)$ is a probability distribution function. In the second expression, Q_a is the normalization of the absorption spectrum of the active atom and τ_s results from the normalization of the emission spectrum of the sensitizer. The second equation follows from the first after some mathematical manipulation.

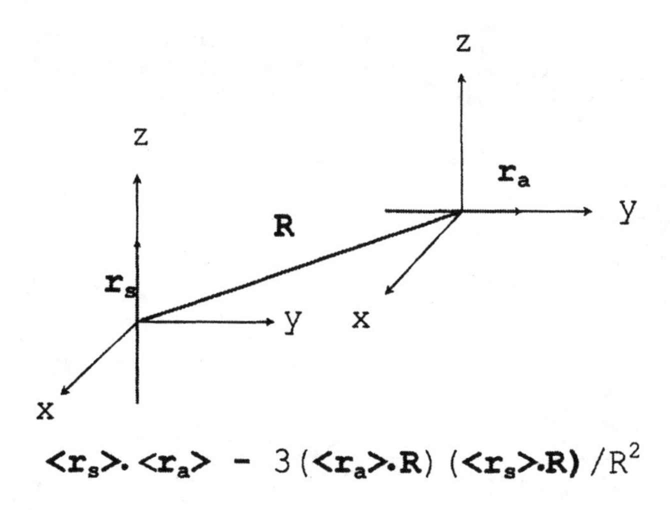

Figure 3. Orientation of dipoles for energy transfer

An examination of the equation describing energy transfer reveals the physics. The strength of the interaction decreases as the separation of the atoms to the inverse 6th power. This is because the energy transfer is usually a dipole dipole interaction and the dipole field decreases as the inverse cube of the distance. The interaction depends of the orientation of the individual dipole moments and the orientation of the dipoles with respect to each other, Figure 3. This dependence appears in the dot product of the vectors. Also, the energy transfer coefficient maximizes when the normalized emission spectrum of the sensitizer, $f_s(E)$, and the normalized absorption spectrum of the active atom, $F_a(E)$, overlap. In this case, there is a resonance between absorption and emission and the integral over the product becomes large.

The effect of the absorption and emission spectra can be seen for the Ho:Tm energy transfer situation. Every energy transfer process has a reverse energy transfer process although they may have substantially different magnitudes. The emission spectrum of the Tm $^3F_4$ to $^3H_6$ transition with the absorption spectrum of the Ho $^5I_8$ to $^5I_7$ transition as well as the emission spectrum of the Ho $^5I_7$ to $^5I_8$ transition with the absorption spectra of the Tm $^3H_6$ to $^3F_4$ transition appear in the Figures 4 and 5. The figures indicate that the former has a much better overlap than the latter. In fact, the calculated and measured
energy transfer parameters reflect this increased energy transfer parameter [4].

The energy transfer parameter depends on the orientation and position of the dipoles that are set by the crystal lattice. The Dexter approach averages over all possible orientations of the dipoles and all possible orientations of the dipoles with respect to each other. This is valid for fluids, to which the energy transfer problem was initially addressed [2]. However, in a crystal material, which most lasers are, the orientations of the dipoles are set by the crystal lattice. In addition, the energy transfer model in fluids needs to determine the minimum separation of the interacting atoms. This is critical because of the inverse 6th power dependence on the separation.

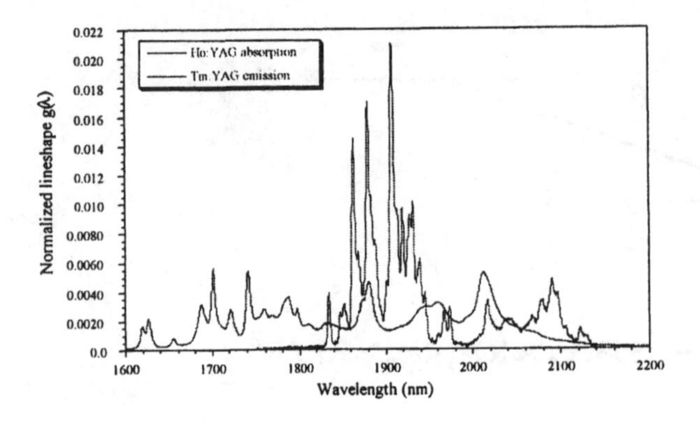

Figure 4. Overlap of Tm:YAG emission and Ho:YAG absorption

The original Forster and Dexter approach was modified to take into account the physics associated with lanthanide series atoms in crystal lattices [5]. Rather than averaging over all possible orientations, the model determines the orientations of the dipoles dictated by the crystal lattice and uses them to calculate the dot products. The separations of the dipoles are also set by the crystal lattice. These separations are used and a summation over all possible locations for the sensitizer and active atoms is then performed. Finally, a quantum mechanical model is used to calculate line to line, rather than manifold to manifold, branching ratios. This allows a summation over all possible initial and final levels for both the sensitizer and active atom to be performed. This summation replaces the need for integrating over the product of the emission and absorption spectra and thus obviates the need to measure these parameters.

Diffusion abets the energy transfer process. Because of the inverse 6th power dependence on the separation, atoms that are close have a far greater probability of achieving a transfer of energy than atoms with greater separation. If the quantum of excitation were frozen in place, atoms at different separations would have different energy transfer rates, complicating the description of the energy transfer process. However, energy transfer can also occur between 2 atoms of the same type and between the same manifolds. For example, a Tm atom in the $^3F_4$ manifold can transition to the $^3H_6$ manifold and simultaneously raise a neighboring Tm atom from the $^3H_6$ manifold to the $^3F_4$ manifold. On the macroscopic scale, nothing appears changed. However, on the microscopic scale, the quantum of excitation has moved or diffused through the laser material. By diffusing through the laser material, the quantum of excitation can spend some time at a Tm atom where the nearest similar neighbor site is occupied by a Ho atom with which energy transfer can occur.

Diffusion is usually orders of magnitude faster than other energy transfer processes. Energy transfer processes are aided by energy resonances. Some transitions have energy resonances in the Tm to Ho energy transfer processes. On the other hand, in the diffusion process, every transition has at least 1 energy resonance. The profusion of resonances greatly enhances the diffusion energy transfer

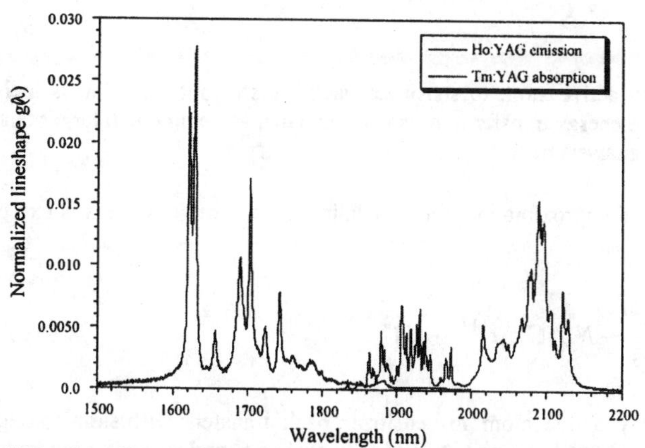

Figure 5. Overlap of Ho:YAG emission and Tm:YAG absorption

processes. In turn, diffusion abets other energy transfer processes by transferring the quantum of excitation to locations where the interacting atoms are in close proximity.

Several models have been developed to describe the energy transfer process. In the static transfer approximation, the population density of sensitizers is expressed as [6]

$$N_S(t) = N_S(0)\exp\frac{[-\frac{t}{\tau} - 4\pi\Gamma(1-\frac{3}{q})N_A(C_{DA}t)^{\frac{3}{q}}]}{3} \quad (12)$$

where N_S is the population of sensitizers, N_A is the population density of active atoms, τ is the natural lifetime, and C_{DA}/R^6 is the energy transfer parameter. In this treatment, sensitizer to sensitizer diffusion is neglected as well as active atom to sensitizer back transfer.

In the diffusion limited approximation, the population density of sensitizers is expressed as [7]

$$N_S(t) = N_S(0)\exp[-\frac{t}{\tau} - \frac{4\pi^{3/2}}{3}N_A(C_{DA}t)^{1/2}F(x)] \quad (13)$$

where

$$F(x) = \left[\frac{(1+10.87x+15.5x^2)}{(1+8.74x)}\right]^{3/4} \quad (14)$$

$$x = DC_{DA}^{-1/3}t^{2/3} \quad (15)$$

This approximation neglects any active atom to sensitizer back transfer but diffusion is included through the x term. However, the energy transfer term is approximated as being much larger than the diffusion term, in contrast to the analysis here.

In the migration assisted approximation, the population density of sensitizers is expressed as [8]

$$N_S(t) = N_S(0)\exp[-\frac{t}{\tau} - \frac{4\pi^{3/2}}{3} N_A (C_{DA}t)^{1/2} - Wt] \quad (16)$$

This approximation neglects any active atom to sensitizer back transfer. Diffusion among the sensitizer atoms is included through the hopping term and is treated as a random walk process that is fast enough to maintain the initial distribution of the excited sensitizers.

In the correlated placement approximation, the population density of the sensitizers is expressed as [9]

$$N_S(t) = N_S(0)\exp[-\frac{t}{\tau} - \frac{4}{3}\pi N_A [X\pi^{1/2}(C_{DA}t)^{1/2}$$

$$+ R_1^3(Y-X)\Phi(z_1)\exp(-z_1) + R_2^3(1-Y)\Phi(z_2)\exp(-z_2)]] \quad (17)$$

$$z_i = \frac{C_{DA}t}{R_i^6} \quad (18)$$

$$\Phi(z_i) = 1 + (\pi z_i)^{1/2}\exp(z_i)\mathrm{erf}(z_i^{1/2}) \quad (19)$$

In this approximation, diffusion among the sensitizers and sensitizer to donor back transfer is neglected. However, a non random distribution of active atoms can occur in a correlated manner. If X is greater than 1, an enhanced probability of sensitizers and active atoms being close exists. If X is less than 1, the probability of sensitizers and active atoms being close is diminished.

Although all of the above approximations can be made to fit experimental data, each with varying degrees of correlation, a somewhat different approach is advocated. The approach assumes that diffusion is sufficiently fast to maintain a nearly uniform distribution of excited sensitizers, an approximation borne out be calculations. It also takes into account both forward and back energy transfer. The differential equations describing the population density of the excited sensitizer, N_2, as well as the population density of the excited active atom, N_7, are [10]

$$\frac{dN_2}{dt} = \frac{N_2}{\tau_2} - P_{28}N_2N_8 + P_{71}N_7N_1 \tag{20}$$

$$\frac{dN_7}{dt} = -\frac{N_7}{\tau_7} - P_{71}N_7N_1 + P_{28}N_2N_8 \tag{21}$$

$$N_1 + N_2 = C_T N_S \tag{22}$$
$$N_8 + N_7 = C_H N_S \tag{23}$$

The first 2 equations describe the natural lifetime as well as the forward and back energy transfer processes. The latter 2 equations imply that the sensitizer and active atoms are either in the ground manifold or the manifold from which the energy transfer will occur.

These 4 equations can be reduced to 2 equations by using the latter 2 equations to eliminate N_1 and N_8, simplifying the differential equations using the light excitation approximation and solving. The results are

$$N_S(t) = N_S(0)\,[\,(\beta/(\alpha+\beta))\exp(-t/\tau) + (\alpha/(\alpha+\beta))\exp(-(\alpha+\beta)t)\,] \tag{24}$$

$$N_A(t) = \alpha N_S(0)\,[\exp(-t/\tau) - \exp(-(\alpha+\beta)t)\,]/(\alpha+\beta) \tag{25}$$

$$\alpha = P_{28}C_H N_S \tag{27}$$

$$\beta = P_{71}C_T N_S \tag{28}$$

$$1/\tau = (\alpha/\tau_7 + \beta/\tau_2)/(\alpha+\beta) \tag{29}$$

The 2 exponential functions define the time constant associated with equilibrium being established between the excited manifolds and the time constant associated with the mutual decay of the excited manifolds. The excited sensitizers initially decay quickly while the excited active atoms increase. After a quasi equilibrium is reached, they both appear to decay at the same rate.

Measurements taken on the fluorescence of a Ho:Tm system supports the sum of exponentials approach [10]. Measurements were obtained by using tunable, Q-switched Co:MgF_2 laser to excite either Ho or Tm to the first excited manifold. Following this, the fluorescence decay of both the Ho $^5I_7$ and $^5I_8$ manifolds was recorded. This information was performed for a variety of Ho and Tm concentrations in several different laser materials. A sample of the data is given in Figures 6 and 7. In this case, the Tm $^3F_4$ manifold was excited. An initial rapid decay of the Tm $^3F_4$ manifold population density is mirrored by an initial rapid rise in the Ho $^5I_7$ manifold population density. After achieving quasi equilibrium, both population densities decay at essentially the same rate, as predicted by the 2 exponential model.

The ratio of the forward and backward energy transfer parameters can be deduced with a knowledge of the energy levels. Consider the case where only a single resonance exists for the energy transfer process. It may be noted that rate equations are written in terms of manifold population densities rather than energy level population densities. Thus the rate of the forward energy transfer parameter, P_{28}, is proportional to the thermal occupation or Boltzmann factors for the interacting levels in manifolds 2 and 8. It is also proportional to the strength of the interaction, represented by a dipole moment. Thus, the forward transfer parameter is proportional to

$$P_{28} \propto \langle \mathbf{r_a} \rangle \langle \mathbf{r_s} \rangle \exp[-(E_2 - E_{2ZL} + E_8)/kT]/Z_{2T}Z_{8T} \qquad (30)$$

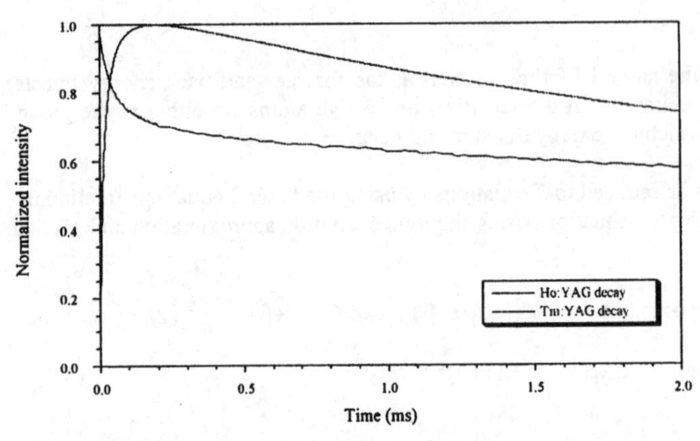

Figure 6. Decay of Tm $^3F_4$ and Ho $^5I_7$ under Tm $^3F_4$ excitation

where E_{2ZL} and E_{7ZL} are the energies of the lowest level in the first excited manifolds of Tm and Ho, k is Bollzmann's constant, T is the temperature, and Z_{2T} and Z_{8T} are partition functions of manifolds 2 and 8, the first excited manifold of Tm and the ground manifold of Ho. Using the same logic, the backward transfer parameter is proportional to

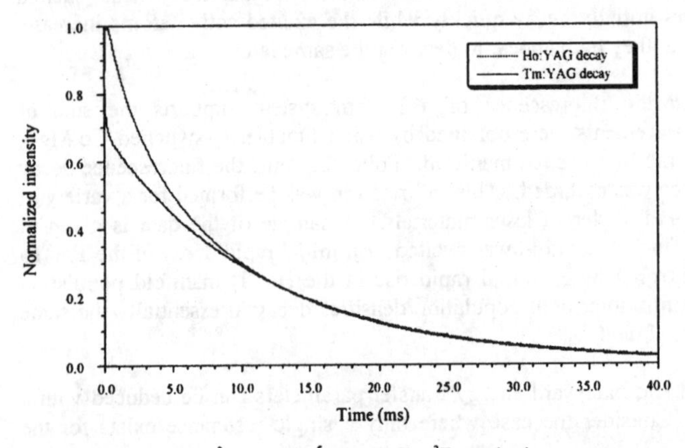

Figure 7. Decay of Tm $^3F_4$ and Ho $^5I_7$ under Tm $^3F_4$ excitation

$$P_{71} \propto \langle r_a \rangle \langle r_s \rangle \exp[-(E_7 - E_{7ZL} + E_1)/kT]/Z_{7T}Z_{1T} \tag{31}$$

By taking the ratio of these energy transfer parameters, the dipole moments cancel because the same transitions are involved in both processes. Because of the resonance, $(E_2 - E_1)$ is equal to $(E_7 - E_8)$. Thus, the ratio of the energy transfer parameters is dependent on the temperature according to

$$P_{28}/P_{71} = [Z_{7T}\, Z_{1T}/\, Z_{2T}\, Z_{8T}]\, \exp[(E_{2ZL} - E_{7ZL})/kT] \tag{32}$$

This implies that the higher energy levels tend to transfer the quanta of energy to lower energy levels much faster than the reverse process.

5. Self Quenching

Self quenching is a special case of generalized energy transfer in which the quanta of energy is exchanged between lanthanide series atoms of the same type. This is likely to be significant if the upper manifold is a metastable manifold. In this case, the self quenching process shortens or quenches the lifetime of the metastable level. Self quenching occurs when a Nd atom in the $^4F_{3/2}$ manifold interacts with a neighboring Nd atom in the ground manifold or the $^4I_{9/2}$ manifold to produce 2 Nd atoms in the $^4I_{15/2}$ manifold.

The differential equations governing the self quenching process are as follows

$$\frac{dN_4}{dt} = -\frac{N_4}{\tau_4} - P_{41}N_4N_1 \tag{33}$$

$$N_1 + N_4 = C_N N_S \tag{34}$$

Although there is a backward process, the Nd $^4I_{15/2}$ manifold has such a short lifetime that the backward energy transfer process seldom occurs. Under this approximation, the solution to this differential equation is approximately [11]

$$N_4 = N_{40}\frac{\exp(-\alpha t)}{[1 - P_{41}N_{40}((1-\exp(-\alpha t))/\alpha]} \tag{35}$$

$$\alpha = (1/\tau_4) + P_{41}C_N N_S \tag{36}$$

The lifetime, τ_4, is therefore shortened by the self quenching process, P_{41} approximately linearly with the concentration, C_N.

A shortening of the Nd lifetime can be observed under high levels of excitation. The fluorescence from the $^4F_{3/2}$ manifold as a function of time appears in Figure 8. An exponential fit to the data also appears. Although the exponential fit to the data is not unreasonable, systematic deviations can be observed. On the other hand, fitting a solution that includes self quenching eliminates almost all of the systematic deviations.

Diffusion also abets the self quenching process. Because the dipole moments between the $^4F_{3/2}$ and the $^4I_{15/2}$ manifolds is small, only Nd atoms is close proximity have a high probability of self quenching. However, diffusion can bring the quantum of excitation to a pair of Nd atoms in close proximity. Because the diffusion process is essentially a random walk process, it requires some time for the quantum of excitation to encounter a pair of Nd atoms. In the interim, the quantum of excitation may decay away by a radiative process. If this were the case, the inverse of the lifetime would depend rather weakly on the Nd concentration until a critical concentration was reached. The critical concentration is defined by the average time interval required for the quantum of excitation to reach a Nd atom pair. After the concentration is well above the critical concentration the inverse lifetime increases linearly with the concentration.

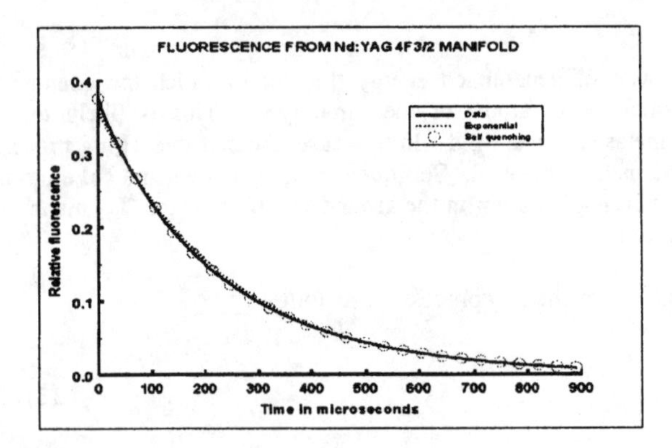

Figure 8. Decay of Nd $^4F_{3/2}$ with curve fits to data

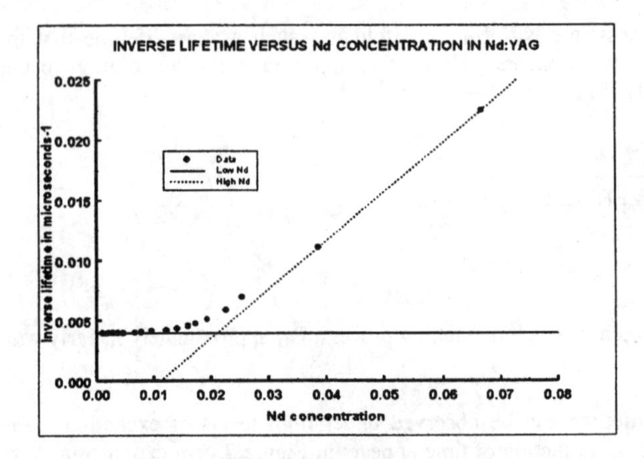

Figure 9. Lifetime of Nd:YAG $^4F_{3/2}$ versus Nd concentration

Data on the inverse lifetime of Nd:YAG supports this model. The inverse lifetime is plotted versus the concentration in the Figure 9. Initially, the inverse of the lifetime is essentially independent of Nd concentration. However, for concentrations above approximately 0.02, the inverse of the lifetime appears to increase approximately linearly with Nd concentration. Both behaviors are consistent with diffusion driven self quenching.

6. Up Conversion

In the up conversion process, 2 atoms both in an excited manifold interact, promoting 1 atom to a higher energy manifold and demoting the other atom to a lower manifold. A familiar example of this involves the Ho:Tm system. A Ho atom in the $^5I_7$ manifold and a Tm atom in the $^3F_4$ manifold interact, process P_{27}. The Ho atom is promoted to the $^5I_5$ manifold while the Tm atom is demoted to the $^3H_6$ manifold. A Ho atom in the $^5I_5$ manifold is likely to decay, usually by a nonradiative process, to the $^5I_6$ manifold before the backward process, P_{51}, occurs. Because the energy transfer parameter involving the Ho $^5I_6$ manifold and the Tm $^3H_5$ manifold is quite large, energy transfer from the Ho $^5I_6$ manifold to the Tm $^3F_4$ manifold is likely, process P_{61}. The backward process, process P_{38}, is not likely because the Tm $^3H_5$ manifold is likely to decay to the Tm $^3F_4$ manifold through a nonradiative process. Thus, a quantum of excitation is lost in the process. This has the effect of shortening the lifetime of the combined Ho $^5I_7$ and Tm $^3F_4$ manifolds.

The differential equations that govern this situation involve 7 different manifolds and can be approximated as

$$\frac{dN_2}{dt} = -\frac{N_2}{\tau_2} - P_{28}N_2N_8 + P_{71}N_7N_1 + \frac{N_3}{\tau_3} - P_{27}N_2N_7 + P_{51}N_5N_1 \tag{37}$$

$$\frac{dN_3}{dt} = -\frac{N_3}{\tau_3} + P_{61}N_6N_1 - P_{38}N_3N_8 \tag{38}$$

$$\frac{dN_5}{dt} = -\frac{N_5}{\tau_5} + P_{27}N_2N_7 - P_{51}N_5N_1 \tag{39}$$

$$\frac{dN_6}{dt} = -\frac{N_6}{\tau_6} + \frac{N_5}{\tau_5} - P_{61}N_6N_1 + P_{38}N_3N_8 \tag{40}$$

$$\frac{dN_7}{dt} = -\frac{N_7}{\tau_7} + P_{28}N_2N_8 - P_{71}N_7N_1 - P_{27}N_2N_7 + P_{51}N_5N_1 \tag{41}$$

$$N_1 + N_2 + N_3 + N_4 = C_T N_S \tag{42}$$

$$N_8 + N_7 + N_6 + N_5 = C_H N_S \tag{43}$$

In these equations, τ_i are lifetimes of the respective manifolds. Approximations to the above rate equations employed by others may neglect some of the participating manifolds and essentially use an effective up conversion parameter, P_{27}. Although this approach does hold some appeal, it approximates away much of the physics of the situation and easily leads to a value for the up conversion parameter which is widely different than actuality.

There is no obvious simple, closed form, solution to the above differential equations. To illustrate the effects of the up conversion process, the population density of the Ho $^5I_7$ and Tm $^3F_4$ manifolds are determined as a function of time utilizing numerical techniques. Results are displayed for 2 in Figures 10 and 11 for Ho:Tm:YAG and Ho:Tm:YLF. In this analysis, the laser is not allowed to oscillate, a situation similar to the pumping phase of a Q-switched laser. The analysis assumes the same Ho and Tm concentrations in both materials and equal pumping rates. Ho:Tm:YAG is subject to up conversion as can be observed in a rapid decrease in the population densities of the Ho $^5I_7$ and Tm $^3F_4$ manifolds after the pump terminates. On the other hand, the decrease in the population densities of these manifolds is not nearly as rapid, indicating there is less detrimental up conversion.

7. Excited State Absorption

Excited state absorption occurs when a metastable manifold accumulates a sufficiently high population density and has an appreciable absorption cross section, σ_{ESA}, that it can compete with other processes. Both pump and probe radiation can fall victim to excited state absorption. For pump beam absorption, the rate equations are given by

$$\frac{dN_p}{dt} + \frac{c}{n}\frac{dN_p}{dz} = -\frac{\sigma_a c}{n} N_1 N_p - \frac{\sigma_{ESA} c}{n} N_2 N_p \tag{44}$$

$$\frac{dN_2}{dt} = -\frac{N_2}{\tau_2} + \frac{\sigma_a c}{n} N_1 N_p - \frac{\sigma_{ESA} c}{n} N_2 N_p + \frac{\beta_{42} N_4}{\tau_4} \tag{45}$$

$$\frac{dN_4}{dt} = -\frac{N_4}{\tau_4} + \frac{\sigma_{ESA} c}{n} N_2 N_p \tag{46}$$

In many cases, the manifold to which the atom is promoted by the excited state absorption process, N_4, decays rapidly, returning a fraction of the quanta it receives, β_{42}, to the metastable manifold. To detect excited state absorption, the pump must be sufficiently intense to generate a significant population in the excited state. It is also necessary to know the excited state population density, N_2, accurately. This is often not a trivial task because the pump beam profile is often not uniform and absorption depletes the pump, thereby causing further nonuniform excitation.

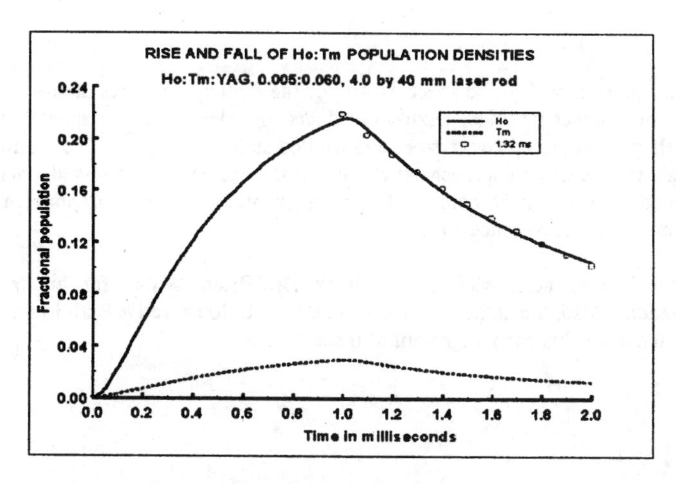

Figure 10. Rise and decay of Tm $^3F_4$ and Ho $^5I_7$ in YA

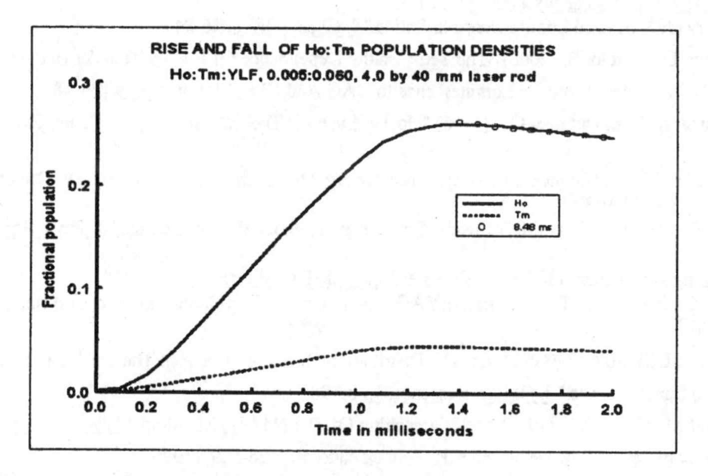

Figure 11. Rise and decay of Tm $^3F_4$ and Ho $^5I_7$ in YLF

The presence of excited state absorption may be detected by observing fluorescence from a manifold that is higher than the metastable manifold [12]. Excited state absorption in Nd can be detected by observing fluorescence from the $^4G_{7/2}$ manifold. In this case, excited state absorption of a probe beam rather than a pump beam is being detected. The probe beam is absorbed from the excited state manifold, $^4F_{3/2}$, and the doubly excited Nd atom relaxes to the $^4G_{7/2}$ manifold from which some fluorescence can be observed. Although this manifold is quenched, there is enough fluorescence to allow the presence of excited state absorption to be deduced.

8. Summary

Extreme effects can be observed in the spectroscopy of the lanthanide series atoms under high levels of excitation or high concentrations. Extreme effects include: amplified spontaneous emission, energy transfer, self quenching, up conversion, and excited state absorption. These effects can often be described using a rate equation approach. If energy transfer effects are involved, energy transfer parameters are needed. These can be computed using a quantum mechanical approach or they can be measured by monitoring fluorescence emission.

The author would like to acknowledge and thank Dr. Brian Walsh for his many contributions to this presentation. Also, the author wishes to thank Dr. Baldassare Di Bartolo for the invitation to address this audience and his encouragement of these studies.

9. References

1. Barnes, N.P., and Walsh, B.M. (1999) "Amplified Spontaneous Emission-Application To Nd:YAG Lasers," IEEE J. Quant. Elect. QE-35 101-109

2. Forster, T.(1949) "Experimentelle Und Theoretische Untersuchung Des Zwishcenmolekularen Ubergangs Von Elektronenregungenergie," Zeitschrift Für Naturforschung. 4A 321-327

3. Dexter, D.L., (1953) "A Theory Of Sensitized Luminescence In Solids," J. Chem. Phys. 836-849

4. Walsh, B.M., Barnes, N.P., and Di Bartolo, B. (2000) "The Temperature Dependence Of Energy Transfer Between The Tm $^3F_4$ and Ho $^5I_7$ Manifolds Of Tm Sensitized Ho Luminescence In YAG And YLF," J. Lumines. 90 39-48

5. Barnes, N.P., Filer, E.D., Morrison, C.A., and Lee, C.J. (1996) "Ho:Tm Lasers I: Theoretical," IEEE J. Quant. Elect. QE-32 92-103

6. Inokuti, M. and Hirayama, F. (1965) "Influence Of Energy Transfer By The Exchange Mechanism On Donor Luminescence," J. Chem. Phys. 43 1978-1989

7. Yokota, M. and Tanimoto, O. (1967) "Effects Of Diffusion On Energy Transfer By Resonance," J. Phys. Soc. Japan 22 779-784

8. Burshtein, A.J.(1972) "Hopping Mechanism Of Energy Transfer," Soviet JETP Physics 35 882-885

9. Rotman, S.R. (1989) "Nonradiative Energy Transfer In Nd:YAG- Evidence For Correlated Placement Of Ions," Appl. Phys. Lett. 54 2053-2055

10. Walsh, B.M., Barnes, N.P., and DiBartolo, B. (1997) "On The Distribution Of Energy Between The Tm $^3F_4$ and Ho $^5I_7$ Manifold In Tm Sensitized Ho Luminescence," J. Of Luminescence, 75 89-98

11. Barnes, N.P., Filer, E.D., and Morrison, C.A. (1996) "Self Quenching Of The Nd $^4F_{3/2}$ Manifold," Proceedings Of The Advanced Solid State Laser Conference, S.A. Payne and C.R. Pollock editors 1, 526-529, (1996)

12. Guyot, Y., and Moncorge, R. (1993) "Excited State Absorption In The Infrared Emission Domain Of Nd^{3+} Doped $Y_3Al_5O_{12}$, $YLiF_4$, And $LaMgAl_{11}O_{19}$," J. Appl. Phys. 826-8530

15. EXCITONIC BOSE-EINSTEIN CONDENSATION VERSUS ELECTRON-HOLE PLASMA FORMATION

CLAUS KLINGSHIRN
Institut für Angewandte Physik
Universität Karlsruhe and
Center for Functional Nanostructures (CFN)
Wolfgang Gaede Str. 1
76131 Karlsruhe
Germany

Abstract: After a short introduction we outline the properties of the electron-hole plasma and give some selected examples how its properties have been verified in direct and indirect gap bulk semiconductors and in quantum wells.

Then we present the basic properties expected for an excitonic Bose-Einstein condensation, claims for its observation and the objections put forward to these claims. The contribution will be finished with a short conclusion and outlook.

1. Introduction

In the first contribution to this book we have introduced the concept of excitons in bulk material and in various types of quantum wells. In this contribution we will follow two aspects of what could happen if the density of excitons or of electron-hole pairs is increased.

We note, that we are throughout this contribution in a time regime longer than the dephasing time of the initially created polarisation (via exciton and / or electron hole pair creation) with the driving laser field. Effects of Rabi-flopping etc. treated e.g. in the contributions by B. Di Bartolo or M. Wegener to this school are here of no relevance. Furthermore the pulses are not so intense that they melt the sample (see the contribution of E. Mazur).

We give in Fig. 1 a very schematic and intuitive sketch of what might happen with increasing generation rate.

B. Di Bartolo and O. Forte (eds.), Frontiers of Optical Spectroscopy, 539-570.

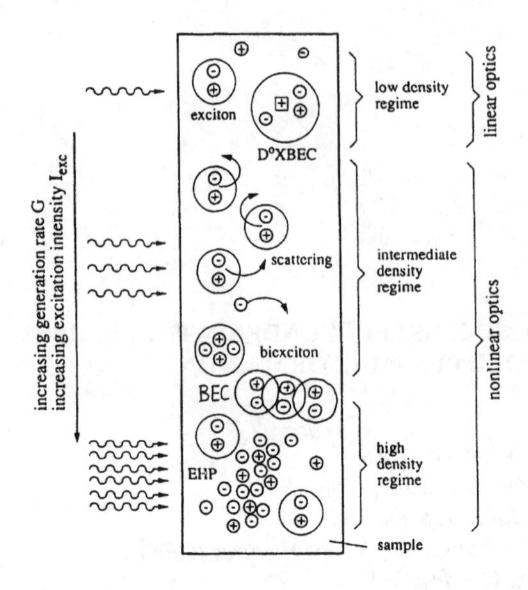

Figure 1: Schematic drawing of various properties of the electron-hole pair system in a semiconductor.

At low excitation density the absorbed incident light may directly form excitons or electron-hole pairs for $\hbar\omega_{exc} > E_g$. The electron-hole pairs may bind to excitons or excitons my be thermally dissociated at higher temperatures. At low temperature the excitons may be also bound to (point-) defects like neutral acceptors or neutral or ionized donors to form bound exciton complexes. After an average lifetime T_1 the excitons and carrier pairs recombine radiatively or radiationless. These processes form the regime of linear optics, this means that the optical properties like the spectra of absorption, transmission, reflection and luminescence do not depend on the light intensity. In other words an expansion of the polarization in a power series of the electric field, as presented in several contributions to this school, can be truncated after the linear term.

With increasing generation rate, we come to a region, where excitons are still good quasi particles, but their density is so high, that they start to interact: there might be elastic and inelastic scattering processes between excitons or excitons and free carriers or two excitons may bind together to form a biexciton or excitonic molecule.

All theses phenomena lead already to excitation induced variations of the optical properties of the semiconductor, i.e. to nonlinear optical effects, which can be described either by higher terms in the power series expansion of **P(E)** including the well know $\chi^{(2)}$ and $\chi^{(3)}$ effects like second harmonic generation, rectification of the light field (the dc effect) or four wave mixing or hyper-Raman scattering, respectively. In the case of incoherent excitation this approach is better replaced by an explicit dependence of the dielectric function $\varepsilon(\omega) = \varepsilon_1(\omega) + i\varepsilon_2(\omega)$ or of the complex index or refraction $\tilde{n}(\omega) = n(\omega) + i\kappa(\omega)$ on the electron-hole pair density $\varepsilon(\omega, n_P)$ or $\tilde{n}(\omega, n_P)$. For such effects see either various contributions to this school or [1] and the references given therein. For even higher excitation densities two different things may happen.

If the density of excitons becomes so high, that the average distance of the excitons and / or carries is comparable to the excitonic Bohr radius, one can no longer say that one electron is bound to one hole. Instead one gets a new, metallic, collective phase of electrons and holes, which is know as electron-hole plasma (EHP). As we shall see later this EHP may occur at low temperature in a liquid phase (electron-hole liquid (EHL)) forming small electron-hole droplets (EHD).

In this situation the Fermi-character of the carriers, i.e. of the constituents of the excitons, dominates.

On the other hand, excitons are to a good approximation Bosons, because they have integer spin. Actually the creation and anniliation operators for excitons obey the commutations rules for Bosons, however with an additional term increasing with increasing exciton density of the order $O(n_P a_{ex}^3)$ where a_{ex} is the excitonic Bohr radius.

For some early theoretical predictions of excitonic BEC see e.g. [2 – 4] and for older or recent reviews or books [5 – 7].

While the properties of the EHP are nowadays well established and used in every commercially available semiconductor laser diode, the excitonic BEC is still an open question as we shall see, in contrast to the one of (alkali-) atoms in a trap. For this latter topic see e.g. the contributions by Ch.J. Pethick and K. Helmerson to this school and the references therein.

We start first to describe the properties of an electron hole plasma, then we discuss the excitonic BEC and give finally a short conclusion and outlook.

2. The electron-hole Plasma

We start with the discussion of the EHP because its properties are well established.

2.1 BASIC PROPERTIES OF THE ELECTRON-HOLE PLASMA

In Fig. 2 we show schematically on the rhs the density dependent variations of electronic properties when an EHP is formed and on the l.h.s. the resulting changes of the optical properties. The upper half gives data for low and the lower half for high temperatures, respectively. The drawings are made for a direct gap semiconductors with dipole-allowed band-to-band transition, but the sketches on the r.h.s. are also valid for indirect gap materials.

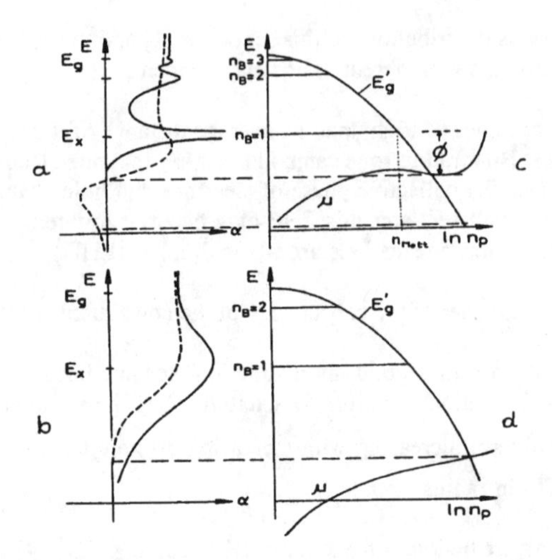

Figure 2: Schematic drawing of the changes of the optical (a, b) and the electronic properties (c, d) when increasing the electron hole pair density n_P and going from a low density exciton gas to an EHP. Low temperatures (a, c), high temperatures (b, d).

At low temperature and density we see the exciton series with main quantum number n_B for $\mathbf{k}_{ex} = 0$ (Fig. 2c) coupling to the radiation field and resulting in the series of exciton absorption peaks followed by the transition into the ionization continuum (Fig. 2a solid line).

If we now start to increase the electron hole-pair (or exciton) density various things happen (for details see e.g. [1, 7] and references therein): The width of the forbidden gap is monotonously decreasing with increasing n_P. This band-gap renormalization (BGR) has two contributions. If the electrons and holes would be randomly distributed the gap would be independent of n_P, because the terms describing mutual electron-hole attraction would cancel on the average with ones for electron-electron and hole-hole repulsion. Actually they are not randomly distributed. At low density e and h bind due to Coulomb interaction to form excitons. Even at densities where excitons do no longer exist as individual quasi-particles there is still a (Coulomb) correlation energy which results from the spatial correlation of the probability to find electrons and holes in the sense that there is in the vicinity of a hole a higher probability to find an electron than to find another hole and vice versa. The resulting average Coulomb attraction lowers the gap with increasing n_P. A further contribution to this BGR comes from the exchange interaction, which describes the fact that equal carriers with parallel spin cannot be at the same place due to the Pauli principle and are thus on the average further apart form each other. Such an increase of the average distance of equal particles reduces their average repulsion and contributes to an overall lowering of the gap as shown in Fig. 2c and d. These effects result in the BGR $Eg'(n_P)$ which is to a good approximation temperature independent.

The next effect is a density dependent decrease of the exciton binding energy caused by a screening of the Coulomb-interaction between an electron-hole pair and especially in structures of reduced dimensionality by phase space filling [1]. The Coulomb potential is then transformed to a screened one according to (1)

$$\frac{1}{4\pi\varepsilon\varepsilon_0}\frac{-e^2}{|\mathbf{r}_e-\mathbf{r}_h|} \Rightarrow \frac{1}{4\pi\varepsilon\varepsilon_0}\frac{-e^2}{|\mathbf{r}_e-\mathbf{r}_h|} \exp\{-|\mathbf{r}_e-\mathbf{r}_n|/l_c\} \quad (1)$$

where the screening length l_c decreases with increasing carrier density. For

$$a_B l_c^{-1} = 1.19 \quad (2)$$

with a_B = excitonic Bohr radius the binding energy of the exciton vanishes and no more bound exciton state exists.

It can be shown that the BGR and the decrease of the exciton binding energy cancel to a good approximation. This means, that the absolute energy of the exciton remains approximately constant over the electron hole pair density as seen in Fig. 2b, but the damping and width increase until the exciton state ceases to exist at a certain density n_M. This density is generally referred to as the Mott density. It describes at low temperatures the transition from an insulating state of neutral excitons (originally of neutral donors) to a metallic state of electrons and holes. Actually it turned out later that this is rather a continuous transition instead of an abrupt one [8]. Obviously the higher exciton states disappear already at lower densities.

The last quantity shown in Fig. 2c, d is the chemical potential of the electron-hole pair system $\mu(n_P,T)$. We introduce this quantity in the following.

In thermodynamic equilibrium the distribution of the electrons in the bands and in defect- or doping centers can be described by one (electro-)chemical potential which is in semiconductor physics called Fermi level E_F for all temperatures.

The densities of electrons in the conduction band n or of holes in the valence band p are then e.g. given by

$$n_e = \int_{E_g}^{\infty} D(E)\cdot f_{FD}(E,E_F,k_BT)\,dE \quad (3a)$$

$$p = \int_{-\infty}^{0} D(E)(1-f_{FD})(E,E_F,k_BT)\,dE \quad (3b)$$

where D(E) is the density of states in the respective band, and f_{FD} (E, E_F, k_BT) the Fermi-Dirac occupation probability (see below). The energy zero is assumed at the top of the valence band.

If we now increase the density of electrons and of holes e.g. by carrier injection in a forward biased pn-junction or by optical excitation, and still assume that the carriers acquire a thermal distribution, which allows to define a temperature T, we can use again (3a, b) but with individual quasi-Fermi levels for electrons and holes $E_F^{e,h}$.

The difference between these quasi-Fermi levels is just the chemical potential of the electron hole pair system

$$\mu(n,p,T) = E_F^e(n,T) - E_F^h(p,T) \tag{4}$$

In thermal equilibrium μ vanishes obviously. It is positive if n and p are increased beyond the equilibrium value and is negative in the opposite case e.g. in a pn junction biased in the blocking direction.

In a similar way the chemical potential of the electron-hole pair system is defined if excitons are still good quasi particles of density n_{ex} at a temperature

$$n_{ex} = \int_{E_{ex(k-0)}}^{\infty} D_{ex}(E_{ex}) f_{BE}(E_{ex}, \mu, k_B T) dE \tag{5}$$

now using the Bose-Einstein occupation probability.

In Figs. 2c and d μ is evidently zero for n_P -> 0 and increases for finite n_P. At low temperatures it may show a non monotonous behaviour (Fig. 2c). At higher temperatures it is a monotonously increasing function of n_P (Fig. 2d).

The energetic distance between the exciton energy and the minimum of $\mu(n_P, T \to 0)$ is called the binding energy of the EHP Φ. It is indicated in Fig. 2b, with $\Phi > 0$ indicating a bound state.

We assume now, that we have excited a direct gap semiconductor with dipole-allowed band-to-band transition to a density n_P which is above n_{Mott} and even so high that μ exceeds the reduced gap E_g'

$$\mu(n_P, T) > E_g'(n_P) \tag{6}$$

Then we have population inversion between conduction and valence band, and optical amplification or gain between μ and E_g' if the band-to-band transition is direct and dipole allowed. For photon energies above $\mu(n_P, T)$ there is band-to-band absorption enhanced by the remaining e-h correlation in the plasma.

At higher temperatures the exciton resonances are thermally broadened (solid line in Fig. 2b) and since $\mu(n_P, T)$ decreases with increasing T for constant n_P it may no longer exceed E_g' for the assumed density n_P resulting just in a bleaching of the absorption tail (dashed line in Fig. 2b) since $\mu(n_P, T)$ gives always the transparency point.

Note that there is no such thing as a simple Burstein-Moss shift i.e. a simple blue shift of the absorption edge by state filling even for an unipolar plasma produced e.g. by strong n-doping. The filling of states is always accompanied by a BGR and the interplay between both effects determines if there is a red- or a blue shift of the absorption edge. At very high densities the state filling dominates always over BGR.

2.2 EXAMPLES FOR BULK SEMICONDUCTORS WITH DIRECT, DIPOLE-ALLOWED GAP

In this section we show some data for direct gap bulk semiconductors with dipole allowed band-to-band transition.

We should note at this point, that there are also excitonic processes in the intermediate density regime [1, 9], yielding optical gain, but all commercially available laser diodes relay on the gain in an EHP, usually in quantum wells of direct gap semiconductors with dipole-allowed band-to-band transition.

In Fig. 3 we show for CdS at various lattice temperature T_L transmission spectra of a CdS crystal platelet which is a few μm thick without optical pump beam and with a 5ns pump pulse of I_{exc} = $4MW/cm^2$ and $\hbar\omega_{exc} = 2.8eV$. Since the lifetime T_1 of the carrier pairs is around ≤ 1ns one has quasi-stationary conditions around the maximum of the pump beam. The temporal and spatial maximum is probed by the weak probe beam.

In the transparent region the transmission is modulated by Fabry-Perot modes of the platelet type sample. At T_L = 5K the transmission minimum around 2.552eV results from the lowest free exciton ($A\Gamma_5$). With pump beam the behaviour corresponds to the expectations of Fig. 2a, c. The exciton resonance disappears both in transmission and reflection [8, 10], there is optical amplification between μ and E_g'. Both quantities are indicated in Fig. 3a. The binding energy Φ of the EHP is about 10meV (see also Figs. 4 and 5). For higher temperatures the excitonic absorption edge is smoother due to the Urbach rule

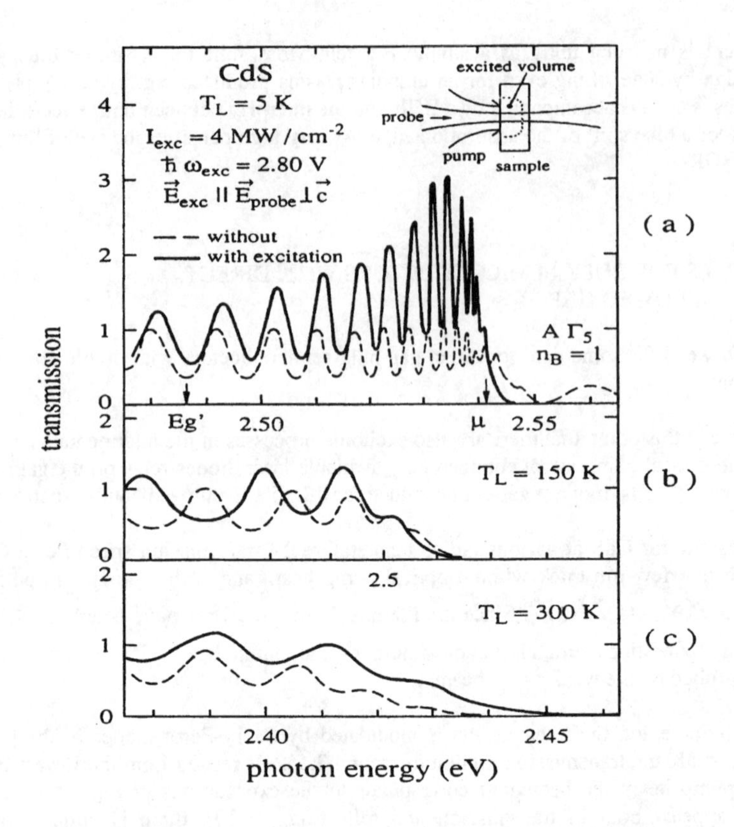

Figure 3: Pump- and probe beam spectra of a thin CdS platelet for various lattice temperatures. From [10].

behaviour [1], the gain is smaller at 150K and vanishes almost at 300K. Only a pronounced bleaching is left.

At all temperatures one observes a blue shift of the Fabry Perot modes, indicating a density dependent decrease of the real part of the refractive index $\Delta n(\omega, n_p)$ as expected from Kramers-Kronig relations [11].

In Fig. 4 we show gain spectra deduced from data like in Fig. 3 now for various values of I_{exc} at a lattice temperature $T_L = 5$K.

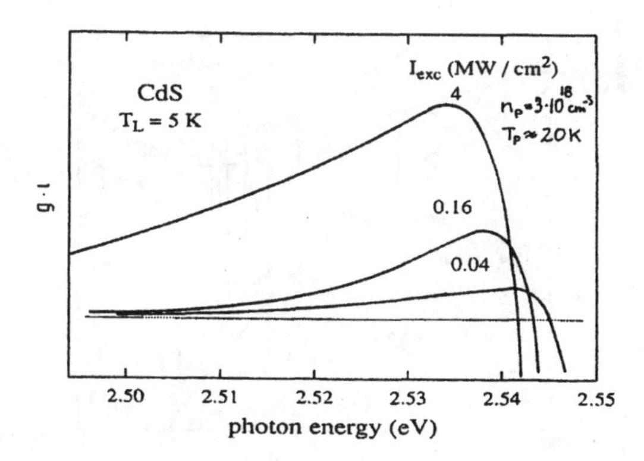

Figure 4: Gain spectra of a CdS crystal for various pump intensities at a lattice temperature $T_L = 5K$. From [8, 10].

The gain spectra become broader with increasing excitation. They can be perfectly fitted with a model taking into account band-to-band recombination under **k**-conservation and including some (final-state) damping and the remaining Coulomb correlation [8, 10, 12]. For the highest excitation the deduced parameters are given. It should be noted that the carrier temperature can be above the lattice temperature. (See also Figs. 6, 7.)

In Fig. 5 we show, that there is good agreement between the calculated and measured dependencies of the chemical potential $\mu(n_P, T_P)$ and the reduced gap $E_g'(n_P)$.

We come back to this Figure in another context later. Before we give some more experimental results, to show that the formation of an EHP is a well established phenomenon in a wide variety of semiconductors.

In Fig. 6a we show an EHP gain spectrum for ZnTe (another II-VI compound) and in b the absorption spectrum for the III-V semiconductor GaAs. While the EHP forms evidently a bound state in ZnTe $\Phi = E_{ex} - \mu = (2.3809 - 2{,}362)\text{eV} = 18.9\text{meV}$ the chemical potential is above the exciton energy in the experiment shown in GaAs, but at lower densities μ falls below E_{ex} resulting again in a positive value of Φ [16].

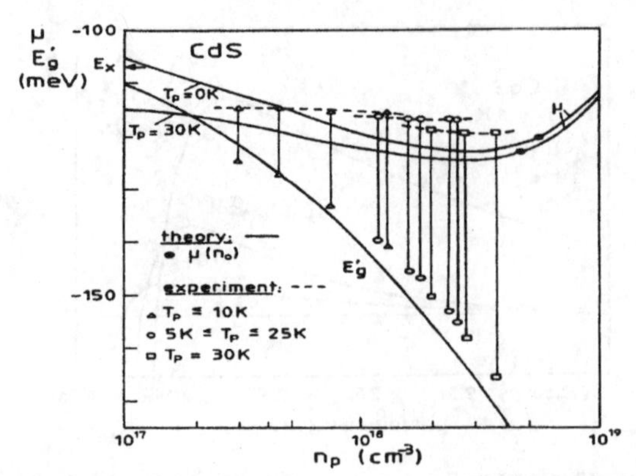

Figure 5: Calculated and measured density dependencies of the reduced gap Eg'(n_P) and the chemical potential $\mu(n_P, T_P)$. According to [10, 12 13].

Even in CuCl an EHP state has been reached under strong pulsed excitation, though the excitonic Bohr radius does hardly exceed the lattice constant [17]. In Cu_2O, a hot candidate to observe excitonic BEC (see below) it was recently possible to bleach the whole exciton series down to the n_B = 2p state completely [18], and it is questionable if the remaining 1s excitons can be still considered as a weakly interacting Bose gas at these densities.

The agreement of the observed BGR with "universal formula" has been demonstrated in [19, 20] for many semiconductors.

2.3 PHASE SEPARATION AND ELECTRON-HOLE LIQUID

We come back to Figs. 2b, d and 5. If the chemical potential $\mu(n_P T)$ shows below a certain critical temperature T_c a non-monotonous behaviour, quasi-equilibrium thermodynamics predict a first order phase transition to occur [5 – 7, 13, 21] in a very similar way as in a real or van der Waals gas. The region with a negative slope in the $\mu(n_P, T_P = \text{const})$ curve is unstable. Instead a phase separation occurs into a liquid like electron-hole plasma state (EHL) given by the high density border of the coexistence region in Fig. 7 and a low

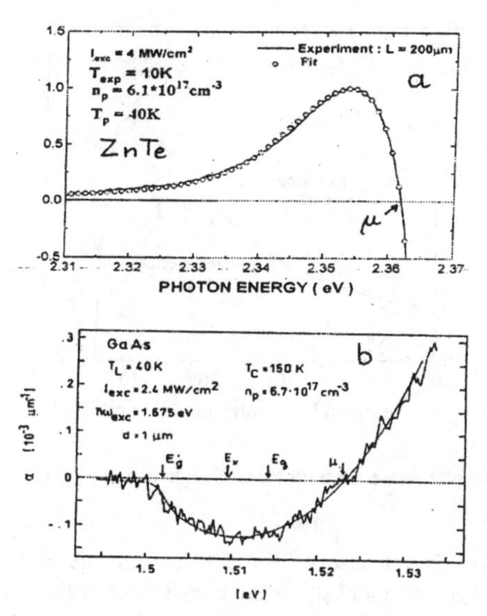

Figure 6: Gain or absorption spectra, respectively, resulting from EHP formation in ZnTe and GaAs. From [14, 15].

density gas phase consisting of excitons and possibly biexcitons and, depending on temperature, of some free carriers.

This means, that under excitation conditions with carrier temperature below T_c and an average density in the coexistence region a phase separation should occur. The low energy side of the coexistence region is given by [7, 22]

$$n_P = g\left(\frac{m_{ex}\,k_B T}{2\pi\hbar^2}\right)^{3/2} e^{-\Phi/k_B T} \tag{7}$$

A certain modification of the phase diagram as indicated by the dashed like may occur at low densities and temperature due to the finite carrier lifetime.

Above T_c a kind of Mott transition will take place as will detailed in 3.4.

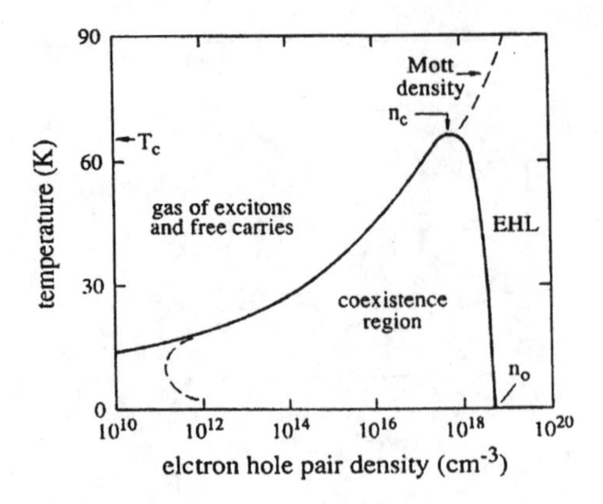

Figure 7: Calculated electron hole pair phase diagram for CdS. According to [13]

Evidently this phase separation is not observed in CdS, though the EHP forms a bound state. The plasma densities could be varied continuously in the coexistence region by a simple variation of the excitation intensity. The reason for this failure is not, that the binding energy of the EHP is too small, but the reason is simply the short lifetime of the electron-hole pairs in the plasma, which is in the sub-ns regime and might be even shorter if stimulated emission sets in.

The formation of a phase separation into liquid like droplets (EHD) surrounded by a low density gas phase requires a spatial motion of the uniformly excited carriers and thus a finite time. If this time is longer than the lifetime of the carrier pairs the phase separation does not take place though $\Phi > 0$, as shown explicitely for CdS [8, 21] and as is very likely true for all similar II-VI and III-V compounds.

2.4 EXAMPLES OF BULK SEMICONDUCTORS WITH INDIRECT GAP

In order to observe the phase transition, one has to investigate according to the above considerations semiconductors with long electron-hole pair lifetime. Ideal candidates are the indirect gap materials Ge and Si. The long lifetime results from the necessity to involve a momentum conserving phonon in the recombination process. This fact has various advantages with respect to EHP research: High densities can be easily reached, under favourable condition even under cw excitation.

Stimulated emission does not occur due to the small optical transition probability though population inversion i.e. $\mu > E_g^{'}$ is easily reached. This latter fact is however a big disadvantage for the application because it prevents the use of Si or Ge as laser or light emitting diodes!

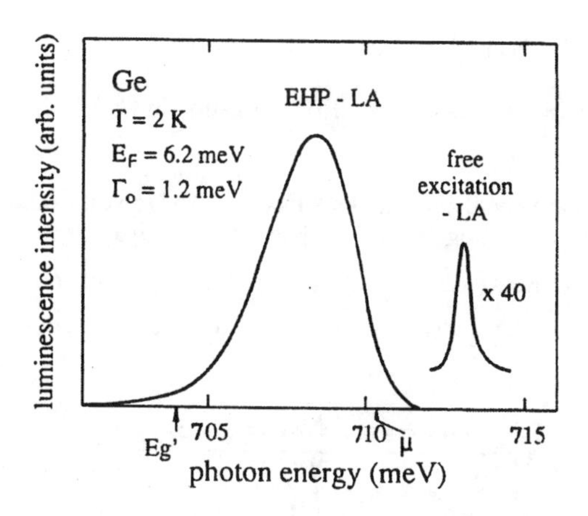

Figure 8: The LA phonon replica of the exciton and EHP emission of Ge. According to [23].

Indeed the formation of an EHL has been observed by many authors. We give below some examples. In Fig. 8 we show the LA phonon assisted luminescence of a Ge crystal showing the free exciton luminescence and the emission from the electron-hole droplets. The fact that a liquid like state has been obtained follows from the fact, that the EHP density remains constant with increasing excitation at constant temperature but the relative weight of the two bands changes as expected from the phase diagram.

The phase diagram itself can be deduced from the luminescence data in the following way.

One increases the generation rate G and thus the average electron-hole pair density at constant carrier temperature. As long as one stays below the low density boarder of the coexistence region only free (and bound) exaction emission is observed. When one crosses this border, EHD emission appears. From the generation rate and the lifetime it is possible to reconstruct the low density boarder while a line shape analysis of the EDH gives the high energy side.

Phase diagrams in very close agreement with the one shown in Fig. 7, have thus been verified experimentally for Ge and Si. The critical densities and temperatures turned out to be ($8 \cdot 10^{16}cm^{-3}$, 6.5K) and ($1.5 \cdot 10^{18}cm^{-3}$, 25K) while the liquid densities for T -> 0 n_0 are $2.3 \cdot 10^{17}cm^{-3}$ and $3.5 \cdot 10^{18}$ respectively, see e.g. [19, 24] and the references therein.

Beautiful experiments have been performed at that time with EHD's under high surface excitation. The droplets are driven in pure samples into the bulk by the phonon wind [25], while macroscopic γ-drops have been observed in potential minima of E_g' created by the application of inhomogeneous stress [26].

The phase diagrams could be varied by the application of homogeneous stress to the samples which reduces if properly applied the degeneracy of the multivally structure of the conduction band and reduces thus the values of Φ, n_c, T_c and n_0 [19, 27].

2.5 EXAMPLES FOR QUANTUM WELLS

To conclude this section we show data for the EHP formation in $GaAs/Al_{1-y}Ga_yAs$ quantum wells in Fig. 9.

The phenomena are qualitatively identical to bulk material. The lowest exciton resonances (hh and lh $n_z = 1$) disappear with increasing excitation intensity, the gap decreases monotonously with increasing n_p and μ shifts above E_g' resulting in population inversion and gain. As mentioned already above, this gain is the underlying process of all presently commercially available laser diodes.

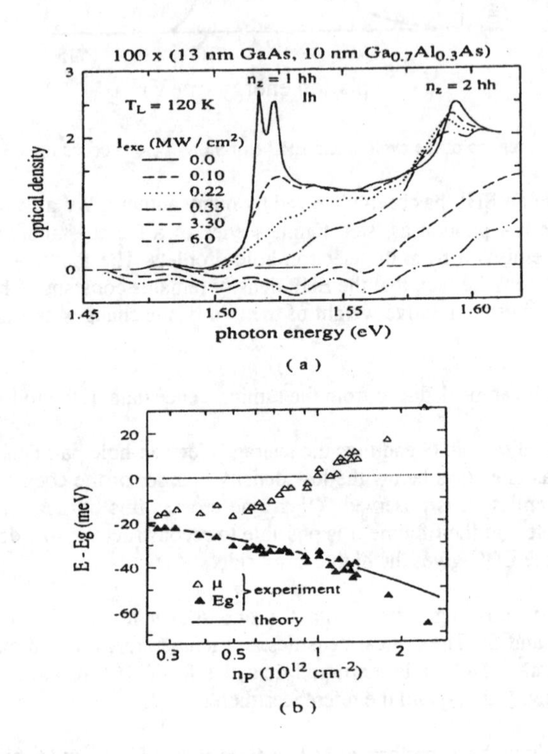

Figure 9: Luminescence spectra of a $GaAs/Al_{1-y}Ga_yAs$ MQW (a) and the density dependence of μ and E_g'.(b). From [28]

It is remarkable, that the higher exciton states ($n_z = 2$) disappear only when these states start to be occupied and show a much smaller shift than the band gap $E_g^{'}$ (n_P), showing a non rigid behaviour of the whole subband structure with increasing n_P [28]. Similar data have been obtained also for $In_{1-y}Ga_yAs/InP$ SQW [29].

No reports of an EHP phase separation are known to the author for this type of materials.

For an EHP in quantum wires see e.g. the contributions by E. Kapon to this and the preceeding school [30].

To summarize we can state that the existence of an EHP and the transition to an EHL under favourable conditions is well established in bulk semiconductors and in quantum structures. The main part of the work for bulk materials has been performed in the seventies and eighties of the last century as can be also seen from the references.

3. Excitonic Bose-Einstein Condensation and Superfluidity

Now we concentrate on the possibility to observe excitonic Bose Einstein condensation (BEC). We start with some general considerations and treat then the attempts to observe it in bulk and quasi two dimensional semiconductors. The chapter will be finished with short comments about excitonic insulators and on so-called "driven BEC".

3.1 BASIC PROPERTIES

A BEC is a macroscopic population of one quantum mechanical state by (ideally non or weakly interacting) Bose particles in thermal (quasi-)equilibrium. It occurs if either the temperature T is lowered below a critical temperature T_c at constant particle density n, or if n is raised above n_c at constant T.

For non-interacting ideal Bosons one finds the following relation between n_c and T_c [7]

$$n_c = 2.612g\left(\frac{mk_BT_c}{2\pi\hbar^2}\right)^{3/2} \tag{8}$$

where m and g are the mass of the particles and the degeneracy of the state, respectively. The condensate can show superfluidity. The facts that the mass of excitons is comparable to the free electron mass while those of e.g. alkali atoms are ten or a hundred times the proton mass and that excitons can be created by pulsed lasers easily in the density range up to $10^{17}cm^{-3}$ allows one to expect values of T_c up to around 10K while the successful experiments to observe atomic BEC in traps required T as low as a few 10μK.

The weak interaction should be slightly repulsive, to avoid condensation of the particles in real space, since a BEC is a condensation in **k**-space.

The only example of a BEC of weakly interacting Bosons is the condensation of alkali and H atoms in a trap, treated in the contributions of Ch. J. Pethick and K. Helmerson to this school.

There are further examples of BEC for Bosons with rather strong interaction like the superfluidity of ^{4}He and systems which involve the pairing of Fermions and the formation of a gap like superconductors or superfluid ^{3}He.

Excitons would evidently be another example for weakly interacting Bosons. Before we start to treat experimental attempts to observe excitonic BEC and the objections against these interpretations, we show how eq (8) can be understood.

We show in Fig. 10a the three occupation probabilities for particles with an effective mass m possible in three dimensions namely the Boltzmann statistics for distinguishable particles and the Fermi-Dirac and Bose-Einstein statistics for indistinguishable particles with half or integer spin, respectively. They are given by

$$f_B(T, E, \mu) = \frac{1}{e^{(E-\mu)/k_B T}} \tag{9a}$$

$$f_{FD}(T, E, \mu) = \frac{1}{e^{(E-\mu)/k_B T} + 1} \tag{9b}$$

$$f_{BE}(T, E, \mu) = \frac{1}{e^{(E-\mu)/k_B T} - 1} \tag{9c}$$

where μ is the chemical potential. Evidently they coincide for $(E - \mu)/k_B T > 2$ and show characteristic differences below.

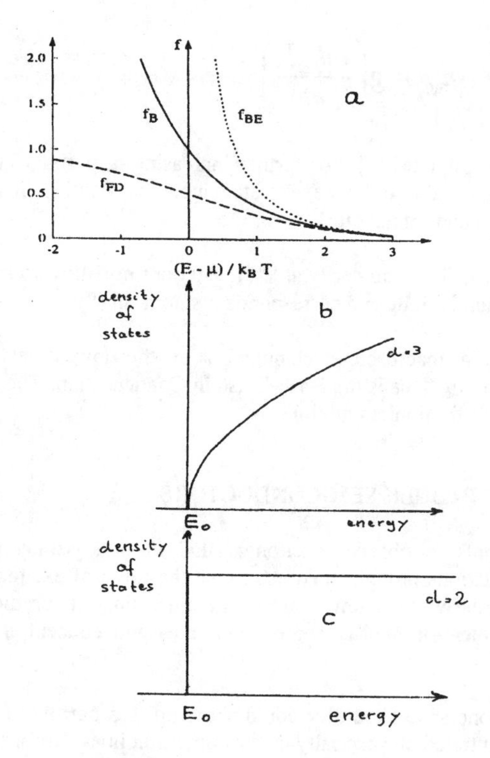

Figure 10: The occupation probabilities according to Boltzmann, Fermi-Dirac and Bose-Einstein as a function of $(E - \mu)/k_BT$ (a) the square root density of states of particles with a finite (effective) mass in three dimensions (b) and the step function in two dimensions (c).

For particles with a well defined density n, μ is defined by eq. (10)

$$n = \int_{E_0}^{\infty} D(E) \cdot f_{B,FD,BE}(T,E,\mu)\, dE \qquad (10)$$

For particles, the density of which is not conserved like photons in black body radiation or phonons in thermal equilibrium one has $\mu \equiv 0$. In Fig. 10b we show the square root density of states of particles with a finite and constant (effective) mass m

$$D(E) \approx \sqrt{E - E_0} \qquad (11)$$

If one performs now the integration (10) for a low density, μ i.e. the origin of Fig. 10a is situated below E_0. If the density increases, μ shifts towards E_0. Eventually E_0 coincides with μ. There is no singularity for Boltzmann particles and Fermions. The density one reaches under this condition for Fermions is given by

$$n_{eff} = g\left(\frac{mk_BT}{2\pi\hbar^2}\right) \tag{12}$$

and is known as effective density of states [1]. For further increasing n, μ shifts into the band. For Fermions, the distribution is then said to be degenerate since the Pauli principle reduces the occupation probability from 1 to ½ and limits it to 1.

For Bosons the situation is different. The singularity at $E - \mu = 0$ does not allow to shift μ beyond E_0. For $\mu = E_0$ the integrated population is finite and corresponds just to n_c in (8).

For further increasing $n > n_c$ a macroscopic population of the lowest state develops and accommodates all particles beyond n_c. This is the Bose-Einstein Condensation. The condensed phase may show superfluidity i.e. loss- or frictionless motion.

3.2 ATTEMPTS TO FIND BEC IN BULK SEMICONDUCTORS

We now present various attempts to observe excitonic BEC in bulk semiconductors and the objections against these interpretations brought forward. Since the topic of excitonic BEC and / or superfluidity appears almost regularly every few years in literature since its prediction in 1962 we cannot treat all examples. We mention shortly some older ones and concentrate then on recent experiments in Cu_2O.

The macroscopic population of one state should or could show up in a narrow luminescence peak. Therefore several attempts concentrated on spectrally narrow emission lines. Unfortunately not only a condensed exciton phase gives rise to narrow emission lines but other processes, too. So the narrow emission or absorption bands from bound exciton complexes (see Fig. 1) (which are by the way also abbreviated by BEC) have been misinterpreted as excitonic BEC e.g. in AgBr, CdS or CuCl [31].

The recombination of biexcitons into a photon and an exciton gives rise to well known and understood emission bands in different semiconductors like the Cu-halides, II-VI compounds or Si [1, 12, 32]. Under resonant two photon excitation of the biexciton narrow emission bands appeared, which have also been interpreted as a BEC of biexcitons [33]. It could be shown however independently by two groups, that these narrow lines result from a cold but non-condensed gas of biexcitons and / or from resonant two-photon or Hyper-Raman scattering [32, 34]. However, it was possible to verify the Bosonic character of biexcitons in CuCl in the sense that they are preferentially scattered into a state which is strongly populated by an external laser pump source [35], but again no spontaneous BEC could be reached. We come back to this type of experiments in 3.6.

For a short while the disappearance of excitonic features from the reflection and transmission spectra of CdSe under high excitation has been considered as an indication for an excitonic BEC [36] but as we know from chapter 2, this is actually an indication for a transition to an EHP.

In Ge the formation of an EHL at low temperatures could be suppressed by the application of stress and magnetic field. Nevertheless it was also in this case not possible to reach an excitonic BEC with increasing pump power, though a Bosonic line narrowing was observed at intermediate densities [37].

After this short overview over older work we concentrate on Cu_2O. This is a direct gap semiconductor with a large exciton binding energy of 150meV but with parity forbidden band-to-band transition. For data see [1] and references therein. The lowest exaction state ($n_B = 1$) is split by exchange interaction by 12meV in a lower para-exciton which is optically forbidden to all orders. Only its Γ_5^- LO phonon satellite is seen very weakly in the luminescence spectra [1, 38]. The ortho-exciton is only quadrupole allowed nevertheless resulting in a beautiful polariton dispersion leading continuously from the ortho exciton branch over the bottleneck and the photon like branch to E -> 0 for k -> 0. [38]. It is weakly seen both in absorption and emission and strongly in two photon absorption [38]. Furthermore its various LO phonon replica show up strongly in luminescence and absorption. The LO-phonon assisted absorption shows nicely the square root dependence of the density of states. This absorption band is superimposed by the n_BP exciton states, which are weakly dipole allowed due to the odd parity of the P envelope function. In good samples the n_BP series can be observed up to $n_B = 7$ and beyond.

The LO assisted luminescence bands, which reflect the distribution of the excitons in their bands, prove that the excitons reach thermal equilibrium with the lattice down to temperatures well below 10K, at least under low excitation. Examples for all of these statements can be found in [1, 7, 39].

Since the relaxation of ortho excitons to para excitons is rather slow at low temperatures [40] there was a hope that ortho excitons might be pumped to beyond n_c at low crystal temperatures. A certain change of the emission line shape from a simple Boltzmann to a narrower Bose type has been found [7, 41], but various authors agree, that the density approaches the $n_c(T_c)$ curve from low densities but never reaches it [7, 42].

The next approach came from transport measurements of presumably para excitons [43]. The experimental set up and the main findings are the following. A brick-shaped Cu_2O sample is exited on one side by intense ns pulses in the LO phonon continuum, producing a cloud of excitons.

Their arrival at the opposite side is monitored by a current pulse produced by their field ionization in a Cu_2O/Cu Schottky contact barrier resulting in a current pulse which is monitored on an oscilloscope. The signal starts after a delay time given by the LA velocity of sound, while the following signal is temporally rather broad, indicating a diffusive transport of the excitons over the sample length of about 3.5mm. It steepens and gets shorter if the density is increased at low temperature or if the temperature is lowered at high excitation.

The interpretation given in [43] is that the excitons undergo a BEC into a superfluid state when the critical values of n_c, T_c are meat, and the condensed cloud propagates with v_{LA} through the sample to the Shottky barrier.

A first small problem in this interpretation is that the signal area saturates with increasing intensity while one would expect rather the opposite behaviour for a BEC.

As is usual in the field of excitonic BEC, various alternative explanations have been put forward and the BEC and superfluidity have seriously been questioned in various ways.

Due to the weak absorption in the 1s ortho exciton state, excitons have to be excited in the LO-phonon or ionization continuum, if one wants to reach high densities. Consequently a large non thermal population of optical and then of acoustic phonons is created by the relaxation of the exciton to the (para-) exciton ground state, and later on by their dominantly non radiative recombination [40]. The phonon cloud created near the surface propagates into the sample and it has been calculated by [44] that an exciton flux very similar to the experimentally observed signal can be expected at the place of the detector caused by the phonon wind, i.e. a ballistic motion of the excitons driven by the expanding non-thermal cloud of LA phonons.

This interpretation does not involve any superfluidity nor BEC of the excitons and explains easily why the signal onset delay at the detector coincides with the time of flight of LA phonons.

Spatially and temporally resolved pump and probe beam spectroscopy of a thin Cu_2O rod has been performed in [45] using the broadening and bleaching of the higher n_BP excitons states caused by a high density of excitons [18, 39] as a measure of the density of excitons along the crystal as a function of distance from the excited surface and time.

The results show exciton transport over distances of around 1mm but the whole propagation process could be simulated within experimental error by classical diffusive propagation for the temperature range from 2 to 30K and the average propagation velocity of the exciton cloud was well below v_{LA}.

Presently work is under way in the group of the author to combine electrical measurements like in [43] with optical ones [45] on the same sample.

Another criticism came from a detailed analysis of the intensity and dynamics of the ortho exciton luminescence [46] and led the authors to the conclusions, that the Bose-Einstein luminescence line shape is due to inhomogenieties of the exciton population and, an artefact more important, that the Auger-recombination of the excitons and the heating of the exciton gas are so dramatic, that one cannot come even close to the conditions of an excitonic BEC.

This huge Auger cross section has in turn been questioned since the Bohr radius is small and the excitons do not even carry an electric dipole moment. Instead an efficient ortho -> para conversion mechanism has been put forward in theoretical investigations [47] for high densities. Experimentally it has been found independently in [45, 48] that the Auger cross section is indeed small.

Even more recently it has been shown [49] that the Schottky barrier exciton detector used in [43] to monitor the arrival of the excitons might be driven under the highest excitation conditions by several orders of magnitude into saturation, explaining both qualitatively the variation of the time dependence of the electrical signal with increasing excitation and the saturation behaviour of its time integral. These arguments are subject of a presently ongoing debate.

First attempts failed to confine excitons in Cu_2O in a potential trap caused by externally applied inhomogeneous stress [50]. New experiments using two photon pumping, partly connected with stress induced wells started recently [51, 52] but did not yet result in an evidence for excitonic BEC.

The same statement is true for excitonic inter subband spectroscopy (i.e. investiagion of the excitonic Lyman series) for which theory predicts features characteristic for excitonic BEC [53, 54].

Though it is not possible to cite all experimental work published during the last forty years, the selection shown here makes clear, that there is until now no clear cut and generally accepted proof for excitonic BEC. On the other hand, there is no generally accepted theoretical result explaining why it should not occur at all [55].

So the search for it will and does continue, including structures of reduced dimensionality. Before we go to such structures in section 3.5 we consider a simple phase diagram in section 3.4.

3.3 A PEDESTRIAN APPROACH TO A PHASE DIAGRAM

We gave in Fig. 7 already a phase diagram for the transition to an EHL. We want to complete this phase diagram including excitonic BEC in Fig. 11 using a simple minded pedestrian approach.

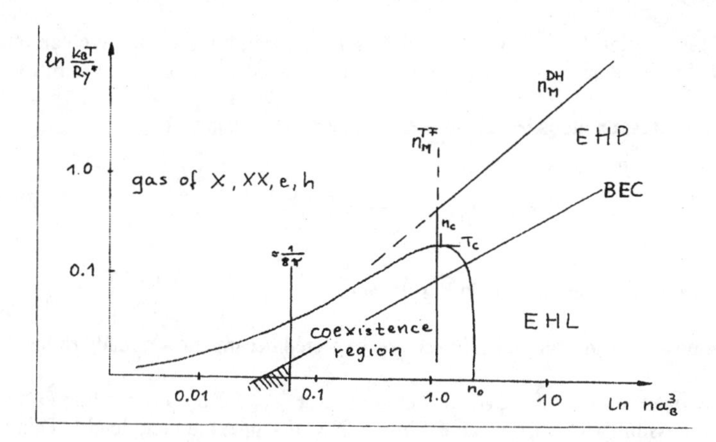

Figure 11: A schematic phase diagram including EHL formation and excitonic BEC. For the meaning of the various lines see text. The units on the x and y axes have to be considered as approximate ones only.

We mentioned in 2.1, that a transition to an EHP occurs beyond a certain density n_M or if the screening length l_c becomes comparable to the excitonic Bohr radius a_B.

In the Debye Hückel approximation l_c is given by

$$l_{DH} = \left(\frac{\varepsilon\varepsilon_0 k_B T}{e^2 n_P} \right)^{1/2} \tag{13a}$$

resulting in

$$n_M^{DH} = (1.19)^2 \frac{\varepsilon\varepsilon_0 k_B T}{e^2 a_B^2} = (1.19)^2 \frac{k_B T}{2 a_B^3 Ry^*} \tag{13b}$$

This approach is only valid as long as the exciton and carrier gases can be described by classical statistics. For very low temperatures it would give the unphysical result that a vanishing density results already in a screening of the exciton. We give therefore n_M^{DH} in Fig. 11 only for higher temperatures. At low temperatures, where the carrier gas is degenerate, the Thomas-Fermi screening length is more appropriate. It gives a temperature independent value of n_M^{TF} given roughly by

$$n_M^{TF}\, a_B^3 \approx 1 \tag{13c}$$

which is also given in Fig. 11. This means than an EHP exists to the right and below these two curves.

Now we give the criterion for a BEC for ideal Bosons according to eq. (8). We can anticipate, that excitons will no longer behave as ideal Bosons close to the Mott transition to a Fermi gas.

In [56] it is claimed that excitons behave as weakly interacting Bosons for

$$n_P\, a_B^3 << \frac{1}{8\pi} \approx 0.04 \tag{13d}$$

The limit 0.04 is shown as vertical line in Fig. 11, too.

The shaded region where an excitonic BEC can be expected became now already rather small.

Now we add the phase diagram for EHL formation from Fig. 7 for a finite binding Ø > 0 of the EHL and see that the possibility to observe an excitonic BEC disappeared completely. This is also stated on p 12 of [7], but is then wavied away rather quickly. The author feels, that some more consideration should be given to this argument.

In Si and Ge the EHL has been clearly observed (see 2.4). Even if Φ was brought close to zero by external fields, no excitonic BEC occurred.

In all III-V and II-VI compounds the plasma forms a bound state. The fact that no phase separation was observed is not due to the fact that the plasma is not bound, but that the carrier lifetime is too short for the phase separation to form. Possibly it is also too short for the BEC condensation to take place, which would explain the failure to observe it in this rather large group of materials.

In CuCl the creation of an EHP has been verified experimentally [17] however our knowledge of its properties is too limited to give a definite statement about Φ. In Cu_2O no report of an EHP is known to the author. Under conditions where it is possible to bleach all n_BP exciton states [18], the 1s exctions are possibly far away from being idealized, weakly interacting Bosons.

To conclude this section, it should be mentioned, that more complex phase diagrams are discussed in theory, which allow for an excitonic BEC pocket and are reviewed e.g. in [7]. However there does not seem to be much of an experimental verification of these diagrams until now.

3.4 BOSE-EINSTEIN CONDENSATION IN (QUASI-) TWO-DIMENSIONAL SYSTEMS

Before discussing recent results on coupled quantum wells in section 3.5, we want to have a short look into two-dimensional systems in general.

Shifting the step-like density of states function of a two-dimensional system of massive particles of Fig. 10c over the Bose-Einstein distribution function of Fig. 10a shows immediately that strictly speaking a BEC is not possible in two dimensions. Either μ is below E_0 then there is no BEC or μ coincides with E_0 then a divergence of the particle density arises from the product of a finite density of states and the divergence in their occupation probability.

Actually the DOS is not step like in the sense of a mathematical heavy side function $\Theta(E\text{-}E_0)$. One has a (often exponential) tail of localized states below E_0. These tail states behave however not like Bosons. They can be empty, they can be occupied by one exciton. If two excitons (i.e. a biexciton) are placed in this state the energy of both shifts to the red. Higher occupancies lead, if possible at all, to a blue shift.

On the other hand it would be also difficult to imagine what a BEC of localized excitons means when every particle sits in another place and at a different energy.

Though there is strictly no BEC in two dimensions for massive particles, a transition to a superfluid state is possible according to Kosterlitz and Touless [7] (KTS) for densities n_c above or temperatures below T_c given by

$$T_c \approx \frac{\left(\hbar^2/2m\right)4\pi n_c}{\ln\ln\left(1/n_c a^2\right)} \tag{14a}$$

resulting to a good approximation in

$$n_c \approx \frac{0.32 g m k_B T_c}{\hbar^2} \tag{14b}$$

In the following we describe to which extend these ideas could be realized in a system not completely different from the one proposed in [57].

3.5 ATTEMPTS TO OBSERVE EXCITONIC SUPERFLUIDITY IN TILTED COUPLED QUANTUM WELLS

According to the above considerations, there were recently attempts by two groups (L.V. Butov et al. and D. Snoke et al.) to observe excitonic superfluidity in the frame of KTS using two coupled and tilted quantum wells [58, 59]. The basic idea of the samples is the some one in both cases and is shown in Fig. 12a. In a n^+in^+ structure two coupled wells with a narrow barrier are incorporated in the intrinsic layer. A voltage applied to the n^+ cladding layers tilts the band structure and separates the lowest electron state from the highest hole state in the way shown in Fig. 12a. Butov et al. use

n^+GaAs and $Al_{1-y}Ga_yAs$ barriers around and between the two GaAs wells in the intrinsic region while Snoke et al. use $In_{1-y}GaAs$ wells.

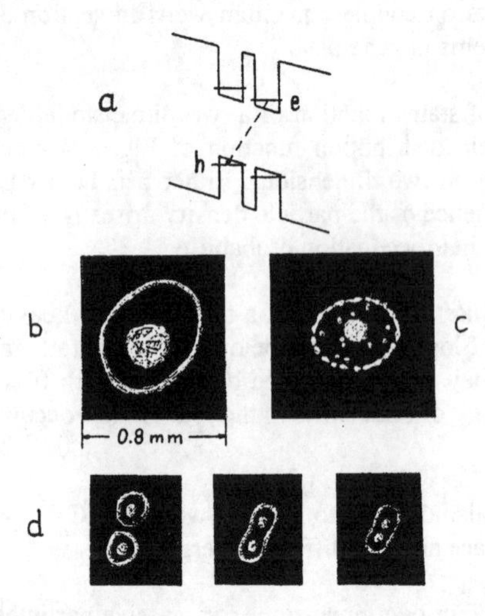

Figure 12: Schematic sketch of the band alignment of the samples used by [58, 59] (a) schematic drawings of the luminescence ring structure observed by [59] (b) and by [58] (c) and the behaviour observed by both groups when to excitation spots are used with decreasing distance (d).

In both cases excitons are formed which carry a permanent dipole moment due to the spatial separation of electrons and holes under the action of an external field. This arrangement increases the exciton lifetime to values in the 100ns regime, but reduces also the exciton binding energy [60], and creates a repulsive interaction which prevents biexciton formation due to the parallel aligned electric dipoles.

At low temperatures and under cw excitation both authors observe rather similar phenomena. With increasing excitation in the barrier luminescence is not only observed from the excitation spot but also from a bright ring with no detectable emission between Fig. 12b. except for some localization centers in the case of [59].The diameter of the ring increases with increasing pump power reaching diameters in the 1mm regime. If the excitation spot is moved on the sample the ring structure follows the excitation spot.

Temporal variations of the excitation intensity result in variations of the luminescence in the ring structure indicating radial velocities of the order of (1 to 5) 10^6cm/s which is considerably larger than the LA velocity of sound in GaAs [59].

Snoke at al. investigate the ring at temperatures of a few K but can follow it up to $T \approx 90K$(!!) Butov et al. works preferentially at lower temperatures and observes a fragmentation of the ring into bright spots which are rather equally spaced along the ring with some bright spots between the directly excited area and the ring resulting from localization sites in the MQW structure [58]. See Fig. 12.

If two widely separated spots are excited, they are surrounded by individual rings. If the excitation spots are brought closer together, both groups find that the rings are deformed and merge to a single structure in the way shown in Fig. 12d.

Though the experimental findings are rather similar, the interpretations of both groups are very different.

Butov et al. claim that the mutual repulsion of the excitons (which is even enhanced by the depolarization field) results in a rapid, presumably ballistic expansion of the excitons with $\mathbf{k}_{\parallel} > \mathbf{k}_{light}$ so that they cannot radiate because $\mathbf{k}_{\parallel}$ is conserved at a plane interface. The velocities given by Snoke are consistent with such arguments. After that expansion, the excitons undergo a BEC transition resulting in the ring and the bright dots along the ring are considered as superfluid vortices.

In contrast Snoke claimes that the excitons are in the KTS superfluid state in the dark range between the excitation spot and the ring. The bright ring results from the transition of the excitons into their normal state due to dilution and subsequent recombination. While being in the superfluid state, emission is forbidden for the excitons for some reason or other, e.g. because of their large value of $\mathbf{k}_{\parallel}$. The argument of Snoke against the interpretation of Butov is, that a purely ballistic propagation of excitons in these structures over distances up to one mm is extremely unlikely. If on the other hand, scattering is allowed, then excitons will be created also at smaller values of $\mathbf{k}_{\parallel}$ and a measurable luminescence should be observed between the exciton spot and the ring.

The mergin of the rings is more difficult to explain in the model of Butov, because the density should be higher in the overlapping region, while the observation of the equally space bright spots is difficult to explain in Snoke's model.

At a recent workshop [61] on "Nonlinear Optics and Excitation Kinetics in Semiconductors" (NOEKS), Snoke came up with a completely different interpretation of the data [62]. He observed, that the ring appears only if the sample is excited in the barrier. If one excites the sample directly in the wells, no ring structure appears, even if the incident intensity is increased to compensate for the reduced absorption.

Snoke's present interpretation is, that the formation of the ring has nothing to do with a superfluid state at all. If excited in the well carriers recombine essentially at the excitation spot. When excited in the barrier, the holes are captured in their well, but electrons partly escape from the wells to the n^+ layer or are not captured at all due to their smaller effective mass. This has the consequence that a two-dimensional "puddle" of holes forms in the well at and around the excitation spot. Apart from the directly excited area, hardly any electrons exist in this "puddle" and the luminescence occurs at its boarder when electrons reach it, coming e.g. from the n^+ layers. This model explains easily the merging of the rings in Fig. 12d by a merging of the two hole "puddles". Butov still explains the bright dots on the ring as superfluid vortices [63].

As a short addendum it can be stated that similar rings have been observed by V. Lyssenko in GaAs single quantum wells and superlattices [64]. The appearance and disappearance of theses rings can be influenced by varying the focussing of the excitation beam.

Luminescence ring structures have been also observed in thin CdS platelets under pulsed excitation and at low temperatures [65] but have been interpreted as light scattering under conditions of stimulated emission.

Evidently we are left in quasi two-dimensional systems with the same situation as in bulk material. There are presently no clear, generally accepted evidences for excitonic BEC nor for superfluidity.

Possibly there are some still unknown reasons why it does not occur [55], which is a challenge to the theoreticians, or, as L.V. Keldysh stated [66] "possibly we had it all the time without noticing it, because its influences on the optical properties are completely unspectacular", which is a challenge to experimentalists.

Furthermore Snoke stated during [61] as a joke that not every circular emission is connected with KTS or BEC as the reader can confirm by entering "Hoag's Object" in a search machine in the Web.

3.6 DRIVEN EXCITONIC BOSE-EINSTEIN CONDENSATIONS

Since it is obviously difficult to find and to uniquely proof excitonic BEC or (KTS) superfluidity in semiconductors, some authors invented the term "driven BEC".

A general property of these suggestions, which are shortly reviewed e.g. in [59] is, that it is no process occurring in thermal (quasi-) equilibrium distribution and thus no BEC.

Instead one is either creating a coherent population by an external laser into which or out of which particles are scattered or it involves lasing on the (photon-like) part of the exciton-polariton dispersion.

It is also a common feature, that similar phenomena have bee found one or a few decades ago under different names.

For example all of the processes shown in Fig. 1 in the intermediate density regime can result in stimulated emission on the lower polariton branch as found theoretically or verified experimentally [1, 12, 32] but at that time nobody had the idea to call such phenomena a BEC.

As mentioned already the scattering of biexcitons into a state which is populated coherently by an external laser source was considered as a proof of the Bosonic character of biexcitons but not as a BEC.

Another phenomenon connected with biexcitons or more generally two polariton states and which has been observed in several I-VII and II-VI compounds [1, 12, 32], has been introduced as two-photon or hyper Raman scattering.

Two polaritons (index i) are created by an incident laser and scatter (possibly after forming an intermediate virtually excited biexciton) in two outgoing particles 01 and 02 under energy and momentum conservation according to

$$2\hbar\omega_i = \hbar\omega_{01} + \hbar\omega_{02} \tag{15a}$$

$$2\mathbf{k}_i = \mathbf{k}_{01} + \mathbf{k}_{02} \tag{15b}$$

Depending on the scattering geometry and on other experimental conditions both or only one of the outgoing quanta can appear in emission. One can also be on the longitudinal exciton branch. This process can occur spontaneously, but it can also show gain and stimulated emission in the sense of an optical parametric amplifier. Actually it is a processes belonging to the group of $\chi^{(3)}$ or of higher order processes like four wave mixing (FWM).

The decay can also be stimulated by sending an external laser beam into one of the outgoing channels resulting in an enhanced scattering into this and via (15a, b) also into the other channel. This would be then a typical example for non degenerate four wave mixing (NDFWM) or electronic coherent (anti-) Stokes Raman Scattering (CARS) [1, 12, 32].

The polariton concept as the quanta of the mixed state of electro-magnetic and polarization wave has been introduced about forty years ago, for excitons e.g. by scientists like Thomas and Hopfield [67] and for quantum wells in micro cavities in the eighties [68]. Now we can read papers speaking about states in a micro cavity which are "half photon and half exciton" and which are scattered under a "magic angle" in a micro cavity to produce populations at $\mathbf{k}_{\parallel} \approx 0$ and $\mathbf{k}_{\parallel} = 2\mathbf{k}_{\parallel i}$ where $\mathbf{k}_{\parallel i}$ is the incident wave vector.

Apart form the fact, that one has transferred the experiments from bulk samples to quasi two-dimensional systems and cavities they are extremely similar to the process described shortly in connection with eq. (15a, b).

3.7 EXCITONIC INSULATORS

To conclude this section on BEC we shortly mention another concept of excitonic BEC, namely the so-called excitonic insulators. There are two scenarios for the occurrence of this phenomenon which will be shortly outlined below. More details and many references can be found in [7].

One scenario occurs in narrow gap semiconductors with E_g^n, or semimetals ($E_g = 0$). If the binding energy of excitons is still finite, the system may lower its energy at sufficiently low temperatures by the spontaneous formation of excitons provided that the following inequalities are fulfilled

$$Ry^* > E_g^n \text{ and } Ry^* >> k_B T \tag{16}$$

Since a gas of excitons is insulating compared to the gas of free carriers usually present in semimetals and narrow gap semiconductors the transition to the excitonic insulator should show up in a characteristic variation of the resistivily.

One of the candidates is or was the semimetal grey tin but no experimental indication of spontaneous exciton formation nor of their condensation is known to the author.

In [69] experiments have been reported in $TmSe_{0.45}Te_{0.55}$, where the resistivity varies roughly three orders of magnitude when tuning the gap by external pressure.

Another system where a transition to an excitonic insulator could occur are wide gap semiconductors in which a degenerate EHP has been created. If the effective masses of electrons and holes are equal (a situation generally not met in real semiconductor) a gap in the sense of a BCS-theory could open simultaneously in the degenerate conduction and valence band populations with decreasing temperature resulting in vanishing total conductivity of the condensates.

4. Conclusion and Outlook

To summarize we have shown that the existence and the properties of the electron-hole plasma in semiconductors are well established both experimentally and theoretically. Concerning excitonic BEC and superfluidity, the situation is still unclear and controversially discussed in literature even 40 years after the first theoretical prediction and it is to some extend a question of taste or of the personal experiences of the scientist if he stresses the "pro" or "contra". The author is obviously presently more on the "contra" side. In any case there is much more theoretical work on the topic than hard experimental facts or as P.B. Littlewood stated [70] the "smoking gun" argument is still missing. Possibly in comes in the future to finish with an optimistic statement.

Acknowledgements:

The author should like to thank the Deutsche Forschungsgemeinschaft for financial support of his work in the above fields, his coworkers for the fine experiments cited partly in this review and many colleagues for stimulating, fruitful and partly controversial discussion especially Profs. Drs. A. Mysyrowicz (Palaiseau), Fortin (Ottawa), D.S. Snoke (Pittsburg), L.V. Keldysh (Moscow), S.W. Koch (Marburg) or H. Haug (Frankfurt am Main).

References:

1. Klingshirn, C. (1997), *Semiconductor Optics*, 2nd printing, Springer, Berlin.
2. Moskalenko, S.A. (1962), *Fiz. Tverd. Tela (Sov. Phys. Solid State)* **4**, 276.
3. Blatt, J.M., Böer, K.W. and Brandt, W. (1962), Bose-Condensation of Exctions, *Phys. Rev.* **126**, 169.
4. Keldyh, L.V. and Kozlov, A.N (1968), Collective Properties of Excitons in Semiconductors, *Sov. Phys. JETP* **27**, 521.
5. Hanamura, E. and Haug, H. (1977) Condensation Effects of Excitons, *Phys. Report* **33C**, 209.
6. Snoke, D.W. and Stringari, S. (1995) *Bose-Einstein Condensation*, Cambridge University Press, Cambridge, UK.
7. Moskelanko, S.A. and Snoke, D.W. (2000) *Bose-Einstein Condensation of Excitons and Biexcitons (and coherent Nonlinear Optics with Excitons)*, Cambridge University Press, Cambridge, UK.
8. Bohnert, K., Anselment, M., Kobbe, G., Klingshirn, C., Haug, H., Koch, S.W., Schmitt-Rink, S. and Abraham, F.F. (1981) Nonequilibrium Properties of Electron-Hole Plasma in Direct-Gap Semiconductors, *Z. Physik B* **42**, 1.
9. See e.g. the contribution of E. Kapon to this book for quantum wires and ~dots.
10. Bohnert, K., Schmieder, G. and Klingshirn, C. (1980), Gain and Reflection Spectroscopy and the Present Understanding of the Electron-Hole Plasma in II-VI Compounds, *phys. stat. sol. b* **98**, 175.
    Majumder, F. A., Swoboda, H.-E., Kempf, K. and Klingshirn, C. (1985) Electron-Hole Plasma in the Direct-Band-Gap Semiconductors CdS and CdSe, *Phys. Rev. B* **32**, 2407.
    Swoboda, H.-E., Majumder, F.A., Lyssenko, V.G., Klingshirn, C. and Banyai, L. (1988) The Electron-Hole Plasma in CdS between 5 K and Room Temperature, *Z. Phys. B* **70**, 341.
11. Kempf, K., Schmieder, G., Kurtze, G. and Klingshirn, C. (1981) Excitation Induced Renormalization Effects of the Excitonic Polariton Dispersion in CdS, *phys. stat. sol. b* **107**, 297.
    Schmitt-Rink, S., Löwenau, J.P., Haug, H., Bohnert, K., Kreissl, A., Kempf, K. and Klingshirn, C. (1983) Theoretical and Experimental Studies of the Transition from the Exciton to the Plasma Phase, *Physica* **117/118B**, 339.
12. Klingshirn, C. and Haug, H. (1981) Optical Properties of Highly Excited Direct Gap Semiconductors, *Physics Reports* **70**, 315.
13. Zimmermann, R., (1988) Many Particle Theory of Highly Excited Semiconductors, *Teubner Texte zur Physik* **18**.
14. Kunz, M., Pier, T., Bhargava, R.N., Reznitsky, A., Kozlovskii, V.V., Müller-Vogt, G., Pfister, J.C., Pautrat, J.L. and Klingshirn, C. (1990), The Electron-Hole Plasma in Cubic ZnTe and ZnSe Crystals, *J. Crystal Growth* **101**, 734.
    Majumder, F.A., Klingshirn, C., Westphäling, R., Kalt, H., Naumov, A., Stanzl, H., and Gebhardt, W. (1994) Gain Processes in ZnTe Epilayers on GaAs, *phys. stat. sol. (b)* **186**, 591.
15. Klingshirn, C. Weber, Ch., Chemla, D.S., Miller, D.A.B., Cunningham, J.E., Ell, C. and Haug, H.'(1989) The Electron-Hole Plasma in Quasi Two-Dimensional and Three-Dimensional Semiconductors, eingel. Beitrag zu NATO Workshop on "Optical Switching in Low Dimensional Systems", Marbella, Spain, Oct. (1988), *NATO ASI Series, B* **194**, 353.
16. Beni, G. and Rice, T.M. (1978) Theory of electron-hole liquid in semiconductors, *Phys. Rev. B* **18**, 768.
17. Hulin, D., Mysyrowicz, Migus, A., A., Antonetti, A. (1985) Subpicosecond time-resolved Mott transition in CuCl, *J. Luminesc.* **30**, 290.
18. Jolk, A., Jörger, M., Klingshirn, C., Franco, M., Prade, P. and Mysyrowicz, A. (2000) Differential Transmission Spectroscopy (DTS) in Cu_2O in the Presence of Cold Excitons, *phys. stat. sol. b* **221**, 295.
19. Forchel, A., Laurich, B., Moersch, G., Schmid, W. and Reinecke, T.L. (1981) Experimental Verification of Scaling Relations for Electron-Hole Liquid Condensation, *Phys. Rev. Lett.* **46**, 678.
20. Klingshirn, C. (1992) Properties of the Electron-Hole Plasma in II-VI Semiconductors, *J. of Crystal Growth* **117**, 753.
21. Koch, S.W. (1984) Dynamics of First-Order Phase Transitions in Equilibrium and Nonequilibrium Systems, *Lecture Notes in Physics* **207** Springer, Berlin.
22. Thomas, G.A., Frova, A., Hensel, J. C., Miller, R. E., and Lee, P. A. (1976) Collision broadening in the exciton gas outside the electron-hole droplets in Ge, *Phys. Rev. B* **13**, 1692.

23. Martin, R.W. (1976), On the mechanism of indirect band to band recombination in germanium electron-hole drops, *Solid State Commun.* **19**, 373 and
Martin, R.W. and Störmer, H.L. (1977) On the low energy tail of the electron-hole drop recombination spectrum, *Solid State Commun.* **22,** 523.
24. Thomas, G.A., Mock, J.B. and Capizzi, M. (1978) Mott distortion of the electron-hole fluid phase diagram, *Phys. Rev. B* **18,** 4250.
Thomas, G.A., Rice, T.M. and Hensel, J.C. (1974) Liquid-Gas Phase Diagram of an Electron-Hole Fluid, *Phys. Rev. Lett.* **33,** 219.
Reinecke, T.L. and Ying, S.C. (1975) Droplet Model of Electron-Hole Liquid Condensation in Semiconductors, *Phys. Rev. Lett.* **35,** 311.
25. Greenstein, M. and Wolfe, J.P. (1980) Formation of the electron-hole droplet cloud in germanium, *Solid State Commun.* **33,** 309.
26. Gourley, P.L. and Wolfe, J.P. (1978) Spatial Condensation of Strain-Confined Excitons and Excitonic Molecules into an Electron-Hole Liquid in Silicon, *Phys. Rev. Lett.* **40,** 526.
27. Brinkman, W.F. and Rice, T.M. (1973) Electron-Hole Liquids in Semiconductors, *Phys. Rev. B* **7,** 1508 and references therein.
28. Weber, C., Klingshirn, C., Chemla, D.S., Miller, D.A.B., Cunningham, J. and Ell, C. (1988) Gain Measurements and Band-Gap Renormalization in GaAs/$AlxGa_{1-x}$ Multiple-Quantum-Well Structures, *Phys. Rev. B* **38,** 12748.
Schlaad, K.-H., Weber, Ch., Cunningham, J., Hoof, C.V., Borghs, G. Weimann, G., Schlapp, W., Nickel, H. and Klingshirn, C. (1991) Many-Particle Effects and Nonlinear Optical Properties of GaAs/(AlGa)As Multiple-Quantum-Well Structures under Quasistationary Excitation Conditions, *Phys. Rev. B* **43,** 4268.
29. Kulakovskii, V.D. et al. (1989) Band-gap renormalization and band-filling effects in a homogeneous electron-hole plasma in $In_{0.53}Ga_{0.47}As$/InP single quantum wells, *Phys. Rev. B* **40,** 8087.
Lach, E., Kulakovskii, V.D., Forchel, A., Reinecke, T.L., Straka, J., Grutzmacher, D., Weimann, G. (1990) Single and many particle effects in the emission spectra of laterally homogeneous 2D plasmas, *phys. stat. sol. b* **159,** 125.
30. Kapon, E. (2003) Self-Ordered Growth and Spectroscopy of Nonplanar Quantum Wires and Quantum Dots, *NATO Sciences Series II* **90,** 243.
31. Czaja, W. and Schwerdtfeger, C.F. (1974), Evidence for Bose-Einstein condensation of free excitons in AgBr, *Solid State Communication* **15,** 87.
Nagasawa, N., Nakata, N., Doi, Y. And Ueta, M. (1975) The Bose condensation of excitonic molecules in CuCl crystals , *J. Phys. Soc. Japan* **38,** 593.
Anzai, T., Goto, T. and Ueta, M. (1975) Zeeman effect of an induced absorption line in highly excited CuCl, *J. Phys. Soc. Japan* **38,** 774.
Weber, J. and Stolz, H., Decay of the proposed Bose-Einstein condensed excitons in AgBr, *Solid State Communication* **24,** 707.
Weber, J. (1976) Indirect near edge emission in pure AgBr, *phys. stat. sol. b* **78,** 699.
32. Hönerlage, B., Levy, R., Grun, J.B., Klingshirn, C., Bohnert, K. (1985) The Dispersion of Excitons, Polaritons and Biexcitons in Direct-Gap Semiconductors, *Physics Reports* **124,** 161.
33. Nagasawa, N., Koizumi, S., Mita, T. and Ueta, M. (1976) Generation of exctionic molecules by giant two-photon absorption in CuCl and CuBr and their Bose condensation, *Journ. of Luminesc.* **12/13,** 587.
34. Levy, R., Klingshirn, C., Ostertag, E., Duy Phach, Vu and Grun, J.B. (1976) Luminescence of a "Cold" Gas of Biexcitons in CuCl, *phys. stat. sol. b* 77, 381.
Duy Phach, Vu., Bivas, A., Hönerlage, B., Grun, J.B., Two-photon resonant Raman Scattering via Biexcitons, *phys. stat. sol. b* **86,** 159.
Hönerlage, B., Duy Phach, Vu. And Grun, J.B. (1978) Polarization properties of biexciton luminescence and two-photon Raman emission in CuCl, *phys. stat. sol. b* **88,** 545.
Ojima, M., Kushida, T., Shionaya, S., Tanaka, Y., and Oka, Y. (1978) Optical studies of resonantly excited excitonic molecules in CuCl, *J. Phys. Soc. Japan.* **45,** 884.
Kushida, T. (1979) Secondary emission under two-photon resonance excitation of excitonic molecule in CuCl, *Solid State Commun.* **32,** 209.
35. Peyghambarian, N., Chase, L.L. and Mysyrowic, A. (1983) Bose-Einstein statistical properties and condensation of excitonic molecules in CuCl, *Phys. Rev. B* **27,** 2325.

36. Akopyan, I.Kh. and Razbirin, B.S. (1970) Bose Einstein condensation of excitons in a CdSe crystal, *JETP Lett.* **12**, 251.
37. Timofeev, V.B., Kulakovskii, V.D. and Kukushkin, I.V. (1983) Spin aligned exciton gas in uniaxially compressed Ge, *Physica B + C* **117/118**, 327.
38. Fröhlich, D., Kulik, A., Uebbing, B., Mysyrowic, A., Langer, V., Stolz, H. and von der Osten, W. (1991)., Coherent Propagation and Quantum Beats of Quadrupol Polaritons in Cu_2O, *Phys. Rev. Lett.* **67**, 2343.
Uihlein, Ch., Fröhlich, D., and Kenklies, R. (1981) Investigation of exction fine structure in Cu_2O, *Phys. Rev. B* **23**, 2731.
39. Jolk, A. and Klingshirn, C. (1998) Linear and Nonlinear Excitonic Absorption and Photoluminescence Spectra in Cu_2O: Line Shape Analysis and Exciton Drift, *phys. stat. sol. b* **206**, 841.
40. Jörger, M., Schmidt, M., Jolk, A., Westphäling, R. and Klingshirn, C. (2001) Absolute External Photoluminescence Quantum Efficiency of the 1s-Orhtoexction in Cu_2O, *Phys. Rev. B* **64**, 113204.
41. Snoke, D.W., Wolfe, J.P. and Mysyrowicz, A. (1990) Evidence for Bose-Einstein condensation of excitons in Cu_2O, *Phys. Rev. B* **41**, 11171.
42. Snoke, D.W. and Wolfe, J.P. (1990) Picosecond dynamics of degenerate orthoexcitons in Cu_2O, *Phys. Rev. B* **42**, 7876.
Naka, N., et al. (1996) A New Aspect of the Bose-Einstein Condensation of 1s Exction system in Cu_2O, *Progr. Crystal Grwoth and Charact.* **33**, 89.
Kavoulakis, G.M. (2001) Bose-Einstein Condensation of excitons in Cu_2O, *Phys. Rev. B* **65**, 035204.
43. Benson, E., Fortin, E., Mysyrowicz, A. (1995) Study of Anomalous Excitonic Transport in Cu_2O, *physica status solidi b* **191**, 345.
Mysyrowicz, Benson, E. and Fortin, E. (1996) Directed Beam of Excitons Produced by Stimulated Scattering, *Phys. Rev. Lett.* **77**, 896.
44. Tikhodeev, S.G., et al. (1997) Comment on "Directed Beam of Excitons Produced by Stimulated Scattering", *Phys. Rev. Lett.* **78**, 3225.
Kopelevich, G.A. et al. (1996), Phonon wind and excitonic transport in Cu_2O semiconductors, *Sov. Phys. JETP* **82**, 1180.
Jackson, A.D. and Kavoulakis, G.M. (2002) Propagation of exction pulses in semiconductors, *Europhys. Lett.* **59**, 807.
45. Jolk, A., Jörger, M. and Klingshirn, C. (2002) Exciton Lifetime, Auger Recombination and Exciton Transport by Calibrated Differential Absorption Spectroscopy in Cu_2O, *Phys. Rev. B* **65**, 245209.
46. O'Hara, K.E., Gullingsrud, J.R. and Wolfe, J.P. (1999) Auger decay of excitons in Cu_2O, *Phys. Rev. B* **60**, 10872 and
O'Hara, K.E., Súillebháin, L.Ó. and Wolfe, J.P. (1999) Strong nonradiative recombination of excitons in Cu_2O and its impact on Bose-Einstein statistics, *Phys. Rev. B* **60**, 10565.
47. Kavoulakis, G.M. and Mysyrowicz, A. (2000) Auger decay, spin exchange, and their connection to Bose-Einstein condensation of excitons in Cu/sub 2/O, *Phys. Rev. B* **61**, 16619.
48. Denev, S. and Snoke, D.W. (2002), Stress dependence of exction relaxation processes in Cu_2O, *Phys. Rev. B* **65**, 085211.
49. Klingshirn, C., Fleck, T. and Jörger, M. (2002) Some Considerations Concerning the Detection of Excitons by Fieldionization in a Schottky Barrier, *phys. stat. sol. (b)* **234**, 23.
50. Trauernicht, D.P., Wolfe, J.P. and Mysyrowicz, A. (1986) Thermodynamics of strain-confined paraexcitons in Cu_2O, *Phys. Rev. B* **34**, 2561.
51. Sun, Y., Wong, G K L, Ketterson, J B (2001) Production of 1s quadrupole-orthoexciton polaritons in Cu2O by two-photon pumping, *Phys. Rev. B* **63**, 125323.
52. Naka, N. and Nagasawa, N. (2002) Two-photon diagnostics of stress-induced exciton traps and loading of 1*s*-yellow excitons in Cu_2O, *Phys. Rev. B* **65**, 075209
Naka, N. and Nagasawa, N. (2002) Nonlinear paraexciton kinetics in a potential trap in Cu_2O under two-photon resonance excitation , *Phys. Rev. B* **65**, 245203.
Naka, N. and Nagasawa, N. (2003) Optical tracking of high-density cooled exctions in potential traps in Cu_2O, NOEKS 7, March 2003 Karlsruhe. *phys. stat. sol. b* **238**, in press.
53. Johnsen, K. and Kavoulakis, G.M. (2001) Probing Bose-Eisntein Condensation of Excitons with Electromagentic Radiation, *Phys. Rev. Lett.* **86**, 858.

54. Jörger, M., Tsitisishvili, E., Fleck, T. and Klingshirn, C. (2003) Infrared Absorption by Excitons in Cu_2O, NOEKS 7, March 2003 Karlsruhe, *phys. stat. sol. b* **238**, in press.
55. Johnston Jr., W.D., Shaklee, K.L. (1974), Considerations relevant to Bose condensation of excitonic molecules in CdSe, *Solid State Communic.* **15**, 73.
56. Fetter, A.L. and Wlecka, J.D. (1971) *Quantum Theory of Many Particle Systems*, Mc Graw Hill, New York.
57. Lozovik, Yu. E., and Yudson, V.I. (1975), Feasibility of superfluidity of paired spatially separated electrons and holes; a new superconductivity mechanism, *JETP Lett.* **22**, 274.
58. Butov, L.V. and Filin, A.I. (1998), Anomalous transport and luminescence of indirect excitons in AlAs/GaAs coupled quantum wells as evidence for exciton condensation, *Phys. Rev. B* **58**, 1980.
Butov, L.V., Ivanov, A. L., Imamoglu, A., Littlewood, P. B., Shashkin, A. A., Dolgopolov, V. T., Campman, K. L. and Gossard, A. C. (2001), Stimulated Scattering of Indirect Excitons in Coupled Quantum Wells: Signature of a Degenerate Bose-Gas of Excitons, *Phys. Rev. Lett.* **86**, 5608.
Butov, L.V., Lai, C. W., Ivanov, A. L., Gossard, A. C., Chemla, D. S. (2002), Towards Bose-Einstein condensation of excitons in potential traps, *Nature* **417**, 47.
Butov, L.V., Gossard, A. C., Chemla, D. S. (2002), Macroscopically ordered state in an exciton system, *Nature* **418**, 751.
59. Snoke, D., Denev, S., Liu, Y., Pfeiffer, L., West, K. (2002), Long-range transport in excitonic dark states in coupled quantum wells, *Nature* **418**, 754.
Snoke, D., (2002), Spontaneous Bose Coherence of Excitons and Polaritons, *Science* **298**, 1368.
60. Szymanska, M.H. and Littlewood, P.B. (2003) Excitonic binding in coupled quantum wells, *Phys. Rev. B* **67**, 193305.
61. Nonlinear Optics and Excitation Kinetics (NOEKS 7), Karlsruhe, February (2003), Proc. to be published in *phys. stat. sol. b* **238**.
62. Snoke, D. (2003), When should we say we have observed Bose condensation of excitons? NOEKS 07, *phys. stat. sol. b* **238**, in press.
63. Butov, L.V. (2003), Bose-Einstein condensation of excitons in semiconductors, NOEKS 07, *phys. stat. sol. b* **238**, in press
64. Lyssenko, V.G. (2003), private communication.
65. Lyssenko, V.G. and Klingshirn, C. (1978), unpublished.
66. Keldysh, L.V. (2003), private communication.
67. Hopfield, J.J. and Thomas, D.G. (1963), Theoretical and Experimental Effects of Spatial Dispersion on the Optical Properties of Crystals, *Phys. Rev. B* **132**, 563.
68. Houdre, R., Weisbuch, C., Stanley, R.P., Oesterle, U., Pellandini P., Ilegems, M. (1994) Measurement of cavity-polariton dispersion curve from angle-resolved photoluminescence experiments, *Phys. Rev. Lett.* **73**, 2043.
69. Bucher, B., Steiner, P. and Wachter, P. (1991), Excitonic insulator phase in $TmSe_{0.45}Te_{0.55}$, *Phys. Rev. Lett.* **67**, 2717.
70. Littlewood, P.B. (2002), private communication.

16. DYNAMICS OF SOLID-STATE COHERENT LIGHT SOURCES

Upconversion Luminescence Dynamics

M. POLLNAU
Advanced Photonics Laboratory
Institute for Imaging and Applied Optics
Swiss Federal Institute of Technology
CH-1015 Lausanne
Switzerland

1. Introduction

This book chapter aims at reviewing in brief the fundamentals of rare-earth-ion spectroscopy in dielectric solids, with special emphasis on energy-transfer upconversion between neighboring active ions in a solid-state host lattice. The energy-level scheme of the $4f$ sub-shell of rare-earth ions is explained and the main intra- and inter-ionic electronic transition processes are introduced. The pump-power dependence of upconversion luminescence and the influence of inhomogeneous active-ion distributions on energy-transfer upconversion are discussed. Examples illustrate how energy-transfer upconversion can impact the performance of solid-state lasers in a negative or positive manner.

2. Spectroscopic Processes of Rare-Earth Ions in Solid-State Laser Materials

2.1 SPECTRA OF RARE-EARTH IONS

The optical transitions of lanthanide (rare-earth) ions in the visible and infrared spectral region occur within the $4f$ subshell. This subshell is shielded by the outer $5s$ and $5p$ subshells and the influence of the host material is relatively small compared to, e.g., the $3d$ transitions in transition-metal ions. The electronic structure of trivalent rare-earth ions derives from the perturbation of the $4f$ energy level in the central-field approximation by the non-centrosymmetric electron-electron interaction, the spin-orbit interaction, and the crystal-field splitting (Stark effect), see the example of the energy-level scheme of Er^{3+} in Fig. 1. The spin-orbit multiplets are commonly denoted by their $^{2S+1}L_J$ terms in Russell-Saunders coupling, although the $4f$ electrons of lanthanide ions exhibit intermediate coupling and the total angular momenta J of the spin-orbit multiplets are linear combinations of the total orbital angular momenta L and total spins S. Single crystal-field (Stark) transitions between two spin-orbit multiplets can be distinguished in many crystalline host materials at ambient temperature. In glasses, inhomogeneous spectral-line broadening occurs due to the local variation of the ligand electric field. Also homogeneous (lifetime) broadening mechanisms are relevant in a number of glasses. This spectral-line

B. Di Bartolo and O. Forte (eds.), *Frontiers of Optical Spectroscopy*, 571-589.

broadening leads to lower absorption and emission cross-sections for the same transition in glasses compared to single-crystalline hosts and, therefore, generally higher pump threshold of laser transitions in glasses.

2.2 INTRAIONIC PROCESSES

Generally, the probability of an allowed electric-dipole transition is seven orders of magnitude larger than that of an allowed magnetic-dipole transition. Since electric-dipole transitions within the 4*f* subshell are parity forbidden, the intensities of radiative transitions

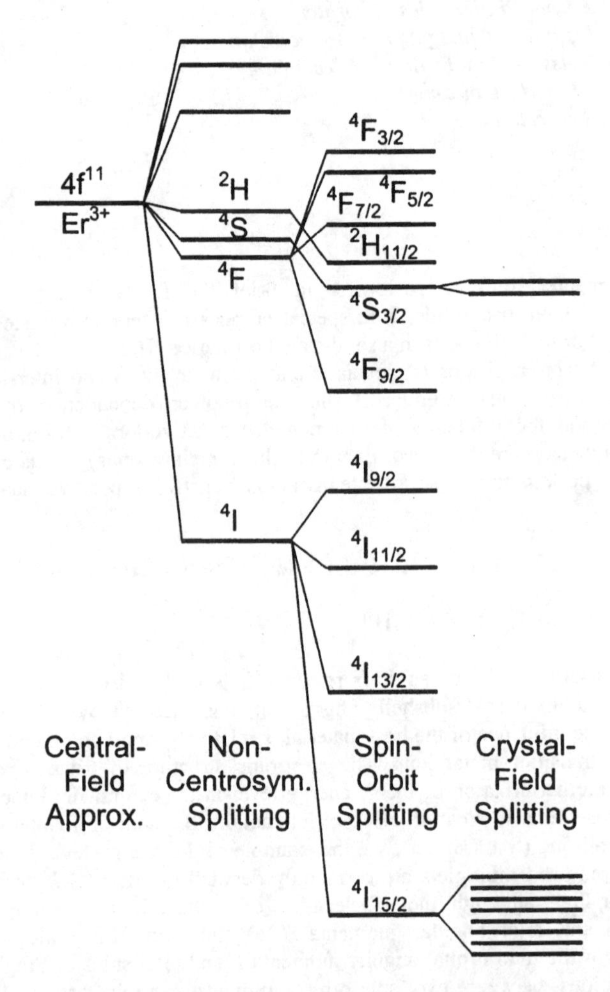

Figure 1. Energy-level scheme of trivalent erbium indicating the splitting of the $4f^{11}$ configuration in the central-field approximation by the non-centrosymmetric electron-electron interaction, the spin-orbit interaction, and the Stark splitting by the local electric field of the host material (indicated only for selected spin-orbit multiplets). (Figure taken from [1]).

in rare-earth ions are weak and the radiative lifetimes of the emitting states are long, typically in the ms range. Mixing of the 4*f* states with higher-lying (typically 5*d*) electronic states of opposite parity at ion sites without inversion symmetry, however, means that electric-dipole transitions become partially allowed and are usually the dominant transitions between 4*f* electronic states. The oscillator strengths f and integrated absorption and emission cross-sections σ of these spin-orbit multiplet-to-multiplet transitions can be calculated with the help of the semi-empirical Judd-Ofelt theory [2, 3]. If the degree of inhomogeneous spectral-line broadening is relatively small and the absorption and emission spectra are structured, the cross-sections $\sigma(\lambda)$ at individual wavelengths that are relevant to pump absorption and stimulated emission of narrow laser lines must be determined experimentally.

Besides ground-state absorption (GSA), also excited-state absorption (ESA) of pump photons [4], see Fig. 2(a), can play a significant role in the excitation mechanisms of rare-earth ions, especially in the case of high-intensity pumping of materials with low dopant concentration [5]. Since the absorption increases exponentially with the absorption coefficient $\alpha(\lambda_P) = N\,\sigma(\lambda_P)$, ESA becomes relevant for the population dynamics of a laser when (a) the ESA and GSA cross-sections $\sigma(\lambda_P)$ are comparable at the pump wavelength λ_P and (b) the population density N of the excited state in which the second pump-absorption step originates becomes a significant fraction of the density of ions in the ground state, i.e., a large degree of ground-state bleaching must be present for ESA to play a significant role [6].

A radiative transition from an excited state i to a lower-lying state j is characterized by the radiative rate constant A_{ij}. If the decay occurs to several lower-lying states, the overall radiative rate constant A_i is the sum of all individual rate constants. The branching ratio of each radiative transition is defined as $\beta_{ij} = A_{ij} / A_i$.

Radiative decay of excited states is in competition with nonradiative decay by interaction with vibrations of the host material, called multiphonon relaxation. For an ordered structure, the vibrational frequency ν of the anion-cation bonds is given by

$$\nu = (1/2\pi)\sqrt{k/M}\,, \qquad (1)$$

where $M = m_1 m_2/(m_1+m_2)$ is the reduced mass for two bodies m_1, m_2 vibrating with an elastic restoring force k. Examples of maximum phonon energies in different host materials are given in Table I. The influence of multiphonon relaxations is stronger in oxides as compared to fluorides or halides [7] because of the smaller atomic mass m_2 of the anion and the larger elastic restoring force k, see (1), due to stronger covalent bonds in oxides [8], both resulting in larger maximum phonon energies in oxides.

The rate constant of a multiphonon-relaxation process, W_i, decreases exponentially with the energy gap ΔE to the next lower-lying state and with the order of the process, i.e., the number p of highest-energy phonons required to bridge the energy gap [9, 10]:

$$W_i = C\exp(-p\beta)\,, \qquad (2)$$

$$p = \Delta E / h\nu\,, \qquad (3)$$

where C and β are constants characteristic of the host material. The rate constant of multiphonon relaxation increases with host temperature. The measurable luminescence lifetime τ_i of an excited state i is the inverse of the sum of the overall radiative rate constant A_i and the rate constant of multiphonon relaxation, W_i. The radiative quantum efficiency is defined as $\eta = A_i / (A_i + W_i)$.

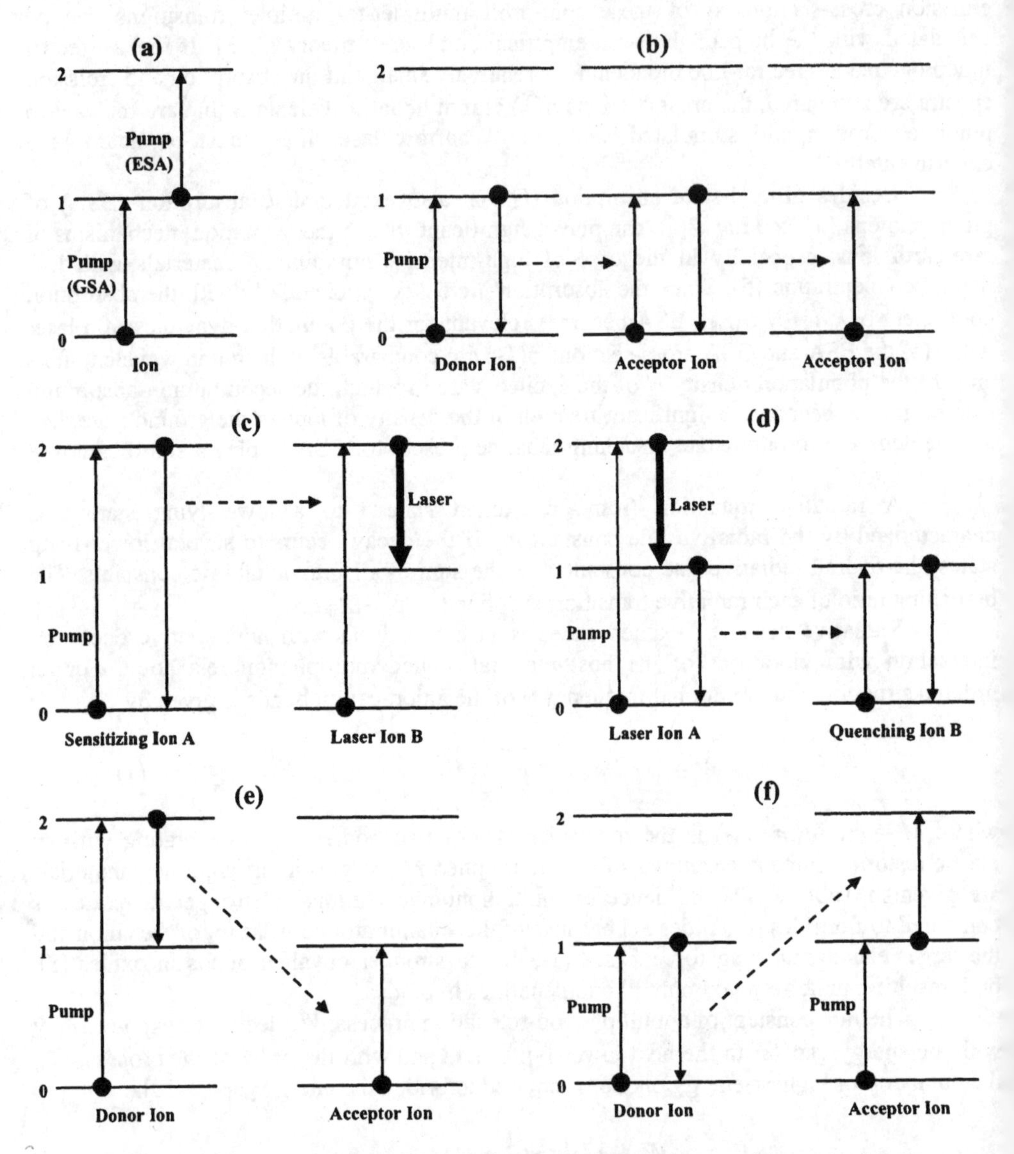

Figure 2. Intra- and interionic processes in fiber lasers: (a) excited-state absorption (ESA); (b) energy migration; (c) sensitization and (d) quenching of a laser ion by an ion of a different type; (e) cross-relaxation and (f) energy-transfer upconversion. (Figure taken from [1]).

TABLE I. Maximum phonon energies in different host materials.

Host Material	Maximum Phonon Energy [cm^{-1}]
Silica Glass	1100
$Y_3Al_5O_{12}$ (YAG)	800
$LiYF_4$ (YLF)	550
ZBLAN Glass	500
$Cs_3Er_2Cl_9$	280
$Cs_3Er_2Br_9$	190
$Cs_3Er_2I_9$	160

A brief example: The luminescence lifetime of the Er^{3+} $^4I_{9/2}$ level is partly quenched by multiphonon relaxation. Typically, nonradiative decay becomes dominant if five or less phonons are required to bridge the energy gap [11]. With an energy gap between the $^4I_{9/2}$ and the next lower lying $^4I_{11/2}$ levels of ~2000 cm^{-1}, radiative decay prevails for phonon energies below ~400 cm^{-1}, i.e., in halides (see Fig. 3).

Like absorption, the strength of a stimulated-emission process is characterized by the emission cross-section $\sigma(\lambda_L)$ of the laser transition. From a simple analysis, for one resonator round-trip of oscillating laser photons, the product τ $\sigma(\lambda_L)$ with τ the luminescence lifetime of the upper laser level, is identified as a "figure of merit" for a possible laser transition. The larger this product, the lower is the expected pump threshold of the laser transition. This "figure of merit", however, does not take into account the numerous parasitic effects that can occur in the population dynamics of a laser system, such

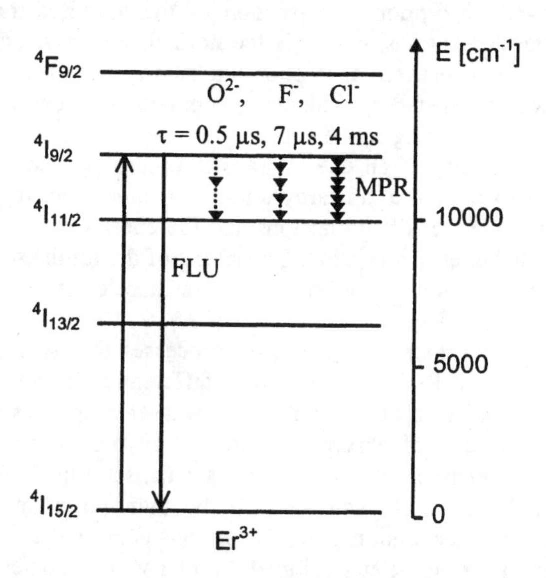

Figure 3. Luminescence lifetime of the Er^{3+} $^4I_{9/2}$ level in different host materials. Multiphonon relaxation requires 2, 4, and 7 phonons in an oxide, fluoride, and chloride host material, respectively, to bridge the energy gap.

as pump ESA, reabsorption of laser photons, and energy-transfer processes. It is often these parasitic processes that lead to surprising performance characteristics - as likely in the negative as in the positive sense - and make the interpretation of rare-earth-doped solid-state lasers challenging. Examples will be discussed in Sect. 4.

2.3 INTERIONIC PROCESSES

In addition to intraionic excitation and decay mechanisms, radiative energy transfer due to reabsorption of emitted photons by other active ions in the sample and nonradiative energy-transfer processes due to multipole-multipole or exchange interactions between neighboring active ions can occur. Radiative energy transfer leads to an increase in the luminescence lifetime. Among the nonradiative energy-transfer processes, most common is the electric dipole-dipole interaction, which can occur as a direct [12] or phonon-assisted [13] energy transfer. A direct energy transfer requires spectral resonance between the involved emission and absorption transitions whereas an indirect transfer can also be non-resonant, i.e., an existing energy gap between the emission and absorption transitions involved in the transfer is bridged by one or several phonons. A process that leads to phonon emission has typically a higher probability than a process requiring phonon absorption. Since the electrostatic field of an electric dipole decreases with distance r as r^{-3}, the probability R_{DA} of an energy transfer between two such dipoles exhibits a strong distance sensitivity of r^{-6} [12]:

$$R_{DA} = \frac{3\hbar^4 c^4}{4\pi n^4} \frac{Q_A}{\tau_D} \frac{1}{r_{DA}^6} \int f_D(E) F_A(E) E^{-4} dE \tag{4}$$

where Q_A is the integral absorption cross-section of the acceptor transition, τ_D is the radiative lifetime of the donor transition, f_D is the normalized emission line shape of the donor transition, F_A is the normalized absorption line shape of the acceptor transition, and r_{DA} is the donor-acceptor distance. Therefore, nonradiative energy transfer occurs predominantly between neighboring active ions.

An obvious possibility of an energy-transfer process is shown in Fig. 2(b). An excited ion transfers its excitation to a nearby ion of the same type. If this process occurs consecutively between a number of similar ions and the energy is thus transferred over a larger distance, it is called energy migration. Quenching of the luminescence lifetime of an excited state by energy transfer to impurities is often accelerated by energy migration among the excited donor ions [14].

Figure 2 displays further energy-transfer processes that typically occur in rare-earth-doped solid-state lasers. Rare-earth ions of a different type can be deliberately co-doped into the host material in order to influence the laser properties of the lasing ions. Efficient excitation by means of absorption and energy transfer of pump light from sensitizing ions to the upper laser level of the lasing ions, see Fig. 2(c), can be exploited when the lasing ions do not sufficiently absorb the pump light at the desired pump wavelength or the dopant concentration of the lasing ions is limited because, e.g., the laser transition terminates in the ground-state multiplet. Similarly, the transfer of excitation from the lower laser level of the lasing ions to nearby quenching ions, see Fig. 2(d), is desirable when the lifetime of the lower laser level is extremely long. The low relaxation rate from the lower laser level would otherwise lead to accumulation of excitation in this level, which can result in self-terminating laser behavior and/or bleaching of the ground-state population density and, consequently, decreased GSA.

Energy-transfer processes that have both ions in excited states before or after the energy transfer are shown in Figs. 2(e) and 2(f). In the former case, an excited ion transfers part of its excitation to a nearby ion in its ground state. This process is called cross-relaxation. Its rate increases with the average number of non-excited neighboring ions, i.e., with dopant concentration. Therefore, cross-relaxation leads to concentration quenching of the measured luminescent decay time of the initial excited state involved in the transfer process. In the inverse process - called energy-transfer upconversion (ETU) - excitation is transferred from one to another excited ion, see Fig. 2(f). After the absorption of two low-energy pump photons, ETU leads to a single excitation of higher energy and a single high-energy photon may be emitted from the second excited state.

In the presence of fast energy migration among the active ions, the excitation is spatially diffused and all these energy-transfer processes can be described by rate-equation analysis using a rate term $W N_d N_a$ that comprises a macroscopic energy transfer probability W and the population densities N_d and N_a of the initial states of the donor and acceptor ions, respectively [15]. This model, however, is usually not applicable at low dopant concentrations where energy migration is weak. In addition, a number of authors reported on active-ion clusters in host materials, see, e.g., [16, 17, 18, 19]. This or any other non-uniform distribution of active ions complicates the analysis of the influence of energy-transfer processes on the performance of rare-earth-doped lasers, as ions within such clusters are more susceptible to interionic processes than isolated ions. In the simplest approach, this distinction defines two different classes of ions that exhibit different population dynamics [20].

3. Upconversion Dynamics

Spectroscopic data such as absorption and emission spectra, luminescent transients, and the pump-power dependence of luminescence intensities are essential to the understanding of excitation mechanisms in luminescent and laser materials and to the improvement of device performance. Special attention has been devoted to the investigation of upconversion-induced luminescence [21, 22], partly because of the availability of near-infrared pump sources for the excitation of visible luminescence [23, 24, 25, 26, 7] and laser emission [27, 28, 29, 30, 31, 32] and partly because these mechanisms can introduce a loss channel for devices emitting in the infrared region [33, 34, 35, 36, 37, 38]. The two most common excitation processes that lead to emission from energy states higher than the terminating state of the first pump-absorption step are ETU and pump ESA.

3.1 POWER DEPENDENCE OF UPCONVERSION LUMINESCENCE

For the interpretation of short-wavelength luminescence, it is often assumed that the order n of the upconversion process, i.e. the number n of pump photons required to excite the emitting state, is indicated by the slope of the luminescence intensity versus pump power in double-logarithmic representation. However, as a consequence of the conservation of energy, a non-linear process which transfers energy from one quantum state to another cannot maintain its non-linear nature up to infinite excitation energy. Consequently, the dependence of an upconversion-luminescence intensity on pump power is expected to decrease in slope with increasing excitation, and a "saturation" of the intensity of an upconversion luminescence for higher pump powers was already observed almost forty years ago [39]. The same saturation can be detected in multiple-step upconversion luminescence [40].

This experimentally observed decrease in the slope of an upconversion-luminescence intensity versus pump power with increasing power is determined by the competition between linear decay and upconversion processes for the depletion of the intermediate excited states. The intensity of an upconversion luminescence which is excited by the sequential absorption of n photons has a dependence on absorbed pump power P which may range from P^n in the limit of infinitely small upconversion rates down to P^1 for the upper state and less than P^1 for the intermediate states in the limit of infinitely large upconversion rates, as will be shown in the following.

The excitation mechanisms in systems with several metastable electronic excited states are usually rather complex. We assume here the simplest possible model:

1) The ground-state population density is constant.
2) The system is pumped continuous-wave (cw) by GSA.
3) Upconversion steps between subsequent excited states take place by either ETU or ESA.
4) The excited states i have lifetimes τ_i and decay with rate constants $A_i = \tau_i^{-1}$ either to their next lower-lying state or directly to the ground state.

In practice, the two different cases of 4) often correspond to a high-phonon-energy material with a predominant multiphonon-induced decay to the next lower-lying state (denoted here by a branching ratio $\beta_i = 1$) and a low-phonon-energy host material with a predominant radiative decay to the ground state ($\beta_i = 0$), respectively. However, it is not important whether the decay mechanism is radiative or non-radiative. The assumptions of $\beta_i = 1$ or $\beta_i = 0$ simplify the solutions of the rate equations and are made here to exemplify the two extreme limiting scenarios.

Since ground-state bleaching is assumed to be negligible, the ground-state population density is

$$N_0 = const. \tag{5}$$

In the presence of ESA, the absorption coefficient α at the pump wavelength for a system with n excited levels is given by the sum of the absorption coefficients $\sigma_j N_j$ of the transitions from states j,

$$\alpha = \sum_{j=0...n-1} \sigma_j N_j , \tag{6}$$

where σ_j is the absorption cross section from state j at the pump wavelength and N_j is its population density. For absorption over a sample length ℓ which is short compared to the absorption length α^{-1}, we expand the exponential function of the Lambert-Beer law for the calculation of absorbed pump power into a Taylor series and approximate it by the leading term:

$$1-\exp(-\ell\alpha)\approx\ell\alpha\,. \tag{7}$$

From (6) and (7), it follows that the pump rate R_i of an individual transition from state i can be written as [41]

$$\begin{aligned} R_i &= \frac{\lambda_P}{hc}\frac{P}{\pi w_P^2 \ell}\left[1-\exp(-\ell\alpha)\right]\frac{\sigma_i N_i}{\alpha} \\ &\approx \frac{\lambda_P}{hc}\frac{P}{\pi w_P^2}\sigma_i N_i \\ &= \rho_P \sigma_i N_i\,, \end{aligned} \tag{8}$$

with λ_p, the pump wavelength, w_p, the pump radius, P, the incident pump power, h, Planck's constant, and c, the vacuum speed of light. The pump constant is

$$\rho_P = \frac{\lambda_P}{hc}\frac{P}{\pi w_P^2}\,. \tag{9}$$

As a consequence of the assumption of small absorption, the pump rate at the transition from state i is independent of absorption at transitions from other states j in (8).

We first demonstrate the relevant effect which leads to a decrease in the slope of an upconversion luminescence with increasing pump power. The simplest system in which upconversion luminescence can be observed is a three-level system as depicted in Fig. 4. Assuming that the system is pumped by GSA and the upconversion step is achieved by ETU with a corresponding parameter W_1, the rate equations describing the excitation mechanisms in this system are (5) and

$$dN_1/dt = \rho_P\sigma_0 N_0 - 2W_1N_1^2 - A_1N_1 \tag{10}$$

$$dN_2/dt = W_1N_1^2 - A_2N_2\,. \tag{11}$$

Under steady-state excitation, this yields

$$A_2N_2 = W_1N_1^2 \tag{12}$$

$$\rho_P\sigma_0N_0 = 2W_1N_1^2 + A_1N_1\,. \tag{13}$$

It follows from (12) that

$$N_2 \propto N_1^2\,. \tag{14}$$

If linear decay (LUM_1 in Fig. 4) is the dominant depletion mechanism of level 1, we can neglect the upconversion term in (13). It follows from (9) and (13) that $N_1 \propto P$ and, consequently, $N_2 \propto N_1^2 \propto P^2$, corresponding to one limit. In contrast, if upconversion (ETU in Fig. 4) is dominant, i.e., we can neglect the linear decay term in (13), then $N_1^2 \propto P$ or $N_1 \propto P^{1/2}$, resulting in $N_2 \propto N_1^2 \propto P$, corresponding to the other limit.

For intermediate pump powers, situations of competition between linear decay and depletion by upconversion are established and, consequently, the slopes of the luminescences are between the two limiting cases. For all pump powers, the slope of the upconversion luminescence is twice that of the direct luminescence because of (14).

We also solve here the simplest case involving ESA as the upconversion mechanism, i.e., the ETU process is replaced by the ESA process in Fig. 4. The rate equations, in this case, are (5) and

$$dN_1/dt = \rho_P \sigma_0 N_0 - \rho_P \sigma_1 N_1 - A_1 N_1 \tag{15}$$

$$dN_2/dt = \rho_P \sigma_1 N_1 - A_2 N_2 . \tag{16}$$

From (9) and (16), we find that $N_2 \propto P\,N_1$. If the linear decay from level 1 is dominant and, thus, the ESA term is negligible in (15), we obtain $N_1 \propto P$ and, consequently, $N_2 \propto P\,N_1 \propto P^2$. For strong ESA, the linear decay term can be neglected in (15), and we derive that N_1 is independent of P, resulting in $N_2 \propto P\,N_1 \propto P$.

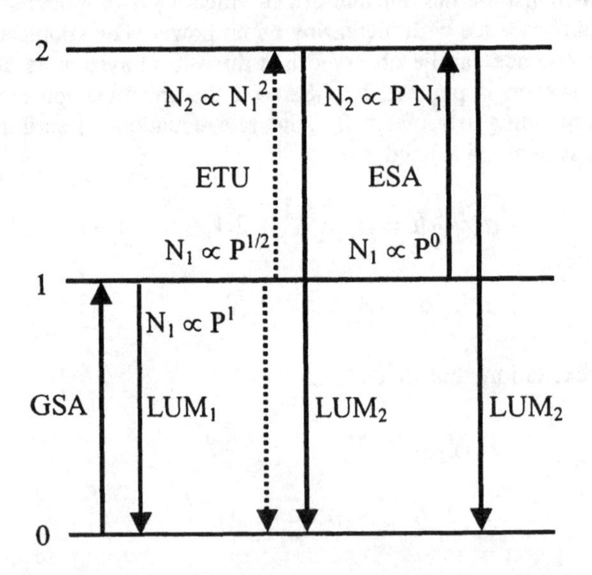

Figure 4. Simple three-level upconversion scheme. Solid and dashed arrows indicate the radiative and nonradiative population and depopulation mechanisms for each level, respectively. The dependence of the population density N_1 on pump power for the corresponding depletion pathways, and the dependence of N_2 on N_1 are indicated for the different cases of ETU and ESA. (Figure taken from [20]).

This shows that, with increasing pump power and the resulting increasing importance of upconversion, the slope of the upconversion luminescence changes from quadratic to linear, whereas the slope of the directly excited luminescence changes from linear to less than linear, with different limits obtained for ETU and ESA. This behavior, therefore, fundamentally derives from the competitive mechanisms of upconversion and direct luminescence for the depletion of the intermediate excited state. A more detailed description of the power dependence of upconversion luminescence can be found in [20].

3.2 INHOMOGENEOUS DISTRIBUTION OF ACTIVE IONS

Participation of only a fraction of ions in energy-transfer processes in general and in ETU in particular has been a major complication in the spectroscopic investigation of these processes, in the interpretation of measured luminescent decay curves, and in the transfer between the micro- and macroscopic pictures of ETU.

There are several scenarios that can result in only a fraction of ions being able to participate in ETU. First, energy transfer is an interionic process whose probability depends strongly on the distance between the participating ions [12]. ETU is, therefore, influenced by the lattice structure of the host material, the dopant concentration, energy migration between the active ions, and the distribution of active ions in the host material. Practically, one can define a critical interionic distance above which ETU is impossible. In a crystalline lattice, situations are often found in which only nearest-neighbor lattice sites for the active ions lie within the critical distance [42]. Consequently, one possible reason for fractional ETU is a non-uniform distribution of ions in the host lattice. A statistical distribution of active ions in the host lattice results in the existence of a range of distances between neighboring ions. Non-statistical distributions have similar consequences. For example, a fraction of ions may be isolated, arranged in clusters [18], or occupy different sites in the host lattice with different nearest-neighbor distances [43, 7].

Second, the probability of an energy-transfer process depends on the spectral overlap of the emission and absorption transitions involved in the energy transfer [12]. Another possible reason for fractional ETU is, therefore, the existence of two spectroscopically different lattice sites for the active ions. Different crystal fields at these sites result in different excited-state energies of the active ions, different spectral overlap of the emission and absorption transitions involved in energy transfer, and, thus, different strengths of the energy-transfer processes [44].

In all these cases, a knowledge of the fraction f of ions that can participate in ETU has important consequences. For example, in the case of a non-uniform distribution of ions in the host lattice, if this distribution is known, the critical distance between neighboring ions for the occurrence of ETU can be derived with the knowledge of f. Vice versa, if the critical distance can be determined by other methods, a knowledge of f allows for testing models of the active-ion distribution in the host lattice.

If only a fraction of ions fulfills the condition to participate in ETU, direct and upconversion luminescence probe different classes of ions. While the former probes all ions, the latter detects only luminescence from those ions that actually participate in ETU. For those ions that fulfill the condition to participate in ETU, there is a quadratic correlation between upconversion and direct luminescent decay at all times, according to (14). However, since the detected overall direct luminescence is composed of the exponential decay arising from the isolated ions and the non-exponential decay arising from those ions that are involved in ETU, the upconversion luminescence does not possess a quadratic dependence on the overall direct luminescence, i.e., a decorrelation from the quadratic dependence of ETU occurs.

Under continuous-wave [45] as well as under pulsed [46] pumping, a decorrelation in the power dependence of upconversion and direct luminescent intensities was observed. Simple models that account for the measured decorrelation were developed [20, 47], however recent investigations show that the description of the observed effects requires a more elaborate model [48].

4. Impact of Energy-Transfer Upconversion on Solid-State Laser Performance

In the following, two examples are presented that illustrate how ETU processes can influence the performance of solid-state lasers in a negative or positive manner.

4.1 UPCONVERSION-INDUCED HEAT GENERATION IN Nd^{3+} 1-μm LASERS

The Nd:YAG (Nd^{3+}:$Y_3Al_5O_{12}$) and Nd:YLF (Nd^{3+}:$LiYF_4$) transitions at 1.064 μm and 1.053 μm, respectively, have been widely used for laser applications, because they offer several advantages over other laser systems: The Nd^{3+} transition at 1 μm involves a four-level scheme with fast multiphonon transitions populating the upper and depleting the lower laser level. Its large stimulated-emission cross-section allows a low laser threshold, while the small quantum defect allows high slope efficiency. YAG has a high fracture limit which is of advantage for use in high-power laser systems. YLF, on the other hand, is an attractive host material because of the wavelength match of the laser transition (1.053 μm) with Nd^{3+} glass amplifiers, the long storage time of the Nd^{3+} upper laser level, its natural birefringence, and its relatively weak thermal lensing on the polarization corresponding to 1.053 μm operation.

However, under conditions of higher excitation density, such as non-lasing conditions, Q-switched operation, or operation as an amplifier, a strong deterioration in the performance of this seemingly simple system is observed. With increasing pump power and intensity, the Nd:YLF system exhibits a significantly reduced storage time under Q-switched operation and a decreasing laser efficiency [49, 50]. This behavior has been explained by lifetime quenching owing to ETU processes involving two neighboring ions in the upper laser level [37, 38, 51], see Fig. 5. As a consequence, Nd:YLF exhibits visible luminescence from energy levels above the pump level. Nevertheless, most of the upconverted excitation decays via multiphonon relaxation back to the upper laser level. The downconverted excitation of the second ion that is involved in each ETU process decays also via multiphonon relaxation to the ground state. Consequently, measurements of the induced thermal lens under non-lasing (high excitation) compared to lasing (low excitation) conditions demonstrate that significant additional heat is generated in the non-lasing situation, with the same pump power [52]. For Nd:YAG, similar effects of luminescence quenching [53, 54], additional heat generation [55, 56], and increased thermal lensing [36] under non-lasing conditions have been observed.

The increased heat load has a number of undesirable consequences, such as spherical aberration in the thermally-induced lens, with consequent degradation in laser-beam quality and higher resonator losses. Ultimately, with sufficient heat load rod fracture will occur. With these effects being particularly pronounced under non-lasing conditions, it follows that Q-switched operation and operation as an amplifier are especially susceptible. A more detailed description of ETU-induced thermal and thermo-optical effects in Nd^{3+} 1-μm lasers can be found in [36].

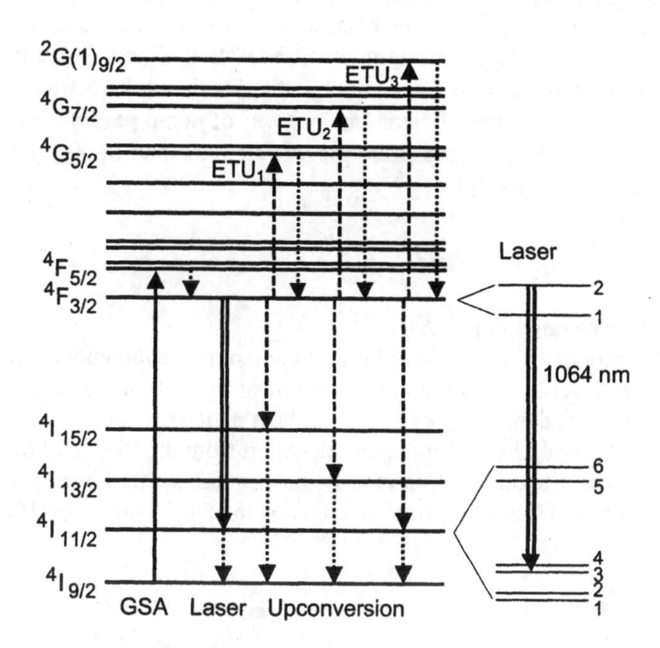

Figure 5. Partial energy-level scheme of Nd^{3+} indicating the dominant processes in the system: GSA, multiphonon relaxation to the $^4F_{3/2}$ upper laser level, laser transition (at 1064 nm in YAG) and ETU processes from the upper laser level, followed by cascaded multiphonon relaxations (dashed lines). (Figure modified from [36]).

4.2 ENERGY RECYCLING IN Er^{3+} 3-μm LASERS

The first observation [57] of coherent emission at 3 μm from erbium ions was reported in 1967. Yttrium aluminum garnet (YAG), today's most widely used solid-state laser material, entered the stage as a host for the erbium 3-μm laser [58] in 1975. In this material with its high phonon energies and strong multiphonon quenching of the $^4I_{11/2}$ upper laser level (leading to an upper-to-lower level lifetime ratio of ~1:50 !), the first CW lasing at 3 μm was observed [59] in 1983.

While erbium-doped fiber lasers have been successfully operated at 3 μm by exploiting ESA [5, 60] or energy transfer from Er^{3+} to co-doped Pr^{3+} ions [35, 61] to deplete the long-living lower laser level, the population mechanisms of the highly erbium-doped crystal laser system are governed by ETU processes between neighboring erbium ions in the lower laser level [62]. In Fig. 6, a partial energy-level scheme of erbium, a suitable pump transition, the laser transition at 3 μm, and the most important ETU process are introduced. The ETU process ($^4I_{13/2}$, $^4I_{13/2}$) → ($^4I_{15/2}$, $^4I_{9/2}$) leads to a fast depletion of the lower laser level and enables CW operation of a laser transition which, otherwise, could be self-terminating owing to the unfavorable lifetime ratio of the upper compared to the lower laser level.

In particular, this ETU process offers another generous advantage. Half of the ions that undergo this process are upconverted to the $^4I_{9/2}$ level and, by subsequent multiphonon relaxation, are recycled to the $^4I_{11/2}$ upper laser level from where they can each emit a second laser photon, for a single pump-photon absorption. For a large number of ions participating in this process, a slope efficiency η_{sl} of twice the Stokes efficiency $\eta_{St} = \lambda_p / \lambda_l$ obtains [6], because the quantum efficiency $\eta_q = n_l / n_p$ of pump photons converted to laser photons increases from 1 to 2 (λ and n are the wavelengths and photon numbers of laser and pump transitions, respectively):

$$\eta_{sl} = \eta_q \eta_{St} = 2\eta_{St} \,. \qquad (17)$$

This mechanism is illustrated in Fig. 6.

The ETU process from $^4I_{13/2}$ can be so dominant that even under direct pumping of the $^4I_{13/2}$ lower laser level and subsequent excitation of the $^4I_{11/2}$ upper laser level by ETU, 3-μm laser operation was demonstrated in several host materials [64].

On the other hand, the multiphonon relaxation from the $^4I_{9/2}$ level that follows each ETU process produces a significant amount of heat, making this laser system even more susceptible to thermal and thermo-optical effects than the Nd^{3+} 1-μm laser [65].

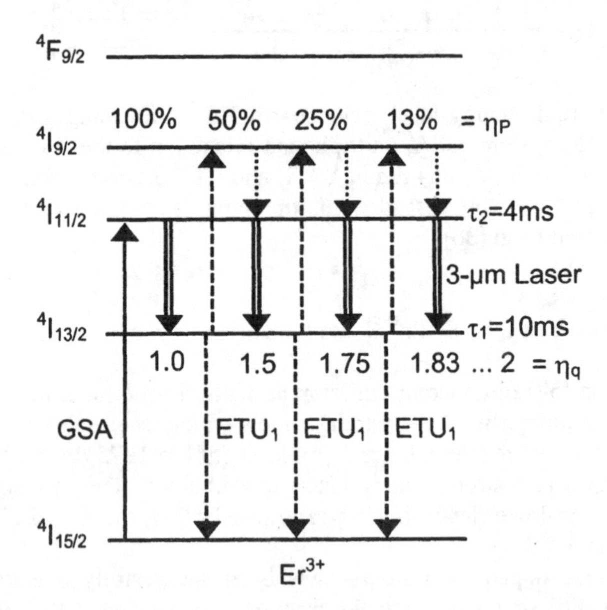

Figure 6. Partial energy-level scheme of erbium illustrating the process of energy recycling from the lower to the upper laser level by ETU. Indicated are the relative pump rate η_P of the upper laser level and the quantum efficiency η_q which increases from 1 to 2 if a large number of ions participate in the process. Lifetimes are given for $LiYF_4$. (Figure modified from [63]).

The pump wavelength that provides the highest Stokes efficiency is 980 nm which corresponds to pumping directly into the upper laser level [66]. For this pump wavelength, the Stokes efficiency is $\eta_{Sl} = \lambda_p / \lambda_l = 35\%$. The highest slope efficiency obtained experimentally [67] is currently $\eta_{sl} = 50\%$ in $LiYF_4:Er^{3+}$. This result shows that energy recycling is indeed efficient and that slope efficiencies far above the Stokes efficiency can be obtained under CW pumping. Under pulsed excitation, the slope efficiency is strongly reduced, because the lower laser level is much less populated than in the steady-state regime and the ETU processes which depend on the square of the population density are less efficient [68, 69]. A detailed review of the Er^{3+} 3-μm laser can be found in [63, 70].

5. Conclusions

In the last decades many research efforts have been undertaken and significant results have been published in the literature that have broadened our knowledge on ETU. Nevertheless, the mechanisms of ETU and its influences on upconversion luminescence dynamics have as yet not been fully understood. Specifically, the presence of inhomogeneous active-ion distributions complicates the understanding and theoretical description of ETU. Solid-state luminescent light sources and lasers often show surprising behavior when influenced by energy-transfer processes. Understanding these processes and exploiting their large potential will contribute to the development of more efficient incoherent and coherent light emitters that will illuminate our future.

Acknowledgements

The author thanks all his colleagues who have contributed to this research over the past ten years. Especially the insight and help provided by W. Lüthy and H.U. Güdel from the University of Bern, D.R. Gamelin from the University of Washington, Seattle, W.A. Clarkson and D.C. Hanna from the University of Southampton, and S.D. Jackson from the University of Sydney have significantly contributed to the work presented here.

References

1. Pollnau, M. and Jackson, S.D. (2003) Mid-infrared fiber lasers, in *Solid-State Mid-Infrared Laser Sources*, Springer Series on Topics in Applied Physics, Vol. **89**, Sorokina, I.T. and Vodopyanov, K.L., eds., Springer-Verlag, Berlin, Heidelberg, pp. 219-253.
2. Judd, B.R. (1962) Optical absorption intensities of rare-earth ions, *Phys. Rev.* **127**, 750-761.
3. Ofelt, G.S. (1962) Intensities of crystal spectra of rare-earth ions, *J. Chem. Phys.* **37**, 511-520.
4. Pollnau, M., Heumann, E., and Huber, G. (1992) Time-resolved spectra of excited-state absorption in Er^{3+} doped $YAlO_3$, *Appl. Phys. A* **54**, 404-410.
5. Pollnau, M., Ghisler, Ch., Bunea, G., Bunea, M., Lüthy, W., and Weber, H.P. (1995) 150 mW unsaturated output power at 3 μm from a single-mode-fiber erbium cascade laser, *Appl. Phys. Lett.* **66**, 3564-3566.

6. Pollnau, M., Spring, R., Ghisler, Ch., Wittwer, S., Lüthy, W., and Weber, H.P. (1996) Efficiency of erbium 3-μm crystal and fiber lasers, *IEEE J. Quantum Electron.* **32**, 657-663.
7. Hehlen, M.P., Krämer, K., Güdel, H.U., McFarlane, R.A., and Schwartz, R.N. (1994) Upconversion in Er^{3+}-dimer systems: Trends within the series $Cs_3Er_2X_9$ (X=Cl,Br,I), *Phys. Rev. B* **49**, 12475-12484.
8. Digonnet, M.J.F. (2001) *Rare-earth-doped fiber lasers and amplifiers*, 2nd ed., Marcel Dekker, Inc., New York, Basel.
9. Riseberg, L.A. and Moos, H.W. (1968) Multiphonon orbit-lattice relaxation of excited states of rare-earth ions in crystals, *Phys. Rev.* **174**, 429-438.
10. Van Dijk, J.M.F. and Schuurmans, M.F.H. (1983) On the nonradiative and radiative decay rates and a modified exponential energy gap law for *4f-4f* transitions in rare-earth ions, *J. Chem. Phys.* **78**, 5317-5323.
11. Güdel, H.U. and Pollnau, M. (2000) Near-infrared to visible photon upconversion processes in lanthanide doped chloride, bromide and iodide lattices, *J. Alloys Compd.* **303-304**, 307-315.
12. Dexter, D.L. (1953) A theory of sensitized luminescence in solids, *J. Chem. Phys.* **21**, 836-850.
13. Miyakawa, T. and Dexter, D.L. (1970) Phonon sidebands, multiphonon relaxation of excited states, and phonon-assisted energy transfer between ions in solids, *Phys. Rev. B* **1**, 2961-2969.
14. Gapontsev, V.P. and Platonov, N.S. (1989) Migration-accelerated quenching of luminescence in glasses activated by rare-earth ions, *Mat. Sci. Forum* **50**, 165-222.
15. Grant, W.J.C. (1971) Role of rate equations in the theory of luminescent energy transfer, *Phys. Rev. B* **4**, 648-663.
16. Delevaque, E., Georges, T., Monerie, M., Lamouler, P., and Bayon, J.-F. (1993) Modeling of pair-induced quenching in erbium-doped silicate fibers, *IEEE Photonics Technol. Lett.* **5**, 73-75.
17. Wagener, J.L., Wysocki, P.F., Digonnet, M.J.F., Shaw, H.J., and DiGiovanni, D.J. (1993) Effects of concentration and clusters in erbium-doped fiber lasers, *Opt. Lett.* **18**, 2014-2016.
18. Quimby, R.S., Miniscalco, W.J., and Thompson, B. (1994) Clustering in erbium-doped silica glass fibers analyzed using 980 nm excited-state absorption, *J. Appl. Phys.* **76**, 4472-4478.
19. Maurice, E., Monnom, G., Dussardier, B., and Ostrowsky, D.B. (1995) Clustering-induced nonsaturable absorption phenomenon in heavily erbium-doped silica fibers, *Opt. Lett.* **20**, 2487-2489.
20. Pollnau, M., Gamelin, D.R., Lüthi, S.R., Güdel, H.U., and Hehlen, M.P. (2000) Power dependence of upconversion luminescence in lanthanide and transition-metal-ion systems, *Phys. Rev. B* **61**, 3337-3346.
21. Auzel, F. (1973) *Proc. IEEE* **6**, 758-786.
22. Wright, J.C. (1976) in Fong, F.K., *Topics in Applied Physics: Radiationless Processes in Molecules and Condensed Phases*, Vol. **15**, Springer, Berlin, Heidelberg, pp. 239-295.
23. Malinowski, M., Jacquier, B., Bouazaoui, M., Joubert, M.F., and Linares, C. (1990) Laser-induced fluorescence and up-conversion processes in $LiYF_4$:Nd^{3+} laser crystals, *Phys. Rev. B* **41**, 31-41.

24. Cockroft, N.J., Jones, G.D., and Nguyen, D.C. (1992) Dynamics and spectroscopy of infrared-to-visible upconversion in erbium-doped cesium cadmium bromide ($CsCdBr_3:Er^{3+}$), *Phys. Rev. B* **45**, 5187-5198.
25. Gharavi, A. and McPherson, G.L. (1992) Up-conversion luminescence from simultaneously excited pairs of Tm^{3+} ions in $CsMgCl_3$ crystals, *Chem. Phys. Lett.* **200**, 279-282.
26. Wermuth, M., Riedener, T., and Güdel, H.U. (1998) Spectroscopy and upconversion mechanisms of $CsCdBr_3:Dy^{3+}$, *Phys. Rev. B* **57**, 4369-4376.
27. Johnson, L.F. and Guggenheim, H.J. (1972) New laser lines in the visible from Er^{3+} ions in BaY_2F_8, *Appl. Phys. Lett.* **20**, 474-477.
28. Silversmith, A.J., Lenth, W., and Macfarlane, R.M. (1987) Green infrared-pumped erbium upconversion laser, *Appl. Phys. Lett.* **51**, 1977-1979.
29. Brede, R., Danger, T., Heumann, E., Huber, G., and Chai, B.H.T. (1993) Room temperature green laser emission of $Er^{3+}:LiYF_4$, *Appl. Phys. Lett.* **63**, 729-730.
30. Thrash, R.J. and Johnson, L.F. (1994) Upconversion laser emission from Yb sensitised Tm in BYF, *J. Opt. Soc. Am. B* **11**, 881-885.
31. Sandrock, T., Scheife, H., Heumann, E., and Huber, G. (1997) High-power continuous-wave upconversion fiber laser at room temperature, *Opt. Lett.* **22**, 808-810.
32. Paschotta, R., Barber, P.R., Tropper, A.C., and Hanna, D.C. (1997) Characterization and modeling of thulium:ZBLAN blue upconversion fiber lasers, *J. Opt. Soc. Am. B* **14**, 1213-1218.
33. Laming, R.I., Poole, S.B., and Tarbox, E.J. (1988) Pump excited-state absorption in erbium-doped fibers, *Opt. Lett.* **13**, 1084-1086.
34. Payne, S.A., Wilke, G.D., Smith, L.K., and Krupke, W.F. (1994) Auger upconversion losses in Nd-doped laser glasses, *Opt. Commun.* **111**, 263-268.
35. Pollnau, M. (1997) The route toward a diode-pumped 1-W erbium 3-μm fiber laser, *IEEE J. Quantum Electron.* **33**, 1982-1990.
36. Pollnau, M., Hardman, P.J., Kern, M.A., Clarkson, W.A., and Hanna, D.C. (1998) Upconversion-induced heat generation and thermal lensing in Nd:YLF and Nd:YAG, *Phys. Rev. B* **58**, 16076-16092.
37. Guyot, Y., Manaa, H., Rivoire, J.Y., Moncorgé, R., Garnier, N., Descroix, E., Bon, M., and Laporte, P. (1995) Excited-state absorption and upconversion studies of Nd^{3+}-doped single crystals $Y_3Al_5O_{12}$, $YLiF_4$, and $LaMgAl_{11}O_{19}$, *Phys. Rev. B* **51**, 784-799.
38. Chuang, T. and Verdún, H.R. (1996) Energy transfer up-conversion and excited state absorption of laser radiation in Nd:YLF laser crystals, *IEEE J. Quantum Electron.* **32**, 79-91.
39. Singh, S. and Geusic, J.E. (1966) Observation and saturation of a multiphoton process in $NdCl_3$, *Phys. Rev. Lett.* **17**, 865-868.
40. Lüthi, S.R., Pollnau, M., Güdel, H.U., and Hehlen, M.P. (1999) Near-infrared to visible upconversion in Er^{3+} doped $Cs_3Lu_2Cl_9$, $Cs_3Lu_2Br_9$, and $Cs_3Y_2I_9$ excited at 1.54 μm, *Phys. Rev. B* **60**, 162-178.
41. Pollnau, M., Graf, Th., Balmer, J.E., Lüthy, W., and Weber, H.P. (1994) Explanation of the cw operation of the Er^{3+} 3-μm crystal laser, *Phys. Rev. A* **49**, 3990-3996.
42. Vasquez, S.O. and Flint, C.D. (1995) A shell model for cross relaxation in elpasolite crystals: application to the $^3P_0$ and $^1G_4$ states of $Cs_2NaY_{1-x}Pr_xCl_6$, *Chem. Phys. Lett.* **238**, 378-386.

43. McPherson, G.L., Varga, J.A., and Nodine, M.H. (1979) Magnetic interactions in exchange-coupled pairs of chromium(III) and molybdenum(III) ions in crystals of $CsMX_3$ halides, *Inorg. Chem.* **18**, 2189-2195.
44. Wenger, O.S., Gamelin, D.R., Güdel, H.U., Butashin, A.V., and Kaminskii, A.A. (2000) Site-selective yellow to violet and near-infrared to green upconversion in $BaLu_2F_8:Nd^{3+}$, *Phys. Rev. B* **61**, 16530-16537.
45. Gamelin, D.R. and Güdel, H.U. (1999) Spectroscopy and dynamics of Re^{4+} near-IR-to-visible luminescence upconversion, *Inorg. Chem.* **38**, 5154-5164.
46. Laversenne, L., Pollnau, M., Bigotta, S., Toncelli, A., and Tonelli, M. (2003) Super-quadratic behavior of luminescence decay excited by energy-transfer upconversion, *International Conference on f-Elements*, Geneva, Switzerland, Final Programme and Abstract Book, p. 64.
47. Pollnau, M. (2002) Decorrelation of luminescent decay in energy-transfer upconversion, *J. Alloys Compd.* **341**, 51-55.
48. Pollnau, M., Laversenne, L., Bigotta, S., Toncelli, A., and Tonelli, M. (to be submitted) Upconversion-luminescence transients in the presence of inhomogeneous ion distributions.
49. Fan, T.Y., Dixon, G.J., and Byer, R.L. (1986) Efficient GaAlAs diode laser pumped operation of Nd:YLF at 1.047 μm with intracavity doubling to 523.6 nm, *Opt. Lett.* **11**, 204-206.
50. Beach, R., Reichert, P., Benett, W., Freitas, B., Mitchell, S., Velsko, A., Darwin, J., and Solarz, R. (1993) Scalable diode-end-pumping technology applied to a 100-mJ Q-switched Nd^{3+}:YLF laser oscillator, *Opt. Lett.* **18**, 1326-1328.
51. Pollnau, M., Hardman, P.J., Clarkson, W.A., and Hanna, D.C. (1998) Upconversion, lifetime quenching, and ground-state bleaching in $Nd^{3+}:LiYF_4$, *Opt. Commun.* **147**, 203-211.
52. Hardman, P.J., Clarkson, W.A., Friel, G.J., Pollnau, M., and Hanna, D.C. (1999) Energy-transfer upconversion and thermal lensing in high-power end-pumped Nd:YLF laser crystals, *IEEE J. Quantum Electron.* **35**, 647-655.
53. Danielmeyer, H.G., Blätte, M., and Balmer, P. (1973) Fluorescence quenching in Nd:YAG, *Appl. Phys.* **1**, 269-274.
54. Devor, D.P., DeShazer, L.G., and Pastor, R.C. (1989) Nd:YAG quantum efficiency and related radiative properties, *IEEE J. Quantum Electron.* **25**, 1863-1873.
55. Fan, T.Y. (1993) Heat generation in Nd:YAG and Yb:YAG, *IEEE J. Quantum Electron.* **29**, 1457-1459.
56. Lupei, V. and Lupei, A. (1996) Emission quantum efficiency and heating effects in YAG:Nd^{3+} lasers, *Opt. Eng.* **35**, 1252-1257.
57. Robinson, M. and Devor, P.D. (1967) Thermal switching of laser emission of Er^{3+} at 2.69 μ and Tm^{3+} at 1.86 μ in mixed crystals of $CaF_2:ErF_3:TmF_3{}^*$, *Appl. Phys. Lett.* **10**, 167-170.
58. Zharikov, E.V., Zhekov, V.I., Kulevskii, L.A., Murina, T.M., Osiko, V.V., Prokhorov, A.M., Savel'ev, A.D., Smirnov, V.V., Starikov, B.P., and Timoshechkin, M.I. (1975) Stimulated emission from Er^{3+} ions in yttrium aluminum garnet crystals at λ=2.94 μ, *Sov. J. Quantum Electron.* **4**, 1039-1040.
59. Bagdasarov, K.S., Zhekov, V.I., Lobachev, V.A., Murina, T.M., and Prokhorov, A.M. (1983) Steady-state emission from a $Y_3Al_5O_{12}:Er^{3+}$ laser (λ = 2.94 μ, T = 300°K), *Sov. J. Quantum Electron.* **13**, 262-263.

60. Pollnau, M., Ghisler, Ch., Lüthy, W., and Weber, H.P. (1998) Cross-sections of excited-state absorption at 800 nm in erbium-doped ZBLAN fiber, *Appl. Phys. B* **67**, 23-28.
61. Jackson, S.D., King, T.A., and Pollnau, M. (1999) Diode-pumped 1.7-W erbium 3-μm fiber laser, *Opt. Lett.* **24**, 1133-1135.
62. Zhekov, V.I., Lobachev, V.A., Murina, T.M., and Prokhorov, A.M. (1983) Efficient cross-relaxation laser emitting at λ=2.94 μ, *Sov. J. Quantum Electron.* **13**, 1235-1237.
63. Pollnau, M. and Jackson, S.D. (2001) Erbium 3-μm fiber lasers, *IEEE J. Select. Topics Quantum Electron.* **7**, 30-40.
64. Pollack, S.A. and Chang, D.B. (1988) Ion-pair upconversion pumped laser emission in Er^{3+} ions in YAG, YLF, SrF_2 and CaF_2 crystals, *J. Appl. Phys.* **64**, 2885-2893.
65. Pollnau, M. (2003) Analysis of heat generation and thermal lensing in erbium 3-μm lasers, *IEEE J. Quantum Electron.* **39**, 350-357.
66. Stoneman, R.C., Lynn, J.G., and Esterowitz, L. (1992) Direct upper-state pumping of the 2.8 μm Er^{3+}:YLF laser, *IEEE J. Quantum Electron.* **28**, 1041-1045.
67. Wyss, C., Lüthy, W., Weber, H.P., Rogin, P., and Hulliger, J. (1997) Emission properties of an optimised 2.8 μm Er^{3+}:YLF laser, *Opt. Commun.* **139**, 215-218.
68. Prokhorov, A.M., Zhekov, V.I., Murina, T.M., and Plantov, N.N. (1983) Pulsed YAG:Er^{3+} laser efficiency (analysis of model equations), *Laser Phys.* **3**, 79-83.
69. Pollnau, M., Spring, R., Wittwer, S., Lüthy, W., and Weber, H.P. (1997) Investigations on the slope efficiency of a pulsed 2.8-μm Er^{3+}:$LiYF_4$ laser, *J. Opt. Soc. Am. B* **14**, 974-978.
70. Pollnau, M. and Jackson, S.D. (2002) Correction to "Erbium 3-μm fiber lasers", *IEEE J. Select. Topics Quantum Electron.* **8**, 956.

17. SOME NOVEL ASPECTS OF INTRAMOLECULAR ELECTRONIC ENERGY TRANSFER PROCESSES

SHAMMAI SPEISER
Department of Chemistry, Technion - Israel Institute of Technology Haifa 32000, Israel

Abstract

Intramolecular electronic energy transfer (intra-EET) is reviewed through examples concerning the bichromophoric naphthalene (**N**) and anthracene (**A**) system as well as the aromatic-dione system.. In addition the effect of the connecting bridge, embedment in stretched films and the possibility of building logic functions based on such interactions are discussed

1.Introduction

Electronic energy transfer (EET) process is one of the most common relaxation mechanisms of an excited chromophore. Its understanding is important for studying the natural photosynthetic processes, light harvesting, polymer photophysics, dye laser operation, light interaction with molecular crystals and photochemical synthesis[1]. EET was studied extensively in condensed systems and its mechanisms are now thought to be well understood [*e.g.* 1-5].

In a simple model each of the two chromophores donor-D and acceptor-A can be found either in its ground or in its excited electronic state, and also possesses several vibrational modes that can be excited. The electronic origin transition energy of donor chromophore E_D is higher than the acceptor energy E_A.

B. Di Bartolo and O. Forte (eds.), Frontiers of Optical Spectroscopy, 591-618.

Within the Born-Oppenheimer approximation the chromophore wavefunctions are

$$\Psi_D = \phi_D(q_i^D, Q_j^D)\chi_D(Q_j^D)$$
$$\Psi_A = \phi_A(q_i^A, Q_j^A)\chi_A(Q_j^A) \quad (1)$$

here ϕ and χ are pure electronic and vibrational functions respectively, q_i and Q_j denote electronic and vibrational coordinate sets. The D and A superscripts mirror the fact that in the zero-order approximation at which there is no interchromophore interaction, the wavefunctions of one chromophore do not depend on the coordinates associated with the other one. Introducing the interaction mixes the wavefunctions, but as the interaction of interest is small compared to the vibronic transition energies, we will neglect this complication.

The chromophore excitations are approximated by one-electron processes. In this case the total two-electron wavefunctions should be antisymmetrized according to the Pauli principle requirement:

$$\Psi_i = 2^{-1/2}[\Psi_{D*}(1)\Psi_A(2) - \Psi_{D*}(2)\Psi_A(1)]$$
$$\Psi_f = 2^{-1/2}[\Psi_D(1)\Psi_{A*}(2) - \Psi_D(2)\Psi_{A*}(1)] \quad (2)$$

here the right hand one-electron wavefunctions are described by eq.(1), the ones with a star corresponding to excited chromophore vibronic states, and the others to ground states. The the zero-order Hamiltonian of the system corresponds to no interchromophore interaction, and the interaction perturbation is

$$V = e^2 / r_{12} \quad (3)$$

where r_{12} is the distance between the interacting electrons. The interchromophore interaction matrix element V_{DA} of perturbation responsible for the EET is

$$V_{DA} = \langle \Psi_i | V | \Psi_f \rangle$$
$$= \langle \Psi_{D*}(1)\Psi_A(2) | V | \Psi_D(1)\Psi_{A*}(2) \rangle - \langle \Psi_{D*}(1)\Psi_A(2) | V | \Psi_D(2)\Psi_{A*}(1) \rangle \quad (4)$$

The first integral in eq. (4) corresponds to the simultaneous donor electron relaxation and acceptor electron excitation and it is known as the Coulomb integral [6]. The second integral describes energy transfer by electron exchange [7] and is known as the Dexter exchange integral. If one substitutes in eq.(4) the initial and final wavefunctions including the spin parts, one obtains different spin selection rules for the two integrals. The exchange integral selection rules are less strict, and thus in some cases (*e.g.* triplet - triplet energy transfer) the first integral vanishes and energy is transferred only via electron exchange. Generally, however, both integrals are non zero and the corresponding mechanisms interfere.

The perturbation affects only electronic wavefunctions, and the nuclear motion can be separated:

$$V_{DA} = \{ < \phi_{D^*}(1)\phi_A(2) | V | \phi_D(1)\phi_{A^*}(2) > - < \phi_{D^*}(1)\phi_A(2) | V | \phi_D(2)\phi_{A^*}(1) > \}$$

$$\times < \chi_{D^*} | \chi_D > < \chi_{A^*} | \chi_A > = V'_{DA} < \chi_{D^*} | \chi_D > < \chi_{A^*} | \chi_A > \quad (5)$$

where in order to separate the Franck-Condon factors an assumption was made that the vibrations of the two chromophores are uncoupled. The quantity V'_{DA} is the pure electronic matrix element.

In solution experiments the initially prepared vibronic states usually undergo vibrational relaxation to a temperature dependent solvent broadened set of states and the energy transfer proceeds from this set. In the supersonic free jet experiments the donor retains with its excitation energy, and EET can be monitored from a specific vibronic state. At sufficiently high vibrational energies the initially prepared state is broadened by interaction with isoenergetic states. Even without this effect the initial state is broadened by the interaction with the radiation field and by the energy transfer process. The acceptor states are broadened by the interaction with the radiation field and by the possible back transfer. The energy profiles of the initial state $P_D(E-E_D)$ and of the final states $P^i_A(E - E^i_A)$ can be defined ($\int P(E - E')dE \equiv 1$). The total EET rate is obtained then from the Fermi Golden Rule:

$$k_{ET} = \frac{2\pi}{\hbar} \sum_i V^2_{DA} \int P_D(E - E_D) P^i_A(E - E^i_A) dE \quad (6)$$

By neglecting the coupling between the vibrations of the two chromophores, we obtain

$$k_{ET} = \frac{2\pi}{\hbar} (V'_{DA})^2 < \chi_{D^*} | \chi_D >^2 \sum_i < \chi^i_{A^*} | \chi^i_A >^2 \int P_D(E - E_D) P^i_A(E - E^i_A) dE \quad (7)$$

Thus within the framework of the employed assumptions the EET rate is proportional to the square of the pure electronic interaction matrix element, Franck-Condon factor for the transition from the donor vibronic state and Franck-Condon weighted energy overlap integral between the donor level and acceptor levels. Usually the acceptor density of states is higher than the donor state width, and the integral in eq.(6) is approximated by the state density ρ or by the so-called Franck-Condon factor weighted state density ρ_{FC}:

$$k_{ET} = \frac{2\pi}{\hbar} V_{DA}^{2}\rho \quad \text{or} \quad k_{ET} = \frac{2\pi}{\hbar}(V'_{DA})^{2}\rho_{FC} \tag{8}$$

This simple picture can be substantially incorrect if the electronic levels are considerably mixed, if the vibrations of the two chromophores are coupled, if other electronic levels of the chromophores are coupled by vibrations to the levels involved in the EET, and so on.

The Coulomb integral can be approximated by two separate integrals corresponding to the electronic transitions in the chromophores. These integrals can then be expanded in series, with the largest term being the dipole - dipole interaction term. If other terms are neglected as well as other contributions to the interaction integral, it can be expressed by the chromophore transition dipole moments:

$$V'_{DA} = \frac{|M_D||M_A|\Gamma}{R_{DA}^{3}} \tag{9}$$

R_{DA} being the interchromophore distance and Γ being the mutual orientation factor of the two transition dipole moments M [1]. Together with the translation of the energy overlap integral and of the Franck - Condon factors into the spectral overlap between donor emission and acceptor absorption it yields

$$k_{ET} = \frac{9000 \ln 10\, \Gamma^{2} \Phi_D}{128\pi^{5} n^{4} N_A \tau_D} \frac{1}{R_{DA}^{6}} \int \overline{F_D}(\bar{\nu}) \varepsilon_A(\bar{\nu}) \bar{\nu}^{-4} d\bar{\nu}^{6} \tag{10}$$

here Φ and τ are fluorescence quantum yield and lifetime, n is refractive index of the medium, F is fluorescence spectrum and ε is absorption spectrum. Thus Foerster critical transfer radius R_0 defined here and the EET rate can be calculated from spectroscopic experimental data.

In cases where the Coulomb mechanism contribution is either small or spin forbidden, the interaction matrix element can be expressed via the exchange integral.

At large enough separations this integral dependence on the interchromophore distance can be approximated by the exponential chromophore wavefunctions spatial overlap dependence [7]:

$$V'_{DA} = K\exp(-R_{DA} / L) \qquad (11)$$

where L is an average decay radius of wavefunctions of the two interacting states.

It can be seen from eqs. (10) and (11) that the Foerster mechanism rate is proportional to R_{DA}^{-6}, and this interaction is most important at intermediate separations.. At close separations were the wavefunctions overlap becomes considerable, the exchange mechanism dominates.

Foerster critical transfer radius for most systems in solution is between 10 and 100A .. Eq. (9) is widely used to describe energy transfer in condensed phase [1]. At distances less than about 5 the Dexter exchange mechanism usually becomes more important. In real systems the exchange integral evaluation by computational methods is extremely complicated, and without it the model mechanism can be employed only in some very limited cases. Still, such computations were performed in a study of series of bichromophoric molecules, yielding good agreement with experimental data on distance and orientation dependence of the EET rate [8].

In addition, more complicated mechanisms than the two considered in eq.(5) can contribute to the pure electronic interaction matrix element. Notably, higher electronic levels for which wavefunctions are localized spatially between the two chromophores, can mediate the transfer. This effect ("superexchange") was detected in bichromophoric molecules, where the rigid bridge connecting the chromophores enhanced the exchange mechanism rate [2, 9].

When the donor chromophore is excited to some specific vibronic state, it can undergo dephasing via intramolecular vibrational redistribution, (IVR), involving other molecular states that have low Franck-Condon factors for excitation (dark states). These states should have the same energy in the sense of eq.(6). It also can undergo energy transfer to the isoenergetic levels of the acceptor chromophore. As all the described processes do not involve energy change, the excitation can be then transferred back, if the fluorescnce is sufficiently slow. Finally, a stationary state is achieved, in which the excitation is spread over the two chromophores in a ratio corresponding to the vibrational state density ratio of the chromophores.

The situation under supersonic jet cooling conditions is different from the well-studied situation in condensed phase experiments. The geometry of the bichromophoric unit is usually well defined. The effect of the relative orientations of the chromophores on the EET rate can thus be conveniently studied under jet conditions. The chromophores are no longer coupled to the low frequency solvent vibrational bath. Consequently, no vibrational relaxation occurrs prior to the EET process. Thus the EET process from a specific vibronic level can be studied, and the density of the coupled states is better defined.

The EET rate depends on the vibronic level choice due to the changes in the density of states. The density of vibronic states of the acceptor chromophore can be so low that the equidistant quasi-continuum model may fail and specific vibronic level structure must be considered. The other source of the rate dependence on the vibronic level is the change in the geometry of the unit that is affected by the vibrations. The assumption of V'_{DA} independence on the vibrations is implied in the Born - Oppenheimer approximation. A better approximation is to use the average of the interaction matrix element over the molecular geometry changes associated with the vibration. Eq. (8) then becomes:

$$k_{ET} = \frac{2\pi}{\hbar}\left[\left(\iint V'_{DA} |\chi_D(Q_j^D)|^2 |\chi_{DA}(Q_j^{DA})|^2 \, dQ_j^D dQ_j^{DA}\right]^2 \rho_{FC} \quad (12)$$

here it is assumed that the initially prepared vibronic state can incorporate donor intrachromophore vibrations (wavefunctions $\chi_D(Q_j^D)$) and interchromophore (or "bridge") vibrations (wavefunctions $\chi_{DA}(Q_j^{DA})$), but no acceptor vibrations. For some vibrations (for example, an interchromophore stretch, or a vibration breaking some symmetry that forbids the EET) such averaged matrix element value may differ significantly from the frozen geometry value.

2. The naphthalene - anthracene bichromophoric molecular system

2.1 GENERAL PROPERTIES

Anthracene first singlet electronic transition origin is at 27695cm^{-1} . The pure electronic transition is polarized along the short axis, and the transition dipole moment in solution is $9.2*10^{-30}$C m, this value can be obtained also from the integrated anthracene absorption spectrum. The fluorescence quantum yield is 0.36 [10]. The second electronic transition origin is at about 30500cm^{-1}, polarized along the long axis. The oscillator strength of this transition is an order of magnitude lower than that of the first one. Consequently, at any excitation energy above 30500cm^{-1} the vibrational mixing between the two electronic states ensures that anthracene stationary states exhibit spectroscopic properties typical of the first transition [10]. The contribution of the second transition to the overall density of states at higher energies can be safely neglected.

The first singlet electronic transition (32020cm^{-1}) of naphthalene is polarized along the long axis, but it has a very low oscillator strength depending on the solvent and on the degree of involvement of the higher transitions in the integration of the corresponding spectral band). The pure S_1 electronic transition dipole moment is $4*10^{-31}$C m [10]. The fluorescence quantum yield is 0.23 [10]. The fluorescence decay lifetime of the anthracene excited at its origin transition in solution is 5ns [10]. In the cold free molecules in the jet this value is ~24ns [10]. The Foerster radius for EET from naphthalene to anthracene in solution, calculated from the spectral data is 23.2A [10]. This implies an EET rate of $7*10^{11}$s^{-1} at a typical sandwich cluster separation of 3.57A . This estimate is, of course, very crude. It does not take into account the exchange mechanism. In addition, at such short distances the dipole-dipole picture is not accurate, and contributions from higher Coulomb terms must be included [10]. Supersonic jet studies of this system revealed unique features for Intra-EET for isolated molecules. [10-14]

2.2 TIME-RESOLVED ANALYSIS OF INTRAMOLECULAR ELECTRONIC ENERGY TRANSFER IN METHYLENES-LINKED NAPHTHALENE-ANTHRACENE COMPOUNDS IN SOLUTION AND IN STRETCHED POLYMER FILMS

Intramolecular electronic energy transfer in methylenes-linked naphthalene and anthracene (referred to as AnN, n =1,3 and 6 denoting number of methylene units) was examined by time-correlated single-photon counting technique in n-hexane and in stretched PVA films. The energy transfer rates in A1N, A3N and A6N in solution were measured and the chain length dependence of these rate constants is discussed in terms of through-bond interaction via the connecting bridge, in addition to direct Coulombic interaction.

The *first* spectroscopic study of bichromophoric molecules containing naphthalene and anthracene chromophore in solution was performed by Schnepp and Levy [15]. In fact, it was the first EET study in a bichromophoric molecule. The three studied molecules contained 9-anthryl and 1-naphthyl linked by a one (**A1N**), two (**A2N**) and three methylene (**A3N**) flexible chain. It was found for all the molecules that all the fluorescence signal following the naphthalene moiety excitation originated from the anthracene moiety. This corresponded to a complete intra-EET process. Several more studies of these molecules followed [1,16,17]. EET was studied also in a bichromophoric molecule containing the two chromophores linked by a rigid bridge consisting of norbornyl and bicyclohexane units [1]. Evidence was found of the bridge mediated superexchange intra-EET mechanism in this molecule.In addition to this flexible system Scholes et al. elucidated the through-bond mechanism of energy transfer using rigidly bridged naphthalene-anthracene moieties, and developed the formulation of exchange interaction through σ-bonds of the connecting bridge [2]. The geometry of flexible-chain linked bichromopholic molecule is determined by the bond length of the chain chemical units, the bond angles, the bond rotation, in addition to the attractive forces between naphthalene and anthracene moieties. The molecular conformation of the moieties connected by a flexible alkane chain exhibits a broad population distribution as a function of the end-group distance as obtained mainly by molecular mechanics calculation [2]. The energy minima with respect to the dihedral angle, rotations about the bond between C-C bonds, are found at three different angles, namely, trans, gauche$^-$ and gauche$^+$.In the present study, the kinetics of intramolecular EET process in A1N, A3N and A6N, are reported both in solution and in stretched PVA films, together with analysis of their conformations. In addition, it is demonstrated that the relaxation of higher vibronic levels of an acceptor molecule depends on the number of intervening methylene units. The results are compared with the study of Schoels et al [6] on rigidly connected anthracene (An) and naphthalene (Np) moieitis and to the intamolecular EET dynamics in AnN observed in supersonic jets.[8]The absorption spectra of the AnN samples in n-hexane, and in the stretched and (fig. 1) non-stretched films was basically consistent with previously reported data which describe AnN spectra as superposition of the spectra of Np and An [1,2,16,17]. The very weak emission intensity in the naphthalene region, from 310nm to 370nm, implies that highly efficient intramolecular EET (>99%) occurred both in A3N and A6N. For A1N only An emission is observed indicative of very efficient intra EET process. The fluorescence intensity of A3N at wavelengths higher than 411 nm, which is assigned as the 0-1 transition (denoted as An(0-1)), was slightly larger than that of A6N. The difference in these emission spectra may be due to the formation of an intramolecular exciplex, followed by the formation of a photocyclomer [16,17].

Basically for all samples two types of experiments were conducted. Time-correlated single-photon counting measurements allowed the determination of the lifetimes of the naphthalene moiety and the rise times of the An signals in A3N and A6N. Direct excitation of An at 281nm could be neglected, since the absorption cross section of Np is 20 times larger than that of An in either A3N or A6N. Lifetime analyses of the fluorescence decay curves with a double exponential function yielded the respective lifetime and normalized pre-exponential. In addition, time resolved fluorescence spectra were recorded where the Np moiety was excited.Figures 2 shows time-resolved fluorescence spectra of A1N, A3N and A6N in stretched PVA films, which were normalized to the maximum intensity of each respective spectrum. These spectra were made from fluorescence decays measured in 1.5nm steps. These spectra show dual fluorescence bands where spectral features at wavelengths shorter than 380nm are due to Np emission while those at longer wavelengths are due to An fluorescence. Such spectra are the manifestation of intramolecular EET [1]. For all AnN molecules the An bands are very similar in their positions and in their relative intensities. They are, however, different at the Np spectral region, where no activity is observed for A1N, some weak intensity at negative times is observed for A3N, and higher intensity in A6N even at 16ps after the peak excitation intensity (corresponding to 0ps). This correlates with the decreasing order of intramolecular EET rates in the series A1N, A3N and A6N. At negative times, An emission is evidently characterized by broad spectral features. Similar spectra were obtained for the AnN samples in solution and in non-streched films, where comparison is made with the fluorescence spectrum of 9-methylanthracene, serving as a reference compound. These results as well as those for the rise and decay of the fluorescence signals associated with the An and Np bands are summarized in Tables 1 and 2.

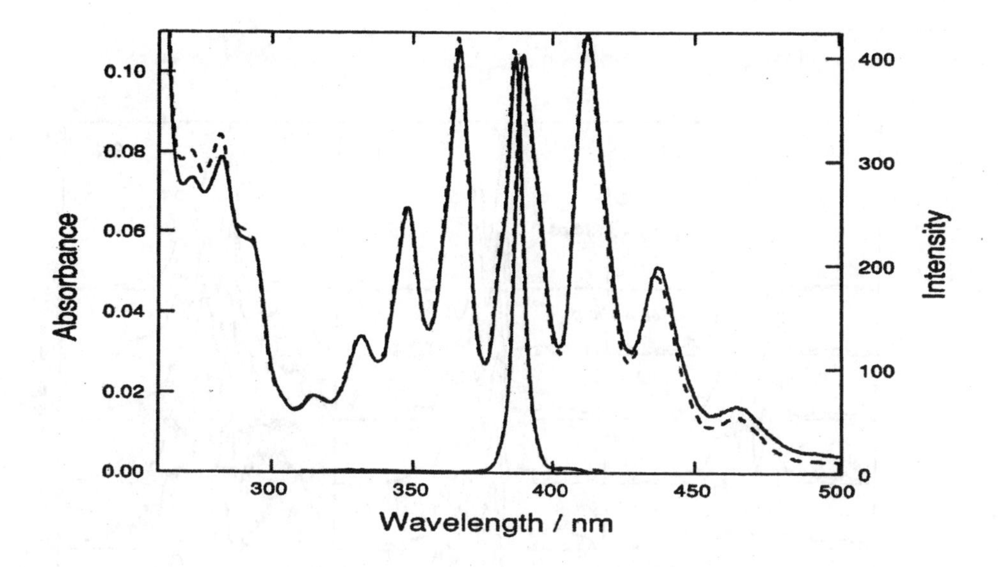

Figure 1. Absorption and fluorescence spectra of A3N in stretched PVA film at 10 K.

Our general conclusions are in agreement with recent observations of inefficient EET in flexibly connected bichromophores which exhibit highly efficient EET processes when they are connected by a rigid bridge, in fact up to 5 order of magnitude enhancement in the EET rate was observed.[18,19] While the distance dependence is the manifestation of the through bond superexchange contribution to the observed EET rates due to the stretched conformations, the direct Coulombic interaction still determines the overall rate for all three AnN molecules in solution for the face to face conformeres.

Table 1 summarizes the results of the analysis for the fluorescence decay curves upon excitation at 280 nm (naphthalene, D) in *stretched* PVA films. The decay curve of D fluorescence in A1N is almost single exponential with very short lifetime at 10 K and 296 K, while those in A3N and A6N are composed of two or three exponential decay functions with short and long lifetimes which gives a distribution of the energy transfer rate even in stretched films. The decay curves of anthracene (A) fluorescence are described simply by a single exponential rise and

decay

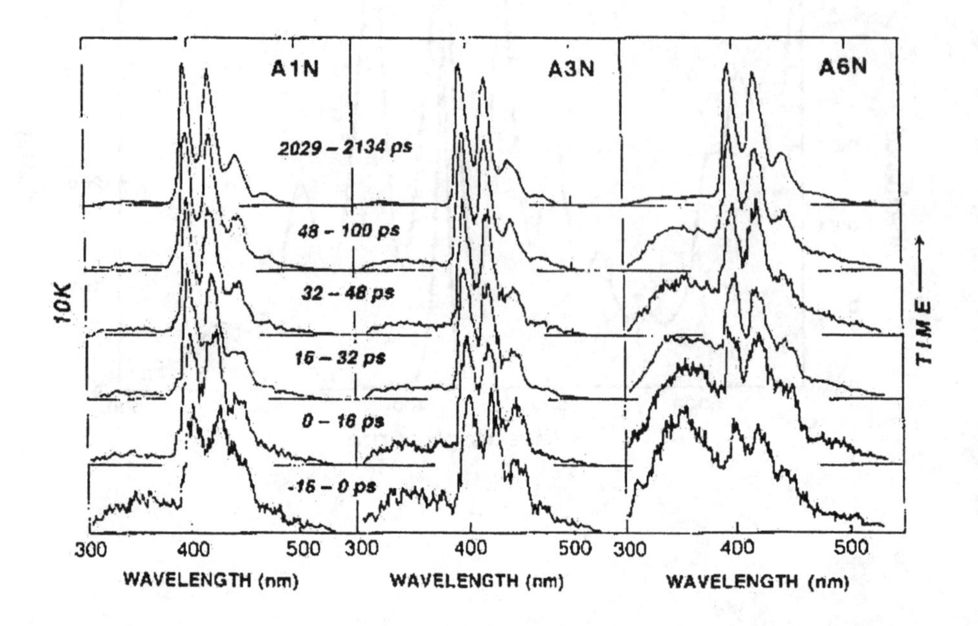

Figure 2. Time-resolved fluorescence spectra of A1N, A3N and A6N in stretched PVA films at 10 K. The excitation wavelength is 280 nm (^1L_a absorption band of naphthalene).

Table 2 summarizes the fluorescence lifetimes for A1N, A3N and A6N in *nonstretched* PMMA films at 10 K and 296 K. All the decay curves can be analyzed phenomenologically in the same way as in the stretched films. In striking contrast to the case of stretched films, the decay kinetics are determined by the faster decays, and as for the component with the dominant amplitude the decay properties show no systematic change on going from A1N to A6N. It then follows that AnN's in *nonstretched* films have a broader distribution of folded conformers, resulting in faster decay kinetics. This can be compared to the solution case, Table 3showing similar behavior.

For this reason, we confine our discussion to the EET dynamics associated with the fluorescence decays with the dominant amplitude in the stretched film at 10 K, which will correlate directly with the D-A distance dependence on the EET rate for the AnN molecules. The fluorescence lifetime is related to the energy transfer rate (k_{EET}) by

$$\tau_F = \frac{1}{k_F + k_{EET}} \cong \frac{1}{k_{EET}} \tag{13}$$

where $k_F = (\tau_F^0)^{-1}$ with τ_F^0 representing the fluorescence lifetime of a reference molecule without an anthracene moiety. In this study we use the lifetime value of 1-methylnaphthalene: i.e., τ_F^0 = 70 ns in PMMA film at 10K. The calculated k_{EET} values are be 47.6 x 10^{10} s^{-1} in A1N, 1.96 x 10^{10} s^{-1} in A3N and 0.145 x 10^{10} s^{-1} in A6N in stretched PVA films at 10K.

According to the Förster's mechanism , eq. 10, the EET rate constant is expressed by the following equation:

$$k_{EET} = \left(\frac{R_0}{R}\right)^6 \frac{1}{\tau_F^0} \tag{14}$$

The transition dipole moment of S_0 to S_1 (^1L_b) transition in naphthalene (D) is in the short molecular axis, while that of anthracene (A) (^1L_a) is in the long molecular axis. The k_{EET} values were calculated by using the literature value R_0 = 25.51 Å for 1-methylnaphthalene (D) and anthracene (A). [16,17]

Taking into consideration that there still exist possible conformers with respect to mutual orientation of the two end groups in stretched films, we calculated for the two extremes of conformations that the molecular planes of D and A are parallel and perpendicular. The calculated k_{EET} values are in fair agreement with the corresponding experimental values, although through - bond interaction in addition to the Förster-type through-space interaction should be considered. In the present case of A1N, a contribution from the similar through-bond interaction via single methylene group is probably responsible for the EET rate larger than the calculated one.

In conclusion, the AnN's molecules incorporated in stretched films prefer a conformation in which the methylene chain is elongated and the two end groups of D and A are separated far apart. For this conformation, the intramolecular EET in AnN's on the whole can be described approximately in terms of the Förster's dipole-dipole interaction mechanism, with some contributions from through-space exchange and through-bond superexchange interactions. On the other hand, the nonstretched films exhibit EET rate which is much faster than the stretched films, indicating that A3N and A6N in films takes predominantly folded conformations similarly to the cases of AnN's in fluid solution.[20]

TABLE 1. Fluorescence decay lifetimes and amplitudes of naphthalene (D) and anthracene (A) , in A1N, A3N (biexponential) and A6N (triexponential) in *stretched* PVA films with excitation of D at 280 nm.

Compound	10K (D/ps)	10K (A/ps)	296K (D/ps)	296K (A/ps)
A1N	2.1	9165	2.9	8379
A3N	51	87	18	38
	173	10920	153	10250
A6N	71	97	56	111
	687	10900	380	10370
	4500			4090

TABLE 2. Bimodal fluorescence decay lifetimes of naphthalene (D) and anthracene (A) in A1N, A3N and A6N in *nonstretched* PMMA films with excitation of D at 280 nm.

Compound	10K (D/ps)	10K (A/ps)	296K (D/ps)	296K (A/ps)
A1N	3.5	15	2.7	14
		12000		11000
A3N	19	14	14	16
	197	12000	140	9300
A6N	37	51	46	80
	188	11000	148	11000

3.Towards molecular scale devices based on controlled intramolecular interactions

As was argued above charge and EET processes are well studied. We have examples whereby different molecules can signal their state from one (the Donor, D) to the other (the Acceptor, A). Here we propose to use this transfer as a way of connecting between logical operations that are implemented on different molecules. With such a concatenation one can begin to think of the construction of larger scale integrated logic circuits, made up of many molecules. In particular, the already demonstrated EET in trichromophoric molecules [21-24] assures us that a fanout operation, that is the communication of a given output as input to more than one circuit, will be possible. While we shall not make use of it, the scheme discussed in this paper can exhibit a bidirectional transfer so that feedback is also possible.

The present work is in molecular electronics [25]. Rather than trying to make molecules that can act as wires [8,9] switches [10-14] and other building blocks of conventional electronic circuits, in this work individual molecules are instructed to implement already non trivial logic tasks. We show that one molecule can communicate its logic output as input to another molecule. This transfer is achieved as an electronic energy transfer from a donor to an acceptor. We do discuss a specific pair for which there is considerable data but the scheme is general enough to allow a wide choice of D and A pairs. The results are for an intermolecular transfer in solution but many similar D – A pairs that are bridged, have been studied [1,9]. Indeed, a rigid bridge will make the energy transfer much more efficient [18,19] so that the rigid concatenation required for a circuit board is an advantage. A bridged pair is a particular example of our general suggestion that a reactive molecular coordinate can act as a bus.

Linked Donor-Acceptor bichromophoric molecular systems of the type Donor-bridge-Acceptor (**D-B-A**), in which the bridge is a saturated hydrocarbon moiety, have been developed into a major tool for studying various aspects of electron donor-acceptor interaction over the past decades. In most cases the intra-EET rate constant k_{EET}, is attributed to two possible contributions. The first is the long range Coulombic contribution which was formulated and by Forster [6] in terms of dipole-dipole interaction. The second contribution to EET is the short range exchange interaction, as formulated by Dexter [7].There are many experiments, in particular photo induced Intra-ELT studies, that indicate donor and acceptor interactions at D and A separations that are much larger than the sum of their van der Waals orbital radii [21-24]. The proposed mechanism is a superexchange interaction operating beyond actual orbital overlap region, usually thought to be mediated by electronic through band (TB) coupling of the interchromophore bridge orbitals.

Following their Intra-ELT work [26], Verhoeven and coworkers examined singlet-singlet intra-EET processes in a series of bichromophoric molecules, DMN-n-ketone, 1(n), containing *rigid* polynorbornyl interchromphore bridge spacers, separating the dimethoxynaphthalene donor from the carbonyl acceptor. When the two chromophores were connected via 4-10 σ bonds in an all *trans* conformation. Intra-EET was observed even for R=11.5Å where direct orbital overlap is not possible. The measured transfer rate constant was much larger than that calculated from the Forster model and depended exponentially on the number of interchromophore σ bonds.

As expected for the TB mechanism, the EET rates display a strong exponential distance dependence of the form shown by eq (1), in which n is the number of bonds in the bridge relay, and β' is an attenuation coefficient with experimentally determined values in the range of 2 - 2.5 per bond [18,19].

$$k_{EET} = A \exp(-\beta' n) = A \exp(-\beta R) \qquad (15)$$

In the background to the discussion is the proposal that entire logic gates and even circuits, can be represented by the spectroscopy of a single molecule. With even as few as 10 molecules, acting independently, one can already achieve acceptable signal to noise.

The logic that can be implemented on a molecule is determined by the number of states of the molecule that can be spectroscopically pumped and probed. High resolution work can make the number of states large and this is one promising direction. The other possible route and the one that is followed here, is to connect different molecules in the manner made familiar by connecting arithmetic units into larger chips. Note that we do not say 'to connect different molecules in the manner made familiar by connecting transistors into larger circuits'. This is because of the considerable versatility of modern molecular spectroscopy which allows the design of non trivial circuits already on individual molecules. For example, either the donor or the acceptor that are here discussed implements by itself both an AND gate and an XOR (eXclusive OR) gate, see figure 2 below. It takes several switches (= transistors) to build any one of these gates.

This section discusses the photophysics of a full adder that is a circuit that receives two binary one digit inputs plus the carry bit from the previous addition and yields as output the sum of the two inputs and a new value for the carry bit (= carry out). There are altogether eight possible inputs to the full adder. It is the (Boolean value of the) intermediate sum that is communicated between the donor and acceptor. The corresponding logic circuit is shown as Figure 3.

The proposed full adder needs two half adders such that one bit can be delivered from the first half adder to the second. A molecular half adder is available for molecules that have a detectable one photon and a detectable two photon absorption. This seems to go against Kasha's rule but in fact there are enough exceptions. Azulene and many of its derivatives provide one class. The emission from the second electronically excited state, S_2, is often as strong or stronger as the fluorescence from S_1 [28,29]. More in general, emission from S_2 is not forbidden, rather, due to competing non radiative processes it often has a low quantum yield but it is definitely detectable particularly so since it is much to the blue as compared to the emission from S_1. If necessary, the emission from S_2 can be detected by photon counting.

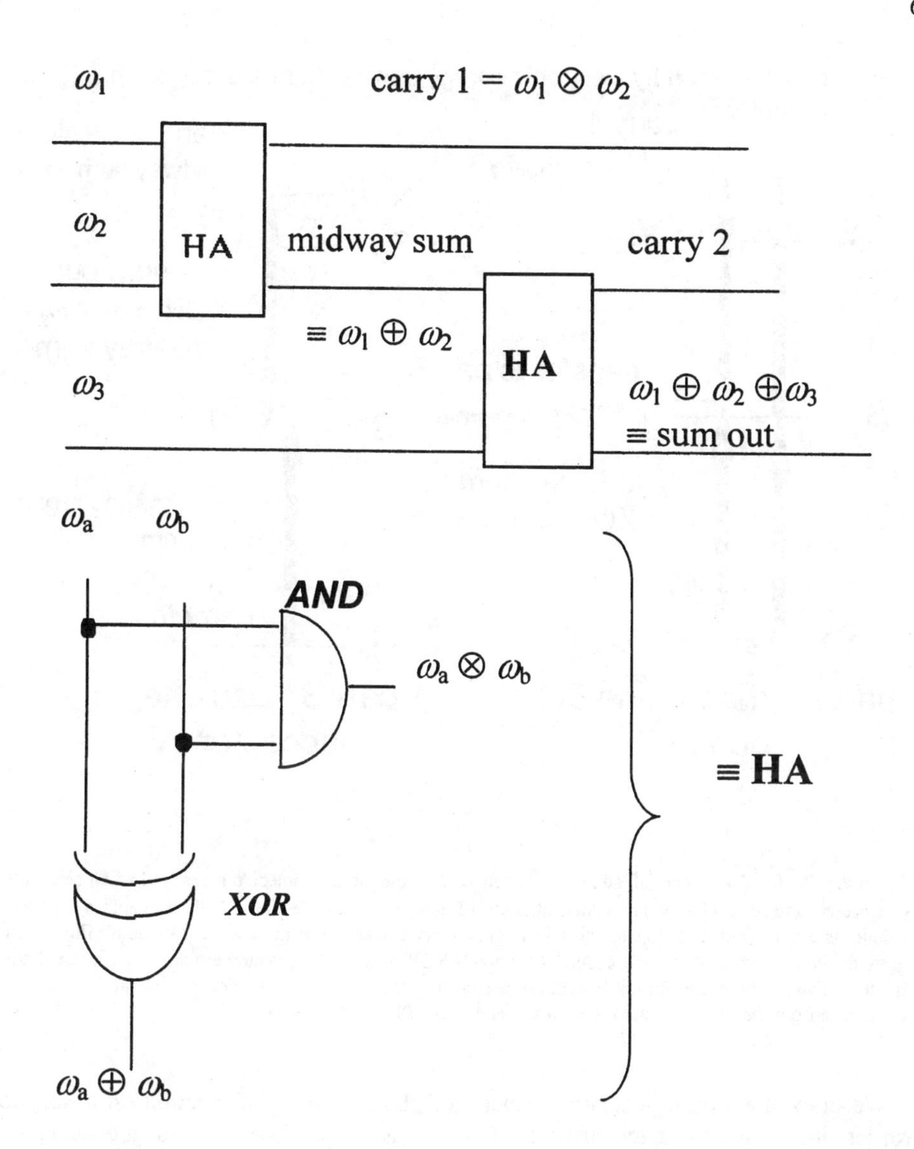

Figure 3. Logic circuits of a full adder, top panel and of a half adder, HA, bottom panel. $\oplus$ is addition of Boolean variables, sometimes The full adder is drawn as if it has three outputs carry1, carry 2 and sum out. Either carry 1 or carry 2 will provide the value of the output carry bit. (If so desired, one can feed them into an OR gate, a one photon broad absorber[18], so as to get an answer that can be communicated to the next circuit). The graphical notation for the AND and the XOR gates is a standard one.

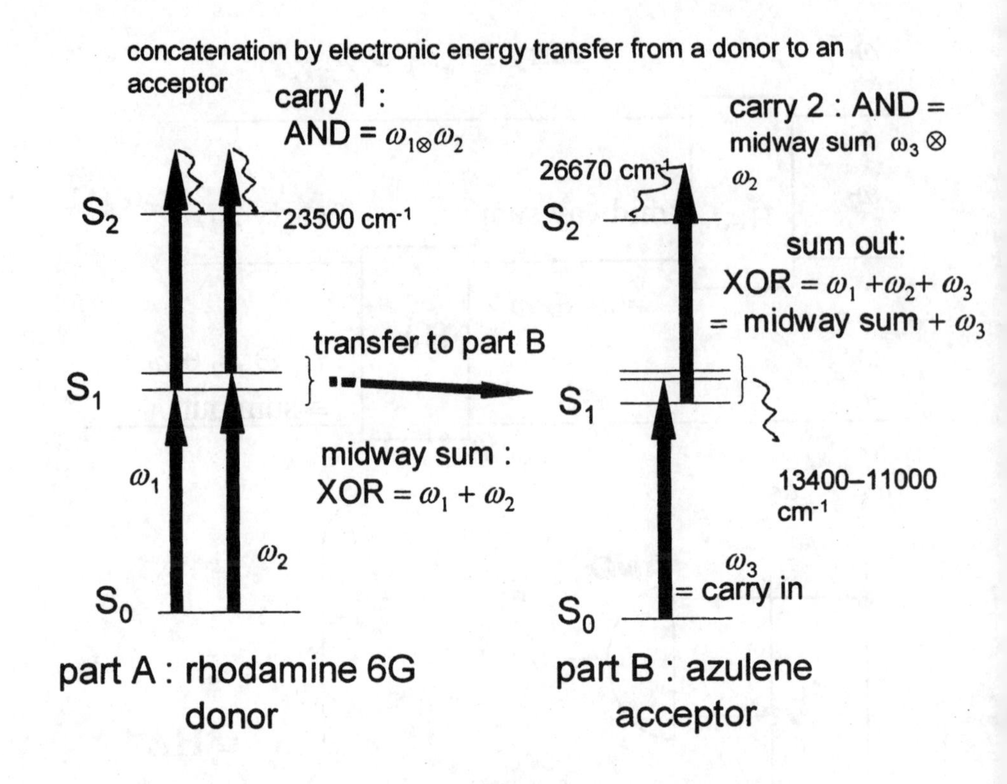

Figure4. The Rh6G-azulene full adder: XOR gate on the Rh6g donor molecule is realized when either ω_1 or ω_2 are present and the excitation is transferred to the acceptor azulene molecule, if both frequencies are present the donor is excited to its second excited state where a carry 1 needs to be detected as 23500cm^{-1} fluorescence to provide the AND gate of the half adder. The sumout XOR needs to be detected either if ω_3 is present alone or if the midway sum is transferred. If both are present the carry 2 needs to be detected as an AND gate via fluorescence from the second excited state of azulene at 26670cm^{-1}.

We draw attention to a relative paucity of photophysical information on absorption profiles and emission characteristics of S_2 levels of potential donors and acceptors. While there is enough data to anchor our proposed scheme on firm observations, it would be useful to have more.

The concatenation between the two half adders is performed by the fairly fast [1] intermolecular electronic energy transfer. Specifically we propose the well characterized transfer from the S_1 level of rhodamine 6G to the S_1 of azulene, figure 2.

There are many possible variations on this theme. While we shall not make use of it, we do mention that the scheme discussed in this Letter can have bidirectional transfer so that the result of the logic implemented on what we call the acceptor can be returned to the donor. For example, the S_2 level of azulene can transfer to a higher excited donor level. See e.g., [33] for a recent discussion of the role of the energy gap in the transfer.

The photophysical details of the full adder are as follows:[21-24] the S_1 level of rhodamine 6G can be readily pumped with photons absorbed within the $S_0 \rightarrow S_1$ band. For example, we take the frequencies $\omega_1 = 18797$ cm^{-1} (the second harmonic of the Nd-YAG laser) and $\omega_2 = 18900$ cm^{-1}.

In fact, since ω_1 and ω_2 are Boolean variables, they can be the same frequency but from two physically distinct laser beams, say using a beam splitter with a beam stop (to do the Boolean variable selection, 0,1 ≡ on , off) in either beam line. We do not need it for the full adder but the absorption to S_1 can be detected through its emission at about 17500 cm^{-1} . This emission is logically equivalent to
$\omega_1 \oplus \omega_2$ because if the intensity is high enough due to two photons being present, the donor will be pumped either directly to S_2 or to higher levels followed by ultrafast nonradiative relaxation to S_2. The large absorption cross-section of 2.5×10^{-18} cm^{2}molecule^{-1} for the $S_1 \rightarrow S_n$ $(n \geq 2)$ of rhodamine 6G insures efficient pumping of S_2. The emission from S_2 is at about 23250 cm^{-1} with a quantum yield of about 10^{-4} [31]. It is this emission which serves to logically implement the left AND gate and it is equivalent to $\omega_1 \oplus \omega_2$ (denoted as carry 1). The S_1 level of the donor transfers the energy, via the Forster mechanism [1] to the azulene acceptor, whose S_1 level is at 14400 cm^{-1} [34]. This level emits in the 13400–11000 cm^{-1} range. The S_2 level of azulene has its absorption origin at 28300 cm^{-1} and so it can be reached from S_1 by a third photon of frequency 14400 cm^{-1} . The same photon can also pump ground state azulene to its S_1 level. Emission (or lack thereof) from S_2 of azulene at 26670 cm^{-1} provides the carry 2 bit while emission from the S_1 level of azulene provides the sum bit. If one wants error correction then the emission from the S_1 level of rhodamine 6G can be used as a check bit, Figure 2.

The energy transfer rate for a solution of 10^{-3} M of azulene, estimated using the S_1 fluorescence spectrum of rhodamine 6G and the absorption spectrum of azulene, is about 10^{10} s^{-1}. This rate can be increased [16] if the two chromophores are incorporated within a single molecule using a short molecular bridge [1]. The increase will be particularly significant (five orders of magnitude) if the bridge is rigid [18]. There are many other couples based on commonly used laser dyes as donors and azulene derivatives [21-24] that can be utilized for the implementation of the logic gate. We have designed a novel donor-acceptor bichromophoric molecule that implement combining rhodamine6G and azulene in a single rigid molecules to test these ideas.

Electron transfer from a donor to an acceptor is another way to achieve concatenation. Rather than a direct donor to an acceptor transfer one can put a bridge in between. One typically thinks of either through-space transfer without any bridge involvement, or of a bridge in terms of the 'super exchange' mechanism . There are however other options, where the bridge itself is actively involved. This allows one to tune the components of the bridge such that the transfer is facile or not.

In order to determine the role played by the interchromophore bridge we compared [18,19] intra-EET between the same two chromophores connected by a *rigid* bridge for which previous studies indicated a through-space exchange interaction controlled EET for a *flexible* bridge [1].

Earlier studies of intra-EET in semiflexible aromatic compounds, of the type shown in scheme 2, established through-space exchange interaction as the process promoting mechanism [1]. This was implemented by the study of the DMN bichromophoric compounds of scheme 1, which were synthesized by Paddon-Row and measured by Ghiggino and coworkers [18].

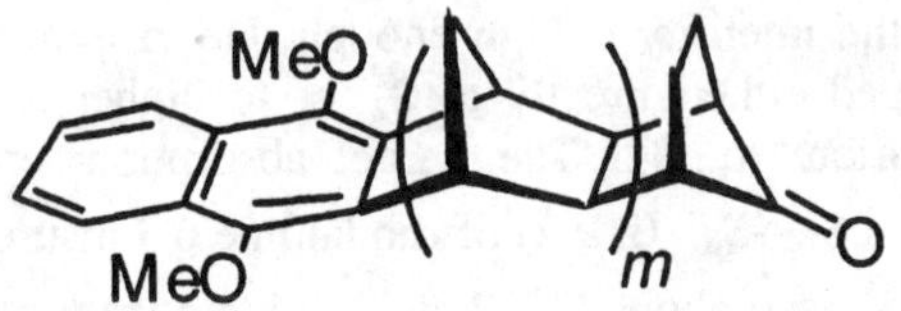

1(4); *m* = 1 1(6); *m* = 2

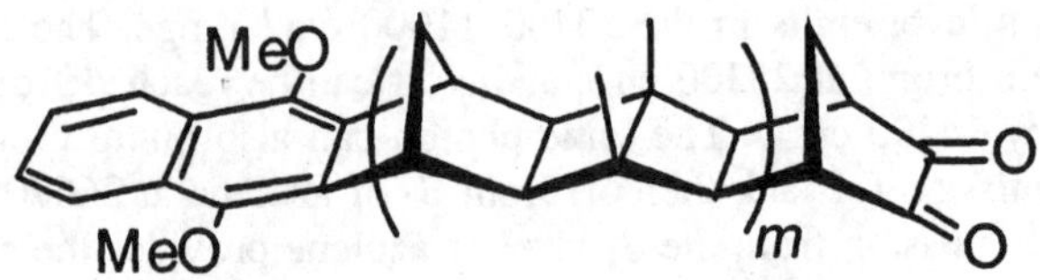

2(6); *m* = 1 2(10); *m* = 2

Scheme 1

The rigid compounds of **scheme 1** were measured utilizing time resolved ultrafast fluorescence spectroscopy to determine the EET rate by measuring the rise time of the donor-dione fluorescence signal. These were compared with the corresponding data obtained for the flexible molecules of **scheme 2** and also to the rigid systems of monoketone rigid system, synthesized by Paddon Row and utilized by Verhoeven and coworkers [26].

This comparison between results pertaining to rigid and flexible bichromophores [12-14] on one hand, and those pertaining to changing of the acceptor's chromophore for rigid molecules on the other hand, is done in Table 1. The β values depend both on the rigidity of the bridge and on the energy gap between bridge electronic energies and those of D* and A*, as probed by comparing EET for monoketone and dione acceptors, as expressed by McConnell's model [27].

1,4-Naph-5,5

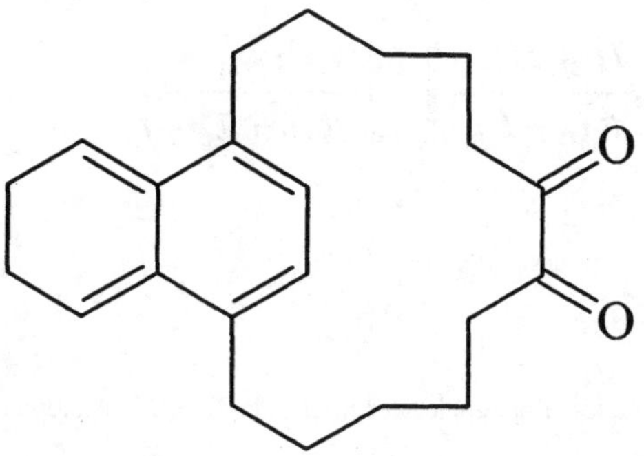

O
O

O-n,n **M-n,n** **P-n,n**

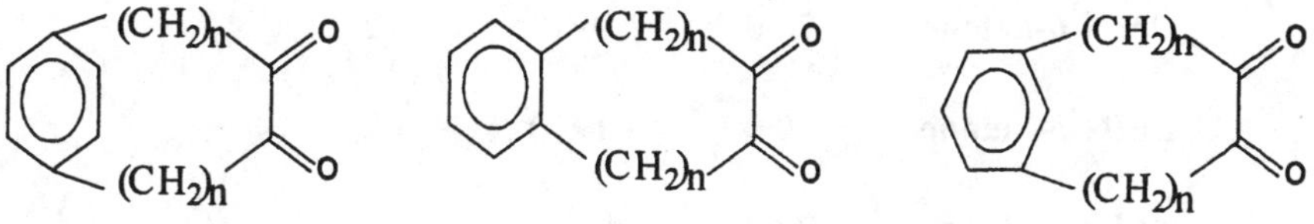

$(CH_2)_n$ O
$(CH_2)_n$ O

n=4,5,6 n=2,3,4 n=3,4

Scheme 2

McConnell suggested that orbital sites intervening between D and A could facilitate ELT. In his superexchange model, an electron is transferred between degenerate D and A orbitals, aided by the presence of empty (not necessarily degenerate) high-lying bridge orbitals. McConnell's expression for the coupling matrix element is given by

$$V_{DA}^{(n)} = \left(\frac{H_{DB_1} H_{B_n A}}{E_{AD} - E_{B_1}} \right) \left(\prod_{i=1}^{n-1} \frac{H_{i,i+1}}{E_{AD} - E_{B_{i+1}}} \right) \tag{16}$$

TABLE 3. Rates of intramolecular singlet-singlet EET in solution.

MOLECULE	*R(A)*	$k_{EET}\,(s^{-1})$	$\beta(A^{-1})$
1,4-Naph-5,5	5.23	6.0×10^{8}	0.9
O-4,4	6.00	5.0×10^{6}	0.9
DMN-4-ketone	5.00	1.2×10^{10}	1.2
DMN-6-ketone	7.50	1.9×10^{8}	1.2
DMN-8-ketone	10.0	3.1×10^{7}	1.2
DMB-6-dione	7.50	$>10^{11}$	
DMN-6-dione	7.50	$>10^{11}$	0.6
DMN-10-dione	12.5	2.5×10^{10}	0.6

where $H_{I,j}$ is the tunneling integral between orbitals i and j, E_{AD} is the degenerate D and A orbitals energy, E_{Bi} is the energy of the *i*th bridge orbital, and n is the number of B orbitals.

In McConnell's model only electronic couplings are taken into account. It complletly ignores vibronic interactions that might affect coupling between the donor and the acceptor to the bridge or intrabridge states coupling. While this may hold for *rigid* bridges it is bound to fail for *flexible* bridges, where bridge vibrartional motion may be coupled to the electron dynamics involved in the transfer process. We provide here preliminary results for such a modified McConnell's model, suitable for dealing with such flexible structures.[24]

The mechanism for electron and electron-energy transfer in donor-bridge-acceptor(**DBA**) systems is studied in the presence of electron-nuclei coupling. The 30 years old McConnell model is generalized in order to account for the flexibility of the molecular bridge and its effect on the electron transfer rate. The original tight-binding model proposed by McConnell [27] is solved numerically, allowing for a non-perturbative coupling between the bridge and the chromophors as well as between the bridge subunits. In the simplest possible correction to this model, which accounts for the flexibility of the bridge, the vibrations of the molecule are described as time-dependent oscillations of the energy level of the bridge subunits. The rate of electron transfer between the donor and the acceptor is calculated by solving the time-dependent Schroedinger equation for the electronic degree of freedom with a time-dependent Hamiltonian, representing the bridge motion. Following the population transfer from the donor to the acceptor in the **DBA** system as a function of time one can obtain the transfer rate. For small electron-nuclei coupling intensities, we find that when the bridge frequency is similar to the energy-splitting between the donor and acceptor levels (the electron transfer rate in the "stiff" bridge), the electron transfer rate is reduced significantly, while for higher or lower vibrational frequencies the transfer rate is hardly affected. This result is consistent with a perturbative treatment of the electron-nuclei coupling and it suggests that the electron transfer rates is sensitive to bridge flexibility when the electron transfer rate is of the order of molecular frequencies (100-1000 cm^{-1}).

Our model is a time-dependent variant of the McConnell Hamiltonian,

$$H(t) = H_0 + V(t) \quad (17)$$

H_0 is the Hamiltonian matrix which represents a "stiff" donor-bridge acceptor system

(see Fig.5) ,

$$H_0 = \begin{bmatrix} 0 & T & 0 & \cdots & & & \varepsilon \\ T & D & t & & & & \\ 0 & t & D & & & & \\ \vdots & & & \ddots & & & \\ & & & & D & t & 0 \\ & & & & t & D & T \\ \varepsilon & & & & 0 & T & 0 \end{bmatrix} \tag{18}$$

The matrix dimension is N+2, where N in the number of the bridge sub-units. The diagonal elements $H_{1,1}$ and $H_{N+2,N+2}$ stand for the electronic energy of the uncoupled chromophors (taken as 0 for both the donor and the acceptor in the present model). The direct coupling between the two chromophors is denoted ε (taken to be negligible in the numerical applications below). The parameter of electronic coupling between each chromophore and its nearby bridge unit is T, and the parameter of coupling between neighboring bridge subunits is t. D is the energy gap between the donor/acceptor levels and the energy levels of the uncoupled bridge subunits.

In order to account for the flexibility of the bridge, a time dependent matrix is added to H_0 (Fig. 6),

$$V(t) = \lambda \sin(\omega t) \begin{bmatrix} 0 & 0 & 0 & \cdots & & & 0 \\ 0 & 1 & 0 & & & & \\ 0 & 0 & 1 & & & & \\ \vdots & & & \ddots & & & \\ & & & & 1 & 0 & 0 \\ & & & & 0 & 1 & 0 \\ 0 & & & & 0 & 0 & 0 \end{bmatrix} \tag{19}$$

λ is the electron-nuclei coupling intensity parameter and ω is the frequency of the bridge oscillations. According to this model, the electronic energy of the bridge subunits oscillates with a typical frequency, corresponding, e.g., to a global nuclear motion (a bridge "phonon") of the bridge subunits.

The electron transfer rate from the donor to the acceptor can be simulated by solving the time-dependent Schroedinger equation, with $H(t)$. The initial state corresponds to 100% population of the donor,

$$\psi(0) = \begin{pmatrix} 1 \\ 0 \\ 0 \\ \vdots \\ 0 \end{pmatrix}. \qquad (20)$$

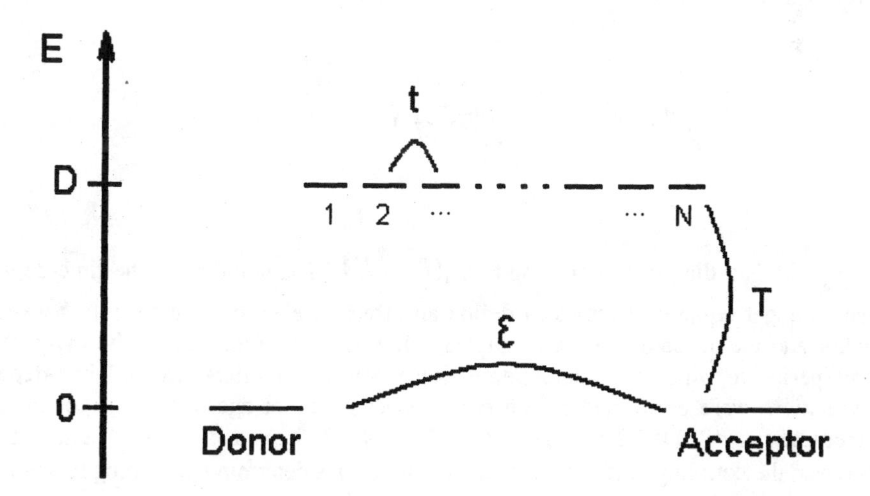

Figure 5. Schematic representation of the McConnell super exchange model

The coupling between the donor and the rest of the system by $H(t)$ leads to population transfer from the donor to the acceptor through the bridge, resulting in a decay of the donor population. This process is reflected by a dynamical change in the wavefunction $\psi(t)$, as a function of time. The numerical procedure for solving the time-dependent Schroedinger equation was based on a split-operator method, in which the time-axis is divided to small time interval, and the propagator at each time interval reads,

$$\psi(t+dt) \cong e^{-i\frac{dt}{2\hbar}V(t+dt)} e^{-i\frac{dt}{\hbar}H_0} e^{-i\frac{dt}{2\hbar}V(t)} \psi(t) \tag{21}$$

The potential matrix $V(t)$ is therefore the operation of $e^{-i\frac{dt}{2\hbar}V(t)}$ is straight forward. The non-diagonal operator $e^{-i\frac{dt}{\hbar}H_0}$ is represented in terms of the diagonalization transformation of H_0,

$$e^{-i\frac{dt}{\hbar}H_0} = Ue^{-i\frac{dt}{\hbar}D}U^{-1} \tag{22}$$

where D is a diagonal matrix, and $H_0 U = UD$. The solution of the time-dependent schroedinger equation enables to define an effective electron transfer rate between the donor and the acceptor. In the examples below we define the rate as the inverse of the time-period in which the donor population reaches its minimal value. Time-dependent simulations were performed for different values of the bridge frequency, for the model parameters: N=8, D=3.2 eV, T=-0.856 eV, t=-0.9795 eV and $\varepsilon=10^{-20}$ a.u.. The ratio between the coupling parameters (t and T) and D was determind according to ab-initio fit by Jordan and Paddon-Row [39] for the molecule in **scheme 1**.

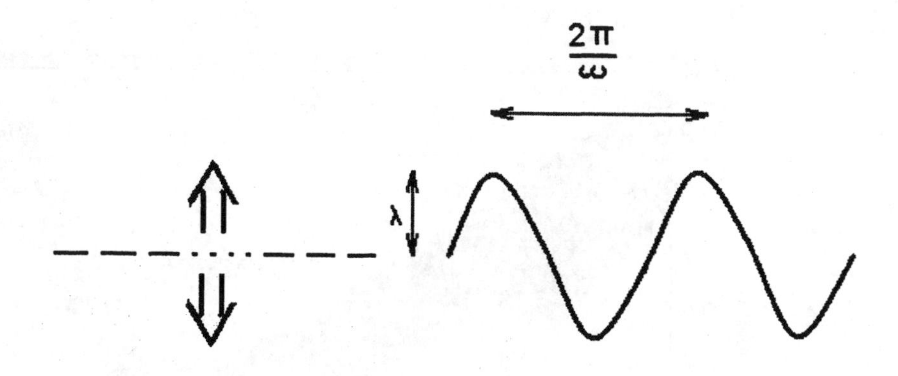

Figure 6. Schematic representation of the oscillatory motion of the molecular bridge

In Figure 7 the population of the donor site $|\psi_1(t)|^2$ is plotted as a function of time (starting from 100% population at t=0) for $\lambda = 0$ and for $\lambda = 0.2D$. The bridge frequency ω was chosen as half of the energy gap between the two lowest energy levels of H_0 (the super exchange rate). Interestingly, the numerical result for $\lambda = 0$, gives a rate of $5.4 \cdot 10^{10}$ 1/sec, which is similar to the electron-energy transfer rate, measured for a bridge with the same length [17,18]. The numerical results for the flexible molecule demonstrate the effect of time delay in the decay of the donor population due to the bridge vibrations. In figure 4, the decay time of the donor population is plotted for the same system with $\lambda = 0.1D$, as a function of the bridge frequency, illustrating the significant effect of the bridge flexibility on the electron transfer rate, when the bridge frequency matches the electron-transfer rate in the stiff system.

In conclusion we note that we have provided evidence that through bond superexchange is significant in promoting Intra-EET in rigidly bridged bichromophoric compounds at interchromophore separations exceeding 12A. This is not the case in flexibly bridged compounds. We believe that the flexibility results in vibronic coupling of bridge modes resulting in loss of coherence needed for efficient bridge mediated through bond coupling, such as invoked in McConnell's model [27]. Similar observations were made for the system composed of naphthalene (donor)-anthracene (acceptor), thus indicating the possibility of controlling EET by judicious molecular engineering.

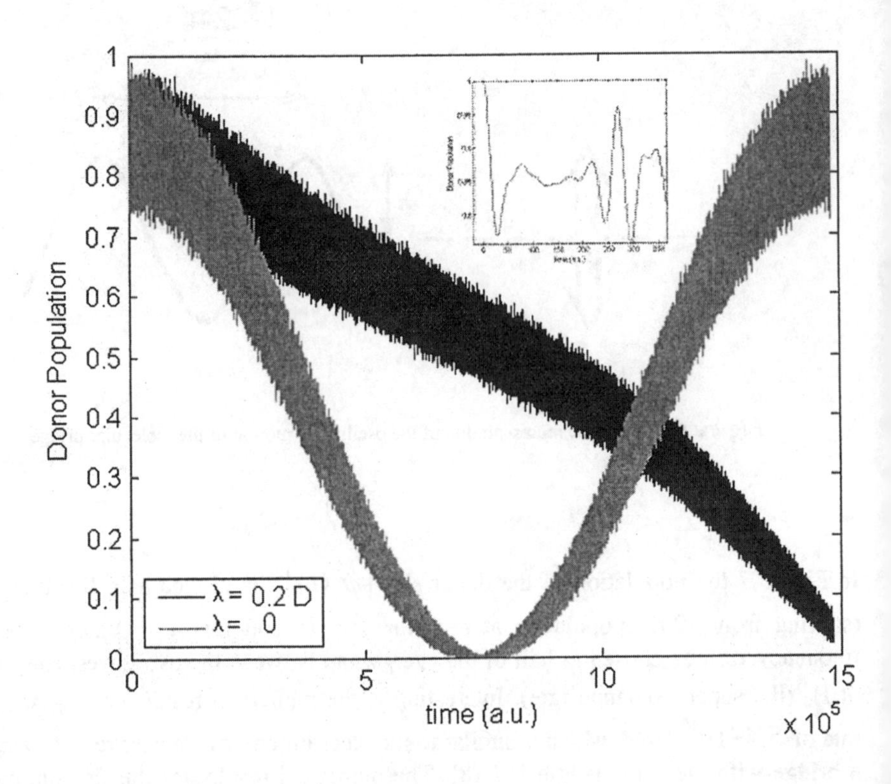

Figure 7. **The time-decay of the donor population for a stiff bridge (grey) and for a flexible bridge (black). The rapid oscillations are due to mixing between the donor/acceptor levels to the bridge levels.Inset: The initial dynamics.**

This is the first direct evidence that the rigidity of the interchromophore bridge is significant for promoting through-bond bridge-mediated intramolecular EET. Such control is needed for implementing molecular logic gates based on Intra EET.The preliminary results outlined in the modified McConnell's model described here show that the electron transfer rates is sensitive to bridge flexibility when the electron transfer rate is of the order of molecular frequencies (100-1000 cm^{-1}).

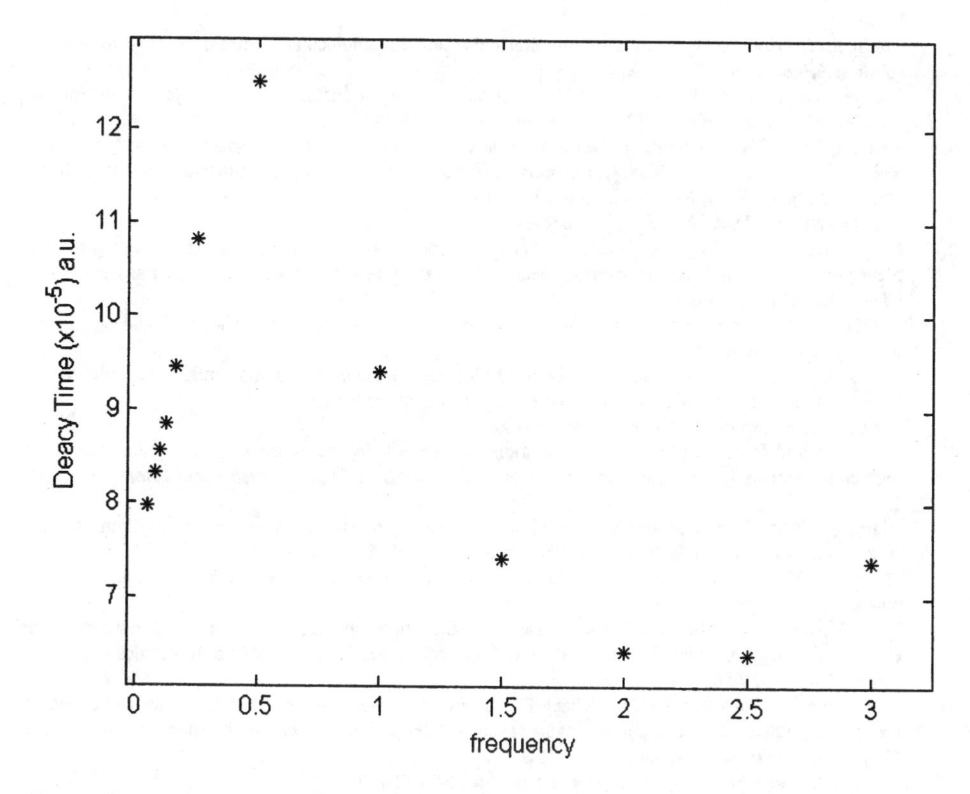

Figure.8. The decay time of the donnor population as a function of the bridge frequency. The frequency is measured in units of the decay rate of the "stiff" bridge.

Acknowledgment

This research was supported by the Fund for the Promotion of Research at the Technion and the Technion VPR Fund.

References

Speiser, S.,(1996) Photophysics and mechanisms of intramolecular electronic energy transfer in bichromophoric molecular systems: solution and supersonic jet studies *Chem. Rev.* **96**, 1953-1976.
Scholes, G.D. (1994) *Electronic interactions and interchromophore energy transfer*, Ph.D. Thesis, the University of Melbourne
Liao D.W, Cheng W.D,. Bigman J,. Karni Y,. Speiser S. and .Lin S.H, (1995) On theoretical treatments of electronic excitation energy transfer *J.Chin.Chem.Soc.* **42**, 177-187.

4. S.Speiser, (1983)Novel aspects of intermolecular and intramolecular electronic energy transfer in solution *J.Photochem.* **22**, 195-211.
5. Speiser S.and.Katriel J. Intramolecular electronic energy transfer via exchange interaction in bichromophoric molecules (1983) *Chem.Phys.Lett..* **102**, 88-94.
6. Foerster, Th. (1968) In *Modern Quantum Chemistry*; O.Sinanoglu, Ed.; Academic Press: New York, Vol.3, 93 ; Forster T., ( )1960in: Comparative Effects of Radiation, eds. M. Burton, J. S. Kirby-Smith and J. L .Magee (Wiley, New York) ,pp. 317-325.
7. D.L. Dexter, (1953). *J.Chem.Phys.* **21**, 836-845.
8. Levy S.T.,. Rubin M.B, and Speiser S., (1992) Photophysics of cyclic α-diketone aromatic ring bichromophoric molecules. Structures, spectra and intramolecular electronic energy transfer *J. Am. Chem. Soc.* **114**, 10747-56.
   S.T.Levy, *Electronic energy transfer in bichromophoric molecules,* M.Sc. Thesis, Technion - Israel Institute of Technology (1992).
9. Toledano E.,. Rubin M.B and Speiser S. (1996). Dependence of intramolecular electronic energy transfer in bichromophoric molecules on the inter-chromophore bridge, *J. Photochem. Photobiol.: Chem. A.*, **94**, 93-100.
10 Speiser S.and Rosenblum G. (1997) Intramolecular electronic energy transfer in bichromophoric molecular systems in supersonic jet expansions", in *Trends in Photochemistry and Photobiology*, **4**, 137-165.
11. Bigman J.,.Karni Y, and Speiser S. (1994) Electronic energy transfer between benzene and biacetyl in a supersonic jet expansion *J.Photochem. Photobiol. A: Chem.* **78**, 101-111.
12. .Bigman J,.Karni Y, and Speiser S., (1993) Electronic energy transfer in bichromophoric molecular clusters *Chem.Phys.* **177**, 601-617.
13. Wang X., Levy D.H.,. Rubin M.B and Speiser S. (2000) Supersonic jet spectroscopy and intramolecular electronic energy transfer in naphthalene-$(CH_2)_n$-anthracene bichromophoric molecules, *J. Phys. Chem.*, **104**, 6558-6565.
14. Rosenblum G., Grosswasser D., Schael F., Rubin M.B. and Speiser S. (1996) Electronic energy transfer in supersonic jet expanded naphthalene - $(CH_2)_n$ - anthracene bichromophoric molecules, *Chem. Phys. Lett.* **263**, 441-446.
15.. Schnepp O. and Levy M., (1962) *J. Am. Chem. Soc.* 84 172-177.
16. Nishimura Y., Yasuda A., Speiser S.,and Yamazaki I. (2000), Time-resolved analysis of intramolecular electronic energy transfer in methylene-linked naphthalene-anthracene compounds", *Chem. Phys. Lett.*, **323**, 117-124 .
17. Hagesawa M., Enomoto S., Hoshi T., Igarashi K., Yamazaki T., Nishimura Y., Speiser S.,and Yamazaki I. (2002) Intramolecular excitation energy transfer in bichromophoric compounds in stretched polymer films", *J. Phys. Chem. B.*, **106**, 4925-4932.
18. Lokan N., Paddon-Row M.N.,. Smith T.A, La Rosa M., Ghiggino K.P. and Speiser.S. (1999).Highly efficient through-bond-mediated electronic excitation energy transfer taking place over 12Å, *J. Am. Chem. Soc.*, **121**, 2917-2918.
19. Speiser S.and Schael F. Molecular structure control of intramolecular electronic energy transfer, (2000).*J. Mol. Liq.*, **86**, 25-35.
20. Schael F., Rubin M.B. and Speiser S. (1998)Electonic energy transfer in solution in naphthalene-anthracene, naphthalene-acridine and benzene-DANS bichromophoric compounds, *J. Photochem. Photobiol. A.*, **115**, 99-108.
21. Remacle F., Speiser S. and. Levine R.D(2001) .Intermolecular and intramolecular logic gates, *J. Phys. Chem. B*, **105**, 5589-5591.
22. Speiser S. (2003) Towards molecular scale devices based on controlled intramolecular interactions", *J. Luminesc.* **102-103**, 267-272.
23. Peskin U., Abu-Hilu M., and Speiser S. (2003)Approaches to molecular devices based on controlled intramolecular electronic energy and electron transfer. Electron transfer through flexible molecular bridges by a time-dependent super exchange model", *Opt. Mat.* (in press).
24. Jortner J. .and Ratner M. A (1997) Molecular Electronics, Blackwell Science.

18. STIMULATED RAMAN SCATTERING SPECTROSCOPY OF FRONTIER NONLINEAR-LASER MATERIALS: ORGANIC CRYSTALS AND NANOCRYSTALLINE CERAMICS

A. A. KAMINSKII

Institute of Crystallography
Russian Academy of Sciences
Leninskii pr. 59, Moscow, 119333 Russia

Dedicated to the 60-th Anniversary
of the Institute of Crystallography
of the Russian Academy of Sciences

Abstract

The stimulated Raman scattering (SRS) phenomenon allows to convert pumping laser emission wavelength of crystalline materials providing suitable molecular or lattice vibration modes which contribute to the third order nonlinear optical susceptibility $\chi^{(3)}$. Renewed interest in this field emerged because of the discovery of efficient SRS in crystals that contain molecular units exhibiting $\chi^{(3)}$-active modes. Particularly, organic nonlinear optical crystals used so far frequency doubling and third harmonic generation seem to have a great potential for SRS applications. This lecture reported same results on an efficient SRS lasing effects that were discovered during last three years in $\chi^{(2)}$- and $\chi^{(3)}$-active organic crystals, as well as recent results on SRS spectroscopy of novel solid-state laser materials – highly nanocrystalline ceramics based on cubic $Y_3Al_5O_{12}$ garnet and RE_2O_3 (here RE = Sc, Y, and lanthanides Ln) sesquioxides.

1. Introduction

Stimulated Raman scattering (SRS) in optical crystalline materials is of topical interest in modern solid-state laser physics. The SRS process allows to shift laser emission wavelength and compress laser pulses, it can improve the spatial quality of laser beams as well as the contrast between peak and background intensities of ultra-short laser pulses, etc. In the last two decides, solid-state SRS science and technology were becoming more wide spread (see, e.g. [1-3]). Growth in the activity has been made possible by the discovery of several new SRS-active inorganic crystals, including a successful application given by nano- and picosecond Raman lasers generating specific and otherwise hard to reach wavelengths in a wide spectral range [3-5]. Among other current applications of new Raman lasers, remote sensing of the atmosphere is of great interest [6]. Furthermore, crystalline lasers using SRS conversion process are very attractive for medical treatments and for laser guide stars in precise astronomical experiments (see, e.g. [7]).

B. Di Bartolo and O. Forte (eds.), *Frontiers of Optical Spectroscopy*, 619-646.

New generation of Raman lasers requires crystalline materials providing a large frequency shifts up to 3000 cm^{-1} or more. Unfortunately, with inorganic crystals such shifts are difficult if not impossible to realize due to their ionic structure. As can be seen from Table 1, among known SRS-active inorganic crystals a largest Raman frequency shift has been measured for calcite ($CaCO_3$) [8] and lithium formate monohydrate ($LiHCOO \cdot H_2O$) [9]. During last three years we have been discovered an efficient SRS effects in several organic crystals many of them as indicated in Table 2 possess a frequency shifts as large 3000 cm^{-1} and a relatively high steady-state Raman gain coefficients for the first Stokes generation. A number of them offer also both nonlinear $\chi^{(2)}+\chi^{(3)}$ susceptibilities, which may give rise to diverse parametric generation acts. It is of interest that in the field of nonlinear optical organic crystals the attention has mainly been directed towards second and third harmonic generation (see, e.g. [25,26]), but not towards SRS. This is rather astonishing, because the bright optical $\chi^{(3)}$ effect, such as the SRS, was discovered in organic liquid (nitrobenzene $C_6H_5NO_2$) in 60´s just in the beginning of laser era [27].

Laser gain inorganic materials based on Ln^{3+}-ion doped (in particular, Nd^{3+} and Yb^{3+}) highly transparent cubic $Y_3Al_5O_{12}$ and RE_2O_3 ceramics (here RE = Sc, Y, and Ln) are the focus of m any research groups (see, e.g. [28-32]). It is enough to note here the results that were achieved concerning of high-power Nd^{3+}:$Y_3Al_5O_{12}$ ceramic laser with laser-diode (LD) pumping. Because of novel ceramic technology, only in the last three yeas has the one-micron (${}^4F_{3/2} \rightarrow {}^4I_{11/2}$ lasing channel) CW output power been increased from a few hundred milliwatts to the 1.5 kW range. This remarkable progress is summarized in Fig. 1. A new generation of laser crystalline ceramics was obtained by the recently developed revolutionary vacuum sintering method, where the raw materials were prepared by nanocrystalline technology [41]. Investigations have shown that the Nd^{3+}:$Y_3Al_5O_{12}$ ceramic lasers demonstrate practically identical efficiency and output power as lasers based on single crystals of the same material (see, e.g. [39,40,42]). However, this new generation of laser ceramics has significant advantages over laser Nd^{3+}:$Y_3Al_5O_{12}$ single crystals. They can be fabricated in large sizes and at low cost and with the required high Ln^{3+} ion concentration, as well as in the form of very large multilayer plates for multifunctional lasers [39,43]. These ceramics provide clear evidence that new ceramic technology is not limited to Nd^{3+}:$Y_3Al_5O_{12}$, but that it is a really universal method for fabricating other new ceramic crystalline laser materials. During last two years efficient CW and femtosecond generation from Nd^{3+}- and Yb^{3+}-ceramic lasers based on refractory $Ia\bar{3}$-cubic Sc_2O_3, Y_2O_3, $YGdO_3$, and Lu_2O_3 sesquioxides was obtained (Table 3). The high optical quality of these undoped and Ln^{3+}-doped laser nanocrystalline ceramics is convincing evidenced by recent experiments on the excitation in them of high-order Stokes and anti-Stokes lasing in the visible and near-IR regions under picosecond pumping [51-53].

Present lecture summarized some main results on SRS spectroscopy of organic crystals and highly transparent nanocrystalline ceramics - newel families of nonlinear laser solid-state materials, and on new self-frequency conversion parametric effects observed in them under ultra-short laser excitation.

2. The steady-state stimulated Raman scattering

The nonlinear frequency conversion effects (SHG, SRS, etc.) are possible in any optically transparent crystals in which the electron cloud of atoms tend to be polarized, i.e. the refractive index n is a function of the electric-field strength $\boldsymbol{E}$ of the propagating laser emission through these crystals (see, e.g. [54-56])

$$n(E) = n_0 + n_1 E + n_2 E^2 + \ldots \qquad (1)$$

here n_0 is the "linear" refractive index, and n_1, n_2 and so on are the higher-order coefficient of $n(E)$. A dielectric polarization vector **P**, defined as the electric dipole moment of the optical crystal can be described phenomenologically in terms of nonlinear susceptibility tensor of a crystal by expressing its polarization as a power series in electric-field strength $\boldsymbol{E}$ as

$$\mathbf{P}(E) = \chi^{(0)} E + \chi^{(2)} E^2 + \chi^{(3)} E^3 + \ldots. \qquad (2)$$

where $\chi^{(0)}$ is the linear susceptibility tensor responsible for linear optical phenomena such as refraction and reflection of the light; and $\chi^{(2)}$, $\chi^{(3)}$, etc. are the nonlinear optical susceptibilities of a crystal. These tensors are related to the linear and nonlinear refractive index as follows

$$\chi^{(0)} \cong \frac{1}{4\pi}(n_0^2 - 1), \quad \chi^{(2)} \cong \frac{1}{2\pi} n_0 n_1, \quad \chi^{(3)} \cong \frac{1}{2\pi} n_0 n_2 \qquad (3$$

and responsible for a large variety of nonlinear optical phenomena. The most important nonlinear frequency conversion effects arise from the second and third terms in Eq. (2), which are connected to electrical polarization as the are quadratic and cubic functions of the electrical field strength. The second terms in Eq. (2) in gives rise to frequency mixing, in particular SHG in acentric crystals, whereas the tensor $\chi^{(3)}$ of the third terms is not subsided to symmetry restrictions. Therefore, in $\chi^{(3)}$ -active crystals several nonlinear processes, such as SRS, third harmonic generation and so on are available in optically isotropic and anisotropic crystals (Table 4).

The Raman lasers based on $\chi^{(3)}$-crystals as mentioned above are extensively growing area in modern laser material science and solid-state laser physics. It is not feasible to present an examination of main theoretical aspects of SRS laser frequency conversion in solids used so far. A few such comprehensive reviews are already present in the literature (see, e.g. [1,3, 254,55,57-60]). Depending on the pump pulse duration τ_p, two temporal SRS regimes, steady-state and transient, can be considered. The main condition for the steady-state pumping condition, which is of more interested for many practical cases and which was realized in most known nano- and picosecond crystalline Raman lasers, is

$$\tau_p >> T_2 = (\pi \Delta \nu_R)^{-1}, \qquad (4)$$

here T_2 is the dephasing (phonon relaxation) time of the SRS-active vibration mode and $\Delta\nu_R$ is the linewidth (FWHM) of the corresponding Raman-shifted line with a frequency ω_{SRS} in the spontaneous Raman scattering spectrum. The condition for the first Stokes steady-state generation regime in Raman lasers based on $\chi^{(3)}$-active crystals [61]

$$R_m \exp\left[2\left(g_{ssR}^{St_1} I_p l_{SRS} - \alpha l_{cr}\right)\right]_{\lambda_{St1}} = 1 \tag{5}$$

is very nearly the same as the condition for stimulated-emission (SE) generation in the usual lasers on the base of activated crystals [62]

$$R_m \exp\left[2\left(\Delta N \sigma_{SE} l_{SE} - \rho l_{cr}\right)\right]_{\lambda_{SE}} = 1. \tag{6}$$

In Eqs. (5) and (6): $g_{ssR}^{St_1} I_p l_{SRS}$ is the Raman gain factor (here: $g_{ssR}^{St_1}$ is the steady-state Raman gain coefficient, I_p is the laser pumping intensity, and l_{SRS} is the SRS-active crystal length), α is the loss coefficient at the first Stokes wavelength λ_{St_1}, l_{cr} is the total crystal length, $R_m = R_{m1} R_{m2}$ is the reflectivity of resonator mirrors, $\Delta N \sigma_{SE}$ is the gain coefficient (here: ΔN is the inversion population of the Stark laser levels and σ_{SE} is the cross-section of inter-Stark laser transition of an activator ions), and ρ is the loss coefficient at the SE wavelength λ_{SE}.

If the intensity of plane-wave fundamental pump-laser radiation is much higher than the intensity of the first Stokes generation ($I_p >> I_{St_1}$), i.e. when the level of pump depletion is very small, the SRS amplification at the first Stokes emission can be written [63] as

$$\frac{dI_{St_1}}{d_{SRS}} = g_{ssR}^{St_1} I_p(l_{SRS}) \cdot I_{St_1}(l_{SRS}) + I_p(l_{SRS}) \cdot \frac{d\sigma}{d\Omega} N_{SRS} \Delta\Omega =$$

$$= g_{ssR}^{St_1} I_p(l_{SRS}) \left[I_{St_1}(l_{SRS}) + I_{St_1}(l_{SRS} = 0)\right], \tag{7}$$

where $I_{St_1}(l_{SRS} = 0)$ is the intensity of the spontaneous Raman scattering at the wavelength λ_{St_1} of the first Stokes generation (in the beginning $l_{SRS} = 0$ of the amplified crystal, i.e. from zero-point fluctuation of spontaneous scattering)

$$g_{ssR}^{St_1} = \frac{2\lambda_{St_1}^2 N_{SRS}}{\pi n_{St_1}^2 h\nu_p} \cdot \frac{d\sigma}{d\Omega} \cdot \frac{1}{\Delta\nu_R} . \tag{8}$$

Clearly, in the first Stokes lasing process, very weak spontaneous Stokes Raman scattering provides the major contribution, because its frequency-shifted emission at ω_{SRS} of the intensive line (s) acts as a "seed" for SRS amplification. This situation is analogous to the luminescence (spontaneous emission) in the laser action in activated crystals. In Eqs. (7) and (8): $\frac{d\sigma}{d\Omega}$ is the Raman scattering cross-section of the vibration transition of the crystal, N_{SRS} is the number (concentration) of SRS-active scattering centers, n_{St_1} is the refractive index of the crystal at wavelength λ_{St_1}, and $\Delta\Omega$ is the small solid angle of SRS lasing. As seen from Eq. (8), the $g_{ssR}^{St_1}$ coefficient is linearly proportional to the Raman scattering cross-section and inversely proportional to the linewidth of the spontaneous Raman scattering transition. The product $\frac{d\sigma}{d\Omega} \cdot \frac{1}{\Delta\nu_R}$ may be considered as the spectroscopic parameter providing a measure for peak intensity of a spontaneous Raman transition. This figure of merit, as shown in [64,65], can be used in a comparative selection for suitable SRS-active crystals. Therefore, high-gain Raman crystals for steady-state SRS laser converters should have a small $\Delta\nu_R$ value and strong spontaneous Raman scattering transition. Solving Eq. (7) yields (see, e.g. [54,55,58,59])

$$I_{St_1}(l_{SRS}) = I_{St_1}(l_{SRS} = 0)\exp(g_{ssR}^{St_1} I_p l_{SRS}) . \tag{9}$$

In many known experimental cases (see, e.g. [3,56,58,59,66]) SRS lasing at the first Stokes wavelength ($\omega_{St_1} = \omega_p - \omega_{SRS}$) becomes with any assurance measurable when the increment in Eq. (9) reaches a value of $g_{ssR}^{St_1} I_p l_{SRS}$ = 25-30, which corresponds to a energy conversion efficiency of approximately 1%. The laser pumping intensity ($I_p = I_{thr}$) providing such an efficiency value is conditionally considered as the first Stokes steady-state threshold pumping intensity ($I_{St_1} / I_{thr} \approx 0.01$). Thus approach makes possible tentatively estimate of the $g_{ssR}^{St_1}$ value for $\chi^{(3)}$-active crystals in rather simple pumping geometries, as in the single-pass SRS experiments (see, e.g. [11,67]).

Due to very strong $\chi^{(3)}$- and $\chi^{(2)}$-nonlinearities of the most used organic crystals (see Table 2), the pumping condition in conducted SRS experiments were slightly different from the model mentioned above. To avoid a manifestation of other possible nonlinear effects (SHG, two-photon absorption and so on) in them, we can make only a comparative estimation of their $g_{ssR}^{St_1}$ coefficients applying several reference $\chi^{(3)}$-active crystals

($PbWO_4$, α-$KY(WO_4)_2$, α-$KGd(WO_4)_2$, and $NaClO_3$ [14,68,69]) and relatively "soft" excitation condition. As a threshold intensity in these comparative experiments we assumed the pumping energy at which the steady-state first Stokes lasing becomes confidently perceptible (usually with signal/noise ratio ≈2). Conducted measurements with our organic crystals showed that in the most cases their first-Stokes pumping "soft" threshold significantly less then the "1%-threshold".

3. Stimulated Raman spectroscopy of nonlinear-laser organic crystals and nanocrystalline ceramics

The spectroscopic single-pass SRS experiments in [17-21,51-53] were done using oriented samples of organic single crystals with different active length (from $l_{SRS} \approx 0.5$ mm for AANP to $l_{SRS} \approx 25$ mm for benzophenone and GuZN-III), as well as several centimetre length nanocrystalline ceramic bars. The reference crystals and measured crystalline matertials were equal in length and their optical faces were polished plane-parallel but not anti-reflection coated. For the excitation steady-state Stokes and anti-Stokes generation in organic crystals, was used a home-made picosecond $Nd^{3+}:Y_3Al_5O_{12}$ laser with ≈30% efficient frequency doubler that generates ≈110 ps pulses (FWHM) at λ_{f1} = 1.06415 μm and an energy up 3 mJ, and ≈80 ps SHG at λ_{f2} = 0.53207 μm wavelength [70]. Pump radiation with Gaussian beam profile, as need, were focused onto the investigated crystal by a lens with a focal distance adjusted (F = 25 cm) such that the SRS lasing was maximum without a surface and volume optical damage sample, resulting in a beam waist diameter of about 160 μm (see used setup in the frame of Fig. 2). The spectral composition of the Stokes and anti-Stokes, as well as self-FD and self SFM generation emission was measured with a CCD-spectroscopic multichannel analyzer (CSMA) consisting of a scanning grating monochromator (with Czerny-Turner arrangement), an analyzer, and a Si-CCD array-sensor (Hamamatsu S3923-1024Q) as a detector. The sensitivity dispersion of this CSMA system is given in the inset of Fig. 3.

3.1. Nonlinear $\chi^{(3)}$- and $\chi^{(2)}$-lasing effects in organic crystals

In addition to very large Raman shift and efficient first Stokes generation in discovered SRS-active organic crystals in same an acentric of them, what are more a polar crystals, were observed combined nonlinear lasing effects, namely self-FD and self-SFM [19-21]. This potential allows to classify these materials a as promising $(\chi^{(3)}+\chi^{(2)})$-medium for a new type of laser-frequency converters.

$C_{14}H_{22}N_8O_{13}Zr$ (GuZN-III) crystal [20], two SRS-spectra (see Fig. 2 and 3) are shown an identification of observed Stokes and anti-Stokes lines related to two SRS-active optical vibration modes of the crystal $\omega_{SRS1} \approx$ 1008 cm^{-1} and $\omega_{SRS2} \approx$ 2940 cm^{-1}. The analysis is shown that for the 62-atomic molecule $C_{14}H_{26}N_8O_{13}Zr$ of a GuZN-III structure with orthorhombic space group D_2^5 and Z = 4 (2 for primitive unit cell) overall degrees of freedom (3N×2) = 372 distributed into (at ***k*** = 0, center of Brillouin zone):

$$\Gamma_N = 92A + 94B_1 + 93B_2 + 93B_3$$

irreducible representations. In accordance with [71], the A modes of a GuZN-III crystal are Raman active only, and those of B_1, B_2, and B_3 are both Raman and IR active. Among them the ($B_1 + B_2 + B_3$) species are acoustic modes. As an illustration, Fig. 4 shows the Raman spectrum of the fully symmetric A species, which was recorded under excitation geometry *≈a(cc)≈a* practically as in the case of the SRS spectrum exhibited in Fig. 3. The assignment of its strongest Raman shifted lines to the respective vibration modes of a GuZN-III crystal yields that the A-symmetry lines at ≈1008 and ≈2940 cm^{-1} are promoting modes of observed SRS lasing components. They correspond to the stretching vibrations of the CH_2 and N-C-O bond systems, respectively.

$C_{13}H_{10}O$ (benzophenone), **α-$C_{14}H_{12}O$** (4-methylbenzophenone), and **$C_{14}H_{10}O_2$** (benzil) crystals [18]. Their Stokes and anti-Stokes spectra are given in Figs. 5 and 6. The analysis conducted in [18] is shown that most of their SRS-active modes (with the frequency of ≈3070, 1650, and ≈1000 cm^{-1}) correspond to the ν(CH) vibrations of the benzene ring, ν(C=O) vibrations of the carbonyl unit, and symmetric ν(CC) vibrations of the benzene ring, respectively. The ≈103 cm^{-1} SRS-mode is lattice vibration;

$C_{12}H_{22}O_{11}$ (sucrose or sugar) crystal [21]. Due to its low symmetry and, hence, the large number of the vibrational modes (3NZ = 270; Γ_N = 133A + 137B, here (A + 2B) are acoustic modes), it is quite difficult at this initial stage of the research to establish the relation of observed SRS-mode $\omega_{SRS} \approx 2960$ cm^{-1} (Fig. 7,a) to the specific C-H vibrational bond (ν[CH] or ν[CH_2]). It is interesting to note here that food sugar glassy caramel offers also very efficient SRS lasing (Fig. 7,b). Besides intensive Stokes and anti-Stokes lasing components, under picocesond pumping in a sugar, which is sufficiently good UV crystals, were also observed rather efficient ($\chi^{(3)}+\chi^{(2)}$)-nonlinear self-frequency conversion effects, namely the self-FD ($\lambda_{self\text{-}FD}$ = 0.3158 μm, i.e. $1/2\lambda_{St1}$, or $\omega_{SHG} = 2\omega_{St1}$) and self-SFM ($\lambda_{self\text{-}SFM}$ = 0.2887 μm, i.e. $\Sigma\lambda_{f2},\lambda_{St1}$ or $\omega_{SFM} = \omega_{f2} + \omega_{St1}$). The SHG- and SRS-potential, availability, very low cost, and various structural modifications of a $C_{12}H_{22}O_{11}$ make this crystal quite attractive for application in modern laser physics and nonlinear optics.

$C_{15}H_{19}N_3O_3$ (AANP) [19] and **$C_{16}H_{15}N_3O_4$** (MNBA) [20] crystals. In these papers has been discovered a great potential for very efficient SRS acting of these two polar organic crystals. To our best knowledge, among all known $\chi^{(3)}$-active crystals they offer the greatest value of the steady-state Raman gain coefficient in near IR. These crystals are promising candidate for a new generation of Raman laser converters, where relatively short their SRS-interaction lengths (less then 1 mm) allow for miniaturization. In AANP and MNBA were observed also several new parametric lasing effects which are illustrated in Table 5 and Fig. 8. According to [19], for the 39-atomic $C_{15}H_{19}N_3O_3$ of a AANP structure overall degrees of freedom 3NZ = 468 distributed into

$$\Gamma_N = 117A_1 + 117A_2 + 117B_1 + 117B_2$$

irreducible representations. The vibration modes can be divided into acoustic $\Gamma_T = A_1 + B_1 + B_2$, internal $\Gamma_i = 111A_1 + 111A_2 + 111B_1 + 111B_2$, and translatory and rotatory $\Gamma_{T'} = 2A_1 + 3A_2 + 2B_1 + 2B_2$ and $\Gamma_R = 3A_1 + 3A_2 + 3B_1 + 3B_2$, respectively. All optical modes are Raman active. Observed SRS spectrum shows (Fig. 8,a) Stokes and anti-Stokes lines which related to the $\omega_{SRS} \approx 1280$ cm^{-1} vibration mode. It connected with strongest vibration of the bond C-N-C which links the pyridine ring and adamantylamino system of AANP crystal. Unfortunately, vibration mode analysis for a MNBA crystal is embarrassing at present due to absence of precise *X*-ray data. It should be done late. The large nonlinearities and hence a very efficient Stokes and anti-Stokes generation related to the $\omega_{SRS} \approx 1587$ cm^{-1} vibration mode and other manifestation of frequency conversion lasing of MNBA with aromatic rings, donor $-OCH_3$ and acceptor group $-NHCOCH_3$, are due to extended π-electron conjugation [25,72].

3.2 High-order Stokes and anti-Stokes generation of $Y_3Al_5O_{12}$ and Y_2O_3 ceramics

The high optical quality of the new generation of laser host ceramics on the base of $Y_3Al_5O_{12}$ and Y_2O_3 is evidenced by recent investigation on the excitation in these materials of high-order Stgokes and anti-Stokes Raman lasing in the visible and near IR regions under picosecond pumping [51-53]. For illustration, several selected spectra of SRS and spontaneous Raman scattering are shown in Figs. 9-11.

$Y_3Al_5O_{12}$ nanocrystalline garnet ceramics. The 80 atoms of the $Ia\bar{3}d$ primitive cell of the $Y_3Al_5O_{12}$ structure have 3N = 240 degrees of freedom, which, according to factor group analysis and symmetry degeneracy [71], give rise to 98 vibration modes belonging to the following irreducible representations (at **k** = 0):

$$\Gamma_N = 3A_{1g} + 8E_g + 14F_{2g} + 5A_{1u} + 5A_{2u} + 5A_{2g} + 10E_u + 14F_{1g} + 16F_{2u} + 18F_{1u}.$$

Of these, 25 (A_{1g}, E_g, and F_{2g}) modes and 17 (F_{1u}) modes should appear in spontaneous Stokes Raman scattering and IR spectra, respectively. The comparison of energy spacing between Stokes and anti-Stokes components in $Y_3Al_5O_{12}$ ceramics (Fig. 9) and single crystals (see Fig. 3 in [53]) in Raman shifted intensive lines in spontaneous Raman scattering spectra makes it possible to attribute their SRS-active vibration modes $\omega_{SRS} \approx 370$ cm^{-1} to the internal A_{1g} and F_{2g} optical vibrations of tetrahedral AlO_4^{5-} and octahedral AlO_6^{9-} groups of the garnet host laser materials.

Y_2O_3 ceramics. The C-modification of the Y_2O_3 sesquioxide with bixbyite mineral structure crystallizes in the $Ia\bar{3}$ cubic system with 16 formulae per unit cell, where 24 Y occupy the C_2 sites and the other 8 Y are locates in centrosymmetric C_{3i} crystallographic positions. Its primitive cell contains 40 atoms and hence 3N = 120 degrees of vibration freedom. The irreducible representations for Y_2O_3 optical and acoustical modes (at **k** = 0) are

$\Gamma_N = 4A_g + 4E_g + 14F_g + 5A_{2u} + 5E_u + 17F_u$.

Factor group analysis predicts 22 Raman active modes ($4A_g + 4E_g + 14F_g$), $17F_u$ modes should appear in IR spectra (among them one Fu is the acoustic mode), and (A_{2u} and E_u) species are both Raman and IR inactive. For $Ia\bar{3}$-structure, spontaneous Raman and IR spectra are mutually exclusive since there are no vibration modes that are both Raman and IR active. The vibration modes ($A_u + E_u + 3F_u$) of C_{3i} octahedral (YO_6^{9-}) are Raman inactive. In much the same manner as for the $Y_3Al_5O_{12}$ crystalline compounds, in the case of Y_2O_3 single crystal and nanocrystalline ceramics the spontaneous Stokes Raman spectra are practically indiscernible (Fig. 10). According to our analysis (see, also [73]) the strongest Raman shifted line ≈ 378 cm^{-1} can be assigned to the totally symmetric A_g- and degenerated F_g- type modes connected with the vibrations of C_2-octahedra (YO_6^{9-}) of these laser host crystalline materials. The observed in [51] and showed in Fig. 11 high-order Stokes and anti-Stokes lasing in nanocrfystalline Y_2O_3 ceramics are governed exactly by these vibration modes.

4. Conclusion

We have demonstrated a great potential for efficient SRS laser acting in several organic and organometallic crystals. These first observation of their large frequency shifts and high steady-state first Stokes Raman gain coefficients, as well as self-FD and self-SFM parametric effects let us hope that these novel materials may be used for a new generation of Raman laser converters, where relatively short their nonlinear $\chi^{(3)}$-interaction lengths allow for very attractive miniaturization. The multiple Stokes and anti-Stokes generation in $Y_3Al_5O_{12}$ and Y_2O_3 nanocrystalline demonstrates good optical quality these novel solid-state laser host materials. Nearly completion the paper will be in order to illustrate the results of our experimental estimations of corresponding value of the gain $g_{ssR}^{St_1}$ coefficients for several investigated crystalline materials. These data a given in Table 6.

Acknowledgments

The grateful duty of the author to stress also here that the reviewed results in this paper were obtained jointly with Professors S.N.Bagayev, H.J.Eichler, J.Hulliger, K.Ueda, H.Klapper, E.Haussühl, T.Kaino, J.Hanuza and their teams within scientific cooperation of the Joint Open Laboratory for Laser Crystals and Precise Laser Systems, as well as Dr. T.Yanagitani and H.Yagi for very useful cooperation in the field of nanocrystalline laser ceramics. Special thanks must go to Professor A.Z.Grasyuk for numerous discussions on SRS problems. The author acknowledges the partial support from the Russian Foundation for Basic Research and the Ministry of Industry, Science and Technology, as well as from Alexander von Humboldt Foundation for Research Prize.

Table 1. Selected easily accessible inorganic SRS-active crystals with laser frequency shift (ω_{SRS}) more than 900 cm^{-1} [2,3,5,8-11] [1)]

Crystal	Space group		Nonlinearity (class)	Lagest SRS-active vibration mode (cm^{-1})
	Notation	Number		
$LiHCOO \cdot H_2O$	$C_{2v}^{9} - Pna2_1$	(No. 33)	$\chi^{(2)}+\chi^{(3)}$ (polar)	≈1372
$NaClO_3$	$T^4 - P2_13$	(No. 198)	$\chi^{(2)}+\chi^{(3)}$	≈936
$NaY(WO_4)_2$	$C_{4h}^{6} - I4_2/a$	(No. 88)	$\chi^{(3)}$	≈914
KH_2PO_4 (KDP)	$D_{2d}^{12} - I\bar{4}_2 2d$	(No. 122)	$\chi^{(2)}+\chi^{(3)}$	≈915
$KAl(SO_4)_2 \cdot 12H_2O$	$T_h^{6} - Pa3$	(No. 205)	$\chi^{(3)}$	≈989
α-$KY(WO_4)_2$	$C_{2h}^{6} - C2/c$	(No. 15)	$\chi^{(3)}$	905
α-$KGd(WO_4)_2$	$C_{2h}^{6} - C2/c$	(No. 15)	$\chi^{(3)}$	901
α-$KYb(WO_4)_2$	$C_{2h}^{6} - C2/c$	(No. 15)	$\chi^{(3)}$	≈907
α-$KLu(WO_4)_2$	$C_{2h}^{6} - C2/c$	(No. 15)	$\chi^{(3)}$	907
$CaCO_3$	$D_{3d}^{6} - R3c$	(No. 167)	$\chi^{(3)}$	≈1085
$Ca_4Gd(BO_3)_3O$	$C_s^{3} - Cm$	(No. 8)	$\chi^{(2)}+\chi^{(3)}$ (polar)	933
$CaWO_4$	$C_{4h}^{6} - I4_2/a$	(No. 88)	$\chi^{(3)}$	≈908
$ZnWO_4$	$C_{2h}^{4} - P2/c$	(No. 13)	$\chi^{(3)}$	907
$Sr_5(PO_4)_3F$	$C_{6h}^{2} - P6_3/m$	(No. 176)	$\chi^{(3)}$	950
$SrWO_4$	$C_{4h}^{6} - I4_2/a$	(No. 88)	$\chi^{(3)}$	922
$Ba(NO_3)_2$	$T_h^{6} - Pa3$	(No. 205)	$\chi^{(3)}$	≈1047
$BaWO_4$	$C_{4h}^{6} - I4_2/a$	(No. 88)	$\chi^{(3)}$	924
β'-$Gd_2(MoO_4)_3$	$C_{2v}^{8} - Pba2$	(No. 32)	$\chi^{(2)}+\chi^{(3)}$ (polar)	960
$PbWO_4$	$C_{4h}^{6} - I4_2/a$	(No. 88)	$\chi^{(3)}$	901

[1)] Most of these crystals are already commercial materials as the laser host-crystals (indicated by bold letters) and crystals for second harmonic generation (SHG), as well as some of them are well known birefringent and scintillator crystals (see, e.g. [12-15]). The diamond is also $\chi^{(3)}$-active crystal with $\omega_{SRS} \approx 1333$ cm^{-1} [16], but it is not easily accessible.

Table 2. SRS-active organic and organometallic crystals [17-21]

Crystal	Space group Notation	Number	Nonlinearity (class)	SRS-active vibration mode (cm^{-1})	Observed nonlinear laser effect [1)]
Organic					
$C_{12}H_{22}O_{11}$ [2)] (sucrose, sugar)	$C_2^2 - P2_1$	(No. 4)	$\chi^{(2)}+\chi^{(3)}$ (polar)	≈2960	SHG, SRS, self-FD, self-SFM
$C_{13}H_{10}O$ (benzophenone)	$D_2^4 - P2_12_12_1$	(No. 19)	$\chi^{(2)}+\chi^{(3)}$	3070, 1650, 998, ≈103	SHG, SRS
$C_{13}H_{10}O_3$ (salol)	$D_{2h}^5 - Pbca$	(No. 61)	$\chi^{(3)}$	≈3150	SRS
α-$C_{14}H_{12}O$ [3)] (4-methylbenzophenone	$C_{2h}^5 - P2_1/c$	(No.14)	$\chi^{(3)}$	3065	SRS
$C_{14}H_{10}O_2$ [4)] (benzil, dibenzoyl)	$D_3^4 - P3_121$	(No.152)	$\chi^{(2)}+\chi^{(3)}$	≈1000	SHG, SRS
$C_{15}H_{19}N_3O_2$ (AANP) [5)]	$C_{2v}^9 - Pna2_1$	(No. 33)	$\chi^{(2)}+\chi^{(3)}$ (polar)	≈1280	SHG, SRS, self-SFM
$C_{16}H_{15}N_3O_4$ (MNBA) [6)]	$C_s^4 - Cc$	(No. 9)	$\chi^{(2)}+\chi^{(3)}$ (polar)	≈1587	SHG, SRS, self-FD, self-SFM
Organometallic					
$C_{14}H_{26}N_8O_{13}Zr$ (GuZN-III) [7)]	$D_2^5 - C222_1$	(No. 20)	$\chi^{(2)}+\chi^{(3)}$	≈1008, ≈2940	SHG, SRS
$C_{13}H_{22}N_5TlZr$ (TlGuZN) [8)]	$C_2^2 - P2_1$	(No. 4)	$\chi^{(2)}+\chi^{(3)}$ (polar)	≈1005, ≈2950	SHG, SRS

[1)] Used abbreviations self-FD and self-SFM are the self-frequency doubling and the self-sum-frequency mixing, correspondingly.

[2)] Strongly shifted Stokes and anti-Stokes picosecond generation ($\omega_{SRS} \approx 2915$ cm^{-1}) was observed also in glassy sugar caramel. Both sugar materials, single crystals and glassy caramel are easily accessible and very cheap. They were bought in pastry shops.

[3)] It is known also the metastable β-$C_{14}H_{12}O$ phase which has trigonal space group $C_3^2 - P3_1$ (No.144) or $C_3^3 - P3_2$ (No.145) [22].

[4)] In accordance with [23] space group could be also $D_3^6 - P3_221$ (No.154).

[5)] Full chemical name is the 2-adamantylamino-5-nitropyridine.

[6)] Full chemical name is the 4'-nitrobenzylidene-3-acetamino-4-methoxyaniline.

[7)] Full chemical name is the bis(guanidinium) zirconium bis(nitrilotriacetate) hydrate.

[8)] Full chemical name is the thallium quanidinium zirconium bis(nitrilotriacetate) dehydrate [24]. Refined data on SRS and SHG will be published soon with Dr. E.Haussühl, who grew and characterized of this crystal.

Table 3. Stimulated-emission (SE) channels of Nd^{3+} and Yb^{3+} lasants in nanocrystalline ceramics fabricated by the vacuum sintering method [41]

Ceramics	Space group [1)]	Ln^{3+} lasants and their SE channels	
		Nd^{3+}	Yb^{3+}
Y_2O_3	T_h^7	$^4F_{3/2} \rightarrow ^4I_{11/2}$ [39,44]	$^2F_{5/2} \rightarrow ^2F_{7/2}$ [47,48]
$YGdO_3$	T_h^7	$^4F_{3/2} \rightarrow ^4I_{11/2}$ [45]	
$Y_3Al_5O_{12}$	O_h^{10}	$^4F_{3/2} \rightarrow ^4I_{11/2}$ [34,39] $^4F_{3/2} \rightarrow ^4I_{13/2}$ [35]	$^2F_{5/2} \rightarrow ^2F_{7/2}$ [49]
Sc_2O_3	T_h^7		$^2F_{5/2} \rightarrow ^2F_{7/2}$ [50]
Lu_2O_3	T_h^7	$^4F_{3/2} \rightarrow ^4I_{11/2}$ [46]	

[1)] Ceramic grains are nano- and(or) micrometer-sized single crystals.

Table 4. Some possible $\chi^{(2)}$- and $\chi^{(3)}$-effects in undoped nonlinear-laser crystals (see, e.g. [57]).

Nonlinear effect [1)]	Frequency		Nonlinear susceptibility	Note
	Incident	Created		
Second harmonic generation [2)]	ω,ω	2ω	$\chi^{(2)}$	UV and visible generation
Sum frequency mixing [2)]	ω_1,ω_2	$\omega_3 = \omega_1+\omega_2$	$\chi^{(2)}$	Up-conversion
Difference frequency mixing [2)]	ω_1,ω_2	$\omega_3 = \omega_1-\omega_2$	$\chi^{(2)}$	IR generation
Third harmonic generation [2)]	ω,ω,ω	3ω	$\chi^{(2)}$	VUV generation
Sum frequency mixing [2)]	$\omega_1,\omega_2,\omega_3$	$\omega_4 = \omega_1+\omega_2+\omega_3$		VUV and UV generation
Stimulated Raman scattering [3)]	ω_1	ω_2	$\chi^{(3)}$	$\omega_2 = \omega_1 \pm \omega_{SRS}$
Two-photon absorption [3)]	ω,ω	–	$\chi^{(3)}$	$\omega_c = 2\omega$

[1)] It is available also self-frequency conversion effects, namely self-FD, self-SRS, etc.
[2)] Phase matching required.
[3)] ω_{SRS} and ω_c are the crystal frequencies,.

Table 5. Parametric lasing effects of nonlinear ($\chi^{(3)}+\chi^{(2)}$)-interaction in organic polar crystals $C_{15}H_{19}N_3O_2$ and $C_{16}H_{15}N_3O_4$ under picosecond $Nd^{3+}:Y_3Al_5O_{12}$-laser excitation at λ_{f1} = 1.06415 μm wavelength [17,19].

$\chi^{(3)}$ and $\chi^{(2)}$ generation component			SRS-ctive crystal vibration mode (cm^{-1})
Wavelength (μm)[1)]	Line	Attribution	
$C_{15}H_{19}N_3O_2$ crystal, $l_{SRS}\approx$ 0.4 mm [2)] (Fig. 8,a)			
0.53207	SHG ($1/2\lambda_{f1}$)	$2\omega_{f1}$	-
0.5710	$\Sigma\lambda_{f1},\lambda_{St1}$ ($\lambda_{self\text{-}SFM}$)[3)]	$\omega_{f1}+(\omega_{f1}-\omega_{SRS})$	≈1280
0.6160	$\Sigma\lambda_{f1},\lambda_{St2}$ ($\lambda_{self\text{-}SFM}$)[3)]	$\omega_{f1}+(\omega_{f1}-2\omega_{SRS})$	≈1280
0.8363	ASt_2 (λ_{ASt2})	$\omega_{f1}+2\omega_{SRS}$	≈1280
0.9366	ASt_1 (λ_{ASt1})	$\omega_{f1}+\omega_{SRS}$	≈1280
1.06415	λ_{f1}	ω_{f1}	-
1.2320 [4)]	St_1 (λ_{St1})	$\omega_{f1}-\omega_{SRS}$	≈1280
1.4626 [4)]	St_2 (λ_{St2})	$\omega_{f1}-2\omega_{SRS}$	≈1280
$C_{16}H_{15}N_3O_2$ crystal, $l_{SRS}\approx$ 1 mm [2)] (Fig. 8,b)			
0.53207	SHG ($1/2\lambda_{f1}$)	$2\omega_{f1}$	-
0.5811	$\Sigma\lambda_{f1},\lambda_{St1}$ ($\lambda_{self\text{-}SFM}$) [3)]	$\omega_{f1}+(\omega_{f1}-\omega_{SRS})$	≈1587
0.6402	SHG ($1/2\lambda_{St1}$)	$2\omega_{St1}=2(\omega_{f1}-\omega_{SRS})$	≈1587
0.7126	$\Sigma\lambda_{St1},\lambda_{St2}$ ($\lambda_{self\text{-}SFM}$) [5)] $\Sigma\lambda_{f1},\lambda_{St3}$ ($\lambda_{self\text{-}SFM}$) [6)]	$(\omega_{f1}-\omega_{SRS})+(\omega_{f1}-2\omega_{SRS})$ $\omega_{f1}+(\omega_{f1}-3\omega_{SRS})$	≈1587
0.7955	ASt_2 (λ_{ASt2})	$\omega_{f1}+2\omega_{SRS}$	≈1587
0.8334	SHG ($1/2\lambda_{St2}$)	$2\omega_{St2}=2(\omega_{f1}-2\omega_{SRS})$	≈1587
0.9104	ASt_1 (λA_{St1})	$\omega_{f1}+\omega_{SRS}$	≈1587
1.06415	λ_{f1}	ω_{f1}	-
1.2804 [4)]	St_1 (λ_{St1})	$\omega_{f1}-\omega_{SRS}$	≈1587
1.6069 [4)]	St_2 (λ_{St2})	$\omega_{f1}-2\omega_{SRS}$	≈1587

[1)] Measurement accuracy is ±0.0003 μm.
[2)] λ_{SRS} is the SRS-lasing length of crystalline element.
[3)] $\lambda_{self\text{-}SFM}$ is the wavelength of the self-sum-frequency mixing generation in which was involved the pumping with the fundamental ω_{f1} frequency and arising in the crystal first or econd Stokes lasing with $\omega_{St1}=\omega_{f1}-\omega_{SRS}$ or $\omega_{St2}=\omega_{f1}-2\omega_{SRS}$ frequency.
[4)] Due to zero sensitivity of used Si-CCD sensor (see Fig. 3) the Stokes lasing at this wavelength is not detectable.
[5)] $\lambda_{self\text{-}SFM}$ is the wavelength of the self-sum-frequency mixing generation in which was involved the pumping with the first and second Stokes lasing emissions with $\omega_{St1}=\omega_{f1}-\omega_{SRS}$ and $\omega_{St2}=\omega_{f1}-2\omega_{SRS}$ frequency.
[6)] Due to strong absorption (optical transparent of this crystal covers the spectral range of ≈0.51-≈2.2 μm, see also Fig. 8,b)it is possible, in general, weak third Stokes lasing at the wavelength λ_{St3} = 2.1570 μm, but this self-sum-frequency mixing generation process is unreal.

Table 6. The steady-state Raman gain coefficients of Raman spectroscopic properties of organic, organometallic crystals and nanocrystalline laser host ceramics[1)]

Crystal	First Stokes lasing characteristics				Raman spectroscopic property		
	λ_{f1} = 1.06415 μm		λ_{f2} = 0.53207 μm		ω_{SRS} (cm^{-1})	$\Delta\nu_R$ (cm^{-1})	T_2 (ps)
	λ_{St_1} (μm)	$g_{ssR}^{St_1}$ (cm/GW)	λ_{St_1} (μm)	$g_{ssR}^{St_1}$ (cm/GW)			
Organic							
$C_{12}H_{22}O_{11}$	-	-	0.6315	>6.5	≈2960	≈7	≈2
$C_{13}H_{10}O$	1.1906	≈2.8	0.5619 0.6360	>10 >10	998 3070	≈3.5 ≈6.5	≈3 ≈1.6
$C_{15}H_{19}N_3O_2$	1.2320	>15 [2)]	-	-	≈1280	≈24 [3)]	≈0.44
$C_{16}H_{15}N_3O_4$	1.2804	>14 [2)]	-	-	≈1587	1.5	≈7
Organometallic							
$C_{14}H_{26}N_8O_{13}Zr$	1.1920 1.5487	≈3.8 ≈3.2[2)]	0.5622 0.6307	>10 >9	≈1008 ≈2940	≈5 ≈14 [3)]	≈2 ≈0.8
Ceramics							
$Y_3Al_5O_{12}$	1.1078	0.1±0.02	-	-	≈370	≈5.7	≈1.9
Y_2O_3	1.1088	0.6±0.08	-	-	≈373	≈4	≈2.6

[1)] Some of listed data are unpublished.
[2)] For this case, using nanosecond $Nd^{3+}:Y_3Al_5O_{12}$ laser ($\tau_p \approx$ 20 ns) and an avalanche Ge detector we can estimated only the lower limiting value of the $g_{ssR}^{St_1}$ coefficient.
[3)] Inhomogeneously broadened line.

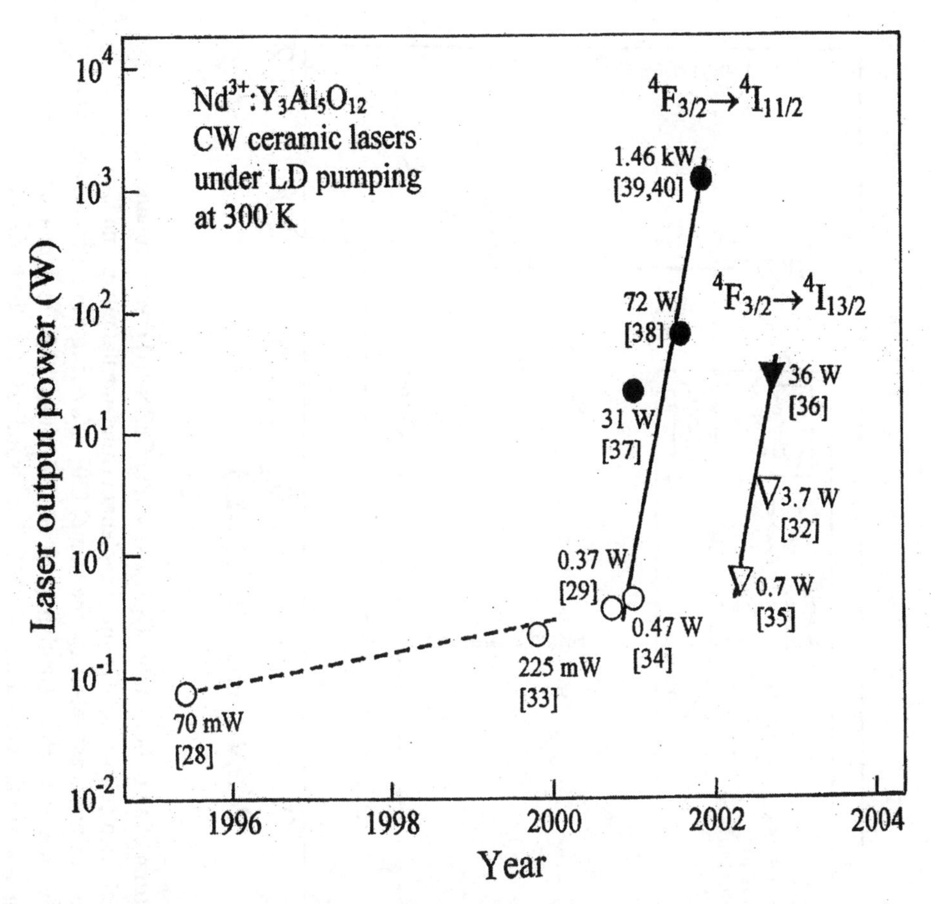

Fig.1. Progress in the development of LD-pumped $Y_3Al_5O_{12}:Nd^{3+}$ nanocrystalline ceramic CW lasers emitting at wavelengths of two $^4F_{3/2} \rightarrow ^4I_{11/2}$ and $^4F_{3/2} \rightarrow ^4I_{11/2}$ generating channels at 300 K under end (longitudinal, open circles and triangles) and side (transverse, filled circles and triangle) pumping geometries. The CW laser output power is given in mW or W

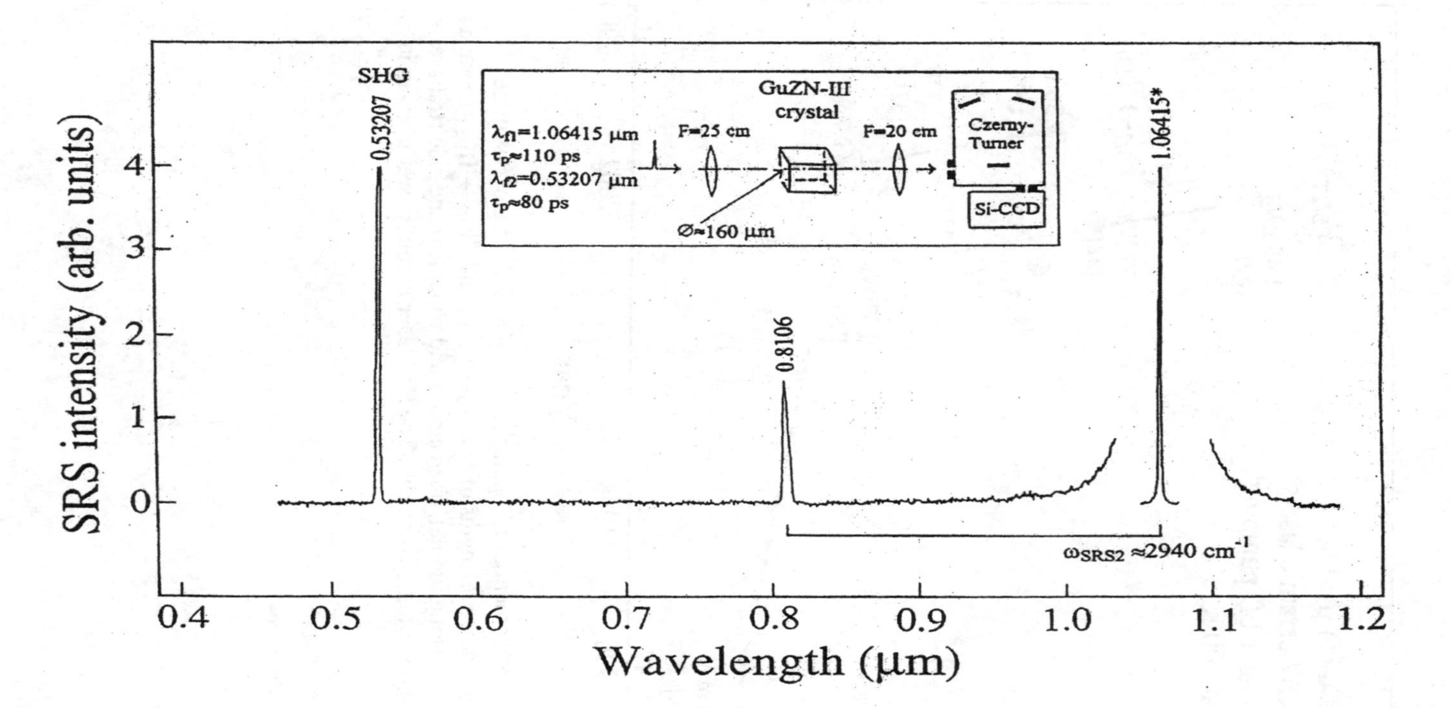

Fig. 2. The orientational SRS and SHG spectrum of an orthorhombic $C_{14}H_{26}N_8O_{13}Zr$ (GuZN-III) crystal obtained in pumping geometry *c(aa)c* under picosecond excitation at λ_{f1} =1.06415 μm wavelength (fundamental pump line is asterisked), as well as a scheme of the experimental single-pass set-up (in the frame) [20]. Wavelengths of all lines are given in μm and their intensity are shown without correction of spectral sensitivity of used analyzing CSMA system (see Fig. 3). Anti- Stokes line related to SRS-active vibration mode of the crystal $\omega_{SRS2} \approx 2940$ cm^{-1} is indicated by the horizontal scale line.

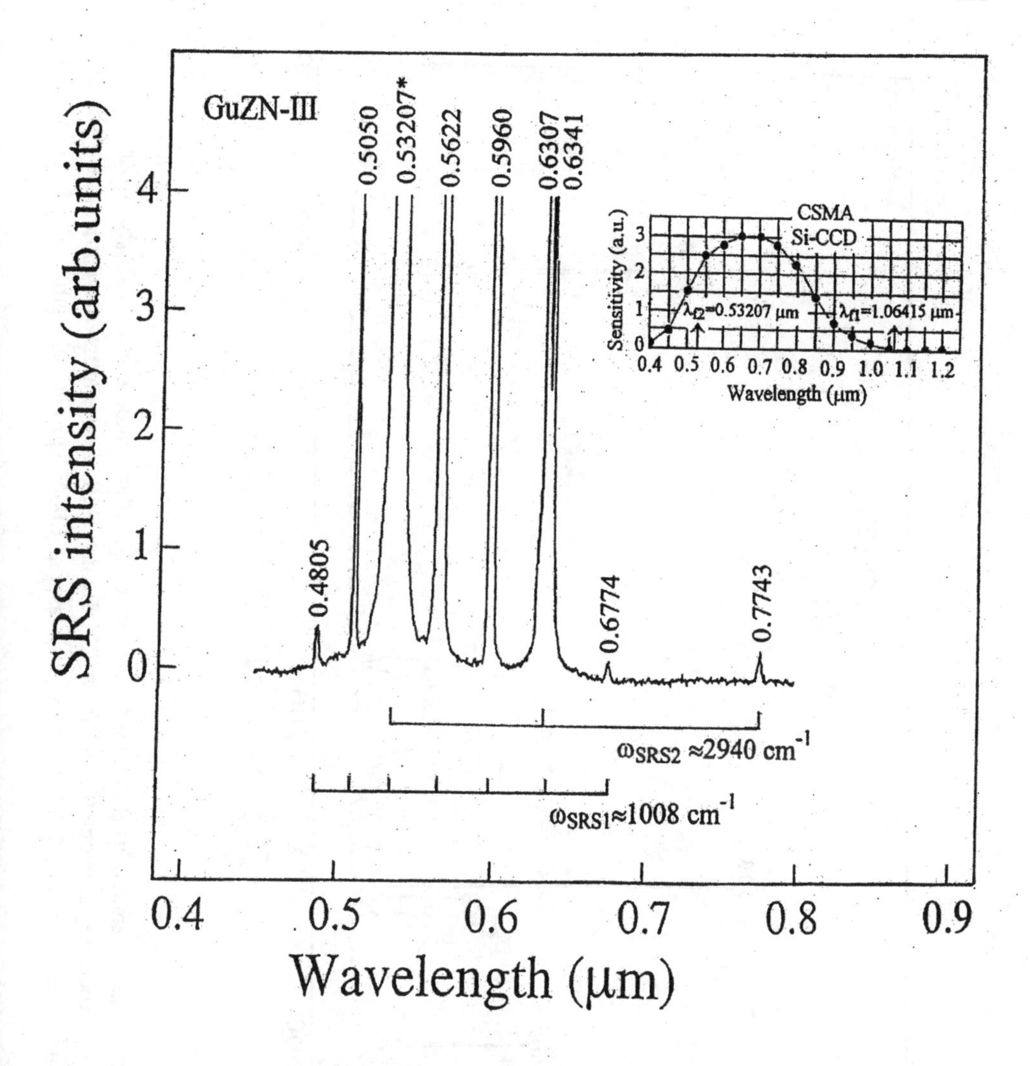

Fig. 3 The orientational SRS and RFWM spectrum of an orthorhombic $C_{14}H_{26}N_8O_{13}Zr$ (GuZN-III) crystal obtained in pumping geometry *a(cc)a* under picosecond excitation at $\lambda_{f2} = 0.53207$ μm wavelength, as well as wavelength dependence the spectral sensitivity of used analyzing CSMA system (in the frame) [20]. Stokes and anti-Stokes lines related to SRS-active vibration modes of the crystal $\omega_{SRS1} \approx 1008$ cm^{-1} and $\omega_{SRS2} \approx 2940$ cm^{-1} are indicated by horizontal scale lines. Other notations as in Fig. 2.

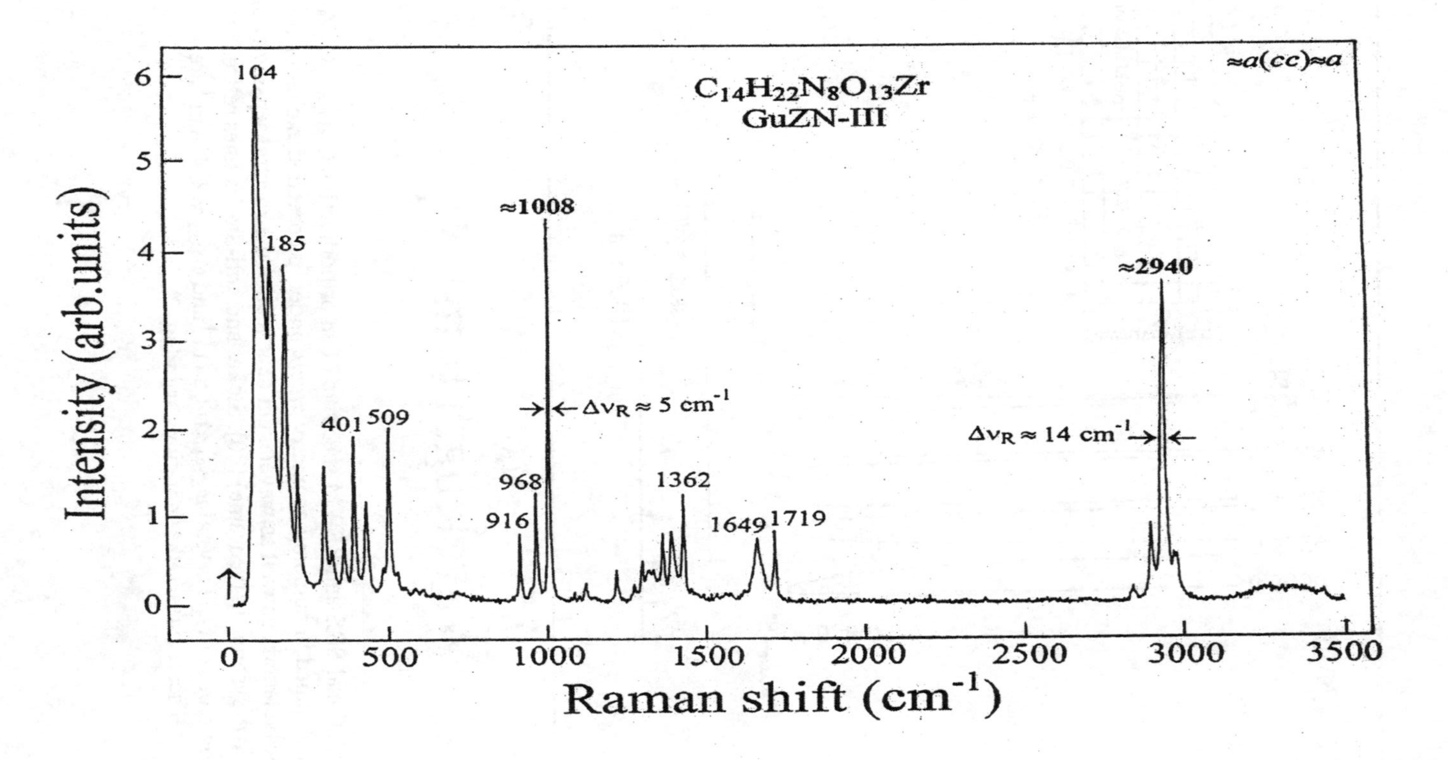

Fig. 4 The room-temperature polarized spontaneous Raman scattering spectrum of an orthorhombic $C_{14}H_{26}N_8O_{13}Zr$ (GuZN-III) crystal registered under experimental geometry *≈a(cc)≈a* [20]. Raman shift of several intensive lines are given in cm^{-1}. The arrow at zero corresponds to excitation by CW Nd^{3+}:$Y_3Al_5O_{12}$ laser at 1.06415 μm wavelength.

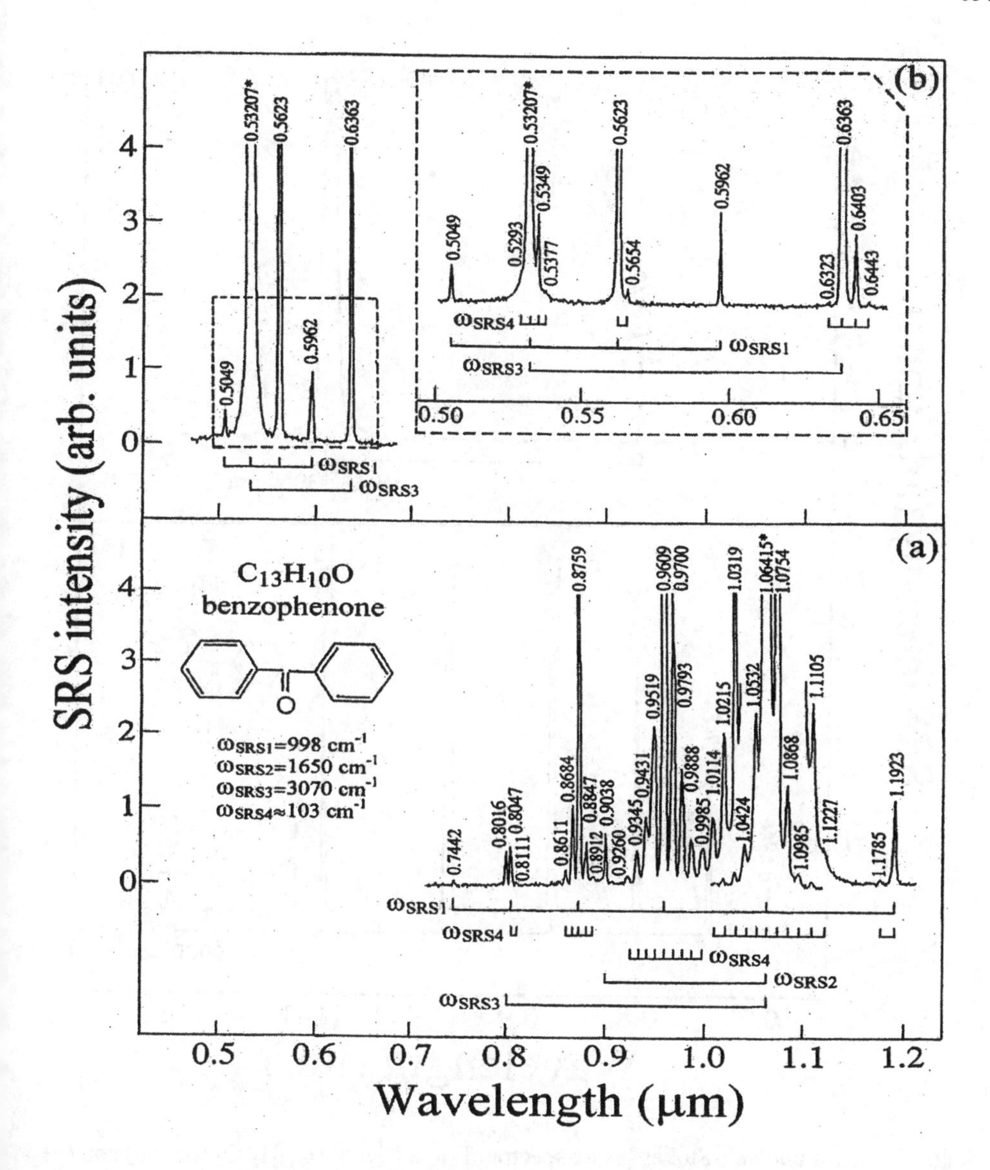

Fig. 5 The orientational SRS and RFWM spectra of an orthorhombic $C_{13}H_{10}O$ (benzophenone) crystal obtained in pumping geometry *≈b(≈c≈c)≈b* under picosecond excitation at (a) $\lambda_{f1} = 1.06415$ μm and (b) $\lambda_{f2} = 0.53207$ μm wavelengths [18]. Stokes and anti-Stokes lasing lines related to SRS-active vibration modes of the crystals $\omega_{SRS1} = 998$ cm^{-1}, $\omega_{SRS2} = 1650$ cm^{-1}, $\omega_{SRS3} = 3070$ cm^{-1}, and $\omega_{SRS4} \approx 103$ cm^{-1} are indicated by horizontal scale lines. Other notations as in Fig. 2.

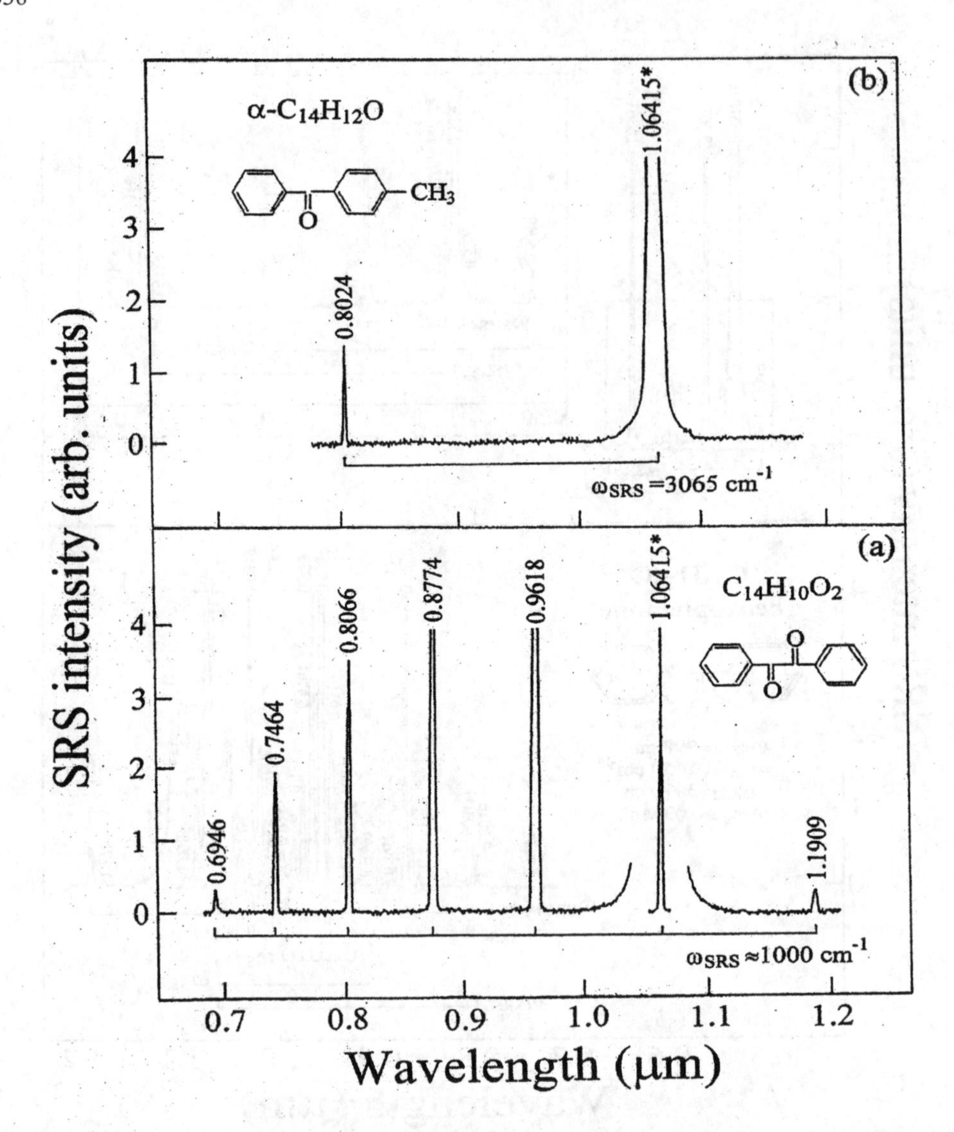

Fig. 6 Stokes and anti-Stokes lasing spectra of (a) a trigonal $C_{14}H_{10}O_2$ (benzil) and (b) a monoclinic α-$C_{14}H_{12}O$ (4-methylbenzophenone) crystals obtained under picosecond excitation at λ_{f1} =1.06415 μm wavelength [18]. Pumping geometry for $C_{14}H_{10}O_2$ crystal was *⊥b(≈b≈b)⊥b* and the α-$C_{14}H_{12}O$ crystal was random oriented. The SRS-active vibration modes of these crystals are indicated by horizontal scale lines. Other notations as in Fig. 2.

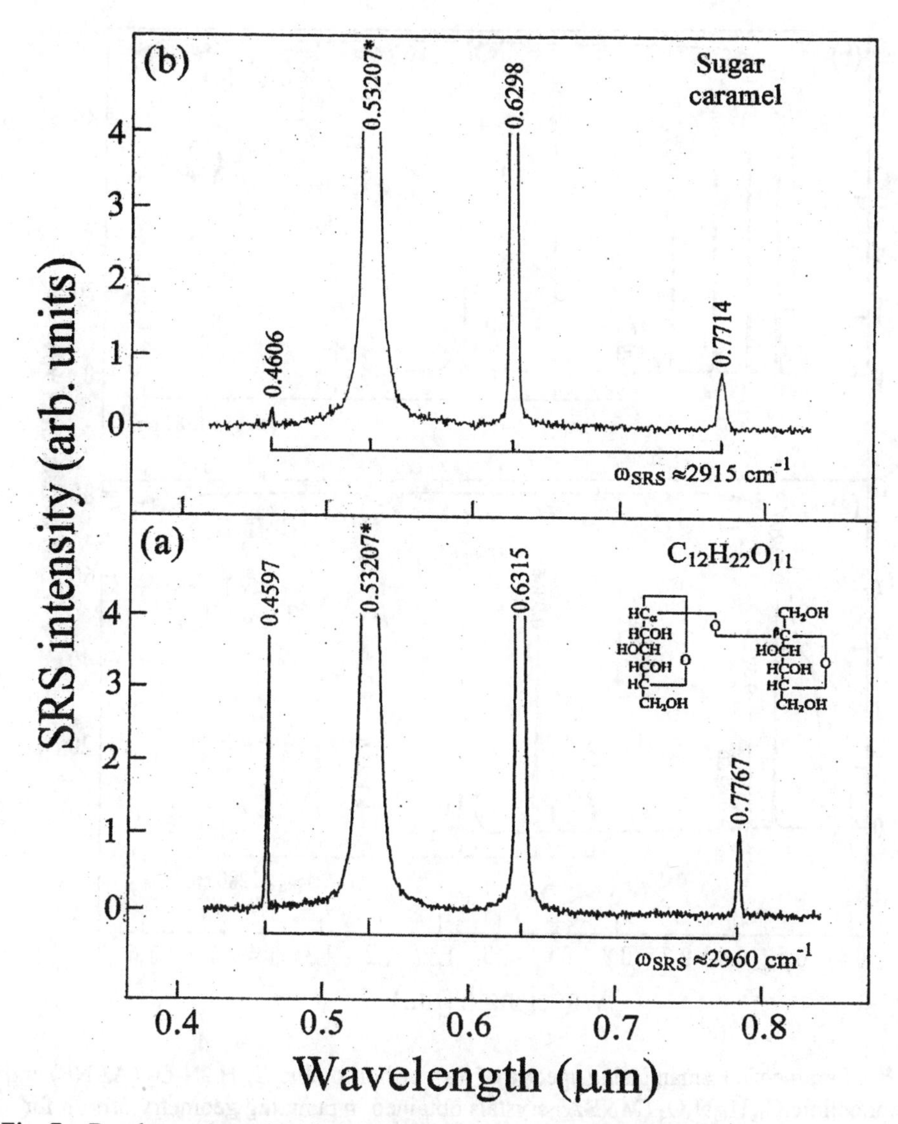

Fig. 7 Random oriented Stokes and anti-Stokes lasing spectra of (a) a monoclinic $C_{12}H_{22}O_{11}$ (sugar) crystal and (b) glassy sugar caramel obtained under picosecond excitation at λ_{f2} = 0.53207 μm wavelength [21]. The SRS-active vibration modes of these organic materials are indicated by horizontal scale lines. Other notations as in Fig. 2.

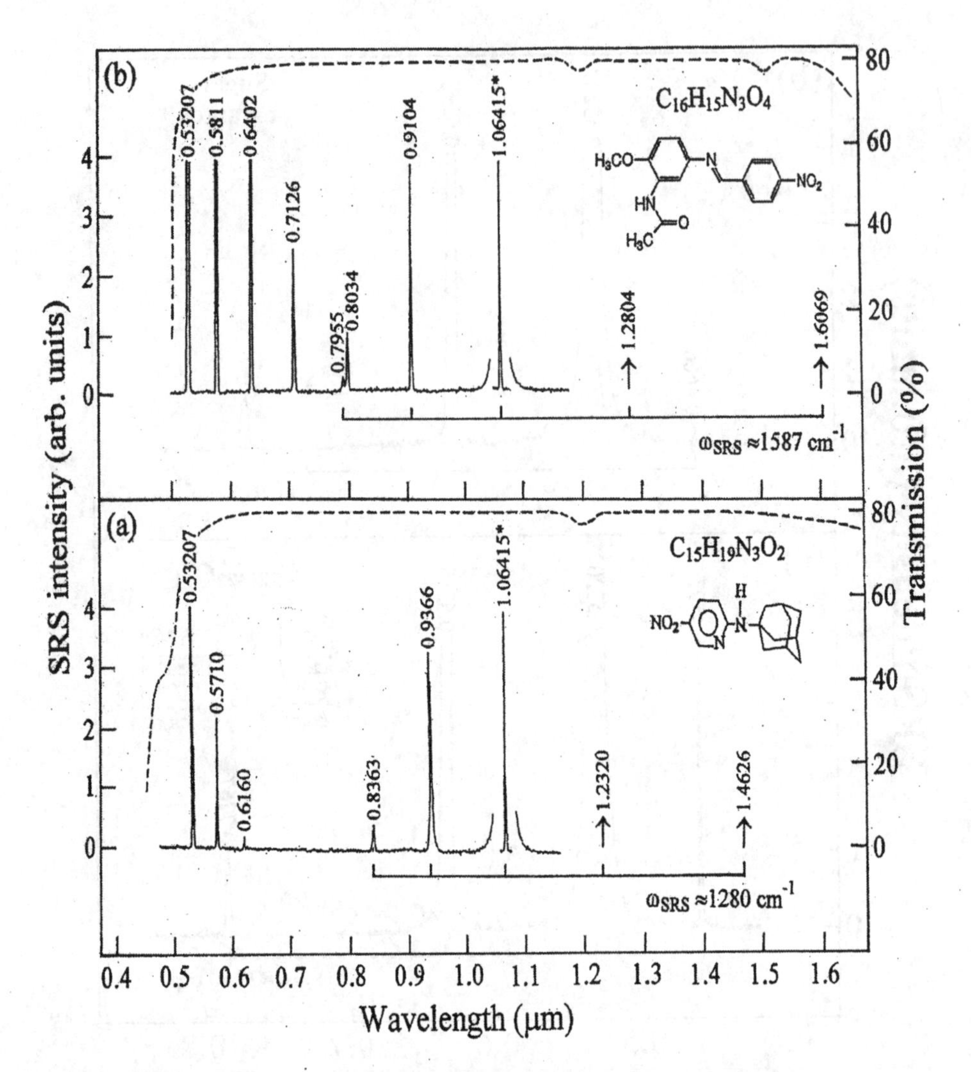

Fig. 8 Parametric Raman lasing spectra of (a) orthorhombic $C_{15}H_{19}N_3O_2$ (AANP) and (b) monoclinic $C_{16}H_{15}N_3O_4$ (MNBA) crystals obtained in pumping geometry *b(aa)b* for AANP and *b(≈a≈a)b* for MNBA under picosecond excitation at λ_{f1} =1.06415 μm wavelength [17,19]. The arrows indicate the spectral positions of the first and second Stokes lines non-detectable by the used Si-CCD sensor (see Fig. 3). Dashed lines are showed the fragments of nonpolarized transmission spectra of 0.4- and ≈1-mm thick samples for AANP and MNBA, respectively. The SRS-active vibration modes of these strongly nonlinear crystal are indicated by horizontal scale lines. Other notations as in Fig. 2.

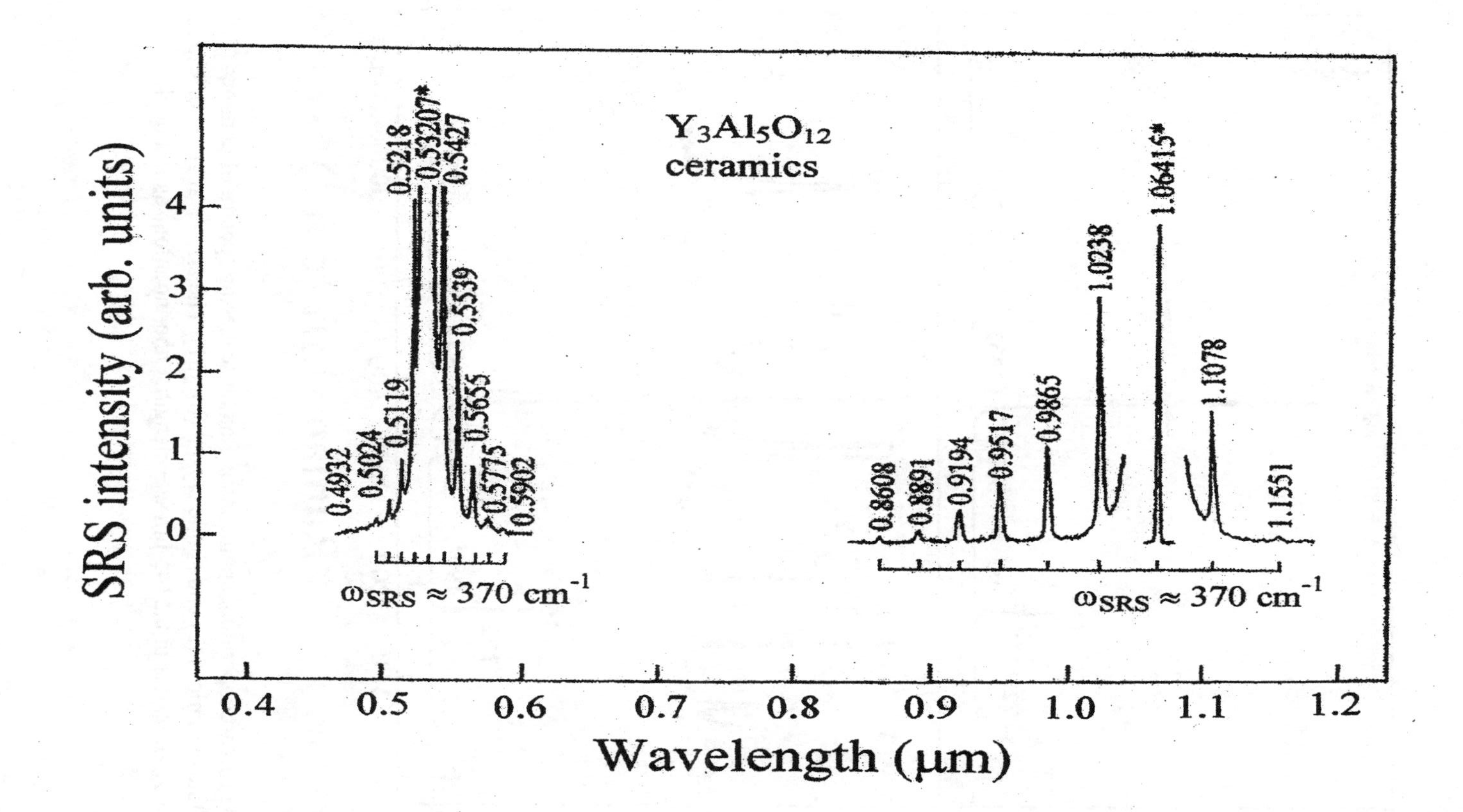

Fig. 9 Stokes and anti-Stokes lasing spectra of undoped $Y_3Al_5O_{12}$ nanocrystalline ceramics recorded at 300 K under picosecond excitation at λ_{f1} = 1.06415 μm and λ_{f2} = 0.53207 μm wavelengths [52,53]. The SRS-active vibration mode $\omega_{SRS} \approx$ 370 cm^{-1} is indicated by horizontal scale lines. Other notations as in Fig. 2.

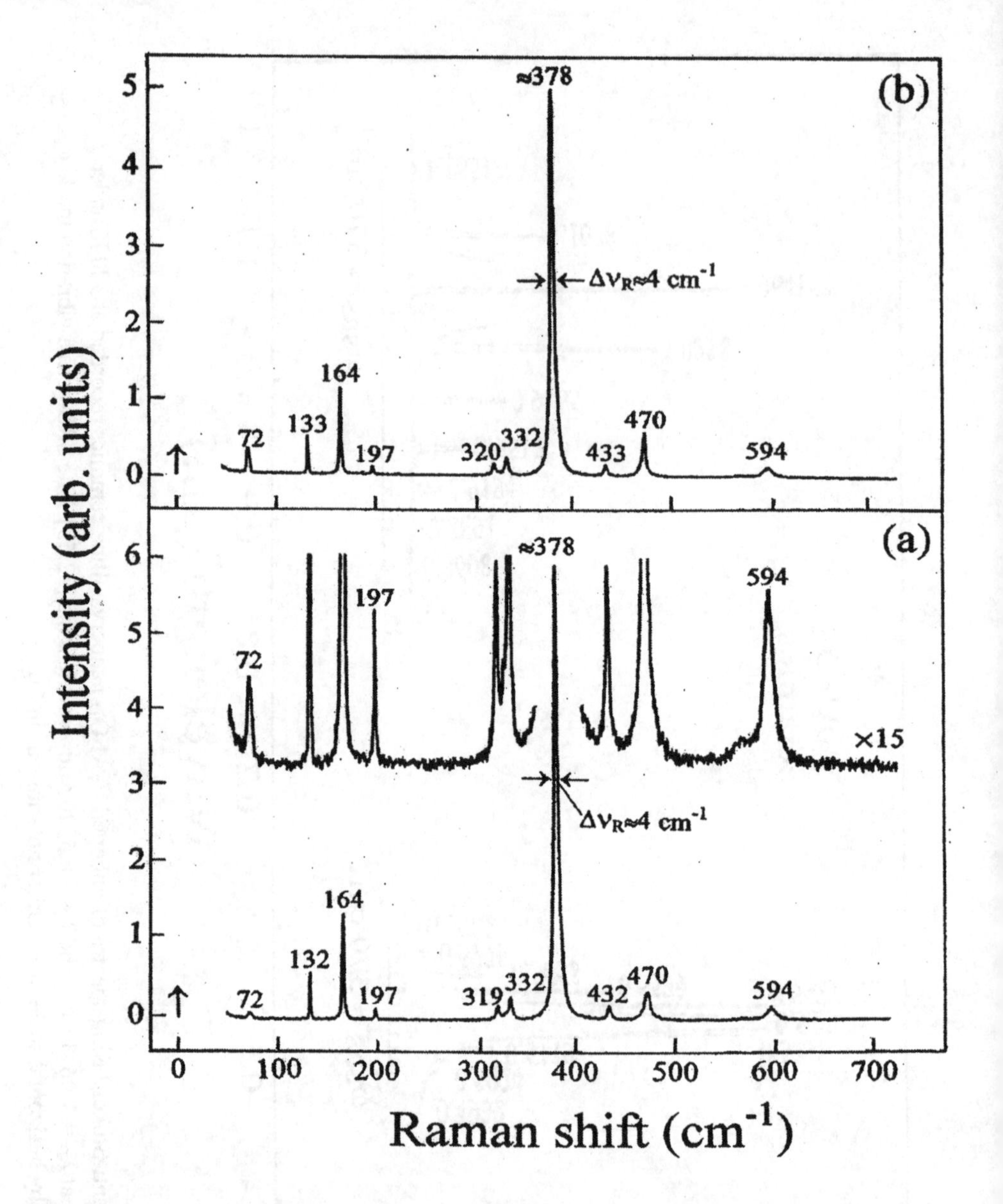

Fig.10. The room-temperature spontaneous Raman scattering spectra of undoped Y_2O_3 nanocrystalline ceramicvs (a) and single crystal (b) [51]. The arrows at zero correspond to excitation by Ar-ion laser at 0.5145 μm wavelength. Other notations as in Fig. 4.

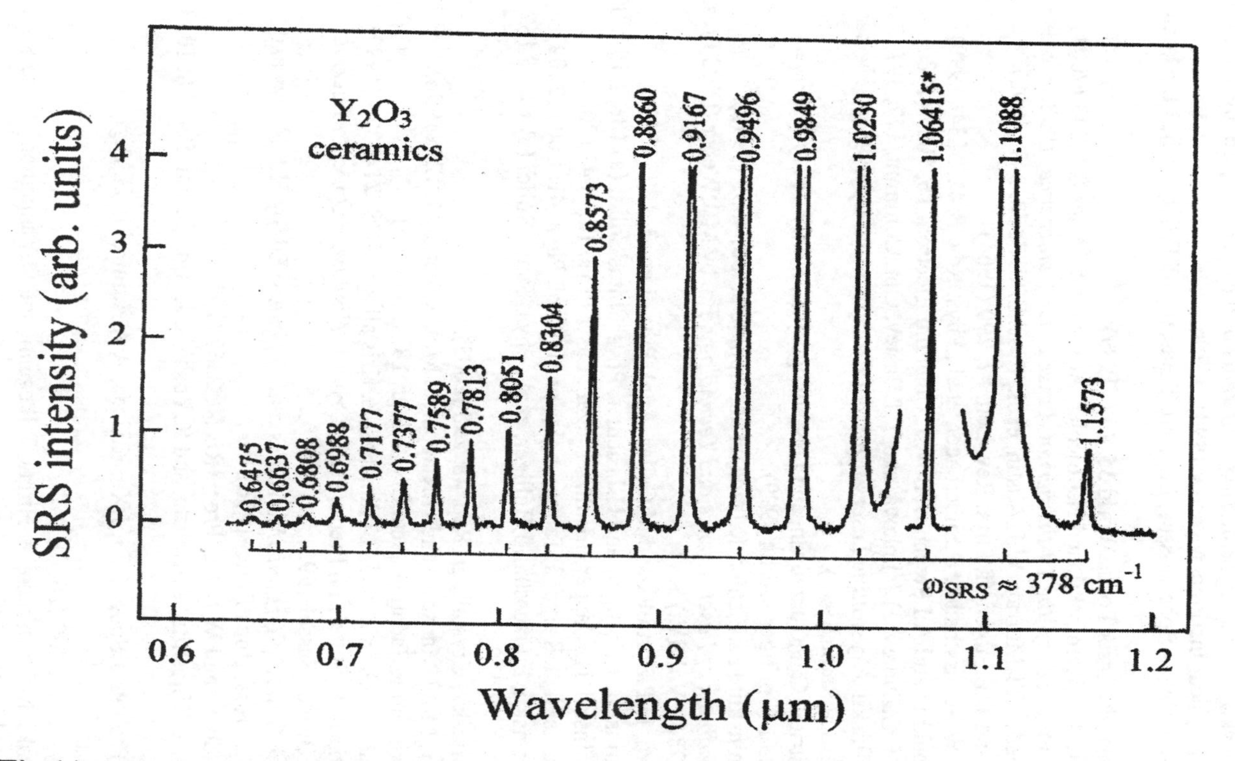

Fig.11. Stokes and anti-Stokes lasing spectra of undoped Y_2O_3 nanocrystalline ceramics recorded at 300 K under picosecond excitation at λ_{f1} = 1.06415 μm wavelength [51]. The SRS-active vibration mode $\omega_{SRS} \approx 378$ cm^{-1} is indicated by horizontal scale line. Other notations as in Fig. 2

References

1. J.T.Murray, R.C.Powell, and N.Peyghambrarian, J. Lumin. **66-67**, 89 (1996).
2. A.A.Kaminskii, in: *Raman Scattering – 70 Years of the Research*, Ed. by V.I.Gorelik, (Lebedev Physical Institute, Moscow 1998), p. 206.
3. Optical Materials – Special issue, Ed. by T.T.Basiev and R.C.Powell, **11**, March 1999.
4. G.A.Pasmanik, Laser Focus World **35**, 137 (1999).
5. J.Hulliger, A.A.Kaminskii, and H.J.Eichler, Adv. Funct. Mater. **11**, 243 (2001).
6. J.Barnes, in: *Proc. of the 19-th Internat. Laser Lidar Conference*, 1998, p. 619.
7. W.T.Roberts, J.T.Murray, W.L.Austin, et al., Proc. SPIE **3353**, 347 (1998).
8. B.Chiao and B.P.Stoicheff, Phys. Rev. Lett. **12**, 290 (1964).
9. K.K.Lai, W.Schusslbauer, H.Silberbauer, et al., Phys. Rev. B **42**, 5834 (1990).
10. S.N.Korpukhin and A.I.Stepanov, Sov. J. Quantum Electron. **16**, 1027 (1986).
11. P.Cerny, P.G.Zverev, H.Jelinkova, and T.T.Basiev, Opt. Commun. **177**, 397 (2000).
12. A.A.Kaminskii, *Crystalline Lasers: Physical Processes and Operating Schemes* (CRC Press, Boca Raton 1996).
13. V.G.Dmitriev, G.G.Gurzadyan, and D.N.Nikogosyan, *Handbook of Nonlinear Optical Crystals* (Springer, Berlin 1999).
14. A.A.Kaminskii, H.J.Eichler, K.Ueda, et al., Appl. Opt. **38**, 4533 (1999).
15. *CRC Handbook of Laser Science and Technology*, Ed. by M.J.Weber (CRC Press, Boca Raton 1986), Vol. IV.)
16. G.Eckhard, D.P.Bortfeld, and M.Geller, Appl. Phys. Lett. **3**, 137 (1963).
17. A.A.Kaminskii, J.Hulliger, and H.J.Eichler, Phys. Status Solidi (a) **186**, R19 (2001).
18. A.A.Kaminskii, H.Klapper, J.Hulliger, et al., Laser Phys. **12**, 1041 (2002).
19. A.A.Kaminskii, T.Kaino, T.Taima, et al., Jpn. J. Appl. Phys. **41**, L603 (2002).
20. A.A.Kaminskii, E.Haussühl, J.Hulliger, et al., Phys. Status Solidi (a) **193**, R167 (2002).
21. A.A.Kaminskii, Crystallogr. Rep. **48**, 295 (2003).
22. H.Kutzke, M.Al-Mansour, and H.Klapper, J. Mol. Struct. **374**, 129 (1996).
23. C.J.Brown and R.Sadanga, Acta Crystallogr. **18**, 158 (1965).
24. E.Haussühl, G.Giester, and E.Tillmanns, Z. Kristallogr. NCS, **214**, 375 (1999).
25. Ch.Bosshard, K.Sutter, Ph.Pretre, et al., *Organic Nonlinear Optical Materials* (Gordon and Beach, Bassel 1995).
26. *Nonlinear Optics of Organic Molecules and Polymers*, Ed. by H.S.Nolwa and S.Miyat (CRC Press, Boca Raton 1996).
27. E.L.Woodbury and W.K.Ng, Proc. IRE **50**, 2367 (1962).
28. A.Ikesue, T.Kinoshita, K.Kamata, and K.Yoshida, J. Am. Ceram. Soc. **78**, 1033 (1995).
29. J.Lu, M.Probhu, J.Song, C.Li, J.Xu, K.Ueda, A.A.Kaminskii, H.Yagi, and T.Yanagitani, Appl. Phys. B **71**, 469 (2000).
30. T.Tachiwaki, M.Yoshinaka, K.Hirota, T.Ikegami, and O.Yamaguchi, Solid State Commun. **119**, 603 (2001).

31. A.Bednarkiewicz, D.Hreniak, P.Deren, and W.Strek, J. Lumin. **102-103**, 438 (2003).
32. A.A.Kaminskii, M.Sh.Akchurin, V.I.Alshits, K.Ueda, K.Takaichi, J.Lu, T.Uematsu, M.Musha, A.Shirakawa, V.Gabler, H.J.Eichler, H.Yagi, T.Yanagitani, S.N.Bagayev, J.Fernandez, and R.Balda, Crystalogr. Rep. **48**, 515 (2003).
33. T.Taira, A.Ikesue, and K.Yoshida, in: *Advanced Solid-State Lasers* (OSA, Washington DC, 1998), p. 430.
34. J.Lu, M.Prabhu, J.Xu, K.Ueda, H.Yagi, T.Yanagitani, and A.A.Kaminskii, Appl. Phys. Lett. **77**, 3707 (2000).
35. J.Lu, J.Lu, A.Shirakawa, K.Ueda, H.Yagi, T.Yanagitani, V.Gabler, H.J.Eichler, and A.A.Kaminskii, Phys. Status Solidi (a) **189**, R11 (2002).
36. J.Lu, J.Lu, T.Murai, K.Takaichi, T.Uematsu, K.Ueda, J.Xu, K.Ueda, H.Yagi, T.Yanagitani, and A.A.Kaminskii, Opt. Lett. **27**, 1120 (2002).
37. J.Lu, J.Song, M.Prabhu, J.Xu, K.Ueda, H.Yagi, T.Yanagitani, and A.Kudryashov, Jpn. J. Appl. Phys. **39**, L1048 (2000).
38. J.Lu, T.Murai, K.Takaichi, T.Uematsu, K.Misawa, M.Prabhu, J.Xu, K.Ueda, H.Yagi, T.Yanagitani, A.A.Kaminskii, and A.Kudryashov, Appl. Phys. Lett. **78**, 3586 (2001).
39. J.Lu, T.Murai, K.Takaichi, K.Misawa, M.Prabhu, J.Xu, K.Ueda,H.Yagi,T.Yanagitani, A.A.Kudryashov, and A.A.Kaminskii, Laser Phys. **11**, 1053 (2001).
40. J.Lu, K.Ueda, H.Yagi, T.Yanagitani, Y.Akiyama, and A.A.Kaminskii, J. Alloys Comp. **341**, 220 (2002).
41. T.Yanagitani, H.Yagi, and Y.Hiro, Japanese patent 10-101411 (1998).
42. J.Lu, J.Lu, T.Murai, K.Takaichi, T.Uematsu, K.Ueda, H.Yagi, T.Yanagitani, and A.A.Kaminskii, in: *Advanced Solid-State Lasers* (OSA, Washington DC, 2002), p.507.
43. K.Ueda, J.Lu, K.Takaichi, H.Yagi, T.Yanagitani, and A.A.Kaminskii, Rev. Laser Eng. **31**, 465 (2003).
44. J.Lu, J.Lu, T.Murai, K.Takaichi, T.Uematsu, K.Ueda, H.Yagi, T.Yanagitani, and A.A.Kaminskii, Jpn. J. Appl. Phys. **40**, L1277 (2001).
45. J.Lu, K.Takaichi, A.Shirakawa, M.Musha, K.Ueda, H.Yagi, T.Yanagitani, and A.A.Kaminskii, Laser Phys. **13**, 940 (2003).
46. J.Lu, K.Takaichi, T.Uematsu, A.Shirakawa, M.Musha, K.Ueda, H.Yagi, T.Yanagitani, and A.A.Kaminskii, Appl. Phys. Lett. **81**, 4324 (2002).
47. J.Lu, K.Takaichi, T.Uematsu, A.Shirakawa, M.Musha, K.Ueda, H.Yagi, T.Yanagitani, and A.A.Kaminskii, Jpn. J. Appl. Phys. **41**, L1373 (2002).
48. A.Shirakawa, K.Takaichi, H.Yagi, J.F.Bisson, J.Lu, M .Musha, K.Ueda, T.Yanagitani, T.S.Petrov, and A.A.Kaminskii, Optics Express **11**, 2911 (2003).
49. K.Takaichi, H.Yagi, J.Lu, A.Shirakawa, K.Ueda, T.Yanagitani, and A.A.Kaminskii, Phys. Status Solidi (a) **200**, R5 (2003).
50. J.Lu, J.F.Bisson, K.Takaichi, T.Uematsu, A.Shirakawa, M.Musa, K.Ueda, H.Yagi, T.Yanagitani, and A.A.Kaminskii, Appl. Phys. Lett. **83**, 1101 (2003).
51. A.A.Kaminskii, K.Ueda, H.J.Eichler, S.N.Bagaev, K.Takaichi, J.Lu, A.Shirakawa, H.Yagi, and T.Yanagitani, Laser Phys. Lett. **1**, N1 (2004).
52. A.A.Kaminskii, H.J.Eichler, K.Ueda, S.N.Bagaev, G.M.A.Gad, J.Lu, T.Murrai, H.Yagi, and T.Yanagitani, Phys. Status Solidi (a) **181**, R19 (2000).
53. A.A.Kaminskii, H.J.Eichler, K.Ueda, S.N.Bagaev, G.M.A.Gad, J.Lu, T.Murrai, H.Yagi, and T.Yanagitani, JETP Lett. **72**, 499 (2000).
54. R.W.Boyd, *Nonlinear Optics* (Academic Press, New York 1992).
55. Y.R.Shen, *The Principles of Nonlinear Optics* (Wiley, New York 1984).

56. J.F.Reintjes, in: *CRC Handbook of Laser Science and Technology, Supplement 2: Optical Materials*, Ed. by M.J.Weber (CRC Press, Boca Raton 1995), p. 346.
57. *Nonlinear Optics*, Ed. by P.G.Harper and B.S.Wherrett (Academic Press, London 1977).
58. W.Kaiser and M.Maier, in: *Laser Handbook,* Ed. by F.T.Arichi and E.O.Schulz-Dubois (North-Holland, Amsterdam 1972), Vol.2, p. 10077.
59. A.Laubereau, in: *Non-linear Raman Spectroscopy and its Chemical Applications*, Ed. by W.Kiefer and D.A.Long (Reidel, Dorderect 1982), p. 183.
60. N.Bloembergen, *Nonlinear Optics* (Benjimin, San Diego 1992).
61. A.Z.Grasyuk, Sov. J. Quantum Electron. **4**, 269 (1974).
62. A.A.Kaminskii, *Laser Crystals, Their Physics and Properties* (Springer, Berlin 1981 and 1980).
63. A.Z.Grasyuk, S.B.Kurbasov, L.L.Losev, et al., Quantum Electron. **28**, 162 (1998).
64. M.Maier, Appl. Phys. **11**, 209 (1976).
65. T.T.Basiev, A.A.Sobol, P.G.Zverev, et al., Appl. Opt. **38**, 594 (1999).
66. D.C.Hanna, D.J.Poiner, and D.J.Pratt, IEEE J. Quantum Electron. **22**, 332 (1986).
67. P.Cerny, H.Jelinkova, T.T.Basiev, and P.G.Zverev, IEEE J. Quantum. Electron. **38**, 1471 (2002).
68. A.A.Kaminskii, S.N.Baghayev, J.Hulliger, et al., Appl. Phys. B **67**, 157 (1998).
69. A.A.Kaminskii, C.L.McCray, H.R.Lee, et al., Opt. Commun. **183**, 277 (2000).
70. H.J.Eichler and B.Liu, Opt. Mater. **1**, 21 (1992).
71. D.L.Rousseau, R.P.Baumann, and S.P.S.Porto, J. Raman Spectrosc. **10**, 253 (1981).
72. *Nonlinear Optics of Organic Molecules and Polymers*, Ed. by H.S.Nalwa and S.Miyat (CRC Press, Boca Raton 1996).
73. G.Schaak and J.A.Konngstein, J. Opt. Soc. Am. **60**, 1110 (1970).

19. STRANGE PROPERTIES OF QUANTUM SYSTEMS AND POSSIBLE INTERPRETATIONS

GIOVANNI COSTA
Dipartimento di Fisica "G.Galilei"
Università di Padova
Istituto Nazionale di Fisica Nucleare, Sezione di Padova
Via F.Marzolo, 8, 35131 Padova, Italy

Abstract. Quantum Mechanics (QM) applies successfully to all phenomena at the atomic, nuclear and subnuclear levels, and all experiments carried out so far have confirmed the validity of the theory. However, various conceptual difficulties are still present and different interpretations have been proposed. In the orthodox formulation, a central role is played by the interaction between the physical system and the observer, that is between the microworld and the macroscopic measuring instruments. In principle, there are two different possibilities: either the domains of validity of QM and classical physics are completely separated, or the properties of the macroscopic objects are reducible to those of their quantum constituents. In the first case, one does not know where to put the boundary between the two domains, while in the second case there is the problem of reconciling the probabilistic character of QM with the deterministic laws of classical physics. After a brief introduction recalling the main principles of QM, the above and related problems will be discussed and different interpretations will be examined.

1. Introduction

The great majority of physicists who apply succesfully Quantum Mechanics (QM) to different sectors of physics (ranging from atomic and condensed matter physics to nuclear and subnuclear physics) adopt, more or less explicitly, the so-called Copenhagen interpretation. This interpretation is based on the idea that the indeterminacy in the states of the physical systems is an essential feature of quantum phenomena. In fact, this prin-

B. Di Bartolo and O. Forte (eds.), *Frontiers of Optical Spectroscopy*, 647-667.

ciple is the starting point of the present formulation which is the result of the work of many physicists: Niels Bohr, Erwin Schrödinger, Werner Heisenberg, Louis de Broglie, Max Born, Wolfgang Pauli and others.

In 1926 wave mechanics was proposed by Schrödinger in terms of wave-packet interpretation. At the same time, Heisenberg adopted a more formal approach based on matrix theory. Born gave the first probabilistic interpretation of the wave function. Probability has a fundamental character: it is not due to incomplete knowledge of a system, but it is an intrinsic property of QM. In 1927 Heisenberg introduced the uncertainty principle and Bohr formulated the complementarity principle; finally Pauli understood that both principles were two sides of the same description.

In his book "Mathematical foundation of quantum mechanics", published in German in 1932 [1], von Neumann presented an axiomatic and complete treatment of QM. The physical observables are described by self-adjoint operators, acting on the vectors in a Hilbert space which represent the physical states of the system. The states evolve in time according to a differential equation, but the act of measurement produces a sudden jump in the system and the reduction to one of the relevant eigenstates.

All the predictions of QM have been tested and confirmed, but discussions about alternative interpretations are going on, in particular on the implications of the Einstein, Podolski and Rosen (EPR) theorem and on the interaction between a quantum system and the instrument employed by the observer who performs a measurement on the system.

In the following sections, after a brief recollection of the main principles of QM, we discuss the properties of the entangled states and the implications of the EPR theorem. After mentioning the Bohm's theory with hidden variables, we examine the Bell's inequality and its experimental tests. As an interesting application, we consider the possibility of teleportation. In the last section, we discuss the measurement problem and some of the paradoxes that one has to face when extending the superposition principle to macroscopic systems. Finally, in connection with the problem of reconciling the laws of QM with the properties of macroscopic systems, we indicate two different approaches, which are based on "spontaneous localization" and on "decoherent histories".

2. Resumé of the main ingredients of Quantum Mechanics

The formulation of quantum mechanics is based on the axioms which are briefly summarized in the following (for more details see e.g. [2]).

- The states of a physical system S are represented by (normalized) vectors in a Hilbert space H, which , following the Dirac notation, will be denoted by $|\Psi>$.

- The physical observables A, B,... are represented by self-adjoint operators $\hat{A}, \hat{B}$ ... acting on the Hilbert space.

- The only possible results of the measurement of the observables A, B,.. are the eigenvalues a_i, b_j ,.. of the operators $\hat{A}, \hat{B}$,...

- The expectation value of the observable A in the generic state $|\Psi>$ is given by $<\Psi|\hat{A}|\Psi>$.

- The time evolution of the state $|\Psi>$ is determined by a linear differential equation

$$i\hbar\frac{\partial}{\partial t}|\Psi(t)>= H|\Psi(t)> \tag{1}$$

With the initial condition $|\Psi(t=0)>= |\Psi_0>$, it can be written as

$$|\Psi(t)>= U(0,t)|\Psi_0> \tag{2}$$

where

$$U(0,t) = exp(-iHt/\hbar) \tag{3}$$

is the unitary operator of time evolution.

- If a measurement of A leads to the eigenvalue a_i, then the state $|\Psi>$ is changed suddenly (quantum jump) into the eigenstate $|a_i>$. This is the famous postulate of the *wave-packet reduction.*

- The maximal information on the states $|\Psi>$ of a system S is determined by measuring a complete set of commuting observables A, B... and it is specified by a complete and orthonormal set of eigenvectors

$$|\Psi>= |a_i, b_j...> \tag{4}$$

where a_i, b_j... are the eigenvalues of the operators $\hat{A}, \hat{B}$, ...
For the sake of simplicity, we assume a discrete spectrum for the eigenvalues, but the generalization to a continuous spectrum can be carried out without difficulties.

- We note that, while the state $|\Psi>$ evolves in a deterministic way, only probabilities can be assigned to the results of a measurement:

$$P(a_i b_j,..) = |<a_i, b_j|\Psi>|^2 \tag{5}$$

- Since the time-evolution equation is linear, if $|\Psi_1(t)>$ and $|\Psi_2(t)>$ are two different solutions, also the linear combination

$$|\Psi(t)>= p|\Psi_1(t)> +q|\Psi_2(t)>, \tag{6}$$

with $|p|^2+|q|^2 = 1$, is a solution. This is the famous *superposition principle*, whose consequences lead to many speculations.

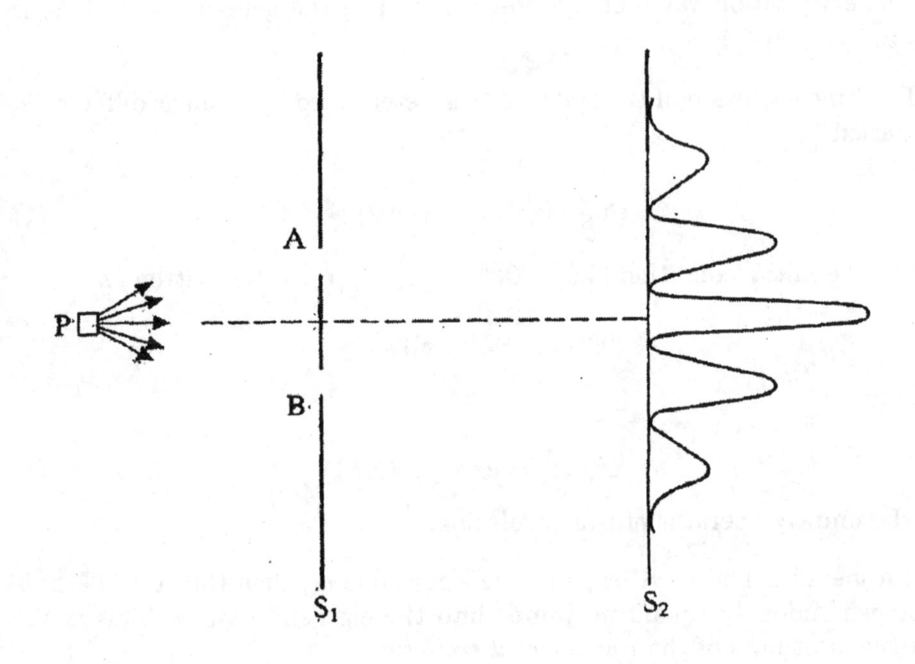

Figure 1. Sketch of the two-slit experiment.

It is instructive to consider explicitly a typical case of superposition, namely the classical example of the two-slit experiment, quantum version of the experiment performed by Thomas Young at the end of 18th century, which indicated the ondulatory nature of light.
The source P (see Fig. 1) emits electrons, with very low intensity: only one electron at a time passes through the two-slit screen S_1 and hits a photographic plate on the screen S_2. If both slits are open, after some time, an interference pattern appears on S_2. The *particle* (electron) behaves as a *corpuscle* when it is emitted and hits the screen S_2, but it behaves as a *wave* when it passes through S_1 . It is meaningless to ask through which slit the electron has passed; in fact, the interference pattern disappears if one tries to answer this question experimentally. According to QM, the state of the electron is the superposition of two distinct states

$$|\Psi(t)>=|\Psi_1(t)>+|\Psi_2(t)> \tag{7}$$

and it is their interference which explains the strange behavior of the electron.

This example characterizes very nicely the content of the principle. On the other hand, if this principle is applied to macroscopic objects, it leads to strange paradoxes. According to Richard Feynman, this experiment is "a phenomenon which is impossible, absolutely impossible, to explain in any classical way, and which has in it the heart of quantum mechanincs". (quoted in [3]).

Another important point which we would like to recall is the distinction between *pure and mixed* states. A pure state is a coherent superposition of states, for example

$$|\Psi>=\sum_i c_i|a_i> \tag{8}$$

where $|c_i|^2$ gives the probability that, by measuring A, one finds the system in the eigenstate $|a_i>$. The expectation value of the observable A in the state $|\Psi>$ is

$$<\hat{A}>=\sum_i |c_i|^2 a_i \tag{9}$$

In statistical terms, a pure state corresponds to a pure ensemble in which all the systems are in the same state. If we have an ensemble with different states $|\Psi_i>$, each of which has a given probability P_i , one speaks of *mixed states* or *mixture.* In this case, the expectation value is

$$<\hat{A}>=\sum_i P_i <\Psi_i|\hat{A}|\Psi_i> \tag{10}$$

with $\sum_i P_i = 1$. It is convenient to introduce the *density matrix*

$$\rho=\sum_i P_i|\Psi_i><\Psi_i| \tag{11}$$

which satisfies the following properties:

$$\rho=\rho^+, \qquad Tr\ \rho=1 \tag{12}$$

and

$$<\hat{A}>=Tr(\rho\hat{A}) \tag{13}$$

We note that for a pure state $|\Psi>, \rho$ reduces to

$$\rho = |\Psi><\Psi| \tag{14}$$

from which one gets immediately:

$$\rho^2 = \rho, \quad Tr\ \rho^2 = 1 \tag{15}$$

Instead, for a mixed state, one has:

$$\rho^2 = \sum_i P_i^2 |\Psi_i><\Psi_i| \neq \rho \tag{16}$$

and

$$Tr\ \rho^2 = \sum_i P_i^2 < 1 \tag{17}$$

For the sake of illustration, we apply the above formalism to the case of spin-$\frac{1}{2}$ states, which we shall employ in the next section.
Let us consider a spin-$\frac{1}{2}$ particle: the three spin components are described by the operators

$$S_i = \frac{1}{2}\hbar\ \sigma_i \tag{18}$$

where

$$[\sigma_i, \sigma_j] = 2i\epsilon_{ijk}\sigma_k \tag{19}$$

Only one component can be *diagonalized*. With the usual choice for the Pauli matrices

$$\sigma_1 = \begin{pmatrix} 0 & 1 \\ 1 & 0 \end{pmatrix} \qquad \sigma_2 = \begin{pmatrix} 0 & -i \\ i & 0 \end{pmatrix} \qquad \sigma_3 = \begin{pmatrix} 1 & 0 \\ 0 & -1 \end{pmatrix} \tag{20}$$

the two eigenstates corresponding to the eigenvalues $+\frac{1}{2}, -\frac{1}{2}$ along the 3rd (z) axis are represented by the two-component vectors

$$|\alpha> = \begin{pmatrix} 1 \\ 0 \end{pmatrix} \qquad |\beta> = \begin{pmatrix} 0 \\ 1 \end{pmatrix} \tag{21}$$

One can easily check that the eigenstates with eigenvalues $+\frac{1}{2}, -\frac{1}{2}$ along the 1st (x) axis are:

$$|\alpha'> = \sqrt{\frac{1}{2}}\{|\alpha> + |\beta>\} = \sqrt{\frac{1}{2}}\begin{pmatrix} 1 \\ 1 \end{pmatrix} \tag{22}$$

$$|\beta' >= \sqrt{\frac{1}{2}}\{|\alpha > -|\beta >\} = \sqrt{\frac{1}{2}}\begin{pmatrix} 1 \\ -1 \end{pmatrix} \tag{23}$$

As an application, let us consider the most general density matrix for the spin-$\frac{1}{2}$ states:

$$\rho = \frac{1}{2}(\mathbf{1} + \mathbf{b}\vec{\sigma}) \tag{24}$$

where $\mathbf{b}$ is a three-vector and $\vec{\sigma} = (\sigma_1, \sigma_2, \sigma_3)$. One can immediately check that the following relations hold

$$Tr\ \rho = 1, \quad Tr\ \rho^2 = \frac{1}{2}(1 + b^2), \tag{25}$$

where $b = |\mathbf{b}|$. Moreover, taking into account the properties of the Pauli matrices:

$$< \vec{\sigma} >= Tr(\rho\vec{\sigma}) = \mathbf{b}, \tag{26}$$

so that b represents the degree of polarization along the direction of $\mathbf{b}$. The case $b = 0$ corresponds to an unpolarized beam, while the case $b = 1$ corresponds to a complete polarization.

3. Entangled states and the EPR argument

Albert Einstein was not satisfied with the probabilistic interpretation of QM. His ideas can be expressed in the following way: each physical variable has a definite value in each state of a physical system and the indetermination is only due to our approximate knowledge of the states . The statistical description does not provide a complete description (ignorance interpretation). Einstein assumed that all physical systems possess intrinsic and well defined properties even when they are not subject to measurement. Then he formulated the following requisites for a physical theory:

a) *completeness*: every element of the physical reality has a counterpart in the theory;

b) *physical reality*: if - without disturbing a system - one can predict with certainty the value of a physical quantity, then there exists an element of physical reality corresponding to this quantity;

c) *locality*: two elements of the physical reality separated by a spatial distance cannot have an instantaneous influence on each other.

These ideas gave rise to a famous "gedanken" experiment which was proposed by Einstein, Podolsky and Rosen (EPR) [4]. In its original form, it was formulated in terms of positions and linear momenta, but it can be expressed (as shown by David Bohm) in an equivalent form in terms of angular momenta (EPRB version) [5].

Let us suppose that a spin-0 particle decays at rest into two spin-$\frac{1}{2}$ particles. The angular part of the state can be described in terms of the eigenstates of S_z.

$$|\Psi>=\sqrt{\frac{1}{2}}\{|\alpha_1>|\beta_2>-|\beta_1>|\alpha_2>\} \tag{27}$$

The total angular momentum is zero ($S=0, S_z=0$), while the two particles have no definite values of S_z.

The state is not factorizable: it is a pure state which is *entangled.* This is the general situation occurring for a system with more than one particle: the system has well-defined quantum states, while the individual particle states are not defined. This situation was pointed out by Schrödinger who wrote in 1935: "the whole is in a definite state, the parts taken individually are not". [6]

When particles 1 and 2 are far apart, a measurement of S_z (1) is performed: there is equal probability for either of the two values : $+\frac{1}{2}, -\frac{1}{2}$; correspondingly, $S_z(2)$ takes the opposite value. Suppose particle 1 is found in $|\alpha_1>$, then the state given in (27) collapses into

$$|\Psi>\longrightarrow|\alpha_1>|\beta_2> \tag{28}$$

After the measurement, the total angular momentum is no longer defined. In fact, one has:

$$|\alpha_1>|\beta_2>=\sqrt{\frac{1}{2}}\{|0,0>+|1,0>\} \tag{29}$$

where the kets on the r.h.s of the equation represent the eigenstates $|S, S_z>$. Moreover, for rotational invariance, the initial state $|\Psi>$ can be expressed also as:

$$|\Psi>=\sqrt{\frac{1}{2}}\{|\alpha_1'>|\beta_2'>-|\beta_1'>|\alpha_2'>\} \tag{30}$$

where the spin S is now quantized along, say, the x-axis. If we measure $S_x(1)$ and find e.g. $+\frac{1}{2}$, particle 2 will be in the state $|\beta_2'>$ with $S_x(2)=-\frac{1}{2}$.

What can we conclude? The spin component $S_z(2)=-\frac{1}{2}$ is an element of reality, since it can be predicted with certainty. Assuming that there are no actions-at-a-distance effects, such element of reality must exist independently of the measurement on particle 1. Then we should attribute a physical reality also to the value $S_x(2) = -\frac{1}{2}$. On the other hand, according to the rules of QM, the eigenvalues of S_x and S_z cannot be determined simultaneously, since S_x and S_z are non- commuting observables.

In conclusion, either QM is not complete, since it is not able to describe all the elements of the physical reality, or the locality assumption is violated.

In 1952 there was an interesting development: David Bohm [7] succeeded in building a theory with hidden variables (the so-called pilot-wave theory), which is complete, in the sense that each state can be defined exactly in terms of a set of variables λ, even if they are not accessible to measurement. In this way, he was able to produce a deterministic theory which gives results equivalent to those of QM. Each component of the system has a specific value for the position $\mathbf{x}$ and the linear momentum $\mathbf{p}$, so that there are trajectories as in classical physics. The possible trajectories are consistent with the probability distributions given by the Schrödinger function ψ $(\mathbf{x})$. This requires the introduction of a quanto-mechanical potential, which has to be related to ψ $(\mathbf{x})$. In a way, the motion of a particle is guided by a wave. (Some more details can be found in [8]).

The weak point in Bohm's theory is that it is *non-local* and *contextual*: the hidden variables cannot be assigned independently of the context, but they depend also on the environment. In fact, the trajectory of a particle can be changed instantaneously by the action performed in a far-away region. On the other hand, it is interesting that a deterministic theory can produce the same results of QM. Moreover, Bohm's theory is built in such a way as to reproduce all the results of QM. But, since the hidden variables are not accessible, the theory cannot be tested, i.e. it cannot be falsified. Then, according to the requirement of the epistemologist Karl Popper, it would not be considered a physical theory!

At the end, the question remains: is it possible to elaborate a model along these lines, in agreement with the results of QM, but which satisfies *both determinism and locality?*

This question was attacked by John Bell in 1963. He tried to investigate whether it is possible to take into account the quantum correlations in a local scheme. He did not succeed in his program: he proved that no -either deterministic or probabilistic- theory which is local can reproduce all the correlations predicted by QM. His research gave rise to an inequality which is able to test, at the experimental level, whether the quantum correlations are satisfied or QM has to be modified taking into account the requirement of locality.

4. Bell's inequality

Before discussing the Bell's inequality [9], we reconsider the EPRB gedanken experiment, formulated in terms of linear polarization states. We define by $|x>$ and $|y>$ the states of a photon with linear polarization along the x-axis and along the y-axis, respectively.

The EPRB experiment can be realized with a source of two photons emitted in opposite directions: if the source is in a state of zero angular momentum and positive parity, the two photons will be in the following state:

$$|\Psi>=\sqrt{\frac{1}{2}}\{|x_1>|x_2>+|y_1>|y_2>\} \tag{31}$$

Then, a measurement of the polarization of one photon, (e.g. x_1), fixes the other photon to have the same polarization (x_2).
In the case of negative parity, the two photons will be in the state

$$|\Psi'>=\frac{1}{2}\{|x_1>|y_2>+|y_1>|x_2>\} \tag{32}$$

but, with obvious modifications, one arrives at the analogous conclusion. The state given in (31) is invariant under rotation of the polarization axes in the x-y plane. In fact, making the rotation:

$$\begin{aligned} x' &= x\ \cos\theta + y\ \sin\theta \\ y' &= -x\ \sin\theta + y\ \cos\theta \end{aligned} \tag{33}$$

one gets

$$|\Psi>=\sqrt{\frac{1}{2}}\{|x'_1>|x'_2>+|y'_1>y'_2>\} \tag{34}$$

Suppose that in a specific EPRB experiment two photons are emitted in two opposite directions along the z-axis and two polarizers measure their linear polarization along the vectors **a** and **b** in the x-y plane. The result of each measurement will be indicated by either (+) or (−) according to the case in which the photon passes or does not pass the test. Suppose that our description is not complete and that, at least in principle, the states of the system can be completely specified by a set of hidden variables λ. We define by

$$P_\lambda(\mathbf{a};\pm) \tag{35}$$

the probability that a single photon (in the state defined by λ) passes (+) or does not pass (−) the test, i.e. that its polarization is along **a**, or orthogonal to **a**. In the case of the two photons, we define by

$$P_\lambda(\mathbf{a}, \mathbf{b}; \pm, \pm) \tag{36}$$

the probability of obtaining (±) for the polarization of one photon with respect to **a** and of the other photon with respect to **b**.
The locality condition can be expressed as

$$P_\lambda(\mathbf{a}, \mathbf{b}; \pm, \pm) = P_\lambda(\mathbf{a}; \pm) \times P_\lambda(\mathbf{b}; \pm) \tag{37}$$

i.e. it is based on the assumption that the measurements on the two photons are independent and exert no influence on each other. We note that these arguments apply both to probabilistic and deterministic theories; in the latter case, all P_λ have to be taken either equal to zero or to 1.

Then we define the correlation function

$$E_\lambda(\mathbf{a}, \mathbf{b}) = P_\lambda(\mathbf{a}, \mathbf{b}; +, +) + P_\lambda(\mathbf{a}, \mathbf{b}; -, -) - P_\lambda(\mathbf{a}, \mathbf{b}; +, -) - P_\lambda(\mathbf{a}, \mathbf{b} : -, +) \tag{38}$$

Making use of the locality condition and of the two relations

$$P_\lambda(\mathbf{a}; +) + P_\lambda(\mathbf{a}; -) = 1 \tag{39}$$

$$-1 \leq P_\lambda(\mathbf{a}; +) - P_\lambda(\mathbf{a}; -) \leq +1 \tag{40}$$

it is easy to prove that the following relation holds

$$|E_\lambda(\mathbf{a}, \mathbf{b}) - E_\lambda(\mathbf{a}, \mathbf{d})| + |E_\lambda(\mathbf{c}, \mathbf{b}) + E_\lambda \mathbf{c}, \mathbf{d})| \leq 2 \tag{41}$$

Assuming the distribution function $\rho(\lambda)$ for the λ-variables, with normalization $\int d\lambda \rho(\lambda) = 1$, we get

$$P(\mathbf{a}, \mathbf{b}; \pm, \pm) = \int d\lambda \rho(\lambda) P_\lambda(\mathbf{a}, \mathbf{b}; \pm, \pm) \tag{42}$$

$$E(\mathbf{a}, \mathbf{b}) = \int d\lambda \rho(\lambda) E_\lambda(\mathbf{a}, \mathbf{b}). \tag{43}$$

Finally one obtains the Bell's inequality

$$S = |E(\mathbf{a}, \mathbf{b}) - E(\mathbf{a}, \mathbf{d})| + |E(\mathbf{c}, \mathbf{b}) + E(\mathbf{c}, \mathbf{d})| \leq 2 \tag{44}$$

Denoting by $|\Psi_a(+)>$ and $|\Psi_a(-)>$ the photon states with polarization along **a** and orthogonal to **a** , respectively, one gets:

$$|\Psi_b(+)>=\sqrt{\frac{1}{2}}\{\cos(\mathbf{a.b})|\Psi_a(+)>+\sin(\mathbf{a.b})|\Psi_a(-)>\} \qquad (45)$$

and it is immediate to show that in QM:

$$P(\mathbf{a},\mathbf{b};+,+)=P(\mathbf{a},\mathbf{b}:-,-)=\frac{1}{2}\cos^2(\mathbf{a.b}) \qquad (46)$$

$$P(\mathbf{a},\mathbf{b}:+,-)=P(\mathbf{a},\mathbf{b};-,+)=\frac{1}{2}\sin^2(\mathbf{a.b}) \qquad (47)$$

so that the QM prediction is

$$E(\mathbf{a},\mathbf{b})=\cos 2(\mathbf{a.b}) \qquad (48)$$

We stress the fact that Bell's inequality is based on the locality requirement, but it applies both to deterministic or probabilistic theories. There are specific orientations of the polarizers for which the QM prediction violate the Bell's inequality. In fact, if one chooses the polarization vectors **a**,**b**,**c**,**d** as indicated in Fig. 2, one gets:

$$S_{QM}=2\sqrt{2} \qquad (49)$$

which is beyond the allowed range.

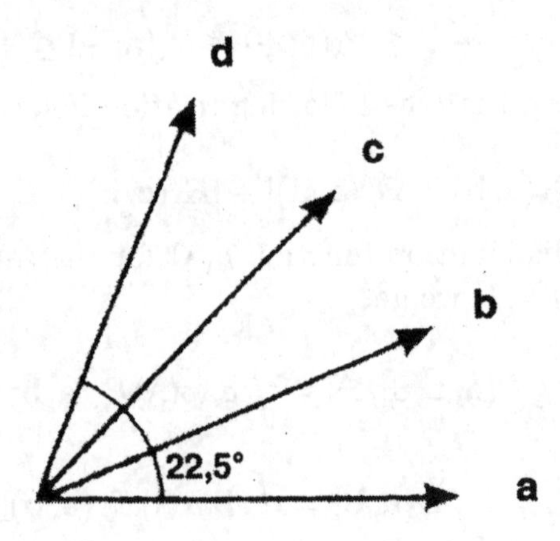

Figure 2. Orientations of the polarizers yielding $S_{QM}=2\sqrt{2}$.

5. Experimental tests

An interesting feature is that the Bell's inequality can be tested by real experiments. Several experiments of increasing accuracy have been performed; here we refer to the experiments by Aspect et al. [10], in which an optical version of the EPRB experiment has been realized.

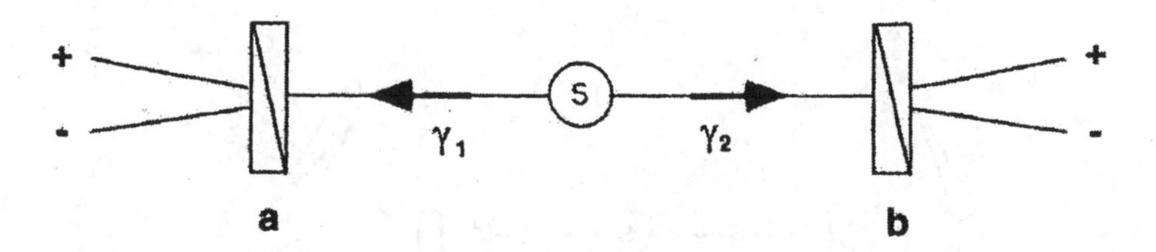

Figure 3. Scheme of the EPRB experiment with two photons γ_1 and γ_2 analized by linear polarizers with orientations **a** and **b**.

As indicated in Fig.3, a pair of photons γ_1 and γ_2 is analysed by two linear polarizers with orientations **a** and **b**, and by photomultipliers; the coincidence rate is monitored. In the experiment, the two photon excitation cascade in ^{20}Ca is employed. The atomic configuration is the following:

$$(1s^2 2s^2 p^6 3s^2 p^6)4s^2 = (Ar)4s^2 \tag{50}$$

From the excited $(Ar)4p^2$ the following transitions occur:

$$\begin{array}{ll} (Ar)4p^2 & (^1S_0) \\ & \downarrow \\ (Ar)4s4p & (^1P_1) \\ & \downarrow \\ (Ar)4s^2 & (^1S_0) \end{array} \tag{51}$$

Two correlated photons (one at each step) are emitted in opposite directions with the same polarization.

The experiment reproduces the configuration of Fig. 2, which corresponds to $S_{QM} = 2\sqrt{2}$. Taking into account the experimental set up, the effective theoretical value is reduced to $S_{th} = 2.70 \pm 0.05$. The experimental value is $S_{exp} = 2.697 \pm 0.015$, while Bell's inequality requires $S \leq 2$. The results are then in agreement with QM.

These experiments have been performed with static setups, i.e. with polarizers which were fixed during the run. One may question the validity of the locality hypothesis, since some time has elapsed, in which some information might have exchanged between the two polarizers. Then it was proposed to modify the settings of the experiment during the flight of the two photons. Such a timing experiment would certainly prevents any faster-than-light influence.

In 1982 an experiment of this type was performed by Aspect et al.[11]. The previous setting has been modified according to the scheme indicated in Fig.4 . Each polarizer has been replaced by a setup involving a switching device followed by two polarizers in two different orientations:

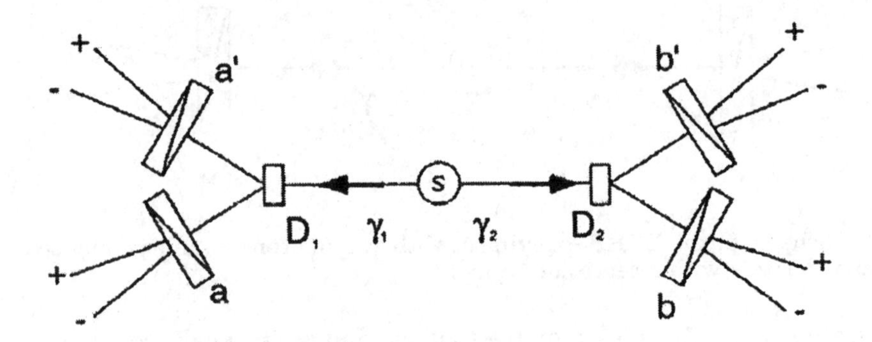

Figure 4. Scheme of the EPRB experiment modified with the inclusion of two switching devices D_1 and D_2.

a, **a**′ on one side, and **b**, **b**′ on the other side. Such an optical switch is able to rapidly redirect the incident light from one polarizer to the other. If the two switches work at random and are uncorrelated, it is possible to make use of a modified Bell's inequality, which requires : $-1 \leq S \leq 0$. The experimental value is $S_{exp} = 0.101 \pm 0.020$, which clearly violates the condition $S \leq 0$.

These results indicate that Nature follows QM. However, lot of discussions are still going on because, from the epistemological point of view, the interpretation is not satisfactory. The formalism of QM tells exactly what are the predictions, but there is a subtle element of non-locality which is not accepted by all the members of the physical community.
It is important to understand whether non-locality can be interpreted as an instantaneous influence of a system on a distant one. Does it mean that we can transfer information faster than the speed of light? This would violate the Einstein's principle of relativity.

6. An application of the EPR theorem

An interesting application of the EPR theorem is teleportation [12, 13, 14]. The dream is to transfer instantaneously something from one place to another distant location. Suppose Alice has a photon in a given state

$$|\Psi_1 >= a|x_1 > +b|y_1 > \tag{52}$$

with $|a|^2 + |b|^2 = 1$, and she wants to transfer to Bob a photon in the same state, but not by sending it directly to him, since the communication may destroy the quantum coherence.

Bennet et al.[12] suggested that it is possible to transfer the quantum state of a particle to another particle (quantum teleportation) making use of the projection postulate.

To get teleportation, the following points have to be realized:

a) A pair of particles (photons 2 and 3) is initially shared by Alice and Bob in the entangled state

$$|\Psi_{23}^- >= \sqrt{\frac{1}{2}}\{|x_2 > |y_3 > -|y_2 > |x_3 >\} \tag{53}$$

This state contains no information on the individual particles 2 and 3; it only indicates that they have opposite polarizations. As we know from the EPR experiment, as soon as a measurement on one particle projects it in a given state, say $|x >$, the other is determined to be in $|y >$, and viceversa.

b) Now Alice has photons 1 and 2, but particle 2 is entangled with 3, which is in the hands of Bob. Next particles 1 and 2 have to be entangled.

Suppose that particles 1 and 2 become entangled in the state:

$$|\Psi_{12}^- >= \sqrt{\frac{1}{2}}\{|x_1 > |y_2 > -|y_1 > |x_2 >\} \tag{54}$$

QM predicts that when particles 1 and 2 are projected into $|\Psi_{12}^- >$, particles 3 is instantaneously projected into the initial state of particle 1. In fact, in the entangled state, whatever the state of particle 1 is, particle 2 must be in the opposite (orthogonal) state. But since particles 2 and 3 are orthogonal to each other, particle 3 must be in the same state of 1.
The final state of particle 3 is then:

$$|\Psi_3 >= a|x_3 > +b|y_3 > \tag{55}$$

which coincides with the state in eq.(52). The transfer of quantum information from particle 1 to 3 can occur over arbitrary distances.

In the above discussion, we have assumed that particles 1 and 2 are in the entagled state $\Psi_{12}^- >$. However, there are four different entangled states in which a two-particle state can be decomposed; the other three are the following:

$$|\Psi_{12}^{+}>=\sqrt{\frac{1}{2}}\{|x_1>|y_2>+|y_1>|x_2>\} \tag{56}$$

$$|\Phi_{12}^{+}>=\sqrt{\frac{1}{2}}\{|x_1>|x_2>+|y_1>|y_2>\} \tag{57}$$

$$|\Phi_{12}^{-}>=\sqrt{\frac{1}{2}}\{|x_1>|x_2>-|y_1>|y_2>\} \tag{58}$$

The projection of an arbitrary state of two particles into the basis of the four states is called a Bell-state measurement. Explicitly, the combination of the two states given in eqs.(52) and (53) can be decomposed as follows:

$$\begin{aligned}&|\Psi_1>|\Psi_{23}^{-}>=\tfrac{1}{2}\{|\Phi_{12}^{+}>(a|y_3>-b|x_3>)\\&+|\Phi_{12}^{-}>(a|y_3>+b|x_3>)-|\Psi_{12}^{+}>(a|x_3>-b|y_3>)\\&-|\Psi_{12}^{-}>(a|x_3>+b|y_3>)\}\end{aligned} \tag{59}$$

The above relations show that particle 1 can be found in each of the four states given in eqs. (55) and (56)-(58) with equal probability of 25%.
Therefore, in order that Bob selects the right state, he needs some extra information; via a classical communication channel Alice has to transmit to him the information about the entangled state. First of all, she has to detect in which of the four states the two photon 1 and 2 have gone. This is done by means of appropriate coincidence after a beam splitter. According to what she finds, she sends the appropriate instruction to Bob. In conclusion, Bob gets a particle in the identical state of the one in Alice's hands; however, the quantum teleportation is not sufficient: he needs also a physical messanger taking the extra information.

The results so far obtained are in agreement with the prediction of QM and, moreover, we can conclude by saying that there is no contradiction with causality, since it is not possible to employ non-local effects by sending signals at superluminal speed. The non-local effects of QM appear to coexist with relativity.

7. Macroscopic quantum superposition and the measurement problem

In the standard interpretation, microsystems obey QM while macrosystems (specifically measuring instruments) obey the laws of classical physics. However, macrosystems are composed by elementary constituents of matter, and their behaviour is described in terms of fundamental interactions which are ruled by QM. For a more general discussion see [15, 16]. But if QM is applied to macrosystems, one can have entanglement for macroscopic states and paradoxes like the famous story of the Schrödinger's cat [6]. In his "gedanken" experiment, a cat, closed in a box, is put in a pure state, which is an even superposition of life and death, through the correlation with a decaying radioactive atom. It is only the observer, when he opens the box, that puts the cat either in a state of a living creature or in a state of a dead one.

The measurement, i.e. the active presence of an observer, is a requisite of the QM postulates. But what is the mechanism that transforms suddenly a pure state into a statistical mixture? Suppose that a quantum system S is in a state $|\Psi_0>$ and we want to measure the observable A. We know from eq. (8) that $|\Psi_0>$ can be decomposed into the eigenstates $|a_i>$:

$$|\Psi_0>=\sum_i c_i|a_i> \tag{60}$$

The interaction of S with the instrument I_1 (in the state $|\Phi_1>$) produces the transition

$$|\Psi_0>|\Phi_1>\longrightarrow |a_i>|\Phi_{1i}> \tag{61}$$

where $|\Phi_{1i}>$ stands for the state of the instrument with its "pointer" indicating the value a_i. However, if we assume that the system $S+I_1$ is governed by a linear evolution equation we should have:

$$|\Psi_0>|\Phi_i>=\sum_i c_i|a_i>|\Phi_{1i}> \tag{62}$$

How can we determine the specific state $|\Phi_{1i}>$? We need a second instrument I_2, and we should get:

$$|\Psi_0>|\Phi_1>|\Phi_2>=\sum_i c_i|a_i>|\Phi_{1i}>|\Phi_{2i}> \tag{63}$$

This procedure can be repeated, and we would obtain an infinite chain (von Neumann's chain). Where can one stop? Can one put a clear separation between a coherent superposition of states and a statistical mixture? We limit ourselves here to mention a couple of hypothetical solutions.

According to Eugene P. Wigner [17], the breaking of the chain occurs at the moment of the conscient perception of the observer (since the consciousness is considered not to be reducible to physical processes).

The many-world interpretation was proposed by Hugh Everett III [18]: all possibilities in the superposition of different states are realized at the same time in different separated branches of the Universe.

These interpretations are rather speculative, and do not provide a satisfactory physical description. The role of the observer appears to be fundamental in QM. However, when QM is applied to astrophysics and cosmology, what is the meaning of the observer? Is it possible a quantum theory without observer? One would like to derive the deterministic laws that approximately govern the domain of familiar experience from the probabilistic properties of QM; how the causal connection among classical events can emerge from the fundamental uncertainty of QM?
Two different approaches have been proposed:

1) *Spontaneous localization* [19, 20].
This approach combines Schrödinger evolution with spontaneous random collapses. It is based on the postulate that all elementary constituents of matter obey the linear dynamics of the Schrödinger equation, but they are subjected also to spontaneous random (non-linear) processes of space localization.

Let us consider the case of a single particle. At a given instant $t = t_0$, the wave funtion ψ $(\mathbf{x})$ is localized around $\mathbf{r_0}$.

$$\psi(\mathbf{r}) \longrightarrow F(\mathbf{r}-\mathbf{r_o}) \sim \exp\{-(\mathbf{r}-\mathbf{r_o})^2/2d^2\} \tag{64}$$

For instance, in the case of a particle with magnetic moment passing through a Stern-Gerlach apparatus, one gets the superposition

$$\psi(\mathbf{r}) = \sqrt{\frac{1}{2}}\{\psi_A(\mathbf{r}) + \psi_B(\mathbf{r})\}, \tag{65}$$

where ψ_A and ψ_B represent the convergence toward two different regions A and B. The stocastic localization occurs either in A or in B, and then it makes a choice between ψ_A and ψ_B.

Let us consider the measurement of this system by means of an instrument: its pointer can indicate that the particle is in one of the two regions A and B. The "state" of the instrument is described by

$$|\Phi(\mathbf{x_1},\mathbf{x_2},\mathbf{x_3}...)> = \sqrt{\frac{1}{2}}\{|\Phi_A(\mathbf{x_1},\mathbf{x_2},\mathbf{x_3}....)> \; + |\Phi_B(\mathbf{x_1},\mathbf{x_2},\mathbf{x_3}...)>\}, \tag{66}$$

where $\mathbf{x_i}$ stands for the position of the i-th particle inside the pointer. It contains a huge number of particles (of the order of the Avogadro number $\sim 6.6 \times 10^{23}$) but the random localization of $\mathbf{x_i}$ inside one of the two regions A and B selects one of the two states Φ_A and Φ_B.
Even if the probability that a single particle is localized is extremely small, it is practically certain that several particles (out of 10^{26}) become localized in a microsecond. According to the authors of [19], taking e.g. $d = 10^{-6}$ cm

and $\lambda = 10^{-16}$ sec^{-1} (frequency of localization), about 10^7 particle will be localized in a second. In conclusion, it appears that the particle localization makes the macroscopic instrument jump into a given position.

2) *Decoherent histories* [21, 22, 23].

A "quantum history h" is defined by the time evolution of a system from the initial state $|\Psi_0>$ at $t = 0$ by applying to it the n-step projection operator

$$E(h) = P(a_n)exp\{-iH(t-t_n)/\hbar\}....P(a_i)exp\{-iH(t_1-t_0)\hbar\} \qquad (67)$$

where $P(a_i)$ is the projection operator relative to the eigenstate $|a_i>$ of $\hat{A}$. We assume that at the time $t = t_i$ the system is in $|a_i>$, but there is no observer performing a measurement.

Different histories have different intermediate steps. To each of them one can assign a probability:

$$P_h =< E(h)\Psi_0|E(h)\Psi_0 > \qquad (68)$$

However, for a given set of histories, it can happen that the sum of the relative probabilities exeeds 1; this means that there is some inconsistency. To get consistent histories, one has to limit one self to a family of histories which are *decoherent.*

For example, if the system is a particle of spin $\frac{1}{2}$, the two histories corresponding to the two possibilities $S_z = +\frac{1}{2}$ and $S_z = -\frac{1}{2}$ at $t = t_i$ are in the same family, but they are incompatible with the two histories relative to $S_x = +\frac{1}{2}$ and $S_x = -\frac{1}{2}$, because S_z and S_x do not commute.

Gell-Mann defines a *decoherence function*

$$D(g,h) =< \Psi_g|\Psi_h > \qquad (69)$$

wich satisfies

$$D(h \text{ or } g,\ h \text{ or } g) = D(h,h) + D(g,g) + D(h,g) + D(g,h) \qquad (70)$$

The last two terms (interference terms), which can be re-written as

$$D(h,g) + D(g,h) = 2Re\{D(h,g)\} \qquad (71)$$

have no definite sign. Only if the sum in eq. (71) vanishes, h and g become consistent histories.

According to Gell-Mann, in the case of a macroscopic system, one has to consider *coarse-grained histories* which contain a very large number of *fine-grained* histories. In the total sum, the interference terms cancel among themselves, and the decoherence function becomes a true probability which gets a value very close to 1.

These two approaches that we have briefly considered (detailed discussions can be found in the quoted references) deserve attention since they are rather promising. On the other hand, they indicate that we have not yet reached a unique and definite formulation.

In conclusion, we may ask: do we really understand QM? We know how to employ the QM formalism and, up to now, all the predictions have been confirmed and no limits appeared for the validity of the theory. On the other hand, some of its features are still argument of discussion. At the end of the 19th century, Heinrich Hertz remarked that "sometimes the equations of physics are more intelligent than the person who invented them" [24]. Maybe this remark could be applied also to the equations of QM.

Acknowledgement

I would like to thank Prof. Rino Di Bartolo for inviting me once more to participate in the stimulating and friendly atmosphere of his School.

References

1. Von Neumann,J.L.(1932) *Matematische Grundlagen der Quantenmechanik*, Springer-Verlag,Berlin; English trans. (1955), *Matematical Foundation of Quantum Mechanics*, Princeton University Press, Princeton, N.J.
2. Schwabl, F. (1992) *Quantum Mechanics*, Springer-Verlag, Berlin Heidelberg.
3. Feynman, R., Leighton, R.B., and Sande, M. (1963) *The Feynman Lectures on Physics*, Vol.I, Addison-Wesley, New York, p.37.
4. Einstein, A., Podolsky, B. and Rosen, N. (1935) Can quantum mechanical description of physical reality be considered complete?, *Phys. Rev.* **47**, 777-780.
5. Bohm, D. (1951) *Quantum Theory*, Prentice Hall, Englewood Cliffs, N.J.
6. Schrödinger, E. (1935) Die gegenwärtige Situation in der Quantenmechanik, *Naturwissenschaften* **23**, 807-812; English trans. (1980), *Proc. Ann. Philos. Soc.* **124**, 323-338.
7. Bohm, D. (1952) A suggested interpretation of the quantum theory in terms of hidden variables, *Phys. Rev.* **85**, 166-179 and 180-193.
8. Costa G. (2003) Entanglement and non-separability in Quantum Mechanics. in B. Di Bartolo (ed.), *Spectroscopy and Systems with Spatially Confined Structures*, Kluwer Academic Publishers, Dordrecht, pp.593-606.
9. Bell, J.S. (1964) On Einstein-Podolsky-Rosen paradox, *Physics* 1, 195-200.
10. Aspect, A., Grangier, P. and Roger, G., (1982) Experimental realization of EPRB Gedankenexperiment: a new violation of Bell's inequalities, *Phys. Rev. Letters* **49**, 91-94.
11. Aspect, A., Dalibard, J. and Roger, G. (1982) Experimental tests of Bell's inqualities using time-varying analyzers, *Phys. Rev. Letters* **49**, 1804-1807.
12. Bennet, C.H., Brassard, G., Crépeau, C., Jozsa, R., Peres, A. and Wooters, W.K. (1993) Teleporting an unknown quantum state via dual classical and

EPR channels,*Phys. Rev. Lett.* **70**, 1895-1899.
13. Bowmeester, D., Pan, J.-W., Mattle, K., Eibl, M., Weinfurter, H. and Zeilinger, A. (1997), Experimental quantum teleportation, *Nature* **390**, 575-579.
14. Boschi, D., Branca, S., De Martini, F., Hardy, L. and Poposcu, S. (1998) Experimental realisation of teleporting an unknown pure quantum state via dual classical and EPR channels, *Phys. Rev. Lett.* **80**, 1121-1125.
15. Laloë, F., (2001) Do we really understand quantum mechanics? Strange correlations, paradoxes and theorems. *Am.J. Phys.* **69**, 655-701.
16. Bassi A. and Ghirardi, G.C. (2003) Dynamical reduction model, preprint arXiv quant-ph/0302164.
17. Wigner, E.P. (1961) Remark on the mind-body question, in I.J. Good (ed.), *The Scientist Speculates*, Heinemann, London, pp. 284-302.
18. Ever III, H. (1975) Relative state formulation of quantum mechanics, *Rev. of Mod. Phys.* **29**, 454-462.
19. Ghirardi, G.C., Rimini, A. and Weber, T. (1986) Unified dynamics for microscopic and macroscopic systems, *Phys.Rev.* D**34**, 470-491.
20. Ghirardi, G.C. , Rimini, A. and Weber, T. (1987) Disentanglement of quantum wave function, *Phys.Rev.* D **36**, 3287-3289.
21. Griffith, R.B. (1984) Consistent histories and the interpretation of quantum mechanics, *J. Stat. Phys.* **36**, 219-272.
22. Omnès, R.(1988) Logical reformulation of Quantum Mchenics, I. Foundations; II. Interferences and the EPR experiments; III. Classical limit and irreversibility, *J.Stat.Phys.* **53**, 893-975.
23. Gell-Mann, M. and Hartle, J. (1993) Classical equation for quantum systems, *Phys.Rev.* D **47**, 3345-3382.
24. Hertz, H.(1896) *Miscellaneous Papers*, Vol.I, MacMillan, London, p.318.

20. MODULATION SPECTROSCOPY REVISITED

GEORGE J. GOLDSMITH
Boston College
Chestnut Hill, , MA 02467

1. Introduction

Modulation Spectroscopy is most widely applied to semiconductors although it has also been used for studies of metals and organic compounds. It involves a straightforward technique in which the derivative of the wavelength dependence of reflectance (or transmittance) or, the dependence under the influence of a periodic perturbation of the dielectric function, by stress, photo excitation, electric field excitation, magnetic field excitation, or thermal excitation. Under these conditions, in regions where optical transitions have a high probability of taking place, *e.g.*, where appropriate pairs of singularities exist exist, specific optical transitions in the band structure can be isolated from the gross spectrum through the application of lock-in techniques. Electro modulation and photo modulation are particularly useful since these perturbations disrupt the translational symmetry of the semiconductor which uncover details of the band structure and energy distribution. Wavelength , and stress (piezo) modulation do not disturb the intrinsic symmetry of the sample. Illustrated in Fig. 1 is a comparison between the total reflection spectrum and the extracted photomodulation reflection spectrum in the vicinity of a singularity near the band edge of GaAs .

Early publications on the application of modulation techniques first began to appear in the mid to late 1960's notably by B. O. Seraphin, and M. Cardona *(1)*.

Here we shall review the background and some of the essentials of the technique, and then present some of the experimental details of electromodulation.

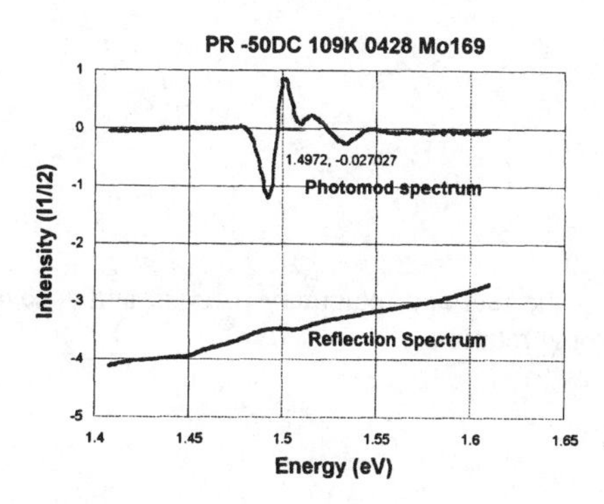

Fig. 1. Comparison of photomodulation and gross reflection spectra at 109K of gallium arsenide in the vicinity of the fundamental band gap.

B. Di Bartolo and O. Forte (eds.), Frontiers of Optical Spectroscopy, 669-685.

2. The Dielectric Function and Reflectivity

We shall examine the relationship the between the measurable parameters, reflectance and absorption, and the dielectric function, and how modulation of the dielectric function affects band-to-band transitions in semiconductors.
When a beam of photons is incident on a substance it is both attenuated and reflected. The intensity of the attenuated beam is $I = I_o e^{(-\alpha x)}$ where I is the intensity of the transmitted light, I_0, the incident intensity , α the absorption coefficient and x, the range in the medium. From the Fresnel relations, the reflectivity is $R = \left|\frac{\hat{n}-1}{\hat{n}+1}\right|^2$ where $\hat{n}$ is the frequency dependent *complex index of refraction*. The region of interest in these studies is usually in the energy range near and above the minimum energy gap. As a consequence of the strong absorption in that range, most studies employ *reflection methods* Reflectivity at normal incidence, in terms of the refractive index, ***n***, **and** the extinction coefficient, **k** , :is:

$$R = \frac{(n-1)^2 + \kappa^2}{(n+1)^2 + \kappa^2} \tag{1}$$

The extinction coefficient is related to the absorption coefficient by the relation,

$$\kappa = \frac{c\alpha}{2\pi\omega}$$, where ω is the frequency.

The reflectivity can thus be written,

$$R = \frac{(n-1)^2 + \left(c\alpha/2\pi\omega\right)^2}{(n+1)^2 + \left(c\alpha/2\pi\omega\right)^2} \tag{2}$$

The quantities, ***n*** and ***k*** are therefore related to the dielectric constant, ε:

$$n^2 - \kappa^2 = \varepsilon. \tag{3}$$

2.1 THE KRAMERS-KRONIG RELATIONS

If we define a complex dielectric constant, $\hat{\mathcal{E}} = \hat{n}^2$, the real and imaginary parts, ε_r and ε_i have an interdependent relationship known as the Kramers-Kronig relation.
The real part of the dielectric constant may be written as:

$$\varepsilon_r(\omega)-1=\frac{2}{\pi}P\int_0^\infty\frac{\omega'\varepsilon_i(\omega')}{\omega'^2-\omega^2}d\omega' \tag{4}$$

P is the Cauchy principal value of the integral: $P\int_0^\infty \equiv \lim_{a\to 0}\left(\int_0^{\omega-a}+\int_{\omega+a}^\infty\right)$

The absorption coefficient, $\alpha(\hbar\omega)$, or $\alpha(E)$, where E is the energy of the photon Eq. (5) can be written in terms of the extinction coefficient, **κ**, and the real part of the refractive index, **n:**

$$n(E)-1=\frac{2}{\pi}P\int_0^\infty\frac{E'k(E)}{E'^2-E^2}dE' \tag{5}$$

and since $k(E')=\frac{hc\alpha}{4\pi E'}$, in terms of the absorption coefficient becomes

$$n(E)-1=\frac{ch}{2\pi^2}P\int_0^\infty\frac{\alpha(E')}{E'^2-E^2}dE'. \tag{6}$$

On integration, this becomes

$$n(E)-1=\frac{ch}{2\pi^2}P\int_0^\infty ln\frac{1}{E'^2-E^2}\frac{d\alpha(E')}{dE'}dE' \tag{7}$$

The absorption coefficient, $\alpha(E')$, depends on the joint density of states and becomes large where $\nabla_k\left[E_c(k)-E_v(k)\right]=0$

E_c and E_v are, respectively, the conduction band and valence band energies at that point in the Brillouin zone.. These sites are called *critical points* or *van Hove singularities*. In three-dimensional space there are four kinds of van Hove singularities (fig. 2.1) *(3)*

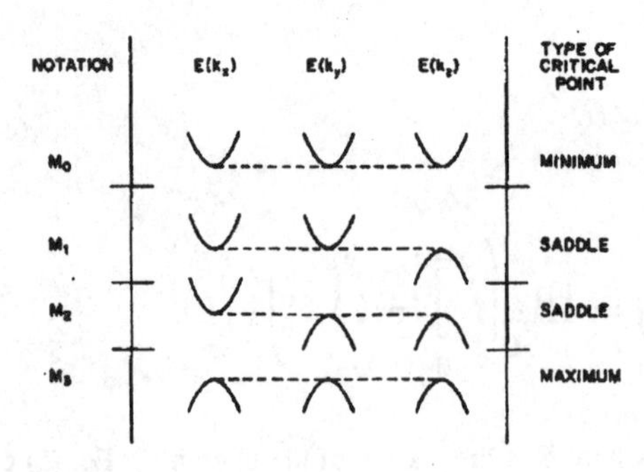

Fig. 2.1 Critical Points for which $\nabla_k E(k) = 0$

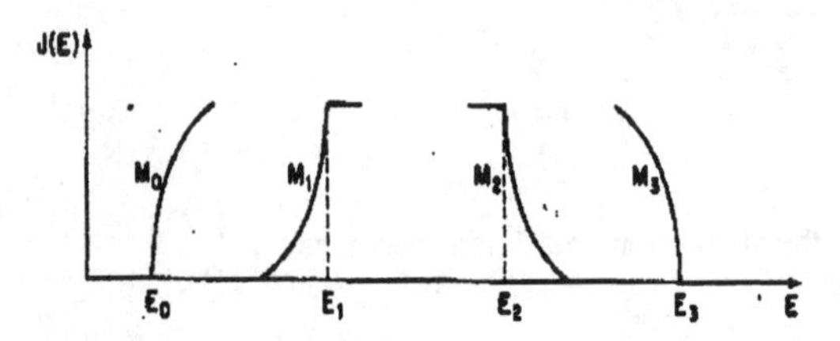

Fig. 2.2 Optical Density of states for the four main types of critical points

Given that α(E') is dependent on the joint density of states, in the vicinity of critical points the logarithmic term in Eq (8) is large and the derivative term, $\frac{d\alpha(E')}{dE'}$ becomes dominant. Thus, because of the joint density of states functional dependence on energy for each of the four van Hove singularities (Fig. 2.2) *(3)*, $\frac{d\alpha(E')}{dE'}$ is large and positive above E_0 and below E_1 and large and negative above E_2 below E_3. Therefore **n** (see Eq 8) will go through a maximum at E_0 and E_1 and a minimum at E_2 and E_3. Further, it can be shown that *direct* transitions between parabolic bands across an energy gap, E_g yield a value of $\alpha(E)$ proportional to $(E—E_g)^{1/2}$. The derivative, $\frac{d\alpha(E)}{dE}$, is proportional to $(E—E_g)^{-1/2}$ which exhibits a singularity at E_g. For forbidden transitions $\alpha(E)$ is proportional to $(E—E_g)^{3/2}$ and for indirect transitions (E) is proportional to $(E—E_g)^2$, and $\frac{d\alpha(E)}{dE}$ goes gradually to zero at the critical point.

Thus the refractive index, **n,** exhibits a structure whenever the derivative of the absorption coefficient with respect to **ω** goes through a maximum or a minimum. This behavior is superimposed on the gross absorption spectrum (or reflection spectrum). If now a periodic perturbation such as an electric field is applied, it produces a corresponding variation in the refractive index (Eq. 7):

$$n(E)-1=\frac{ch}{\pi}P\int_0^{\infty}\frac{\alpha(E',0)}{E'^2-E^2}dE'+\frac{ch}{\pi}P\int_0^{\infty}\frac{\alpha(E',E)}{E'^2-E^2}dE' \qquad (8)$$

so that the change is: $$\Delta n(E,E)=\frac{\mathrm{ch}}{\pi}\int_0^{\infty}\frac{\Delta\alpha(E',E)E}{E'^2-E^2}dE'. \qquad (9)$$

The band structure of silicon at room temperature, and two electromodulation spectra, one at 88K and the other at 294K are shown in Fig. 3. They correspond to a transition along the **Γ** point in the band structure between a pair if singularities at $\Gamma_{25\infty}$ and what appears to be $\Gamma_{2\infty}$. They are separated by ~3.38 eV, at room temperature, The minima in the Fig. 3 spectra locate a band gap wh;ch, at room temperature, corresponds to the separation of the singularities.
Hamakawa, et. al. *(4)* have calculated the line shape of the dielectric constant, ε_1 and ε_2 for the four types of critical points with and without electric field (fig. 4). The quantities, **Δn** and **Δα** would show the same functional dependence.

Seraphin *(5)* extracted a useful relation on applying the Kramers-Kronig relation to the change of the dielectric function with energy, in terms of the change in the values of the refractive index and the extinction coefficient,

$$\frac{\Delta R}{R}=A\frac{ch}{\pi}\int_0^{\infty}\frac{\Delta\alpha(E')dE'}{E'^2-E^2}+\frac{ch}{2E}B\Delta\alpha(E) \qquad (10)$$

where

$$A=\frac{n^2-k^2-1}{[(n+1)^2+k^2][(n-1)^2+k^2]}$$

$$B=\frac{2nk}{[(n+1)^2+k^2][(n-1)^2+k^2]}$$

A and **B** are known as the "Seraphim Coefficients".

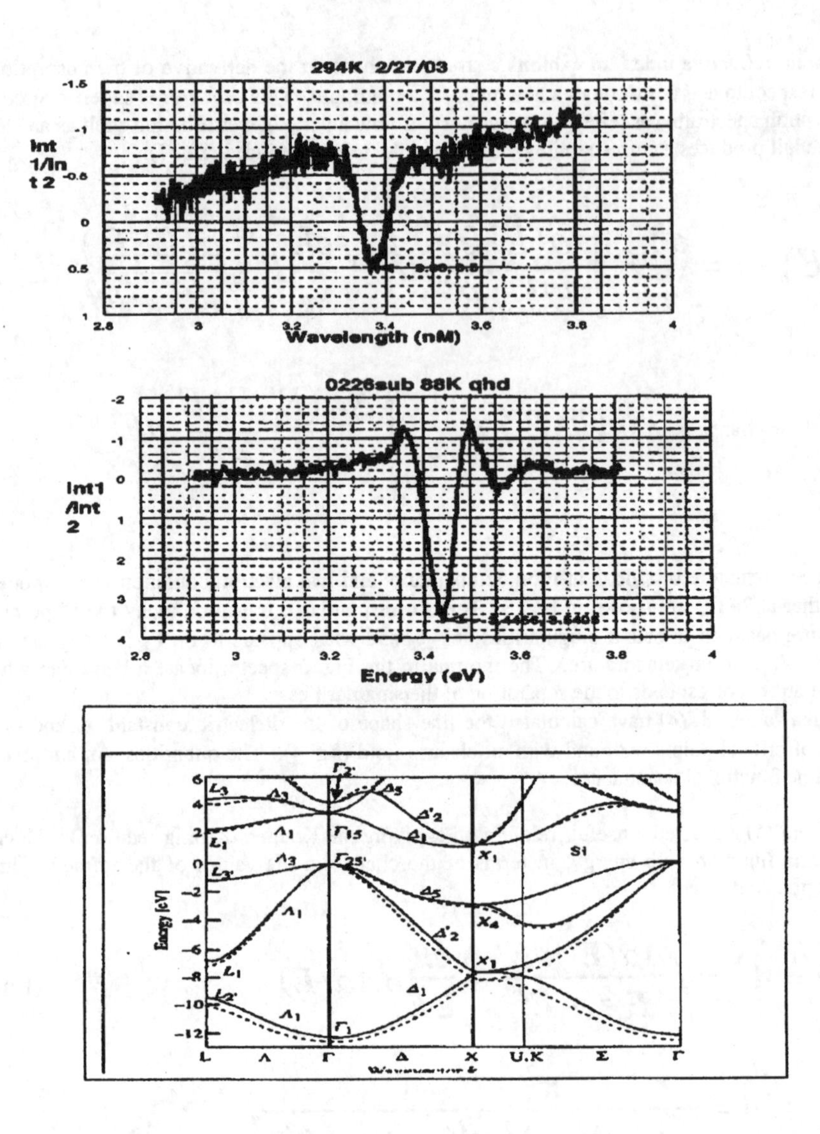

Fig. 3 Band Structure and 294K and 88K EM spectra at □ point

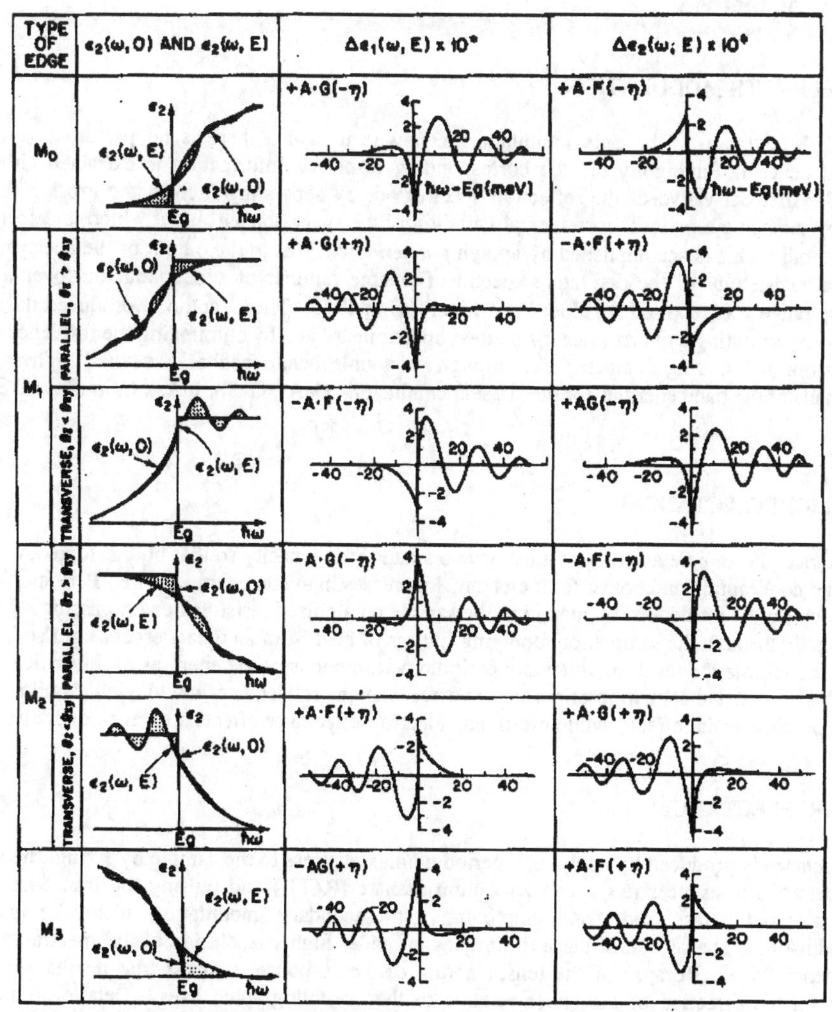

*LINE SHAPES OF $\Delta\epsilon_1(\omega, E)$ AND $\Delta\epsilon_2(\omega, E)$ CALCULATED AT THE CONDITION: $\hbar\theta = 10$ meV, $E_g = 0.8$ eV AND $B = 1$, HERE $\eta = (\hbar\omega - E_g)/\hbar\theta$, $A = (B - \theta^{1/2})/\omega^2$

Fig. 4 Field induced changes in the dielectric constant at the various types edges *(4)*

3. Modulation Methods

3.1 WAVELENGTH MODULATION:

Wavelength Modulation is the only modulation technique in which there is no perturbation of the sample and hence no ambiguity in the corresponding response. Information is extracted either by forming the first derivative of the reflection spectrum or by separating it from the gross reflection spectrum by imposing a periodic incremental variation of the spectrometer output which is detected by lock-in methods. This is accomplished by applying a periodic $\Delta\lambda/\Delta t$ to the output of the spectrometer. A very thorough study of the derivative spectra of a large number of semiconductors over a wide temperature range was reported by Zucca and Shen *(6)* in 1979. These authors produced the $\Delta\lambda/\Delta t$ perturbation by vibrating a quartz plate inside their spectrometer and by eliminating the influence of the lamp spectrum and grating characteristics through a double beam feedback system. Their study yielded details of the band energies for the six semiconductors, GaAs. GaSb, InAs,, InSb, Si, and Si.

3.2 THERMOREFLECTANCE

In this instance, the sample and mount must have a small heat capacity so that both sample and heater can heat and cool rapidly and not suffer incremental increases in average temperature. Periodic thermal pulses, which can be introduced by mounting the sample on a pulsed resistive heater, passing a pulse of current directly through the sample, or supplying pulses of heat from an incandescent light source. are applied to the sample the result of which are periodic perturbations of the energies of the critical points It would be expected that thermoreflectance and wavelength reflectance would yield similar results since both produce scalar effects, while piezo- and electro- reflectance effects are tensor quantities.

3.3 PIEZOREFLECTANCE

Piezoreflectance is produced by applying a periodic uniaxial stress to the sample by fixing the sample to a piezoelectric transducer (e.g., lead zirconium titanate [PZT]) and pulsing the transducer or by applying a fixed stress and then employing a secondary modulation technique such as electromodulation. The second of these techniques provides higher resolution of the spectral features. This is accomplished. Because of the tensor nature of the response to stress, the results are highly dependent on the direction of the stress relative to the crystallographic axes. Details of the band structure such as the nature of excitonic states can be elucidated.

3.4 ELECTROMODULATION

Of the several modulation spectra systems electromodulation, while it is sometimes difficult to interpret, yields the sharpest spectra . Of the several different possible configurations we will discuss only electrolyte electroreflectance (EER) and Schottky Barrier electroreflectance (SBR) and photoreflectance (PR). During the decades of the 80's and the 90's the technology of GaAs synthesis improved. This II-VI alloy provides much improve electron mobility, a direct minimum gap, and structural compatibility with other similar II-VI materials such as GaAlAs. These developments along with the introduction of Molecular Beam Epitaxi (MBE) and Chemical Vapor Deposition (CVD) techniques for fabrication of precisely structured materials gave a large impetus to much improved device performance. Electromodulation is one of the most useful techniques for characterizing these new structures.

3.4.1 *Electrolyte Electroreflectance*

In this technique, the sample is immersed in a weak electrolyte along with an inert electrode. In one configuration the sample is cemented to a microscope slide using silver conducting paint, A thin line of silver paint is continued to the top of the slide where a contact wire is attached. The entire face of the slide excepting the front of the sample is coated either with paraffin or a solution of an inert plastic ("Microstop")(Fig. 5) The electrolyte may be either a dilute aqueous or ethylene glycol .solution of some electrolyte such as KOH. A periodic voltage pulse is applied between the sample and the inert electrode forming a Helmholtz double layer at the face of the sample. The spectrum is measured by directing the output of a monochromator at the face of the sample and scanning through an appropriate wavelength range while measuring the reflected light intensity (**R**) and the output of the lock-in amplifier (**ΔR**) to arrive at **ΔR/R(λ)**.

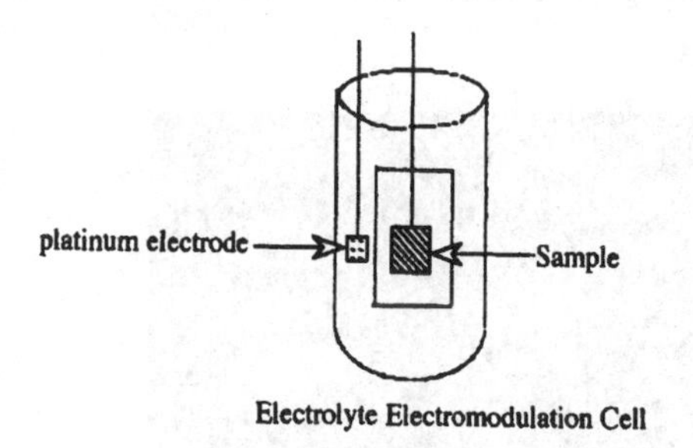

Fig. 5. Electrolyte Electromodulation Cell

3.4.2 *Schottky Barrier Electroreflectance (SBR)*

A suitable Schottky barrier may be deposited on the polished surface of a semiconductor by evaporating a transparent layer of a metal which has a work function greater than that of the semiconductor (e. g., ~5 nM of Au). An ohmic contact is applied to the back surface (e. g. In-Ga eutectic). Modulation is accomplished by applying a square wave to the front contact and a dc bias if required by the particular study. Probe excitation is applied to the transparent front electrode. A sample mounted on the cold finger of the cryostat is shown in fig. 6.

3.4.3 *Photoreflectance (PR):*

Photoreflectance is a variation on electroreflectance in which the electric field is generated by pulses of strongly absorbed monochromatic light to the front of the sample. The electric field is created by the optical separation of electron-hole pairs at the surface of the sample and their subsequent drift caused by trapped surface charges formed by pinning of the Fermi level at the semiconductor-air interface.

This technique requires no special preparation of the reflecting surface of the sample unless a dc bias is required in which case a transparent metal front electrode is required.

3.4.4 *Experimental Details*

The experimental apparatus (Fig. 8) consists of a light source (either quartz-halogen 100 watt incandescent or a 75 watt Xe discharge depending on the wavelength range of interest). The lamp illuminates a monochromator, the wavelength drive of which is activated by a stepping motor. The sample is mounted in the vacuum space of a temperature controlled cryostat. The detector of the light reflected from the sample is either a Si or GaAsP photodiode the output of which (**R**) is fed to a lock-in amplifier which provides the **ΔR** signal. Modulation is provided either by ahe square wave output of a function generator of, in the case of photomodulation, by the chopped beam from the modulating source.

The collected data are processed by a LabVIEW program. The program collects data over the selected wavelength range displaying **ΔR /R** in real time as a function of the number of data points. At the end of the selected run the data are displayed as a function of wavelength. The following settings can be entered: beginning wavelength, ending wavelength, the delay before recording a reading (relevant to the time constant of the lock-in amplifier), the wavelength step to track the wavelength scan of the monochromator, number of steps per reading. On completion of the wavelength range,

Fig. 6 Electromodulation sample and mount

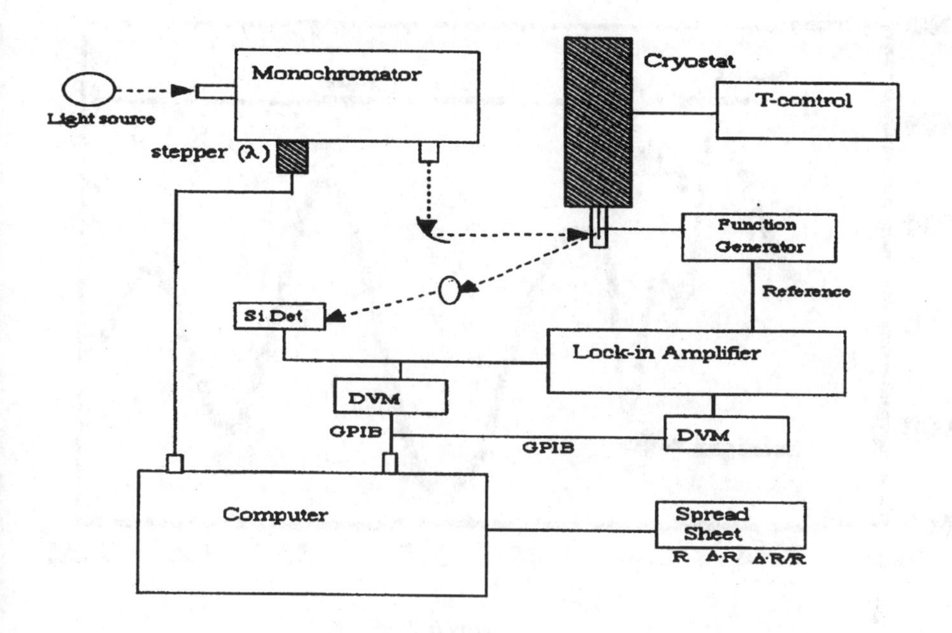

Details of Block Diagram (*Electromodulation*):

Monochromator: : McPherson (model 218) 0.3 M Plane grating scanning monochromator.
1–range 105–1000 nM. Built–in adjustable wavelength drive, fitted with auxiliary stepper drive.
Cryostat: Janis custom cryostat. ~0..5 liter coolant capacity.
Temperature control sensor, Si diode,sample T—range, 85–300K
(as measured by Cu–Constantan thermocouple on sample mount).
Temp. control: Lake Shore T- Controller
Function Gen.: Laboratory instrument (provides modulating pulse train and reference source.)
Lock-in amp: Signal Recovery DSP
DVM: Kiethley 197A multimeter with GPIB capability
Detector: UV enhanced Si photodiode
Computer: MacIntosh PowerMac

Fig.7 Block diagram of modulation spectroscopy experiment

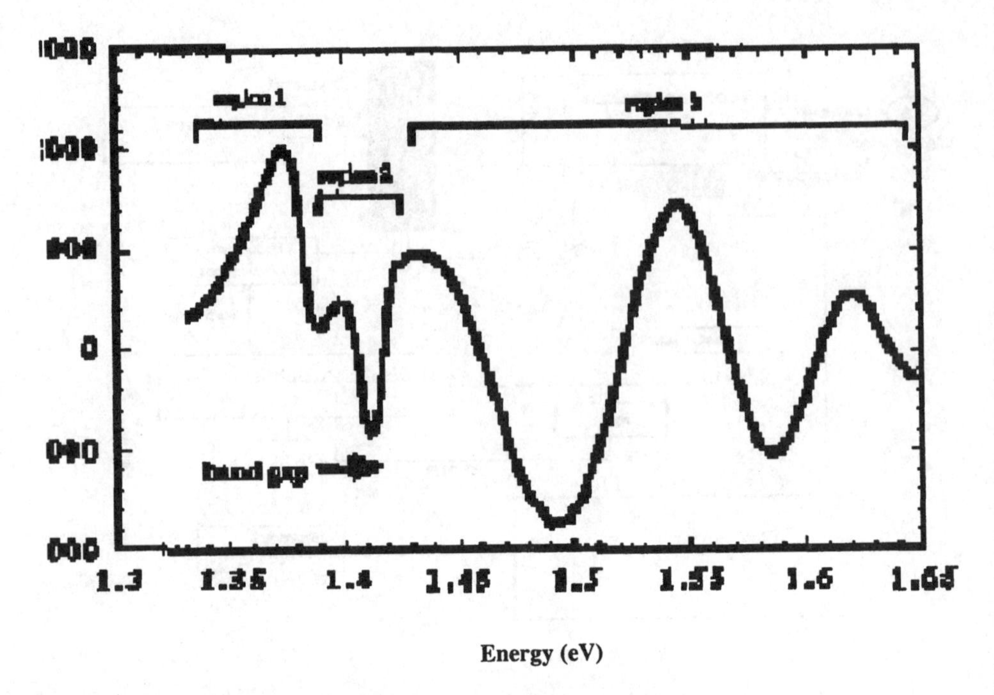

Fig. 8 Electromodulation spectrum of epitaxial GaAs

the data are recorded graphically on the front panel as **ΔR/R** as a function of wavelength. All the significant parameters are preserved and stored independently. The reading number, the wavelength and the corresponding energy in eV, the value of **R, ΔR**, and the ratio **ΔR/R** are all stored independently*.

4. Some Applications

4.1 SINGLE HOMOJUNCTION

Gallium arsenide device fabrication is frequently carried out ON a substrate platform of semi-insulating GaAs (i.e., doped in such a way as to compensate the free carrier concentration) onto which is deposited by CVD or MBE a thin layer of the material in which the desired device is fabricated. Such "homojunction" devices present unique problems of field distribution and optoelectronic behavior. H. Poras *(7)* carried out a comprehensive study of a series of such GaAs structures using several different variations on photo– and electro–modulation techniques._ The basic electromodulation spectrum is shown in Fig.8.

---

* Photomodulation. The same basic block diagram is applicable to the photomodulation application except for the following modifications: The function generator is eliminated. to be substituted by dc bias if it is required. Modulation is accomplished by directing the chopped beam of a laser onto the front surface of the sample, while providing a reference signal from the chopper control to the lock-in amplifier.

4.1. 1 *Energy range below the band gap (Region 1)*

It will be noted that there is considerable structure at energies both above and below the bandgap. Poras demonstrated that the source of the below bandgap structure was the presence of carbon impurities in the epitaxial layer originating from the trimethyl gallium source material in the CVD process. Because these structures are observed in an energy range where GaAs is somewhat transparent, the spectral phenomena are the result of photoabsorption processes and reflections at the back surface and at the epitaxial layer–interface.

4,1,2 *Energy Range at the band gap (Region 2)*

The structure sclose to the bandgap arises from field inhomogeneities at the interface caused by inhomogeneous distribution of impurities near the surface. It was also possible to arrive at a very precise value for the room temperature minimum band gap

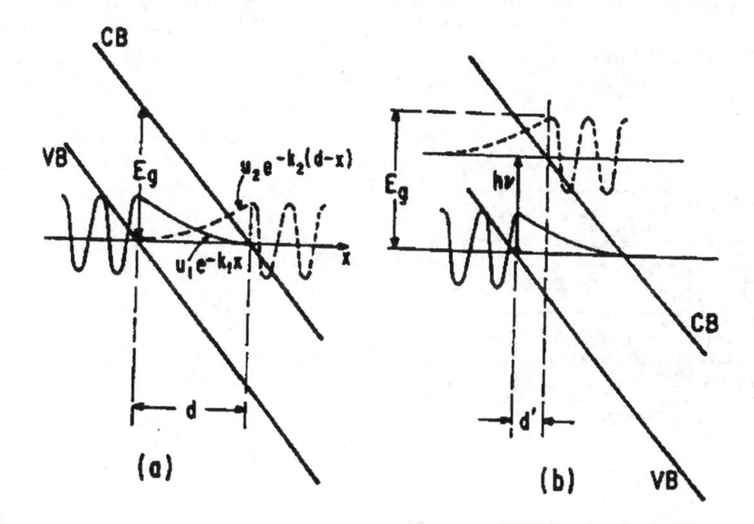

Fig. 9. Electron tunneling, (a) in the dark, (b) with incident photons

4.1.3 *Region above the bandgap (Region 3)*

The oscillations on the high energy side if the band gap are caused by the
Franz–Keldysh effect resulting from t acceleration of optically excited electrons across the tilted boundaries by the electric field (Fig. 9).

4.2 FRANZ-KELDYSH OSCILLATIONS

Indexing of the location of the Franz–Keldysh oscillations (FKO) (, Fig. 10) *(8)* will yield the value of the surface field and the carrier concentration. The position of thenth extremum in the FKO oscillation is given by the relation, $n\pi = \frac{4}{3}\frac{(E_n - E_g)^{\frac{3}{2}}}{(\hbar\theta)^{\frac{3}{2}}} + C$.A plot of $\frac{4}{3\pi}(E_n - E_g)^{\frac{3}{2}}$ *vs.* the index number, **n**, is linear with a slope of $(\hbar\theta)^{\frac{-3}{2}}$. the surface field can then be calculated from the relation, $(\hbar\theta)^3 = \frac{q^2\hbar^2E^2}{2m_\|}$. $(\hbar\theta)$ is the "elecro-optic energy" and $m_\|$, the effective mass of the electron.

Finally the carrier concentration is calculated from the relation $E_s{}^2 \infty(\frac{2q}{\kappa\varepsilon_o}N_D)V_{bias}$. In this instance the surface field was 0.516 V/μV.

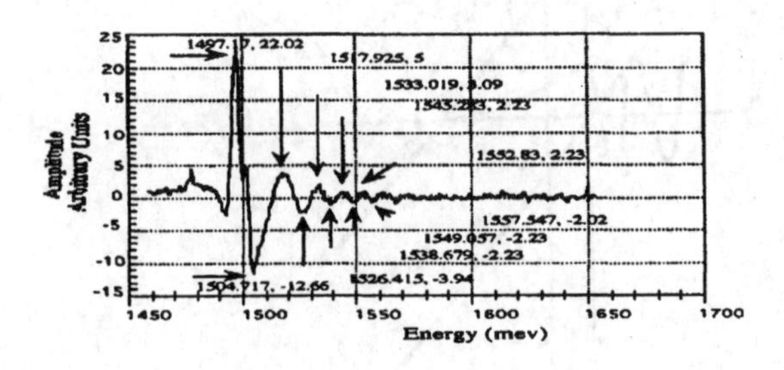

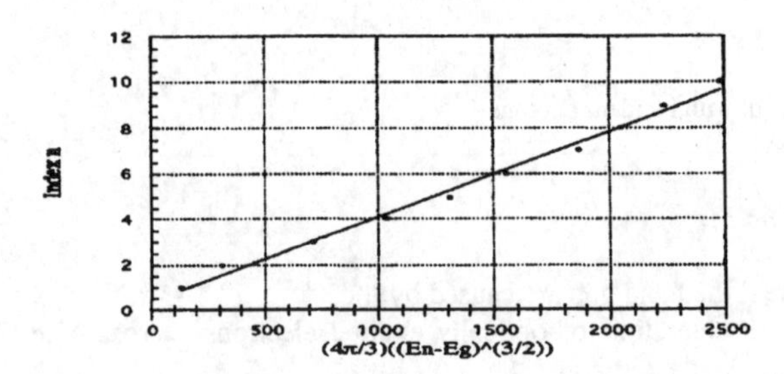

Fig.10 F-K oscillations in GaAs at 100K; (lower) $(\frac{4\pi}{3})(E_n - E_g)^{\frac{3}{2}}$ *vs.* index number

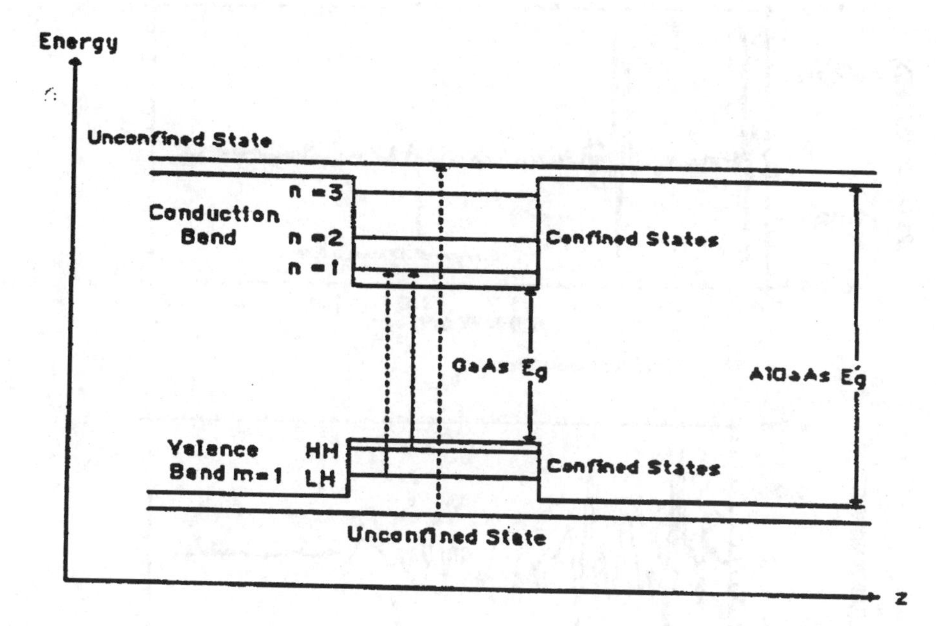

Fig. 11. Structural composition and energy band diagram of a superlattice

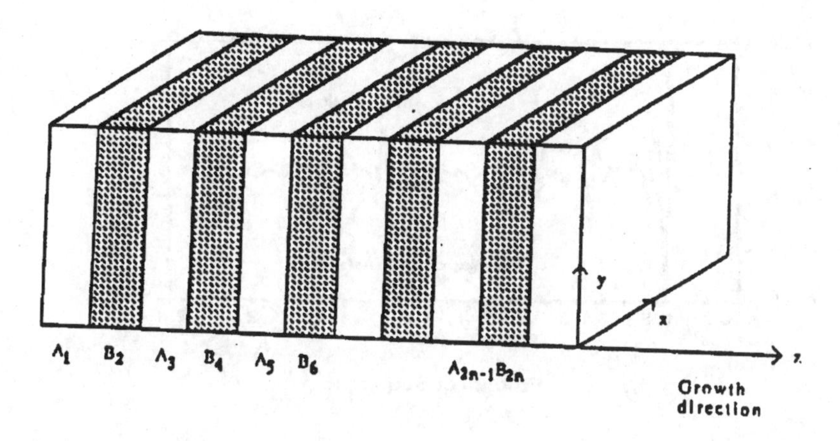

Fig. 12 Multiple quantum well or suuperlattice

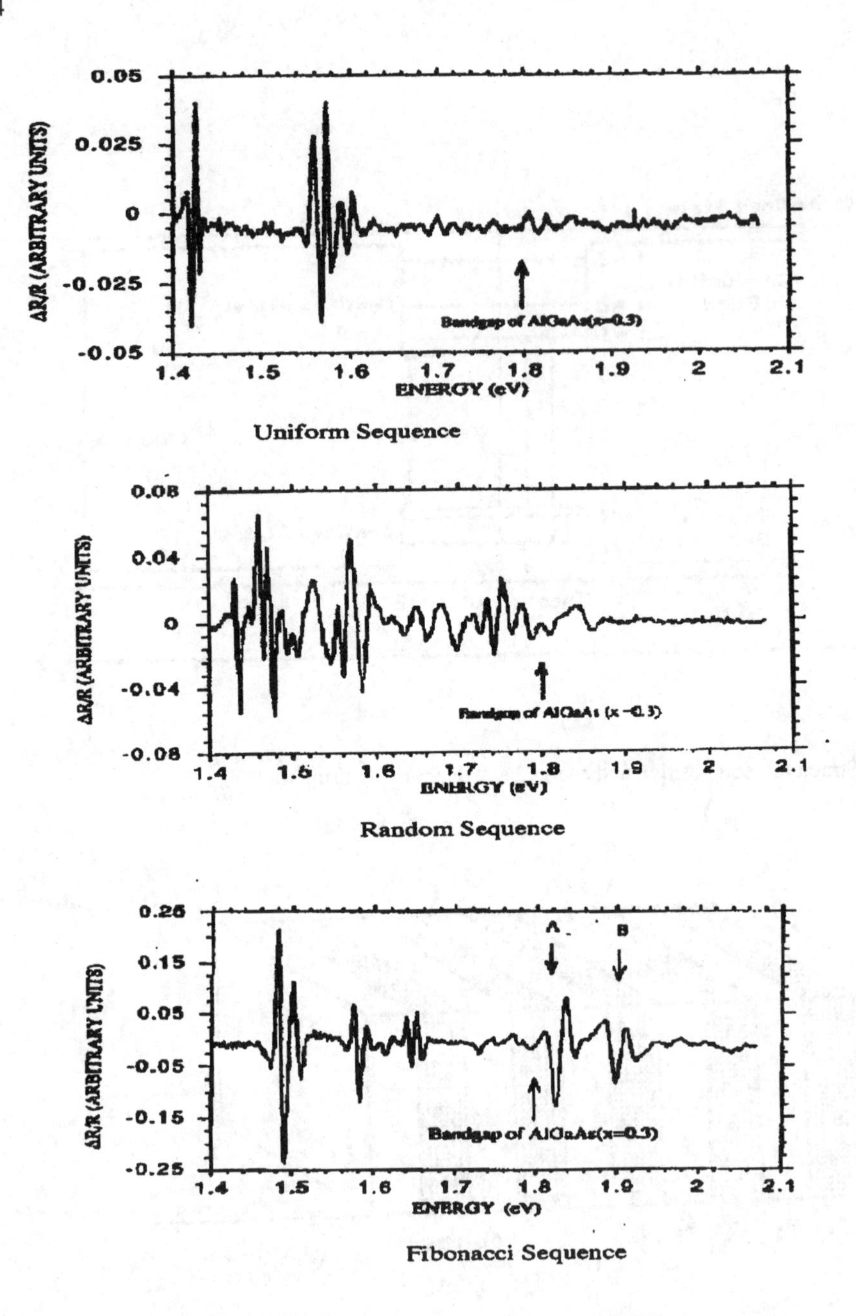

Fig. 13 Photomodulation spectra of three differently composed superlatt

References

1. B. O. Seraphin, *Proceedings of the International Conference on the Physics of Semiconductors, Paris, 1964*, Dunod Cie (1964); M. Cardona, *Modulation Spectroscopy:* F. Seitz, D. Turnbull, and H. Ehrenreich, Academic Press (1969);

See also: F. Pollak,, H. H. Shen, *Modulation spectroscopy of semiconductors: Bulk/thin films microstructures, surfaces,/interfaces, and devices*. Mater. Sci. Eng**R**10, 275 91993)

2. Jacques I. Pankove, *Optical Processes in Semiconductors*, p.393, Dover Publications, New York (1971)

3· L. van Hove: Phys. Rev. **89,** 1189 (1953).

4. Y. Hamakawa, P. Handler, and F. A. Germano, *Phys. Rev.* **167,**709 (19680

5. B. O. Seraphim, *Electroreflectance*, Semiconductors and Semimetals, eds. R. K. Willardson and A. C. Beer, Academic Press, New York pp. 10-17 (1972).

See also: D. Aspnes: Modulation spectroscopy/electric field effects on the dielectric function of semiconductors. in *Handbook of Semiconductor*ed. by M. Balkanski (North Holland, Amsterdam 1980) vol. 2, pp. 115-116.

6. R. R. Zucca and Y. Shen; *Wavelength Modulation Spectra of Some Semiconductors*, Phys. Rev. B, **1,** 2668, (1970).

7. Henry Poras, PhD Thesis, *Boston College*, (1993)

8. Danni Liu, MS Thesis, *Boston College,* (1996)

9. Alan Tai, PhD Thesis, *Boston College,* (1991)

15. ADVANCES IN SOLID STATE LASERS AT NASA LANGLEY RESEARCH CENTER

JAMES BARNES
NASA Langley Research Center - MS 430
Hampton, VA 23681-0001, USA

Abstract

An overview of NASA's vision for 21st Century using space-based solid-state lasers for Earth observations was presented. Current and future usage of lasers for observing Earth processes, such as climate systems caused by man-induced forces upon the environment were highlighted. For some projects mission details and measurements systems using lasers were discussed, covering also the laser systems development processes.

NASA international collaborations with missions such as CALIPSO and ICESAT were stressed to invigorate thoughts from others regarding the benefits of joint missions.

B. Di Bartolo and O. Forte (eds.), Frontiers of Optical Spectroscopy, 687.

18. COMBINATORIAL CHEMISTRY TO GROW SINGLE CRYSTALS AND ANALYSIS OF CONCENTRATION QUENCHING PROCESS: APPLICATION TO Yb^{3+}-DOPED LASER CRYSTALS.

G.BOULON
Physical Chemistry of Luminescent Materials,
Claude Bernard / Lyon1 University,
CNRS UMR 5620, Bat. A.Kastler, 10 rue Ampère,
69622 Villeurbanne Cedex,
France
e-mail: georges.boulon@pcml.univ-lyon1.fr

Abstract

LHPG (Laser Heated Pedestal Growth) technique which is suitable to grow crystalline fibres has been successfully applied for the study of a lot of laser crystals. Our general approach on the research of diode-pumped Yb^{3+}-doped host crystals is presented through three typical examples of oxides and fluorides having cubic structure which are considered among important crystals either for basic or applied reasons. Spectroscopic characterizations were carried out. Especially, Yb^{3+} $^2F_{5/2}$ excited level experimental decay time dependence on Yb^{3+} concentration is analyzed by using our own approach on the synthesis of a concentration gradient fibre developed in the Laboratory. Our main objective is to contribute to have a better understanding of concentration quenching mechanisms in laser crystals and more generally in luminescent materials.

1.Introduction

Within the past decade, the development of high-power InGaAs laser diodes used as pump sources between 900 and 980 nm has lead to a strong interest in Yb^{3+}-doped materials like crystals, bulk glasses and more recently glassy and crystalline fibers which are emitting around 1030 nm. Relative to their Nd^{3+}-doped counterparts, Yb^{3+}-doped materials are much more likely to yield high efficiencies at high powers. Indeed, the simple electronic structure of Yb^{3+} ions implies *de facto* the absence of parasitic effects such as excited-state absorption or up-conversion and makes high doping rates achievable in most host matrices. In addition, the small quantum defect between absorption and emission wavelengths, 11% relative to 30-40% in Nd^{3+}-doped laser hosts, contributes to weak thermal effects so that Yb^{3+}-doped materials have hence turned out to be relevant for efficient high power continuous-wave lasers, up to the kilowatt class. Another advantage of Yb^{3+}-doped materials is to bring new advances in diode-pumped ultra-short sources in the femtosecond scale of time by playing with the Yb^{3+} wide fluorescence spectrum. The understanding and the optimization of the optical properties of Yb^{3+} –doped materials is then necessary. Among different parameters which are influencing the emission properties, Yb^{3+} concentration dependence is probably the most important since it leads to optimize laser materials. This is a huge task due to multiple influences in the optical materials. Finally, since the use of luminescent materials for phosphors, amplifiers, lasers, scintillators and detectors the detailed study of so-called concentration quenching processes has not

B. Di Bartolo and O. Forte (eds.), Frontiers of Optical Spectroscopy, 689-714.

been too much understood. Our objective is to bring a new look on such processes by taking benefit of the Yb^{3+} simple electronic structure in the near infra-red in several well characterized Yb^{3+} -doped laser crystals due to recent applications. Indeed, such investigations may be considered like something playing with extreme physical and chemical conditions, topics of this Summer School. Up to now, several oxides were deeply investigated leading to general methods of evaluation [1,2].

Among such hosts, we are dealing with the main families of laser isotropic cubic crystals as oxides ($Y_3Al_5O_{12}$ or YAG garnet, Y2O3 yttria sesquioxide) and fluoride (CaF_2). They are considered to be among the best laser crystals due to unusual combination of following favourable properties:
-optical :large transparency and low phonon energy which is in favor of high radiative transition probabilities between electronic levels, especially in Y_2O_3 and CaF2 with the dominant phonon energies of 380cm^{-1} and 321cm^{-1} respectively.
-crystallographic structure: both of them have cubic structure and then are isotropic,
-thermal: refractory crystals characterized by high melting points (1850°C for YAG, 2380°C for Yttria and 1360°C for CaF_2). This is obvious that the high melting point of yttria is an obstacle for classical crystal growth techniques as the growth by the Czochralski method. However, the Bridgman process has been performed in expensive rhenium crucible. Un-doped CaF_2 crystals are now growing with large diameter mainly to be used as window materials for excimer lasers which are needed in the semiconductor technology as optical lithography.
-the highest thermal conductivities of oxides, excepted sapphire: 12.8 W m^{-1} K^{-1} in undoped-Y2O3, 7.4 W m^{-1} K^{-1} in 3%Yb-doped Y_2O_3; 9.8 W m^{-1} K^{-1} in undoped-YAG and 6.7 W m^{-1} K^{-1} in 3%Yb-doped YAG; 10 W m^{-1} K^{-1} in undoped CaF2.
-mechanical (robust).
The first step of this lecture will be to show *a new combinatorial chemistry*, original method for the synthesis of sample having a continuous longitudinal concentration gradient for two extreme conditions, from the lowest to the highest Yb^{3+} concentrations which has been recently developed [3] allowing the obtaining of crystalline samples doped in an extremely wide range of concentration. Based on the Laser Heated Pedestal Growth method (LHPG), which does not need the use a crucible, it allows the study of many types of crystals, especially new class of high melting temperature laser crystals like garnets and sesquioxides [4]. These advances in crystal growth are ideally suited for the proposed study of quenching processes.

The second step will be to *analyze quenching processes* in Yb^{3+}-doped crystals depending of the extreme chemical conditions of the host purity [4]. Gradient concentration fibre is an unique tool to make the correlation between Yb^{3+} concentration and lifetime measurements which have been measured *in situ* in the same sample in relation to the distance from the top of the crystallized rod.This fast and simple combinatorial method allows:
-to measure the Yb^{3+} intrinsic radiative lifetime which is very important to measure stimulated emission cross-section,
-the influence of radiation trapping by the presence of the resonant $^2F_{5/2} \leftrightarrow {}^2F_{7/2}$ transition in Yb^{3+} near 980 nm,
-the detection of Yb^{3+} pairs detecting visible emission spectrum by up-conversion mechanism under Yb^{3+} IR pumping which is controlled by convoluting the IR emission spectrum,
-at last, the nonradiative energy transfer to uncontrolled impurities in doped hosts giving a notable contribution to the quenching processes.

All these quenching mechanisms are useful for estimate potential development as optical materials.

2. Fibre crystal growth

2.1. PULLING FIBRE USING FLOATING ZONE BY THE LASER HEATED PEDESTAL GROWTH (LHPG) METHOD

Poplawski [5] is the first one who has initiated crystal growth using a pedestal growth design process based on melting materials by the energy created by an image furnace. Then Haggerty et al [6] have developed the LHPG technique and has been improved by Feigelson [7] at the Stanford University. The LHPG technique seen in "Figure1." has been installed at the Claude Bernard/Lyon1 University recently[8], is based on the utilization of a CO_2 laser beam at 10.6 μm wavelength which is focused on the end of a source rod (typically 0.5 to 1mm in diameter) containing the desired host and dopant materials, by means of circularly symmetric laser optics producing a homogeneous circular distribution of radiation on the rod. The source rod materials can be used from oriented fibre single crystal or polycrystalline reacted materials prepared by solid state reaction. A seed crystal, once dipped into the molten zone, is withdrawn at some rate faster than the source material is fed in. By conservation of melt volume, this leads to the crystalline fibre growing at some constant fraction of the source rod diameter. Pictures shown in "Figure 2." are illustrating the types of samples which can be grown.

2.2. CONCENTRATION GRADIENT FIBRE

The study of a laser system requires the measurement of several variables and particularly the dopant content effect, needing the growth of quantities of samples. LHPG method allows to obtain a continuous variation in composition and a single rod constitutes a "library" involving an infinite number of samples in a selected range of composition. With such a technique a "concentration gradient fibre" where composition changes continuously from one end to the other between two well-defined compositions C_A and C_B is investigated [3]. In first, the gradient is created in the feed ceramic rod as illustrated in "Figure 3(a).". So during the melting zone process the amounts of the solubilized species vary in the liquid phase. Then with high freezing rate, the crystallized solid keep the variation of concentration. Each point of the sample can be considered as a single crystal where composition and physical properties like fluorescence lifetime can be correlated by means of in situ measurements.

Such an example, "Figure 3(b)." shows the ytterbium concentration gradient between 0 to 15 % obtained in Yb^{3+}-doped CaF_2 crystalline fibre (pulling rate : 30 mm/h). This "combinatorial chemistry method" has been successfully applied to the study of Yb^{3+}-Er^{3+} co-doped and Yb^{3+} mono-doped sesquioxides [9-11] and YAG [4]. GGG garnet and other fluorides are under progress in our team [12].

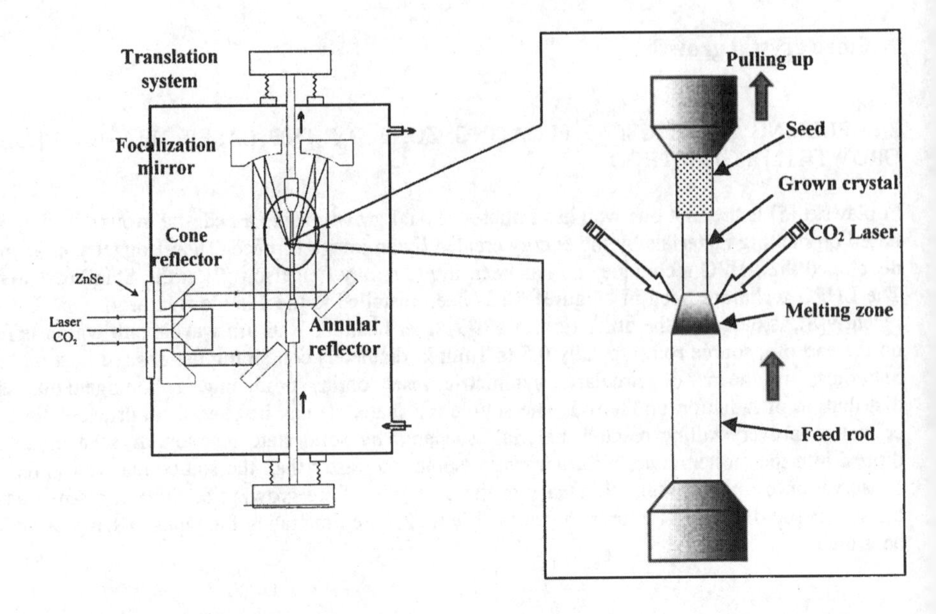

Figure 1. Schematic diagram of the laser optics used in the LHPG technique

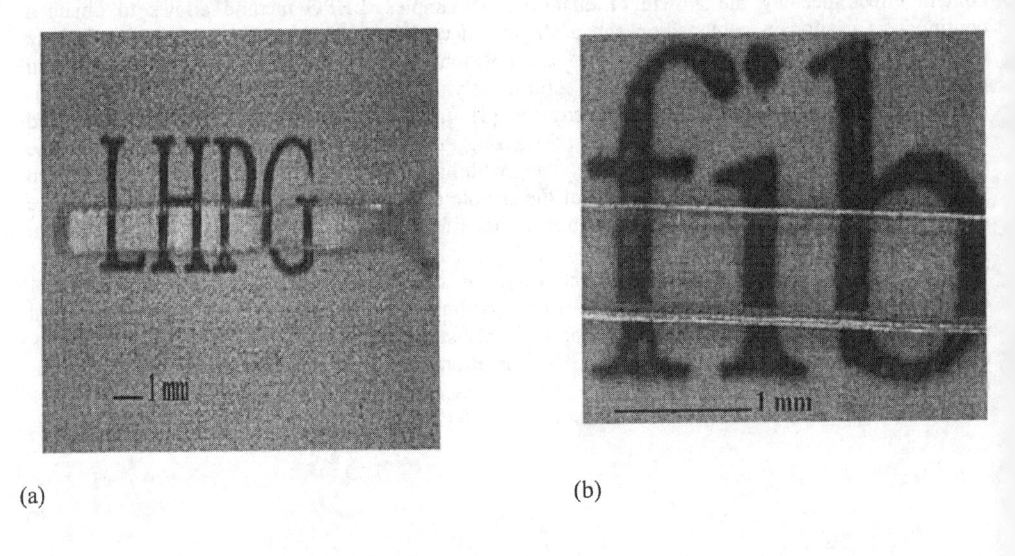

(a) (b)

Figure 2. (a)Photograph of a YAG fiber grown by the LHPG method and (b)Microscopic view of the LHPG grown YAG fiber.

3. Illustration of our approach for Yb^{3+}-doped crystals

3.1. Yb $^{3+}$ SPECTROSCOPY IN INORGANIC CRYSTALS

The Yb^{3+} ion has been selected since high average power lasers have gained much attentions because of the availability of IR high power laser diodes. This topic is under study in our group since few years in the framework of the CNRS Research Group (GDR 1148 on Laser Materials) as can be seen in the recently published issue of Optical Materials [13]. Yb^{3+} activator ion possesses many advantages because of its simple electronic structure as can be seen in "Figure 4.".

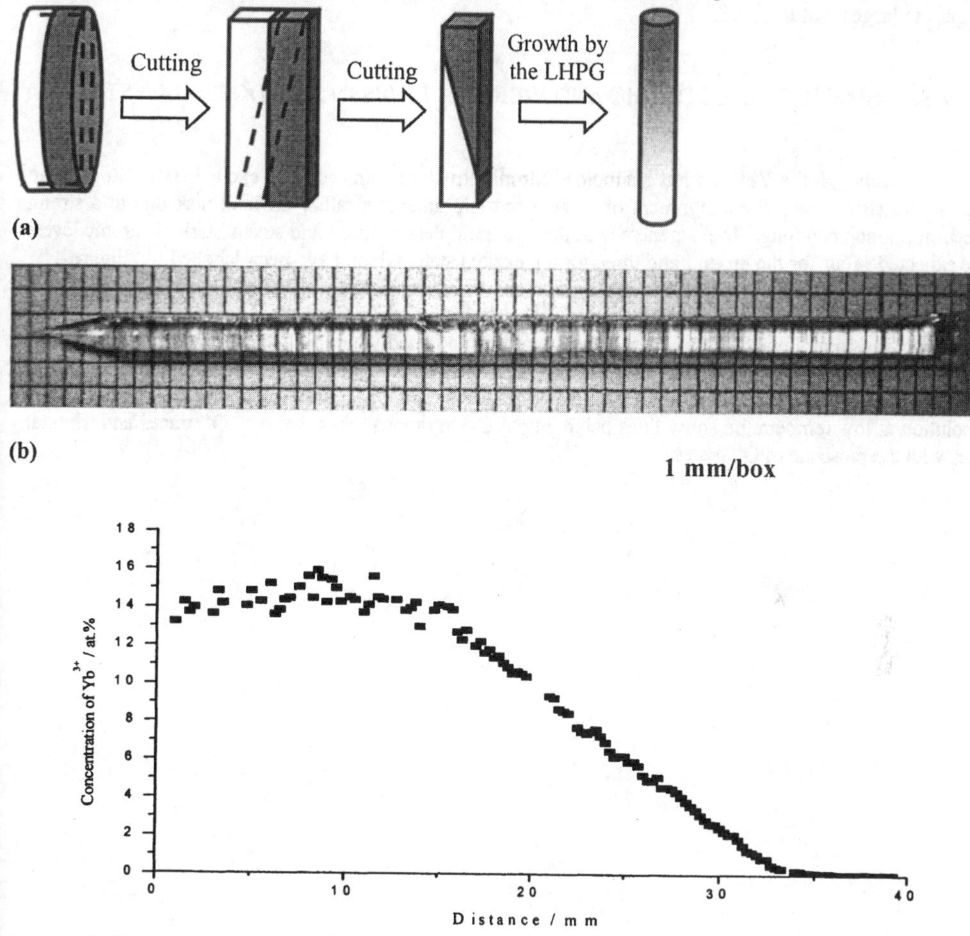

Figure 3. (a) Scheme of the concentration gradient fibre growth process.
(b) As-grown single crystal of $Ca_{1-X}Yb_XF_{2+X}$ concentration gradient fibre (with x vary from 0 to 15 at% in the crystal). The distribution of Yb^{3+} along the fibre was measured by electron probe microanalysis (EPMA)

* There is no excited state absorption, no cross-relaxation process and no more up-conversion internal mechanism able to reduce the effective laser cross-section.
* The intense and broad Yb^{3+} absorption lines are well suited for IR InGaAs diode laser pumping between 900 and 980 nm.
* No absorption in the visible range.
* Small quantum defect, much smaller than in Nd^{3+} ion, between absorption and emission wavelengths leads to a low thermal load (11% relative to 30-40% in Nd^{3+}-doped laser hosts).

With respect to Nd^{3+} ion, which is an ideal 4-level scheme for laser operation, Yb^{3+} ion is only a quasi-four level scheme and then "Figure 5." shows that a high crystal field is required in Yb^{3+} sites to get the largest value of ΔE.

3.2. ASSIGNMENT OF ELECTRONIC AND VIBRONIC LINES IN Yb^{3+}–DOPED CRYSTALS

Although the Yb^{3+} ion has a simple electronic structure with only one excited state ($^2F_{5/2}$) above the ground state ($^2F_{7/2}$), the assignment of pure electronic lines is a rather difficult task due to a strong electron-phonon coupling. The degeneracy of the two multiplets is raised and seven Stark electronic levels are expected : four for the ground and three for the excited state, which have been labelled in "Figure 4.". Room temperature and low temperature absorption and emission spectra of Yb^{3+}-doped garnet ("Figure 6. and Figure 7.") or sesquioxide crystals show clearly many more lines than can be expected for an electronic transition alone. The problem is even more complicated when multisites are occurring which is encountered in CaF_2 crystals with the occurrence of several types of multisites. The assignment shown in Figures is only a first approximation one. In addition of the main square anti-prism symmetry site, higher resolution at low temperature show lines belonging to C_{4v} tetragonal sites without O^{2-} traces and trigonal sites with the presence of O^{2-} traces.

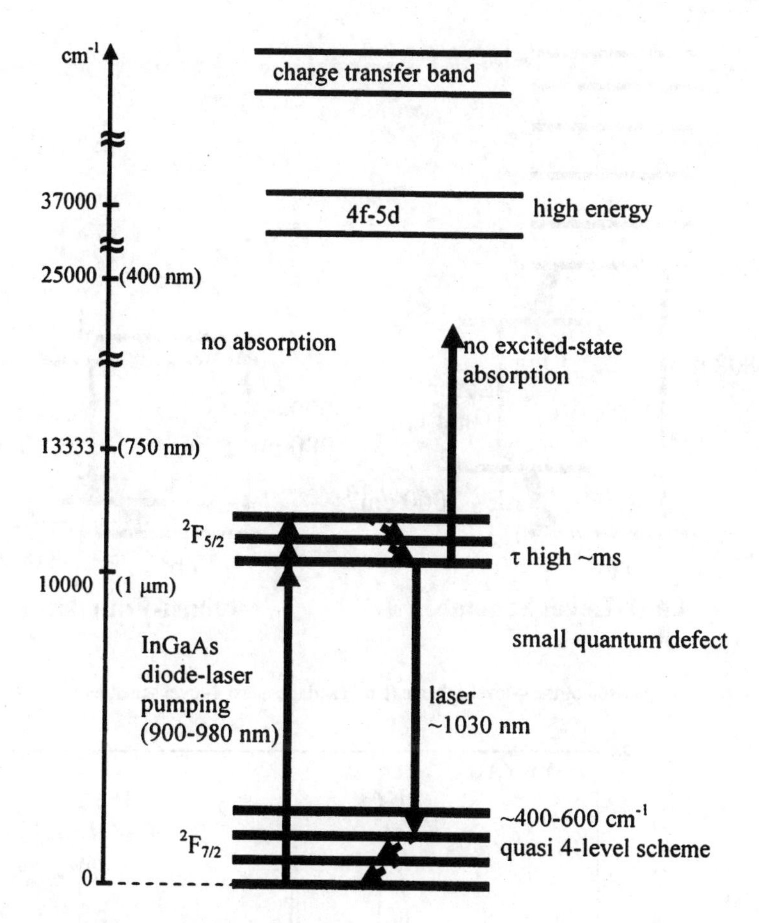

Figure 4. Main spectroscopic data of Yb^{3+}-doped laser materials

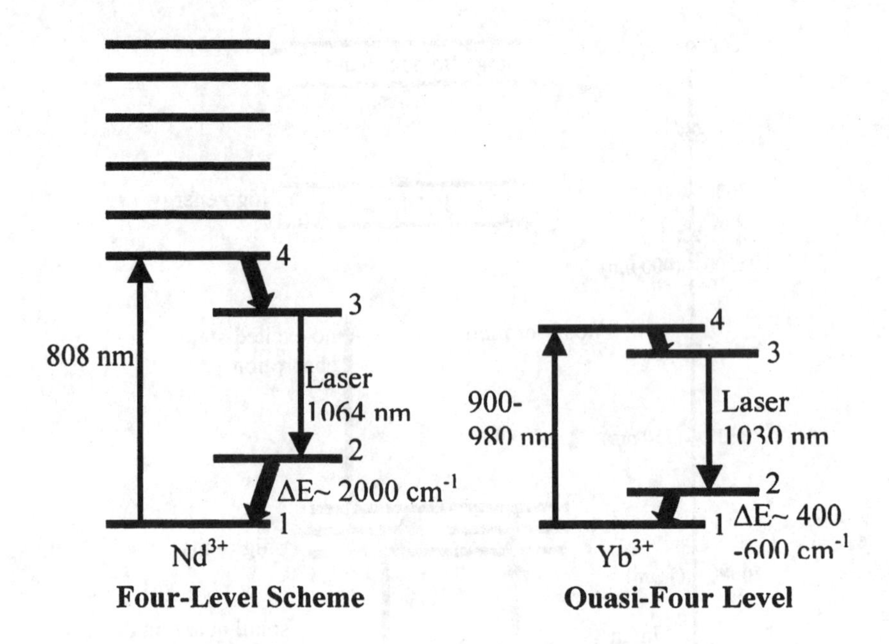

Figure 5. Comparison of the 4-level scheme (Nd^{3+}) and the quasi 4-level scheme (Yb^{3+}).

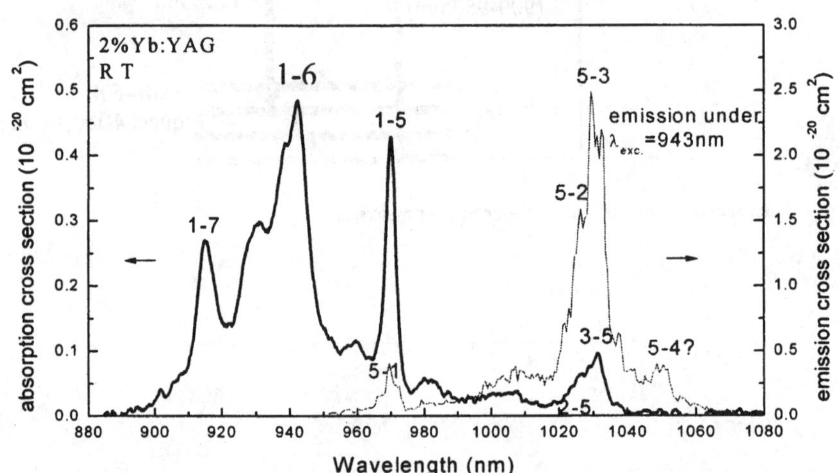

Figure 6. Absorption spectrum and emission spectrum of 2% Yb^{3+}-doped YAG, under 943 nm laser pumping at room temperature. The absorption cross-sections (left) and emission cross-section (right) are given.

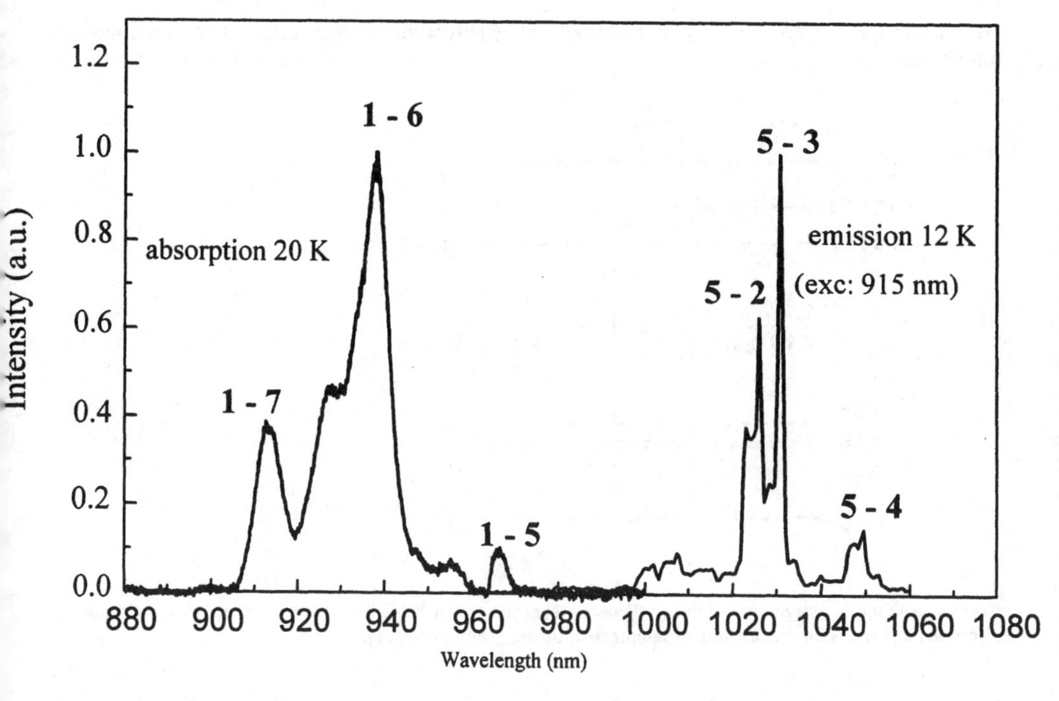

Figure 7. Absorption spectrum and emission spectrum of 2% Yb^{3+}-doped YAG under 915 nm laser pumping at low temperature.

3.2.1.Assignment of the spectroscopic data by comparing absorption, emission and Raman spectra.

-Yb^{3+}-doped YAG [14]

Since the published data concerning the energy levels of Yb^{3+} in YAG are contradictory, mainly due to a strong vibronic coupling, we have tried to contribute to this debate by again analyzing results in the experimental following approach. We compare absorption, emission and Raman corrected spectra which have been both drawn in wave number scale, to have at a glance, a quick evaluation on the difference between electronic and vibronic spectroscopic properties. In so, we are admitting the hypothesis that Raman spectrum should reflect vibronic structure accompanying the main 0-phonon line electronic resonant transition as this is the case with $^2E \rightarrow ^4A_2$ transition of Cr^{3+}-doped YAG. These spectra were then adjusted to the same energy scale, by taking the origin of the absorption and the emission at the 0-phonon line energy, in coincidence with the Rayleigh line of the Argon laser (514.5 nm) used to record the Raman spectrum. In Yb^{3+}-doped oxide crystals, only the lowest level of the $^2F_{5/2}$ excited state is emitting due to fast non-radiative relaxation processes between the 3 Stark components separated by energy gap of the same order of magnitude as the phonon energy. Hence, this is the reason why the resonant transition is the best adapted to give vibronic side band, on one hand, in the highest frequency side (shorter wavelengths) of the absorption spectrum, between 968 nm and 900 nm, and, in the other hand, in the lowest frequency side (longer wavelengths) of the emission spectrum, between 968 nm and 1100 nm ("Figure 6." and "Figure 7."). In such a way, symmetric distributions of vibronic lines are expected around $1 \leftrightarrow 5$ resonant electronic transition. By rotating the absorption spectrum around the origin, we get a direct comparison with the emission and the Raman spectra, which are drawn to the lowest frequency side. Therefore, we should be able to distinguish more clearly electronic and vibronic lines. This is actually our general approach on several Yb^{3+} -doped crystals under analysis in our Laboratory applied here especially on Yb^{3+}:YAG for which a first analysis has been done in "Figure 8." but by only using expected symmetry of absorption and emission spectra.

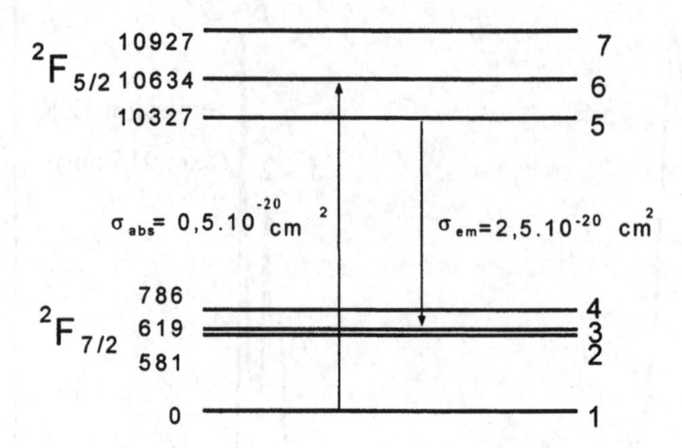

Figure 8. Attempt of assignment of the Yb^{3+} energy level scheme in YAG from the spectra at low temperature for absorption and emission spectra and the approach of the barycentre law [14].

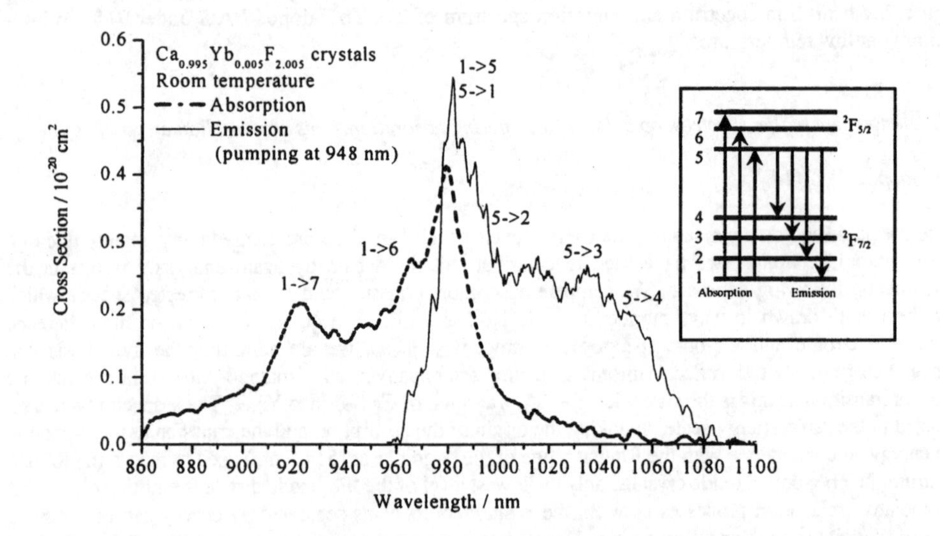

Figure 9. Absorption and emission spectra of 0.005% Yb-doped CaF_2 fibre at room temperature. The main electronic transitions have been assigned by using the energy diagram

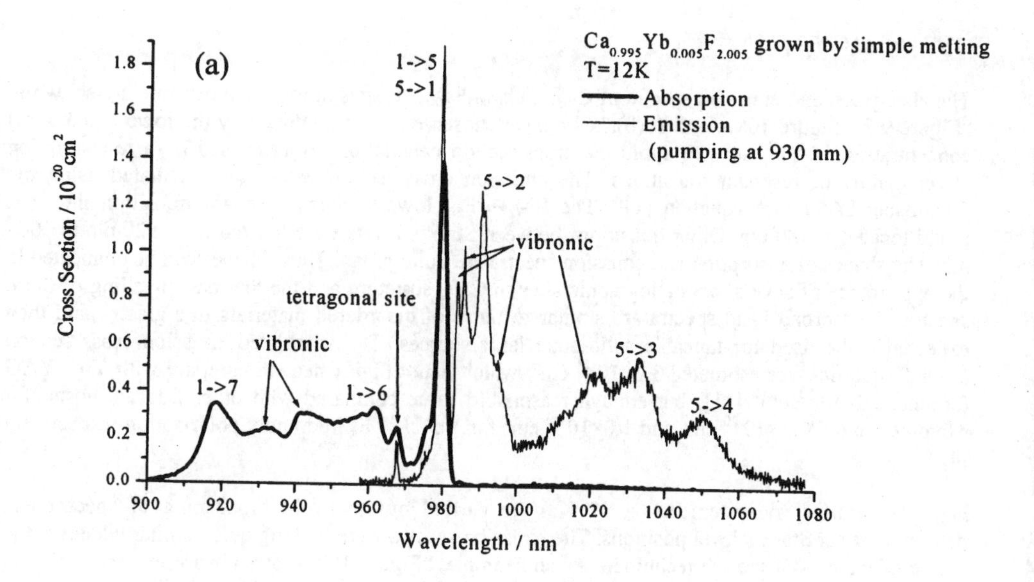

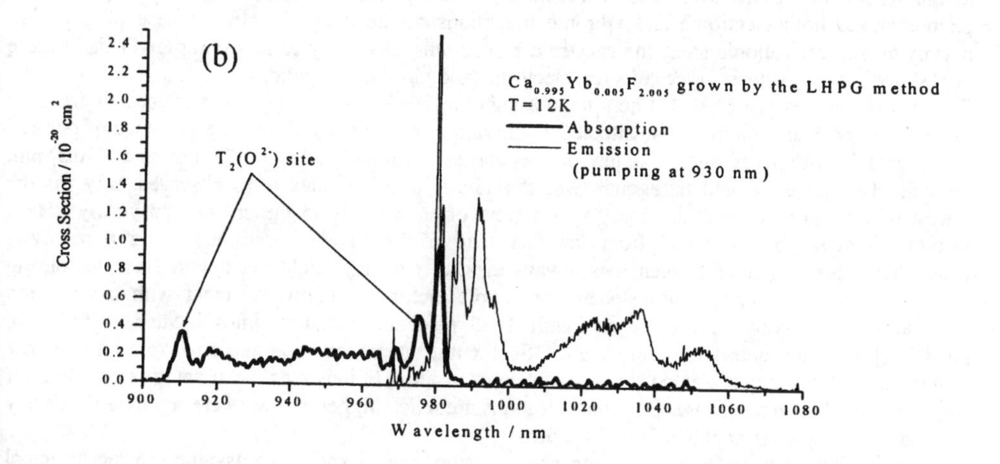

Figure 10. Absorption and emission spectra of $Ca_{0.995}Yb_{0.005}F_{2.005}$ crystal at low temperature (T=12 K) prepared (a) by the simple melting and (b) by the LHPG method. The arrows indicate the expected vibronic lines shifted from the electronic line of the phonon energy of 321 cm^{-1}.

-Yb^{3+}-doped CaF2 [15]

The absorption and emission spectra of $Ca_{0.995}Yb_{0.005}F_{2.005}$ crystals at room temperature are shown in "Figure 9.", "Figure 10(a). and 10(b)." .We have chosen one sample doped by the lowest (0.5 at.%) concentration of Yb^{3+} ions to avoid the cross-section calculation errors caused by a re-absorption effect mainly of resonant transitions. The emission cross-section value was calculated using the Füchtbauer-Ladenburg equation [11]. The $^2F_{7/2}\leftrightarrow^2F_{5/2}$ lowest energy resonant transition line was found located at 980 nm. Other transitions between Stark's levels were located from 920 nm to 1060 nm. The shapes of absorption and emission spectra are quite broad. These shapes can be connected to the appearance of several crystallographic sites in CaF_2 structure and the thermal spreading of these spectra. Such broad band spectra are similar to those of disordered materials like glasses and then expected to be used for tuneable solid state laser sources. The stimulated emission cross section around 1030 nm was estimated 3.0×10^{-21} cm^2, which is not high when we compare with Yb^{3+}:YAG (around 2.0×10^{-20} cm^2 [11]) but enough reasonable value compared with other fluoride hosts (for example around 0.3×10^{-20} cm^2 and 1.0×10^{-20} cm^2 for Yb^{3+}:$LiYF_4$ in σ and π polarization respectively [4]).

Low temperature measurements at 12 K were carried out to avoid broadening of spectra and determine exact Stark's level positions. The emission spectra were looking quite similar whatever the nature of the crystal growth technique. As an example, "Figure 10." shows absorption and emission spectra of $Ca_{0.995}Yb_{0.005}F_{2.005}$ crystal (a) by simple melting method and (b) by LHPG method. The assignment of Yb^{3+} Stark levels is known to have problems since the appearance of a strong electron-phonon mixes both electronic and vibronic transitions of the main site. In the case of Yb^{3+} ion occupying several cationic sites, the spectra are becoming obviously further complex. This "Figure 10." shows clearly different types of sites, electronic and vibronic transitions.
Previous researches [16,17,18,19] help us to assign the 1$\leftrightarrow$5 resonant transition of, both, tetragonal C_{4V} site at 968.5 nm, rhombic C_{2V} site at 977 nm and also $T_2(O^{2-})$ site with the presence of one O^{2-} anion in substitution of one F^- anion in the corner of anion cube at 975 nm and 910.5 nm, respectively. These isolated tetragonal C_{4V}, rhombic C_{2V} defect sites were observed only for the lowest (0.5 at.%) concentration. $T_2(O^{2-})$ sites were observed only in the crystals grown by LHPG method. It could be explained from the fact that in the LHPG conditions using pure argon atmosphere, ppm order of oxygen ions always existed and they could react with fluorides during crystal growth, whereas CF_4 atmosphere for simple melting is known to react with oxygen and eliminate it effectively. Cubic O_h site with 1$\leftrightarrow$5 resonant transition lines located at 963 nm, [16,17,18], were not detected in our samples. Such cubic site observation needs to work on very low concentration (less than 0.2 at.%) since charge compensating F^- ions must be situated somewhere far from the Yb^{3+} site in such case. Our concentrations are much higher and we were not able to see any cubic site in samples from 0.5 at.% to 30 at.%.
Other main lines seen in both absorption and emission spectra should be assigned to the principal Yb^{3+} site that is the set of square-antiprism sites. The electronic absorption transitions (1$\rightarrow$5, 1$\rightarrow$6, 1$\rightarrow$7) and the resonant 5$\rightarrow$1 and non-resonant (5$\rightarrow$2, 5$\rightarrow$3, 5$\rightarrow$4) electronic emission transitions are observed. In addition, vibronic lines can be also seen. One of the main difficulty of assignment is to separate such electronic and vibronic lines. To identify vibronic peak, Buchanan et al. [20] have already proposed the comparison between absorption and emission symmetric spectra around 0-phonon line. This method is however difficult to apply in the case of crystals accepting multi-sites because the broad peaks make difficult to identify the coincidence of peaks with the existence of the several sites. Our own experimental approach has been used both absorption, emission spectra and Raman spectra.

First, vibronic transitions were assigned by the help of Raman spectrum. The Raman spectrum has a very simple structure: only one sharp peak situated at 321 cm^{-1} in lowest doped sample, this unique peak becoming lightly broader in the 30% doped sample. Thus, the vibronic lines should be found in this frequency from the different 0-phonon electronic transitions. Some pairs of signals were found to keep in this frequency as can be seen in "Figure 10.". The 1012 nm line could be assigned as the vibronic transition from the 5→1 resonant transition line. The symmetric of 1012 nm line is 950 nm, this position is in obscurity of the broad absorption. Nevertheless, we have estimated there is a vibronic transition, which is difficult to resolve due to the overlapping between 962.5 nm and 942.5 nm lines. The 1024 nm line could be also assigned as the vibronic overtone line of the 991.5 nm line. Consequently, the 991.5 nm line could be an electronic transition which has been assigned as 5→2 transition. In the similar approach, the 933.5 nm line could be the vibronic line of the 1→6 (962.5 nm) transition.

4. Analysis of concentration quenching processes

Gradient concentration fibre is an unique tool to make the correlation between Yb^{3+} concentration and lifetime measurements which have been measured in situ on the same sample in relation to the distance from the top of the crystallized rod. By using a reference made of a homogeneous single crystal fibre of well-defined composition, the two curves were correlated. The lifetimes have been directly measured on each point of the concentration gradient fibre of 800 μm diameter, with a beam laser of about 500 μm size. Because the volume of materials excited is steady, radiative trapping due to geometrical effect can be supposed to be characterized by a constant value for each concentration, which should be weak due to the small size of the excited sample. Experimental lifetime values have been fitted to an unique exponential profile with an excellent agreement. Results are shown in "Figure11." as a function of the Yb^{3+} concentration for the three samples Y_2O_3, YAG and CaF_2.

4.1. RADIATIVE LIFETIME MEASUREMENT FROM THE CONCENTRATION GRADIENT FIBRE

The measurement of intrinsic lifetimes has received attention since a long time in the literature [21-24].The determination of the intrinsic radiative lifetime of Yb^{3+} in crystals requires a lot of precautions. Especially in YAG, measurements of the room temperature effective stimulated emission cross-section have ranged from $1.6 \cdot 10^{-20}$ cm^2 upon $2.03 \cdot 10^{-20}$ cm^2 [22]. Depending to the concentration, the self-trapping process is more or less involved. The intrinsic lifetimes can be read by following the concentration dependence to the lowest values. The value has been estimated 0.950 ms ± 0.001 in "Figure 11.", that is to say in the same range as the two others 0.951 and 0.9489 ms respectively mentioned previously [22]. In yttria, the radiative lifetime of 0.720 ± 0.001 ms has been measured and in CaF_2 the value is 2.05 ms in fibre samples grown by the LHPG technique and 2.14 ms in bulky samples which were grown by a simple melting process [15]. The last measurements show the strong dependence of the crystal growth method on the optical parameters.The intrinsic lifetime can be read by following the concentration dependence to the lowest values in "Figure 11." . The value has been estimated 2.05 ±0.01 ms in CaF_2.
The direct calculation of the spontaneous emission probability from the integrated absorption intensity [25] is adapted to confirm the radiative lifetime value. Radiative lifetime can be deduced from the absorption spectrum according to the formula (1):

$$1/\tau_{rad} = A_{if} = \frac{g_f}{g_i}\frac{8\pi n^2 c}{\lambda_0^4}\int \sigma_{fi}(\lambda)\,d\lambda \qquad (1)$$

Where the g's are the degeneracies of the initial and final states (g_f=4 for $^2F_{7/2}$ and g_i=3 for $^2F_{5/2}$), n is the refractive index (n=1.43), c is the light velocity, λ_0 is the mean wavelength of the absorption peak (980 nm), $\sigma(\lambda)$ is the absorption cross-section at wavelength λ. The radiative lifetime in CaF_2 was calculated as $\tau_{rad\ (theo.)}$=2.0 ms by the equation (2) quite close to the experimental value of $\tau_{rad\ (exp.)}$=2.05 ms measured in "Figure11.". The lower value of $\tau_{rad\ (theo.)}$ with respect of the $\tau_{rad\ (exp.)}$ might be related to the over estimated value of the integrated absorption cross-section including all types of sites at room temperature and not only the highest population of the main square-antiprism sites.

4.2. SELF-TRAPPING AND SELF-QUENCHING PROCESSES

Identifying the origin of the concentration dependence of the observed experimental lifetime allows understanding the excited state dynamics. The curves can be divided into two regimes in "Figure 11." :

(i) in the lowest concentration range (up to 1.3% in YAG, 5% in yttria and 10% in CaF_2) the experimental lifetime increases as the doping concentration increases.
(ii) for the higher concentration range, the measured lifetime decreases when the doping rate increases up to 4% in YAG, 25% in yttria and 30% in CaF_2. This concentration range has been taken as an example in our experiments, but other concentration ranges can be extended up to 100% if solid solution of Yb^{3+}-doped oxides or fluorides exists.
The first regime is the indication of the radiative energy transfer whereas the second one corresponds to the usual quenching process by a non-radiative energy transfer to defects and other impurities in the host. In the intermediate region, competition occurs between the two trends and consequently they compensate each other, leading to a constant value of the measured lifetime.

4.2.1.Radiative energy transfer

The first regime is an indication of fluorescence re-absorption, the so-called self trapping process, between levels 1 and 5, by the 1↔5 resonant transition, giving radiative energy transfer and then migration of the energy on a long distance. Resonant radiative energy transfer is well known for acting as a radiation trap. Actually resonant transitions permit a long-range energy diffusion between identical ions by successive re-absorption/reemission processes. This serial mechanism affects the experimental excited state decay since, after the initial emission occurs, each subsequent event acts as a time reset on the relaxation of this excitation. Consequently, a lengthening of the fluorescence lifetime measured over the volume of the sample relative to the lifetime of a single isolated ion should be observed.

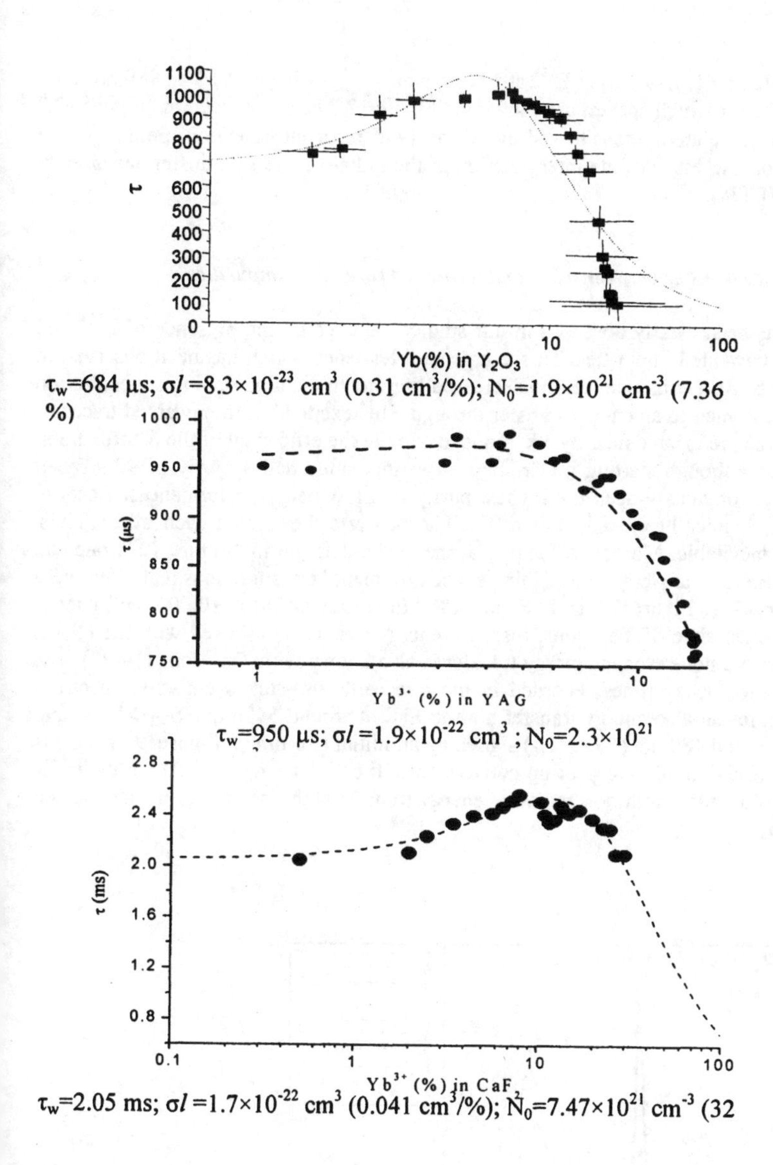

Figure 11. Transversally measured lifetime of the Yb^{3+} $^2F_{5/2} \rightarrow$ $^2F_{7/2}$ transition in fibres of Y_2O_3, YAG and CaF_2 versus molar % concentration along the fibre. Continuous curve is from Eq. (6).

Strong spectral overlap between the fluorescence and absorption spectra enhances fluorescence re-absorption. The most relevant examples are those of the resonant laser transitions such as $^2E \rightarrow {}^4A_2$ of Cr^{3+} in ruby at 694 nm, and several rare earth ions in laser host as $^5I_7 \rightarrow {}^5I_8$ in Ho^{3+} at 2100 nm,

$^3F_4 \rightarrow ^3H_6$ in Tm^{3+} at 2000 nm, $^4I_{13/2} \rightarrow ^4I_{15/2}$ in Er^{3+} at 1540 nm and $^2F_{5/2} \rightarrow ^2F_{7/2}$ in Yb^{3+} at 980 nm. The case of Yb^{3+}-doped Y2O3 for which spectra are shown in "Figure 12." is clearly seen in "Figure 13." where the decreasing of the relative intensities of the of the $1 \leftrightarrow 5$ resonant transition points out the self-trapping phenomenon. An essay of the interpretation of the radiative energy transfer has already been published [26], [27], [28].

4.2.2.Non-radiative energy transfer between Yb^{3+} and unwanted rare earth impurities

An important feature has been clearly observed in our studied samples by the presence of Er^{3+} and Tm^{3+} rare earth ions as unwanted impurities. This is a general tendency which has been observed in all Yb^{3+}-doped laser hosts. As a consequence, the decreased lifetime in the high doping concentration region can be naturally assigned to an energy transfer through Yb^{3+} excited ion to unwanted traces of rare earth impurities which are located near by Yb^{3+} ions to explain the efficiency of the recorded non radiative energy transfer, although starting materials used in this study had a purity of 99.99%. It means that laser materials obviously need the highest purity to allow better performances. Because rare earth elements are chemically related, it is difficult to separate them from each other. Thus, traces of impurities are inevitable. Moreover, looking at the Dieke diagram in "Figure 10.", one can see that many resonant energy transfers are possible between trivalent lanthanide ions and effectively observed both in "Figure15.", "Figure17." and "Figure 18.". In particular in the 10,000 cm^{-1} energy range matching with excited state of Yb^{3+} ions, resonant energy transfer is allowed with the ($^4I_{11/2}$) excited level of Er^{3+} ions and non-resonant energy transfer is also known with Tm^{3+} ions [14-27]. The profiles of the fluorescence decay times recorded in the rare earth anti-stokes emission spectrum under Yb^{3+} IR pumping, reveal also energy transfer both to Er^{3+} at around 540 nm ($^4S_{3/2} \rightarrow ^4I_{15/2}$) and 650 nm and to Tm^{3+} at around 480 nm ($^1G_4 \rightarrow ^3H_6$) signed by an initial rise-time ("Figure19."). This is remarkable to see in Yb:CaF2 another step of up-conversion in the violet range at 410 nm under IR pumping at 940 nm by Yb^{3+} rare earth non radiative energy transfer with only traces of unexpected rare earth impurities [15].

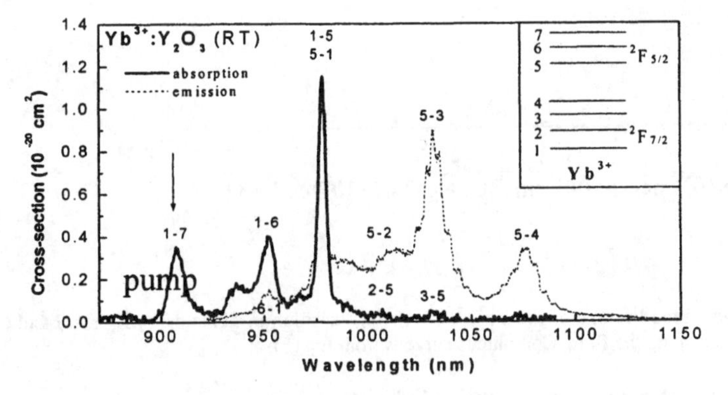

Figure12. Absorption and emission spectra of Yb^{3+}-doped Y_2O_3.

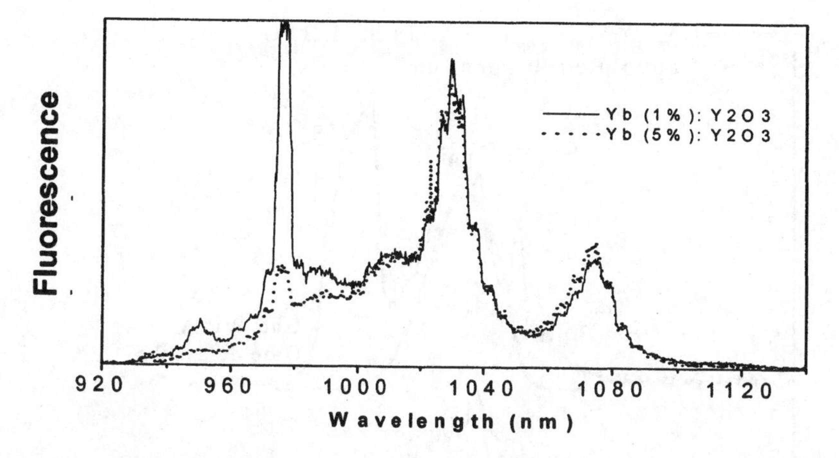

Figure 13. Evidence of the self-trapping process on the Yb^{3+} $^2F_{5/2} \rightarrow$ $^2F_{7/2}$ resonant transitions in Yb^{3+}-dopedY_2O_3. The re-absorption of 1<->5 and 1<->6 transitions near 975 nm and near 950 nm, respectively, is observed.

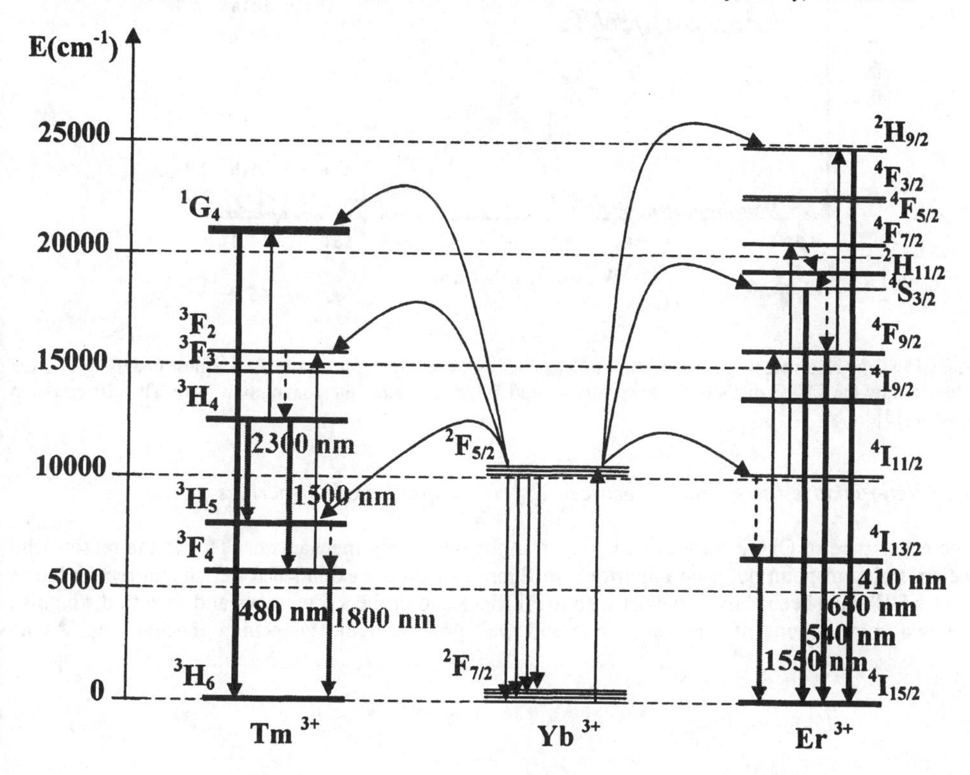

Figure 14. Energy level diagrams of Tm^{3+}, Yb^{3+} and Er^{3+} respectively. The main visible transitions observed by up-conversion emission are indicated. The IR down-conversion emissions are also mentioned.

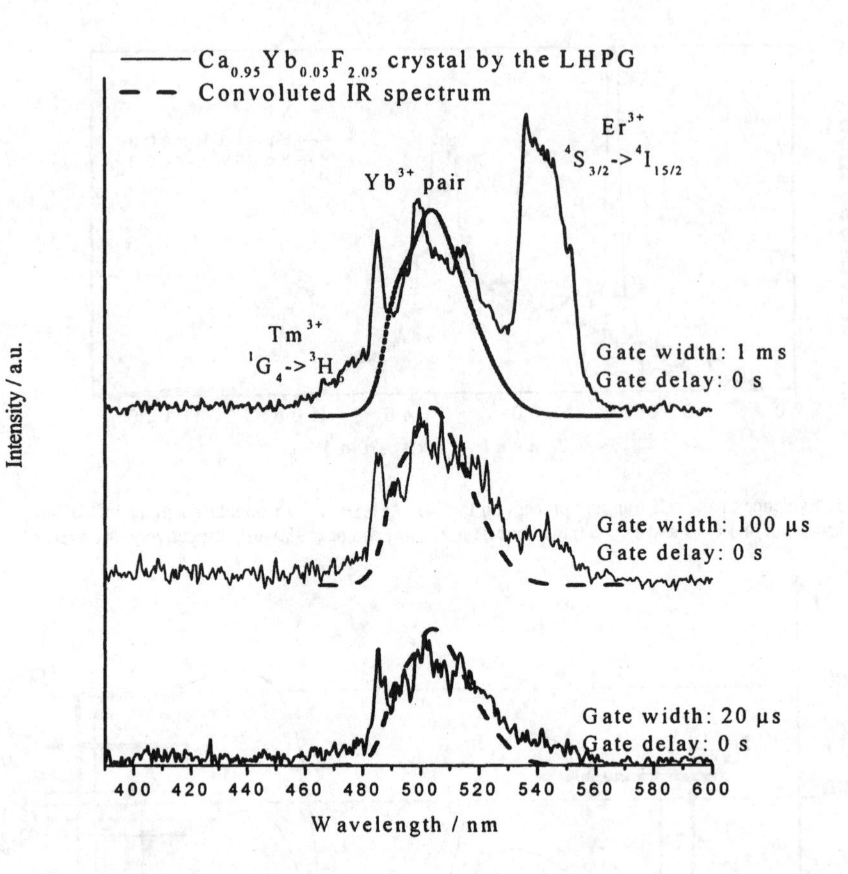

Figre 15. Time resolved emission spectra of $Ca_{1-X}Yb_XF_{2+X}$ (x=0.05) crystals in visible region under $\lambda = 932$ nm excitation by the LHPG method. Clear spectra around 500 nm indicate the convolution from Yb^{3+} IR emission spectra [15].

4.2.3.Non-radiative energy transfer between Yb^{3+} and unwanted OH^- impurities

The occurrence of OH^- is very often involved in the quenching mechanisms. This is the reason why we are trying to point out such impurities. In fluorides OH- are excluded for chemical reason but in oxides OH^- have been shown to exist with the difficulty to probe a few traces and then to distinguish the two contributions of rare earth ions and OH^- groups in the quenching mechanisms. As an example, “Figure 16.” shows clearly the presence of OH^- groups in YAG characterized by an absorption coefficient of 3.5 cm^{-1}. Consequently, the contribution of OH^- cannot be excluded but only in oxide samples. The strongest influence of self-trapping in fluorides for a wider concentration range seen in “Figures 11.” by the increasing of the experimental lifetimes can be explained by the weakest effect of self-quenching for higher Yb^{3+} concentrations. As among impurities rare earth ions are always present in both oxides and fluorides, we may say that this important effect might be associated to the absence of OH- groups in fluoride crystals with respect of oxide crystals.

4.3. Yb^{3+} PAIR AND COOPERATIVE EMISSION

In the case of Yb^{3+}-doped crystals, because of the presence of only one excited level, we cannot expect an excited state mechanism inside Yb^{3+} ions which could reach other rare earth ions by an up-conversion process. Only pair emission from two neighbour ions or cooperative emission from aggregates which are observed in all our samples can occur with a weak probability, depending on the shortest distances between two Yb^{3+} neighbour crystallographic sites :

-Y_2O_3 (C_2 site – C_{3i} site distance = 3.51 Å and C_2 site – C_2 site distance = 3.53 Å),
-YAG (site distance = 3.67 Å),
-CaF_2 (site distance = 3.84 Å).

As an example, “Figure 15.”, “Figure 17.” And “Figure 18.” “ show the visible pair emission around 500 nm in CaF_2, Y_2O_3 and YAG as compared with the convoluted IR emission spectrum obtained under IR pumping. An additional argument in favour of the pairing effect is provided by the value of the green decay time of pairs, half- value of the $^2F_{5/2}$ IR emitting excited level without any initial rise-time in the three hosts (“Figure 19.”). Such observation of pairing effect is a signature of the inhomogeneous distribution of Yb^{3+} in hosts.

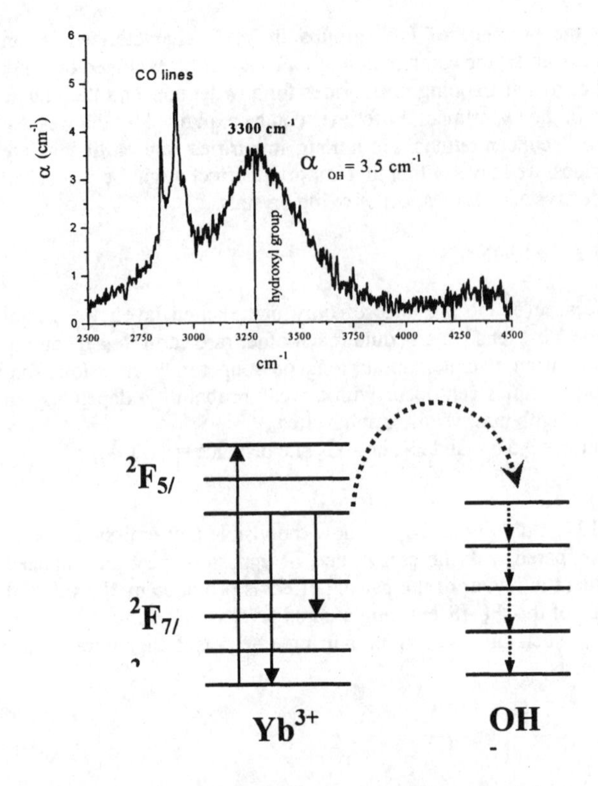

Figure 16. IR absorption spectrum of Yb^{3+}-doped YAG in the spectral area of 2500-4500 cm^{-1}. OH^- groups may be involved as quenchers in Yb^{3+}-doped garnets.

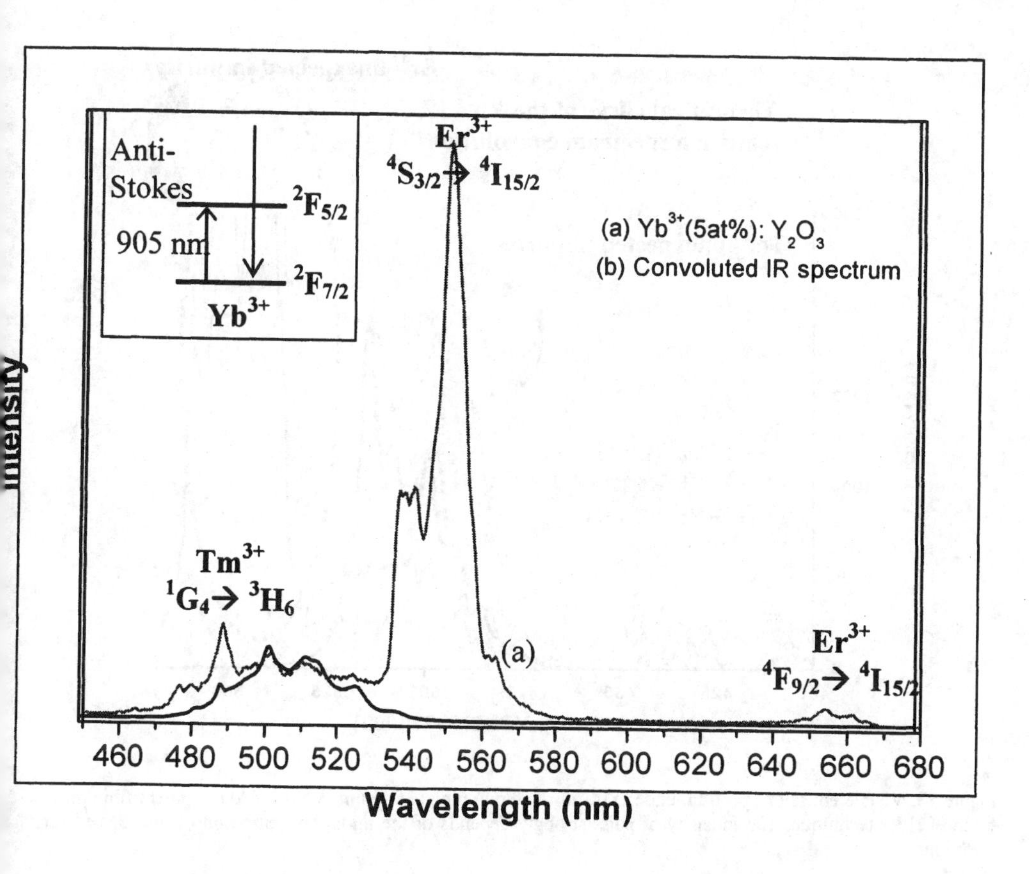

Figure 17. Up-conversion emission spectrum of Yb^{3+}-doped Y_2O_3 under 905 nm IR pumping due to both unexpected Tm^{3+} and Er^{3+} rare earth ions in the host and the presence of pairs ("Yb^{3+}coop "for cooperative emission).

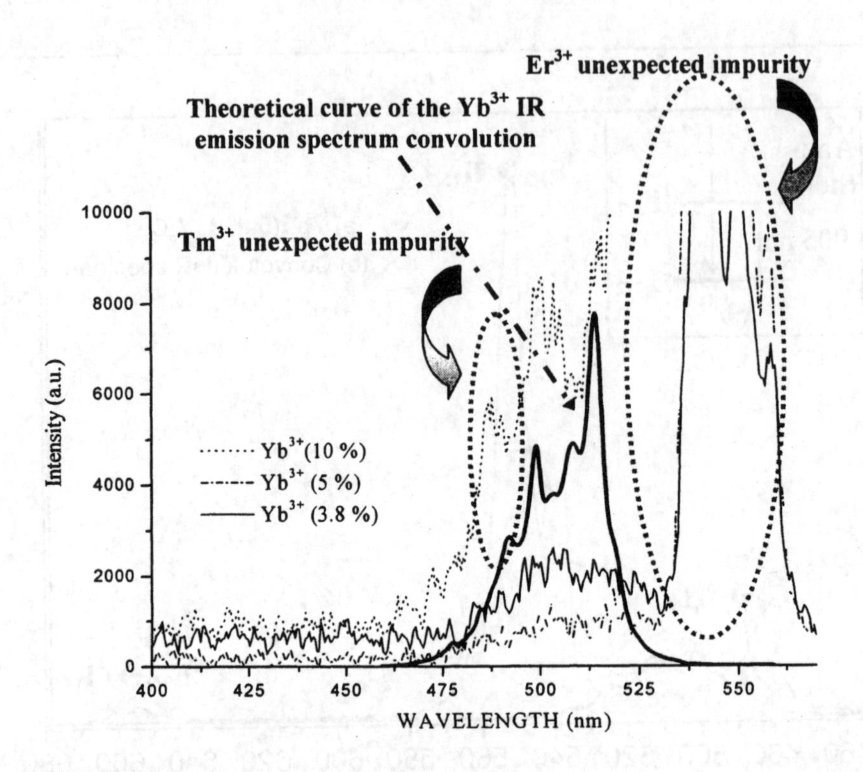

Figure 18. Visible emission spectrum under Yb^{3+} infrared pumping at 940 nm in Yb^{3+}:YAG gradient fibre grown by the LHPG technique. The intensity of pairs strongly depends of the initial concentration of unexpected rare earth ions.

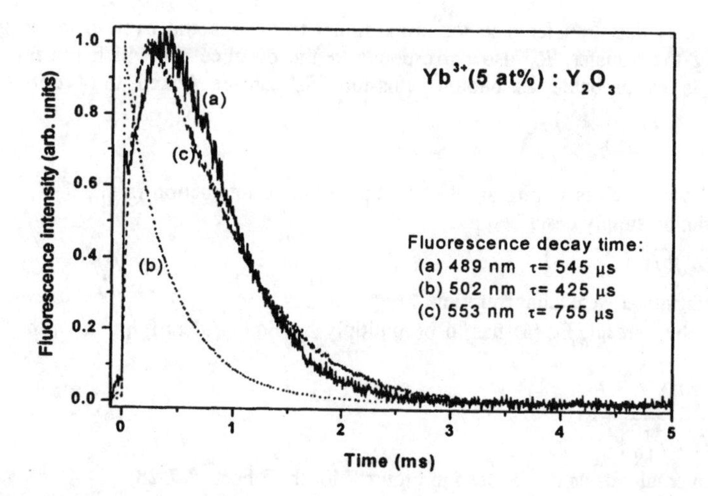

Figure 19. Fluorescence decay curves of the anti-Stokes emissions in 5% Yb^{3+}-doped Y2O3 under 906 nm pumping in the excited state of Yb^{3+} ions. (a) and (c) represent decays of rare earth impurities whereas (b) represents decay of pairs without any initial rise-time.

5.Model to interpret radiation self-trapping and self-quenching mechanisms in Yb^{3+}.

For radiation trapping, a simple method already applied to solids is used on a Compton type spatial diffusion equation containing the Einstein relationships between spontaneous and induced emission and absorption but neglecting of the population inversion. In the case of "weak opacity", which is the only one which can be analytically solved, the following formula has been deduced:

$$\tau_t = \tau_i\left(1 + \sigma N l\right) \tag{2}$$

where τ_t is the measured lifetime in the trapping conditions, τ_i is the intrinsic lifetime without trapping, σ is the transition cross-section, N the ion doping concentration, and l is the average absorption length in the lifetime measurement experiment.

When the quenching centre is an impurity analogous to the active centre and having levels in resonance or quasi resonance with the considered ion first excited state, the quenching probability can be of the same order as the transfer probability for diffusion within the considered ion sub-system. The strong quenching situation prevails as well as the limited diffusion process. In that case, because both probabilities for quenching and diffusion are about equal, they can be assumed to be ruled by an equivalent critical transfer distance R_0 defined by :

$$C = \left(R_0 / R\right)^s / \tau_s \tag{3}$$

where C is Dexter parameter for energy transfer probability P as given by $P = C/R^s$; R is the distance between the two ions in interaction; τ_s is the ion lifetime before interaction; s is the multipolar index for the interaction; R_0 is the critical transfer distance proportional to the spectral overlap insuring energy conservation during the transfer; R_0 also corresponds to the distance for which the non-radiative energy transfer is as probable as photon emission. R_0 can be linked to a critical concentration by :

$$R_0 = (3/4\pi N_0)^{1/3} \quad (4)$$

In such case it has been shown that, assuming an electric dipole-dipole interaction (s=6), the self-quenching behaviour can be simply described by:

$$\tau(N) = \tau_w / [1 + (9/2\pi)(N/N_0)^2] \quad (5)$$

where τ_w is the measured lifetime at weak concentration.

In case of photon trapping be present, Eq.(4) has to be multiply by Eq.(1) identifying τ_w with τ_i, giving respectively:

$$\tau(N) = \frac{\tau_w (1 + \sigma N l)}{1 + (9/2\pi)(N/N_0)^2} \quad (6)$$

This model fits well experimental data as can be seen in Figure 7 for the 3 hosts [27-28].

The theoretical fittings of Yb-doped crystals are shown in "Figure 11.", as continuous line for the LHPG method.

The fitting parameters of Yb:CaF2 were found as :

$\sigma l = 1.7\times10^{-22}$ cm^3 (0.041 cm^3/%), $N_0 = 7.47\times10^{21}$ cm^{-3} (32 %) for the LHPG samples, respectively.

N_0 being the parameter of the self-quenching, hence much higher N_0 values than that of $Yb:Y_2O_3$ ($N_0 = 1.9\times10^{21}$ cm^{-3}) and Yb:YAG ($N_0 = 2.3\times10^{21} cm^{-3}$) mean weaker self-quenching probability as compared with oxide laser crystals. Such weak self-quenching probability is agreeable property as laser application.

It is also interesting that two curves for the two crystal growth techniques, the simple melting and the LHPG ones, show same N_0 value. It means two curves have theoretically same self-quenching probability, therefore it confirms that the difference of decay time dependence between the two growth methods mainly come from the self-trapping process and not from the self-quenching processes.

Finally, it has been shown that self-quenching, for a rather large doping range, is well described by a limited diffusion process within the doping ion subsystem towards impurities analogous to the doping ions themselves.

Fast diffusion towards intrinsic non-radiative centers cannot explain the observed results. In addition, the intrinsic center, cluster-like pair of rare earth active centers, which should yield a particular multiphonon-assisted energy transfer between them has not been observed from the fitting of figures. Being due to the vibronic properties of the host it cannot be however suppressed but seem be much weaker than the limited diffusion process in the three examples of this study[28].

As an application, a simple quantitative method for optimizing the gain material concentration for amplifiers and lasers has also been proposed and performed [28].

6. Conclusion

The LHPG method have been described in order to be applied for the growth and the characterization of optical parameters of Yb^{3+}-doped host crystals, which might be used as high average power solid-state lasers under pumping by high power laser diode. We have chosen $Y_3Al_5O_{12}$ (YAG) yttrium aluminium garnet, Y_2O_3 yttria sesquioxide and CaF_2 fluoride as typical host examples which are considered among important crystals due to their unusual combination of favourable properties. Moreover, a combinatorial chemistry approach has been applied on concentration gradient crystal fibres. Gradient concentration fibre is an unique tool to make the correlation between Yb^{3+} concentration and lifetime measurements which have been measured *in situ* in the same sample in relation to the distance from the top of the crystallized rod. This fast and simple combinatorial method allows to measure:

-the Yb^{3+} intrinsic radiative lifetime at the lowest Yb^{3+} concentrations, which is one of the basic parameters for laser application for measuring stimulated emission cross-section,

-the radiation trapping by the decreasing intensity of the resonant $^2F_{5/2} \leftrightarrow {}^2F_{7/2}$ transitions (1<->5 and 1<->6) between 930 nm and 980 nm when Yb^{3+} concentration increases,

-the non-radiative energy transfer to uncontrolled rare earth impurities in doped hosts giving a notable contribution to the quenching processes,

-the non-radiative energy transfer to OH- impurities in doped hosts giving also a notable contribution to the quenching processes mainly in oxides and probably less in fluorides explaining why the self-trapping is much higher than the self-quenching in a wide concentration range above 10%,

-Yb^{3+} pairs detected by the up-conversion visible emission spectrum under Yb^{3+} IR pumping, signature of the inhomogeneous distribution of ions in crystalline hosts but also possible signature of intrinsic quenching, due to the electron-phonon coupling itself. This is an intrinsic self-generated quenching center for lanthanides. This center, in fact a cluster-like pair of active centers, is shown to come from a particular multiphonon-assisted energy transfer between them. Being due to the vibronic properties of the host it cannot be suppressed. However his occurrence was not detected in our crystals.

Finally, a limited diffusion process model within the doping ion subsystem towards impurities analogous to the doping ions themselves has been given to interpret such mechanisms. Research is in progress to detect any new unexpected impurities which could contribute to quenching mechanisms.

Acknowledgements

We would like to thank all researchers of the Laser Materials group working in the Physical Chemistry of Luminescent Materials of the Claude Bernard/Lyon1 University who are involved in the Yb^{3+}-doped crystals project: Alain Brenier, Marie-Thérèse Cohen-Adad, Christelle Goutaudier, Yannick Guyot, Kheirredine Lebbou, Gérard Métrat as permanent members and Amina Bensalah, Helena Canibano, Abdes El Hassouni, Masahiko Ito, Laetitia Laversenne, as PhD students. We would like also to thank Prof.Fukuda, Dr.Yoshikawa and all the team of the Tohoku University in Japan and, in addition, Dr.François Auzel from CNRS Laboratory of Paris- Meudon, for the fruitful co-operation of this program.

References

1. Brenier, A. Boulon, G. (2001) Overview of the best Yb^{3+}-doped laser crystals, J.of Alloys and Compounds **323-324**, 210-213
2. Brenier, A. Boulon, G. (2001) New criteria to choose the best Yb $^{3+}$ -doped laser crystals, Europhysics Letters **55**, 647-652
3. Cohen-Adad,M.T. Laversenne, L. Gharbi, M. Goutaudier, C. Boulon, G. Cohen-Adad, R. (2001) New combinatorial chemistry approach in material science, Journal of Phase Equilibria **22 n°4**, 379-385
4. Boulon, G. Laversenne, L. Goutaudier, C. Guyot, Y. and Cohen-Adad, M. T. (2003) Radiative and non-radiative energy transfers in Yb^{3+}-doped sesquioxide and garnet laser crystals from a combinatorial approach based on gradient concentration fibers , J.of Luminescence **102-103**, 417-425
5. Poplawsky, R. P. (1962) J. Appl. Phys. **33,** 1616
6. Haggerty J.S. (1972) NASA Report CR-120948
7. Feigelson, R.S. (1986) J. Crystal Growth **79,** 669-680.
8. Foulon, G. Brenier, A. Ferriol, M. Cohen-Adad, M.T. Boulon, G. (1995)Laser heated pedestal growth and optical properties of Yb-doped LiNbO3 single crystal fibers, Chem..Phys.Lett. **245,** 555-560
9. Laversenne, L. Guyot, Y. Goutaudier, C. Cohen-Adad, M.T. Boulon, G. (2001) Optimization of spectroscopic properties of Yb 3+-doped refractory sesquioxides :cubic Y2O3, Lu2O3 and monoclinic Gd2O3, Optical Materials **16,** 475-483.
10. Laversenne, L. Kairouani, S. Guyot, Y. Goutaudier, C. Boulon, G. Cohen-Adad, M. T. (2002) Correlation between dopant content and excited-state dynamics properties in Er-Yb-co-doped Y2O3 by using a new combinatorial method, Optical Materials **19,** 59-66
11. Laversenne L., Goutaudier, C. Guyot, Y. Cohen-Adad, M. Th. Boulon, G. (2002) Growth of rare earth (RE) doped concentration gradient crystal fibers and analysis of dynamical processes of laser resonant transitions in RE-doped Y2O3 (RE=Yb, Er,Ho), J. Alloys Compounds **341,** 214-219
12. A.Bensalah, H.Canibano, M.Ito, (2002-2004) PhD's under progress at the LPCML, Claude Bernard/Lyon1 University
13. Boulon, G. (Editor) (2003) Optical Materials Scientific Committee of the French Research Group:GDR 1148 CNRS "LASMAT" Research on Laser Materials, **22** 81-176
14. Yoshikawa, A. Boulon, G. Laversenne, L. Canibano, H. Lebbou, K. Collombet, A. Guyot, Y. Fukuda T. (2003) Growth and spectroscopic analysis of Yb^{3+}-doped $Y_3Al_5O_{12}$ fibre single crystals, J. of Applied Phys. **94,** 5479-5488
15. Ito, M. Goutaudier, C. Lebbou, K. Guyot, Y. Fukuda, T. Boulon, G. (2003) Crystal Growth, Yb^{3+} Spectroscopy, Concentration Quenching Analysis and Potentiality of Laser Emission in $Ca_{1-X}Yb_XF_{2+X}$,J. of Physics : Cond. Matter (submitted)
16. Kirton, J. McLaughlan, S. D. (1967) Physical Review **155,** 279
17. Kirton, J. White A. M., (1969) Physical Review **178,** 543
18. Yu. K. Voron'ko, V. V. Osiko, I. A. Shcherbakov, (1969) Soviet Physics JETP **29,** 86
19. Falin, M. L. Gerasimov, K. I. Latypov, V. A. Leushin, A. M. Bill, H. Lovy, D. (2003) EPR and optical spectroscopy of Yb ions in CaF2 and SrF2, J. of Lumin. **102-103,** 239-242
20. Buchanan, R. A. Wickersheim, K. A. Pearson, J. J. Herrmann, G. F. (1967) Physical Reviw **159,** 245
21. Sumita, D. S. Fan, T. Y. (1994) Optics Letters **19,** 1343-1345
22. Uehara, N. Ueda, K.Y. Kubota, (1996) Jpn. J. Appl. Phys. **35,** L499
23. Hehlen, M. P.(1996) in OSA TOPS on Advanced Solid-State Lasers, S. A. Payne and C. Pollock, Eds. **1,** 530
24. Christensen, H. P. Gabbe, D. R. Jenssen, H. P. (1982) Phys. Rev. B **25,** 1467
25. DeLoach, L. D. Payne, S. A. Chase, L. L. Smith, L. K.. Kway, W. L Krupke, W. F. (1993) IEEE J.Quantum Electronics **29,** 1179
26. Boulon, G. Brenier, A. Laversenne, L. Guyot, Y. Goutaudier, C. Cohen-Adad, M.T. Métrat, G. Muhlstein, N. (2002) Search of optimized trivalent ytterbium doped-inorganic crystals for laser applications, J. Alloys Compounds **341,** 2-7
27. Auzel, F. Bonfigli, F. Gagliari, S. Baldacchini, G. (2001) The interplay of self-trapping and self-quenching for resonant transitions in solids, role of a cavity, J. Lum. **94-95,** 293-297
28. Auzel, F. Baldacchini, G. Laversenne, L. Boulon, G. (2003) Radiation trapping and self-quenching analysis in Yb, Er and Ho-doped Y2O3, Optical Materials **24,** 103-119

23. TABLE-TOP SOFT X-RAY LASERS AND THEIR APPLICATIONS

GIUSEPPE TOMASSETTI
Dipartimento di Fisica, Università dell'Aquila,
and g.c. LNGS-INFN,INFM,
Coppito (AQ), 67010, ITALY

Abstract. A review of the developments of the table-top soft X-ray lasers, which have been carried on in the last years utilizing different experimental techniques, is reported. A comparison between the performances of the different devices is illustrated.

1. Introduction

Research on soft X-ray lasers - more properly, at the moment, X-UV single pass, high gain (g >10) ASE radiation sources - has made great progresses since 1985, when lasing in Ne-like Se (Se^{24+}) ions, at $\lambda = 20.6$ and 21 nm was first achieved by D.L. Mathews et al. [1]. In that experiment x-ray lasing was achieved by focussing a high power, 0.5 ns, Nd-glass laser on a Se target via energy levels population by electron collisional excitation, according to a theoretical proposal first put forward by L.I. Gudzenko and L.A.Shelepin in the far 1964. Today, the main challenge in improving X-ray lasers is in building table-top lasers, i.e. low cost, small size, high repetition rate devices, as alternative to X-ray sources based on large research facility as, for instance, storage rings. In the last two decades many experiments have shown the excellent features of such type of "lasers" in respect to monochromaticity, collimation, very high peak spectral brightness, and fairly good coherence.
Table-top X-ray lasers open new possibilities in smart investigations on high density plasma dynamics, in material science, for example, measurements of surface reflectivity and optical properties of materials, testing multi-layer mirrors and gratings. Important industrial applications will be nanobiology, i.e. single shot imaging of biological cells, X-ray holography and phase-contrast imaging of microstructures, phase variation in samples by interference methods.
Soft X-ray lasers actually cover the electromagnetic spectrum from the 50 nm down to 3.56 nm [2], an energy/pulse of 0.1μJ ÷3 mJ, and some devices can operate at a repetition rate up to 10 Hz [3-4]. The population inversion is created between external levels of highly ionized atoms inside high temperature plasmas. Because of the short duration of the population inversion (1 ps-1 ns) and the lack (today) of high reflectivity mirrors the resonator is missing. Due to the short length of the plasma columns, which can be experimentally created (typically 0.5 -5 cm) and the mirror-less amplification, the saturation of the active medium requires very high gain values. This means that the

B. Di Bartolo and O. Forte (eds.), Frontiers of Optical Spectroscopy, 715-720.

pumping power density required to pump a soft X-ray laser scales more strongly than λ^{-4}.
The physical processes involved in creating a population inversion are mainly or the collisional excitation of the electrons from the ground state or the electron collisional recombination. However, other processes, like photo-ionization [5] or resonant photo-absorption [6] have been investigated.

2. Pumping Techniques

From the experimental point of view, different schemes have been tested to create the plasma columns with the proper conditions for XUV amplification. Up to 1980/90 high powerful lasers [1] (able to drive energies of $\approx$ 500 J in $\approx$ 1 ns) were mainly used. Focused on the surface of a target they can produce hot and dense cylindrical plasmas along the focusing line, where the population inversion may occur under quasi-stationary conditions of density and temperature. With this approach X-ray laser pulses of few mJ [1,2] of energy have been obtained at many wavelengths using different materials. However, today, all the efforts are toward using small-scale, high power, ultra-fast ($\Delta T \approx$ 1 ps) optical laser pulses as drivers or alternatively, compact, high power discharges in a gas or metal vapors filled capillaries.
In recent years a deep understanding of the factors influencing the gain has suggested using a laser pumping by a multi-pulse irradiation to obtain lasing in Ne-like or Ni-like ions. By this technique one irradiates a solid target with several pulses of few joule and short time duration. In a double pulse arrangement [7], for instance, a first pulse ($\approx$ 1 Joule, $\approx$ 1 ns) forms low temperature plasma with the required ionization stage and a small density gradient for the laser propagation. About 1 ns later, the second, high intensity pulse, ($\approx$ 1–10 J in 1 ps), quickly heats the free electrons to reach much higher temperatures. In this way it is possible to optimize the transient collitional excitation of the upper laser level keeping fixed the optimal ionization stage. With such pumping mechanism (so called Transient Collitional Excitation) the gain can be improved of one or two orders of magnitude respect to the quasi-steady state excitation in the same transition.
The TCE utilizes chirped pulse amplification (CPA) technology to create the laser pulses of picosecond duration. Advantages of this technique are the possibility of a drastic reduction of the energy of the pumping laser pulses (to few Joules) and the possibility of scaling to shorter wavelengths, because the increase of the pumping power produces higher temperature plasmas. Disadvantages of the technique are the low efficiency (10^{-8}), the short length of the plasma (< 2 cm), the complicated and expensive optical systems and the low energy/pulse of x-ray laser due to the shorter duration of the gain. This scheme has been successfully operated in many laboratories. For instance, at λ = 13.9 nm in Ni-like Ag at Lemail [8] and at λ = 14.7 nm in Ni-like Pd at LLNL [9].
The development of CPA technique for pulse shortening has opened a successful way to pump both collisional and recombination soft X-ray lasers. The Optical Field Ionization (OFI) represents another promising approach [3] using fast pumping laser pulses. In [4], a polarized, 20-50 femtosecond optical laser pulse, driving a relatively low energy (70–600mJ), is focused into a <50 µm spot, to reach an intensity of 10^{15}–$5 \cdot 10^{17}$ W/cm^2 across a low density gas. Along the path, the quantum tunnelling process ionizes the atoms to

the high-charged states. The ionization stages and the energy distribution of electrons depend on the pumping laser characteristics (OFI being a threshold process), but the largest effect on the free electron kinetic energy distribution is due to the polarization. A linear polarization minimizes the mean electron distribution and it is suitable for strong recombination. At contrary, when the pumping laser is circularly polarized, the released electrons will maintain a kinetic energy equal to the ponderomotive potential of the laser at the time of ionization and the hot electron distribution is apt to the collisional excitation. In recent investigations on collisional excitation OFI x-ray lasers, by focusing a 300–600 mJ circularly polarized, 35 femts, 10 Hz Ti:sapphire laser in a few mm-long cell filled with Xe or Kr, a gain of 67cm^{-1} in Pd like Xe at $\lambda = 41.8$nm and a gain of 78 cm^{-1} in Ni-like Kr at $\lambda = 32.8$ nm have been demonstrated [10]. Amplification has been also observed in the 2-1 transition in H-like LiIII by collisional recombination at $\lambda = 13.5$ nm [11]. However, in spite of the relatively high repetition rate, which can be reached due to the low energy of the pumping laser pulse the energy/pulse of the X-ray beam results, up to now, to be modest ($\approx$ 10–20 nJ) [3-4].
In 1994, the demonstration of a capillary discharge pumped soft X-ray by J.J. Rocca et al. [12] at 46.9 nm in Ne-like Ar, really opened the possibility of a concrete realization of efficient, compact and high repetition rate soft x-ray lasers. In this efficient pumping scheme, a current flowing through a few mm in diameter and 10–30 cm long capillary channel, initially filled with gas or a metal vapor creates the active medium. The high (several tens of kA) and fast (rise time lower than 50 ns) current pulse, compresses the plasma column towards the capillary axis (z-pinch), causing the plasma heating and the increasing of the electron density. The laser developed by Rocca et al. in Ne-like Ar has been demonstrated to operate at a repetition rate up to 7 Hz with a mean power of 1 mW and can give up to $\approx$ 1 mJ/pulse. Very recently, also other Laboratories [13-15] reported capillary discharge pumped soft x-ray lasing in Ne-like Ar. Fig. 1 and fig.2 report respectively the 46.9 nm, $\approx$ 2 ns long x-ray laser pulse and the saturation of the active medium obtained in L'Aquila experiment. With this experimental technique, however, it results still difficult to determine the proper working point to scale to the shorter wavelength region. Lasing has been reported only at the wavelengths of 52.9 nm and at 60.8 nm respectively in Ne-like Cl and S. The realization of capillary discharge soft X-ray laser operating at shorter wavelength is still under investigation.
The generation of population inversion by photoionization ie ionizaton by incoherent X-ray photons has been considered as a possible pumping system to achieve wavelengths shorter than 1 nm. This scheme (ISPI) requires photon energies at least high enough to photoionize preferentially the K shells electrons of a considered element while photons with lower energies must be removed by appropriate filters to avoid pumping of lower laser level.

Figure 1. Intensity of the laser line at two different initial gas pressures: 330 mTorr (full circles) and 250 mTorr (triangles) as a function of the capillary length. The gain resulted to be respectively of 1 and 0.7 cm^{-1}.

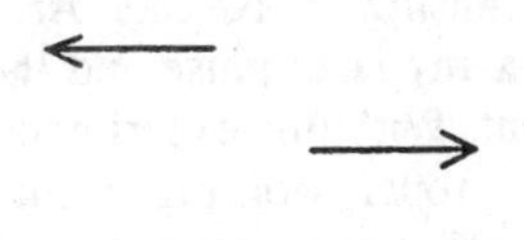

tim e (2 0ns/div)

Figure 2. Time evolution of soft x-rays emitted by the discharge in a 3-mm in diameter, 20-cm long capillary channel initially filled with 0.32 Torr of Ar.

In the ISPI scheme a broad band X ray pumping source must be used with very high photon density to get high gain in ASE operation regime as our. The scheme has the advantage of avoiding a large pumping rate of the lower level and therefore the necessity of a metastable upper level but unfortunately many competitive processes deplete the population inversion as auto-ionization, auger effect and collisional ionization of outer shells. These competitive processes can rapidly quench the gain. Thus the pumping method must be faster than those competitive processes. This is the reason why ISPI X ray lasers never have been demonstrated but only proposals have been forwarded in

particular for low Z elements as, for instance transition of Ne at 1.5 nm by photoionization of 1s electrons of Ne atoms creating a population inversion and gain.
In conclusion, concerning this process we are waiting for ultra-fast very intense laser or chirped amplification to use the necessary hundred Terawatt ligth pulses as pumping tools.
Measurements on the spatial coherence of soft X-ray lasers have been done in few Laboratories. Recently, for a TCE silver 20 μJ, λ = 13.9 nm soft X-ray laser, a spatial coherence length l_t = 100 μm was found [16]. Also a 10 μJ output energy soft X-ray Ni-like Sn laser, at λ = 12.0 nm, has been also extensively studied by a double slit experiment (as almost usually) with a total coherence length l_t = 35 μm at 21cm from the source. Since 1997 on a capillary discharge pumped table-top laser J.J. Rocca and coworkers observed a certain degree of coherence of the X-ray beam. Recently, with a plasma column of 36 cm, they reported [17] a fully coherent (coherence length as big as the dimension of the source), milliwatt average power laser, with a peak brightness ≈ 210^{25}ph/s $mm^2 mrad^2$0.01%bw.
High order harmonic generation appears to be a promising method for the production of ultra-short coherent XUV pulses from a compact laser system. In spite of relatively small number of photons/ harmonic pulse (typically ≈ 10^5 at 31.8 nm) the spectral brightness can be very high, due to the extremely short pulse duration and the good spatial coherence 10^{22}-10^{23} photons/(A s ster).

References

1. Matthews D.L. et al. (1985) Demonstration of a Soft X-ray Amplifier, *Phys. Rev. Lett.* **54**, 110
2. MacGowan B.J. et al. (1987) Demonstration of Soft X-Ray Amplification in Nichel-like Ions, *Phys. Rev. Lett.*,59, 2157
3. Lemoff B.E. et al. (1995) Demonstration of a 10 Hz Femtosecond-Pulse-Driven XUV Laser at 41.8 nm in Xe IX, *Phys. Rev. Lett.*, **74**, 1574-1577
4. Sebban S. (2001) Saturated Amplification of a Collisionally Pumped Optical-Field-Ionization Soft X-Ray Laser at 41.8 nm, *Phys. Rev. Lett.* **86**, 3004
5. Moon S. J. et al. (2002), Advances toward inner-shell photo-ionization x-ray lasing at 45 A, *Proc. of the 8th International Conf. on X-ray lasers, Aspen*
6. Pretzler G. et al. (2000) Hot.electron generation in copper and photopumping of cobalt, *Phys. Rev.E*, 62, 5618
7. Nickels J. et al. (1997), Short Pulse X-ray laser at 32.6 nm Based on transient Gain in Ne-like Titanium, *Phys. Rev. Lett*, **78**, 2748-2751
8. Kuba J. et al. (2000), Two color transient pumping in Ni-like silver at 13,9 nm and 16.1 nm, *Phys. Rev. A*, **62**
9. Dunn J. et al. (2000) Gain saturation regime for laser-driven tabletop, transient Ni-like ion X ray-laser, *Phys. Rev. Lett.*, **84**, 4834–4837
10. Sebban S. (2001), Investigations on collissional optical field ionization soft x-ray lasers, *Proceedings of SPIE, 4505, 195*
11. Avitzour Y. et al. (2002) Experimental and Theoretical simulation for conditions for lasing at 13.5 nm in LiIII, *Proc. of the 8th International Conf. on X-ray lasers, Aspen*
12. Rocca J.J. et al. (1994), Demonstration of a table-top soft-xray laser, *Phys. Rev. Lett.*, **73**, 2192
13. Ben-Kish A. et al.(2001), Plasma Dynamics in Capillary Discarge Soft X-Ray Lasers, *Phys. Rev. Lett.* **87**,1
14. Niimi G. et al. (2001) Observation of multi-pulse soft x-ray lasing in a fast capillary discharge, *J. Phys. D, Appl.Phys.* **34**, 1-4
15. Tomassetti G. et al. (2002) Capillary discharge soft x-ray lasing in Ne-like Ar pumped by long current pulses, *Eurpean Phys. J. D* **19**, 73-77

16. Ros D. et al. (2002) Recent Progress on the understanding of the transient Ni-like Ag X-ray laser at 13.9 nm at LULI facilities, *Proc. of the 8th International Conf. on X-ray lasers, Aspen*
17. Liu Y. et al. (2001) Achievement of essentially full spatial coherence in a high-average-power soft-x-ray laser, *Phys. Rev. A*, 63, 1-5

24. RARE EARTH ION DOPED CERAMIC LASER MATERIALS

XUESHENG CHEN
Department of Physics & Astronomy
Wheaton College
Norton, MA 02766, USA

1. Introduction

Rare-earth ion doped ceramic laser materials have recently shown great potential for efficient microchip or high power lasers. Contrast to popular single-crystal laser materials such as the single crystal Nd:YAG, ceramic laser materials are easy to be fabricated into large size at potentially very low cost and readily to be doped with high rare-earth ion concentrations. Mass production of ceramic laser materials is possible because no sophisticated technique and expensive equipment are needed, compared with what are involved in the single-crystal growth and subsequent fabrication. The first part of this lecture reviews some recent work done by others on rare-earth ion doped ceramic laser materials, and the second part reviews some work done by the author and her collaborators on a new kind of ceramic materials which have great potentials and unique features in laser applications.

2. Nd doped Ceramic YAG

The average grain size of the Nd:YAG ceramics was about 10 μm. Highly efficient solid-state lasers are very important for many technologically important applications, such as remote sensing, communications, and target recognition. Since the single crystal Nd-doped $Y_3Al_5O_{12}$ (Nd:YAG) laser material was first fabricated by Guesic et al.[1] in 1964, progress in the fabrication technique, the Czochralski (Cz) method, has rapidly improved its optical quality. Single crystal Nd:YAG lasers have been applied with remarkable success to various fields. But it is extremely difficult to grow large size and dope more than 1 at.% of Nd homogeneously as a luminescent element in a single crystal YAG, because the effective segregation coefficient of Nd for the host material is 0.2, thus limiting its possibility for high power and highly efficient lasers. However, a recent breakthrough in the solid-state laser material made of transparent polycrystalline YAG ceramics brought a new hope to the community of high-power and high-efficiency lasers and microchip lasers.

In 1995 the first transparent Nd:YAG ceramic laser was obtained [2] with a slope efficiency of 28%, pumped by a 600 mW laser diode (end pumping scheme). In this work, fabrication of transparent Nd:YAG ceramics was started with powders of Al_2O_3, Y_2O_3, and Nd_2O_3 of less than 2μm particle size, followed by hot press.

Since then, the ceramic formation process and sintering process have been optimized. The first high-power CW ceramic Nd:YAG laser [3] was demonstrated in 2000, with a laser output power of 31 W because of the successful fabrication of highly transparent, high-quality ceramic Nd:YAG. The Nd:YAG ceramic rod was pumped by 808nm laser diodes of a total power with a grain-boundary width of less than 1nm. The porosity level was at 1ppm level.

B. Di Bartolo and O. Forte (eds.), Frontiers of Optical Spectroscopy, 721-731.

Such low porosity level and narrow boundary width ensured very low scattering loss inside the rod. Recently, in December 2003, ceramic Nd:YAG lasers achieved high output power of more than 1 kilowatt. [4]

Laser characteristics of Nd:YAG ceramics were shown to be equivalent or far superior to those of high-quality Nd:YAG single crystals. Ceramic lasers have a combination of properties of both glass and single-crystal. Compared to single-crystal YAG's, ceramics are much easier to fabricate, less expensive, readily increased in size, and suited for mass production. Most importantly, the rare-earth-ion doping concentration, hence the lasing efficiency in ceramics, could be much higher than that in a single crystal. Ceramic lasers with as high as 5 at.% Nd have been successfully demonstrated. Shown in Fig. 1 and Fig. 2 are comparisons of absorption coefficients and laser characteristics between the 1.0 at.% Nd:YAG single crystal and 4.8 at.% Nd:YAG polycrystalline ceramic. Clearly, the highly doped ceramic Nd:YAG has higher

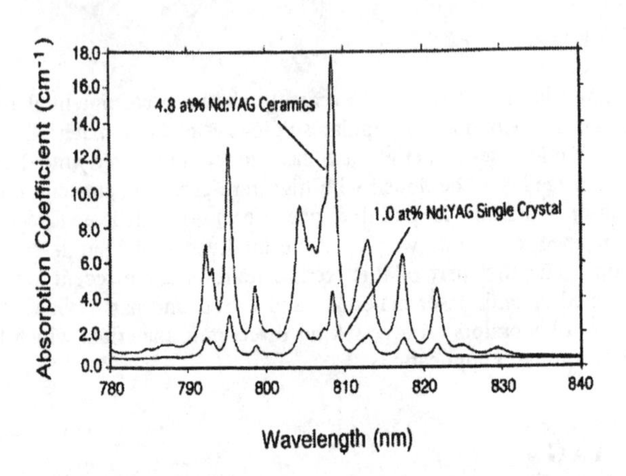

Fig.1. Absorption spectra for (a) the single crystal 1.0 at.% Nd:YAG and (b) the ceramic 4.8 at.% Nd:YAG between 780 and 840 nm.

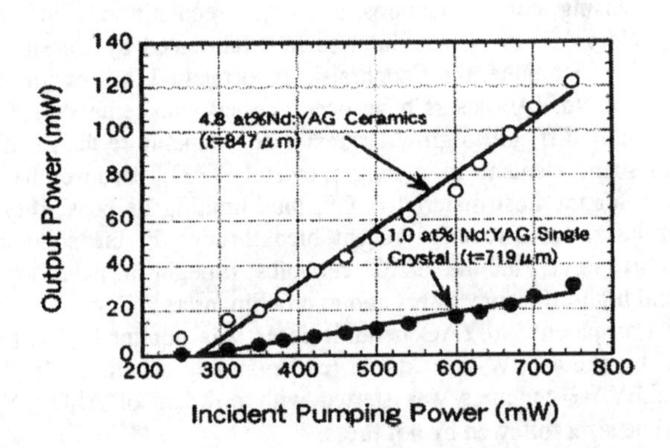

Fig.2. Microchip laser characteristics of the single crystal 1 at.% Nd:YAG and the ceramic Nd heavily-doped YAG.

TABLE 1. Comparison of physical properties of polycrystalline YAG ceramics sintered at 1800°C with YAG single crystal grown by the Czochralski method.

	Polycrystalline	Single Crystal
Bulk density (g/cm^3)	4.55	4.55
Vickers hardness (Gpa)	12.8	12.6 a)
Refractive index c)	1.81	1.81 a)
Thermal conductivity (J/cm·C·s) b)		
at 20 °C	0.105	0.107 a)
at 200 °C	0.067	0.067 a)
at 600 °C	0.046	0.046 a)

a) Measured at <111> orientation.

b) Measured by laser flash method.

c) Measured by ellipsometer.

TABLE 2. Comparison of absorption coefficients of Nd:YAG ceramics at different Nd concentrations, along with the 0.9% Nd:YAG single crystal.

Material	Nd concentration (at.%)	Absorption Coefficient (cm^{-1})	Wavelength (nm)
Single Crystal YAG	0.9	3.45	808.3
Ceramic YAG	1.1	2.45	808.3
	2.4	4.9	808.3
	4.8	11.7	808.3
	9.1	32.6	808.3

absorption coefficient and higher overall laser efficiency. Table 1 shows comparative data for both ceramic YAG and single crystal YAG on bulk density, Vickers hardness, refractive index, and thermal conductivity. These values show excellent agreement with each other. [5] Table 2 shows the comparisons of absorption coefficients at 808nm of the ceramic Nd:YAG at different Nd concentrations ranging from 1 to 9% along with corresponding information of a single crystal Nd:YAG. The maximum output power of the 4.8% Nd:YAG ceramic laser was 3 times higher than the conventional single crystal Nd:YAG because the ceramic YAG has four times higher Nd ion concentration than the 0.9% Nd:YAG single crystal.[5]

3. New Lead-based Ceramic Laser Materials

The author and her collaborators have been working on a new, highly transparent electro-optic ceramic material, Er^{3+}-doped $Pb_{1-x}La_xZr_yTi_{1-y}O_3$ (Er:PLZT) fabricated by Boston Applied Technologies, Inc. Since the host material (PLZT of some special compositions) has excellent transparency in a wide optical window as shown in Fig.3 (29% of the optical loss in the visible region is resulted from surface reflection which can be eliminated by proper anti-reflection coating). PLZT has exceptionally high electro-optic (EO) effect, which has been successfully used for a variety of optical devices in telecommunications. The EO-based laser material would have unique features

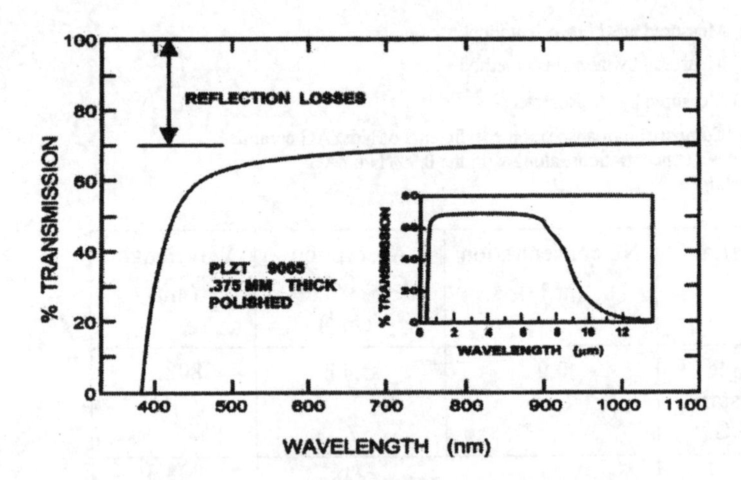

Fig.3. Transmission spectral of PLZT, where 29% of optical loss in the visible region is resulted from surface reflection, which can be eliminated by proper anti-reflection coating.

in phase and mode self-modulation that will lead to a revolutionary laser system of higher efficiency, more compactness, and integrated multi-functions. Our optical property study of this new electro-optical ceramic material Er-doped PLZT is reviewed in this second part of the lecture. Our work provides crucial information in its promising applications in new optical devices including microchip and high-power ceramic lasers.

3.1. MATERIAL

PLZT is fabricated into a transparent ceramic based on the perovskite structure of ABO_3 [6], formulated as $(Pb,La)(Zr,Ti)O_3$, where Pb^{2+} and La^{3+} ions occupy A sites while Zr^{4+} and Ti^{4+} are at B sites. From our research, it is believed that the Er^{3+} and Yb^{3+} doping ions take B-sites in the 2%Er doped PLZT and 0.5%Er-2.5%Yb co-doped PLZT – two transparent ceramic samples reported here. This kind of ceramic materials shows polycrystalline structure with the average grain size in micrometer range and grain boundary width in nanometer. Fig.4 is a picture of a typical sample, which shows an average grain size of 10μm. This work was focused on optical and photoluminescence studies of the Er^{3+} ions in these two new materials with a particular attention to the effect of Er^{3+} concentration and addition of the Yb^{3+} ion. Co-doping Yb^{3+} is for the purpose of increasing the absorption of pump light. The wavelength of the pump light has to match the Yb^{3+} energy level in near infrared and to one of the Er^{3+} levels in order for Yb^{3+} to be a useful ion to increase absorption strength.

The Er doped and Er-Yb co-doped PLZT were fabricated by means of pressure assisted sintering (PAS). Powders with proper Er or Er-Yb / PLZT stoichiometry were cold-pressed into a pre-form of a 2 - 4 inch diameter. The cold pressed slug was then subjected to a high temperature (1100-1300°C) and high pressure (up to 2500 psi), sintering under an oxygen atmosphere in an HP22-0614 SC hot press system. The completed slug, which was transparent, was then cut and polished into wafers or cubes for various studies.

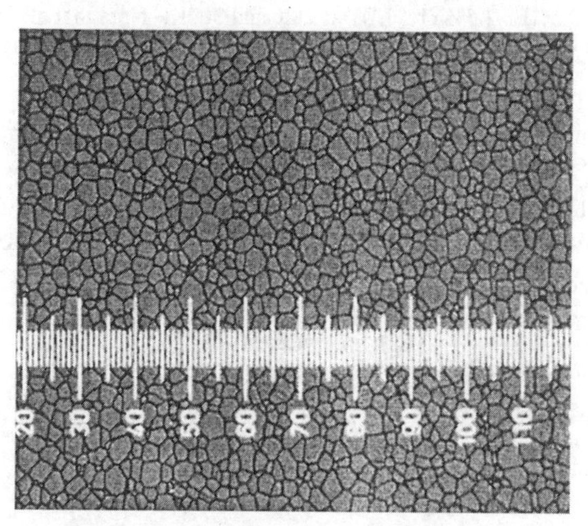

Fig.4. Picture of a typical Er-doped PLZT sample.
The average grain size is 6μ.

3.2. EXPERIMENTAL

For absorption measurement, a Perkin Elmer Spectrophotometer and Ocean Optics Fiber Optic Spectrometer were used. For photoluminescence measurement, the sample was excited by a diode laser at ~800 nm or ~970 nm or an Ar laser at ~488 nm (all in multimode). Luminescence emitted by the sample was observed at 90° with respect to the axis of the excitation laser beam and analyzed by a 1-m McPherson Model 2051 (f/8.7) high resolution, coma free, scanning monochromator with a grating of 600/mm, blazed at 1.25μ. The monochromator provided a resolution of 0.8 Å at 50 μm slit widths and had a wavelength reproducibility of 0.1 Å.

An appropriate long-pass filter was used between the sample and the monochromator entrance to prevent scattering of the pump laser light from getting into the monochromator. Photoluminescence (PL) from the sample was modulated at a frequency of 250Hz before entering the entrance slit. A PbS detector was used at the exit of the monochromator to convert the PL signal to electrical, which was then amplified by a lock-in amplifier and recorded by a computer. The sample was mounted on a cold finger of a Closed Cycle Refrigerator in order to change the temperature from 25 to 300K, controlled by a Lake Shore Model 805 temperature controller.

3.3. EXPERIMENTAL RESULTS AND DISCUSSIONS

Absorption and photoluminescence (PL) spectra were investigated under different conditions, for example, different pump laser wavelengths, for the 2%Er:PLZT and 0.5%Er-2.5%Yb:PLZT ceramics. Shown in Fig.5 are absorption spectra of these two samples at room temperature. In order to separate the spectra for better viewing, the curve of the Er-Yb co-doped sample is shifted up by 1 unit. Shown in Fig.6 are PL spectra of the samples in the 1550 nm region under a ~800 nm diode laser pump at. The PL in the same spectral region pumped by a ~970 nm laser diode is shown in Fig.7. PL spectra at different temperatures ranging from 25 to 300K were also measured. Fig.8 shows PL of the 2%Er:PLZT from 940 to 1760 nm at 25K and 300K, while Fig.9 shows PL spectra of 0.5%Er-2.5%Yb:PLZT at 28K and 300K. Note that all the PL spectra shown here have not been corrected for equipment's non-uniform intensity response to different wavelengths.

From results shown in Figs.5-9, some observations, conclusions, and comments can be made as follows.

1) The Er^{3+} and Yb^{3+} energy levels in solids are shown in Fig.10 as a reference. The bottom curve in Fig.5 shows absorption lines that are due to following Er^{3+}'s energy manifolds:

$^4S_{3/2}$ and $^2H_{11/2}$ -------- ~520 nm (strongest)
$^4F_{9/2}$ -------------------- ~660 nm (weak)
$^4I_{9/2}$ -------------------- ~800 nm (weakest)
$^4I_{11/2}$ ------------------- ~990 nm (weak)
$^4I_{13/2}$ ------------------- ~1560 nm (medium-strong)

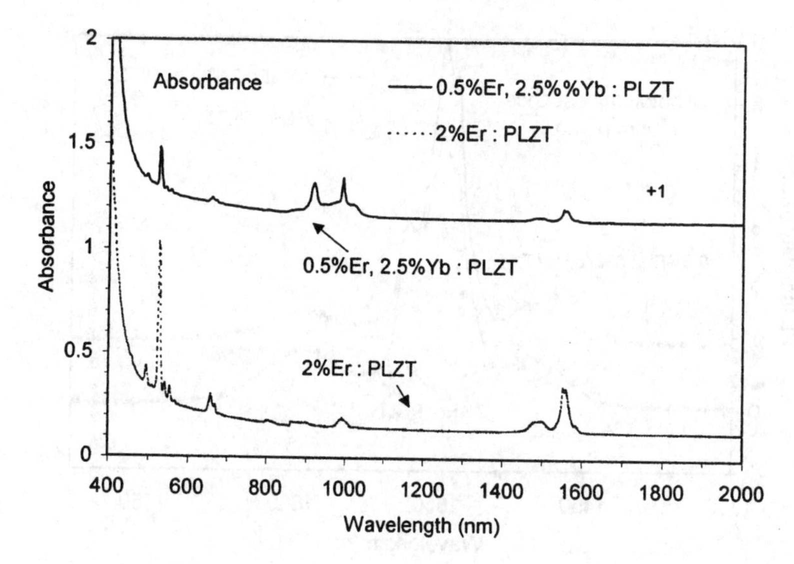

Fig.5. Absorbance spectra of 2%Er:PLZT and 0.5%Er-2.5%Yb:PLZT at room temperature.

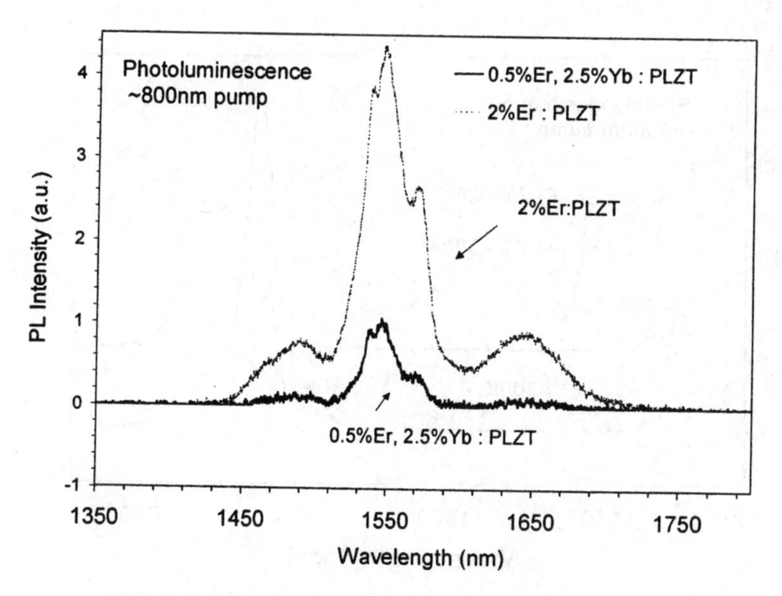

Fig.6. Photoluminescence spectra of 2%Er:PLZT and 0.5%Er-2.5%Yb:PLZT under ~800 nm excitation.

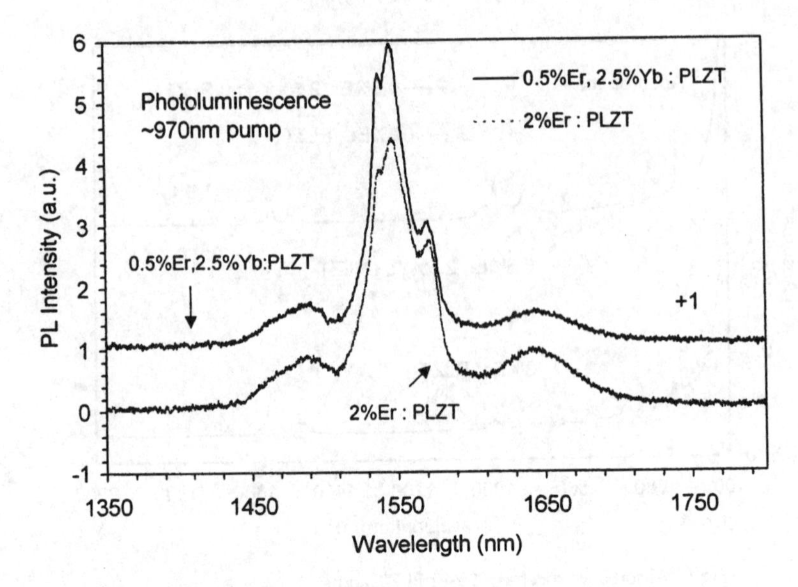

Fig.7. Photoluminescence spectra of the same samples as in Fig.6 but under ~970 nm excitation.

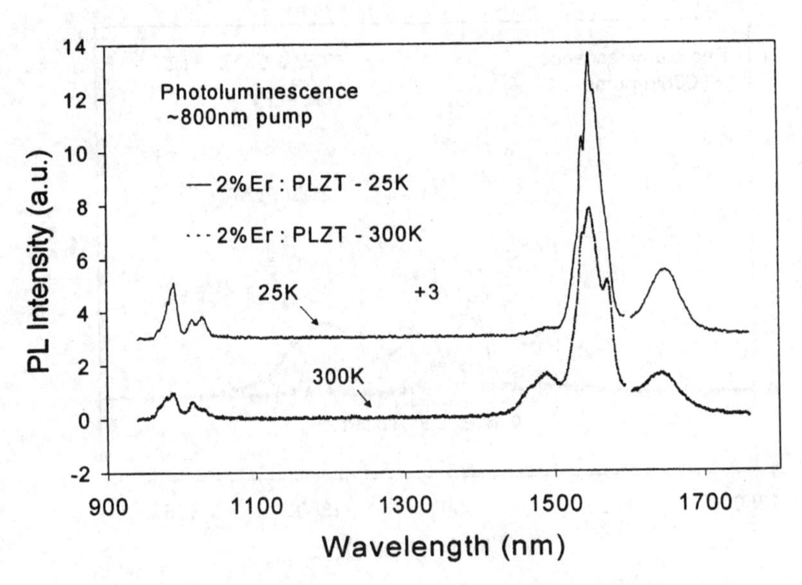

Fig.8. Photoluminescence spectra of 2%Er:PLZT at 25K and 300K under diode pump at ~800nm.

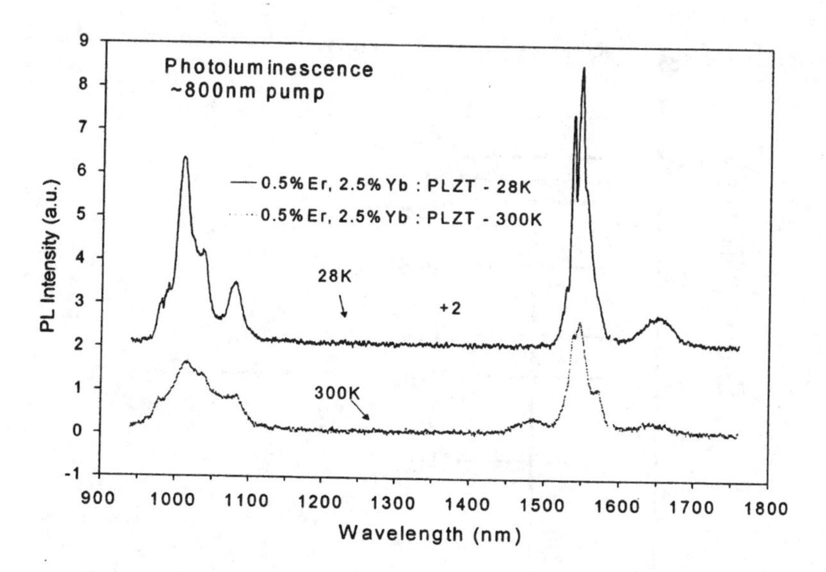

Fig.9. Photoluminescence spectra of 0.5%Er-2.5%Yb:PLZT at 28K and 300K under ~800 nm excitation.

Comparing the two absorbance spectra in Fig.5, one can see absorbance at ~520 nm, ~660 nm, and ~1560 nm is higher in 2%Er:PLZT (the lower curve) than in 0.5%Er, 2.5%Yb:PLZT. It is obviously due to higher Er^{3+} concentration in the 2%-Er sample. One exception is that the absorbance in the 900-1100nm region is higher in the sample of lower Er concentration (0.5%), as shown in the top curve. One logic explanation is that, in the absorbance spectrum of 0.5%Er-2.5%Yb:PLZT, Yb^{3+} contributes to the absorption lines at ~920nm and ~990nm groups due to its $^2F_{5/2}$ manifold, together with weak Er^{3+}absorption in the two groups.

2) PL in the 1550 nm region is due to transition from $^4I_{13/2}$ to $^4I_{15/2}$ in Er^{3+}. We believe that Yb^{3+} results in similar PL intensity for both the 2%Er and 0.5%Er doped samples, as shown in Fig.7, when pumped by the ~970 nm diode laser. Yb^{3+} in 0.5%Er-2.5%Yb:PLZT can absorb the pump photon energy efficiently through its $^2F_{5/2}$ manifold (because its energy matches the pump photon energy), and then this absorbed energy can be transferred to the Er^{3+} ion's $^4I_{11/2}$ manifold for it has very similar energy to the $^2F_{5/2}$ of Yb^{3+} (see Fig.10).

3) Since $^2F_{5/2}$ is Yb^{3+} 's only manifold, corresponding to the 900-1100nm wavelength range, it cannot assist in absorbing pump energy if the sample is pumped by a ~800 nm diode laser because of a large energy mismatch. In this case, the Er^{3+} concentration has to be the key factor to determine PL intensity. As shown in Fig.6, the PL spectrum of the 2%Er-doped PLZT's has about four times the intensity of the 0.5%Er-doped sample, matching the ratio of the two Er concentrations (2%/0.5% = 4). We also observed similar behavior when the samples were pumped by an Ar laser at ~488nm. (But it should be keep in mind that doping Er^{3+} beyond a certain concentration level would cause PL quenching, instead of increasing, due to too much energy migration.)

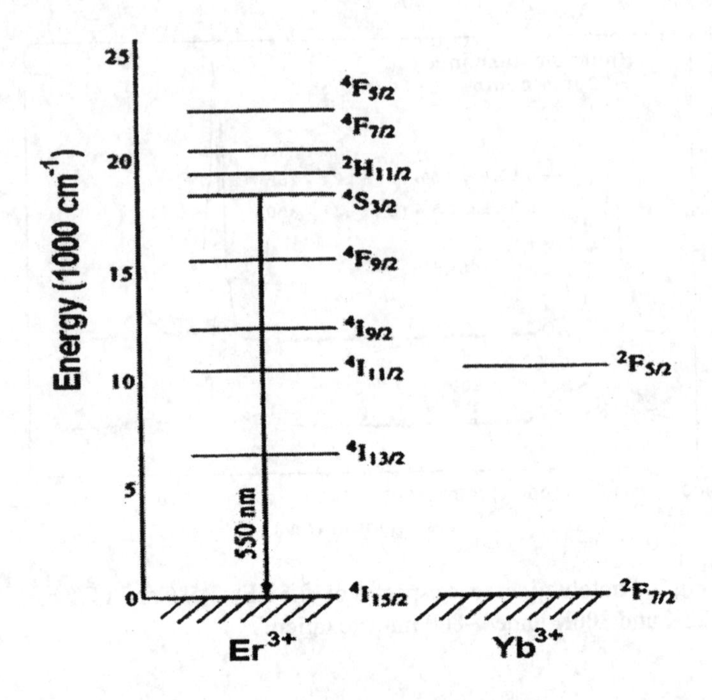

Fig.10. Energy level diagrams of Er $^{3+}$ and Yb^{3+} in solids.

4) In Figs.8 and 9, two groups of PL peaks are due to $^4I_{11/2}$ -> $^4I_{15/2}$ (~1000 nm) and $^4I_{13/2}$ -> $^4I_{15/2}$ (~1550 nm) transitions of the Er ions, pumped by a ~800 nm diode laser. We can see that the overall PL intensity for both samples decreases with temperature. There is, however, one difference between Fig.8 and Fig.9: the ratio of PL intensity at ~1000 nm to that at ~1550 nm in the 0.5%Er-doped PLZT (Fig.9) increases with temperature; this behavior does not show up in the 2%Er-doped PLZT sample. It is probably due to faster increase with temperature in upconversion [7] [8] in the 0.5% sample from the $^4I_{13/2}$ level to higher levels than from the $^4I_{11/2}$ level.

5) Strong green luminescence (~550nm) was observed in both the 2%Er-doped PLZT and the 0.5%Er-2.5%Yb-codoped PLZT under either ~800 nm or ~970 nm pump. The observed green luminescence is due to upconversion energy transfer processes.[9] Efficient upconversion in these materials is good for their possible applications in diode-pumped upcoversion lasers, but is undesirable for their potential applications in the 980 or 1550 nm laser. In order to make these materials efficient for the 980 or 1550 nm laser, the upconversion processes need to be greatly reduced either from material engineering or by selecting a pump laser at other wavelengths, such as 1480 nm.

3.4. CONCLUSIONS

We have investigated optical properties of the novel, highly transparent, electro-optic ceramics, 2%Er:PLZT and 0.5%Er-2.5%Yb:PLZT, developed by Boston Applied Technologies, Inc. Strong photoluminescence in these two ceramics in the 1550nm region was obtained when pumped by diode lasers either at ~970 nm or ~800 nm. We also observed efficient green upconversion luminescence. Absorption and PL spectra show the advantage of using Yb^{3+} for enhancing Er^{3+} emission in the 1550nm region. This work shows promising future for developing microchip and high power ceramic lasers using Er-doped PLZT ceramics.

3.5. ACKNOWLEDGEMENT

The author would like to thank National Science Foundation for supporting her research on optical property studies of the novel electro-optic ceramic materials: Er doped PLZT ceramics.

4. Summary

This lecture reviewed an exciting area of developing transparent ceramic laser materials for microchip and high-power lasers with high efficiency, focusing on recent breakthrough of ceramic Nd:YAG lasers and a new type of highly transparent ceramic material – Er-doped PLZT. Er-doped PLZT has exceptionally high electro-optic effect that can allow unique features in phase and mode self-modulation and lead to a revolutionary laser system of higher efficiency, more compactness, and integrated multi-functions.

References:

[1] E. Guesic, et al, *Appl. Phys. Lett.*, 4 (10)(1964) 182.
[2] A. Ikesu, et al, *J. of Am. Ceram. Soc.*, 78 (1995) 1033.
[3] Jianren Lu, et al, *Jpn. J. Appl. Phys.* 39 (2000) L1048-1050.
[4] J. Lu, et al., *CLEO*, PR 2003, F2I-(1)-3, December 19, 2003.
[5] K. Yoshida, et al, *CLEO*, Pacific Rim'99, 963.
[6] G.Haertling, Ferroelectrics, 51, 116, (1991).
[7] Xuesheng Chen, et al., *Journal of Luminescence*, 85, 295, (2000).
[8] Xuesheng Chen, et al., *Journal of Luminescence*, 83-84, 471, (1999).
[9] *Advances in Energy Transfer Processes*, eds. B. Di Bartolo and Xuesheng Chen, World Scientific Publishing Company, (2001).

25. SHORT SEMINARS

CONFOCAL FLUORESCENCE AND RAMAN MICROSCOPY OF FEMTOSECOND LASER-MODIFIED FUSED SILICA

Wilbur J. Reichman, James W. Chan, and Denise M. Krol
Lawrence Livermore National Laboratory, Livermore, CA 94550, USA

Confocal fluorescence and Raman microscopy were used to probe for spatial variations in defect concentration and glass structure across femtosecond (fs) laser-modified lines in fused silica. Modified lines were written with 130 fs laser pulses from an amplified Ti-sapphire laser operating at 800 nm at a repetition rate of 1 KHz. The sample was scanned at 20 μm/s with laser pulse energies ranging from 1 to 35 μJ, resulting in modified lines with diameters ranging from 8 to 40 μm. The fluorescence intensity decreased with increasing distance from line center, indicating a greater concentration of non-bridging oxygen hole center defects near the line center. The Raman intensity increased with increasing distance from line center. No significant variations in the concentration of 3- and 4-membered ring structures were observed.

OPTICAL CHARACTERIZATION OF QUANTUM DOTS

Francisco Rafael Leon, Denis M. Krol, and Thomas Huser
Univ. of California at Davis, Department of Applied Science, Livermore, CA 94550 USA

Semiconductor nanocrystals have optical properties, which depend on their composition, size and surface structure. In such structures, exciton confinement leads to size-dependent excitation and emission spectra and they are therefore often called "quantum dots." Because of this size-tunable optical behavior, colloidally-grown quantum dots have possible applications as fluorescent markers for biological imaging and tags for the detection of biomolecules. Fluorescence lifetime and quantum efficiency are also important optical characteristics. By using a scanning confocal microscope with chromatic and temporally-resolved detection systems, we will observe how the properties of single quantum dots vary with synthesis technique. In addition, we hope to correlate these observations with size measurements obtained via AFM and bond resonance data obtained using SERS. In collaboration with a quantum simulation team and crystal growers, our goal is to understand how to tune the optical properties of quantum dots in order to make them a viable tool for biologists.

B. Di Bartolo and O. Forte (eds.), Frontiers of Optical Spectroscopy, 733-740.

NEW PHOSPHORS FOR ULTRAVIOLET EXCITATION

P. Vergeer and A. Meijerink
Debye Institute, Utrecht University, Princetonplein 1 Utrecht, THE NETHERLANDS

Finding new phosphors for vacuum ultraviolet excitation sources is important for the development of efficient mercury free fluorescent tubes and plasma display panels [1]. The main drawback of these phosphors is the energy loss when one vacuum ultra violet (VUV) photon is converted into one visible photon. To solve this problem the process of down conversion [2], where for each VUV photon absorbed two visible photons are generated, is being studied. Here the possibility of down conversion via energy transfer in the Pr^{3+}-Mn^{2+} and the Pr^{3+}-Eu^{3+} couples is reported. Contrary to what is predicted theoretically, down conversion between Pr^{3+} and Mn^{2+} is not observed. Also Zachau et. al. [3] observed no energy transfer for Pr^{3+} and Eu^{3+}. Upon excitation of the $^1S_0$ level of Pr^{3+} indeed no Eu^{3+} emission is observed. This is however not due the absence of energy transfer, but to quenching. To investigate the quenching mechanism samples of YF_3:Pr^{3+} co-doped with Yb^{3+} were studied and found to also show quenching of the $^1S_0$ emission. This indicates that the quenching occurs via a metal-to-metal charge transfer state.

1. T. Jüstel, H. Nikol and C. Ronda, Angewandte Chemie International Edition 37 (1998) 3084.
2. R. T. Wegh, H. Donker, K. D. Oskam, and A. Meijerink, Science 283 (1999) 663.3. M. Zachau, F. Zwaschka and F. Kummer, Proceedings – Electrochemical Society 97-29, edited by C. Ronda and T. Welker, (1998) 314.

MAIN TOPICS OF INTERESTS IN THE AREA OF LUMINESCENCE MATERIALS

Artur Bednarkiewicz and Dariusz Hreniak
Inst. for Low Temp. and Structure Research, Polish Acad. of Sci. , Wroclaw, POLAND

This talk had the purpose of presenting to the participants the activities that are taking place in the Institute for Low Temperature and Structure Research of the Polish Academy of Sciences. The Institute is located in Wroclaw, Poland.

INTERACTION OF FEMTOSECOND PULSES WITH TRANSPARENT MATERIALS

Rafael Gattass and Iva Maxwell
Harvard University - Gordon McKay Lab, Cambridge, MA 02138 USA

In this talk we presented an overview of femtosecond microstructuring of transparent materials. Bulk structuring of transparent materials can be achieved by focusing high-intensity femtosecond pulses. The morphology of the structures depends on the incident energy per pulse and on the focusing conditions. At high focusing conditions the damage threshold in silicate glasses is just a few nanojoules. This energy range is available from an oscillator.

We demonstrated laser writing of embedded waveguides in silicate glasses with a femtosecond oscillator. Laser machining at high laser repetition rate results in a cumulative thermal mechanism of material modification which leads to structural index of refraction changes beyond the focal volume. The cross-sectional index profile of the fabricated structure depends on the number laser pulses irradiated.

We presented a parametric study of the role of the laser repetition rate in the size of the machined structures. We identify two distinct regimes of processing in the kHz to MHz range. As the time interval between pulses is reduced, we observe a transition from a repetitive modification process (identical to what is frequently called multiple shot damage) to a cumulative thermal mechanism. In the repetitive regime, each pulse acts independently and the energy deposited diffuses out of the focal volume before the next pulse arrives. In the cumulative thermal regime, the heat deposited is additive and leads to structures of significantly different morphology whose size increases dramatically with the decrease of the time interval between pulses.

ULTRAFAST PHASE TRANSITIONS IN SOLIDS

Maria Kandyla
Harvard University - Gordon McKay Lab, Cambridge, MA 02138 USA

In this talk we presented the coherent excitation and coherent control of the A1 phonon mode in Te. First, the underlying theory about the excitation of the A1 phonon mode and only this in a certain class of materials is discussed. The theory, called Displacive Excitation of Coherent Phonons (DECP), predicts the excitation of the A1 phonon mode as a result of electronic excitation following absorption of an ultrashort laser pulse by the material. Since there is no symmetry breaking mechanism in the electronic excitation through absorption the effect can only be encountered in materials which possess the symmetry preserving, breathing A1 phonon mode. Earlier experiments which demonstrate the effect on Te, Sb, Ti_2O_3 and Bi are presented.

Next, our results on the excitation of A1 phonons in Te were discussed. We excite Te with an amplified femtosecond laser pulse and therefore promote a significant amount of electrons from bonding to antibonding states. The A1 phonon mode is excited with a large amplitude of ion oscillations and the bandstructure of the material is altered considerably. Our results exhibit a band crossing taking place in Te hundreds of femtoseconds after the excitation. The indirect band gap of Te collapses and the material undergoes a semiconductor to semimetal transition. As the oscillations of the ions continue and the ions move toward their equilibrium position the material recovers its bandgap. The oscillations have a frequency of about 3.6 THz, too fast to give time to the material to exhibit a metallic behavior while it is in the crossed bands state. We have been working on a double pump experiment which will allow us to stabilize the bandstructure in the semimetallic state for times much grater than before and therefore study the response of the material under this transition. Our most recent results on this coherent control were presented.

RELAXATION PATHWAYS FROM ELECTRONIC EXCITED STATES OF OXYGEN DEFICIENT CENTERS IN GE-DOPED SILICA

A. Cannizzo and M. Leone
Dipartimento di Scienze Fisiche ed Astronomiche, Univ. di Palermo, Palermo, ITALY

Ge-doped amorphous silica (a-SiO_2) is a material of large interest due to its application in technological devices. Much interest arises from its enhanced photosensitivity that enables to realize on fiber optical devices as Bragg gratings, filters and mirrors [1]. This photosensitivity has been connected to the presence of oxygen deficient centers (ODC). They are characterized by the so-called optical B-type activity, which is constituted by an absorption band centered at 5.1-5.4 eV and two related emission bands: a singlet-singlet transition at 4.2-4.3 eV and a triplet-singlet transition at 3.0-3.2 eV named α_E and β, respectively [2, 3]. The involved excited singlet and triplet states are strongly connected via an intersystem crossing (ISC) process with a typical rate of 2×10^9 sec^{-1} at 300 K, which shows a great dependence on temperature, [1, 4] revealing a strong defect-lattice coupling. This luminescence activity can be excited also in the vacuum ultraviolet (v-UV) range around 7.4 eV. Many works were dedicated to the study of the ISC process in the ultraviolet (UV) excitation range, but poor attention was devoted to the v-UV excitation range, in order to investigate the defect-lattice coupling in higher excited electronic states. We reported an experimental investigation of the emission spectra and of fluorescence time decay of the B-type activity exciting in the v-UV and in the UV range, at room temperature and at 10 K. We have found that, exciting in the v-UV range, the triplet population is 100 times greater than exciting in the UV. We rationalize this result hypothesizing the presence of ISC processes characterized by rates faster than 10^{10}~10^{11} sec^{-1}.

1. G. Pacchioni, L. Skuja and D.L. Griscom, *Defects in SiO_2 and related dielectrics: Science and Technology* (Kluwer Academic Publishers, Dordrecht, 2000).
2. V.B.Neustruev, J. Phys. Cond. Matter 6, 6901 (1994).
3. L. Skuja, J. Non-Cryst. Solids 239, 16 (1998).
4. M. Leone, S. Agnello, R. Boscaino, M. Cannas and F. M. Gelardi Phys. Rev. B 60, 11475 (1999).

DETECTING QUANTUM SIGNATURES IN THE DYNAMICS OF TRAPPED IONS

A. Militello, A. Napoli and A. Messina
Dipartimento di Scienze Fisiche ed Astronomiche, Univ. di Palermo, Palermo, ITALY

Trapped ions provide the possibility of observing interesting quantum dynamical features of a few bosonic-fermionic degrees of freedom quantum system. Such a kind of confined particles may indeed be thought of as a composition of a one-, two- or three-dimensional harmonic oscillator and a one-half spin system. Laser-driving the ion with suitable configurations of classical fields, it is possible to implement a wide class of spin-boson-like Hamiltonians. Moreover, suitable vibronic interactions and atomic population measurements via quantum jump techniques provide tools to actually realize some thomographic procedures [1]. We present here two new proposals for extracting information concerning the trapped ion centre of mass motion.

The first one allows *direct* measurement of (in principle) any bosonic operator mean value [2]. As direct we mean that the expectation value is immediately obtained without complete reconstruction of the ion motion state. The method is based upon the idea of implementing a vibronic coupling that induces atomic transitions modulated by the mean value of the vibrational operator of interest. The procedure, interesting by in its own, reveals to be useful in detecting some specific nonclassical behaviours of the system.

Our second proposal concerns a technique aimed at *directly* detecting centre of mass motion density operator matrix elements, i.e. Fock populations and coherences [3]. The method is based upon the possibility of implementing a suitable class of unitary transformations, realizable as sequences of vibronic interactions.
The experimental feasibility of both the methods was briefly discussed.

1. D.J. Wineland et al., J. Res. Natl. Inst. Stand. Technol. 103, 259 (1998); L.G.Lutterbach and L.Davidovich, Phys. Rev. Lett. 78 2547 (1997);
2. B.Militello, A.Napoli, A.Messina, Phys. Rev. A 66 23402 (2002)
3. B.Militello, A.Napoli, A.Messina, Proceedings of the Wigner Centennial Conference, Hungary, Pecs (2002)

NON-EQUILIBRIUM POLARIZATION IN DIELECTRICS AND RELATED PHENOMENA

V.A. Trepakov
A.F. Ioffe Physical-TechnicalInstitute, Russian Acad. of Science –St.-Petersburg, RUSSIA

The purpose of this report is to give a short overview of some relatively new and not widely known phenomena and accompanying effects that have been recognized and studied in Ioffe Institute. These phenomena regard a macroscopic or microscopic polarization appearance under photo- or non-uniform thermal excitation conditions.

As an example of the thermally induced effects, a family of so-called "dielectric thermoelectric effects" is considered (M. Marvan, V. Gurevich, A. Tagantsev, G. Smolensky, V. Trepakov, A. Kholkin, see [1] and references therein). These effects are present in materials of all symmetry classes. They reveal themselves in experiments at nearly the same conditions as classical thermoelectric effects in conventional semiconductors and metals, but are caused by polarization responses of the phonon system to the temperature gradients and related displacement currents. These are thermopolarization effect ($P \sim b\ \partial T/\partial x$, which can be re-written also as $P = \varepsilon_0 \chi b_0 \nabla T = b \nabla T$), kinetic "reverse thermopolarization effect" (dielectric Peltier), and dielectric Thomson effects. The dielectric thermoelectric effects appeared to be pronounced, especially in highly polarizable media with soft modes and in ferroelectrics near transition points. The equivalence of the conventional and displacement currents in respect thermal effects production, which follows from the Maxwell equations, has been shown to be universal due to dielectric thermoelectric effect observations.

Polarization can also appear under photoexcitation of impurity centers with internal degrees of freedom (V. Vikhnin, V. Trepakov, [2,3] and references therein). Presence of the degrees of freedom lead to local configuration instability, and possibility to the reconstruction of the impurity or defect structure in the excited state. This is accompanied by very unusual optical properties controlled by defects. As an example, it is shown that the local configuration instability of the octahedral Cr^{3+} centers in the ^2E excited degenerated state in perovskite-like materials with soft TO phonon modes ($KTaO_3$, $KTaO_3$:Li,Nb, $SrTiO_3$ and $SrTiO_3$:Mg) leads to very strong and unusual changes in position and shape of the R zero-phonon ($^2E \rightarrow ^4A_2$) line of Cr^3 v.s. changes in temperature and electric filed due to the quadratic multimode Jahn-Teller effect in the excited ^2E state. It has recently been found [3] that not only soft lattice, but also relaxation modes can give a pronounced contribution and controlled temperature shift of the zero-phonon optical line shift.

1. V.A. Trepakov, E.T. Rafikov, M. Marvan, L. Jastrabik, N.P. Divin, Europhys.Lett. 21 (1993) 891.
2. V.A. Trepakov, L. Jastrabik, S. Kapphan, E. Giulotto, A. Agranat Optical Materials 19 (2002) 13.
3. V. A. Trepakov, , I. B. Kudyk, S. E. Kapphan, M. E. Savinov, A. Pashkin, L. Jastrabik, A. Tkach, P.M. Vilarinho, A.L. Kholkin, J. of Luminescence, in press.

OPTICAL TRANSITIONS IN QUANTUM NANOSTRUCTURES BASED ON IONIC MATERIALS

O. Proshina, I. Ipatova, and A. Maslov,
A.F. Ioffe Physical-Technical Institute, Russian Acad. of Science –St.-Petersburg, RUSSIA

We investigate theoretically optical transitions in quantum nanostructures based on materials with high ionicity. In case of the strong electron–phonon interaction with longitudinal optical phonons, the polaron effect results in multiple phonon replicas of the exciton optical transition line [1], [2]. It was shown that the maximum influence of electron–phonon interaction on optical spectra occurs in quantum dots.
The exciton states in many semiconductors are degenerate due to the valence band degeneration. In the approach of Luttinger Hamiltonian the degeneration is classified according to the angular momentum of hole $J^{(h)} = 3/2$. We take into account the electron and hole polarization of the medium in quantum dot based on semiconductor with high ionicity. The allowed optical transition occurs into the polaron exciton state with the angular momentum J = 1. The forbidden transition corresponds to J = 2. It is known that exciton levels differ due to the exchange interaction.
It is shown in our study that in case of a spherical quantum dot, the polaron exciton in the strong confinement regime creates the anisotropic polarization of the medium. Nevertheless the polarization does not split the degenerate ground state of the polaron exciton. The average value of angular momentum component J_z vanishes, $\langle J_z \rangle = 0$. The optical light emitted from the spherical dot is not polarized.
The picture is different in a case of ellipsoidal quantum dot. The reduction of symmetry results in the splitting of size quantization levels of the hole [3]. This splitting depends on the properties of the material and on the specific of the quantum dot shape. The ground state of the hole in nonspherical quantum dot is determined by the angular momentum component $J_z^{(h)} = \pm 3/2$ or $J_z^{(h)} = \pm 1/2$. When the hole with an angular momentum component $J_z^{(h)} = \pm 3/2$ contribute to the exciton transition, the average value of the polaron

exciton angular momentum component $< J_z >= \pm 1$. The light emitted from nonspherical dot has the specific polarization. The study enables us to make a conclusion on the quantum dot geometry.

1. I.P.Ipatova, A.Yu.Maslov, O.V.Proshina. *Europhys. Lett.,* 53(6), pp. 769–775 (2001)
2. I.P.Ipatova, A.Yu.Maslov, O.V.Proshina. *Surface Science,* 507–510, pp. 598–602 (2002)
3. Al.L.Efros, A.V.Rodina. *Phys.Rev. B,* 47, p. 10005 (1993)

LEDS MAKE THINGS BETTER

Cees Ronda
Philips GmbH - Forschungslaboratorien - Aachen, GERMANY

The advent of efficiently emitting LEDs in the blue and UV part of the spectrum has induced quite some work on luminescent materials. In this contribution, I will deal with a formulation of the requirements, LED phosphors have to fulfill and translate these requirements into suitable ions and host lattices. Color rendering will be discussed as well.

Application fields and the current state of the art were reviewed in terms of materials choice and efficiency. New developments were touched upon. The contribution ended with an outlook.

26. POSTERS

PHOTOREFLECTANCE AND LUMINESCENCE MEASUREMENTS OF GAINNAS/GAAS MULTIPLE QUANTUM WELL STRUCTURES

Andreas Grau, Michael Hetterich, and Claus Klingshirn
Institut für Angewandte Physik, Universität Karlsruhe, Karlsruhe, GERMANY

In recent years GaInNAs/GaAs has developed to one of the most promising material systems for the realization of optoelectronic devices emitting in the near infrared at telecom wavelength. The optical properties of GaInNAs are quite unusual.

Incorporation of only a few percent of nitrogen as for e.g. in InGaAs leads to a strong bandgap reduction and an increased effective electron mass.

It is now really interesting to find out more about the fundamental parameters of this material system (e.g. the behaviour of the effective mass with different nitrogen or indium contents). In this work the results of photoreflectance (PR) and luminescence measurements at various temperatures were shown.

First the effect of nitrogen on the bandstructure was discussed. In addition to that we described how one can extract information from PR spectra by fitting procedures and compare it with theoretical calculations. In particular one can get information about the coupling parameter C_{MN} , which is important in material systems with nitrogen.

SELF-CONSISTENT CALCULATION OF GROUND AND EXCITED ENERGY LEVELS OF A DOPED QUANTUM DOT BY A QUANTUM GENERIC ALGORITHM

Mehmet Sahin, Ulfet Atav, and Mehmet Tomak
Selcuk Üniversitesi Fen-Edebiyat Fakülltesi Fizik Bolümü Kampüs, Konya – TURKEY

In this study, we have theoretically calculated ground and excited energy levels of a doped quantum dot self-consistently. For this purpose, we have assumed that there are effectively two energy levels in a modulation doped spherical semiconductor quantum dot (QD) and we have determined the subband energy levels, corresponding wavefunctions, chemical potential, and potential profile of the QD. Electrical charge and thermodynamical equilibrium were also taken into consideration.

THE WIRES DIRECTION PHOTOCONDUCTIVITY OF GAAS/ALGAAS QUANTUM WIRES MEASURED ALONG

Miroslav Saraydarov
Dept. of Solid State Physics and Microelectronics, Sofia University Sofia, BULGARIA

A new photoconductivity (PC) study of undoped GaAs/AlGaAs quantum wires (QWRs) is carried out measuring the PC along the wires direction. The PC spectrum reveals several peak structures, related to the QWRs. This suggestion is confirmed by the observed dependence of the spectrum on the exciting light polarisation and by photoluminescence and photoluminescence excitation measurements on a similar sample. A long pre-illumination of the sample with infrared light (hn = 1.18 eV) is found to reduce considerably the substrate related background in the PC spectrum, which makes the QWR structures better manifested.

HIGH EXCITATION SPECTROSCOPY OF ZnO

H. Priller, J. Brückner, Th. Gruber, C. Klingshirn, H. Kalt, and A. Waag
Institut für Angewandte Physik, Universität Karlsruhe, Karlsruhe, GERMANY

We investigated ZnO epitaxial layers grown by MOVPE (Metal Organic Vapor Phase Epitaxy) techniques. The sample was grown on sapphire with a GaN buffer layer (60 nm) and a ZnO layer with a thickness of 400 nm.

The cw (continuous wave) luminescence experiments were carried out under excitation with the 325 nm line of a HeCd laser. For the reflection measurements we used a 150 W Xenon lamp. High excitation measurements were carried out with an Excimer Laser (308 nm, 15 ns pulse duration).

At medium excitation intensities exciton-exciton scattering with a super linear increase of the luminescence was observed. At higher intensities a new, broad band appeared between the bound exciton emission and the P band emission. This band strongly shifts to lower energies with the excitation intensity even below the P band. This band is attributed to the formation of an electron hole plasma, which leads to stimulated emission at a lower energy with a super linear increase, and a spectral narrowing of the emission.

Measurement of the gain with the variable stripe length method also showed stimulated emission.

PROPERTIES OF PECVD a-SiO_x:H FILMS

A. O. Kodolbas, A. Bacioglu, and O. Oktu
Hacettepe University, Faculty of Engineering, Dept.of Physics Eng. Ankara,TURKEY

Hydrogenated amorphous silicon-oxygen alloy (a-SiO_x:H) films were deposited by RF glow-discharge decomposition of SiH_4+CO_2 gas mixture at a substrate temperature of 300°C. Optical band gap of the samples can be tuned between 1.65 and 2.73 eV by varying the oxygen content
from 4.5 to 64.2 at. %. While, both room temperature photo and dark conductivity decreases with oxygen alloying, for oxygen content below 15 at. % measured conductivities are comparable to those of unalloyed a-Si:H. In contrast, deep defect density and Urbach parameter continuously increases with oxygen alloying.

OPTICAL INVESTIGATION OF SPIN INJECTION INTO OPTICALLY ACTIVE NANOSTRUCTURES

Daniel Troendle, Robert Hauschild, Hendrik Burger, and Heinz Kalt
Institut für Angewandte Physik, Universität Karlsruhe, Karlsruhe, GERMANY

In the past years the spin degree of freedom has become subject of substantial interest. Especially spin-based quantum information processing (spintronics) appears to be quite promising. Major challenges on the way to make a spin-based electronic device possible are the preparation of spin-polarized carriers and their injection into semiconductors, the transport of spins, the storage of spin information, and its coherent manipulation.

In our project we plan to use a micro-Photoluminesence setup inside a 14T Magnet to investigate GaInNAs/GaAs quantum dots with a ZnMnSe spin aligner. In this poster we presented the recently obtained first spin dependent measurements done on a well characterized ZnCdSe/ZnSe quantum well sample which is used for confirmation purpose of the system functionality. Besides we presented information about other experimental techniques available in our labs.

ULTRAFAST PHASE TRANSITIONS IN SOLIDS

Maria Kandyla

Harvard University - Gordon McKay Lab, Cambridge, MA 02138 USA

In this talk we presented the coherent excitation and coherent control of the A1 phonon mode in Te. First, the underlying theory about the excitation of the A1 phonon mode and only this in a certain class of materials was discussed. The theory, called Displacive Excitation of Coherent Phonons (DECP), predicts the excitation of the A1 phonon mode as a result of electronic excitation following absorption of an ultrashort laser pulse by the material. Since there is no symmetry breaking mechanism in the electronic excitation through absorption the effect can only be encountered in materials which possess the symmetry preserving, breathing A1 phonon mode. Earlier experiments which demonstrate the effect on Te, Sb, Ti_2O_3 and Bi were presented.
Next, our results on the excitation of A1 phonons in Te were discussed. We excited Te with an amplified femtosecond laser pulse and therefore promoted a significant amount of electrons from bonding to antibonding states. The A1 phonon mode was excited with a large amplitude of ion oscillations and the bandstructure of the material was altered considerably. Our results exhibited a band crossing taking place in Te hundreds of femtoseconds after the excitation. The indirect band gap of Te collapses and the material undergoes a semiconductor to semimetal transition. As the oscillations of the ions continue and the ions move toward their equilibrium position the material recovers its bandgap. The oscillations have a frequency of about 3.6 THz, too fast to give time to the material to exhibit a metallic behaviour while it is in the crossed bands state. We have been working on a double pump experiment which will allow us to stabilize the bandstructure in the semimetallic state for times much greater than before and therefore study the response of the material under this transition. Our most recent results on this coherent control were presented.

STIMULATED EMISSION OF $Nd_{0.5}La_{0.5}Al_3(BO_3)_4$ RANDOM LASER AND THE THRESHOLD CONDITIONS FOR LARGE AND SMALL PUMPING REGIMES

K. J. Morris, M. Bahoura, G. Zhu, and M. A. Noginov
Norfolk State University, Center for Materials Research, Norfolk, VA 23504, USA

We have studied stimulated emission in $Nd_{0.5}La_{0.5}Al_3(BO_3)_4$ ceramic random laser in a broad range of pumped spot diameters *d*. The developed heuristic model adequately describes the dependence of threshold pumping energy density versus d at $d \geq 150\mu m$. At small pumping beam diameter ($d<100\mu m$), a very bright and strongly localized emission was observed in the center of the pumped area. At the appearance of the bright-spot, no lasing could be achieved. The spectrum of the bright-spot emission is not the spectrum of Nd^{3+} emission.

SPECTROSCOPY AND OPTICAL MICROSCOPY WITH NANO-LOCAL LIGHT SOURCES

Jinquan Liu, Andrea Callegari, Jerome Morville, Dino Tonti, Awos Alsalman, and Majed Chergui
LSUPC, ICMB, Faculté des Sciences de Base, BSP Ecole Polytech. Féd. de Lausanne
Lausanne-Dorigny, SWITZERLAND

Instead of using only near-field light coming out from the fiber aperture as the source in normal scanning near-field optical microscopy (SNOM), we are developing two version of apertureless scanning near-field microscopy (A-SNOM) with a probe (a molecule or a nanocrystal) attached to the fiber tip.

By using the probe attached on the fiber tip, we can measure the fluorescence spectra induced by resonant energy transfer from the donor probe to the acceptor (sample). Furthermore, gating fluorescence with pulsed laser, we can reduce the scattered light substantially, get rid of donor fluorescence, and even study time resolved spectra. By attaching a sharp metal nanocrystal to the fiber tip, we can implement a tunable light source with high enhancement and spatial confinement of optical field. Spectroscopic investigation of the sample using sum frequency generation with pulsed lasers offers the possibility of time-resolved pump-probe experiments.

THE SIZE-EFFECT AND PHASE TRANSITIONS-EFFECT ON LUMINESCENCE PROPERTIES OF $BaTiO_3:Eu^{3+}$ NANOCRYSTALLITES PREPARED BY THE SOL-GEL METHOD

D. Hreniak, W. Strek, G. Boulon, and R. Pązik
Inst. for Low Temp. and Structure Research Polish Acad. of Sci. , Wroclaw, POLAND

The effect of sintering temperature on optical properties of $Eu^{3+}:BaTiO_3$ nanocrystallites was investigated. The emission spectra and luminescence decays measurements were performed. The strong attention was paid on the effects of grain size and concentration of active ions on emission properties of $Eu^{3+}:BaTiO_3$ nanocrystallites and strong correlation of grain sizes and luminescence properties of Eu^{3+} was found. To explain these differences a detail analysis of luminescence spectra has been performed. It is well known that the intensity parameters Ω_2,Ω_4 and Ω_4 of the Judd-Ofelt theory [1,2] well characterize the emission yields of RE ions. The intensity parameters Ω_2 and Ω_4 have been determined from photoluminescence spectra following the method described recently by Kodaira et al [3].

1. B. R. Judd, Phys. Rev. 127 (1962).
2. G. S. Ofelt, J. Chem. Phys. 37 (1962) 511.
3. C.A. Kodaira, H.F. Brito, O.L. Malta, O.A. Serra, J. Luminescence 101 (2003)11.

ENERGY TRANSFER IN Nd^{3+} and Yb^{3+} DODOPED NANOMETRIC YAG CERAMICS

Artur Bednarkiewicz and Wieslaw Strek
Inst. for Low Temp. and Structure Research Polish Acad. of Sci. , Wroclaw, POLAND

Spectroscopic properties of Yb^{3+} and Nd^{3+} co-doped $Y_3Al_5O_{12}$ nanocrystallites were studied. The mechanisms of cooperative interaction responsible for the $Nd^{3+} \rightarrow Yb^{3+}$ energy transfer, which was observed for all grains sizes, were discussed. In the case of smaller grains (with diameter ~15nm) $Yb^{3+} \rightarrow Nd^{3+}$ energy transfer, orange in color 'hot' anti-Stokes emission of Nd^{3+} ions was also observed. Regarding energy level scheme of the ions, the anti-Stokes emission is not obvious. However this luminescence was observed and bands were prescribed to energy levels of neodymium ions. The power dependence measurements carried, gave sequentially increasing order of the anti-Stokes process numbers vs. the energy of emitting level of Nd^{3+} ion. Moreover the relation was linear. This suggests $^4F_{3/2}$ level of neodymium to be the starting level for multiphonon assisted energy transfer. Due to low probability of the processes with high orders, another hypothesis was put considering sequential excitation of lower laying Nd^{3+} levels (^4I_J where J=11/2, 13/2, 15/2). Also long rise-times (of the order of seconds) of the observed Nd^{3+} emission stand behind the second hypothesis. The larger grains doped with Nd and Yb ions do not exhibit Nd→Yb energy transfer. We explain the observed behavior due to differences in thermal energy dissipation abilities of grains with different sizes during CW pumping. The experiments allowed us to support the thesis about influence of grain size onto the luminescence properties of rare-earth doped materials.

ENVIRONMENT AND SHAPE EFFECTS ON DYNAMICS OF CdSe NANOCRYSTALS: COMPARING QUANTUM DOTS AND RODS

Camilla Bonati, Mona Mohamed, Dino Tonti, Jinquan Liu, and Majed Chergui
LSUPC, ICMB,Faculté des Sciences de Base, BSP Ecole Polytech. Féd. de Lausanne
Lausanne-Dorigny, SWITZERLAND

Semiconductor nanocrystals present tunable optical properties due to the effects of quantum confinement, and therefore are appealing structures both for fundamental studies and applications such as biological labeling, local imaging and designing optoelectronic devices.

CdSe quantum dots and rods with low size distribution (less than 5%) and high quantum yield (more than 50%) have been prepared by chemical methods in colloidal solution, and analyzed in this work by means of fluorescence up-conversion and pump-probe techniques.

Comparing the dynamics of the relaxation processes of these nanometer-size structures in different environments (changing both the capping material and the solvent) show that there is an interaction between the crystals and the surrounding media, depending both on the surface properties of the semiconductor particle and the polarity of the solvent.

The lack of a complete geometrical symmetry in quantum rods reveal the presence of states that are not observed in dots of the same size, and that can be related to the degeneracy of energy levels in the perfectly spherical quantum dots, as confirmed by the dynamics of the relaxation processes in the two different geometries.

GAMMA AND PROTON IRRADIATION EFFECTS ON KU1 QUARTZ GLASS

M. R. Nemtanu and B. Constantinescu
National Inst. for Lasers, Plasma and Radiation Physics, Bucharest-Magurele, ROMANIA

Studies on gamma and high-energy proton irradiation-induced modifications in ultraviolet transmission properties on KU1 quartz glass, known to be radiation-resistant, were presented. It was irradiated at high doses with gamma and high-energy protons (similar to thermonuclear reactor conditions). The optical transmission components of the future fusion devices will be expected to maintain their transmission properties under high levels of ionizing radiation during hundreds of hours. The results confirmed that KU1 quartz glass is highly resistant and show the modifications in UV transmission varying with irradiation dose.

FEATURES OF FEMTOSECOND LASER ABLATION OF SOLID TARGETS

M. Vitiello, S. Amoruso, R. Bruzzese, N. Spinelli, R. Velotta,
X. Wang, C. Altucci, and C. de Lisio
Dipartimento di Scienze Fisiche, Università di Napoli "Federico II", Napoli, ITALY

Progress towards the characterization of laser ablation and cluster production by fs laser pulse irradiation of solid targets of different materials was reported. In particular, we have employed optical emission spectroscopy (OES) and ion probes for the diagnostics of fs laser-ablation plume of different materials, in high vacuum. The OES analysis showed the presence of temporally separated populations in the material blow-off. A first temporal component was characterized by spectral line emission of the atomic constituents of the target material, and was observed just after the irradiation of the laser pulse to the target. A second component , which is much more delayed with respect to the first one, was characterized by the emission of a structureless continuum spectrum. The continuum emission has been ascribed to clusters and particles of nanometric size in the plume. Moreover, the ion probe measurements also showed the presence of a third component due to high energy ions.

STUDY OF THE SURFACE OF $SrTiO_3$ SINGLE CRYSTALS BY OPTICAL SECOND HARMONIC GENERATION

G. Cerrone, L. Marrucci, F. Miletto, D. Paparo, and U. Scotti di Uccio
Dipartimento di Scienze Fisiche, Università di Napoli "Federico II", Napoli, ITALY

A detailed characterization of the surface properties of SrTiO3 (STO) crystals is of fundamental importance in many topical research fields that range from catalysis to thin-film growth [1].

The most used crystallographic orientations of STO surfaces are (100) and (110), but vicinal cuts are also frequently considered. In an idealized description, the (100) STO is terminated by two undisturbed atomic layers, i.e., SrO or TiO2 plane. Most applications of STO demand an extremely accurate control of surface structure and chemistry. For instance, deposition on (100) STO substrates of high quality HTc superconductive thin-films is undoubtedly connected to the kind of atomic layer that terminates the STO structure. Despite the relevance of this problem, an effective technique suitable for recognizing *in situ* the two terminations as well as other surface reconstructions is still missing. Indeed standard probing techniques either are not suitable for *in situ* monitoring (AFM, STM) or their use is restricted to evacuated environments (LEED, RHEED, etc.).

From this point of view, linear optical techniques offer invaluable advantages, but, generally, are not surface specific. However, in the last decade, the nonlinear optical technique of surface second harmonic generation (SSHG) has proved to be a powerful tool for surface analysis [2]. Its strength relies on the fact that the second-order optical nonlinear processes are forbidden in the bulk of centrosymmetric materials. However, at their interfaces, this symmetry is automatically broken and SSHG may take place. This symmetry breaking occurs in the first few atomic layers in the case of ionic or covalent crystals, making second harmonic generation highly surface specific.

In this work we reported experimental investigations on STO surfaces with different terminations and orientations by means of SSHG. To date very few applications of SSHG to STO surfaces can be found in the literature [3]. In any case, they have been focussed only on (110)-oriented surfaces. Compared to ref. [3], we have improved the basic SSHG technique by employing an interferometric experimental scheme that allows measuring both the amplitude and phase of the SSHG signal for exploiting all the potentialities of this technique.

In order to gain the maximum information on the surface symmetries, we measured the SSHG signal at different polarization combinations of both the input pump-beam and output second-harmonic beam, and at different azimuth angles with respect to the surface normal. We have found that the second-harmonic amplitude and phase show different symmetries between (100)- and (110)-oriented surfaces. Experiments are in progress on (100) STO surfaces with different terminations aimed at studying their influence on the second-harmonic optical field.

1. C. Noguera "Condens. matter" 12, 367-410 (2000)
2. C. T. Williams, D. A. Beattie "Surface science" 500, 545-576 (2002)
3. E. D. Mishina et al. "Journal of experimental and theoretical physics", 94, 3, 552-567 (2002)

DLS MEASUREMENT OF NANOMETRIC CARBON CLUSTERS PRODUCED IN LAMINAR PREMIXED FLAMES

A. Bruno, D. Cecere, P. Minutolo, and A. D'Alessio
Dipartimento di Scienze Fisiche, Università di Napoli "Federico II", Napoli, ITALY

There is an increasing demand for diagnostics to determine the physical and the chemical characteristics of very small particles or molecular clusters in the size range of 1-10 nm. The greatest fraction of long-lived particles of this dimension are produced in combustion process are suspected of human toxicity. Epidemiological results have also showed a causal relationship between particulate matter concentrations in ambient air and increased deaths in several cities around the world (Wichmann et al. 2000).

Because of their size, optical properties, and partial water solubility, combustion generated d=1-10 nm particles may also have significant effects on climate acting as cloud condensation nuclei.

This Dynamic Light Scattering [1,2] is based on the study of the intensity fluctuations of the light diffused by nanoparticles suspended in a medium, which are due to Brownian motion. These fluctuations are random and related to the translational diffusion coefficient D through the Stoke-Einstein relation and so to the diameter of the particles.

In the Rayleigh limit, the light intensity scattered by the particles is dependent on the six power of the diameter and so to the particle diameter distribution function , N(d), and it depends from a form factor that we assume to be spherical [4,5].

Laminar, premixed, sooting ethylene/air flat flames were studied using extra-situ sampling technique. Combustion products were collected by means of an isokinetic water-cooled, stainless steel probe placed at different height above the burner. The isokinetic property guarantees that the velocity of the gases in the flame is not perturbed by the presence of the probe. From the probe this material was transported through a sampling line and bubbled in 4ml of DI water. This type of sampling technique based on bubbling of combustion products in water showed the ability to isolate nanoparticles from other flame products.

Size distribution function obtained with DLS of samples collected in water for different C/O ratio and different height above the burner (z). A narrow range of particle size distribution from 2nm up to 4 nm was observed in two flames. Good agreement was found for the DLS distribution function of the hydrodynamic diameter and the distribution function of the equivalent volume diameter obtained with the Atomic Force Microscopy for the flame C/O =0,77 at z=3.5mm also using the mean diameter obtained by DLS measurements, the imaginary part of the refractive index at λ=266nm for these particle was evaluated [7].

1. Finsy,R., Particle Sizing by Quasi-Elastic Light Scattering, Elsevier,1994
2. Berne, B.J., Pecora,R., Dynamic light scattering Wiley,New York,1976
3. Wyn Brown, Dynamic Light Scattering, Clarendon Press,1993
4. S.W.Provencher, Comput. Phys. Commun. 27, 229-242 (1982)
5. Gousbet A.,Applied Optics, 16, 222, (1986)
6. Minutolo P.,Gambi G., D` Alessio A., Proc.27 symposium on Combst.1461 (1998)
7. Borghese A. and Merola S., Proc.27 Symposium on Combst. 2101 (1998)

INDEX